Electron Microscopy in Mineralogy

Coordinating Editor: H.-R. Wenk

Editorial Board: P. E. Champness
J. M. Christie · J. M. Cowley · A. H. Heuer
G. Thomas · N. J. Tighe

With 272 Figures

Springer-Verlag Berlin Heidelberg New York 1976

ISBN 3-540-07371-X Springer-Verlag Berlin Heidelberg New York
ISBN 0-387-07371-X Springer-Verlag New York Heidelberg Berlin

Library of Congress Cataloging in Publication Data. Main entry under title: Electron microscopy in mineralogy. Includes bibliographical references and index. 1. Mineralogy, Determinative. 2. Electron microscope. I. Wenk, H.-R., 1941– . II. Thomas, Gareth. QE369.M5E35. 549'.133. 75-23431.
This work is subject to copyright. All rights are reserved, whether the whole or part of the material is concerned, specifically those of translation, reprinting, re-use of illustrations, broadcasting, reproduction by photocopying machine or similar means, and storage in data banks.
Under § 54 of the German Copyright Law where copies are made for other than private use, a fee is payable to the publisher, the amount of the fee to be determined by agreement with the publisher.
© by Springer-Verlag Berlin · Heidelberg 1976.
Printed in Germany.
The use of registered names, trademarks, etc. in this publication does not imply, even in the absence of a specific statement, that such names are exempt from the relevant protective laws and regulations and therefore free for general use.
Typesetting, printing and bookbinding: Universitätsdruckerei H. Stürtz AG, Würzburg.

Preface

During the last five years transmission electron microscopy (TEM) has added numerous important new data to mineralogy and has considerably changed its outlook. This is partly due to the fact that metallurgists and crystal physicists — having solved most of the structural and crystallographic problems in metals — have begun to show a widening interest in the much more complicated structures of minerals, and partly to recent progress in experimental techniques, mainly the availability of ion-thinning devices. While electron microscopists have become increasingly interested in minerals (judging from special symposia at recent meetings such as Fifth European Congress on Electron microscopy, Manchester 1972; Eight International Congress on Electron Microscopy, Canberra 1974) mineralogists have realized advantages of the new technique and applied it with increasing frequency. In an effort to coordinate the growing quantity of research, electron microscopy sessions have been included in meetings of mineralogists (e.g. Geological Society of America, Minneapolis, 1972, American Crystallographic Association, Berkeley, 1974). The tremendous response for the TEM symposium which H.-R. Wenk and G. Thomas organized at the Berkeley Conference of the American Crystallographic Association formed the basis for this book. It appeared useful at this stage to summarize the achievements of electron microscopy, scattered in many different journals in several different fields and present them to mineralogists. A group of participants as the Berkeley symposium formed an Editorial Committee and outlined the content of this book. We decided to review the present state of the art of transmission electron microscopy in general chapters which would be followed by short original research contributions.

In this way we hope to produce a "living textbook". It would be inappropriate to base a book in this rapidly expanding field entirely on established facts without taking into account present efforts and pointing out new directions. In the selection of contributors we tried to be representative rather than comprehensive, but accommodated contributions to the Berkeley conference as far as they fitted into the narrower framework of the book.

We hope that "Electron-Microscopy in Mineralogy" will serve a multiple function as (1) a summary of present achievements with fairly complete references to previous work, thus an indispensable reference handbook for the electron microscopist, (2) an introduction for the classical mineralogist to the field of electron microscopy and (3) a textbook for students in an advanced mineralogy and crystallography course.

The various chapters of the book were edited by section editors (G. Thomas accepted manuscripts 1, 2.2, 2.4, 2.5 and 3; P. Champness 4.2, 4.3, 4.4, 4.5 and 4.11; A. Heuer 5.2, 5.3, 5.4, 5.5, 5.8 and 5.9; J.M. Christie 6.2, 6.4, 6.5, 6.6 and 6.7 and H.-R. Wenk 2.1, 2.3, 4.1, 4.6, 4.7, 4.8, 4.9, 4.10, 5.1, 5.6, 5.7, 6.1, 6.8, 7.1, 7.2, 7.3, 7.4, 7.5, 7.6, 7.7 and 7.8) but individual authors are responsible for text and illustrations. Section editors used the services of outside referees in their evaluation of the manuscripts and apart from contributors to the book, A.E. Bence, J. Boland, E.P. Butler, M.S. Hampar, M. Korekawa, F. Laves, W.S. MacKenzie, J.D.C. McConnell, A.C. McLaren, M. Muir, P.H. Ribbe, P. Robinson, H. Schulz, and J. Zussmann have been kind enough to review manuscripts. H.-R. Wenk coordinated the project, to integrate the individual manuscripts into a uniform format. All manuscripts were received by the end of February 1975.

Our main thanks go to the contributors of articles for helping, often with great enthusiasm, to make this project possible. We are also grateful to Springer Verlag Inc. and especially to Dr. K.F. Springer for his personal interest. The publisher's care and high-quality reproduction contributed to make this book attractive. H.-R. Wenk is obliged to the Research School of Earth Sciences at ANU, Canberra and to the Institut für Kristallographie, Universität Frankfurt a.M. for their hospitality and help with clerical work. They served as editorial offices during his sabbatical leave.

We hope that this volume of electron microscopy will make many earth scientists aware of the new technique and give them access to its use in solving geological problems. We also hope that the book will attract students to pursue quantitatively one of the many fruitful research projects that have reviewed or introduced here.

November, 1975   The Editors

Contents

Section 1   **Introduction**
H.-R. WENK 4

Section 2   **Contrast**

2.1   Fundamentals of Electron Microscopy
O. VAN DER BIEST and G. THOMAS 18
2.2   Interpretation of Electron Diffraction Patterns
J.A. GARD 52
2.3   Contrast Effects at Planar Interfaces
S. AMELINCKX and J. VAN LANDUYT 68
2.4   Computer Simulation of Dislocation Images in Quartz
J.W. MCCORMICK 113
2.5   The Direct Imaging of Crystal Structures
J.M. COWLEY and S. IIJIMA 123
2.6   A Comparison of Bright Field and Dark Field Imaging of Pyrrhotite Structures
L. PIERCE and P.R. BUSECK 137

Section 3   **Experimental Techniques**
N.J. TIGHE 144

Section 4   **Exsolution**

4.1   Exsolution in Silicates
P.E. CHAMPNESS and G.W. LORIMER 174
4.2   Coarsening in a Spinodally Decomposing System: TiO_2-SnO_2
M. PARK, T.E. MITCHELL, and A.H. HEUER 205
4.3   Magnetite Lamellae in Reduced Hematites
N.J. TIGHE and P.R. SWANN 209
4.4   Precipitation in the Ilmenite-hematite System
J.S. LALLY, A.H. HEUER, and G.L. NORD JR. 214
4.5   Pigeonite Exsolution from Augite
G.L. NORD JR., A.H. HEUER, and J.S. LALLY 220
4.6   The Transformation of Pigeonite to Orthopyroxene
P.E. CHAMPNESS and P.A. COPLEY 228

4.7	On the Detailed Structure of Ledges in an Augite-enstatite Interface D.L. Kohlstedt and J.B. Vander Sande	234
4.8	The Phase Distributions in Some Exsolved Amphiboles M.F. Gittos, G.W. Lorimer, and P.E. Champness	238
4.9	Physical Aspects of Exsolution in Natural Alkali Feldspars C. Willaime, W.L. Brown, and M. Gandais	248
4.10	Analytical Electron Microscopy of Exsolution Lamellae in Plagioclase Feldspars G. Cliff, P.E. Champness, H.-U. Nissen, and G.W. Lorimer	258
4.11	Exsolution in Metamorphic Bytownite T.L. Grove	266

Section 5 Polymorphic Phase Transitions

5.1	Polymorphic Phase Transitions in Minerals A.H. Heuer and G.L. Nord Jr.	274
5.2	Direct Observation of Iron Vacancies in Polytypes of Pyrrhotite H. Nakazawa, N. Morimoto, and E. Watanabe	304
5.3	Rutile: Planar Defects and Derived Structures B.G. Hyde	310
5.4	High-resolution Electron Microscopy of Unit Cell Twinning in Enstatite S. Iijima and P.R. Buseck	319
5.5	Polytypism in Wollastonite H.-R. Wenk, W.F. Müller, N.A. Liddell, and P.P. Phakey	324
5.6	High-resolution Electron Microscopy of Labradorite Feldspar H. Hashimoto, H.-U. Nissen, A. Ono, A. Kumao, H. Endoh, and C.F. Woensdregt	332
5.7	Origin of the (c) Domains of Anorthite A.H. Heuer, G.L. Nord Jr., J.S. Lally, and J.M. Christie	345
5.8	On Polymorphism of $BaAl_2Si_2O_8$ W.F. Müller	354
5.9	The Submicroscopic Structure of Wenkite F. Lee	361

Section 6 Deformation Defects

6.1	Deformation Structures in Minerals J.M. Christie and A.J. Ardell	374
6.2	Work Hardening and Creep Deformation of Corundum Single Crystals B.J. Pletka, T.E. Mitchell, and A.H. Heuer	404
6.3	Dislocation Structures in Synthetic Quartz D.J. Morrison-Smith	410

	6.4	The Microstructure of Some Naturally Deformed Quartzites
		N.A. LIDDELL, P.P. PHAKEY, and H.-R. WENK 419
	6.5	Defects in Deformed Calcite and Carbonate Rocks
		D.J. BARBER and H.-R. WENK 428
	6.6	Plasticity of Olivine in Peridotites
		H.W. GREEN II 443
	6.7	The Role of Crystal Defects in the Shear-induced Transformation of Orthoenstatite to Clinoenstatite
		S.H. KIRBY 465

Section 7 Special Techniques and Applications

	7.1	Amorphous Materials
		M.L. RUDEE 476
	7.2	Signals Excited by the Scanning Beam
		R. BLASCHKE 488
	7.3	Analytical Electron Microscopy of Minerals
		G.W. LORIMER and G. CLIFF 506
	7.4	X-ray Microanalysis Using a Scanning Electron Microscope
		M. SUTER, H.-U. NISSEN, R. WESSICKEN, and P. WIEDERKEHR 520
	7.5	Quantitative X-ray Microanalysis of Thin Foils
		R. KÖNIG 526
	7.6	Particle Track Studies
		I.D. HUTCHEON and J.D. MACDOUGALL 537
	7.7	Stony Meteorites
		J.R. ASHWORTH and D.J. BARBER 543
	7.8	Microcracks in Crystalline Rocks
		L. DENGLER 550

Subject Index 557

List of Contributors

AMELINCKX, S.	S.C.K.-C.E.N., 2400 Mol, Belgium
ARDELL, A.J.	Department of Geology, University of California, Los Angeles, CA 90024, USA
ASHWORTH, J.R.	Department of Physics, University of Essex, Colchester, CO4 3SQ, Great Britain
BARBER, D.J.	Department of Physics, University of Essex, Colchester, CO4 3SQ, Great Britain
BLASCHKE, R.	Institut für Medizinische Physik der Universität Münster, 4400 Münster, FRG
BROWN, W.L.	Laboratoire de Géologie-C.O No 140, Université de Nancy I, 54000 Nancy Cédex, France
BUSECK, P.R.	Departments of Geology and Chemistry, Arizona State University, Tempe, AZ 85281, USA
CHAMPNESS, P.E.	Department of Geology, University of Manchester, Manchester, MI3 9PL, Great Britain
CHRISTIE, J.M.	Department of Geology, University of California, Los Angeles, CA 90024, USA
CLIFF, G.	Department of Metallurgy, University of Manchester, Manchester, MI3 9PL, Great Britain
COPLEY, P.A.	Department of Mineralogy and Petrology, Cambridge, CB2 3EW, Great Britain
COWLEY, J.M.	Department of Physics, Arizona State University, Tempe, AZ 85281, USA
DENGLER, L.	Department of Geology and Geophysics, University of California, Berkeley, CA 94720, USA
ENDOH, H.	Department of Applied Physics, Faculty of Engineering, Osaka University, Osaka 565, Japan

GANDAIS, M.	Laboratoire de Minéralogie-Cristallographie, Université Pierre et Marie Curie, 75230 Paris Cédex 05, France
GARD, J.A.	Department of Chemistry, University of Aberdeen, Old Aberdeen, AB9 2UE, Great Britain
GITTOS, M.F.	The Welding Research Institute, Research Laboratory, Abington, CB1 6AL, Great Britain
GREEN II, H.W.	Department of Geology, University of California, Davis, CA 95616, USA
GROVE, T.L.	Department of Earth and Space Sciences, State University of New York at Stony Brook, Stony Brook, Long Island, NY 11790, USA
HASHIMOTO, H.	Department of Applied Physics, Osaka University, Osaka 565, Japan
HEUER, A.H.	Department of Metallurgy and Materials Science, Case Western Reserve University, Cleveland, OH 44106, USA
HUTCHEON, I.D.	Enrico Fermi Institute, University of Chicago, Chicago, IL 60637, USA
HYDE, B.G.	Gorlaeus Laboratories, Rijksuniversiteit Leiden, Leiden, The Netherlands
IIJIMA, S.	Department of Physics, Arizona State University, Tempe, AZ 85281, USA
KIRBY, S.H.	U.S. Geological Survey, Menlo Park, CA 94025, USA
KÖNIG, R.	Battelle-Institut e.V., 6000 Frankfurt/Main 90, FRG
KOHLSTEDT, D.L.	Department of Materials Science and Engineering, Cornell University, Ithaca, NY 14853, USA
KUMAO, A.	Physics Laboratory, Kyoto Technical University, Kyoto, Japan
LALLY, J.S.	U.S. Steel Research Laboratory, Monroeville, PA 15146, USA
LEE, F.	Department of Geological and Physical Sciences, California State University, Chico, CA 95929, USA
LIDDELL, N.A.	Department of Physics, Monash University, Clayton, Victoria 3168, Australia
LORIMER, G.W.	Department of Metallurgy, University of Manchester, Manchester, MI3 9PL, Great Britain

List of Contributors

MACDOUGALL, J.D.	Geological Research Division, Scripps Institution of Oceanography, La Jolla, CA 92093, USA
MCCORMICK, J.W.	Department of Geology, University of California, Los Angeles, CA 90024, USA
MITCHELL, T.E.	Department of Metallurgy and Materials Science, Case Western Reserve University, Cleveland, OH 44106, USA
MORIMOTO, N.	Institute of Scientific and Industrial Research, Osaka University, Osaka 565, Japan
MORRISON-SMITH, D.J.	Hurstpierpoint College, Hassocks, BN6 9JS, Great Britain
MÜLLER, W.F.	Institut für Kristallographie der Universität, 6000 Frankfurt/Main 1, FRG
NAKAZAWA, H.	National Institute for Researches in Inorganic Materials, Sakura-Mura, Niiharigun, Ibaragi-ken, 300-31, Japan
NISSEN, H.-U.	Labor für Elektronenmikroskopie II, ETH Zürich, 8006 Zürich, Switzerland
NORD JR., G.L.	U.S. Geological Survey, Reston, VA 22092, USA
ONO, A.	Electron Optics Division, JEOL Ltd., Tokyo 196, Japan
PARK, M.	NL Industries, Bridge Station, Niagara Falls, NY 14305, USA
PHAKEY, P.P.	Department of Physics, Monash University, Clayton, Victoria 3168, Australia
PIERCE, L.	Department of Chemistry, Arizona State University, Tempe, AZ 85281, USA
PLETKA, B.J.	Department of Metallurgy and Materials Science, Case Western Reserve University, Cleveland, OH 44106, USA
RUDEE, M.L.	Department of Applied Physics and Information Science, University of California, La Jolla, CA 92093, USA
SUTER, M.	Laboratorium für Kernphysik, ETH Zürich, 8049 Zürich, Switzerland
SWANN, P.R.	Metallurgy Department, Imperial College, London, SW7 2BP, Great Britain
THOMAS, G.	Department of Materials Science and Engineering, University of California, Berkeley, CA 94720, USA

TIGHE, N.J.	Physical Properties Section, United States Department of Commerce, National Bureau of Standards, Washington, D.C. 20234, USA
VAN DER BIEST, O.	Department of Materials Science and Engineering, University of California, Berkeley, CA 94720, USA
VANDER SANDE, J.B.	Department of Materials Science and Engineering, Massachusetts Institute of Technology, Cambridge, MA 02139, USA
VAN LANDUYT, J.	Rijksuniversitair Centrum Antwerpen, Faculteit der Wetenschappen, 2020 Antwerpen, Belgium
WATANABE, E.	Electron Optics Division, JEOL Ltd., Tokyo 196, Japan
WENK, H.-R.	Department of Geology and Geophysics, University of California, Berkeley, CA 94720, USA
WESSICKEN, R.	Labor für Elektronenmikroskopie II, ETH Zürich, 8049 Zürich, Switzerland
WIEDERKEHR, P.	Labor für Elektronenmikroskopie II, ETH Zürich, 8049 Zürich, Switzerland
WILLAIME, C.	Laboratoire de Minéralogie-Cristallographie, Université Pierre et Marie Curie, 75230 Paris Cédex 05, France
WOENSDREGT, C.F.	Geologisch en Mineralogisch Instituut, Rijksuniversiteit Leiden, Leiden, The Netherlands

Section 1 Introduction

◄ Clinopyroxene from lunar basalt 15058. Complex structure with lamellar intergrowth of pigeonite and augite on (100) (SE-NW) and (001) (SW-NE). Pigeonite lamellae which appear bright in the right part of the figure show contrast of antiphase domains with the displacement vector 1/2 ($a+b$). Note the change in volume ratio pigeonite to augite across the (100) interfaces. Dark field, magnification 62000×. (Photograph by W.F. Müller)

CHAPTER 1

Introduction

H.-R. WENK

Until recently mineralogists have been mainly accustomed to two classical tools i.e. the polarizing microscope and X-ray diffraction techniques. With the *optical microscope* we are able to determine the morphology and optical properties; we can see twins and exsolution lamellae which are larger than the wavelength of light. From *X-ray diffraction* data we can derive accurately the position of atoms in the unit cell on the scale between 1 and 100 Å. However such a crystal structure determination gives us an average over many thousands of unit cells, and assumes that all cells are identical.

Increasingly evident though are the important structures contained in minerals on the scale 100–10000 Å, which could not be directly imaged with conventional techniques. Diffuse reflections on X-ray photographs were attributed to small domains; asterism in Laue patterns or small extinction coefficients in structure refinements indicated that crystals are not perfect but contain defects. The *electron microscope* is the ideal instrument to investigate this range of inhomogeneities in minerals which also contains important geological information on parameters describing the cooling history of minerals and rocks and on conditions of deformation. The purpose of this book is to give mineralogists access to this new field by showing them advantages of electron microscopy, pointing out important applications and illustrating with examples how the microstructure can be interpreted in terms of the geological history. Emphasis is on transmission electron microscopy (TEM) of thin crystal foils, the technique which has had so far the greatest impact on mineralogical research. But additionally a few examples illustrate recent advances in scanning electron microscopy (SEM) which may well become an equally important technique.

Contrary to X-ray diffraction which was adopted in mineralogy immediately after its discovery, the electron microscope had its most important application in metallurgy. It took more than 30 years after the first commercial instruments were available in 1939 for electron microscopy to become an established tool in mineralogy and petrology. There are reasons for the delay.

Specimen preparation techniques such as electro-polishing or chemical thinning which are used in metallurgy were unsuitable for silicate minerals. Analyses were restricted to replicas and powder fragments. Only since ion thinning devices have become available (see Tighe, Chapter 3 in this book) can thin foils be easily prepared from a specific area in a standard thin section, thus permitting the use of the electron microscope as a petrographic microscope for very small objects. Another advantage of the ion beam thinner is that one no longer depends on cleavage planes, but can prepare foils of any crystallographic orientation.

An additional limitation to quantitative microscopy has been the fact that the *contrast theory* of electron diffraction, developed first by metallurgists and physicists for simple structures (among the pioneers are Bollmann, 1956; Hirsch et al., 1956), did not take into account some of the complications which arise in minerals, such as the difficulty of obtaining two beam conditions and elastic anisotropy. As will be shown in the following chapters, research in these fundamentally important fields is by no means concluded.

Scattered studies of minerals using replica methods commenced several decades ago (e.g. Eitel et al., 1939). Early direct TEM research of minerals includes studies of mica (Amelinckx, 1952) and clay minerals (Honjo and Mihama, 1954). There is an extensive literature on clay minerals mainly concerned with morphology of particles and mineral identification. EM-work on clays has already been reviewed (for example Beutelspacher and Van der Marel, 1968) and is therefore excluded from this book. Ribbe (1962) and McConnell and Fleet (1963) were among the first *mineralogists* to use the electron microscope. McLaren and Phakey (from 1965 onwards) and Nissen (from 1967 onwards) influenced the development through their programs which were entirely devoted to EM studies of minerals. In 1970 TEM work on lunar material stimulated a tremendous boom in which mineralogists were joined by materials scientists and physicists. Their results, which quickly changed the aspect of modern mineralogy over the past five years, demonstrated electron microscopy as a powerful tool. Until now new data have had most significant impact on the understanding of feldspars and pyroxenes but in almost every mineral group EM analysis revealed some unexpected features.

The importance of electron microscopy may best be illustrated by some examples. The brief preview of some classical applications which will be expanded in later chapters is intended to convince the reader that electron microscopy should be included in the curriculum of every modern mineralogist.

Determination of the atomic structure from X-ray data relies on measurements of the amplitude of the scattered waves. Since the wavelength of X-rays is only 0.5–2 Å, the phase shift during diffraction by the atoms cannot be determined experimentally, and therefore the electron density distribution in the unit cell cannot be calculated directly as a Fourier transform of the structure factors. In the electron microscope a diffraction pattern is obtained in the back focal plane of the objective lens (see Frontispiece to Section 2). This is — similar to the case of X-rays — the Fourier transform of electrostatic potential differences in the specimen which corresponds to the electron density distribution (Fig. 1a). But in contrast to X-rays, the electron microscope, under ideal conditions, can be used to obtain the inverse transform of the diffraction pattern experimentally in the image plane without losing the phase information (Fig. 1b). Through application of the delicate method of "high resolution electron microscopy" it has recently become possible to record two-dimensional images of the electron density with a resolution of about 3 Å in which the contrast from single atoms can be recognized (see Chapter 2.4 and, for example, Buseck and Iijima, 1974). Resolution is limited by several factors, whereby spherical aberration is the most serious experimental constraint (e.g. Allpress and Sanders, 1973). Good images of the structure are only obtained in very thin foils. The use of high voltages—if stability can be achieved—is advantageous since it produces short wavelengths and therefore diffraction patterns with many reflections, thus giving a large number of terms in the Fourier summation.

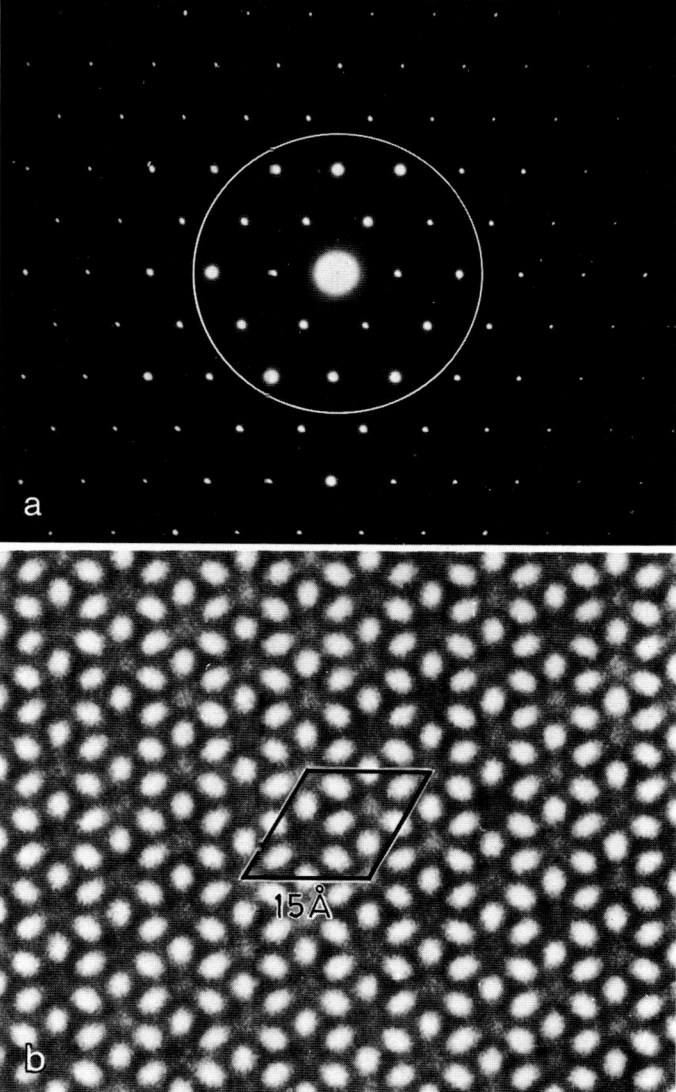

Fig. 1 a–c. Structure of tourmaline. (From Iijima et al., 1973.) (a) Electron diffraction pattern. hk0 plane of the reciprocal lattice. Size of the objective aperture is indicated. (b) High-resolution electron microscope image for diffracting conditions shown in (a). Unit cell is indicated. Due to short-range order there are variations in darkness from unit cell to unit cell. (c) z-projection of the structure of tourmaline. Notice that the regions with low electron density correspond to the white regions in (b)

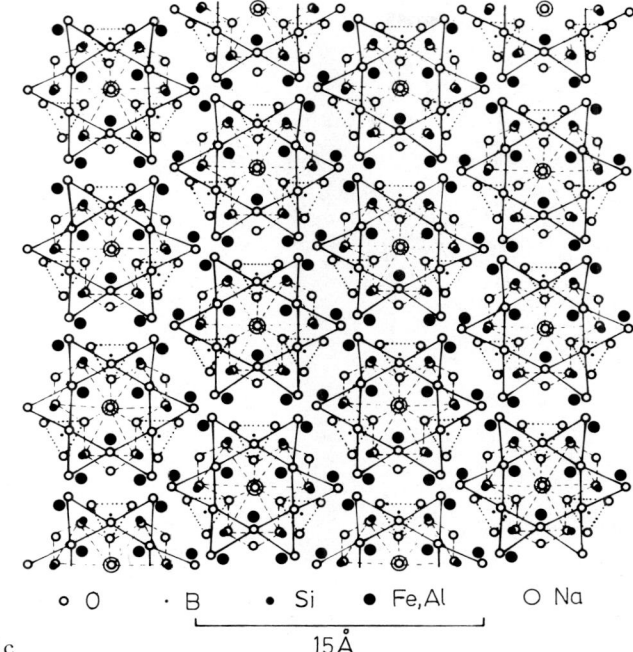

Fig. 1c

In Fig. 1b (Iijima *et al.*, 1973) — displaying a z-projection of the electrostatic potential differences in tourmaline — we observe a close correspondence to the crystal structure as determined by X-ray techniques (Fig. 1c). There are intensity variations from unit cell to unit cell which were attributed by Iijima *et al.* (1973) to short-range order which cannot be seen with X-rays. This example shows the close correspondence between X-ray and electron diffraction. X-ray data permit a more precise determination of the average electron density in the unit cell of the lattice, while electron microscopy resolves imperfections of the lattice such as local site occupancies.

This field, the goal of which is to resolve single atoms, is at present one of the frontiers of the science of electron microscopy. Some more conventional examples will also illustrate the close relation between the two techniques and some of the unique advantages of the electron microscope.

X-ray crystallographers demonstrated that certain minerals contain *submicroscopic twins and exsolution lamellae*. X-ray precession photographs of an optically homogeneous alkali feldspar (Fig. 2a), for example, could be interpreted as a mixture of albite and microcline, both twinned according to the albite twin law. Volume fractions can be estimated from the intensity of the X-ray reflections. Using the electron microscope and especially darkfield techniques, we are now in a position to study the morphology and texture of these domains (Fig. 2 from Brown *et al.*, 1972). TEM-work of McConnell (1965) and Nissen (1967) confirmed the earlier hypothesis of Laves (1950) that orthoclase contains very small domains of different crystallographic orientation which McConnell (1965)

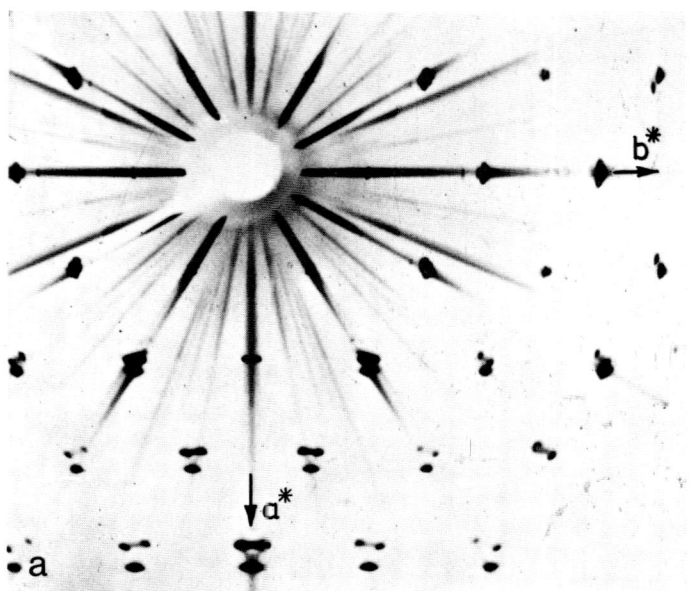

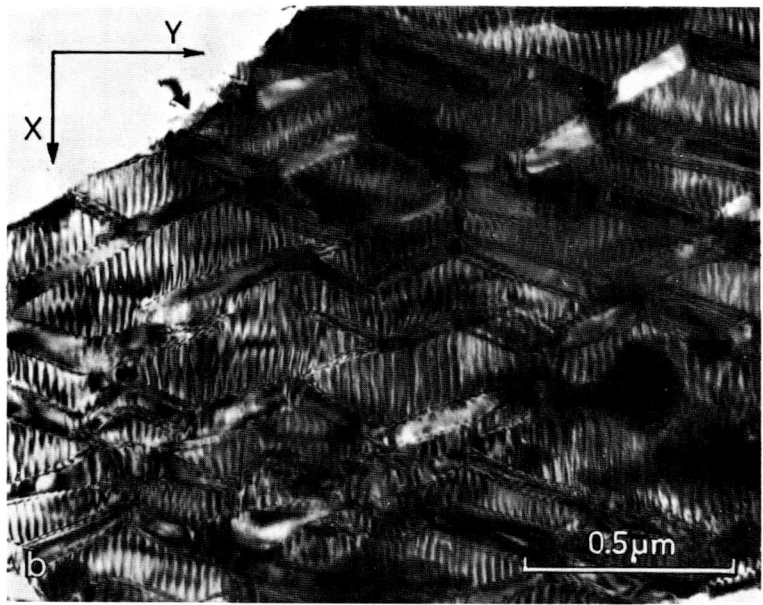

Fig. 2a and b. Exsolution in cryptoperthite. (From Brown et al., 1972.) (a) hk0 precession photograph (CuKα radiation). Double spots connected by diffuse streaks are from albite-twinning (010) in microcline, elongated reflections with larger a^* are from albite also twinned on (010). (b) TEM image of the same crystal. The narrow zigzag bands correspond to microcline in two twin orientations, the striated part is twinned albite

interprets as "waves of distortion". If these domains are larger, then they correspond to the twin texture of microcline.

Exsolution lamellae have been studied in plagioclase, in pyroxenes and in many other minerals (Section 4) which were assumed until recently to be homogeneous. These findings are of immediate consequence for geochemistry which has treated them as single phases and ideal solid solutions. The chemical composition of lamellae which are only 1 000 Å wide can now be quantitatively determined with the electron microscope microanalyzer (see review by Lorimer and Cliff in Chapter 7.3 of this book). The results from these studies are the basis of present discussions which center on the mechanisms of exsolution. Depending on the cooling history, either heterogeneous nucleation and growth or spinodal decomposition dominated.

Another structural feature which was long assumed to exist in silicate minerals though it could not be proven with X-rays is the *antiphase domain* (Chapter 2.3 and Section 5). Laves and Goldsmith (1954) and Megaw (1962) suggested domains in anorthite and bytownite which have a reverse Al/Si order, i.e. they are separated by a boundary along which adjacent units are displaced with respect to each other (Fig. 3a). They assumed the presence of these domains from the variable diffuseness and streaking of a particular group of reflections (b-reflections) in X-ray diffraction patterns. Such antiphase domain boundaries (b-APB's with a displacement vector $1/2\ [a+b]$) were imaged first by Christie et al. (1971) using bright field and dark field techniques. Fig. 3b shows c-APB's in anorthite which are attributed to positional order of Ca (Czank, 1973). APB's are obvious in lattice resolution micrographs. Fig. 3c (McLaren and Marshall, 1974) shows (111) fringes and a c-boundary in anorthite. The displacement vector is $1/2\ [a+b+c]$.

Formation of APB's is usually considered to be the result of ordering and antiphase domains are found in crystals with a superstructure. Therefore their occurrence, size and shape may give information about the cooling history of rocks. Müller et al. (1972) observed a variation in size of c-domains in anorthite from rapidly quenched volcanic rocks and annealed metamorphic crystals.

The frontispiece to this chapter shows a complex microstructure in a lunar pyroxene with two sets of exsolution lamellae and antiphase domains. A possible interpretation is that during cooling in the basaltic melt a homogeneous pyroxene crystal decomposed into augite (Ca-rich) and pigeonite (Ca-poor) thereby producing a large lamellar structure parallel to (100). At a lower temperature a second stage of exsolution occured in both the augite and the pigeonite lamellae. Augite exsolved pigeonite and pigeonite exsolved augite on a fine lamellar system parallel to (001). At an even lower temperature pigeonite underwent a symmetry change from C2/c to P2$_1$/c through positional ordering of the [SiO$_3$]-chains and this gave rise to antiphase domains. Such a complex microstructure is evidence that cooling was moderately slow and occured in several distinct stages.

The study of *dislocations* is of particular interest in deformed crystals and rocks. From the geometry of dislocations, particularly the direction of the Burgers vector, glide systems can be determined. The dislocation microstructure with such features as tangles, networks and loops gives information on the movement of dislocations which is a function of physical conditions which prevailed during

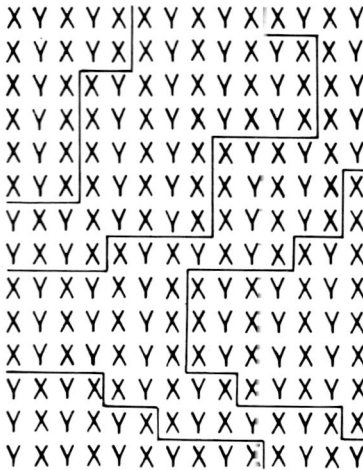

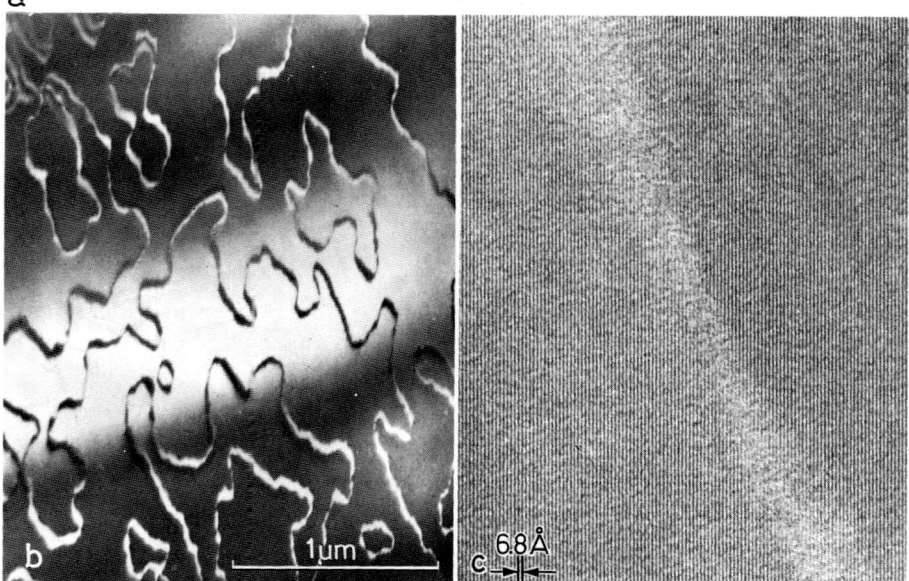

Fig. 3a–c. Antiphase domains in anorthite. (a) Diagram illustrating the antiphase relationship in feldspar. (From Megaw, 1962.) (b) c-APB's with a displacement vector $1/2\,[a+b+c]$ in a metamorphic anorthite An 94–95 (Mera 79 from the Bergell Alps) imaged in darkfield using a c-reflection. (c) Direct lattice resolution of $(1\bar{1}\bar{1})$ planes in anorthite (An 100) showing the displacement $1/2[a+b+c]$ across a c-APB. (From McLaren and Marshall, 1974)

the deformation process. The most important variables are temperature, stress, strain and time.

Metallurgists have established the mechanisms of dislocation movement (a good brief introduction is given by Hull, 1965). At low temperature dislocations move "conservatively" on slip planes, requiring only small shearing stresses. They multiply, their density increases and due to interacting forces the strain energy of the deformed crystals augments. Fig. 4a shows a high dislocation density in a low carbon steel. Such structures are typical of materials deformed plastically at low temperature ("cold-work"). On annealing or plastic deformation at higher temperature ("hot-work"), dislocations leave the slip planes and climb into positions which are closer to equilibrium such as networks (Fig. 4b). Climb is achieved by diffusion of vacancies and as a diffusion process it needs an activation energy and is temperature-controlled. During further annealing dislocations arrange on well-defined low-angle boundaries breaking up a deformed crystal into subgrains (Fig. 4c). This "recovery" is one mechanism to release strain energy which has been accumulated during deformation. Another

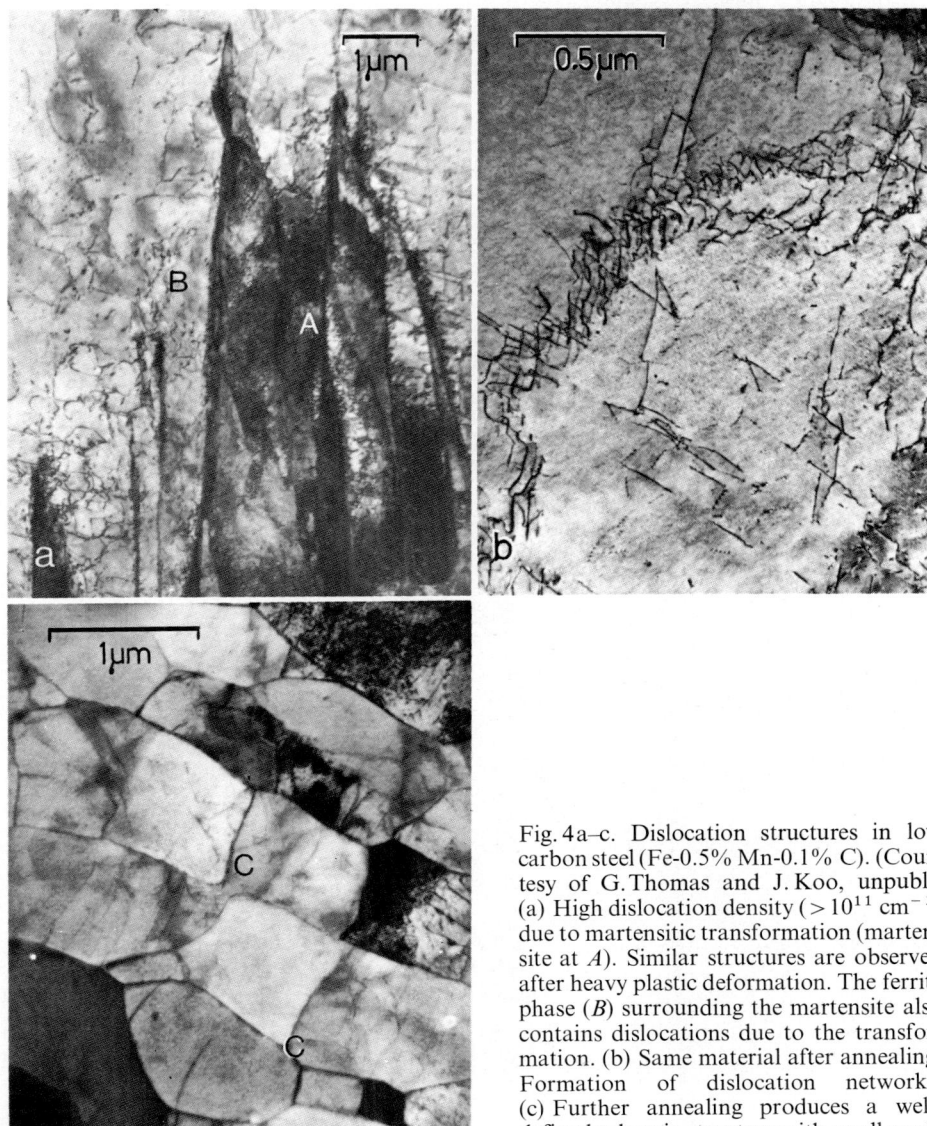

Fig. 4a–c. Dislocation structures in low carbon steel (Fe-0.5% Mn-0.1% C). (Courtesy of G. Thomas and J. Koo, unpubl.) (a) High dislocation density ($>10^{11}$ cm^{-2}) due to martensitic transformation (martensite at A). Similar structures are observed after heavy plastic deformation. The ferrite phase (B) surrounding the martensite also contains dislocations due to the transformation. (b) Same material after annealing. Formation of dislocation networks. (c) Further annealing produces a well-defined subgrain structure with small-angle boundaries. Changes in contrast at C indicate small misorientation across the walls

mechanism is by recrystallization i.e. nucleation of new dislocation-free grains particularly along grain boundaries and in heavily strained areas.

Similar processes take place in minerals (see Chapter 6.1) and the microstructure of a metamorphic quartzite from Central Australia (Frontispiece to Section 6) looks remarkably similar to a recovered steel. Analogy to metals has been successfully applied to interpret the geological history of rocks based on the deformation microstructure (see for example the papers by Liddell *et al.* on quartz, Chapter 6.4, Barber and Wenk on calcite, Chapter 6.5, and Green, Chapter 6.6 on olivine in this volume). Glide systems often change with temperature (see e.g. Fig. 1 in Green, Chapter 6.6 of this volume) but in most minerals they are inadequately known and only for a very few (e.g. olivine), have Burgers vectors been determined. Any quantitative interpretation of texture and preferred orientation in deformed rocks relies on these parameters which can best be determined with the electron microscope.

Electron microscopy has also been used to determine the age of terrestrial, meteoritic and lunar rocks. This is based on the principle that during radioactive decay nuclei release particles which penetrate the surrounding material with a high velocity and leave a visible *"track"* in the crystal which can be seen with the electron microscope, using either the scanning or transmission mode. The density of these fission tracks around radioactive inclusions is proportional to the age of the crystal and the length is a function of the particle energy. Long tracks indicating high energy have been found around inclusions of whitlockite from lunar rocks and Hutcheon and Price (1972) attribute a particularly long track to fission of ^{244}Po an element which is extinct today due to its short halflife but may still have existed 4 billion years ago. Tracks in material from the moon's surface and meteorites (Fig. 5) give information about the evolution

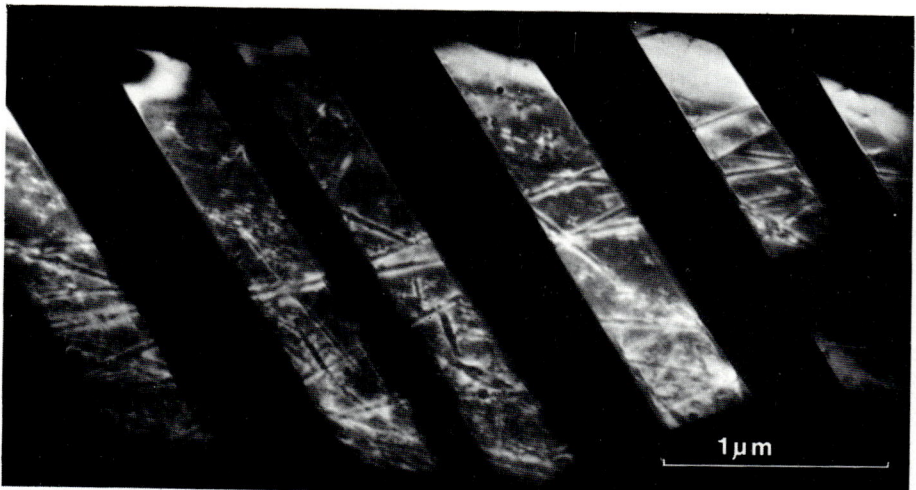

Fig. 5. Dark field TEM micrograph of a pyroxene grain from the Pesyanoe meteorite. The high track density is produced by energetic heavy nuclei (Mainly Fe) in solar flare before compaction of the meteorite. Notice lamellar structure due to exsolution. (Courtesy of J.D. Macdougall)

Introduction

of cosmic radiation and enable conclusions about the age and composition of the universe (see Hutcheon and Macdougall, Chapter 7.6 of this volume).

Recently geophysicists have paid a lot of attention to microcracks in rocks in their effort to explain changes in elastic properties shortly before earthquakes occur, and use these observations in their theories to predict earthquakes. Microcracks can be quantitatively studied with a scanning electron microscope (SEM, see Dengler, Chapter 7.8 of this volume). There is no need to mention the importance of the SEM in modern paleontology where it was instrumental in the classification of microfossils.

Many papers in this book emphasize the importance of EM in petrology. Description of domains, lamellae, and dislocations provide geological information which cannot be obtained with standard mineralogic and geochemical methods. Of main concern is the history of rocks after the crystallization of minerals and determination of such parameters as rate of cooling in igneous rocks, time and temperature of annealing, and temperature and strain-rate during deformation in metamorphic rocks.

Thus the electron microscope is applied to a wide variety of geological problems ranging from a study of the crystal structure to the dimension of the universe. Because of this it has become a standard instrument in such diverse disciplines as crystallography, mineralogy, petrology, geophysics and paleontology, linking together all the earth sciences in an even more general way than the optical microscope.

In our few examples we have emphasized comparisons between X-ray diffraction and electron diffraction. An advantage of electron microscopy is that structures and textures can be imaged in *real space* and the results are therefore easier to visualize than *diffraction patterns*. Yet at this point a call for caution is appropriate. Unlike the optical microscope, structures cannot be seen directly with the electron microscope. We merely observe contrast originating, for example, from a strain field around dislocations, and this is transformed into an image in the instrument. The interpretation of contrast has to be based on the theory of electron diffraction and cannot be accomplished by simple associations with macroscopic structures. For all quantitative calculations the crystal structure must in principle be known first and therefore electron microscopy does not replace X-ray diffraction studies. On the other hand there are many examples where electron microscopic evidence was instrumental in the interpretation of X-ray data. The two techniques complement each other ideally.

The following articles will summarize in some detail recent progress in electron microscopy concentrating in mineralogic applications. Reviews of important subjects provide a general sketch of basic achievements while original research papers illustrate the various directions in which this new science is expanding.

References

Allpress, J.G., Sanders, J.V.: The direct observation of the structure of real crystals by lattice imaging. J. Appl. Cryst. **6**, 165–190 (1973).
Amelinckx, S.: Screw dislocations in mica. Nature **169**, 580 (1952).

Beutelspacher, H., Van der Marel, H.W.: Atlas of electron microscopy of clay minerals and their admixtures. Amsterdam: Elsevier 1968.
Bollmann, W.: Interference effects in electron microscopy of thin crystal foils. Phys. Rev. **103**, 1588–1589 (1956).
Brown, W.L., Willaime, C., Guillemin, C.: Exsolution selon l'association diagonale dans une crytoperthite: étude par microscopie électronique et diffractions des rayons X. Bull. Soc. Franç. Minéral. Crist. **95**, 429–436 (1972).
Buseck, P.R., Iijima, S.: High resolution electron microscopy of silicates. Am. Mineralogist **59**, 1–21 (1974).
Christie, J.M., Lally, J.S., Heuer, A.H., Fisher, R.M., Griggs, D.T., Radcliffe, S.W.: Comparative electron petrography of Apollo 11, Apollo 12 and terrestrial rocks, Second Lunar Sci. Conf. Geochim. Cosmochim. Acta, Suppl. 2, Vol. 1, 69–89 MIT Press (1971).
Czank, M.: Strukturuntersuchungen von Anorthit im Temperaturbereich von 20 °C bis 1430 °C. Ph.D. Diss. ETH, Zürich, 74 p. (1973).
Eitel, W., Müller, H.O., Radczewski, O.E.: Übermikroskopische Untersuchungen an Tonmineralien. Ber. Deutsch. Keram. Ges. **20**, 165–180 (1939).
Hirsch, P.B., Horne, R.W., Whelan, M.J.: Direct observations of the arrangement and motions of dislocations in aluminium. Phil. Mag. Ser. 8, **1**, 677–684 (1956).
Honjo, G., Mihama, K.: A study of clay minerals by electron diffraction diagrams due to individual crystallites. Acta Cryst. **7**, 511–513 (1954).
Hull, O.: Introduction to dislocations. Oxford: Pergamon Press 1965.
Hutcheon, I.O., Price, P.B.: Plutonium-244 fission tracks: Evidence in a lunar rock 3.95 billion years old. Science **176**, 909–911 (1972).
Iijima, S., Cowley, J.M., Donnay, G.: High resolution electron microscopy of tourmaline crystals. Tschermaks Mineral. Petrog. Mitt. **20**, 216–224 (1973).
Laves, F.: The lattice and twinning of microcline and other potash feldspars. J. Geol. **58**, 548–571 (1950).
Laves, F., Goldsmith, J.R.: Long-range short-range order in calcic plagioclases as a continuous and reversible function of temperature. Acta Cryst. **7**, 465–472 (1954).
McConnell, J.D.C.: Electron optical study of effects associated with partial inversion in a silicate phase. Phil. Mag. **11**, 1289–1301 (1965).
McConnell, J.D.C., Fleet, S.G.: Direct electron-optical resolution of antiphase domains in a silicate. Nature **199**, 586 (1963).
McLaren, A.C., Marshall, D.B.: Transmission electron microscopy study of the domain structure associated with the b-, c-, d-, e- and f-reflections in plagioclase feldspars. Contrib. Mineral. Petrol. **44**, 237–249 (1974).
McLaren, A.C., Phakey, P.P.: Dislocations in quartz observed by transmission electron microscopy. J. Appl. Phys. **36**, 3244–3246 (1965).
Megaw, H.D.: Order and disorder in feldspars. Norsk Geol. Tidsskr. **42**, 104–137 (1962).
Müller, W.F., Wenk, H.-R., Thomas, G.: Structural variations in anorthite. Contrib. Mineral. Petrol. **34**, 304–314 (1972).
Nissen, H.U.: Direct electron-microscopic proof of domain texture in orthoclase (K Al Si$_3$ O$_8$). Contrib. Mineral. Petrol. **16**, 354–360 (1967).
Ribbe, P.H.: Observations on the nature of unmixing in peristerite plagioclases. Norsk. Geol. Tidsskr. **42**, 138–151 (1962).

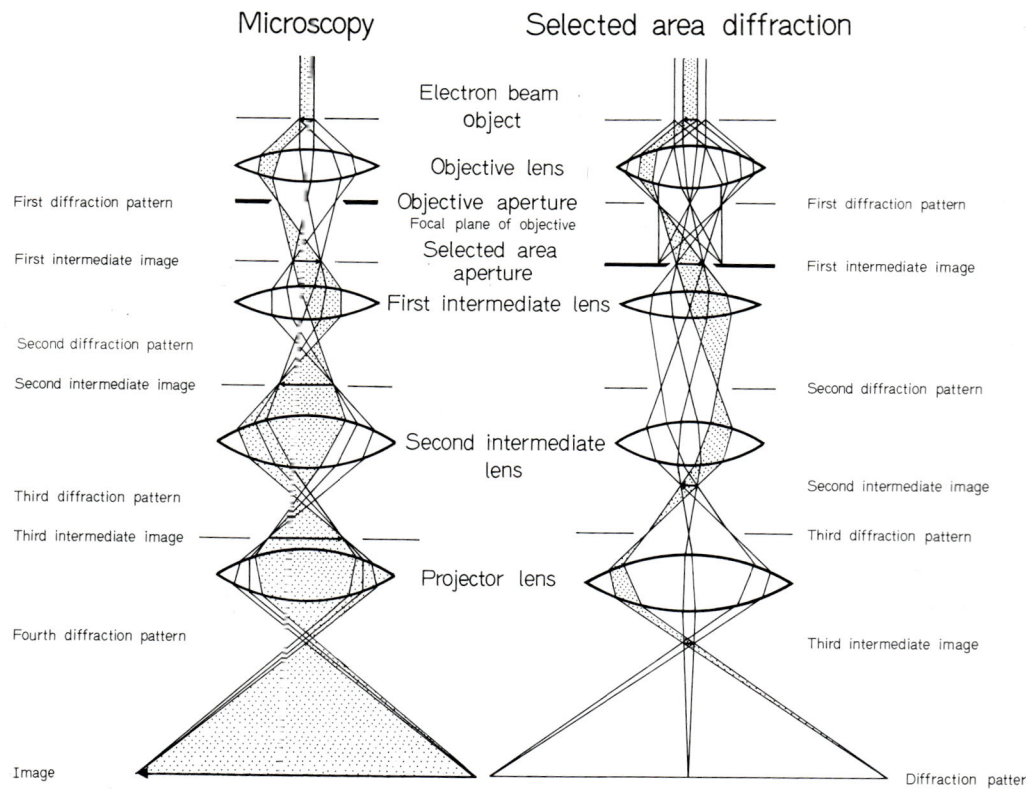

Section 2 Contrast

◀ Raypath in a transmission electron microscope. Microscopy and diffraction are compared. Dotted area indicates equivalent raypaths in the two modes. (Courtesy of Siemens)

CHAPTER 2.1

Fundamentals of Electron Microscopy

O. VAN DER BIEST and G. THOMAS

1. Introduction

It would be impossible to review adequately within one chapter the basics of electron microscopy and the reader is urged to consult the bibliography at the end for more details on the topics discussed here. It will be shown in this chapter what kind of information can be gained by using electron microscopy techniques which have become standard in metals research. The principles of these techniques will be reviewed and they will be illustrated with examples drawn from current research on ceramics and minerals in our laboratory. Wherever minerals and ceramics may present a problem because of their more complex structure, this will be indicated. Some examples of what are often called "non-conventional techniques" will also be presented.

2. Preliminaries

The information that is obtained by electron microscope methods is derived from the scattering processes that take place when the electron beam travels through the specimen. There are two main types of scattering (a) elastic—the interaction of the electrons with the effective potential field of the nuclei—involving no energy losses and which can be coherent or incoherent (poor phase relationships) (b) inelastic—the interaction of the electrons and the electrons in the specimen involving energy losses i.e. absorption. It is the elastic scattering that produces a diffraction pattern; and if the scattering centers in the specimen are arrayed in an orderly, regular manner such as in crystals, the scattering is coherent and results in spot patterns, Kikuchi patterns and, if the sample is a fine-grained polycrystal, ring patterns.

The basic reason for the utilization of the electron microscope is its superior resolution resulting from the very small wavelengths of electrons compared to other forms of radiation for which an optical system can be constructed. The resolution is given by the Rayleigh formula which is derived from considering the maximum angle of electron scattering (α) which can pass through the objective lens. This formula is:

$$R = \frac{0.61 \lambda}{\alpha} \tag{1}$$

Fundamentals of Electron Microscopy

where R is the size of the resolved object, λ is wavelength, and α is identical to the effective aperture of the objective lens.

In the electron microscope, the effective aperture is limited chiefly by spherical aberration. The spherical aberration error is:

$$\Delta S = C_s \alpha^3 \tag{2}$$

where C_s is the coefficient of spherical aberration of the objective lens ($\simeq$ focal length e.g. 3 mm).

Thus R increases with decreasing α; whereas ΔS decreases with decreasing α. As a result, in electron optics one arrives at an optimum aperture and minimum resolution given by:

$$\alpha_{opt} = A(\lambda^{1/4}) C_s^{-1/4}, \tag{3}$$

$$\Delta R_{min} = B(\lambda^{3/4}) C_s^{1/4} \tag{4}$$

where A, B are constants of order 1.

The relativistic wavelength of electrons depends on the accelerating voltage and is given by the modified DeBroglie wavelength:

$$\lambda = \frac{h}{\left[2m_0 eE \left(1 + \frac{eE}{2m_0 c^2}\right)\right]^{1/2}} \tag{5}$$

$$\lambda = \frac{12.26}{E^{1/2}(1 + 0.9788 \times 10^{-6} E)^{1/2}} \text{ (Å)} \tag{6}$$

where h = Planck's constant, m_0 is rest mass and e the charge of the electron, E is the accelerating potential (volts), c is the velocity of light, v is the electron velocity, and thus decreases with energy E. Some values pertinent to electron microscopy are given below in Table 1.

Table 1

E (volts)	λ (Å)	λ^{-1} (Å)$^{-1}$	$(v/c)^2$
100 kV	0.037	27.02	0.3005
500 kV	0.0142	70.36	0.7445
1 MeV	0.0087	114.7	0.8856
10 MeV	0.0012	846.8	0.9976

Another advantage of the small wavelength of electrons is that the depth of field and depth of focus are very large in electron microscopes.

At 100 kV $\alpha_{opt} \simeq 6 \times 10^{-3}$ rad. and $\Delta R_{min} \simeq 6.5$ Å for $C_s = 3.3$ mm. Other factors which affect resolution are astigmatism and chromatic aberration of the imaging system, and chromatic aberration resulting from energy losses in the specimen. These errors produce poor resolution for non-axial illumination.

$$\Delta C = C_c \alpha \frac{\Delta E}{E} \tag{7}$$

where C_c is the coefficient of chromatic aberration of the lens.

ΔE arises from voltage fluctuations in the incident beam, current fluctuations in the lenses, and absorption processes in the specimen. Whilst α_{opt} and ΔR_{min} change slowly with E (through λ) ΔC rapidly decreases with E and this is one of the main advantages of high-voltage operation.

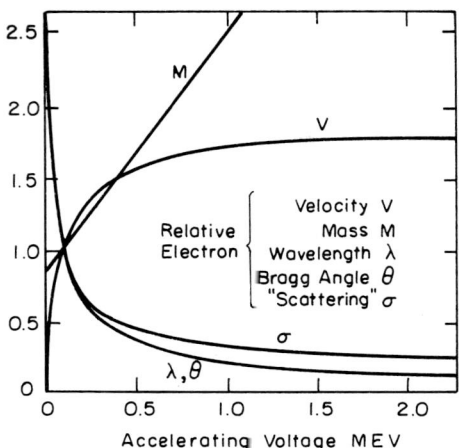

Fig. 1. Properties of electrons as a function of voltage, relative to those at 100 kV. (After Fisher, 1968)

High-voltage electron microscopy is now well established in many laboratories in different countries. High voltages are useful for several reasons (Bell and Thomas, 1972; Cosslett, 1970; Fisher, 1968; Fisher, 1972; Howie, 1970; Humphries, 1972; Lacaze and Thomas, 1973). The effective scattering cross sections decrease with energy as indicated in Fig. 1. This predicts an improvement in specimen penetration for a given level of resolution, and is especially important for the study of inorganic materials which are difficult to thin. Experimentally (Fig. 2), it is found that the most significant gain in penetration is for light elements (Si, Al) which are important in minerals. The reduction in inelastic scattering with increasing voltage implies a reduction in ionization and other damaging processes, and this has been observed for several biological and polymeric solids. However, knock-on damage occurs above a threshold energy which is roughly proportional to atomic number (e.g. $\simeq 500$ keV for copper; 1 MeV for gold). A further gain from increasing the voltage is the rapid reduction in spherical aberration. This can be seen from Eq. (2), i.e., $\Delta S \simeq C_s \alpha^3$ and since $C_s \simeq \lambda^{-1}$, $\Delta S \simeq \lambda^2$.

When an image is formed with the scattered beams, two main mechanisms of contrast arise. If the transmitted and scattered beams can be made to recombine, thereby preserving their amplitudes and phases, then a lattice image of the planes which are diffracting may be resolved directly (phase contrast). This technique is discussed in detail in this book by Cowley and Iijima (Chapter 2.5 of this volume).

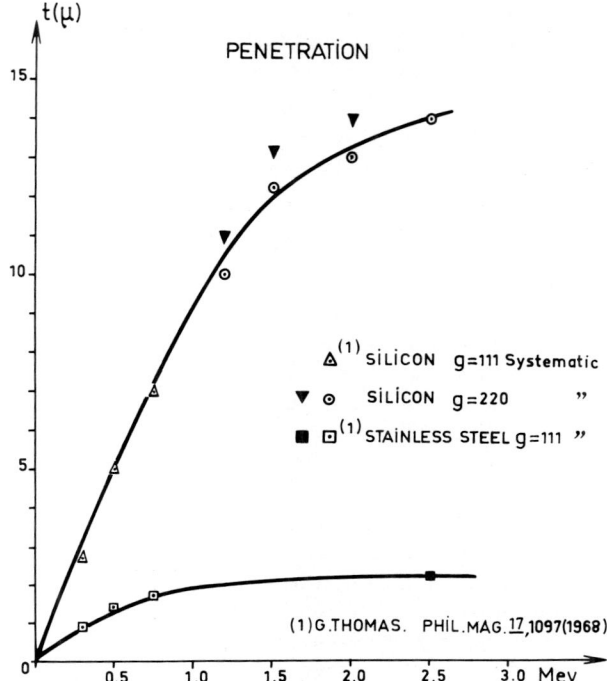

Fig. 2. Experimental data on penetration in silicon and stainless steel (Lacaze and Thomas, 1973)

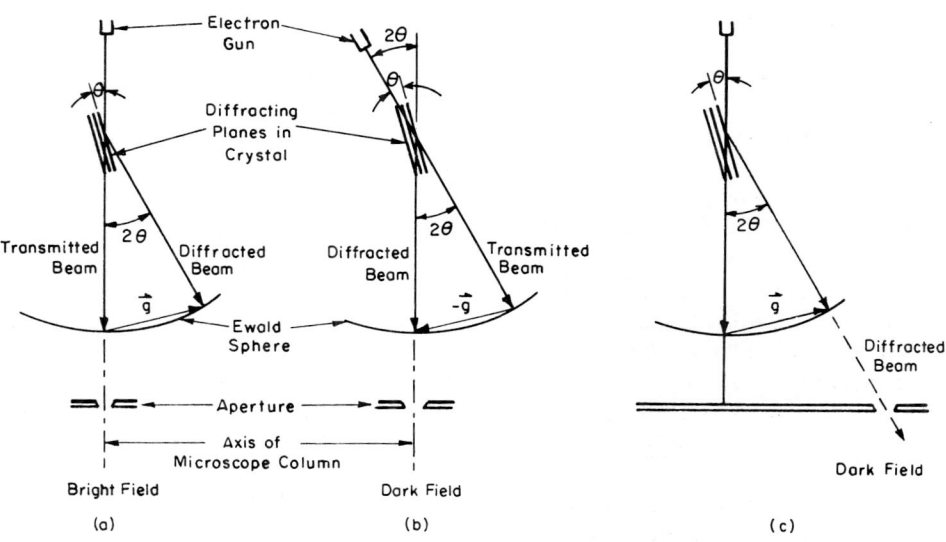

Fig. 3a–c. Amplitude contrast imaging: the phases of the transmitted and diffracted beams do not recombine. Objective apertures are used to stop off the diffracted beams to form a bright field image (a). Dark field images are obtained by gun-tilting or beam deflection (b) or with an off-axis aperture (c)

Although lattice imaging is a very powerful method with a resolution of approximately 2 Å, the more general technique of imaging is by amplitude contrast (i.e., without recombination of the phases of the transmitted and scattered waves) as is illustrated in Fig. 3 and in the Frontispiece of this Section.

Amplitude contrast is achieved in one of two ways, either (a) formation of the bright field image by removing all diffracted beams or (b) formation of the dark field image by allowing only one strong diffracted beam to form this image. These operations are carried out by means of the objective aperture which is inserted at the back focal plane of the objective lens (Fig. 3). The dark field image is best obtained by gun tilting or by deflection so as to allow the beam to pass along the optic axis (Fig. 3b), thereby reducing errors from chromatic and spherical aberrations which occur if the objective aperture is moved off the optic axis (Fig. 3c). An important point to realize is that for axial dark field the gun translation or beam deflection must be done such that the direction of g is reversed (i.e., if g is excited for bright field the corresponding gun-tilt dark field should be obtained in $-g$). This must be remembered when the correct alignment between images and diffraction patterns is made.

Fig. 4 shows that the image is a magnified picture of a diffraction spot selected by the aperture in the back focal plane of the objective lens. The intermediate lens can be focused at the back focal plane of the objective thereby allowing the diffraction pattern to be observed on the screen and photographed. If an aperture with diameter D is placed in the image plane (Fig. 4) only electrons passing through an area D/M on the specimen can reach the final screen. (M is the magnification of the objective.)

However, because of spherical aberration, the error in selection of the area for analysis can be appreciable. The minimum selected area ΔA is given by Eq. (2),

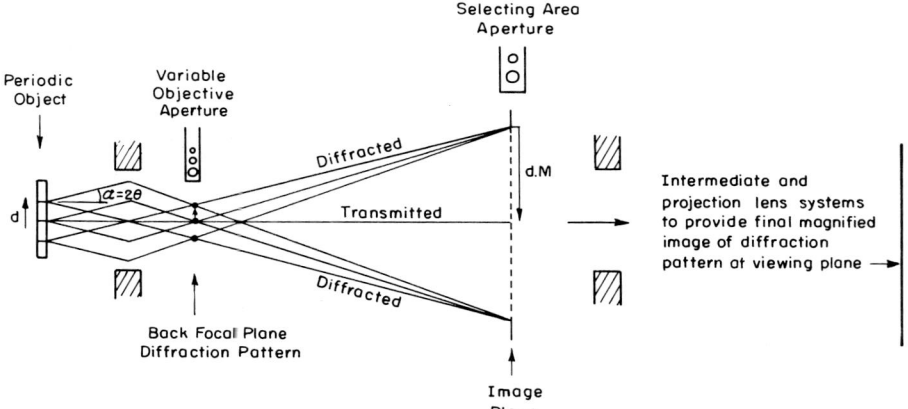

Fig. 4. Phase contrast imaging from a periodic object. The diffraction pattern is formed in the back focal plane. The period d is imaged as magnified fringes if the diffracted and transmitted beams recombine at the image plane (when $2\Theta < \alpha$). Notice the inversion between diffraction pattern and image relative to the object

replacing α by the Bragg angle Θ, i.e.,

$$\Delta A \simeq C_s \Theta^3 \simeq C_s (\lambda/d)^3 \quad \text{for small angles.}$$

Since C_s varies approximately as λ^{-1} then ΔA varies as λ^2 and thus decreases rapidly with increasing voltage. For example, $\Delta A \simeq 2\,\mu m$ at 100 and $0.02\,\mu m$ at 1 MeV. This illustrates another advantage of high voltage microscopy. The smallest size aperture D normally used is $5\,\mu m$ and considerable care must be exercised so as to align the microscope correctly with proper focusing.

3. Electron Diffraction

3.1 Geometry of Electron Diffraction

The geometry of electron diffraction can easily be understood from the familiar Ewald's sphere construction. There are two important differences between the X-ray case and electrons (1) the radius of the reflecting sphere λ^{-1} is very large and increases with accelerating voltage (see Table 1, Section 2) (2) Bragg's law does not strictly apply to electron diffraction. Alternatively, one can say that the Laue conditions for diffraction are relaxed.

For a particular reciprocal lattice vector, which is customarily denoted g, the orientation is fixed by the vector s, which is often called the excitation error (Fig. 5). When used as a scalar, s is conventionally taken positive when the reciprocal lattice point (relpoint) is inside the sphere and negative outside. As we are aiming only at a qualitative understanding of electron diffraction effects, we will make the kinematical approximations. This means that the diffracted intensity is always much smaller than the intensity in the transmitted beam and is the result of a number of independent scattering events. The kinematical approximation is a reasonable one in very thin foils and when the excitation error is large.

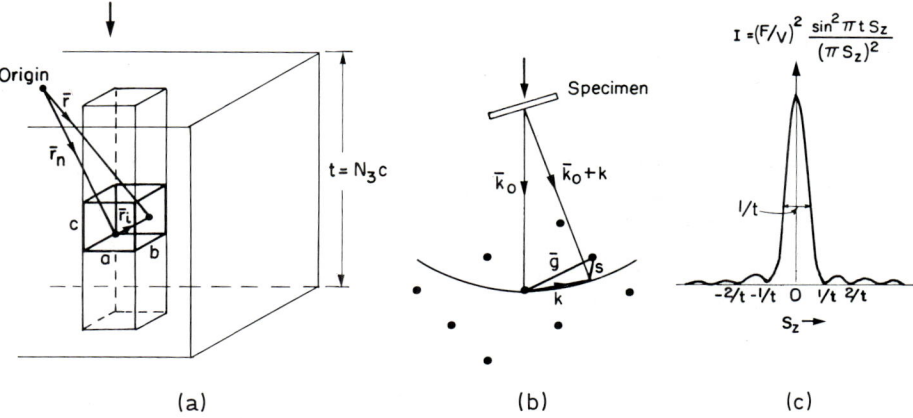

Fig. 5. (a) Sketch of a column of unit cells in a parallelpiped crystal for calculating the interference function. (b) Showing relation between the reciprocal lattice and reflecting sphere when Bragg's Law is not exactly satisfied. (c) The interference function along s_z direction showing the kinematical intensity distribution for this foil

The amplitude of the scattered wave ψ is then proportional to the atomic scattering amplitude f_i times the phase factor summed over all atoms.

$$\psi \sim \sum_{\text{atoms}} f_i \exp(2\pi i \mathbf{k} \cdot \mathbf{r}). \tag{8}$$

Consider a parallel-piped crystal made up of unit cells each $\mathbf{a} \cdot (\mathbf{b} \times \mathbf{c})$ in volume with N_1, N_2, N_3 as the number of unit cells scattering along the principal axes a, b, c. Fig. 5a.

Let $\mathbf{r}$ = the position of an atom with respect to the origin of the crystal; $\mathbf{r}_n$ = the position of a unit cell with respect to the origin of the crystal; $\mathbf{r}_i$ the position of an atom with respect to the origin of the unit cell and, in reciprocal space, $\mathbf{k} = \mathbf{g} + \mathbf{s}$ (Fig. 5b).

Thus

$$\psi \sim \sum_{\text{atoms}} f_i \exp[2\pi i (\mathbf{g}+\mathbf{s}) \cdot (\mathbf{r}_i + \mathbf{r}_n)] \tag{9}$$

now

$$\sim \sum_{\text{all unit cells}} \left[\sum_{\text{all atoms per cell}} f_i [\exp 2\pi i (\mathbf{g}+\mathbf{s}) \cdot \mathbf{r}_i] \exp 2\pi i (\mathbf{g}+\mathbf{s}) \cdot \mathbf{r}_n \right] \tag{10}$$

$$\sum_{\text{all atoms/cell}} f_i \exp[2\pi i (\mathbf{g}+\mathbf{s}) \cdot \mathbf{r}_i]$$

does not depend on the shape of the crystal, and at $s \approx 0$, this becomes the structure factor F_g (Eq. 10). Since $|\mathbf{s}| \ll |\mathbf{g}|$ the dependence on $\mathbf{s}$ is not very strong, hence

$$F_g \simeq \sum f_i \exp[2\pi i (\mathbf{g}+\mathbf{s}) \cdot \mathbf{r}_i]$$

thus

$$\psi \sim \sum_{\text{all unit cells}} F_g \exp[2\pi i \mathbf{g} \cdot \mathbf{r}_n] \exp[2\pi i \mathbf{s} \cdot \mathbf{r}_n]. \tag{11}$$

Since $\mathbf{g} \cdot \mathbf{r}_n$ = integer, $\exp 2\pi i (\mathbf{g} \cdot \mathbf{r}_n) = 1$.

Also the quantity $\mathbf{s} \cdot \mathbf{r}_n$ does not change appreciably from cell to cell. Thus by approximating the sum by an integral

$$\psi = F_g \int_{\text{crystal}} \frac{1}{V_c} \exp(2\pi i \mathbf{s} \cdot \mathbf{r}_n) \cdot dV \tag{12}$$

where V_c = volume of the unit cell.

By definition $\mathbf{r}_n = u\mathbf{a} + v\mathbf{b} + w\mathbf{c}$

$$\mathbf{s} = s_x \mathbf{a}^* + s_y \mathbf{b}^* + s_z \mathbf{c}^*$$

$$\psi \cong \frac{F_g}{V_c} \int_0^{N_1 a} \int_0^{N_2 b} \int_0^{N_3 c} \exp(2\pi i(s_x x + s_y y + s_z z)) \, dx \, dy \, dz \tag{13}$$

which approximates to:

$$\psi \sim \frac{F_g}{V_c} \frac{\sin(\pi s_x N_1 a)}{(\pi s_x)} \frac{\sin(\pi s_y N_2 b)}{(\pi s_y)} \frac{\sin(\pi s_z N_3 c)}{(\pi s_z)}. \tag{14}$$

This is known as the interference function.

Fundamentals of Electron Microscopy

For a crystal in the form of a thin plate $N_3 \ll N_2, N_1$ and for $s_x = s_y = 0$

$$\psi \sim \frac{F_g}{V_c} \frac{\sin(\pi s_z N_3 c)}{(\pi s_z)}, \tag{15}$$

$N_3 c = t$ the thickness of the plate, thus the intensity is

$$I \sim \left(\frac{F_g}{V_c}\right)^2 \frac{\sin^2(\pi t s_z)}{(\pi s_z)^2}. \tag{16}$$

This well known function is shown in Fig. 5c.

For values of $|s|$ smaller than $\frac{2}{t}$ the diffracted intensity will still be quite high.

One can adapt the Ewald's sphere construction by replacing each relpoint with a reciprocal lattice rod (relrod). These relrods have an axis perpendicular to the surface of the foil with length $\frac{1}{t}$. As the part of the crystal which contributes to the diffraction pattern is not infinite but limited by the selected area aperture, the dimensions of the rods in the direction parallel with the foil surface are also finite. Whenever the reflecting sphere cuts through a rod, diffraction will occur in the direction of the points of intersection. This simple geometrical model is very useful when interpreting diffraction patterns (see Thomas et al., 1965).

One of the consequences is that if the crystal is very thin, it is possible for relrods from the upper and lower levels of the reciprocal lattice to extend sufficiently to cut the reflecting sphere giving spots at positions which do not correspond to allowed reflections. In order to check for this effect, one should merely move into thicker parts of the foil, when such spots will disappear. Other diffraction effects and the identification of diffraction patterns will be discussed in the paper by Gard (Chapter 2.2 in this volume).

3.2 Identification of Phases Using Selected-area Diffraction

Compared with conventional metallographic and petrographic microscopes, the electron microscope has not only the advantage of increased resolution, but it also allows one to identify uniquely the phases present in complex microstructures; for example through an analysis of their diffraction pattern and the matching of lattice constants with published values (e.g. ASTM files). In ceramics and minerals, coexisting phases often have a similar anion arrangement but different cation distributions. The difference in lattice parameter between two phases may then be very small and the spots due to second phases will be close to the matrix spot. An extreme case is shown in Fig. 6 where matrix and precipitate spots are so close together that they cannot be separated even at high order reflections. The matrix has the spinel structure with lattice parameter $a = 8.33$ Å and the precipitate has the sodium chloride type structure with a lattice parameter very close to one half the value for the spinel structure. In the diffraction pattern the precipitate spots coincide with every other spot along a reciprocal lattice direction in the spinel matrix.

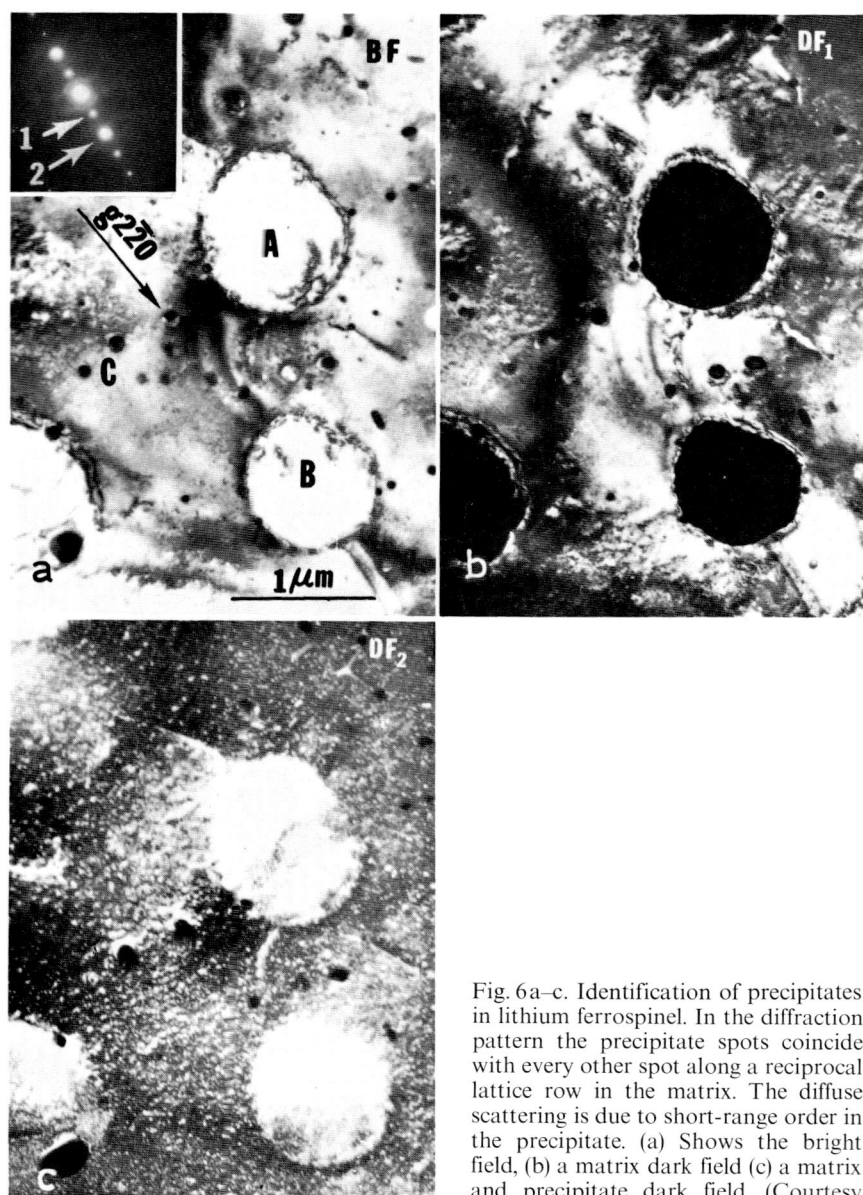

Fig. 6a–c. Identification of precipitates in lithium ferrospinel. In the diffraction pattern the precipitate spots coincide with every other spot along a reciprocal lattice row in the matrix. The diffuse scattering is due to short-range order in the precipitate. (a) Shows the bright field, (b) a matrix dark field (c) a matrix and precipitate dark field. (Courtesy R. K. Mishra)

The selected-area diffraction technique allows one to isolate a diffracted beam using an aperture in the back focal plane of the objective lens. Magnification of this diffracted beam permits to observe those areas in the foil which contribute to the diffracted intensity. Fig. 6b is a dark field of spot 1 in which the precipitates A, B, C, show up dark because they do not contribute to this diffracted

Fundamentals of Electron Microscopy

beam (the diffraction pattern is inset in Fig. 6a). Spot 2 is common to the spinel matrix and the precipitate hence in a dark field image of this spot precipitate and matrix show up bright.

From the diffraction pattern one can establish the orientation relationship (if any) between the matrix and the precipitate. This orientation relationship is specified by listing the crystallographic planes and directions parallel in matrix and precipitate e.g. in the case of precipitation (exsolution) of hcp in an fcc matrix $(0001)_{hcp} \| (111)_{fcc}$ and $[2\bar{1}\bar{1}0]_{hcp} \| [1\bar{1}0]_{fcc}$ is normal. When the matrix has a high symmetry, many orientation variants may occur e.g. the (0001) plane of hcp may be parallel with any of four {111} planes in fcc. The presence of several orientation variants may complicate the analysis of a diffraction pattern. Here again the dark field technique has to be used in order to determine from which variant different spots originate.

Second phases can also be detected through the strain they cause in the matrix or through the characteristics of the interface with the matrix e.g. in Fig. 6a interface dislocations are visible. These contrast mechanisms will be discussed in Section 4.

3.3 Kikuchi Patterns

Kikuchi patterns are produced in relatively thick and perfect crystals. They are formed as a result of coherent scattering of the inelastically scattered electrons in the upper parts of the foil and appear as pairs of parallel lines in the diffraction image (Fig. 7).

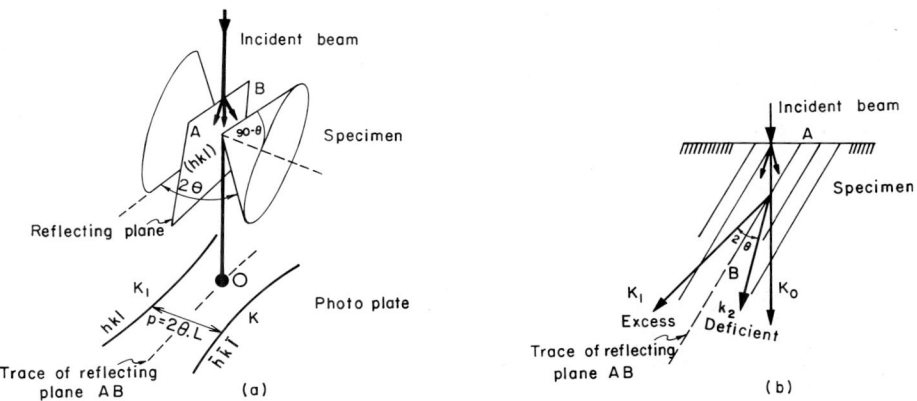

Fig. 7a and b. Geometry of formation of Kikuchi lines. (a) Incident beam is inelastically scattered. Inelastic electrons are then rescattered coherently by plane *AB* to produce cones of radiation with intersect the reflecting sphere as slightly curved lines, which bisect the reflecting plane. Each Kikuchi pair belongs to a unique reflecting plane. (b) If the reflecting plane is inclined with respect to the incident beam more electrons are scattered into the direction K_1 at the expense of the electrons scattered in the direction K_2. (Courtesy N. Holland publishers; Thomas, 1970)

When the reflecting plane is inclined with respect to the incident beam, the process of coherent scattering is to change those electrons which have been initially inelastically scattered in the direction K_2, into the direction K_1. As a result, a dark line will be present near the incident beam and a bright line will be observed near the diffraction spot (i.e. on the screen or a positive print). If the reflecting plane is parallel with the incident beam, the two lines have about equal intensity. Since the energy losses of the inelastically scattered electrons are typically of the order of 10–100 eV, the change in wavelength is negligible. Hence the pairs of Kikuchi lines always subtend an angle of exactly 2Θ, irrespective of the foil position. The Kikuchi lines lie normal to the particular g and are related to the spot pattern as shown in Fig. 8. In thin specimens no Kikuchi lines will be observed as the number of inelastically scattered electrons generated is too small.

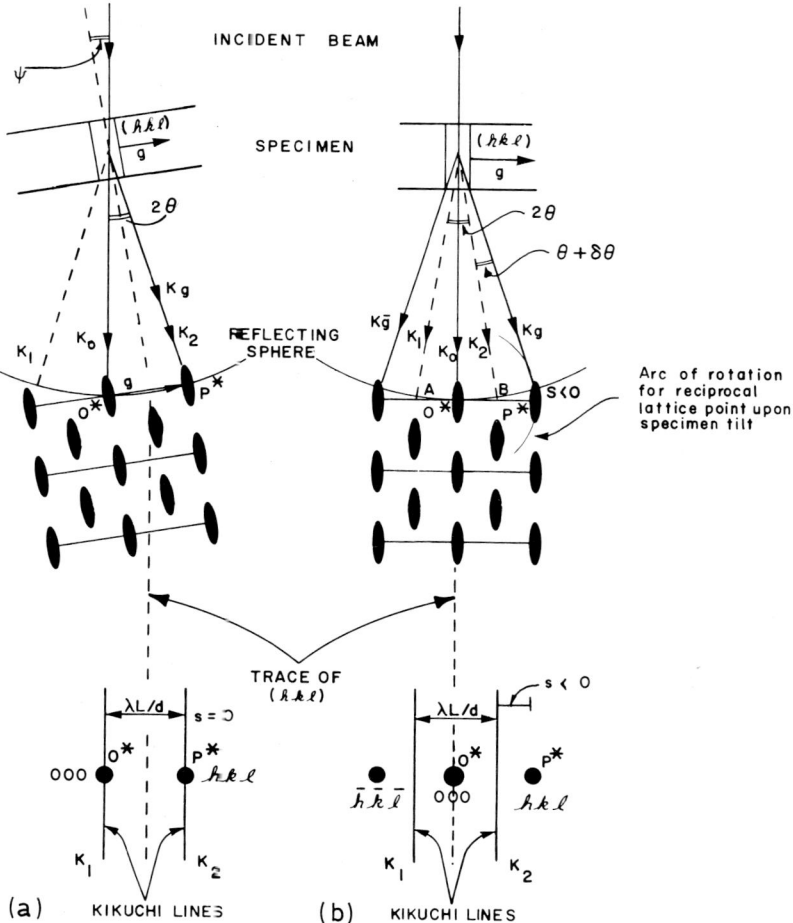

Fig. 8a and b. Reciprocal lattice-reflecting sphere construction showing relation of Kikuchi pattern to spot pattern for (a) exact Bragg two-beam orientation and (b) the symmetrical orientation. (Courtesy Physica Status Solidi **12**, 354, 1965)

As the thickness of a specimen is increased, Kikuchi line patterns are observed in addition to spots; with further increase in thickness, only Kikuchi patterns will be observed until eventually in very thick specimens no coherent diffraction will occur because of complete absorption.

For electron microscopy, there are two important orientations of the specimen (Fig. 8): (a) the systematic Bragg reflection, where the Bragg angle for one g is satisfied and (b) the symmetrical orientation, where the reflecting planes are parallel with the incident beam.

The former is a prerequisite for contrast work and the latter for orientation determinations. Both are easily achieved if a goniometric tilt specimen holder is used. As can be seen from Fig. 8, these two positions are related simply by a tilt of Θ if in (a) the first order Bragg reflection is excited. The position of the Kikuchi line, with respect to the spot pattern, determines the exact relationship between diffracting planes in the crystal and the direction of the incident beam. Thus, it is possible to obtain the orientation of a foil to better than 0.1°, which is more accurate than by the X-ray Laue method. Also, the position of the Kikuchi pattern relative to the spot pattern determines the sign and magnitude of the excitation error and gives the direction and tilt of the specimen with respect to the beam. The tilt angles can thus be calibrated by measurements of Kikuchi line shifts. For an electron microscope with $\lambda L \simeq 2$ Å cm a tilt of 1° of the foil corresponds to a shift of Kikuchi lines of about 1 cm on the plate.

The Kikuchi lines can be easily indexed from the symmetry of the patterns or from measurements of their widths since the latter are related to the d-spacings through the camera constant equation,

$$\lambda L = p_1 d_1 = p_2 d_2 = p_n d_n$$

or $\quad \dfrac{p_1}{p_2} = \dfrac{d_2}{d_1}$

where p_1, etc., is the width of a parallel hkl Kikuchi pair corresponding to a spacing $d(hkl)$. Therefore, the ratios of the spacings enable the lines to be indexed just as for a spot pattern. As can be seen from Fig. 7, the center of the Kikuchi pair is the trace of the reflecting plane in the diffraction pattern and these planes must make the precise angles to each other as required by the crystallographic relations for the particular crystal. The indexing is thus easily checked by measuring these angles. An example is given in Fig. 9 for silicon. The solution is given in Fig. 9b. Measurement from the plates yield

$$\frac{p_1}{p_2} = 1.73; \quad \frac{p_1}{p_3} = 1.2.$$

The indexing given in Fig. 9b is consistent with these ratios and with the interplanar angles indicated in the figure. The poles A, B, C, can then also be indexed. These indices can be used to determine the scale factor of the pattern, e.g. the angle between the A pole [215] and the B pole [114] is measured to be 4.05 cm. The calculated angle is 8.5° hence the scale factor is 2.1°/cm.

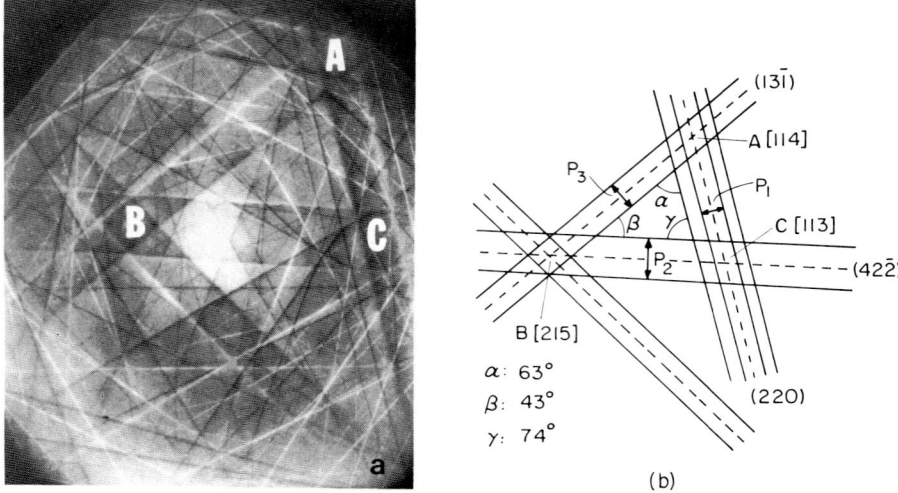

Fig. 9. (a) Kikuchi pattern from a thick silicon single crystal. (b) Solution of the Kikuchi pattern in (a)

Indexing and orientation analysis is greatly facilitated by the use of Kikuchi maps as described in great detail elsewhere (see Thomas, 1970 for review). Orientations are identified by comparing the unknown pattern to the relevant map.

Because orientations can be determined very accurately, one of the most obvious advantages of using Kikuchi patterns is in determining orientation relationships, misorientations due to sub-boundaries, and other detailed crystallographic information (Okamoto and Thomas, 1968; Thomas et al., 1965; Thomas, 1965).

4. Contrast of Perfect Crystals

4.1 Kinematical Theory

Qualitatively, the contrast from perfect crystals can be understood from the kinematical theory. As explained earlier, the kinematical theory is applicable only to thin specimens and for conditions away from the exact Bragg position ($|s| \gg 0$). In Section 2 the kinematical intensities were calculated assuming all the energy is conserved:

$$\text{Diffracted:} \quad (\psi_d)^2 = \left(\frac{F}{V_c}\right)^2 \frac{\sin^2 \pi t s}{(\pi s)^2}.$$

$$\text{Transmitted:} \quad |\psi_t| = 1 - |\psi_g|^2. \tag{17}$$

The kinematical theory predicts that bright and dark field images are complementary. In practice, absorption occurs and this symmetry property is modified.

The effect of absorption on image contrast can only be adequately treated by the dynamical theory of diffraction. However, the kinematical theory is very useful in a qualitative way and as shown in the following can account for many effects.

4.1.1 Bragg Contours, Bend Contours (s Fringes)

A crystal may be oriented at the Bragg angle (for a certain set of planes) with respect to the optic axis of the microscope but, unless the beam is perfectly parallel, the entire irradiated area will not be at the Bragg angle. For a divergent or convergent incident beam, there will be a continuous range of incident angles from one side of the Bragg angle to the other. From (17) it follows then that in a foil of constant thickness periodic intensity oscillations will occur with changing angle of incidence (changing s). The resulting contours (loci of constant s) are called Bragg contours. A variation of s will also occur when the foil is bent or strained, a common occurrence in the electron microscope. The resulting contours are called bend or extinction contours.

4.1.2 Thickness Fringes

In an analogous way, for a constant angle of incidence, the intensity of the diffracted and transmitted beam will vary periodically with thickness. The period $t_0 = 1/s$ is called the kinematic extinction distance. The resulting contours (loci of constant t) are called thickness fringes.

4.2 Dynamical Theory

The kinematical theory of electron diffraction breaks down near the exact Bragg condition ($s \simeq 0$) and for thick foils. These two conditions usually prevail in routine electron microscopy work.

In the dynamical theory, electrons in the diffracted beam can be scattered back into the forward direction. As the electron travels down the crystal there is a continuous "dynamical" interaction between the two (or more) beams. The scattered intensity can become equal in magnitude to the incident intensity.

The contrast features described in kinematical theory can also be explained using the dynamical theory. The dynamical theory will not be reviewed here as there are numerous extensive reviews available in the literature (e.g. Howie, 1970). We will summarize some important results, valid for the two-beam case unless otherwise indicated.

In the dynamical theory the extinction distance ξ_g is defined as $\pi V/\lambda F_g$ where V is the volume of the unit cell and F_g the structure factor for the particular reflection; the deviation parameter $w = \xi_g s$ is used to measure departures from the exact Bragg condition. For thickness fringes the period of the intensity oscillation with thickness at the exact Bragg condition is equal to the extinction distance. At every integral number of extinction distances all the electrons are moving in the forward direction (transmitted) and in every odd half multiple extinction distances, all electrons are moving in the diffracted direction. The exchange of electron intensity between the transmitted and diffracted beams is exactly analogous to the motion of two coupled harmonic oscillators which periodically exchange all the vibrational energy of the system. This forms the basis of the dynamical theory. For Bragg or extinction contours, the theory shows that subsidiary maxima occur when $(s^2+(\xi_g)^{-2})t^2$ is integral. Absorption can be formally introduced into the dynamical theory by the substitution of

$$\frac{1}{\xi_g} \to \frac{1}{\xi_g} + \frac{i}{\xi'_g}$$

ξ'_g is called the absorption length. One of the important results of the two-beam dynamical theory for perfect crystals is that the transmitted intensity $|\psi_0|^2$ has a maximum for w slightly positive. In the dark field $|\psi_g|^2$ is maximum for $w=0$.

From the foregoing we expect to produce contrast effects in an otherwise perfect crystal due to the following:

1. Changes in thickness t – wedge fringes, fringes at inclined defects,
2. Changes in excitation error s – Bragg contour fringes,
3. Changes in orientation – changes in s and g.

Thus, e.g in polycrystals the intensity varies from grain to grain because of differences in diffracting conditions. Since the contrast is very orientation-sensitive, it is essential to use a goniometric specimen holder, preferably with as large a tilting range as possible so that the diffracting conditions can be varied in a systematic manner. Without such a stage, quantitative characterization of microstructure is almost impossible.

5. Contrast in Imperfect Crystals

5.1 General Comments

In real materials we have to consider and characterize complicated microstructures including the following effects:

1. Changes in orientation with or without change in structure or composition, e.g., grains, twins, precipitates.
2. Lattice defects: point defects, line defects, planar defects, volume defects (effects due to elastic displacements).
3. Phase transformations: (a) changes in composition but not structure (e.g. spinodals), (b) changes in composition and structure (general precipitation) (c) changes in structure but not composition (e.g. martensites), (d) interphase interfaces (coherent, partially coherent, incoherent).

The contrast from these will arise from such effects as changes in the local diffracting conditions {changing s and g (d-spacings)}, changes in phase factor on crossing interfaces, structure factor changes, changes in effective thickness (changing ξ_g). The situation can become quite complex especially when the defect density is high and strain fields overlap as in the case of heavily deformed crystals, or crystals containing large volume fractions of precipitated particles.

Fundamentals of Electron Microscopy 33

The combination of bright and dark field imaging techniques and diffraction pattern analysis is essential in the characterization procedure. Analysis should always start from the diffraction pattern and most of the interpretation will be carried out *at the microscope*. For contrast work it is recommended that two-beam orientations be used, i.e. that only the transmitted beam and one strong diffracted beam are visible in the diffraction pattern (see, however, Section 5.5). For this reason it becomes essential to recognize orientations by inspection so that particular reflections of interest can be brought into operation. It is convenient to start by tilting the foil into a recognizable symmetrical orientation and then tilting from there. The use of Kikuchi maps greatly facilitates this process. It cannot be emphasized too strongly how important it is to have a strong working knowledge of diffraction patterns and three-dimensional crystallography.

5.2 Information Requirements for Analysis (Two-beam Conditions)

Generally we need to know the foil orientation, direction of the diffraction vector, sign of s, and foil thickness.

Because of the 180° ambiguity in spot patterns, the spot pattern by itself does not give the unique foil orientation and hence the geometry of defects in the foil are not known since the image is a two-dimensional projection of the object. Thus the top of the foil is not distinguishable from the bottom, nor up from down. Special absorption contrast effects such as the asymmetry in the dark field image when s is not quite zero, can be utilized to solve this problem. At $s>0$ the bottom of the foil is in stronger contrast than the top (Fig. 10); the reverse is true at $s<0$ (see Bell and Thomas, 1965). Alternatively, large angle tilting experiments can be performed to observe the change in projected size of an object in the foil.

Once the orientation is known, the sense of slope of crystallographic planes and directions can be derived. This information is needed for quantitative analyses involving the determination of the sense of strain fields (e.g. vacancy or interstitial loops or faults).

Stereomicroscopy is also useful, because by this technique one can obtain information on the depth distribution of defects. Stereo pairs can be obtained by tilting 8–10° along a Kikuchi band so that the diffracting conditions are not altered.

In orienting the diffraction pattern with the image, due to the inversion between the image and the diffraction pattern, rotate the diffraction pattern 180° (plus magnification rotation) with respect to the image, with the negatives both emulsion-side up [1]. Crystallographic data can then be transferred directly from the pattern to the image. It is recommended that this be done directly on the non-emulsion side of the negative where it can be wiped off later after prints are made. In this way the geometry is preserved with minimum confusion, and the correct sense of the diffraction vector *g* in the image is retained. *g* is of course identified from the diffraction pattern and the region in the specimen corresponding to this *g* will reverse contrast in the dark field image of this reflection.

---

[1] This rotation is clockwise for Siemens Elmiskops, it is anticlockwise in the JEM 7. Thus it must be checked for each type of microscope.

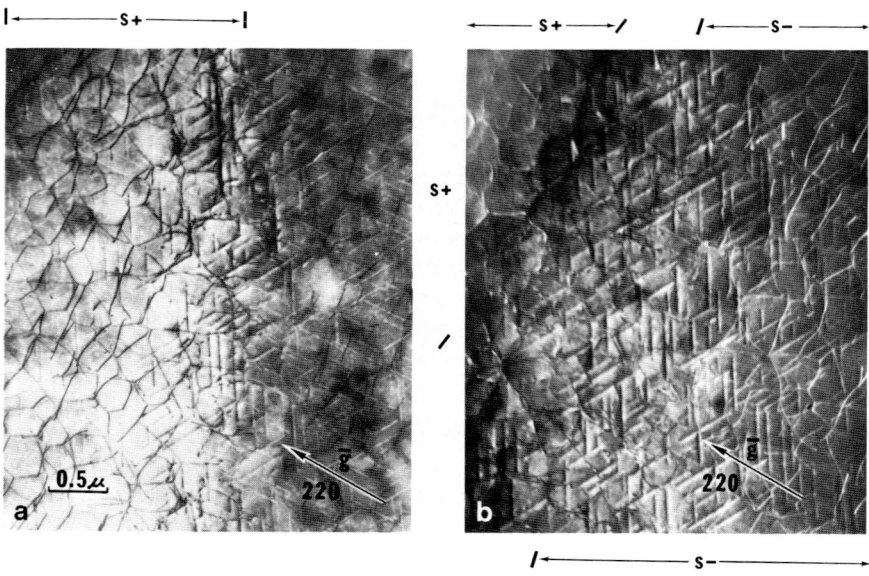

Fig. 10a and b. Bright (a) and dark field (b) images of boron diffused silicon, containing dislocations near the bottom of the foil and precipitates in the form of small rods near the top. In bright field maximum intensity is at $s>0$ and in dark field at $s=0$. The contrast is also maximum at $s>0$ in bright field but is depth dependent with s in dark field. The precipitates are more visible at the top of the foil $s<0$ and the dislocations more visible at the bottom of the foil $s>0$ in dark field. Thus dark field images at $s \neq 0$ are very useful for analyzing the depth dependence of defects

The sign of the deviation parameter s is important in several instances. In a two-beam absorbing case the intensity in bright field is a maximum for $s>0$ and in dark field at $s=0$. These conditions are thus readily seen directly in the image (Fig. 10). In the diffraction pattern for $s>0$ the Kikuchi line will lie to the outside of the corresponding spot since the reciprocal lattice point will be lying inside the reflecting sphere. For $s<0$ the Kikuchi lines lie inside the spot as for example in symmetrical orientations (Fig. 8).

The foil thickness can be found in several ways, e.g. trace analysis of projected defects that go completely through the foil (faults, twins, precipitates, or from measurements of subsidiary fringes either in the convergent beam pattern or from Bragg contour fringes (Siems *et al.*, 1962). Recently a general method has been developed by Von Heimendahl (1973) in which one or two latex balls of known diameter are applied to both foil surfaces in the area viewed. Changes in dimensions are observed after a known tilt and from the geometry the thickness can be calculated with approximately 4% accuracy.

5.3 Visibility of Lattice Defects: General Criteria

Defects can be described in terms of translational vectors which represent displacements of atoms from their regular positions in the lattice. If the general displacement vector is $R(x, y, z)$, the kinematical amplitude scattered from the crystal as a whole becomes

$$\psi \sim \int_{\text{crystal}} \exp[2\pi i(g+s) \cdot (r_n+R)] \, dt \tag{18}$$

or

$$\psi \sim \int_{\text{crystal}} \exp[2\pi i s \cdot r_n] \exp[2\pi i g \cdot R] \, dt \tag{19}$$

since $g \cdot r_n =$ integer, and neglecting $s \cdot R_n$ [refer to Eq. (11)].

Thus the amplitude scattered by the perfect crystal is modified by the phase factor $2\pi g \cdot R = n 2\pi$ and n can be integral, zero or fractional. The case $g \cdot R = 0$ is particularly important in contrast work. It means that R lies in the reflecting plane (hkl), hence d (and thus $|g|$) is unaltered, so that the path difference between transmitted and diffracted waves is unaffected by R. Since g is normal to the lattice plane (hkl), $g \cdot R = 0$, is the condition for no contrast at a point (x, y, z) due to a displacement R. It should be emphasized that R for a general defect varies with position.

The magnitude of $g \cdot R$ must be sufficient to change the intensity such that contrast is detectable over background (about 10% is enough). For example, for dislocation line defects in crystals $g \cdot b > \frac{1}{3}$ if the lines are to be detectable.

Examples of defects which are of general interest in the study of crystals and which can be described in terms of such displacements are: dislocations, dislocation loops, coherent volume defects (e.g. point defect clusters, voids, coherent particles), planar defects such as stacking faults, domain boundaries due to chemical, magnetic or electric order, twin and grain boundaries, interphase interfaces etc.

The visibility of these defects can all be understood in terms of the simple $g \cdot R$ criterion. However, the detailed interpretation of contrast behavior, such as the intensities, and variations in contrast with depth in the foil, influence of excitation errors especially in dark field (Fig. 10), influence of other reflections, and the analysis of the sense of the displacements associated with the defects, require the application of the dynamical theory.

5.4 Planar Defects

The simplest type of crystalline defect is the stacking fault. This case also covers antiphase domain boundaries, which are stacking faults in an ordered structure. For this type of defect $R(x, y, z)$ is a step function i.e. $R = 0$ for parts of the crystal above the stacking fault and below the fault R is a constant vector, independent of position, called the displacement vector of the fault. A fault inclined with respect to the surface of the foil will show up as a series of fringes, parallel with the line of intersection of the fault plane or surface with the foil surface. The depth periodicity of these fringes is one half the extinction distance ξ_g. If a reflection is used for which ξ_g is very large, as is often the case for antiphase domain boundaries, then only one fringe may be observed. The properties of these fringes are determined by the value of the phase angle $\alpha = 2\pi g \cdot R$ which modifies the transmitted and diffracted intensities at the bottom of the crystal relative to their values for a perfect crystal at the top. When $\mathbf{g} \cdot \mathbf{R} = n$ with n zero or an integer, the fault will be invisible. However, due to the translational symmetry of the crystal, one can always add a lattice vector to the displacement vector, without altering the physical configuration around the fault. The integer is thus in fact not uniquely determined. The displacement vector can be determined if three linearly independent reflections $\mathbf{g}_i$ can be found for which the fault is invisible. Then we can write:

$$\mathbf{g}_i \cdot \mathbf{R} = n_i \quad i(1, 2, 3) \tag{20}$$

where n_i must be zero or an integer.

In order to keep the integers on the right-hand side completely arbitrary, it is necessary to restrict the $\mathbf{g}_i$ used in these equations to the lowest order $\mathbf{g}$ along a systematic row for which a fault goes out of contrast. Otherwise, one restricts the values of the integers e.g., if $\mathbf{g}_i \cdot \mathbf{R} = n_i$ then $(2\mathbf{g}_i) \cdot \mathbf{R} = 2n_i$ and the integer on the right-hand side can only be even.

At high voltages, it is not always simple to determine along a systematic row of reflections the lowest order reflection for which the fault goes out of contrast. Because of dynamical interaction between lower and higher order reflections one may observe a fault, although the phase angle $\alpha = 2\pi \mathbf{g} \cdot \mathbf{R}$ is zero for the reflection which is used to form the dark field image (see Section 6).

One can predict the following properties of the solutions of Eq. (20), where the n_i are now arbitrary integers: (1) Every lattice vector will be a solution of these equations. This solution is of course a trivial one. (2) If $\mathbf{R}$ is a solution of the system then $-\mathbf{R}$ is also a solution, and these two solutions are not necessarily equivalent unless the difference between the two vectors (i.e. $2\mathbf{R}$) is also a lattice vector. (3) A linear combination of two solutions is also a solution. An example is provided in Fig. 11, where the fault A is invisible for $\mathbf{g} = 1\bar{3}1, 11\bar{3}$ and $1\bar{1}\bar{1}$. Hence we can write:

($\mathbf{R} = [uvw]$ and p, q, r are zero or integers)

$$u - 3v + w = p$$
$$u + v - 3w = q$$
$$u - v - w = r.$$

The solution of these equations yields:

$$\mathbf{R} = \left[\frac{-p - q + 2r}{2}, \frac{-2p - q + r}{4}, \frac{-q - r}{4} \right].$$

For $p = 1, q = 1$ and $r = 0$ this yields:

$$\mathbf{R} = \left[\bar{1}, \frac{\bar{3}}{4}, \frac{\bar{1}}{4} \right] = [\bar{1}00] + \frac{1}{2} [0\bar{1}\bar{1}] + \frac{1}{4} [0\bar{1}1].$$

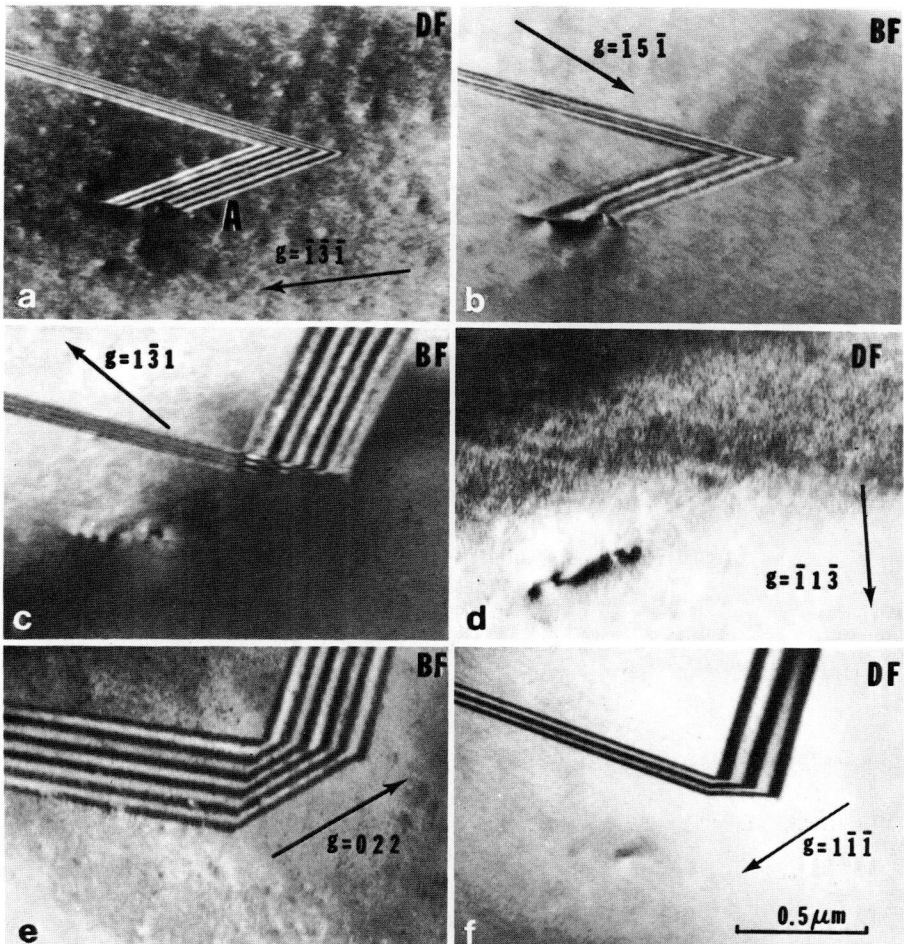

Fig. 11a–f. Three faults meeting at a junction in fully ordered lithium ferrite. Only spinel reflections have been used in these photographs. These diffraction experiments allow one to determine the displacement vector of these faults and the Burgers vector of the dislocation confining fault A (Van der Biest and Thomas, 1974)

The first two vectors at the right-hand side are lattice vectors (The crystal has a spinel structure with an fcc lattice). Thus the smallest displacement vector describing the fault is $1/4[0\bar{1}1]$. In this case R is equivalent to $-R$ as $2R$ is a lattice vector of the structure. This is not generally the case and one will be called an intrinsic fault, due to removal of a plane of atoms and the other will be called extrinsic due to insertion of an extra plane of atoms. The sign of R will be opposite in both cases and hence the sign of $\alpha = 2\pi g \cdot R$ will be reversed for the same g. By convention the top half of the crystal is at rest and the bottom half is displaced by the fault displacement vector.

The dynamical theory predicts the dependence of the intensity of the fringes on the sign of α (for $\alpha \neq \pi$). In bright field for α positive the first fringe is light whereas for α negative the first fringe is dark. The reverse is true for dark field images although the effect of absorption modifies the symmetry such that the fringes are complementary only at the lower surface of the foil. These rules can be used to determine whether a fault is intrinsic or extrinsic. In general, it is

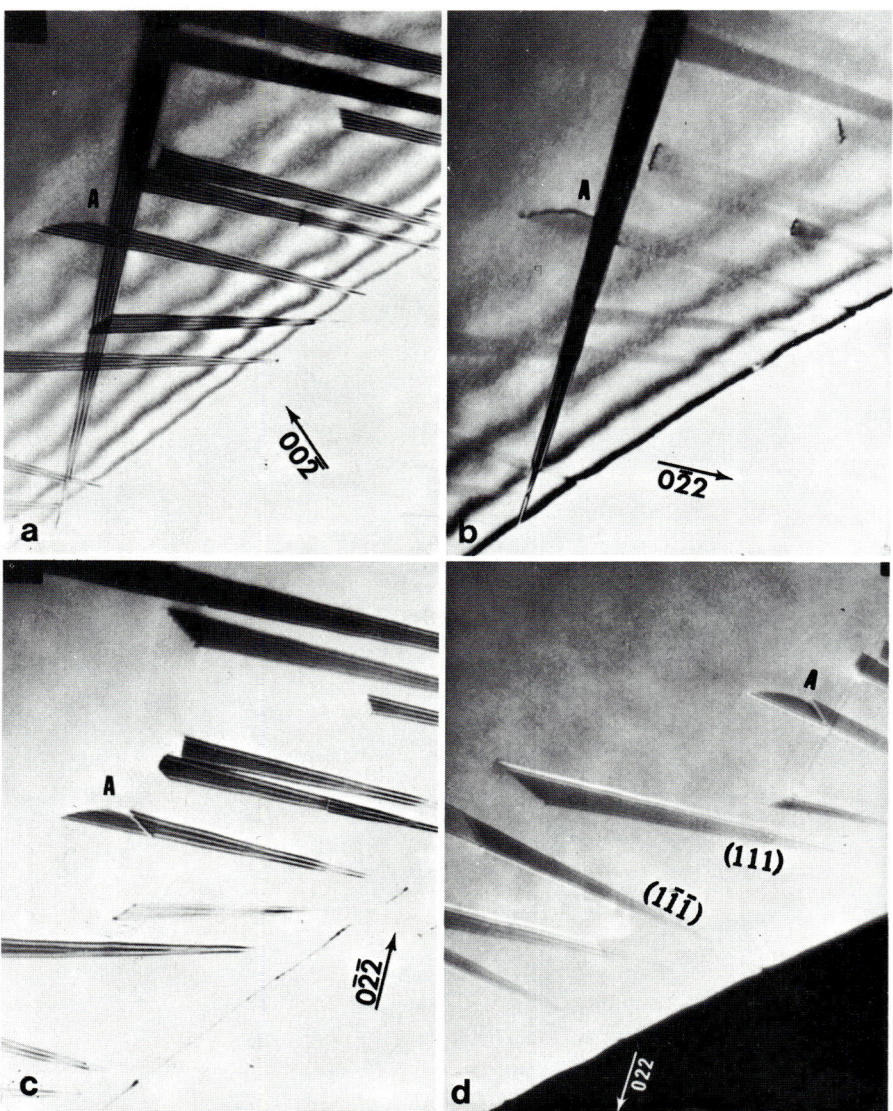

Fig. 12a–d. Contrast experiments for faults in $TaC_{0.8}$ (a–c) bright field images showing the shear nature of the faults $\{b=\langle 112 \rangle\}$. The faults do not completely vanish in (b–d), indicating local changes in composition (structure factor contrast). The dark field image in (d) shows that (111) and (1$\bar{1}\bar{1}$) are intrinsic and extrinsic respectively

essential to know the sense of slope of the fault. For structures with an fcc lattice and faults with displacement vector $\pm\frac{a}{3}[111]$, rules have been worked out so that it is not necessary to know the sense of slope (Gevers et al., 1963). The type of stacking fault can be determined from a dark field picture and the direction of *g* (allowance has to be made for optical rotation). For reflections of type {111}, {220} and {400} the rule is as follows: if the origin of the *g* vector is placed at the center of the dark field image of the fault, *g* points away from the light fringe if the fault is intrinsic and towards it if the fault is extrinsic. In reflections of type {200}, {222} and {440} the opposite rule applies.

An example is given in Fig. 12, showing stacking faults of type $\frac{1}{3}\langle 111\rangle$ in TaC, which has a sodium chloride type structure. Application of this rule to Fig. 12 shows that the faults on (111) are intrinsic and the fault on (1$\bar{1}$1) is extrinsic. The figure also illustrates the symmetry properties of α fringes in bright and dark field.

Consider two overlapping faults of the same kind (Fig. 13). If these are close together the phase factors add, giving a net phase shift of $2\pi/3 + 2\pi/3 = -2\pi/3$, i.e. the color of the first fringe changes at the point of overlap. If three faults overlap, the phase change is $2\pi =$ zero and no contrast occurs. If the faults are far apart, the outer fringes will be of the same color (Fig. 13).

Similarly if two overlapping intrinsic-extrinsic faults exist, they are close together $\alpha = +2\pi/3 + (-2\pi/3) = 0$ and no contrast occurs. However if they are far enough apart, the outer fringes will be of opposite color and the center part of the overlap will have weak or zero contrast where the two phase shifts cancel (Fig. 13).

Other types of planar boundaries are possible. Boundaries between enantiomorphic pairs, or more generally inversion boundaries, are also imaged as α

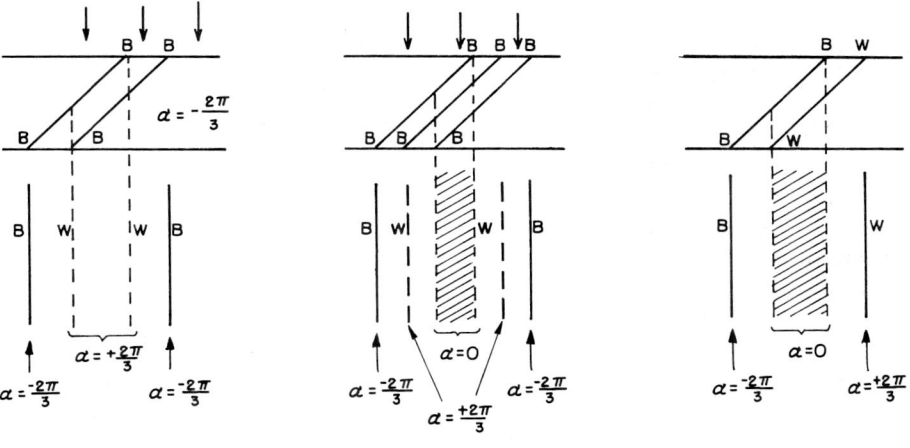

Fig. 13. Scheme showing contrast expected from overlapping faults in fcc crystals with $\alpha = \pm 2\pi/3$. *B* and *W* are dark and light fringes respectively

fringes. The value of α is then given by $\alpha = \alpha^1 - \alpha^2$ where α^1 and α^2 are the phase angles in the structure factor expressions $F_g^1 = |F_g^1| \exp(i\alpha_g^1)$ and $F_g^2 = |F_g^2| \exp(i\alpha_g^2)$, calculated with respect to the same origin (see also Van der Biest and Thomas, 1975). The superscripts 1 and 2 refer to the structures respectively above and below the plane of the boundary. This procedure to evaluate α can be applied whatever the nature of the geometrical operations characterizing the boundary may be. In order to have a pure α boundary it is of course necessary that $|F_g^1| = |F_g^2|$. In the specific case of an enantimorphous pair $|\alpha| = |\alpha_g^l - \alpha_g^r|$. An example of the analysis of boundaries between enantiomorphs in ordered lithium ferrite is given in Fig. 14. The ordered structure can exist in both a left-handed ($P4_332$) and a right-handed ($P4_132$) arrangement. In addition, four translation variants exist for each enantiomorph. As a result one may expect seven different boundaries: three pure translation boundaries, a pure inversion boundary (the point of inversion is considered to be fixed) and three boundaries involving an inversion and a translation. The values of $|\alpha|$ for the reflections used in Fig. 14 are given below in Table 2.

Table 2

g	T_1	T_2	T_3	I	$I+T_1$	$I+T_2$	$I+T_3$
$0\bar{1}2$	π	0	π	$\pi/2$	$\pi/2$	$\pi/2$	$\pi/2$
$\bar{1}10$	0	π	π	π	π	0	0
$0\bar{1}1$	π	π	0	π	0	0	π
$\bar{1}01$	π	0	π	π	0	π	0

The boundaries can be identified from their visibility or invisibility in different diffraction conditions using Table 2 (Fig. 14a–d). The results are indicated in Fig. 14a. The translational variants are distinguished by a number, the enantiomorphs by a letter L or R. The atomic configuration of one domain was assumed. The configuration of all the others follow from the character of their boundaries, in a self-consistent manner. Inversion boundaries can be recognized quickly in dark field when one takes advantage of violations in Friedel's law in non-centrosymmetric crystals in certain multi-beam orientations (Serneels et al., 1973).

Fig. 14e and f were taken with the diffraction conditions shown in Fig. 14g, and illustrate this method of analysis. The fringes in Fig. 14e running from right to left are thickness fringes. They remain continuous across the boundaries in the bright field picture. In dark field, however (Fig. 14f) these fringes change contrast at some boundaries (e.g. at the boundaries marked a and c, but remain continuous across others (e.g. at j and g). Comparison with Fig. 14a shows that the latter are translation boundaries, whereas the former are inversion boundaries.

If the orientation of g changes by a small amount $\Delta s = g_1 - g_2$ across the boundary, where g_1 and g_2 are the g vectors above and below the boundary, δ fringes are produced which have symmetry properties different from α fringes (we owe this terminology to Amelinckx, 1970). Specifically in bright field, the fringe profile will not be symmetrical whereas in dark field it is. The properties of these fringes have been reviewed by Amelinckx (1970, 1972). If Δs becomes very large so that only one g is excited, then the boundary will be visible as thickness fringes such as those occurring at grain boundaries.

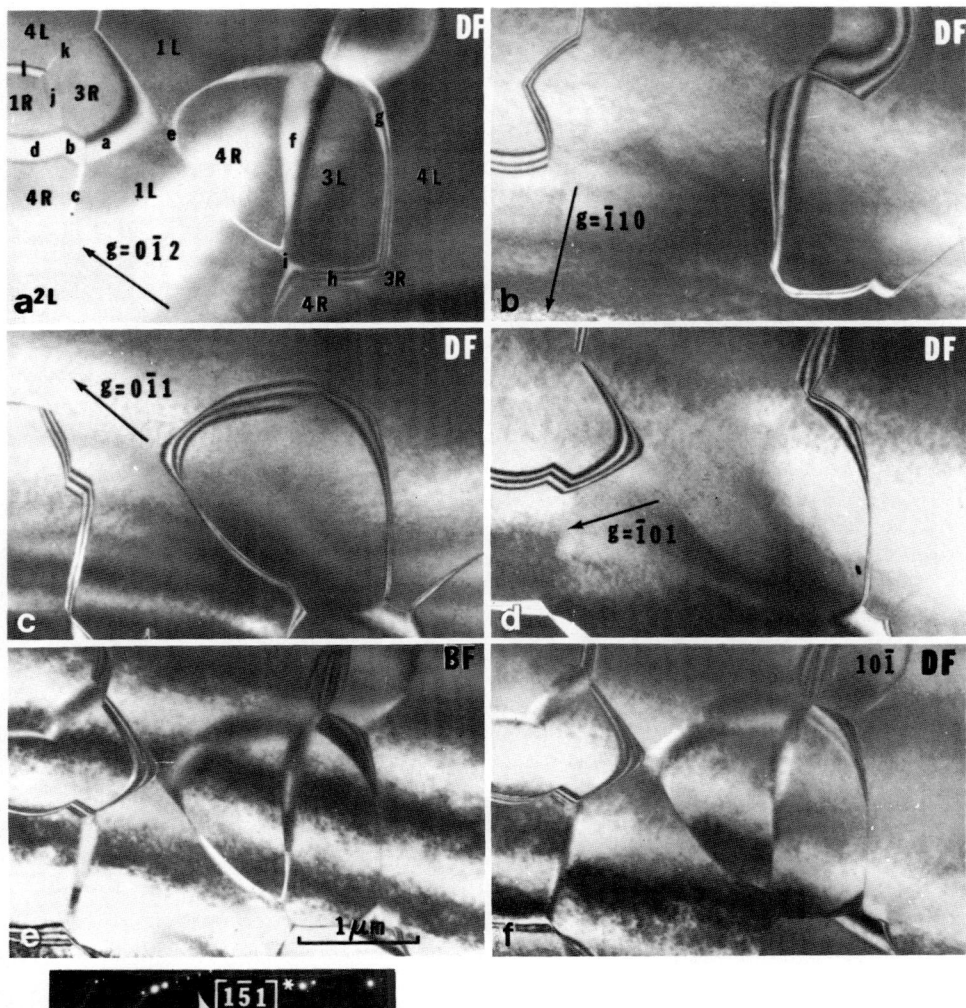

Fig. 14a–g. An identical area of crystal photographed under five different diffraction conditions. The operating reflections in Fig. 4a–d are indicated by vectors. The diffraction pattern corresponding to Fig. 4e and f as shown in Fig. 4g (BF bright field; DF dark field with reflection used indicated)

5.5 Dislocations

The strain field associated with a segment of dislocation line is determined by the direction of the line, parallel with a unit vector u, and the Burgers vector of the dislocation b. For a screw dislocation u and b are parallel and the displacement field $R(x, y, z)$ has only components parallel with b. Hence when for a screw dislocation $g \cdot b = 0$, $g \cdot R = 0$ and the dislocation is out of contrast. For a pure edge dislocation, u and b are at right angles. The displacement field has two components, one (R_b) parallel with b, the other component (R_n) parallel with the vector product of u and b i.e. normal to the glide plane determined by u and b. The latter component is the smaller one. An edge dislocation may not be completely out of contrast when $g \cdot b = 0$. There will be some residual contrast, unless g is parallel with u so that $g \cdot R_n$ is also equal to zero. However, the $g \cdot b = 0$ criterion can still be used to determine the direction of the Burgers vector by determining those g vectors for which contrast is much weaker than in other reflections.

For perfect dislocations b is a lattice vector and $g \cdot b = n$ where n is an integer or zero. For $g \cdot b = 1$ the dislocation will show up as a line, parallel with but not coinciding with the core of the dislocation. For $g \cdot b = 2$ (two-beam conditions) the dislocation will show up as two lines parallel with the core of the dislocation so that in principle studies of the image in different reflections enable the magnitude of b to be determined as well as its direction. If two-beam conditions cannot be realized, the magnitude of $g \cdot b$ can be determined using computer simulation (see Section 6 and Fig. 18). In minerals with complex structures the image contrast will also be affected by the elastic anisotropy. Again this will have to be evaluated using detailed computer calculations (see McCormick, Chapter 2.4 of this volume).

For partial dislocations, $g \cdot b$ can take non-integral values and some complexities arise due to dynamical effects.

In the fcc structure especially when anisotropy is considered it has been shown that whereas $g \cdot b = \pm \frac{1}{3}$ is always an invisibility criterion, the case for $g \cdot b = \mp \frac{2}{3}$ leads to visibility or invisibility depending on the sign of s and the position in the foil (thickness dependence). These difficulties have been discussed by Clarebrough (1971). The Burgers vector of a partial dislocation is restricted by the displacement vector of the fault that it confines and this will aid in its determination.

An example of a Burgers vector determination is given in Fig. 11 for the dislocation confining fault A. The dislocation is out of contrast for $g = 1\bar{1}\bar{1}$ and $g = 022$. Hence b is parallel with the direction given by the cross product of these two vectors i.e. the $[0\bar{1}1]$ direction. Previously it was found that the displacement associated with fault A is $\frac{1}{4}[0\bar{1}1]$ plus a lattice vector. From the $g \cdot b$ analysis this can be reduced now to $\frac{1}{4}[0\bar{1}1] + \frac{n}{2}[0\bar{1}1]$. The Burgers vector of the dislocation will be the shortest one of these, hence $b = \pm\frac{1}{4}[0\bar{1}1]$.

The position of the dislocation image with respect to the actual line depends on the sign of s. This can be seen geometrically by considering Fig. 15. In (a) we set $s > 0$ and assume an edge dislocation oriented as shown. We see that if g points to the right the sense of tilt of the planes on the LHS cause greater deviation $s \gg 0$ from diffraction whereas on the RHS the reverse is true. In this case the image will thus appear to the RHS. Similarly, if the foil is tilted slightly to make s negative, the image flips to the LHS. Thus on crossing an extinction contour (s changing sign) the image changes position. The same effect is true when g is changing sign (for s invariant).

Thus without pre-defining the sense of the Burgers vector, as long as the sign of s and the direction of g are known, it is possible to reconstruct in the foil the orientation of the dislocation which gives rise to the observed image shift when

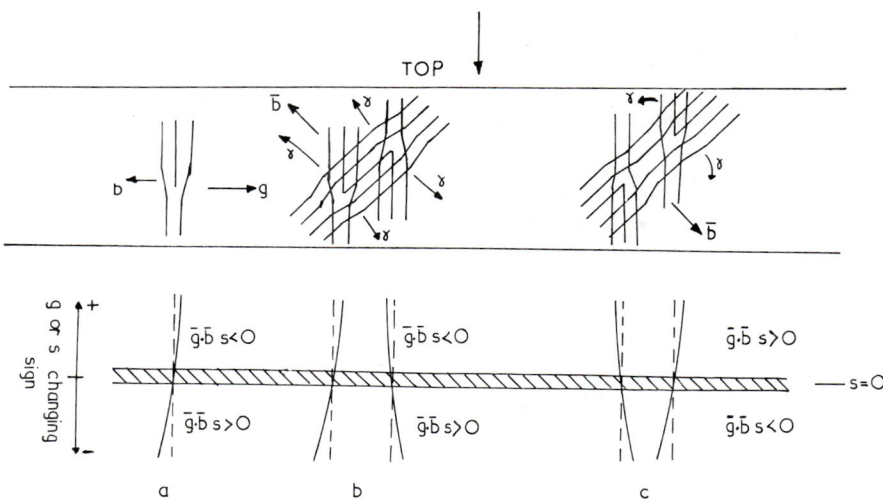

Fig. 15a–c. Scheme showing that a dislocation locally tilts reflecting planes closer to or further away from the Bragg condition on opposite sides of the extra half plane. The images are therefore to one side of the true position of the dislocation. (b) and (c) show the contrast behavior for an interstitial type (b) and a vacancy type (c) loop

g or s changes sign. This result has useful applications e.g. to distinguish between dislocation dipoles (two closely spaced dislocations of opposite Burgers vector) and a super-dislocation (two closely spaced dislocations with the same Burgers vector).

The dislocation images in the dipoles will shift in opposite directions whereas the dislocations comprising the super dislocation will shift in the same direction when the sign of s is changed. The same technique can be used to determine whether a dipole or dislocation loop has interstitial or vacancy character (Fig. 15b and c).

5.6 Small Volume Defects

In the early stages of a phase transformation, second phases in the form of small particles will often show up through the coherency strains that are set up in the matrix. The strain around each particle depends on its shape and the elastic properties of the matrix. Small dislocation loops, smaller in diameter than the image width of a dislocation under the conditions used, will show a similar contrast. A feature common to all strain contrast images is "the line of no contrast" (LC). In the simple case of a spherically symmetric strain field, LC is perpendicular to the operating reflection passing through the center of the defects as $\boldsymbol{g} \cdot \boldsymbol{R} = 0$ for this line. For a spherically symmetric strain field, this line of no contrast will rotate as $\boldsymbol{g}$ changes and this provides a means of checking if the strain field is symmetrical. In other cases, the line of no contrast tends to follow the symmetry of the strain field. An example of this type of contrast is shown in Fig. 16. These are strain contrast images of coherent cuboid particles of wustite

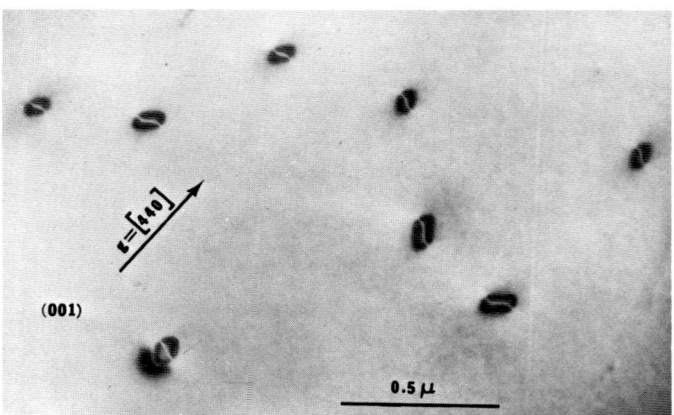

Fig. 16. (Co,Fe)O precipitates as small square prisms, bounded by {100} planes, when cobalt ferrite is heated for 10 min at 1230 °C in air. Note the particular shape of the line of no contrast when $g = 440$. (Courtesy L.C. DeJonghe)

in a cobalt ferrospinel matrix. Contrast of this type of particle was calculated for a limited number of diffracting conditions by Sass et al. (1967).

The details of this type of contrast, even in the simple case of spherically symmetrical strain fields, is rather involved, and the reader is referred to the original papers (Ashby and Brown, 1963 a, b). Quantitative results regarding the sign of the strain field and the amount of strain are possible.

Recently, the lattice image technique has been successfully employed actually to resolve small particles (Phillips and Tanner, 1973). This technique is very valuable for the analysis of complex situations in which strain fields overlap. The resultant strain patterns in amplitude-contrast images are very complex and in many cases completely mask the individual particles. Such contrast is referred to as tweed or basket-weave. Progress is being made in understanding these images by computational techniques (Fillingham et al., 1972) and lattice imaging (Phillips and Tanner, 1973). Very recently lattice imaging has been successfully applied to the study of complex inorganic crystals (see Chapter 2.5 of this volume by Cowley and Iijima) and to early stages of phase transformations in alloys (Schneider et al., 1974).

6. Many Beam Effects and Contrast at High Voltages

In the previous discussion of contrast in electron microscope images it was assumed that a two-beam condition was realized, characterized by the reciprocal lattice vector g and its deviation parameter s or w. It may not be possible to achieve this situation for two reasons: (1) Ceramics and minerals have often rather large lattice parameters, of the order of 10 Å or more. Hence diffraction spots are spaced very close together and even at 100 kV accelerating voltage, a two-beam condition cannot be realized except for a few selected directions in reciprocal space. (2) Because ceramics and minerals are often difficult to thin, thick foils have to be used and examined at high voltage (300 kV and higher). At these voltages the sphere of reflection becomes much flatter (Section 2) so that many beams are excited. Furthermore the electron-scattering factors increase with voltage, due to the increase of the relativistic mass (Fig. 1), so that the diffracted intensities are higher. These factors mean that two-beam theory no longer applies and it is more difficult or impossible to predict contrast from simple geometrical arguments. However, if cautiously applied, the invisibility criteria for stacking faults and dislocations in $g \cdot R = 0$ and $g \cdot b = 0$ can still be useful if one chooses systematic orientations. This means that only reflections along a reciprocal lattice row are excited i.e. ... $-2g$, $-g$, 0, g, $2g$, $3g$ ... so that the invisibility condition is valid for every reflection along the row. Using these systematic orientations one can still determine Burgers vectors, displacement vectors and the symmetry of strain fields. However the details of the contrast cannot be accounted for by simple theories. Many-beam dynamical theory has to be used. This usually involves computer calculations if possible coupled with image simulation techniques (Head et al., 1973). Fig. 17 shows an example of a six-beam calculations for a $\frac{1}{2}[111]$ type antiphase boundary in anorthite. The calculations show that the boundary is also visible in the $22\bar{2}$ dark field even though the phase angle $\alpha = 2\pi$ for this reflection. This has also been observed (Müller et al., 1973; see also Fig. 20). One may conclude that at high voltages it will be difficult to find the first order g for which a fault is invisible, when the fault is visible for another reflection along the row. This is due to dynamical interactions between the reflections excited.

Although the interpretation of electron micrographs becomes less straightforward, due to these dynamical effects, controlled tilting experiments in the electron microscope, coupled with computer calculations and (or) image simulation, becomes an even more powerful technique to extract quantitative information out of electron micrographs.

Dislocation image characteristics under systematic diffracting conditions have also been found useful for determining the magnitude of the product $g_1 \cdot b$ where g_1 is the first-order reflection of the systematic set and thus can be of tremendous help when it is necessary to choose between a number of different possibilities, and for examining crystals with a complex structure. Fig. 18 shows an example of the determination of the magnitude of the Burgers vector of a dislocation in cobalt ferrite. The analysis is based on comparing observed and calculated image profiles for different values of $g_1 \cdot b$. 12 beams along the systematic row have been taken into account. Other parameters which have to be determined or estimated are the deviation parameters, the absorption lengths, the foil thickness and position of the dislocation in the foil. The calculated intensity profiles are compared with a microphotometer tracing across the image. The microphotometer signal is proportional to the logarithm of the intensity rather than the intensity. In addition, a number of factors have been ignored in the calculation, the most prominent one being the diffuse scattering. These are not expected to influence the maxima and minima in the intensity profiles, so that when comparing calculated and experimental intensity profiles, instead of focusing on intensities, it is better to match the number and spacings of the fringes. In Fig. 18 the direction of the Burgers vector was determined to be parallel with $[10\bar{1}]$ and under

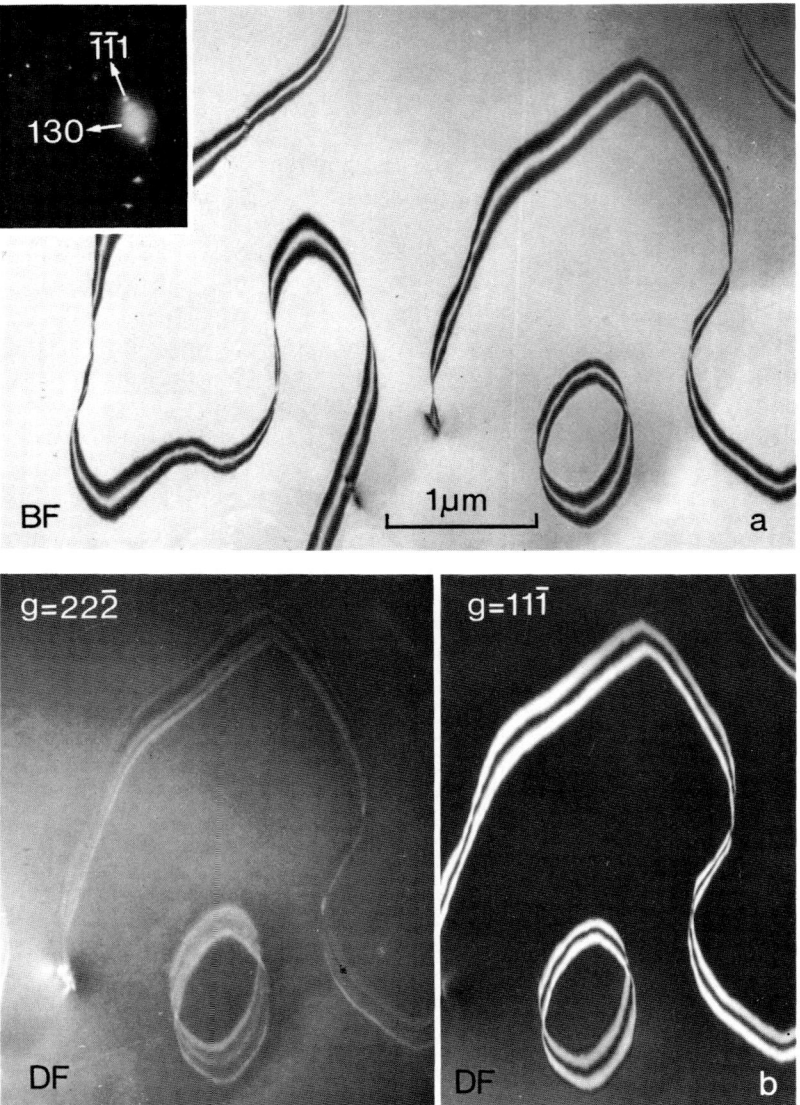

Fig. 17a and b

these diffraction conditions only $g_1 \cdot b = 1$ and $g_1 \cdot b = 2$ are possible. From the fringe spacings of the image, i.e. separation of the subsidiary minima, it can be concluded that the value of $g_1 \cdot b = 2$ and hence $b = \frac{1}{2}[10\bar{1}]$.

Another area where detailed computer calculations will be necessary is the contrast in non-centro-symmetric crystals. It has been shown that in these crystals and in multibeam conditions Friedel's law is generally not valid for the dark field image (see Serneels et al., 1973). One useful application of this is that

Fig. 17a–c. Fringes across c-APB's in anorthite. (a) Bright field, (b) dark field, (c) computed 6 beam intensity profiles for c type APB's in anorthite. Profiles for bright field ($g=000$) and $11\bar{1}$ and $22\bar{2}$ dark field images (W.F. Müller et al., 1973)

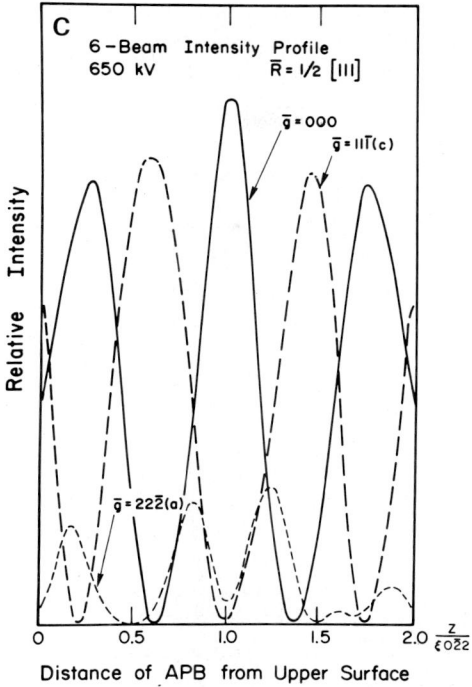

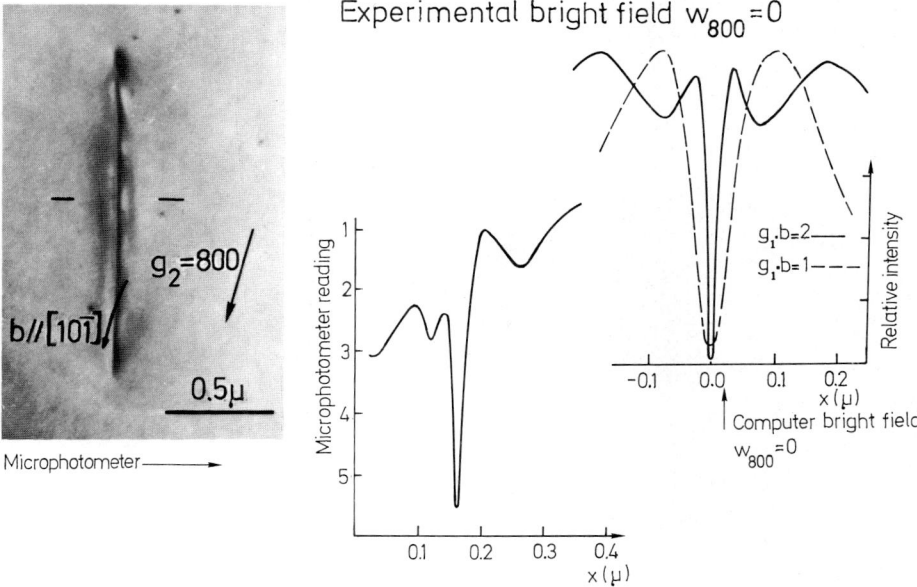

Fig. 18. Comparison between calculated and actual dislocation images in $CoFe_2O_4$. The magnitude of the Burgers vector is found from the fringe spacings, not from their intensity. $g_1 \cdot b = 2$ for this image. (Courtesy L.C. DeJonghe)

in the case of enantiomorphous pairs, the presence of both enantiomorphs can be quickly established (see Section 5.4). In principle it should be possible to determine the absolute configuration of the structure from the contrast in dark field using a multibeam dynamical theory for non-centro-symmetric crystals.

7. Non-conventional Techniques

In a two-beam orientation ($s \approx 0$) the width of the image of a dislocation is roughly equal to $\xi_g/3$ i.e. of the order of 100Å for most materials. Dislocation pairs whose projected separation is of the same order of magnitude will not be resolvable. This will be the case for closely spaced partials, and dipoles, for specimens with high dislocation density e.g. due to plastic deformation or at interphase interfaces.

The image width can be reduced using two techniques (1) the weak beam dark field (2) the high order g bright field. By weak beam dark field is generally meant that a dark field is formed of a beam with large deviation parameter s. In a two-beam case this means that the incident beam is tilted towards $-g$ instead of towards $+g$, as explained earlier for the high resolution dark field (Fig. 3b). The weak beam dark field is most useful at low voltages although it can also be applied at high voltages. It is based on the observation that the image width of the dislocation for larger values of the deviation parameter roughly drops off as $1/s$. The theoretical aspects of the method have been discussed by Cockayne (1972). An example is shown in Fig. 19, where the conventional high resolution dark field ($s=0$) (Fig. 19a) is compared with the weak beam dark field (Fig. 19b). The specimen shows the radiation damage due to ion implantation in silicon. The geometry of the defects is much better defined e.g. the small strain centers can be seen to be hexagon-shaped dislocation loops in the weak beam image.

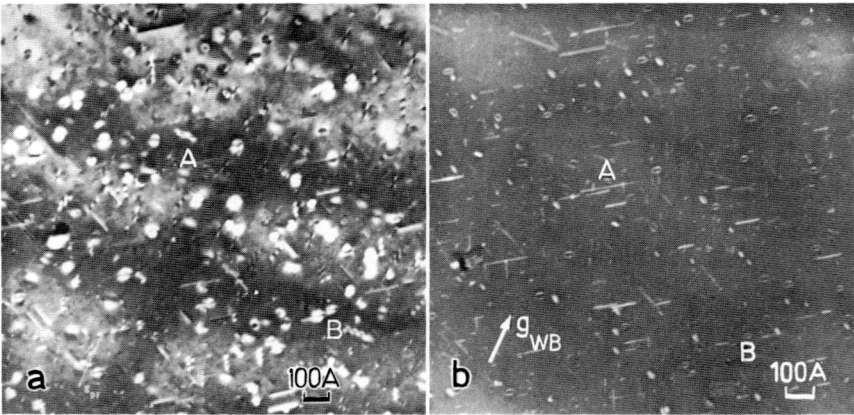

Fig. 19a and b. Conventional dark field (a) and weak beam dark field (b) of defects in silicon due to ion implantation. (Courtesy K. Seshan and J. Washburn, 1972)

Fig. 20a–c. Contrast conditions are shown for a translation fault (a), a (101) fault (b) and 1/4[101] partials (c). Weak π contrast occurs in (b) due to the influence of the superlattice reflections in the systematic set. (c) shows the advantage of using high resolution images in $n\mathbf{g}$ ($n=6$) in which the two partials are well resolved (Van der Biest and Thomas, 1974)

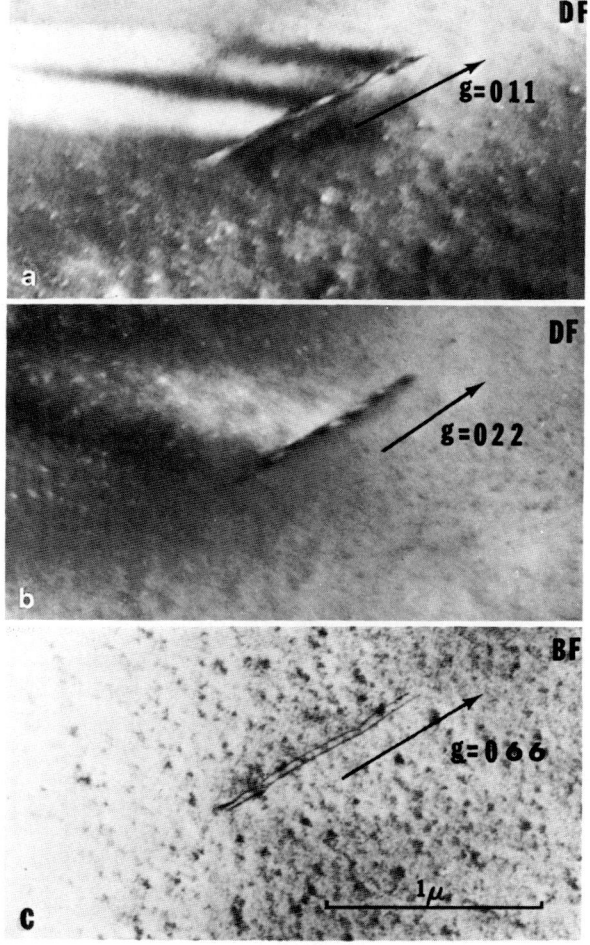

At high voltages a high-order bright field technique can be used to decrease the image width of a dislocation. The specimen is oriented so that a systematic set of reflections is operating and the Bragg condition is satisfied for a high-order reflection ($n\mathbf{g}$, with n typically 6 or 8). An example is shown in Fig. 20 where a dissociated dislocation, which in turn confines an antiphase boundary, is studied using increasing order of $\mathbf{g}$. For the first-order reflection which is a superlattice reflection, the antiphase boundary is in good contrast but the decomposition of the dislocation is not evident. Using the second-order reflection, the stacking fault in between the partial dislocations is visible. The antiphase boundary should be invisible, but is observed because of dynamical interaction with the first-order beam. When the sixth-order reflection is used, the partial dislocations are sharply resolved. This imaging technique has been discussed by Bell and Thomas (1972) and Goringe et al. (1972).

A Representative General Bibliography for Transmission Electron Microscopy

Transmission electron microscopy of metals. G. Thomas. New York: J. Wiley & Sons 1962.
Electron microscopy and strength of crystals. G. Thomas and J. Washburn (eds.). New York: Wiley-Interscience 1963.
The direct observation of dislocations. S. Amelinckx. Oxford: Academic Press 1964.
Fundamentals of transmission electron microscopy. R. D. Heidenreich. New York: Wiley-Interscience 1964.
Thin Films. Am. Soc. Metals, Chaps. 3, 8–10 (1964).
Electron microscopy of thin crystals. P. B. Hirsch, A. Howie, R. B. Nicholson, D. W. Pashley and M. J. Whelan. London: Butterworths 1965.
Techniques for electron microscopy. D. H. Kay (ed.). Oxford: Blackwell Sci. Pubs. 1965.
Interpretation of electron diffraction patterns. K. W. Andrews, D. J. Dyson and S. R. Keown. London: Hilger & Watts 1967.
Einführung in die Elektronenmikroskopie. M. von Heimendahl. Braunschweig: F. Vieweg 1970.
Electron optical applications in materials science. L. E. Murr. New York: McGraw-Hill 1970.
Modern diffraction and imaging techniques in materials science. S. Amelinckx, R. Gevers, G. Remaut and J. van Landuyt (eds.). Amsterdam: North Holland Press 1970.
Durchstrahlungs-Elektronenmikroskopie fester Stoffe. E. Hornbogen. Weinheim: Verlag Chemie 1971.
Electron microscopy in materials science. U. Valdre (ed.). New York: Academic Press 1971.
Modern metallographic techniques and their application. V. A. Phillips. New York: Wiley-Interscience 1971.
Electron microscopy and structure of materials. G. Thomas, R. Fulrath and R. Fisher (eds.). Univ. Calif. Press 1972.
Electron optics and electron microscopy. P. W. Hawkes. London: Taylor & Francis Ltd. 1972.
Computed electron micrographs and defect identification. A. K. Head, P. Humble, L. M. Clarebrough, A. J. Morton, and C. T. Forwood. Amsterdam: North Holland Press 1973.

References

Amelinckx, S.: The study of planar interfaces by means of electron microscopy. In: Modern Imaging and Diffraction techniques (eds. S. Amelinckx, R. Gevers, G. Remaut and J. van Landuyt), p. 257–294. Amsterdam: North Holland Press 1970.
Amelinckx, S.: The geometry and interfaces due to ordering and their observation in transmission electron microscopy and electron diffraction. Surface Science **31**, 296–354 (1972).
Ashby, M. F., Brown, L. M.: Diffraction contrast from spherically symmetrical coherency strains. Phil. Mag. **8**, 1083–1103 (1964).
Ashby, M. F., Brown, L. M.: On diffraction contrast from inclusions. Phil. Mag. **8**, 1649–1676 (1964).
Bell, W. L., Thomas, G.: Useful properties of dark-field electron images. Phys. Stat. Sol. **12**, 843–852 (1965).
Bell, W. L., Thomas, G.: Applications and recent developments in transmission electron microscopy. In: Electron microscopy and structure of materials (eds. G. Thomas, R. Fulrath, R. Fisher), p. 23–59. Berkeley: University of California Press 1972.
Clarebrough, L. M.: Contrast from shockley partial dislocations. Austr. J. Phys. **24**, 79–96 (1971).
Cockayne, D. J. M.: A theoretical analysis of the weak-beam method of electron microscopy. Z. Naturforsch. **27**a, 452–460 (1972).
Cosslett, V. E.: Recent progress in high voltage electron microscopy. In: Modern diffraction and imaging techniques in materials science (eds. S. Amelinckx, R. Gevers, G. Remaut and J. van Landuyt), p. 341–375. Amsterdam: North Holland Press 1970.

Fillingham, P.J., Leamy, H.J., Tanner, L.E.: Simulation of electron transmission images of crystals containing random and periodic arrays of coherency strain centers. In: Electron microscopy and structure of materials (eds. G.Thomas, R. Fulrath and R. Fisher), p. 163–170. Berkeley: University of California Press 1972.

Fisher, R.M.: High voltage electron microscopy, Proc. 26th EMSA Meeting, p. 324. Baton Rouge, La.: Claitors Publs. 1968.

Fisher, R.M.: Fundamental aspects and applications of high voltage electron microscopy. In: Electron microscopy and structure of materials (eds. G.Thomas, R. Fulrath and R. Fisher), p. 60–84. Berkeley: University of California Press 1972.

Gevers, R., Art, A., Amelinckx, S.: Electron microscopic images of single and intersecting stacking faults in thick foils. Phys. Stat. Sol. **3**, 1563–1588 (1963).

Goringe, M.J., Hewat, E.A., Humphreys, C.J., Thomas, G.: Defect contrast and resolution in high voltage electron microscopy. Proc. 5th Int. Congress on Electron Microscopy, p. 538. London: Institute of Physics 1972

Head, A.K., Humble, P., Clarebrough, L.M., Morton, A.J., Forwood, C.T.: Computed electron micrographs and Defect identification. Amsterdam: North Holland Press 1973.

Heimendahl, M. von: Specimen thickness determination in transmission electron microscopy in the general case. Micron **4**, 111–116 (1973).

Howie, A.: The theory of high energy electron diffraction. In: Modern imaging and diffraction techniques in materials science (eds. S. Amelinckx, R. Gevers, G. Remaut and J. van Landuyt), p. 295–339. Amsterdam: North Holland Press 1970.

Humphreys, C.J.: The optimum voltage in very high voltage electron microscopy. Phil. Mag. **25**, 1459–1472 (1972).

Lacaze, J.-C., Thomas, G.: Transmission electron microscopy at 2.5 MeV. J. Mic. **97**, 301–308 (1973).

Müller, W.F., Wenk, H.-R., Bell, W.L., Thomas, G.: Analysis of the displacement vectors of antiphase domain boundaries in anorthites. Contrib. Mineral. Petrol. **40**, 63–74 (1973).

Okamoto, P.R., Thomas, G.: On the four-axis hexagonal reciprocal lattice and its use in the indexing of transmission electron diffraction patterns. Phys. Stat. Sol. **25**, 81–91 (1968).

Phillips, V.A., Tanner, L.E.: High resolution electron microscope observation on GP zones in an aged Cu-1.97 wt. % Be crystal. Acta Met. **21**, 441–448 (1973).

Rowcliffe, D.J., Thomas, G.: Structure of non-stoichiometric TaC. J. Materials Science (1974).

Sass, S.L., Mura, T., Cohen, J.B.: Diffraction contrast from non-spherical distortions – in particular a cuboidal inclusion. Phil. Mag. **16**, 679–690 (1967).

Schneider, K., Sinclair, R., Thomas, G.: Lattice imaging of ordered alloys. Proc. 8th Int. Congress on Electron Microscopy, Canberra, p. 520 (1974).

Serneels, R., Snykers, M., Delavignette, P., Gevers, R., Amelinckx, S.: Friedel's law in electron diffraction as applied to the study of domain structures in non-centrosymmetrical crystals. Phys. Stat. Sol. (b), **58**, 277–292 (1973).

Seshan, K., Washburn, J.: On precipitation of phosphorous in ion implanted silicon. Radiation Effects **14**, 267–270 (1972).

Siems, R., Delavignette, P., Amelinckx, S.: The buckling of a thin plate due to the presence of an edge dislocation. Phys. Stat. Sol. **2**, 421–438 (1962).

Thomas, G.: Kikuchi electron diffraction and dark field techniques in electron microscopy studies of phase transformations. Trans. AIME **233**, 1608–1619 (1965).

Thomas, G.: Kikuchi electron diffraction and applications. In: Modern diffraction and imaging techniques in materials science (eds. S. Amelinckx, R. Gevers, G. Remaut, J. van Landuyt), p. 159–185. Amsterdam: North Holland Press 1970.

Thomas, G., Bell, W.L., Otte, H.M.: Interpretation of electron diffraction patterns from thin platelets. Phys. Stat. Sol. **12**, 353–366 (1965).

Van der Biest, O., Thomas, G.: Cation stacking faults in lithium ferrite spinel. Phys. Stat. Sol. **24**, 65–77 (1974).

Van der Biest, O., Thomas, G.: Cation stacking faults in lithium ferrite spinel. Phys. Stat. Acta Cryst. **31** A, 70–75 (1975).

Chapter 2.2

Interpretation of Electron Diffraction Patterns

J.A. Gard

1. Introduction

Many minerals and synthetic inorganic phases of low symmetry occur only as fine powders or fibre aggregates. In such cases, determination of the unit cell from X-ray data alone can be difficult or impossible, because of failure to resolve multiple reflections or to detect weak ones on polycrystalline X-ray photographs, together with uncertainties about systematically absent reflections. However, electron diffraction of small single crystals can be used for determination of the unit cell, enabling the X-ray lines to be indexed. Alternatively, electron diffraction of polycrystalline platy textures can sometimes be used, both to determine the unit cell and to provide indexed intensity data for use in structure analysis. d-spacings and intensities measured on X-ray patterns are, however, generally more accurate and reliable than electron-diffraction data. Correlation of both sources of data can therefore yield information not available from either technique alone.

The principles of electron diffraction are similar to those of X-ray diffraction, but there are at least three major differences. (1) The wavelengths of electron beams are much shorter than those of X-rays (e.g. 0.037 Å at 100 kV compared with 1.54 Å for Cu $K\alpha$); this affects both the geometry and the types of diffraction pattern that can be obtained. (2) Electrons are much more strongly diffracted than X-rays; electron-diffraction patterns can therefore be viewed, and photographically recorded with short exposures, from either a single crystal as small as 0.1 μm diameter, or from a single layer of thin crystals. (3) The specimen may be affected by the high vacuum of the instrument, bombardment by electrons, or contamination.

Two types of instrument are used for electron diffraction. (1) An electron diffraction camera has one or two lenses above the specimen to focus the electron beams on the photographic plate; the "general-area" diffraction pattern is produced either by transmission through a thin, usually polycrystalline specimen, or by "reflection" at an oblique angle from the surface of a bulky specimen. Reflection electron diffraction has few applications in mineralogy, and only transmission diffraction will be dealt with here. (2) An electron microscope can be quickly adjusted to display either an enlarged image of the specimen, or a diffraction pattern of a small area of it, selected by means of an aperture at the level of the primary image. An additional specimen stage can be inserted below the projector lens of some electron microscopes, permitting the use of both selected- and general-area diffraction with the same instrument. Selected-area diffraction (sometimes called microdiffraction) and its applications will be described first.

2. Selected-area Electron Diffraction

A good description of the principles and operation of the electron microscope has been given by Agar et al. (1974). Agar (1960) has devised a procedure for bringing the objective lens into correct focus so that only the selected area contributes to the diffraction pattern. The selected area can usually be varied between 0.5 and 2 µm diameter by choice of aperture, enabling a single particle to be isolated for diffraction. This technique has a number of uses, including the following. The shape of a crystal may be compared with its diffraction pattern (after allowing for rotation due to the different excitations of the lenses), so that faces, growth steps, twin lamellae and other features can be indexed; relative orientations of initial material and product of topotactic reactions may be determined; unknown unit cells can be determined and the Bravais lattice and space group deduced from the systematically absent reflections; stacking polytypes and twinning may be indicated by reflections too weak or too diffuse to appear on X-ray photographs; systematically strong reflections can indicate sub-lattices. For many of these purposes some reconstruction of the reciprocal lattice must be attempted. The procedures employed will depend on the nature of the crystals and the facilities available for orienting the specimen.

3. Calibration and Measurement of Diffraction Patterns

As has been shown in the preceeding chapter the wavelength of an electron beam may be calculated from $\lambda = 12.26\{E(1+0.979 \cdot 10^{-6} E)\}^{-1/2}$ Å, where E is the potential of the electron gun in volts. It is so short that the angle 2θ between the direct and diffracted beam does not normally exceed 5°. Using small angle approximations, $d \cdot 2r = 2\lambda L$ mm Å, where L is the effective distance from the specimen to the photographic plate, and $2r$ the diameter of a powder ring (both in mm), or the distance between a pair of spots equidistant from the direct beam spot, e.g. 300 and $\bar{3}00$. λL is known as the "camera constant", and it may be estimated from a powder pattern of a standard substance; it is therefore not often necessary to determine λ and L independently. In the electron microscope, the powder rings are often slightly elliptical due to lens aberrations, and L is sensitive to small changes in specimen height. The use of an internal standard is therefore strongly recommended, especially if a tilting stage is used. For example, aluminium may be evaporated from a filament in vacuum on the substrate before or after adding the specimen. $2\lambda L = 2r_s \cdot 2.338$ mm Å for the inner (111) ring of aluminium, and $d_{100} = 2r_s \cdot 2.338\, h(2r_h)^{-1}$ Å, where $2r_h$ is the distance between the $h00$ and $\bar{h}00$ spots. A measuring microscope, preferably with two orthogonal scales, may be used to measure spacings along the row of spots and the ring diameter in the same direction. A simple turntable for aligning the row of spots with the measuring scale is a useful accessory.

The use of an internal standard has two other advantages. Firstly, the d-spacings may be roughly estimated directly on the viewing screen, and the zone that the pattern represents provisionally identified. Secondly, as an aid to rapid focusing of the objective lens, the brightness of the ring pattern is minimal when only the selected area contributes to it.

4. Determination of Unit Cells

4.1 Use of Laue Zones

An electron-diffraction pattern represents an almost planar section of the reciprocal lattice through the origin, as it is cut by an Ewald sphere of reflection with a large radius, equal to the camera length L (Vainshtein, 1964, p. 29) on the scale of the pattern. A crystal usually lies with its thinnest direction parallel to the electron beam, and resolution of reciprocal lattice points is poorest in this direction (Pinsker, 1953, p. 79). The points are therefore effectively extended into short spikes, as shown in Fig. 1. This represents real and reciprocal lattices of a monoclinic crystal with its principal (001) plane in contact with the substrate, approximately normal to the electron beam in the untilted position shown in Fig. 1a. The angle between the c-axis and the beam is ϕ ($\approx \beta$-90°). The Ewald sphere cuts several layers of the reciprocal lattice, giving rise to circular Laue zones with their common center at the point of intersection of the c-axis with the plane of the diffraction pattern. The zero-order zone, which passes through the direct beam spot, comprises $hk0$ reflections; other Laue zones carry successively increasing l-indices as shown. The a- and b-dimensions of the unit cell can be estimated from the distances between the relevant parallel rows of spots. The spacing of the "third row", in this case the c-period, may be calculated from the equation:

$$c = L(2\lambda L) l \{(r_0 + r_l)(r_0 - r_l)\}^{-1},$$

where r_0 and r_l are the radii of the zero- and l-order Laue zones, respectively; L may be estimated from $\lambda L/\lambda$, λ being derived from the nominal cathode voltage. Many specimen stages have tilt facilities of ca. $\pm 5°$, enabling suitable Laue zones to be obtained. Unless the zones are well defined, however, their common center cannot be accurately located on a single diffraction pattern. The author has described elsewhere a procedure for locating these centers on a series of diffraction patterns recorded from the same crystal (Gard, 1956), and has discussed deduction of the Bravais lattice type from the relative positions of spots in adjacent Laue zones. Using the Laue zone method, a provisional unit cell can often be obtained and refined by correlation with X-ray powder data. Some examples of its use are FeCr (σ-phase) and $Cr_{23}C_6$ (Brown and Clark, 1952), foshagite (Gard and Taylor, 1958), and tricalcium silicate hydrate (Buckle et al., 1958).

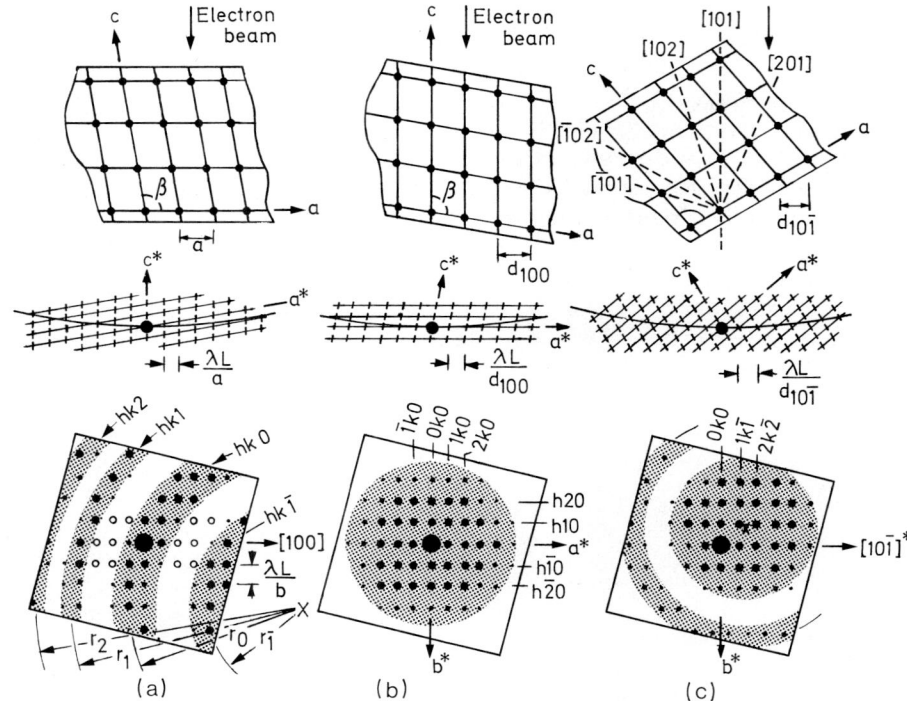

Fig. 1a–c. Real and reciprocal lattices, and their electron-diffraction patterns, for a monoclinic crystal with its (001) face on the substrate. (a) With the specimen untilted; a and b may be determined directly as shown, and c estimated from the radii, r_0, r_1, r_2 and $r_{\bar{1}}$, of Laue zones. Continuation of rows of spots (open circles) shows that the lattice is primitive. (b) With the crystal tilted to place c parallel to the beam; d-spacings in the $hk0$ zone may be measured. (c) With the crystal tilted to place e.g. [101] parallel to the beam, so that the $hk\bar{h}$ zone is central. Broken lines indicate other zone axes that may be brought in turn parallel to the beam for measurement of d_{h0l} spacings, from which a projection of the reciprocal lattice down [010] may be constructed

5. Use of a Goniometric Specimen Stage

Greater scope is possible with a goniometric stage, i.e. one that permits tilt exceeding $\pm 45°$ about any chosen axis; some stages are capable of $\pm 60°$. An unknown unit cell and many or all of the space group absences can be determined by use of electron diffraction alone. Two types of goniometric stage are in general use. (1) *Rotation-tilt*, with tilt about a fixed axis normal to the electron beam, and rotation about an axis normal to the specimen plane (see e.g. Mackay, 1968). (2) *Double-tilt*, with tilt about two orthogonal axes, one normal to the beam, while the other is in the specimen plane and tilts with it (see e.g. Lucas, 1970). The present author has described procedures for operation of both types of stage, and has displayed graphs for rapid estimation of the effective tilt angles during rotation round a chosen axis, usually a row of spots common to a series of diffraction patterns (Gard, 1971, pp. 45–48).

The first step in determination of a unit cell is to orient the crystal with a zone axis parallel to the electron beam. The crystal giving the pattern shown in Fig. 1a, for example, should be tilted so that the center of the Laue zones is moved towards the direct beam spot, eventually giving the pattern shown in Fig. 1b, representing the $hk0$ zone. The crystal may then be rotated round the b^*-axis, in each direction in turn, recording each zone as it appears when the various $[u0w]$ axes, shown in Fig. 1c, are placed parallel to the beam. Angles of tilt should be carefully noted for all patterns. As the monoclinic b^*-axis is common to all the zones, the spots fall on a series of orthogonal nets. The reciprocal spacing $(d_{h0l})^{-1}$ is measured for each pattern, then, using these values and the corresponding tilt angles, a projection of the reciprocal lattice down b is constructed. After slight adjustments, a regular array of indexable spots should appear, from which c^* and β^* can be measured, or calculated from the equations

$$2(d_{00l})^{-2} = (d_{10l})^{-2} + (d_{10\bar{l}})^{-2} - 2(d_{100})^{-2}$$

and

$$4\cos\beta^* = d_{00l}\, d_{100}\{(d_{10l})^{-2} - (d_{10\bar{l}})^{-2}\}.$$

Spots on the first zone encountered may not, of course, fall on an orthogonal net. The crystal should be tilted around one or other of the closely spaced rows of spots until an orthogonal or nearly orthogonal zone appears.

If the lattice is not primitive, some of the diffraction patterns will be centered orthogonal nets, indicating systematically absent reflections that must be taken into account while allocating indices. Reflections not in the plane of the projection (e.g. with odd k-indices for projections down b) may be inserted as open circles, as in Fig. 3. The Bravais lattice type is readily deduced from projections of orthogonal unit cells, bearing in mind that an F-centered reciprocal lattice indicates an I-centered real cell, and *vice versa*. Several crystals should be examined, and preferably rotated round more than one row of spots, with the object of recording all three faces of the reciprocal cell, as direct measurement of a^*, b^* and c^* is preferable, and further absences may indicate the possible space groups. For example, roggianite was shown to be I-centered tetragonal, with odd-l reflections absent from the $h0l$ zone, due to the presence of a c-glide (Gard, 1969); orthorhombic $FeAlO_3$ has space group $Pc2_1n$ or $Pcmn$ because $l=2n$ in the $0kl$ zone, and $h+k=2n$ in the $hk0$ zone, indicating c- and n-glide planes, respectively (Dayal *et al.*, 1965). Other examples are given by Gard and Bennett (1966). Interpretation of space group absences presents some difficulties, which will be discussed later. Unfortunately, some planes of the reciprocal lattice are virtually inaccessible, for example the zone normal to a lath axis, without application of special preparation techniques, such as ultramicrotomy or argon ion bombardment thinning (e.g. the c-periods of muscovite and chlorite: Oertel *et al.*, 1973).

Projections of the reciprocal lattice are more difficult to construct and interpret if the crystal is triclinic, or if it cannot readily be rotated round a monoclinic b^*-axis. Obviously, tilting the reciprocal lattice shown in Fig. 1b round the a^*-axis would give progressively more oblique diffraction patterns. The pseudo-cell of tacharanite ($Ca_{12}Al_2Si_{18}O_{69}H_{36}$) was determined in this way (Cliff et al., 1975). The procedure is described below as a general example of construction of an unknown reciprocal lattice from electron-diffraction patterns.

Most of the material in the electron microscope specimens occurred as aggregates of thin laths. Isolated crystals suitable for electron diffraction were very small. Tilt of about 22° round the long axis, subsequently identified as monoclinic b, gave the pattern shown in Fig. 2a, which has a row of spots spaced $(15.56 \text{ Å})^{-1}$. This row was called a^*, and the reciprocal pseudo-cell (referred to the sharp spots, ignoring the streaks) was determined by rotating crystals round this axis. The $[0\bar{1}1]$, $[0\bar{3}1]$ and $[0\bar{5}1]$ zone axes were in turn placed parallel to the electron beam, giving the diffraction patterns shown in Fig. 2b, c and d, respectively. The corresponding [011], [031] and [051] zones were also recorded by tilting in the opposite direction. Each pair of diffraction patterns $[0v1]$ and $[0\bar{v}1]$ appeared to be identical mirror images, confirming the monoclinic symmetry. d_{020} measured

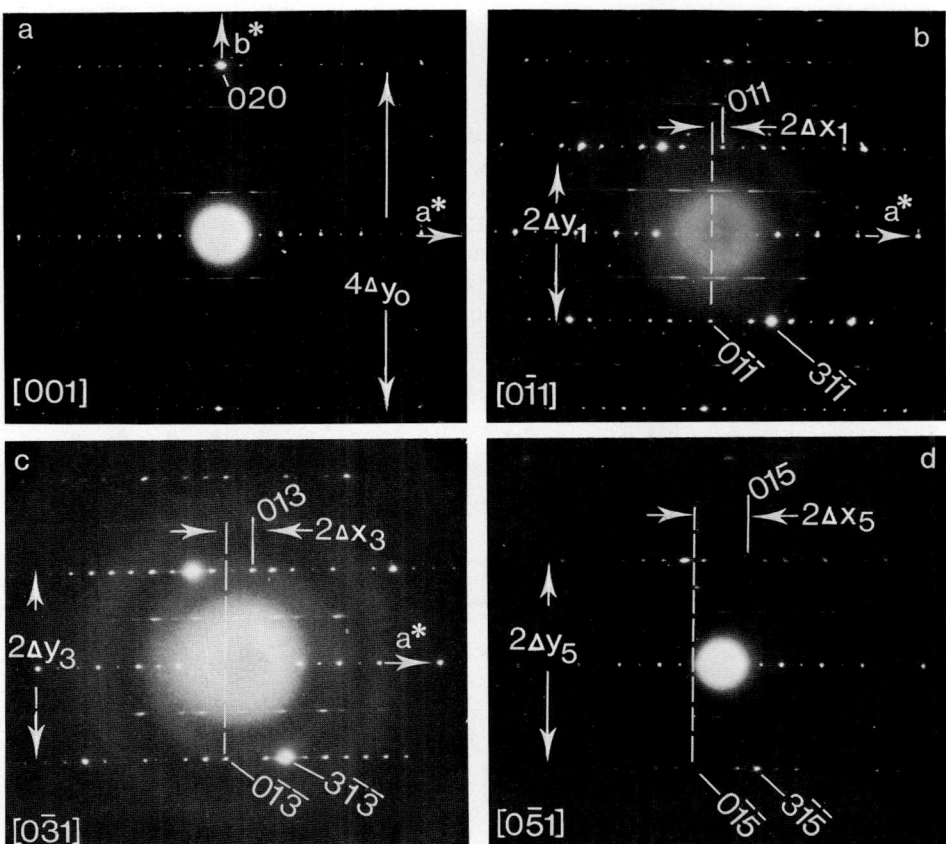

Fig. 2a–d. Electron-diffraction patterns of tacharanite from a crystal tilted round a^*. The measurements $2\Delta x_1$ and $2\Delta y_1$ are explained in the text. For clarity, no internal standard was used

on Fig. 2a gave $b = 3.65$ Å. Each pattern was placed in turn on a Pye two-way measuring microscope with a^* exactly parallel to the x-direction of traverse. Taking the $[0\bar{3}1]$ pattern of Fig. 2c as an example, the following measurements were made on each pattern.

The x- and y-scale readings were noted with the crosswires of the microscope set on the 013 and $0\bar{1}\bar{3}$ spots in turn; the differences of readings will be called $2\Delta x_3$ and $2\Delta y_3$ (in mm), respectively, as shown in Fig. 2c. The diameters of the inner aluminium ring, $2r_x$ and $2r_y$, were also measured. The distances δx_3 and y_3, shown in Fig. 3, were then calculated from the equations.

$$\delta x_3 = 2\Delta x_3 (2.338 \cdot 2r_x)^{-1} \text{ Å}^{-1};$$
$$y_3 = 2\Delta y_3 (2.338 \cdot 2r_y)^{-1} \text{ Å}^{-1}.$$

It may be assumed that the slight distortion due to astigmatism of the electron microscope lenses will be the same for all patterns taken from the same crystal. The average value of $(\delta x_5 - \delta x_3)$, $(\delta x_3 - \delta x_1)$, $(\delta x_1 + \delta x_{\bar{1}})$ etc. is called δx_2, and is used in the following equations to calculate c and β^*,

$$c = 5(y_5^2 - y_0^2)^{-1/2} \text{ Å}; \quad \cot \beta^* = \tfrac{1}{2} c \cdot \delta x_2.$$

These values of c and β^* were used to construct the [100] and [010] projections of the reciprocal lattice shown in Fig. 3, which was obviously A-face-centered. The accuracy of the [010] projection was checked with $h0l$ spacings measured on diffraction patterns from crystals rotated round b^*. A row of spots spaced $(12.4 \text{ Å})^{-1}$, at first thought to be c^*, was indexed $[10\bar{2}]^*$; the $0kl$ zone was not observed because of the good (100) cleavage. The pseudo-cell with the best fit for the data was A-centered monoclinic, with $a = 16.85$, $b = 3.64$, $c = 27.36$ Å,

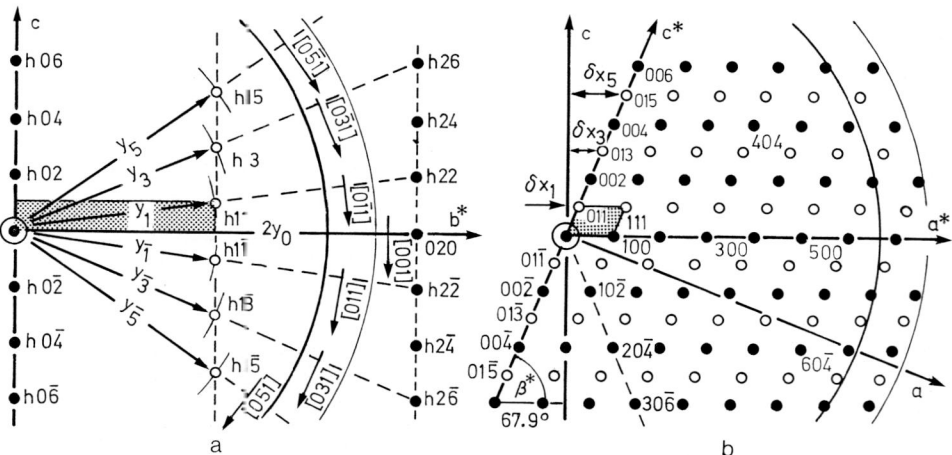

Fig. 3a and b. Projections down a^* and b^* of the reciprocal lattice of tacharanite, constructed from measurements shown in Fig. 2. Arcs represent 2.338 and 2.024 Å powder rings from aluminium (not present in Fig. 2). Reflections with k even are shown as closed circles, and with k odd as open circles. The pseudo-cell is shaded

$\beta = 112.1°$. A slightly larger cell was found to fit the X-ray data; evidently tacharanite shrinks slightly due to dehydration in the vacuum.

The methods described above may also be applied to triclinic crystals, which would give an oblique [100] projection. They may also be used to locate reflections that are too weak or diffuse to appear on X-ray powder photographs, but which indicate stacking polytypes (for example, in xonotlite and foshagite: Gard, 1960, 1966; Gard and Taylor, 1960), or complex twinning relationships (e.g. in a synthetic anorthite with excess Al_2O_3: Bhatty et al., 1970).

6. Multiple Diffraction: Rules for Space-group Extinctions

A strongly reflected beam, generated in the first part of a crystal, will act as a primary beam and will be diffracted as it traverses the rest of the crystal, adding its own contribution to each diffraction spot. Such contributions due to secondary scattering can be visualized by displacing the origin of the pattern to each strong spot in turn. The general result is that weak, and even "forbidden" spots gain intensity at the expense of the strong ones. A weak reflection is therefore suspect if its indices are the sum of those of any two strong reflections. Fig. 4 shows that the extinction rules remain valid for all Bravais lattice types, and for both axial and diagonal (n-) glide planes parallel to the plane of the diffraction pattern, but not for extinctions along a single row of spots in the pattern, such as a screw axis. For example, the $00.l$ diffuse spots and streaks in Fig. 7a are "forbidden" reflections inserted along a 6_3 screw axis. "Forbidden" reflections disappear if the row of spots is isolated by tilting the crystal round that row (Vainshtein, 1964, p. 63). Coherent multiple diffraction is, of course, equivalent

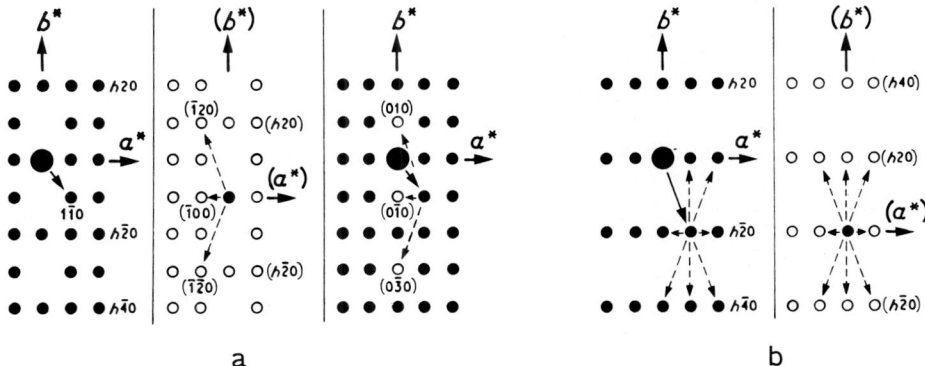

Fig. 4a and b. Extinction rules for electron diffraction. (a) Left: Primary diffraction pattern from the upper layer of a crystal lying on (001); $0k0$ reflections with k odd are absent. Center: secondary diffraction pattern from the $1\bar{1}0$ primary beam in subsequent layers of the crystal. Right: Sum of these patterns showing presence of "forbidden" odd $0k0$ spots. (b) Left: Primary diffraction pattern from a crystal with all odd-k reflections absent, due to a b-glide plane parallel to (001). Right: Secondary pattern from the $\bar{1}20$ primary beam; no "forbidden" spots with k odd are inserted, as all secondary spots fall on existing primary spots

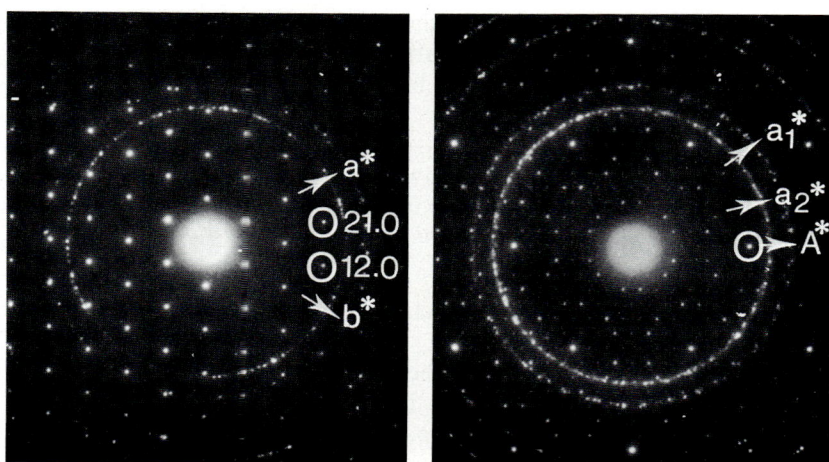

Fig. 5. Electron-diffraction patterns from particles of zinc hydroxysulphate. Left: From a single crystal; right: from overlapping crystals with 21.0 of one coinciding with 12.0 of the other. Secondary spots form a net with $A = a \cdot \sqrt{7}$

to dynamical scattering, the effects of which on the extinction conditions are discussed more fully by Gjønnes and Moodie (1965).

Where double diffraction occurs from two separate overlapping crystals at different orientations, the secondary spots are additional to the two primary patterns. Where an $hk.0$ spot of one hexagonal crystal coincides with a $kh0$. of the other, these extra spots form a hexagonal superlattice with $A = a(h^2 + k^2 + hk)^{1/2}$. Fig. 5 shows a primary pattern with $a = 8.35$ Å, and a double diffraction pattern in which 21.0 and 12.0 coincide; a weak hexagonal net is present with $A = 22.1$ ($= 8.35 \cdot \sqrt{7}$) Å. Both patterns are from crystals of zinc hydroxysulphate. Other examples have been given by Eckhardt (1958), Stabenov (1959a, b), Zvyagin and Gorshkov (1966) and Gard (1971, p. 32).

Where the angle α between two overlapping crystals is small, a secondary reflection occurs close to the main beam. Interference between the two beams causes a set of parallel "Moiré" fringes to appear on a micrograph, with spacing $D = d(2 \sin \frac{1}{2} \alpha)^{-1} \approx d\alpha^{-1}$, where d is the lattice spacing for both primary and secondary reflections (see Bassett et al., 1958). The fringes can often be seen on micrographs of layer lattice minerals.

7. "Two-dimensional" Electron-diffraction Patterns

If a crystal is very thin or disordered, the reciprocal lattice spikes shown in Fig. 1 join to form continuous streaks. Resolution of reciprocal lattice points along the streaks may then be too poor for application of the procedures described above. However, the a- and b-dimensions of the unit cell can be accurately determined from diffraction patterns of crystals with good (001) cleavage, using

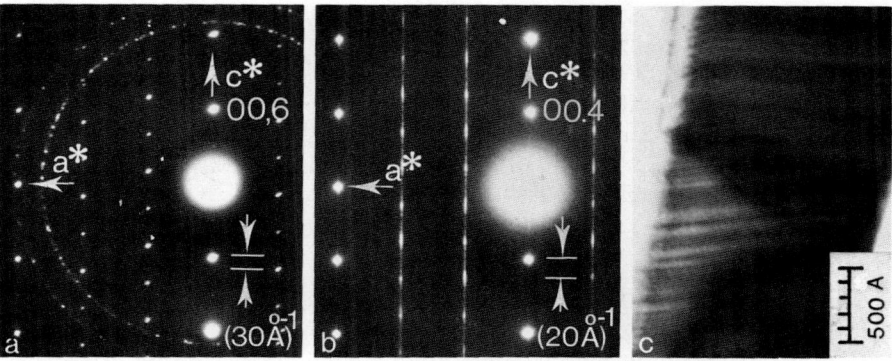

Fig. 6a–c. Electron-diffraction patterns of SrSiO$_3$ prepared at (a) 1600 °C, (b) 1150 °C. (c) Lamellae parallel to (00.1), multiples of 20 Å wide, which give the diffraction pattern (b)

an internal standard. These values may help in indexing the longest X-ray spacings, such as d_{010} and d_{100}; if b and d_{010} are identical, the crystals must be monoclinic, or nearly monoclinic. β can sometimes be estimated from $\sin \beta = d_{100} a^{-1}$. Okenite (Gard and Taylor, 1956) is an example of a unit cell determined in this way.

In some disordered structures, the "thin" direction of the domains may be placed normal to the electron beam, permitting direct observation of the streaks. One example is SrSiO$_3$, in which sheets of Si$_3$O$_9$ rings may be stacked in various ways. X-ray photographs of samples prepared at 1500 °C display more lines than those prepared at 1150 °C (Moir et al., 1975). Fig. 6 compares electron-diffraction patterns of the $h0.l$ zones for the two samples. All reflections were sharp for the 1500 °C sample, but spots with $h-k \neq 3n$ were streaked parallel to c^* for the 1150 °C sample. Maxima on the streaks corresponded to a c-period of 20.24 Å, but weak spots were also present, suggesting periodic structures with c up to 4×20 Å. Lamellae between 20 and 80 Å wide were visible on micrographs (Fig. 6c). Evidently each lamella has a regular stacking sequence that changes at its boundaries.

Fig. 7. (a) Electron-diffraction pattern of the $2kk.l$ zone of a hexagonal crystal of disordered offretite. The streaks are due to thin lamellae of erionite; those between the even $00.l$ spots are inserted by multiple diffraction. (b) Paths of the electron beam through the crystal, showing the splitting of reflections by refraction; the refraction angles are greatly exaggerated

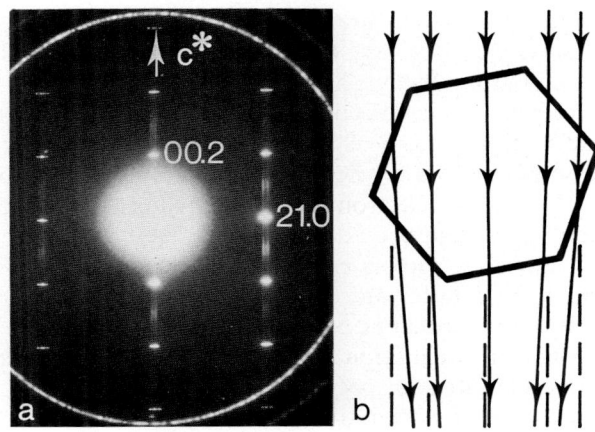

Another example is shown in Fig. 7a, a diffraction pattern from a hexagonal prism of the zeolite offretite, containing stacking faults present as lamellae of erionite (Bennett and Gard, 1967; Gard and Tait, 1971). The streaking of odd-l reflections shows that the lamellae are only a few c-periods wide. All reflections are split into several components, due to refraction of the electron beam through angles of up to 2'. Fig. 7b shows the various paths that the electron beam can take through the crystal.

8. Topotactic Reactions

Microdiffraction can be used to study topotactic reactions, by indicating relative orientations before and after heat or chemical treatment. Where the product is a good pseudomorph, the reaction can be conducted outside the microscope, as in the transformations of foshagite at 700 °C to wollastonite (Gard and Taylor, 1958) and calcite to fluorite in NaF solution (Gard, 1964). Relative orientations can be derived more directly by heating the specimen in the microscope, either with the electron beam, as with the conversion of brucite to periclase (Goodman, 1958), or with a specimen heating stage (e.g. Cartz and Tooper, 1964).

In addition to the works previously quoted, various aspects of single-crystal electron diffraction are discussed by Ross and Christ (1958), Reimer (1959), Alderson and Hallicay (1965), Hirsch *et al.* (1965), McConnell (1967), Andrews *et al.* (1971) and Beeston *et al.* (1972).

9. Intensities of Reflections in Electron-diffraction Patterns

X-rays are scattered only by the electrons in an atom, but the nuclear charge Z, the atomic number, is also involved in electron scattering. Thus

$$f(\theta) = 2.38 \times 10^{-10} \, (\lambda/\sin\theta)^2 \, (Z - f_x),$$

where $f(\theta)$ and f_x are atomic scattering amplitudes for electrons and X-rays, respectively (see, e.g. Hirsch *et al.*, 1965, p. 90). Reflections with hkl odd are absent from X-ray photographs of KCl, because both K^+ and Cl^- have 19 electrons; such reflections are, however, clearly visible on the $\langle 110 \rangle$ electron-diffraction pattern of KCl shown in Fig. 8. Tetrahedral Al and Si atoms can also be resolved for Zeolite A (Gard, 1960) and anorthite (see e.g. Bhatty *et al.*, 1970, Fig. 4), electron-diffraction patterns of which display spots doubling a and c, respectively.

Structure factors can rarely be reliably calculated from electron-diffraction intensity data for any but extremely thin single crystals, because the intensities are strongly modified by dynamical scattering effects that vary greatly with crystal thickness, orientation and perfection. Sturkey (1962) and others have shown that intensities on polycrystalline patterns are much more reliable, because of the range of thickness of the crystals; the question of misorientation does not

Fig. 8. Electron-diffraction pattern of KCl with [0$\bar{1}$1] parallel to the electron beam. Weak reflections with odd hkl values are clearly visible

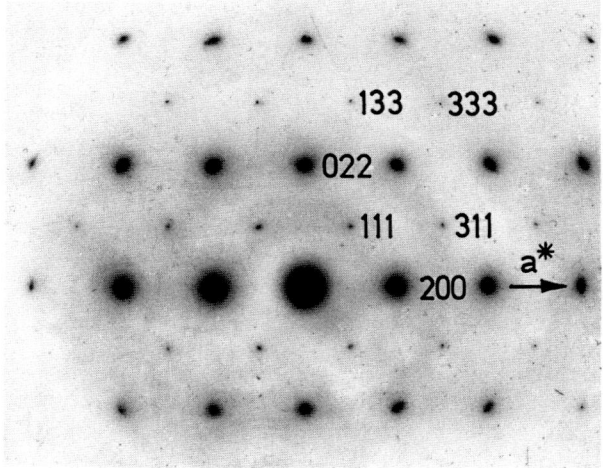

arise. Electron diffraction of specimens with preferred orientation, usually platy "textures" at oblique incidence, have been extensively used for structure analysis, particularly in the Soviet Union (see Pinsker, 1953; Vainshtein, 1964; Zvyagin, 1967). The general principles will be described in the next Section.

9.1 Electron Diffraction of Platy Textures

Specimens are often prepared by allowing a drop of a dilute suspension of the mineral to dry on a thin collodion film, stretched over holes, between 2 and 4 mm diameter, in a copper plate. The platy crystals should lie flat on the substrate without appreciable overlapping, but with random orientation around the "texture axis", which is c^* for a mineral with (001) cleavage. The specimen is illuminated with a weak electron beam, usually several hundred μm diameter. With the substrate normal to the beam, a "normal texture" ring pattern is obtained, as in Fig. 9a. It becomes an "oblique texture" pattern on tilting the specimen through an angle ϕ, usually between 45 and 70° (e.g. Fig. 9b).

9.2 Indexing Oblique Texture Patterns

The reciprocal lattice points lie on sets of "rods" parallel to c^*. Rotation round c^* generates a set of cylinders, with radius m'_{hk} Å^{-1}, each carrying a set of ring nodes with equal hk values, as shown in Fig. 10a. For the pseudohexagonal clay minerals, for example, the radius of each cylinder is $(3h^2+k^2)\ b^{-1}$, and successive rings bear the hk-indices (02,11), (13,20), (04,22) etc. If the crystals are sufficiently thin or disordered, the cylinders may have some intensity at the level $Z=0$, and the normal texture pattern displays rings for all the hk-indices, as in Fig. 9a. When the specimen is tilted through angle ϕ, the cylinders are

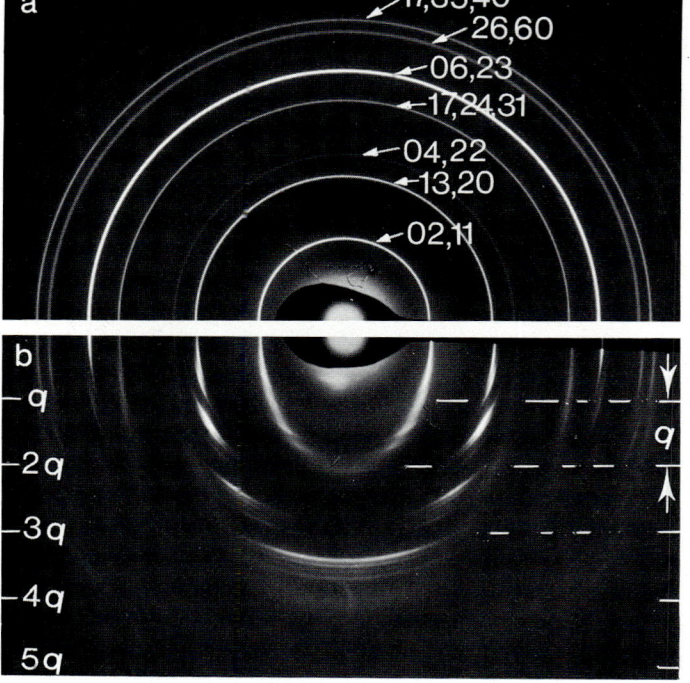

Fig. 9a and b. Texture electron-diffraction patterns of kaolinite (Morbihan, Brittany). (a) Substrate normal to the electron beam; hk-indices are shown. (b) Substrate tilted 50° from normal, showing arcs on the ellipses; the lq-levels, from which c^* is calculated, are shown

intersected by the almost planar Ewald sphere as ellipses, for which the radius of the major axis is $M'_{hk} = m'_{hk}/\cos \phi$. Points along the ellipse may be calculated from $y^2 = (M^2 - x^2 M^2) m^{-2}$. Each ring node is "smeared" by misorientation, so it appears on the ellipse as a powder arc centered on the direct-beam spot (see Figs. 9b and 10b). Fig. 11 shows that the height of any ring node above the (001) plane is $D'_{hkl} = ha^* \cos \beta^* + kb^* \cos \alpha^* + lc^*$ Å^{-1}. On the diffraction pattern, the height of the maximum of the arc above the minor (tilt) axis is

$$D_{hkl} = (\lambda L/\sin \phi)(ha^* \cos \beta^* + kb^* \cos \alpha^* + lc^*) = hp + ks + lq$$

where p, s and q are measured in mm on the diffraction pattern. This equation applies to *all* textures of triclinic crystals lying on (001). It is much simpler for textures in which c coincides with c^*, such as orthogonal or hexagonal crystals on a unit cell face, or monoclinic on (010); interchange of another zone axis for c may be necessary. D then becomes lq, and the arcs fall on layer lines intersecting the ellipses (see, e.g. Fleischmann and Thirsk, 1963, for texture patterns of hexagonal $Cd(OH)_2$). Such patterns are relatively easy to index. Allocation of indices to triclinic and monoclinic textures is more difficult; for the latter, $D = hp + lq$. The hk-indices of each ellipse can be deduced by comparison of a normal texture pattern with a selected-area pattern of a single crystal. The oblique texture pattern should next be examined to estimate the periodic repeat q that is present in all the ellipses; a series of patterns with increasing angles

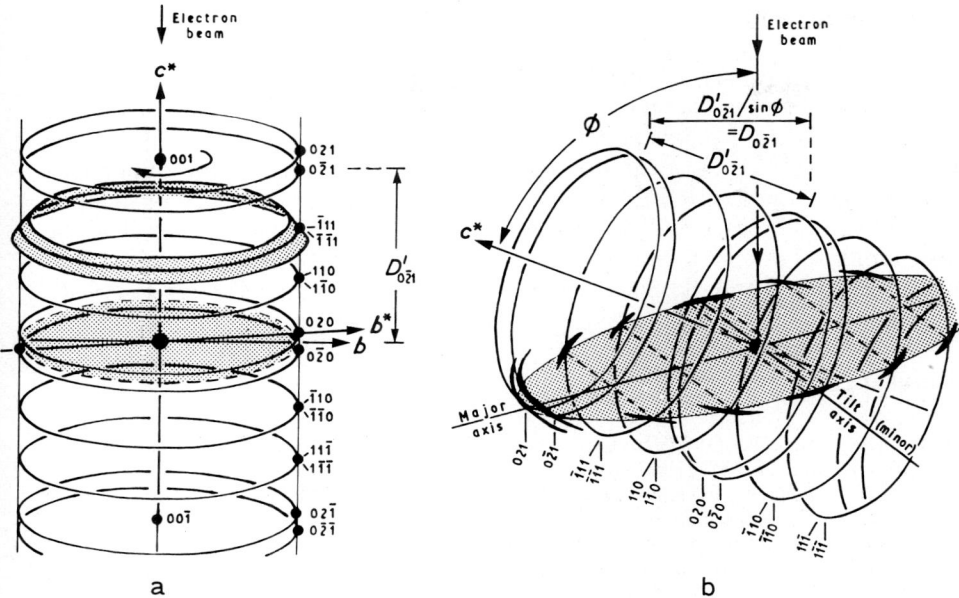

Fig. 10a and b. Schematic diagram of the (02l, 11l) reciprocal lattice cylinder of a platy texture of kaolinite: (a) "normal texture"; (b) "oblique texture", showing the arcs on the ellipse formed by intersection of the ring nodes by the Ewald sphere

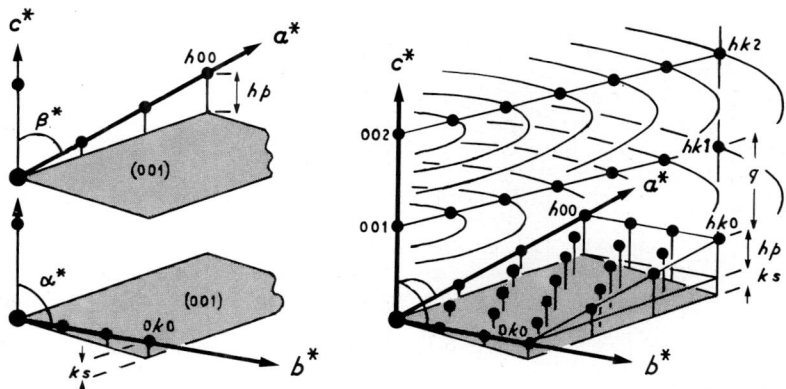

Fig. 11. The reciprocal lattice of a C-centered triclinic crystal (e.g. kaolinite), showing that the height above the (001) plane of an $hk0$ reflection is $(hp+ks)$, and that of an hkl reflection is $(hp+ks+lq)$. Some of the ring nodes formed by rotation round c^* are shown

of tilt may help. Heights of arcs are symmetrically disposed about each lq level, and all combinations of (hk) values for each ellipse are represented in each q-period, but overlapping may reduce the number of arcs that are visible. Indexing is dealt with more fully by Vainshtein (1964), Zvyagin (1960, 1967) and Gard (1971, pp. 66–73).

9.3 Structure Analysis

When the arcs are indexed, intensities may be estimated, usually by visual comparison of the maximum at the center of each arc with a logarithmic exposure strip. The structure factor is calculated from $|F_g|^2 = IR'mn^{-1}$, where I is the intensity, R' the distance of the arc from the major axis, m the minor axis radius, and n the multiplicity factor. The methods used for structure analysis from X-ray intensities can be applied to electron diffraction, except that the distribution of potential ϕ_{xyz} is plotted instead of ϱ_{xyz}. The subject is discussed by Cowley (1967) and the general procedures are described by Vainshtein (1964).

References

Agar, A.W.: Accuracy of selected-area microdiffraction in the electron microscope. Brit. J. Appl. Phys. **11**, 185–189 (1960).
Agar, A.W., Alderson, R.H., Chescoe, D.: Principles and practice of electron microscope operation. In: Practical methods in electron microscopy, vol. 2 (ed. A.M. Glauert). Amsterdam-London: North Holland Press 1974.
Alderson, R.H., Halliday, J.S.: Electron diffraction. In: Techniques for electron microscopy, 2nd ed., Chap. 15 (ed. D. Kay). London: Blackwell 1965.
Andrews, K.W., Dyson, D.J., Keown, S.R.: Interpretation of electron diffraction patterns, 2nd ed. London: Hilger 1971.
Bassett, G.A., Menter, J.W., Pashley, D.W.: Moiré patterns on electron micrographs. Proc. Roy. Soc. (London) Ser. A **246**, 345–368 (1958).
Beeston, B.E., Horne, R.W., Markham, R.: Electron diffraction and optical diffraction techniques. In: Practical methods in electron microscopy, vol. 1 (ed. A.M. Glauert). Amsterdam-London: North-Holland Press 1972.
Bennett, J.M., Gard, J.A.: Non-identity of the zeolites erionite and offretite. Nature **214**, 1005–1006 (1967).
Bhatty, M.S.Y., Gard, J.A., Glasser, F.P.: Crystallization of anorthite from glasses. Mineral. Mag. **37**, 780–789 (1970).
Brown, J.F., Clark, D.: Three-stage electron microscope in crystal structure analysis. Acta Cryst. **5**, 615–619 (1952).
Buckle, E.R., Gard, J.A., Taylor, H.F.W.: Tricalcium silicate hydrate. J. Chem. Soc. **1958**, 1351–1355.
Cartz, L., Tooper, B.: Dehydration of phlogopite micas in the electron microscope. Proc. 3rd Europ. Congress on Electron Microscopy, Prague, **A**, 335–336 (1964).
Cliff, G., Gard, J.A., Lorimer, G.W., Taylor, H.F.W.: Tacharanite. Mineral. Mag. **40**, 117–130 (1975).
Cowley, J.M.: Crystal structure determination by electron diffraction. Prog. Materials Sci. **13**, 267–321 (1967).
Dayal, R.R., Gard, J.A., Glasser, F.P.: Crystal data on $FeAlO_3$. Acta Cryst. **18**, 574–575 (1965).
Eckhardt, F.J.: Elektronoptische Untersuchungen an tonigen Sedimenten. Neues Jahrb. Mineral., Monatsh. **1958**, 1–17.
Fleischmann, M., Thirsk, H.R.: Growth of thin passivating layers. J. Electrochem. Soc. **110**, 688–698 (1963).
Gard, J.A.: Use of stereoscopic tilt device in unit cell determinations. Brit. J. Appl. Phys. **7**, 361–367 (1956).
Gard, J.A.: Weak reflections in electron-diffraction patterns. Proc. 2nd Europ. Congress on Electron Microscopy, Delft, **1**, 203–206 (1960).
Gard, J.A.: Reaction between calcite and sodium fluoride. Proc. 3rd Europ. Congress on Electron Microscopy, Prague, **A**, 333–334 (1964).

Gard, J.A.: A system of nomenclature for the fibrous calcium silicates, and a study of xonotlite polytypes. Nature **211**, 1078–1079 (1966).
Gard, J.A.: Electron microscope and diffraction study of roggianite. Clay Minerals **8**, 112–113 (1969).
Gard, J.A.: Interpretation of electron micrographs and diffraction patterns. In: The electron optical investigation of clays, Chap. 2 (ed. J.A. Gard). London: Mineralogical Society 1971.
Gard, J.A., Bennett, J.M.: A goniometric specimen stage, and its use in crystallography. Proc. 6th Int. Congress on Electron Microscopy, Kyoto **1**, 593–594 (1966).
Gard, J.A., Tait, J.M.: Structural studies on erionite and offretite. Advan. Chem. Ser. **101**, 230–236 (1971).
Gard, J.A., Taylor, H.F.W.: Okenite and nekoite (a new mineral). Mineral. Mag. **31**, 5–20 (1956).
Gard, J.A., Taylor, H.F.W.: Foshagite: composition, unit cell and dehydration. Am. Mineralogist **43**, 1–15 (1958).
Gard, J.A., Taylor, H.F.W.: The crystal structure of foshagite. Acta Cryst. **13**, 785–793 (1960).
Gjønnes, J., Moodie, A.F.: Extinction conditions in the dynamic theory of electron diffraction. Acta Cryst. **19**, 65–67 (1965).
Goodman, J.F.: Decomposition of magnesium hydroxide in an electron microscope. Proc. Roy. Soc. (London) Ser. A **247**, 346–352 (1958).
Hirsch, P.B., Howie, A., Nicholson, R.B., Pashley, D.W., Whelan, M.J.: Electron microscopy of thin crystals. London: Butterworth 1965.
Lucas, J.H.: Proc. 7th Int. Congress on Electron Microscopy, Grenoble **1**, 159–160 (1970).
Mackay, A.L.: Diffraction measurements with the electron microscope. J. Sci. Instrum. (J. Phys. E) Ser. 2, **1**, 907–910 (1968).
McConnell, J.D.C.: Electron microscopy. In: Physical methods in determinative mineralogy, Chap. 7 (ed. J. Zussman). London: Academic Press 1967.
Moir, G.K., Gard, J.A., Glasser, F.P.: Crystal chemistry and solid solutions amongst pseudowollastonite-like polytypes of $CaSiO_3$, $SrSiO_3$ and $BaSiO_3$. Z. Krist. **141**, 437–450 (1975).
Oertel, G., Curtis, C.D., Phakey, P.P.: Transmission electron microscope and X-ray diffraction study of muscovite and chlorite. Mineral. Mag. **39**, 176–188 (1973).
Pinsker, Z.G.: Electron diffraction. (Transl. J.A. Spink and E. Feigl.) London: Butterworth 1953.
Reimer, L.: Elektronenmikroskopische Untersuchungs- und Präparationsmethoden. Berlin-Göttingen-Heidelberg: Springer 1959.
Ross, M., Christ, C.L.: Mineralogical applications of electron diffraction. I. Theory and techniques. Am. Mineralogist **43**, 1157–1178 (1958).
Stabenov, J.: Elektroneninterferenzen an übereinanderliegenden Kristallschichten. Z. Krist. **112**, 289–311 (1959a).
Stabenov, J.: Elektroneninterferenzen an übereinanderliegenden Kristallschichten. Z. Physik **156**, 503–521 (1959b).
Sturkey, L.: Practical considerations in interpretation of electron diffraction patterns. Symposium on techniques in electron metallography. Spec. Tech. Publs. Am. Soc. Test. Mater. **339**, 31–46 (1962).
Vainshtein, B.K.: Structure analysis by electron diffraction. (Transl. E. Feigl and J.A. Spink.) Oxford: Pergamon 1964.
Zvyagin, B.B.: Electron diffraction determination of the structure of kaolinite. Soviet Phys. Crystallogr. **5**, 32–42 (transl. from Kristallografiya **5**, 40–50) (1960).
Zvyagin, B.B.: Electron-diffraction analysis of clay minerals. (Transl. S. Lyse.) New York: Plenum Press 1967.
Zvyagin, B.B., Gorshkov, A.I.: Secondary diffraction in mineral crystals superimposed with rotation. Proc. 6th Int. Congress on Electron Microscopy, Kyoto, **1**, 603–604 (1966).

Chapter 2.3

Contrast Effects at Planar Interfaces

S. Amelinckx and J. Van Landuyt

1. Introduction

Minerals are often known to be to a certain extent disordered structurally as can be deduced from the X-ray diffraction data and occasionally from high-resolution optical microscopy.

This structural disorder is usually related to structural or compositional defects which are more or less confined to or limited by crystallographic planes, depending on the nature of the defects.

Possible types of defects are antiphase boundaries, stacking faults, inversion boundaries, and boundaries between twins, ordering domains, and exsolution domains. Transmission electron microscopy is a particularly useful technique for the study of such interfaces as is apparent from the vast amount of studies in the materials science field. An increasing number of studies on minerals by electron microscopy and diffraction have also become available recently.

2. The Origin of Domain Structures

"Single crystals" are often fragmented into domains separated by planar interfaces of different kinds (Fig. 1). Such fragmentation is usually the consequence of a phase transformation.

Many phase transformations can be considered as being the result of the creation of order; others are displacive, i.e. result from small cooperative displacements of atoms, whilst the atoms remain in the same interstices of the structure.

The distinction is not always straightforward especially since the ordering of magnetic moments or of electrical dipoles is often accompanied by a displacive transformation; this is in general the case in ferroelectric or magnetic transitions. It is then not possible to distinguish cause and effect.

Ordering of atoms occurs in a large number of alloys. Whereas in the disordered form the lattice sites are occupied at random, they will be occupied by atoms of a given chemical species in the ordered form. This usually results in the loss of symmetry elements in such a way that the pointgroup of the ordered phase is a subgroup of the pointgroup of the disordered phase. Usually but not always, this decrease in symmetry of the structure is also reflected in a decrease in symmetry of the lattice.

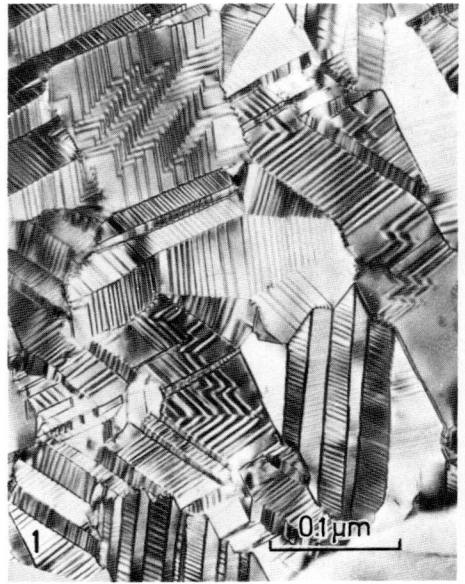
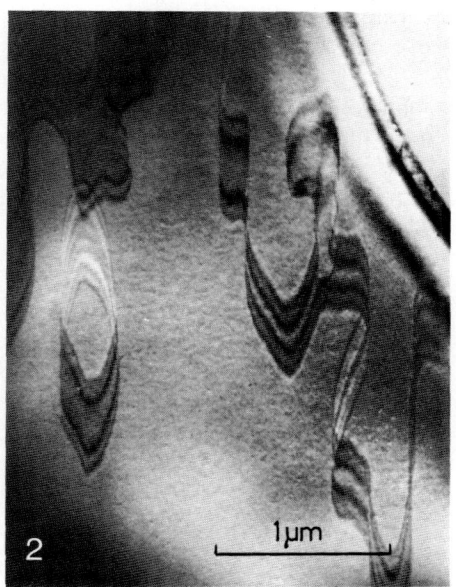

Fig. 1. Fragmented structure of rutile film obtained by the recrystallization of a nearly epitaxic oxide layer. Twins as well as antiphase boundaries are visible

Fig. 2. Domain structure in a ferroelectric crystal (Ba Na(NbO$_3$)$_5$)

In a number of crystals the anions form a close-packed arrangement, whereas the sublattice of octahedral or tetrahedral interstices of this anion lattice is only partly filled with cations. Within the same framework of large anions the cations may then order according to different variants, giving rise to a domain structure (Van Tendeloo and Amelinckx, 1974a) this is for example the case in rutile (Van Landuyt et al., 1964a, 1964b) and in pyrrhotite (Van Landuyt and Amelinckx, 1972). Also in this case the pointgroup of the ordered structure is a subgroup of the disordered one. The latter case can either be considered as the ordering of "interstitials" or as the ordering of "vacancies", whereas the former can be called substitutional atomic order.

The ordering of magnetic dipoles that occurs when paramagnetic solids become either ferromagnetic, ferrimagnetic or antiferromagnetic is usually accompanied by a small lattice deformation as a result of magnetostriction. Also in this case a domain structure results.

Similarly the ordering of electric dipoles on transition from the paraelectric to the ferroelectric or anti-ferroelectric phase causes a decrease in symmetry and as a result a domain structure. Ferroelectric domains in BaNa(NbO$_3$)$_5$ are shown in Fig. 2. The phase transformations of this type are usually of the displacive type, the structure being a "frozen in" soft-mode configuration.

The cooperative Jahn-Teller effect is similarly accompanied by domain formation (Van Landuyt et al., 1972). It was shown recently that also the Verwey ordering in Fe$_3$O$_4$ may lead to the formation of domains (Yamada et al., 1968).

It is possible that Verwey ordering is in fact the result of a cooperative Jahn-Teller effect (Chakraverty, 1974).

In general one can state that any transition that leads to a *decrease* in symmetry will lead to a domain structure. In a large number of cases one finds that the pointgroup of the ordered phase is a subgroup of the pointgroup of the disordered phase; in such cases one can make use of simple group theoretical considerations to formulate a number of general statements; this is the subject of the following paragraph.

2.1 Orientation Variants

In this paragraph we only consider phase transitions for which the pointgroup H of the ordered phase is a subgroup of the pointgroup G of the disordered phase or prototype phase, i.e. $H \subset G$.

Three questions are of interest in the study of domain structures by means of electron microscopy

1. How many different orientation variants are formed on ordering?
2. How are these variants related one to the other?
3. How many crystallographically different planar interfaces between such variants occur?

The answer to these questions have been derived by Van Tendeloo and Amelinckx (1974a); they will be summarized here.

The number of variants is simply given by $n = p/q$, where p and q are the orders of the pointgroups G and H respectively. This can be shown to be a direct application of Lagranges theorem. These variants are related one to the other by operations of G which are not in H. A number of operations of G may produce the same variant starting from a given initial variant. In most cases the set of operations that produces *all* different variants, starting from a given initial variant, forms also a pointgroup: the *variant generating group* V of order $n = p/q$. The group V is also a subgroup of G and it has no element in common with H other than the identity. For a given pair of a G and a H group there are in general several V groups, which, however, generate the same variants. The variant-generating groups for all possible pairs of pointgroups G and H have been tabulated.

Crystallographically different interfaces separate variants related by operations of V which belong to different classes of G. If the group H is non-centro-symmetric, whereas G is centro-symmetric, the V-group contains the inversion operation and domains related by an inversion operation occur.

We shall comment as an example the ordering alloy Ni_4Mo. The disordered structure is face-centered cubic and the pointgroup G is $\frac{4}{m}\overline{3}\frac{2}{m}$ of order 48. The pointgroup H is $4/m$ of order 8; the number of orientation variants is therefore 6. Acceptable V-groups are $3m$ and 32; the 6 elements of these groups belong to *two* different classes of G. There are therefore two different types of interfaces which have been called *perpendicular* twins and *anti-parallel* twins (Ruedl et al., 1968).

Contrast Effects at Planar Interfaces 71

The perpendicular twins are interfaces such that the tetragonal axis in the domains on either side are along mutually perpendicular directions parallel to cube directions of the face-centered cubic parent structure. The anti-parallel twins separate domains which have their z-axis parallel to the same cubic direction of the fcc structure, however, the x-axes are rotated in opposite senses.

2.2 Translation Variants

We shall now look for the number of different ways the ordered structure can be built within a given orientation variant of the structure. We shall call these *translation variants*.

The number of different translation variants for a given orientation variant is given by the number of different translation vectors τ which are lattice vectors for the disordered structure but not for the ordered structure. All different τ-vectors lead therefore from the origin of the unit cell of the superlattice to all lattice points within the primitive unit cell of the superlattice. This number can be obtained by dividing the volume of the primitive unit cell of the superstructure V_0 by the volume of the primitive unit cell of the disordered structure V. Let **M** be the transformation matrix relating the primitive base of the disordered structure (a_1, a_2, a_3) with that of the ordered structure $(a_{1.0}, a_{2.0}, a_{3.0})$ then:

$$\begin{pmatrix} a_{1.0} \\ a_{2.0} \\ a_{3.0} \end{pmatrix} = \mathbf{M} \begin{pmatrix} a_1 \\ a_2 \\ a_3 \end{pmatrix} \tag{1}$$

and $V_0 = |\mathbf{M}| V$; the number t of translation variants is then given by $t = |\mathbf{M}|$. If non-primitive unit cells are used one should correct for the multiplicity; let m_1 and m_2 be respectively the multiplicities of the unit cells of the disordered and ordered structures respectively, then one has:

$$t = (m_1/m_2)|\mathbf{M}|. \tag{2}$$

3. Geometry of Interfaces

The boundaries between different domains are planar interfaces. From the contrast point of view it is of interest to distinguish between three categories:

1. Translation interfaces: between translation variants
2. Twin interfaces separating orientation variants
3. Inversion boundaries.

We shall discuss briefly the geometrical characteristics of each of them.

3.1 Translation Interfaces

If the structure in part II of the crystal (Fig. 3) can be derived from that in part I of the same crystal by means of a parallel translation, the interface can be charac-

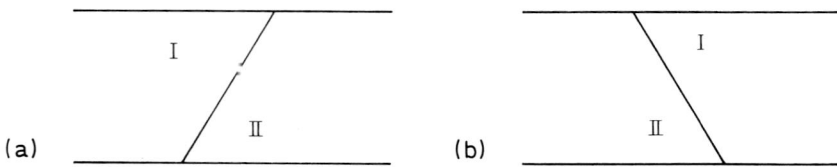

Fig. 3a and b. Inclined interface in a crystal illustrating the definition of part I and part II

terized by a constant displacement vector **R**, always representing the displacement of the crystal part last met by the electrons with respect to the front part; it is only determined modulus a lattice vector.

If **R** is a lattice vector for the disordered structure but not for the ordered one, the interface is called an *antiphase boundary* (APB). If on the other hand **R** is not a lattice vector for the disordered structure it is called a *stacking fault* (SF). The occurrence of stacking faults is common in structures consisting of close-packed layers, because on such layers there are well-defined positions which correspond to relative minima in stacking-fault energy.

In the cubic close-packed stacking sequence, represented by the stacking symbol $abcabc..$ two types of elementary faults can be distinguished: the *intrinsic fault* described by the stacking symbol

$$... abca\overset{\uparrow}{\underline{bc}}abc ... \qquad (3)$$

and the *extrinsic fault* represented by the symbol

$$... abca\overset{\uparrow}{\overline{b}}\overline{ac}abc .... \qquad (4)$$

The faults are in fact triplets of hexagonal stacking within a cubic matrix; there are two of these in each of the faults, they are indicated by brackets but differ in their arrangement.

Two such faults only differ in the sign of **R**, which can be written either as $\pm\frac{a}{3}[111]$ or as $\mp\frac{1}{6}[11\bar{2}]$, as shown by Gevers *et al.* (1963).

In the hexagonal close-packed structure, two main types of faults can be distinguished: *single* faults i.e. with one cubic triplet:

$$... ab\overline{abc}bcb ... \qquad (5)$$

or *double* faults i.e. with two cubic triplets:

$$... abab\overline{abc}aca .... \qquad (6)$$

A schematic view of the lattice planes in the vicinity of a stacking fault with an indication of the displacement vector is shown in Fig. 4.

Fig. 4. Schematic view of a stacking fault, indicating the displacement vector **R**

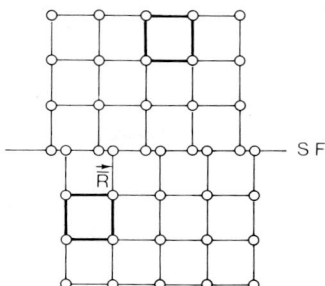

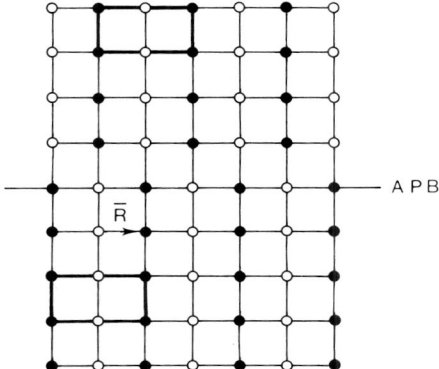

Fig. 5. Two-dimensional model for *APB* in an AB-alloy. (*APB* = antiphase boundary)

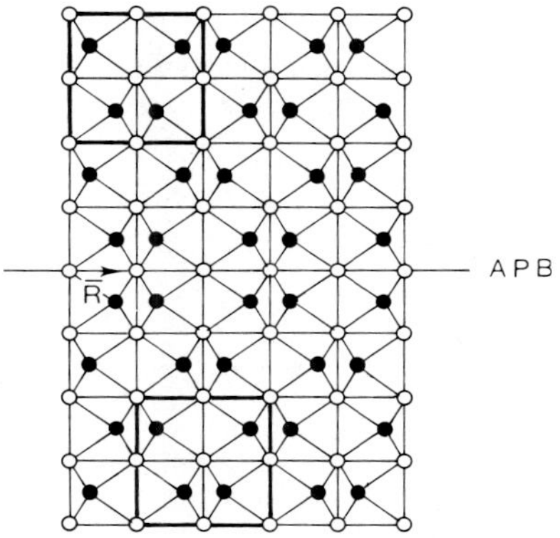

Fig. 6. *APB* resulting from a displacive transformation. The **R**-vector is indicated

Stacking faults can be generated either by growth or by glide of partial dislocations.

The periodic arrangement of stacking faults gives rise to *polytypes* i.e. compounds with the same chemical composition but different stacking sequences of the same layers.

Whereas stacking faults are *geometrical* faults, APB's are rather *chemical* faults.

Antiphase boundaries are common in ordered alloys. A two-dimensional schematic view of an APB in an AB alloy is shown in Fig. 5. APB's may also result from displacive transformations as shown by the two dimensional model of Fig. 6. APB's are *conservative* if the displacement vector is in the boundary plane (Fig. 8). If this is not the case, a slab of material with thickness $t = \mathbf{R} \cdot \mathbf{n}$ ($\mathbf{n}$ is unit normal on boundary plane) is either removed or inserted. If this slab has a chemical composition which is different from that of the crystal, the APB is termed *non-conservative* since its presence causes a change in chemical composition. The name *crystallographic shear plane* is also given to such interfaces in non-metallic compounds such as oxides. So called *non-stoichiometric* compounds often contain regular sequences of crystallographic shear planes. They are then called *shear-structures* (Magnéli, 1953 and Wadsley, 1964) (Fig. 7). Similarly the periodic

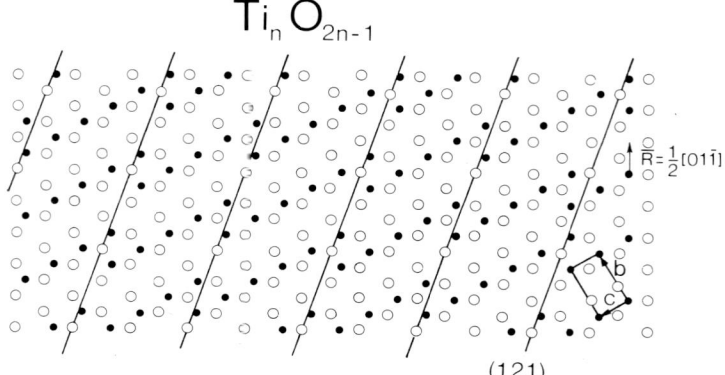

Fig. 7. Two-dimensional model of a shear structure of the type occurring in rutile. Open circles represent oxygen ions, closed circles represent titanium ions

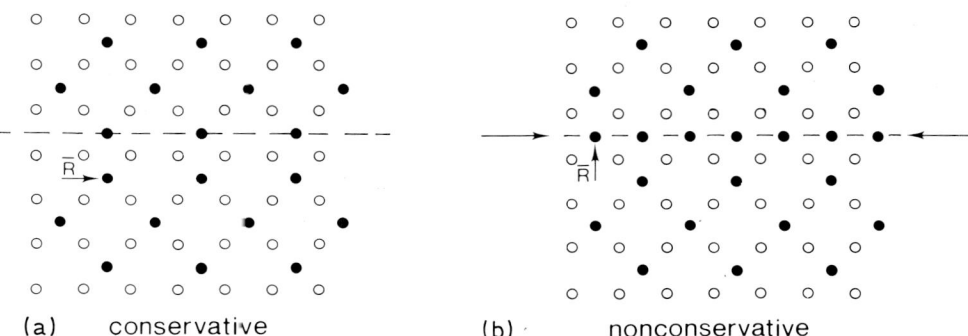

Fig. 8a and b. Two-dimensional model of APB resulting from the ordering of vacancies (or interstitials). (a) Conservative, (b) non-conservative

arrangement of APB's in alloys gives rise to *long-period superstructures* (Ogawa et al., 1958; Glossop and Pashley, 1959; Van Tendeloo et al., 1974b). In ordered alloys a planar interface may be simultaneously a SF and an APB, such interfaces are called *complex faults* (Kear et al., 1969).

APB's may be of thermal nature, i.e. result from ordering, either substitutional ordering or ordering of vacancies (Fig. 8) or they may be generated by glide. Glide often generates *complex faults*.

Certain complex compounds, so-called *mixed layer compounds*, consist of the stacking of lamella say A and B with different chemical compositions in varying succession and proportions, leading to compounds of the type $A_m B_n$. The best-known example is the series of hexagonal ferrites, which contains so-called $M(\equiv BaFe_{12}O_{19})$ and $Y(\equiv Ba_2MeFe_{12}O_{22})$ blocks in regular sequences leading to compounds with a general formula $M_n Y_m$ (Van Landuyt et al., 1973a, 1973b, 1974). At constant composition, crystals with different stacking sequences can still occur e.g. $MYMY_4$ and MY_2MY_3; they are called *mixed layer polytypes*. Such compounds may contain so called *sequential faults* which *do not* perturb the chemical composition, but *do* change the stacking sequence, e.g.

$$MY_2\,MY_3\,MY_2\,MY_3\,MY\underset{\uparrow}{\,}MY_4\underset{\uparrow}{\,}MY_2\,MY_3\,MY_2\,MY_3\,\ldots. \qquad (7)$$

They may also contain *compositional faults*, which do change the composition e.g.

$$\ldots MY_2\,MY_3\,MY_2\,MY_3\,\underset{\uparrow}{MY_3}\,MY_3\,MY_2\,MY_3\,MY_2\,\ldots. \qquad (8)$$

Examples of mixed-layer compounds among minerals are the bastnaesite-synchisite-vaterite series (Van Landuyt and Amelinckx, 1975) and the chlorites.

3.2 Twin Interfaces

From the electron microscopic point of view it is convenient to describe a twin boundary as resulting from the displacement field pictured in Fig. 9. The displacement field now consists of a vector with constant direction and sense, but with a magnitude which increases linearly with the distance to the contact plane between the two crystal parts.

We shall consider two extreme situations which are both of practical interest:

1. the displacement per atom plane is a very small fraction of the interatomic distance, i.e. the "obliquity" of the twin is small (Cahn, 1954);

2. the displacement per atom plane is a large fraction of the interatomic distance.

In the second case, the two crystal parts have to be considered as separate, although related by a symmetry operation of the disordered crystal. From the diffraction point of view they behave as two different crystals, i.e. diffracting separately.

In the first case the crystal can still be considered as "single", the lattices of the two crystal parts being only slightly different. The points in the reciprocal lattice

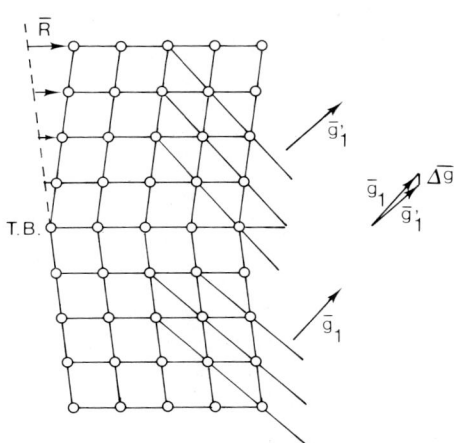

Fig. 9. Displacement field associated with a twin interface. The magnitude of $\boldsymbol{R}$ increases linearly with the distance from the interface. The homologeous diffraction vectors $\boldsymbol{g}_1$ and $\boldsymbol{g}'_1$ in the two crystal parts differ by $\varDelta \boldsymbol{g}$, which is perpendicular to the contact plane

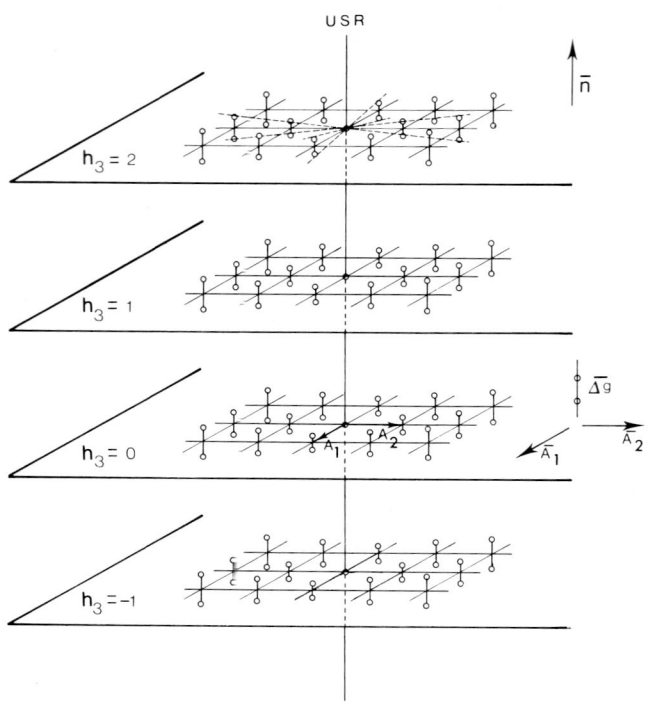

Fig. 10. Reciprocal lattice of a reflection twin; Δg is parallel to the row of unsplit spots; it increases in magnitude with the distance from the unsplit row (USR)

are then "practically" coincident in such a way that the two crystal parts still diffract simultaneously, although with a different excitation error (i.e. deviation from the Bragg position). Such interfaces will be termed *domain boundaries*.

One has to distinguish between reflection twins and rotation twins.

For reflection twins the mirror plane is a common lattice plane and the two individuals are related by a mirror operation with respect to this plane which is "almost" a symmetry plane for the parent structure. When coherent, the composition or contact plane coincides with the mirror or twin plane (Fig. 9). The reciprocal lattice of such a twin is shown in Fig. 10. The difference between two "almost coincident" points Δg is normal to the mirror plane.

For rotation twins the two crystals have a common lattice row and one individual is derived from the other by a rotation over 180° about this common row, which is "almost" a two-fold symmetry operation for the parent structure. The contact plane can now be either perpendicular or parallel to the rotation axis; in the first case the twin is *normal*, in the second case it is *parallel*. The reciprocal lattice of such a twin is shown in Fig. 11. The difference Δg between "almost coincident" points is now perpendicular to the twin axis.

It is not a simple matter to distinguish from electron diffraction evidence between these two types of twins.

Coherent reflection twins, which are the most common type, are fortunately easy to identify. When the interface is seen edge on, the diffraction pattern will

Fig. 11. Reciprocal lattice of a rotation twin; Δg is perpendicular to the twin axis; its magnitude increases with the distance from the plane of unsplit spots (*USP*)

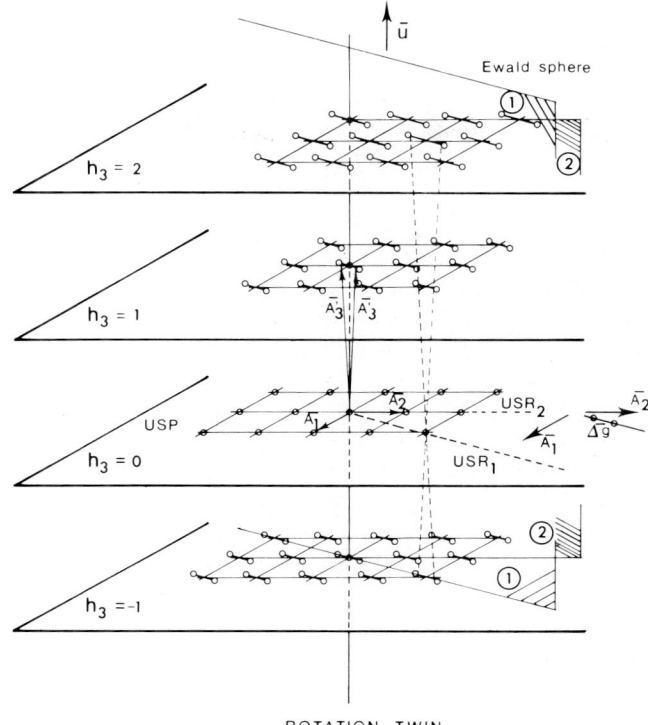

ROTATION TWIN

exhibit a row of unsplit spots, of which the direction is perpendicular to the mirror plane. All other spots are split in a direction parallel to the unsplit row (U.S.R.) and by an amount which is proportional with their distance to the unsplit row. From the amount of splitting the twinning vector can be deduced (Figs. 12, 13).

In all cases Δg can be determined by means of extinctions; for those reflections g for which $\Delta g = 0$ there will be no contrast. For coherent reflection twins, this will be the case for the family of planes which is parallel to the mirror plane, as well as for the family of lattice planes which is perpendicular to the mirror plane, and parallel to the twinning vector.

3.3 Inversion Boundaries

On transforming from a centrosymmetrical into a non-centrosymmetrical structure, boundaries between domains in which the structures are related by an inversion operation are often formed. Such a boundary is shown schematically in two dimensions in Fig. 14. The lattice is now the same in both domains but the structure is *not*.

Such an inversion can sometimes be combined with a parallel translation as in Fig. 15. The structures in the two domains can of course again be deduced one

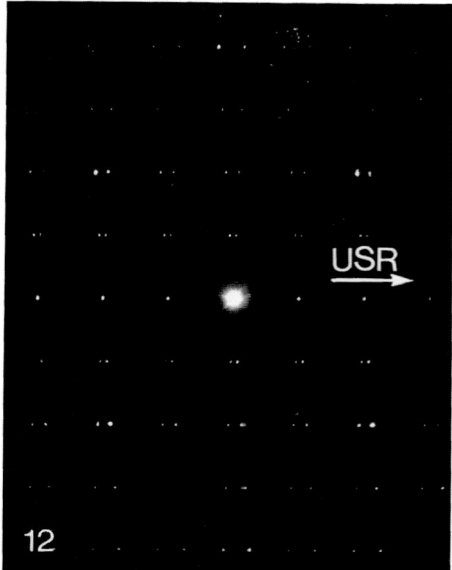

Fig. 12. Example of a diffraction pattern from a reflection twin in leucite. The pattern was taken across the boundary XY in Fig. 13

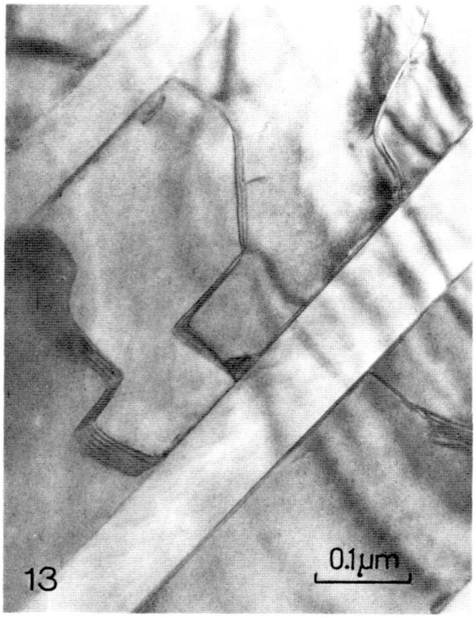

Fig. 13. Twin boundaries and APB's in leucite; the diffraction pattern taken across the twin boundary is shown in Fig. 12

from the other by means of a single inversion. Nevertheless it is meaningful to introduce the concept of a displacement vector R also in this case (Snykers et al., 1972, Serneels et al., 1973) because the presence of such a displacement can be revealed for reflections which are not affected by the non-centrosymmetrical nature of the structure (g_2 in Fig. 15). Boundaries of this nature have been found in the alloy δ-NiMo (Van Tendeloo and Amelinckx, 1973), in the χ-phase $Fe_{36}Cr_{12}Ti_5Mo_5$ (Meulemans et al., 1970, Snykers et al., 1971), and in Li-ferrite (Van der Biest and Thomas, Chapter 2.1 of this volume).

4. Two-beam Dynamical Theory

4.1 Perfect Crystal

We now develop the necessary formalism to describe the contrast effects at single and overlapping planar interfaces in the framework of two-beam dynamical theory (Amelinckx, 1964). The following assumptions are systematically made:

1. the column approximation is valid

2. anomalous absorption is adequately taken into account by the use of a complex lattice potential or, stated otherwise, of a complex extinction distance (Yoshioka, 1957)

3. the symmetrical Laue case of diffraction is realized.

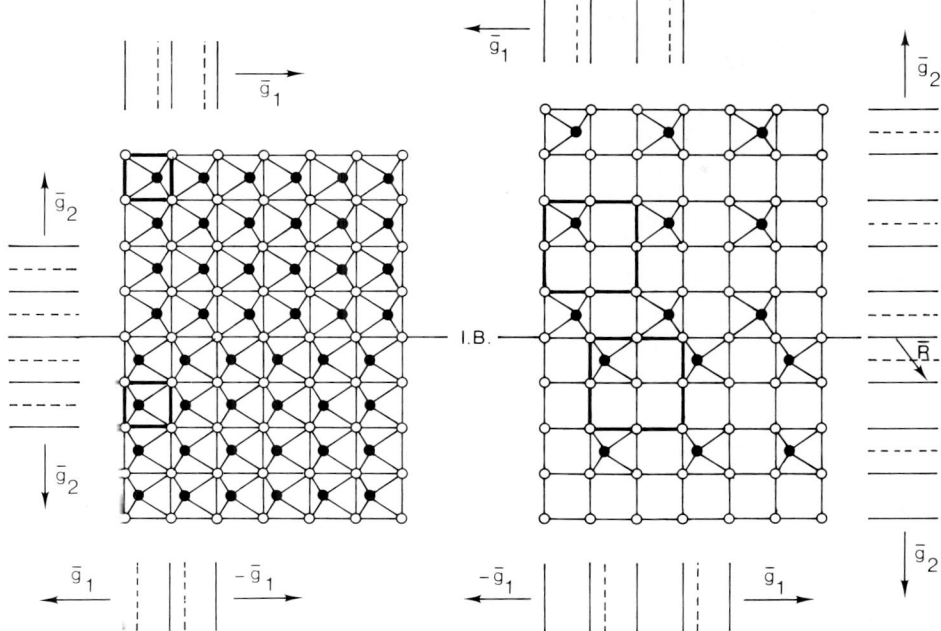

Fig. 14. Two-dimensional model for an inversion boundary in a non-centrosymmetrical crystal. The boundary can be revealed by using reflections such as g_1, but not by using reflections such as g_2

Fig. 15. Two-dimensional model representing a combination on an inversion boundary and an antiphase boundary. The translation can be revealed by reflections such as g_2, whereas reflections of the type g_1 reveal the inversion boundary

If the crystal is non-centrosymmetrical we shall moreover assume that the real and the imaginary parts of this lattice potential may have different phases for the same Fourier-coefficient. Let the complex lattice potential in that case be represented by:

$$V(\mathbf{r}) = v_0 + i w_0 + \sum_{g} (v_g + i w_g) e^{2\pi i \mathbf{g} \cdot \mathbf{r}} \tag{9}$$

where $\mathbf{g}$ is a reciprocal lattice vector.

The Fourier-coefficients v_g and w_g will in general be complex; we shall write:

$$\begin{aligned} v_g &= |v_g| e^{i\Theta_g}; & v_{-g} &= |v_{-g}| e^{-i\Theta_g} \\ w_g &= |w_g| e^{i\Theta_g}; & w_{-g} &= |w_{-g}| e^{-i\Theta_g}. \end{aligned} \tag{10}$$

We shall furthermore put:

$$(v_g + i w_g)/k_g = \frac{1}{\xi_g} e^{i\Theta_g} + \frac{i}{\xi'_g} e^{i\phi_g} = \frac{1}{q_{-g}} e^{i\Theta_g}$$

$$(v_{-g} + i w_{-g})/k_g = \frac{1}{\xi_g} e^{-i\Theta_g} + \frac{i}{\xi'_g} e^{-i\phi_g} = \frac{1}{q_{-g}} e^{-i\Theta_g}$$

(11)

where:

$$\frac{1}{q_g} = \frac{1}{\xi_g} + \frac{i}{\xi'_g} e^{i\beta_g} \qquad \frac{1}{q_{-g}} = \frac{1}{\xi_g} + \frac{i}{\xi'_g} e^{-i\beta_g}$$

(12)

with:

$$\beta_g = \phi_g - \Theta_g.$$

(13)

Also we have $w_0/k_0 = 1/\xi'_0$; $k_0 \cong k_g$ is the wave vector of the incident and scattered electron.

With this notation the generalized Howie-Whelan system of coupled differential equations becomes (Howie and Whelan, 1961):

$$\frac{d\psi_0}{dz} = \frac{\pi i}{q_{-g}} e^{-i\Theta_g} \psi_g - \frac{\pi}{\xi'_0} \psi_0$$

$$\frac{d\psi_g}{dz} = \frac{\pi i}{q_g} e^{i\Theta_g} \psi_0 + 2\pi i s_g \psi_g - \frac{\pi}{\xi'_0} \psi_g$$

(14)

where s_g describes the deviation from the exact Bragg condition; ψ_0 and ψ_g are the amplitudes of transmitted (ψ_0) and scattered (ψ_g) beams respectively, z measures the depth in the crystal, the origin being in the entrance face. This system of equations describes the dynamical interplay between incident and diffracted beam.

The substitution:

$$\psi_0 = T_g e^{-\frac{\pi z}{\xi'_0}} e^{\pi i s z}, \qquad \psi_g = S_g e^{-\frac{\pi z}{\xi'_0}} e^{i\Theta_g} e^{\pi i s z}$$

$$\begin{pmatrix} \psi_0 \\ \psi_g \end{pmatrix} = e^{-\frac{\pi z}{\xi'_0}} e^{\pi i s z} \begin{pmatrix} 1 & 0 \\ 0 & e^{i\Theta_g} \end{pmatrix} \begin{pmatrix} T_g \\ S_g \end{pmatrix}$$

(15)

transforms the system of Eq. (14) into: (16)

$$\frac{dT_g}{dz} = \frac{\pi i}{q_{-g}} S_g - \pi i s_g T_g$$

$$\frac{dS_g}{dz} = \frac{\pi i}{q_g} T_g + \pi i s_g S_g.$$

(16)

The normal absorption, described by the parameter ξ'_0 is clearly accounted for by the decreasing exponential factor in the Eq. (15). Since this factor is common for ψ_0 and ψ_g we shall ignore it in the rest of the discussion. The Eq. (15) also demonstrates that due to the presence of the factor $e^{i\Theta_g}$, a phase difference is introduced between diffracted and transmitted beam. This phase difference is of opposite sign for the $+\boldsymbol{g}$ and the $-\boldsymbol{g}$ reflections operating; this has direct consequences on the contrast behavior of inversion boundaries as we shall see.

In the centrosymmetrical case we can choose the origin in a center of symmetry and then $V(\boldsymbol{r}) = V(-\boldsymbol{r})$ (17) as a result $\beta_g = 0$ or π and $\Theta_g = 0$. Assuming $\beta_g = 0$ and simplifying the notation, the set of Eq. (16) reduces to:

$$\frac{dT}{dz} + \pi i s T = \frac{\pi i}{q_{-g}} S$$

$$\frac{dS}{dz} - \pi i s S = \frac{\pi i}{q_g} T \qquad (17)$$

where $\quad \dfrac{1}{q_g} = \dfrac{1}{\xi_g} + \dfrac{i}{\xi'_g}$.

Putting $\sigma = \dfrac{1}{q_g}\sqrt{1+(sq_g)^2}$ we obtain for the real and imaginary parts of σ

$$\sigma = \sigma_r + i\sigma_i \quad \text{with} \quad \sigma_r \simeq \frac{1}{\xi_g}\sqrt{1+(s\xi_g)^2}\,; \quad \sigma_i \simeq \frac{1}{\xi'_g\sqrt{1+(s\xi_g)^2}} \qquad (18)$$

where higher order terms in ξ_g/ξ'_g have been neglected.

The solution of Eq. (17) with the initial values $T=1$, $S=0$ at $z=0$ is now:

$$T(z,s) = \cos\pi\sigma z - i(s/\sigma_r)\sin\pi\sigma z$$
$$S(z,s) = [i/(\sigma_r \xi_g)]\sin\pi\sigma z \qquad (19)$$

where we have used the further approximation that σ has been replaced by σ_r in the coefficients but not in the exponentials or goniometric expressions. The solution of the same system of equations, but for the initial values $T=0$ and $S=1$ at $z=0$, is on the other hand:

$$T = S(z, -s) \equiv S^-$$
$$S = T(z, -s) \equiv T^-. \qquad (20)$$

This follows by noting that the initial values $T=1$, $S=0$ reduce to the initial values $T=0$ $S=1$ if one interchanges T and S. Performing the same interchange in the set of Eqs. (16) or (17) and simultaneously changing the sign of s leaves the set of equations unchanged. The solution is therefore obtained by performing this substitution in the expressions Eq. (19) which leads immediately to the proposed solution Eq. (20).

In matrix notation we thus obtain:

$$\begin{pmatrix} T \\ S \end{pmatrix}_{out} = \begin{pmatrix} T & S^- \\ S & T^- \end{pmatrix} \begin{pmatrix} T \\ S \end{pmatrix}_{in} = \mathbf{M}(z,s) \begin{pmatrix} T \\ S \end{pmatrix}_{in} \tag{21}$$

where the subscripts in and out refer respectively to the incoming and outgoing wave for the slab of perfect crystal.

The square matrix $\mathbf{M}$ is the *response* matrix of the crystal slab. Knowing the amplitudes of the wave incident on the crystal in the directions of the incident beam and of the Bragg scattered beam it allows calculation of the amplitudes of the outgoing waves in the same directions. If the considered slab is the first met by the electrons one has:

$$\begin{pmatrix} T \\ S \end{pmatrix} = \mathbf{M}(s,z) \begin{pmatrix} 1 \\ 0 \end{pmatrix}. \tag{22}$$

Evidently $\mathbf{M}(s,z)$ must have the property:

$$\mathbf{M}(z_1+z_2+\cdots+z_n,s) = \mathbf{M}(z_n,s)\ldots\mathbf{M}(z_2,s)\mathbf{M}(z_1,s) \tag{23}$$

which is easy to verify. Also the property is physically evident, it expresses the fact that it is immaterial whether or not the perfect crystal is thought to be divided into a number of lamellae by introducing interfaces parallel to the foil surfaces. For the non-centrosymmetrical crystal one cannot ignore the phase factor $e^{i\Theta_g}$ in the scattered amplitude even when $\beta_g = 0$.

The response matrix must therefore be written as:

$$\begin{pmatrix} T_g & S_g^- e^{-i\Theta_g} \\ S_g e^{i\Theta_g} & T_g^- \end{pmatrix}. \tag{24}$$

For a single perfect slab it leads to the correct expressions.

This response matrix can be written as a product of matrices:

$$\begin{pmatrix} 1 & 0 \\ 0 & e^{i\Theta_g} \end{pmatrix} \begin{pmatrix} T_g & S_g^- \\ S_g & T_g^- \end{pmatrix} \begin{pmatrix} 1 & 0 \\ 0 & e^{-i\Theta_g} \end{pmatrix} \begin{pmatrix} 1 \\ 0 \end{pmatrix}, \tag{25}$$

$$\begin{pmatrix} T_g \\ S_g e^{i\Theta_g} \end{pmatrix} = \mathbf{S}(\Theta_g) \mathbf{M}(z,s) \mathbf{S}(-\Theta_g) \begin{pmatrix} 1 \\ 0 \end{pmatrix} \tag{26}$$

where

$$\mathbf{S}(\alpha) = \begin{pmatrix} 1 & 0 \\ 0 & e^{i\alpha} \end{pmatrix}. \tag{27}$$

Note that the $\mathbf{S}$ matrix has the property that:

$$\mathbf{S}(\alpha_1)\mathbf{S}(\alpha_2) = \mathbf{S}(\alpha_2)\mathbf{S}(\alpha_1) = \mathbf{S}(\alpha_1+\alpha_2) \tag{28}$$

and further that

$$\mathbf{S}^{-1}(\alpha) = \mathbf{S}(-\alpha). \tag{29}$$

4.2 Faulted Crystals

4.2.1 General

The deformation of the crystal is characterized by means of the displacement field $\boldsymbol{R}(\boldsymbol{r})$. The atom which would be at $\boldsymbol{r}$ in the perfect crystal is now to be found at $\boldsymbol{r} + \boldsymbol{R}(\boldsymbol{r})$ in the faulted crystal. Since the crystal is described by its lattice potential one can write:

$$V_F(\boldsymbol{r}) = V_P(\boldsymbol{r} - \boldsymbol{R}) \tag{30}$$

where F and P refer to faulted and perfect crystal respectively. The lattice potential of the faulted crystal then becomes:

$$V_F(\boldsymbol{r}) = \sum_{\boldsymbol{g}} v_{\boldsymbol{g}} e^{2\pi i \boldsymbol{g} \cdot (\boldsymbol{r} - \boldsymbol{R})} = \sum_{\boldsymbol{g}} (v_{\boldsymbol{g}} e^{-i\alpha_{\boldsymbol{g}}}) e^{2\pi i \boldsymbol{g} \cdot \boldsymbol{r}} \tag{31}$$

where $\alpha_{\boldsymbol{g}} = 2\pi \boldsymbol{g} \cdot \boldsymbol{R}$. The deformation is thus taken into account by replacing the Fourier-coefficient $v_{\boldsymbol{g}}$ of the perfect crystal by $v_{\boldsymbol{g}} e^{-i\alpha_{\boldsymbol{g}}}$

In view of the inverse proportionality between the extinction distance $\xi_{\boldsymbol{g}}$ and the Fourier-coefficient $v_{\boldsymbol{g}}$ (Hirsch et al., 1965) one concludes that the following substitution must be performed in the system of equations:

$$\frac{1}{q_{\boldsymbol{g}}} \to \frac{1}{q_{\boldsymbol{g}}} e^{i\alpha_{\boldsymbol{g}}} \tag{32}$$

and since moreover $\alpha_{-\boldsymbol{g}} = -\alpha_{\boldsymbol{g}}$, we also have:

$$\frac{1}{q_{-\boldsymbol{g}}} \to \frac{1}{q_{-\boldsymbol{g}}} e^{i\alpha_{\boldsymbol{g}}}. \tag{33}$$

The system of Eq. (17) therefore becomes:

$$\begin{aligned} \frac{dT}{dz} + \pi i s T &= \frac{\pi i}{q_{-\boldsymbol{g}}} e^{i\alpha_{\boldsymbol{g}}} S \\ \frac{dS}{dz} - \pi i s S &= \frac{\pi i}{q_{\boldsymbol{g}}} e^{-i\alpha_{\boldsymbol{g}}} T. \end{aligned} \tag{34}$$

The substitution:

$$T = T' e^{\pi i \alpha'}, \quad S = S' e^{-\pi i \alpha'} \tag{35}$$

with $\alpha' = \alpha/2\pi$ reduces the set of equations to the following one:

$$\begin{aligned} \frac{dT'}{dz} + \pi i \left(s + \frac{d\alpha'}{dz}\right) T' &= \frac{\pi i}{q_{-\boldsymbol{g}}} S' \\ \frac{dS'}{dz} - \pi i \left(s + \frac{d\alpha'}{dz}\right) S' &= \frac{\pi i}{q_{\boldsymbol{g}}} T'. \end{aligned} \tag{36}$$

The deformation is clearly described by the variation of the effective s value; when α' varies with z the effective s-value:

$$S_{\text{eff}} = s + \frac{d\alpha'}{dz} \tag{37}$$

is in general also a function of z.

For a stacking fault $\alpha' = $ constant and $\frac{d\alpha'}{dz} = 0$; the system of equations then reduces to that for a perfect crystal. This expresses the physically evident fact that a parallel displacement of the crystal does not affect the amplitudes of transmitted or scattered beam.

For a domain boundary α' is a linear function of z and therefore $\frac{d\alpha'}{dz} = k = $ constant. The presence of the domain boundary causes the second part of the crystal to have an s-value: $s_2 = s_1 + k$ differing by a constant from that of the first part (s_1).

4.2.2 Crystal with a Planar Interface: Response Matrices

The transmitted and scattered beams which emerge from the first crystal part are now incident on the second part. For a crystal containing a planar interface both crystal parts are of course perfect, but related one to the other in some definite way, depending on the type of interface.

1. Inversion boundary: whereas $+g$ is active in the first part $-g$ is active in the second part. The deviation from the Bragg position is the same in both parts.
2. Translation boundary: $\alpha = $ constant, $s_1 = s_2$, i.e. the orientation is again the same in both parts.
3. Domain boundary (twin): the s-value in the second part (s_2) is constant and different from that in the first part (s_1).

Transmission through the crystal must now be described by the product of a succession of response matrices, one relating to each part. A parallel displacement such as that caused by the presence of a translation interface has to be taken into account by the introduction of an appropriate matrix, as we shall see further.

4.2.2.1 Inversion Boundary

The response matrix of the crystal is now described by the product of response matrices:

$$\underbrace{\mathbf{S}(-\Theta_g)\,\mathbf{M}(z_2, s)\,\mathbf{S}(\Theta_g)}_{\text{part II}}\,\underbrace{\mathbf{S}(\Theta_g)\,\mathbf{M}(z_1, s)\,\mathbf{S}(-\Theta_g)}_{\text{part I}}. \tag{38}$$

Where the phase angle for reflection g is Θ_g in the first part, it is $-\Theta_g$ in the second part, referred to the same origin.

Fig. 16. Two-dimensional model for a general interface, which is simultaneously a twin boundary and a translation interface

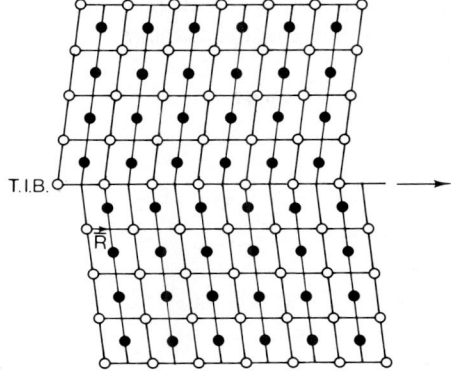

Making use of the properties of the **S** matrices one finds, apart from irrelevant phase factors,

$$\begin{pmatrix} T \\ S \end{pmatrix} = \mathbf{M}(z_2, s)\, \mathbf{S}(2\Theta_g)\, \mathbf{M}(z_1, s) \begin{pmatrix} 1 \\ 0 \end{pmatrix}. \tag{39}$$

We shall see further below that this is equivalent to the response of a crystal containing a pure translation interface with $\alpha = 2\Theta_g$.

4.2.2.2 Domain Boundary

The response of the crystal is given by the product of the response matrices of the two crystal parts, which have now a slightly different orientation:

$$\begin{pmatrix} T \\ S \end{pmatrix} = \mathbf{M}(z_2, s_2)\, \mathbf{M}(z_1, s_1) \begin{pmatrix} 1 \\ 0 \end{pmatrix}. \tag{40}$$

Note that the s values are different for the two matrices.

4.2.2.3 Translation Boundary

We shall now consider for generality an interface which is simultaneously a translation interface and a domain boundary i.e. $\alpha \neq 0$ and also $s_1 \neq s_2$ (Fig. 16). The response matrix for the first part is clearly $\mathbf{M}(z_1, s_1)$. The transmitted and scattered beams of the first part are now incident on the second part. Let the response matrix of the second part be represented by:

$$\begin{pmatrix} A & B \\ C & D \end{pmatrix}. \tag{41}$$

The amplitudes of scattered and transmitted beam for part II are solutions of the set of Eq. (34). The substitution:

$$\begin{pmatrix} T'' \\ S'' \end{pmatrix} = \begin{pmatrix} 1 & 0 \\ 0 & e^{i\alpha} \end{pmatrix} \begin{pmatrix} T \\ S \end{pmatrix} \tag{42}$$

transforms the Eq. (34) into the set for a perfect crystal. It does not affect the initial values. For the initial values:

$$\begin{pmatrix} T'' \\ S'' \end{pmatrix}_{in} = \begin{pmatrix} 1 \\ 0 \end{pmatrix} \tag{43}$$

the solution is:

$$T'' = T(z_2, s_2), \quad S'' = S(z_2, s_2). \tag{44}$$

For the original system of Eq. (34) the solution is therefore:

$$T = T(z_2, s_2), \quad S = S(z_2, s_2) e^{-i\alpha} \tag{45}$$

i.e. the elements A and C are:

$$A = T(z_2, s_2) \equiv T_2 \quad \text{and} \quad C = S(z_2, s_2) e^{-i\alpha} \equiv S_2 e^{-i\alpha}. \tag{46}$$

In order to determine B and D we write the solution of the set of Eq. (34) for the initial values:

$$\begin{pmatrix} T \\ S \end{pmatrix}_{z=0} = \begin{pmatrix} 0 \\ 1 \end{pmatrix}. \tag{47}$$

Interchanging T and S clearly reduces these initial values to the previous ones. The following substitution leaves the system of Eq. (34) invariant:

$$T \rightleftarrows S; \quad \alpha \to -\alpha; \quad s \to -s. \tag{48}$$

The solution thus becomes:

$$B = S(z_2, -s_2) e^{i\alpha} \equiv S_2^{(-)} e^{i\alpha}; \quad D = T(z_2, -s_2) \equiv T_2^{(-)} \tag{49}$$

and the complete matrix therefore is:

$$\begin{pmatrix} T_2 & S_2^{(-)} e^{i\alpha} \\ S_2 e^{-i\alpha} & T_2^{(-)} \end{pmatrix}. \tag{50}$$

By analogy with Eq. (24) this can be written as a product of matrices:

$$\begin{pmatrix} 1 & 0 \\ 0 & e^{-i\alpha} \end{pmatrix} \begin{pmatrix} T_2 & S_2^{(-)} \\ S_2 & T_2^{(-)} \end{pmatrix} \begin{pmatrix} 1 & 0 \\ 0 & e^{i\alpha} \end{pmatrix} \tag{51}$$

or finally for the faulted crystal:

$$\begin{pmatrix} T \\ S \end{pmatrix} = \mathbf{S}(-\alpha) \mathbf{M}_2 \mathbf{S}(\alpha) \mathbf{M}_1 \begin{pmatrix} 1 \\ 0 \end{pmatrix}. \tag{52}$$

Contrast Effects at Planar Interfaces

This expression can readily be generalized for overlapping interfaces. Let the phase angles of the interfaces be α'_j, when referred to the front part of the crystal and let the s values be s_j, we then obtain:

$$\begin{pmatrix} T \\ S \end{pmatrix} = \ldots \mathbf{S}(-\alpha'_2) \mathbf{M}_3 \mathbf{S}(\alpha'_2) \mathbf{S}(-\alpha'_1) \mathbf{M}_2 \mathbf{S}(\alpha'_1) \mathbf{M}_1 \begin{pmatrix} 1 \\ 0 \end{pmatrix}. \tag{53}$$

Introducing the angles $\alpha_j = \alpha'_j - \alpha'_{j-1}$ and putting $\alpha'_0 = 0$, we can rewrite this:

$$\begin{pmatrix} T \\ S \end{pmatrix} = \ldots \mathbf{M}_3 \mathbf{S}(\alpha_2) \mathbf{M}_2 \mathbf{S}(\alpha_1) \mathbf{M}_1 \begin{pmatrix} 1 \\ 0 \end{pmatrix}. \tag{54}$$

Where now the α_j represent the phase angles resulting from the relative displacement of two successive lamella, each lamella always being displaced with respect to the previous one.

Note that Eq. (52) is the same as that for an inversion boundary, provided $s_1 = s_2$ and $\alpha = 2\Theta_g$ and ignoring an irrelevant phase factor which does not affect the intensities.

4.2.3 Interpretation of Response Matrix [Eq. (52)]

For a single interface the explicit expressions for T and S can easily be written down, they are:

$$\begin{cases} T = T_1 T_2 + S_1 S_2^{(-)} e^{i\alpha} \\ S = T_1 S_2 e^{-i\alpha} + S_1 T_2^{(-)}. \end{cases} \tag{55}$$

The amplitude of the transmitted beam thus results from the interference between the doubly transmitted $(T_1 T_2)$ and the doubly scattered beams $(S_1 S_2^{(-)} e^{i\alpha})$ (Fig. 17). One has to remember that for the double scattering event the beam is incident on part II in the direction of the primary scattered beam; this is accounted for

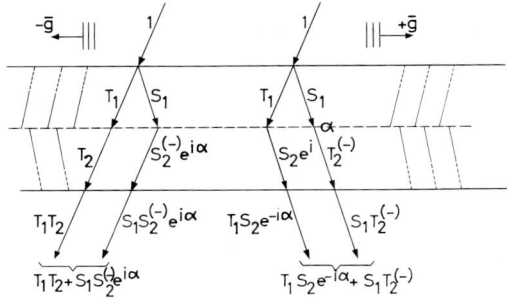

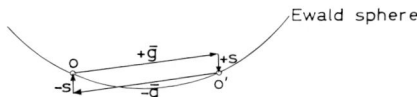

Fig. 17. Schematic representation of the different contributions to the transmitted and scattered amplitudes for a crystal containing an interface

by the use of $-s$ for the deviation parameter. Furthermore the wave scattered by the second part suffers a phase change α as a result of the stacking fault, this introduces the phase factor $e^{i\alpha}$.

The amplitude of the scattered beam similarly results from the interference between the beam scattered by part I and subsequently transmitted by part II in the scattered direction $(S_1 \, T_2^{(-)})$ and the beam transmitted by part I and subsequently scattered by part I suffering a phase change $-\alpha$ because scattering by part II now occurs from the other side as compared to the term $S_1 \, S_2^{(-)} \, e^{i\alpha}$ in T.

This type of analysis is also valid in the case of overlapping interfaces. However the number of terms becomes rapidly large.

4.2.4 Vacuum Matrix

Crystal parts which are far from any reflecting position behave with respect to scattering as a vacuum layer. We therefore need the response matrix for a layer of non-scattering crystal. We can express this by noting that the extinction distance for that layer becomes infinite. For an infinite extinction distance the set of Eq. (17) reduces to:

$$\frac{dT}{dz} + \pi \, i \, s \, T = 0; \qquad \frac{dS}{dz} - \pi \, i \, s \, S = 0. \tag{56}$$

The solution is clearly:

$$T = T_0 \, e^{-\pi \, i \, s \, z}; \qquad S = S_0 \, e^{\pi \, i \, s \, z} \tag{57}$$

where T_0 and S_0 are the amplitudes of the beams incident on the vacuum layer respectively in the incident and in the scattered directions. We can write:

$$\begin{pmatrix} T \\ S \end{pmatrix} = \begin{pmatrix} e^{-i s z} & 0 \\ 0 & e^{\pi i s z} \end{pmatrix} \begin{pmatrix} T_0 \\ S_0 \end{pmatrix} \equiv \mathbf{V}(z, s) \begin{pmatrix} T_0 \\ S_0 \end{pmatrix}. \tag{58}$$

When $\mathbf{V}$ is the vacuum matrix, which depends on the s-value of the crystal part that *preceeds* the vacuum layer, whereas z is the thickness of the layer.

4.3 Discussion of Fringe Profiles

The Eq. (55) yields expressions for the amplitudes T and S. The intensities are obtained from

$$I_T = TT^* \quad \text{and} \quad I_S = SS^* \tag{59}$$

where the asterisk means complex conjugate.

The explicit analytical expressions cannot easily be discussed in the general case; such a discussion can be found in Amelinckx (1970) and Gevers et al. (1965).

We shall limit ourselves to two special cases

(1) $\alpha \neq 0$ $\quad s_1 = s_2 = 0 (\delta = 0)$ $\quad$ (α-fringes)
(2) $\alpha = 0$ $\quad s_1 \xi_{1,g} - s_2 \xi_{2,g} = \delta \neq 0$ $\quad$ (δ-fringes)

where $\xi_{1,g}$ and $\xi_{2,g}$ are the extinction distances in the crystal parts on either side of the boundary.

4.3.1 α-Fringes

The symmetry properties can be derived directly from Eqs. (59) and (55); one has:

$$I_T(z_1, z_2, s, \alpha) = I_T(z_2, z_1, s, \alpha) \tag{60}$$

$$I_S(z_1, z_2, s, \alpha) = I_S(z_2, z_1, -s, -\alpha)$$

which means that the bright field fringe pattern will be symmetrical with respect to the centre of the foil; the dark field image on the other hand will be asymmetrical. In particular the nature of the outer fringes will be the same in the bright field (BF) image, whereas their nature will be different in the dark field (DF) image.

Explicitly one can write this expression as the sum of three terms

$$I_{T,S} = I_{T,S}^{(1)} + I_{T,S}^{(2)} + I_{T,S}^{(3)} \tag{61}$$

with

$$I_{T,S}^{(1)} = \tfrac{1}{2} \cos^2(\alpha/2) [\cosh 2\pi \sigma_i z_0 \pm \cos 2\pi \sigma_r z_0], \tag{62a}$$

$$I_{T,S}^{(2)} = \tfrac{1}{2} \sin^2(\alpha/2) [\cosh 4\pi \sigma_i u \pm \cos 4\pi \sigma_r u], \tag{62b}$$

$$I_{T,S}^{(3)} = \tfrac{1}{2} \sin \alpha [\sin 2\pi \sigma_r z_1 \sinh 2\pi \sigma_i z_2 \pm \sin 2\pi \sigma_r z_2 \sinh 2\pi \sigma_i z_1]. \tag{62c}$$

The upper sign applies to I_T, whereas the lower sign applies to I_S; z_0 is the total thickness ($z_0 = z_1 + z_2$) whereas, $2u = z_1 - z_2$. In fact u is the distance of the interface from the foil center.

For $\alpha = 0$ or a multiple of 2π the first term is the only non-vanishing one; this means that this term represents the contribution from the perfect crystal; it cannot represent fault fringes.

In sufficiently thick foils, i.e. for

$$2 \sinh 2\pi \sigma_i z_0 \gg tg\, \alpha/2$$

and for $\alpha \neq \pi$ the general behavior of the fringe pattern is described by the third term, which itself consists of two terms.

The first term is large at the entrance face since $\sinh 2\pi \sigma_i z_2$ is larger there ($z_2 \simeq z_0$). It represents a damped sinusoid which disappears at the back surface where $z_2 \simeq 0$. The first extremum will be a maximum for $\sin \alpha > 0$ i.e. the first fringe will then be bright.

The second term is similarly large at the exit face since there $\sinh 2\pi \sigma_i z_1$ is large because $z_1 \simeq z_0$, it disappears at the front face since then $z_2 \simeq 0$. This term also represents a damped sinusoid; the first extremum, which now corresponds to the last fringe, is a maximum or a minimum for $\sin \alpha > 0$, depending on whether we consider I_T or I_S. If $\sin \alpha < 0$ the maxima become minima and *vice versa*.

In conclusion we find the results summarized in Table 1 for the nature of the outer fringes.

Table 1. α-Fringes

	B.F.		D.F.	
	F	L	F	L
$\sin \alpha > 0$	B	B	B	D
$\sin \alpha < 0$	D	D	D	B

B: bright; D: dark; F: first fringe; L: last fringe

It should be noted that these results are only valid for *thick* foils. For thin foils, as well as for a discussion of the detailed behavior in the central part of the fringe pattern one has to use the complete expression.

Fig. 18 shows a number of computed fringe profiles for $\alpha = 2\pi/3$, $s=0$, $\xi_g/\xi'_g = 0.035$ with varying thickness.

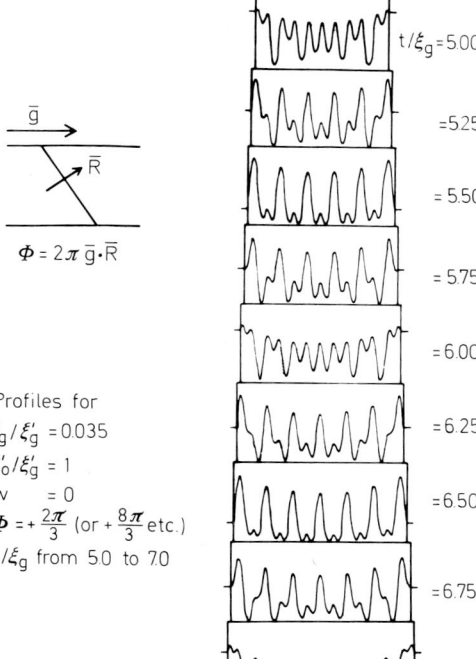

Fig. 18. Two-beam computed profiles for stacking fault fringes with $\alpha = 2\pi/3$, $s=0$, $\xi_g/\xi'_g = 0.035$ and varying thickness. (Courtesy Booker)

Contrast Effects at Planar Interfaces

For $\alpha = \pm 2\pi/3$ and in sufficiently thick foils, the fringes have the following properties.

1. New fringes are created in the center of the foil with increasing thickness.
2. The fringes are parallel to the closest surface.
3. The BF fringe pattern is symmetrical with respect to the foil center, whereas the DF image is similar to the BF image to the front surface, but pseudo-complementary close to the back surface.

These properties can be verified on Fig. 19.

The singular case $\alpha = \pi$ gives rise to an extremely simple pattern if $s=0$. The profile is now described completely by the second term:

$$I_{T,S} = 1/2\,(\cosh 4\pi\sigma_i u \pm \cos 4\pi\sigma_r u) \tag{63}$$

where $\sigma_r = 1/\xi_g$. The following properties of the fringe pattern are immediately apparent from this formula:

1. Bright and dark field image are complementary with respect to the background given by:

$$1/2 \cosh 4\pi\sigma_i u.$$

2. The central fringe ($u \simeq 0$) is bright in the BF image and dark in the DF image.
3. The fringes are parallel to the foil center (i.e. determined by u) rather than to the closest surface (i.e. determined respectively by z_1 or z_2); their depth period is $1/2\xi_g$.
4. With increasing foil thickness, new fringes are created at the surface, rather than in the center.

Examples of π-fringes are visible in Fig. 20.

4.3.2 δ-Fringes

We shall limit ourselves to the discussion of the simplest case $s_1 = -s_2 = s$ i.e. the *symmetrical* case. For a general discussion we refer to Gevers et al. (1965). Assuming that the foil is sufficiently thick one finds that the profile is adequately described by the third term.

$$w^4 I_{T,S}^{(3)} = -(1/2)\delta\,\{\cos 2\pi\sigma_{1r}z_1 \sinh[2(\pi\sigma_{2i}z_2 \pm \phi_2)]$$
$$\pm \cos 2\pi\sigma_{2r}z_2 \sinh[2(\pi\sigma_{1i}z_1 + \phi_1)]\} \tag{64}$$

with

$$w^2 = 1 + (s\xi_g)^2; \quad \delta = s_1\xi_{g,1} - s_2\xi_{g,2}; \quad 2\phi_j = \operatorname{argsinh}(s\xi_g)_j \tag{65a-c}$$

where now $\xi_{g,1}$ and $\xi_{g,2}$ are the extinction distances in the first and second part respectively. The upper sign applies to I_T and the lower to I_S. Close to the front surface the first term again dominates, whereas close to the back surface the second term is the dominating one. One concludes that close to the front surface BF and DF are similar, whereas they are pseudo-complementary close to the back surface. This is in fact a general property of two-beam defect images in thick foils; it is a result of anomalous absorption.

The fringes have furthermore the following properties:

1. The depth period may be different close to front and back surface if $\xi_{g,1}$ is significantly different from $\xi_{g,2}$.

2. The pattern is symmetrical in the DF as far as $\xi_{g,1}$ is equal to $\xi_{g,2}$.

3. Fringes are parallel to the closest surface and therefore new fringes are added in the central part.

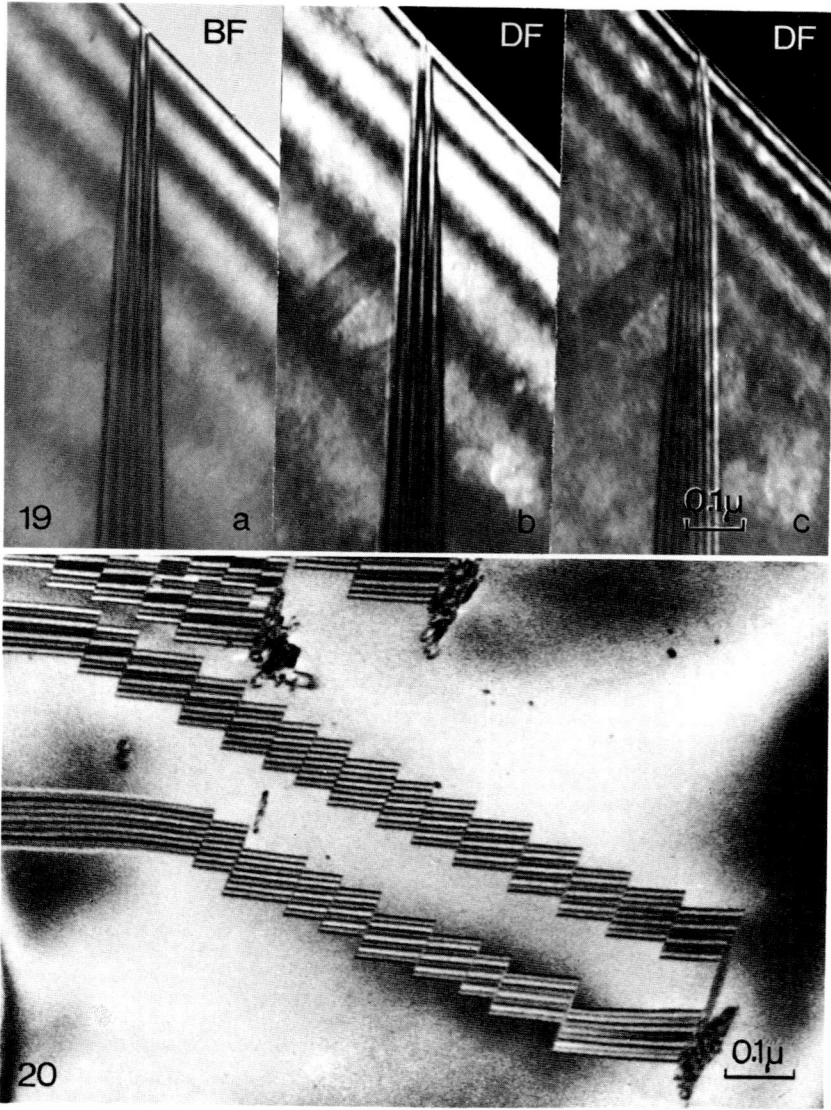

Fig. 19 a–c. Stacking fault fringes in stainless steel. (a) BF image with $s \simeq 0$; (b) DF image with $s \simeq 0$ (corresponding to (a)); (c) DF image for large s-value

Fig. 20. Antiphase boundaries in rutile. For the operating reflection $\alpha = \pi$

4. The nature of the outer fringes depends on the sign of δ in the manner shown in Table 2; they have the same nature in the DF, but opposite nature in the BF.

Table 2. δ-Fringes (B: bright; D: dark; F: first fringe; L: last fringe)

	B.F.		D.F.	
	F	L	F	L
$\delta > 0$	B	D	B	B
$\delta < 0$	D	B	D	D

5. Extinction occurs for $\delta = 0$.

6. The domains on either side of the boundary have a different contrast in the BF image, whereas it may be the same in the DF image (symmetrical position).

These properties can be verified in Fig. 21.

Also from Fig. 21 it is evident that in the BF image the contrast of the fringes is best at the end of the pattern which is close to the bright fringe. This behavior follows from Eq. (64). A model for this pair of interfaces is shown in Fig. 22. The central part is bright in the BF image because of the anomalous transmission effect, i.e. $s_1 > 0$. On the other hand $s_2 < 0$ in the outer regions. As a result $\delta_1 < 0$ for interface (1) and $\delta_2 > 0$ for interface (2). This is consistent with the nature of

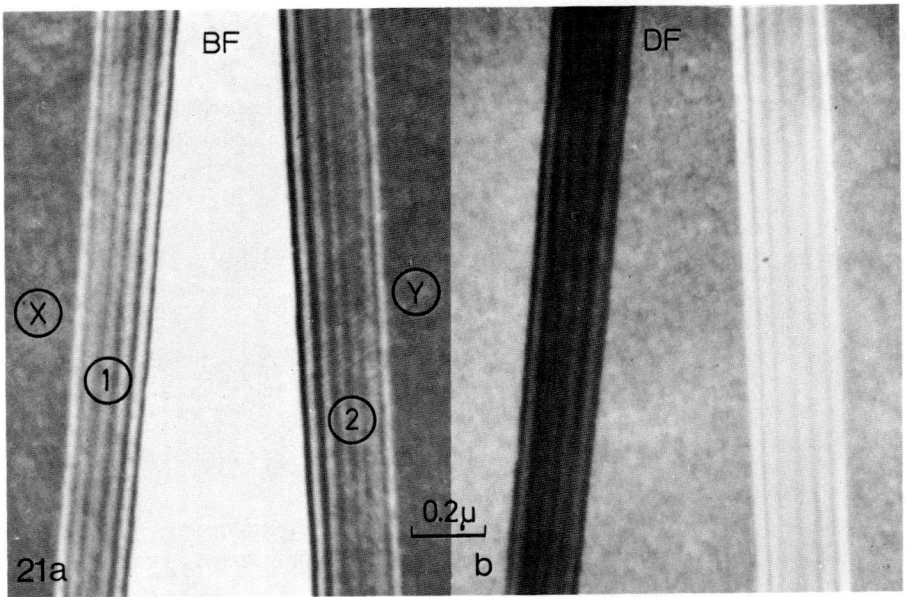

Fig. 21 a and b. δ-Fringes at twins in the intermetallic compound Zr_3Al_3. (a) BF-image. Note the difference in background in the different domains. The boundaries cause tilts of opposite sign leading to outer fringes of opposite nature but in a different succession at the two interfaces. (b) DF-image. The outer fringes have the same nature. In this particular case $s_1 = -s_2$ (symmetrical situation); the background is the same in both domains, but the overall intensity of the fringe pattern is different. (Courtesy Delavignette, Nandedkar)

the outer fringes, as follows from Table 2. Since furthermore $\phi_1 < 0$ and $\phi_2 > 0$ for interface (1) the magnitude of $\sinh[2(\pi\sigma_{2i}z_2 + \phi_2)]$ is larger than that of $\sinh[2(\pi\sigma_{1i}z_1 + \phi_1)]$ both taken close to the surfaces; i.e. for $z_1 = z_2 = t$. This explains the better contrast for the fringes close to the top part. For interface (2) on the other hand z_1 and z_2 have to be interchanged since part I and part II are interchanged and now the contrast is better close to the bottom end of the pattern, as observed.

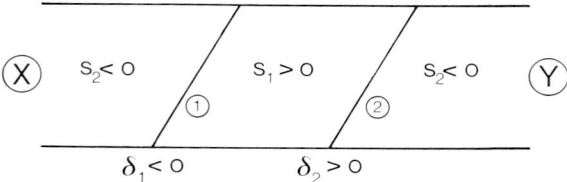

Fig. 22. Model for pairs of domain boundaries of Fig. 21

Another striking effect in the DF image is the difference in average intensity of the fringe pattern for both interfaces. This difference can only be explained by taking into account the other terms of the complete expression given by Gevers et al. (1965). It results mainly from the difference in magnitude of the term $I_S^{(1)}$ for both interfaces. In particular the term

$$\tfrac{1}{2} A^2 C^2 \cosh 2[(\pi\sigma_{1i}z_1 + \pi\sigma_{2i}z_2) + (\phi_1 - \phi_2)]$$

with

$$C^2 = \tfrac{1}{2}(1 + x_1 x_2 + w_1 w_2)$$

determines the general level of intensity. For interface (1) $\phi_1 - \phi_2$ is negative whereas it is positive for interface (2). As a result the hyperbolic cosine will be much larger for interface (2) than for interface (1) i.e. the background intensity will be larger than for (1).

In the BF image the average intensity level is slightly higher for interface (1) than for interface (2). Also this is consistent since now the term which determines the average intensity level is given by the first part of $I_T^{(1)}$ i.e.

$$1/2 A^2 C^2 \cosh[2(\pi\sigma_{1i}z_1 + \pi\sigma_{2i}z_2) + (\phi_1 + \phi_2)]$$

where for interface (1) $\phi_1 + \phi_2$ is slightly positive if $(s_2) \gtrsim (s_1)$ and slightly negative for interface (2). The average intensity level should be the same if $|s_1| = |s_2|$.

4.3.3 Mixed Fringes

It is possible that for certain interfaces as well α as δ are different from zero. Such interfaces produce fringes which have properties which are intermediate between those of α and δ-fringes. (Gevers et al., 1965).

4.4 Use of Fringe Pattern

The information contents of two-beam fringe patterns and of the corresponding diffraction patterns can be exploited to obtain the following information.

4.4.1 Translation Interfaces

Such interfaces produce fringe patterns of the α-type.

4.4.1.1 Determination of the R-vector

One can look for a number of reflections g_i for which the extinction condition $g_i \cdot R$ = integer is satisfied. For each such reflection one obtains an equation of the type $g_i \cdot R = n$. The solution for R is not always unambiguous; because of the ambiguity on n. Examples are given by Van Landuyt (1966) and Hayashi et al. (1970). In any case the vector R is only defined modulus a lattice vector.

If R is not a simple fraction of a lattice vector e.g. $R = R_0 + \varepsilon$, extinction, even in two-beam conditions, will in general not be perfect: a so-called *residual fringe pattern* is left for certain reflections for which $g \cdot R_0$ = integer due to $g \cdot \varepsilon \neq 0$. Such residual contrast can be used to determine the deviation ε from a simple vector R_0. In this case one knows that $g_i \cdot \varepsilon$ is smaller than 1 since ε is small, and a determination of ε is possible; examples can be found in Van Tendeloo and Amelinckx (1974c and 1975).

For periodic arrays of translation interfaces, giving rise to diffraction effects due to the superperiod, one can use the diffraction pattern to determine R (Van Landuyt et al., 1970, and De Ridder et al., 1972).

4.4.1.2 Plane of the Interface

In order to determine the plane of the interface the best procedure is to orient the specimen so as to bring the plane parallel to the incident electron beam. The diffraction pattern then allows the immediate finding of the direction normal to the plane. In a general orientation the sense of inclination of the interface can be found from the nature of the outer fringes using Table 1.

4.4.1.3 Nature of the Interface

One can distinguish stacking faults from antiphase boundaries by noting that stacking faults produce fringe contrast for certain fundamental reflections, but are out of contrast for *all* superlattice reflections. On the other hand antiphase boundaries produce fringe patterns for certain superlattice reflections and are out of contrast for all fundamental reflections.

One can also determine the *sense* of R from the nature of the outer fringes. For faults in the fcc structure the knowledge of the sense of R allows conclusion as to whether a fault is *intrinsic* or *extrinsic* (Gevers et al., 1963, and Van Landuyt et al., 1966). The simple rule, which is demonstrated by Art et al. (1963) is as follows. If in a "two-beam situation" the active diffraction vector points towards a bright fringe in the dark field image the fault is intrinsic for a reflection belonging to class A and extrinsic if the reflection belongs to class B. The class A reflections are 200, 222, 440, whereas the class B reflections are 400, 111, 220. One should take care to orient the image correctly with respect to the diffraction pattern. It is evident that only a dark field image is required.

In hexagonally closed packed structures one can similarly determine whether a fault is single or double (Blank et al., 1963). One now has to produce two images;

making use of reflections with $h+k$ values of the same type i.e. both with $h+k=3n$, $h+k=3n+1$ or $h+k=3n-1$, but with an l-value of different parity i.e. one with l=even and the other with l=odd. If the nature of the outer fringes is different in the two images the fault is single; if it remains the same the fault is double. Examples of application can be found in Blank *et al.* (1963).

4.4.1.4 Domain Contrast

The background intensity on both sides of the interface is always the same as well in BF as in DF whatever the diffraction conditions. This is perhaps the simplest criterion that allows the conclusion that an interface has purely translational character. In case the fault plane is parallel to the foil plane no fringe pattern results, but the faulted area will exhibit a shade which is different from that of the perfect part; except for faults close to the surface the faulted area will show up darker than the perfect part of the crystal. In Fig. 23 the faulted parts, limited by partial dislocations, show up as dark regions.

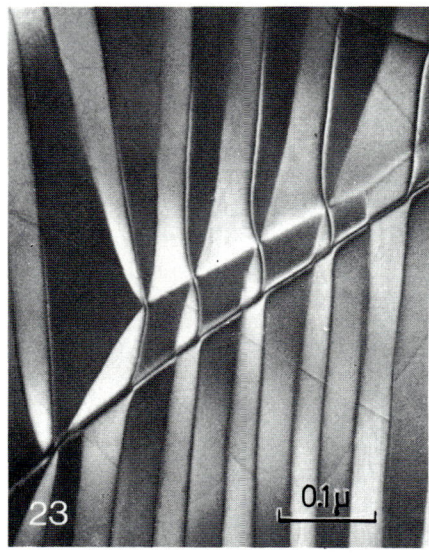

Fig. 23. Stacking faults limited by partial dislocations in a plane parallel to the foil plane in molybdenite showing up as dark areas

4.4.2 Twin Interfaces

Combined use of the fringe pattern, which is now of the δ-type, and of the diffraction pattern allows deduction of detailed information on twin interfaces.

4.4.2.1 Type of Twins

In order to distinguish rotation twins from reflection twins one has to find whether the reciprocal lattice looks like Fig. 10 or like Fig. 11. This is not always straightforward since one has to perform tilting experiments to explore a large part of reciprocal space. Most twins occurring in practice, can be described as reflection twins.

4.4.2.2 Determination of the Twinning Vector and of the Twin Plane (Gevers et al., 1964b, 1964c)

We shall consider in particular the case of a reflection twin (Fig. 9). Let the twin vector be τ such that $R = \tau z$. The vector $\Delta g = g_1 - g_2$ is then perpendicular to the coherent twin interface. The condition for the disappearance of contrast is then $g \cdot R = 0$ or $g \cdot \tau = 0$, which allows determination of the direction of τ. The magnitude of τ can be deduced from the diffraction pattern. The simplest orientation is such that the twin plane is parallel to the electron beam. One can then deduce directly the plane of the interface; the twin plane is perpendicular to the row of unsplit spots (Fig. 12), whereas the magnitude of the spot doubling is determined by the twinning vector.

4.4.2.3 Domain Contrast

If for certain diffraction conditions the background intensity in bright and dark field images is different in the domains on either side of the boundary, one concludes that the interface has twin character. Also spot-splitting is usually apparent, especially for the spots far from the center of the diffraction pattern.

4.4.2.4 Fringe Contrast

If the two crystals always diffract simultaneously because the twin vector is small, the interface gives rise to δ-fringes. The δ-fringes disappear for $\delta = 0$ i.e. if reflections are used for which $g \cdot \tau = 0$. The sense of sloping of the interface can be deduced from the nature of the outer fringes, using Table 2. This Table also allows deduction of the sign of δ which is a useful piece of information for studying the twin geometry (Gevers et al., 1964b, c). For large twin vectors only one crystal part reflects at a time and the interface produces wedge fringes for the reflecting part.

4.4.3 Interphase Interfaces

We shall briefly discuss the following topics:
    a) interfaces between different polytypes
    b) interfaces between matrix and coherent precipitates with a slightly different lattice constant
    c) interfaces between two crystal parts with the same lattice but different structure factors (Ashby and Brown, 1963).
    Since we have in fact two different crystals in all cases, there will in general be differences in contrast within the domains: moreover the interface will in general be imaged by fringes.

4.4.3.1 Polytypes

In crystals which are only partly transformed or a result of a shear transformation, interfaces occur between two different polytypes of the same material. This is for instance the case in cobalt (Fourdeux et al., 1967) and in zinc sulfide (Secco et al.,

1969). When imaging the boundary by means of reflections which only belong to *one* crystal type, wedge fringes are produced. However when reflections "common" to both types are used there is in general a small difference in diffraction vector Δg. For instance for the interface cubic-hexagonal, Δg is due to the deviation of c/a from the ideal value in the hexagonal part; Δg is then along the c-direction (corresponding to the [111] direction in the cubic phase). The interface is therefore imaged as δ-fringes; from the sign of δ one can conclude whether c/a is larger or smaller than ideal (Fourdeux et al., 1967). The contrast due to the interface between the $2H$ and $4H$ polytypes of ZnS was discussed by Secco et al. (1969).

4.4.3.2 Coherent Precipitates

Certain coherent precipitates such as γ' in Ni, have a slightly different lattice constant from that of the matrix. This results in a small difference in diffraction vector for precipitate and matrix (Ardell, 1967). There is also a small difference in structure factor but the main effect is the difference in lattice parameter. The interface is therefore imaged by means of δ-fringes. From the sign of δ one can conclude which lattice parameter is the larger one.

4.4.3.3 Exsolution Lamellae

Exsolution lamellae in feldspars have very approximately the same lattice constant as the matrix, but a different structure amplitude. The extinction distances are therefore slightly different in the two regions. This gives rise to different intensities in the two crystal parts as well as to fringe patterns which have different spacings at both ends. The thickness extinction contours will be shifted at the interface, because the extinction distances are different in lamella and matrix (Fig. 24).

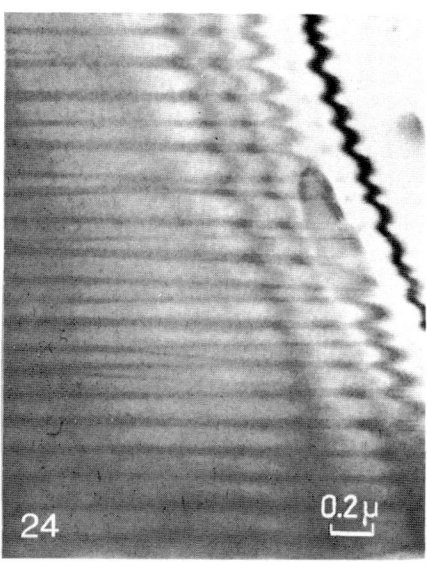

Fig. 24. Exsolution lamellae in moonstone. Note the shift of the thickness extinction contours

Presumably also the lattice constant will be slightly different and this will contribute to the contrast, but the main effect is thought to be the difference in extinction distance in this case.

If the lattice constants are sufficiently different it is evident that other contrast mechanisms will be responsable for the image: orientation contrast, wedge fringes or possibly δ-fringes if the lattices are twin-related.

4.4.4 Inversion Boundaries

Contrast at inversion boundaries requires a separate discussion since it implies the violation of Friedel's law.

In the framework of two-beam theory for non-centrosymmetrical crystals, contrast at inversion boundaries will only be found if corresponding Fourier-coefficients of the real and the imaginary parts of the lattice potential have different phase factors. Even then a contrast difference in the domains will only be found in the DF and never in the BF image (Gevers *et al.*, 1966). In any case the contrast will be weak unless the absorption distance was very small.

The boundaries themselves are imaged by α-like fringes i.e. fringes which are symmetrical in the BF and asymmetrical in the DF (Fig. 2). The phase angle α for *pure* inversion boundaries is $\alpha = 2\Theta_g$ where Θ_g is the phase angle of the Fourier-coefficient of the real part of the lattice potential corresponding to the active reflection g. This result is only true if there is *no* phase difference between real and imaginary part of the potential but it remains approximately true in other cases.

However the largest intensity differences between inversion domains are found in multiple-beam situations (Snykers *et al.*, 1972 and Serneels *et al.*, 1973). Multiple beam theory does predict in fact that Friedel's law will be violated in the DF image but will still hold in the BF image. Of course if only reflections operate which belong to a zone along which the projection is centrosymmetrical the crystal behaves as if it was centrosymmetrical, and the contrast between domains disappears. Inversion boundaries which still show an α-type fringe pattern under these circumstances, are of the type shown in Fig. 15, i.e. contain a parallel "displacement" as well.

The absence of contrast in the BF image under all circumstances against the presence of contrast in the DF image under certain conditions, is the most typical characteristic of inversion domains. The diffraction pattern is clearly not affected since the lattice is the same in the two domains.

Table 3 summarizes the characteristic contrast features of different interfaces.

4.5 Application of the Response Matrix Method to Combination of Interfaces

Using the response matrix formalism it is a simple matter to describe, at least formally, the contrast effects for a number of situations. We shall only give a few examples; more can be found in Amelinckx (1970).

4.5.1 Microtwin in a Face-centered Cubic Crystal

We assume that the matrix is in a reflecting position, whereas the thin twin is far from any reflecting position and therefore behaves as a vacuum layer.

Table 3. Identification of interfaces in the electron microscope

Type of interface	Image contrast effect	Extinction criterium	Diffraction pattern
1. Antiphase boundaries (also shear planes) **R** = Lattice vector of basic lattice but *not* of superlattice	Imaged as α-fringes No intensity difference between domains Visible in superlattice reflections only Long depth period	**g · R** = 0 or integer No contrast for fundamental reflections	No effect (apart from fine structure)
2. Stacking faults **R** = Not a lattice vector	Imaged as α-fringes No contrast difference between both parts Visible in fundamental reflections Small depth period	**g · R** = 0 or integer	No effect (apart from fine structure)
3. Domain boundaries **R** = Displacement field = $k\, z\, \tau$ (z = Distance from interface) (k = Small constant) (τ = Twinning vector)	δ-Fringes at interface Intensity difference in domains for most reflections Imaged by fundamental reflections Depth period is small; may be different on both ends	$\delta = 0$ for imaging reflections **g · R** = 0 in general	If $\Delta\mathbf{g}$ is large enough spot splitting at high order spots (in fact superposition of two slightly different patterns) All reflections are simultaneously excited in two parts
4. Twins **R** = $k\, z\, \tau$ (z = Distance from interface) (k = Large constant) (τ = Twinning vector)	One part only is diffracting in general Interface imaged by thickness fringes	**g · R** = 0; i.e. no contrast when imaged with common reflection	One row of common reflections (unsplit spots) perpendicular to the twin plane Other spots are split parallel to row of unsplit spots Amount of splitting increases with distance from unsplit row In fact superposition of two diffraction patterns
5. Inversion boundaries	Interface imaged as α-type fringes Intensity difference between domains only in DF, and for certain multiple beam situations	No contrast in BF No contrast in DF if **g** belongs to a zone producing in projection a center of symmetry	No effect (apart from possible fine structure)

Depending on the exact thickness of the twin lamella, the displacement between the two crystal parts on either side of the microtwin gives rise to a phase shift: $\alpha = 0$, or $\pm \dfrac{2\pi}{3}$. The amplitudes of transmitted and scattered beams can therefore be found from the following relation:

$$\begin{pmatrix} T \\ S e^{i\alpha} \end{pmatrix} = \mathbf{M}_2 \, \mathbf{V}(z_2, s) \, \mathbf{S}(\alpha) \, \mathbf{M}_1 \begin{pmatrix} 1 \\ 0 \end{pmatrix} \qquad (66)$$

where the succession of **V** and **S** is immaterial since it is not determined unambiguously by the physical situation; this is reflected by the fact that the matrices **V** and **S** commute. On multiplying out one finds:

$$\begin{pmatrix} T \\ S e^{i\alpha} \end{pmatrix} = \mathbf{M}_2 \begin{pmatrix} 1 & 0 \\ 0 & e^{i\beta} \end{pmatrix} \mathbf{M}_1 \begin{pmatrix} 1 \\ 0 \end{pmatrix} \qquad (67)$$

with $\beta = \alpha + 2\pi i s z_2$. The contrast will thus be the same as that due to a pure translation interface with an s-dependent phase angle β. Furthermore it should be noted that the thickness that matters for anomalous absorption is $z_1 + z_3$, whereas the thickness that causes normal absorption is $z_1 + z_2 + z_3 = z_0$ (z_0 = total thickness). Similar contrast effects occur in cases where a thin lamella of a second phase is included within a crystal in such a way that the matrix only is reflecting. The regions of overlap then exhibit stacking fault-like fringes (Fig. 25), whereas the edge parts produce wedge fringes.

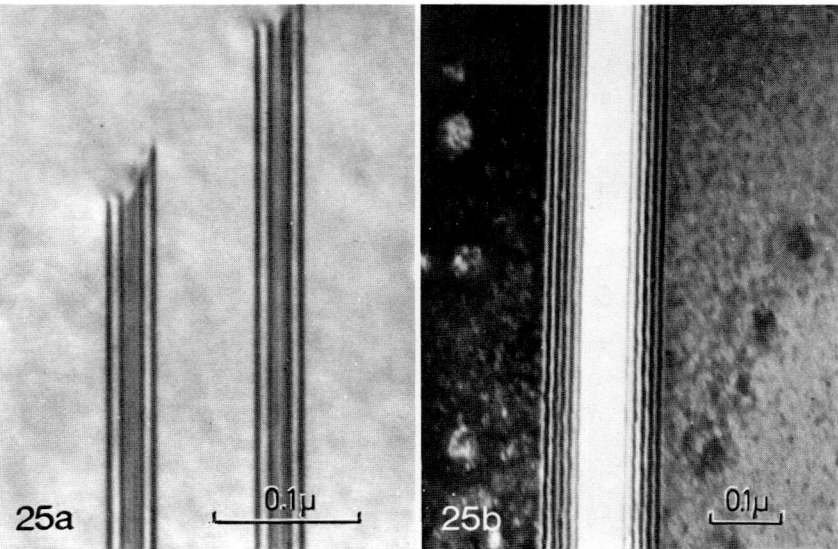

Fig. 25a and b. Thin lamella of a second phase: (courtesy Delavignette-Secco d'Aragona). (a) Thin lamella of wurtzite in a sphalerite matrix; notice the presence of α-fringes in the overlapping parts; (b) narrow strip of sphalerite in a wurtzite matrix, here the interfaces are imaged as wedge fringes

4.5.2 Overlapping Translation Interfaces

The matrix expression giving transmitted and scattered amplitudes for a crystal containing overlapping interfaces with phase angles α_1 and α_2 is

$$\binom{T}{S} = \mathbf{M}_3 \mathbf{S}(\alpha_2) \mathbf{M}_2 \mathbf{S}(\alpha_1) \mathbf{M}_1 \binom{1}{0}. \tag{68}$$

For very small separations between the faults $\mathbf{M}_2 \simeq \begin{pmatrix} 1 & 0 \\ 0 & 1 \end{pmatrix}$ and one has

$$\mathbf{S}(\alpha_2) \mathbf{M}_2 \mathbf{S}_1(\alpha_1) \simeq \mathbf{S}(\alpha_1 + \alpha_2) \tag{69}$$

which means that the phase angles add.

In the case of faults with phase angles $\alpha_1 = \alpha_2 = \pi$, one obtains in the case $s = 0$:

$$\binom{T}{S} = \mathbf{M}_3(z_3, 0) \mathbf{S}(\pi) \mathbf{M}_2(z_2, 0) \mathbf{S}(\pi) \mathbf{M}_1(z_1, 0) \binom{1}{0}. \tag{70}$$

Working out leads to:

$$\binom{T}{S} = \mathbf{M}_3(z_3, 0) \mathbf{M}_2(-z_2, 0) \mathbf{M}_1(z_1, 0) \binom{1}{0} = \mathbf{M}(z_1 + z_3 - z_2, 0) \binom{1}{0}. \tag{71}$$

The crystal then behaves as a perfect crystal with a thickness $z_1 + z_3 - z_2$; the contrast is uniform in the overlap region.

4.5.3 Overlapping Domain Boundaries

We consider in particular the case where a thin wedge-shaped domain is formed within a large domain. This situation occurs frequently on the formation of twins with a small twin vector, such as in ferroelectrics and antiferromagnetics. The s-values in the three crystal parts are now such that $s_1 = s_3 \neq s_2$. In the overlap region the fringe pattern is then described by

$$\binom{T}{S} = \mathbf{M}_3(z_3, s_1) \mathbf{M}_2(z_2, s_2) \mathbf{M}_1(z_1, s_1) \binom{1}{0}. \tag{72}$$

One can deduce symmetry properties of the fringe pattern from this relation. It is for instance easy to show that interchanging z_1 and z_3, does not change T but changes S; one has in particular

$$T(z_1, z_2, z_3\; s_1, s_2, s_3) = T(z_3, z_2, z_1, s_3, s_2, s_1)$$

and

$$S(z_1, z_2, z_3\; s_1, s_2, s_3) = S(z_3, z_2, z_1, -s_1, -s_2, -s_3). \tag{73}$$

This means that in the overlap region the bright field image will exhibit a symmetrical fringe pattern, whereas the dark field image will not, in accordance with the observation.

5. Planar Interfaces and Lattice Fringes

Recently lattice resolution has been used extensively for the study of defects in crystals. Translation interfaces cause a shift, whereas twins cause an orientation change of the lattice fringes. Depending on whether the translation interface intersects the fringes or is parallel with them, a shift of the fringes, or a spacing discontinuity results. From the fringe shift or the discontinuity one can deduce a value for α and hence determine $\mathbf{R}$.

Within the framework of two-beam theory, the formation of lattice fringes is considered to result from the interference between the direct beam and the scattered beam, i.e. the wave function can be written as:

$$T + S e^{2\pi i \mathbf{g} \cdot \mathbf{r}} = T + S e^{2\pi i g x} \tag{74}$$

where $\mathbf{r}$ is the position vector and $\mathbf{g}$ the active diffraction vector; choosing the x-axis parallel to $\mathbf{g}$ one can write $\mathbf{g} \cdot \mathbf{r} = gx$. The intensity distribution in the lattice fringes is then

$$I = I_T + I_S + 2\sqrt{I_T I_S} \sin(2\pi g x + \phi) \tag{75}$$

where $tg\phi = (s/\sigma) tg\pi\sigma t$ (t = foil thickness).

If the second part of the crystal is displaced over $\mathbf{R}$, the scattered beam emerging from this part of the crystal suffers a phase shift $\alpha = 2\pi \mathbf{g} \cdot \mathbf{R}$ with respect to the incident beam. The wave function thus becomes

$$T + S e^{2\pi i g x} e^{i\alpha} \tag{76}$$

and the intensity distribution is

$$I = I_T + I_S + 2\sqrt{I_T I_S} \sin(2\pi g x + \phi + \alpha). \tag{77}$$

The fringe pattern in the second part is thus shifted with respect to that in the first part over a fraction $(\alpha/2\pi)$ of the interfringe distance.

Lattice fringes in a crystal of silicon carbide containing a stacking fault are shown by Van Landuyt and Amelinckx (1971). Examples of sequential faults in "parisite" revealed by anomalies in the fringe spacing will be discussed below.

6. Case Studies

Rather than being inevitably incomplete in reviewing electron microscopy work from other laboratories on minerals, this chapter is intended to illustrate applications of the described techniques and image characteristics by a number of studies to which the authors have contributed in recent years.

We shall briefly report some studies on minerals in which planar defects occur, associated with

   a) ordering: pyrrhotite, tealite
   b) displacive transformations: leucite
   c) exsolution: moonstone, labradorite
   d) layer stackings: parisite-series, molybdenite.

6.1 Ordering Systems: Antiphase Boundaries and Ordering Twins

6.1.1 Pyrrhotite (Van Landuyt and Amelinckx, 1972)

Pyrrhotite has a defective NiAs structure from which different nonstoichiometric compounds such as Fe_7S_8 can be derived by the incorporation of iron vacancies. Ordering of vacancies gives rise to various possible planar defects some of which we want to illustrate here (compare also Nakazawa *et al.*, Chapter 5.2 and Pierce and Buseck, Chapter 2.6 of this volume).

The pyrrhotite structure can be described with respect to the NiAs structure as follows (Fig. 26): The sulphur ions replace the anions in NiAs, the iron ions replace the cations but alternating layers are full or defective, i.e. they contain iron vacancies. The vacancies in the defective layers are ordered as shown in Fig. 27. By changing the origin these Fe-layers are stacked in four successive different positions I, II, III and IV in such a way that a c-axis of four times the original repeat results as compared with NiAs (Fig. 28). Furthermore the lattice is slightly

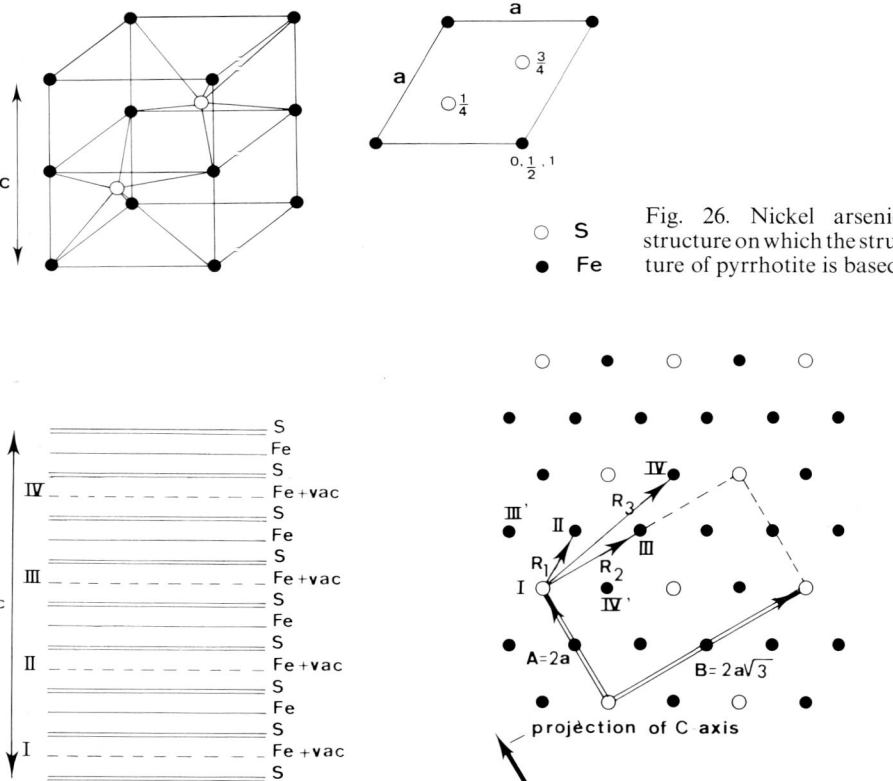

Fig. 26. Nickel arsenide structure on which the structure of pyrrhotite is based

Fig. 27. Pyrrhotite: arrangement of full (full line) and defective (dashed line) Fe-planes

Fig. 28. Pyrrhotite: arrangement of atoms within the defective planes. The displacement vectors R_i are indicated

monoclinically deformed. As a consequence of mishaps in the ordering of the vacancies in the Fe-layers different types of interfaces can be predicted.

a) Antiphase boundaries: if different parallel orientations occur within the defective layers. These have displacement vectors of the type R_1, R_2 and R_3 as indicated in Fig. 27.

b) Antiphase boundaries where defective layers join full Fe-layers.

c) Stacking faults associated with errors in the stacking sequence of the layers I to IV.

d) Twin domains may arise since the monoclinic deformation is associated with the succession of the layers which can occur in six possible senses.

The α-type fringe pattern as observed in pyrrhotite (Fig. 29) is due to one of the types of antiphase boundaries as described above. The same crystal region also exhibits twin domain boundaries.

6.1.2 Teallite (Marinkovic and Amelinckx, 1964)

Teallite is a mixed sulfide of tin and lead, $PbSnS_2$. The structure was found by Hofmann (1935) to be isomorphous with that of SnS; in particular it had not been possible to determine whether Pb and Sn formed an ordered arrangement. This example is particularly interesting since electron microscopy and diffraction have been able to elucidate this point and thus contribute substantially to the structure determination.

Antiphase boundaries were observed (Fig. 30), which already indicated an ordered arrangement for the Pb and Sn atoms since no antiphase boundaries

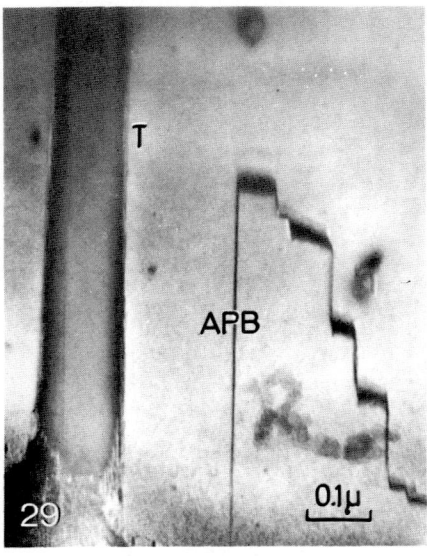

Fig. 29. Antiphase boundaries and twin boundaries in pyrrhotite

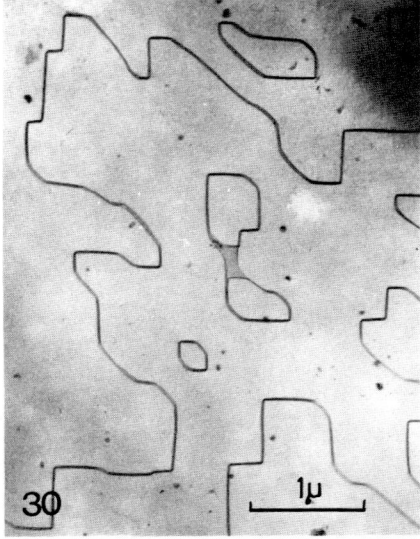

Fig. 30. Antiphase boundaries in teallite. (Courtesy Marinkovic)

were ever detected in SnS. An answer to the question whether the order was within the c-layers or perpendicular to them was given by the electron diffraction evidence that the unit mesh in the c-plane is conserved with respect to the SnS-structure. Ordering is thus expected in the c-direction. Since the c-repeat contains four layers only two possible sequences can be formed:

... Pb Sn |Pb Sn Pb Sn| Pb Sn ... or Sn Sn |Pb Pb Sn Sn| Pb Pb ...

(for simplicity the presence of sulphur in the layers has been omitted in the symbols Pb and Sn).

For both types of ordered stackings, antiphase boundaries are possible; only the occurrence of APB's is thus not sufficient to decide between both stacking sequences. Contrast calculations, cross checked by diffraction experiments and image observations, allowed to decide upon the ... Pb Pb Sn Sn ... sequence as the only one able to give rise to the antiphase boundaries as observed in tealite specimens.

The antiphase boundaries (APB) thus have the arrangement of layers:

$$\text{glide plane} \rightarrow \begin{array}{c|c} \text{Pb} & \text{Sn} \\ \text{Sn} & \text{Pb} \\ \hline \text{Sn} & \text{Pb} \\ \text{Pb} & \text{Sn} \end{array}$$
$$\text{APB}$$

A particularly interesting observation supporting this model is shown in Fig. 31. It illustrates the "interaction" of a dislocation ribbon with an antiphase boundary. From the model for the antiphase boundary it can be predicted that a glide dislocation, e.g. between two Sn-layers changes to a glide plane between two Pb-layers. If now the dislocation is split in partials, it is expected that the stacking fault energies will be different for both situations (see the graphic representation above). Consequently the width of a ribbon will change upon crossing an APB as actually occurs in Fig. 31. This would not have been the case for stacking sequences of the type: ... Pb Sn Pb Sn ... where the stacking fault energies would be the same on both sides of the boundary.

6.2 Displacive Transformations: Antiphase Boundaries and Twins

Leucite ($KAlSi_2O_6$)

Above 625 °C leucite is cubic with $a_0 = 13.43$ Å, below this temperature it is tetragonal with a 16 molecule cell having dimensions (Wyart, 1940) $a_0 = 12.98$ Å; $c_0 = 13.68$ Å. The positions of the oxygen ions remain unchanged through this transformation; only the metal ions are displaced so as to produce the tetragonal symmetry. As illustrated in Section 2.1 it is expected that this type of displacive transformation will give rise to the formation of twin boundaries.

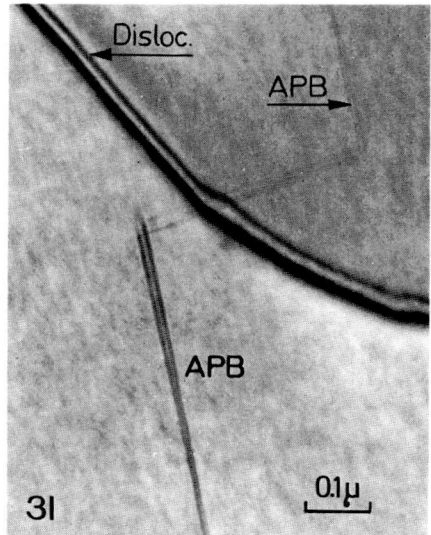

Fig. 31. Dislocation ribbons intersecting antiphase boundaries in teallite. Note the "refraction" and change in width of the ribbons. (Courtesy Marinkovic)

Fig. 32. Two sets of intersecting twin domains in leucite. One set is parallel with the electron beam. Notice also the presence of antiphase boundaries as curvilinear defects

Fig. 13 shows an example where also antiphase boundaries are present. The straight fine lines represent twin boundaries parallel with the electron beam. The boundary plane is (101) as evidenced by the diffraction pattern where the characteristic spot splitting is clearly visible (Fig. 12).

The intersection of two families of {101} twins is shown in Fig. 32. One family is inclined with respect to the beam and is imaged as a fringe pattern (large obliquity → wedge fringes), the other family as a fine line.

The twin domains are often crossed by curvilinear two-dimensional defects as in Figs. 13 and 32. These are antiphase boundaries as furthermore evidenced by the identical background intensity on both sides of the boundary. The antiphase boundaries are most probably already present in the cubic high-temperature phase. This is moreover substantiated by the fact that these boundaries are often continuous across the twin boundaries.

6.3 Feldspars: Exsolution Domains

Moonstone and Labradorite

Since chapters of this book (Champness and Lorimer, Chapter 4.1, Willaime *et al.*, Chapter 4.9, Cliff *et al.*, Chapter 4.10, and Grove, Chapter 4.11 this volume) are devoted to exsolution we shall be rather brief on this example and only stress the types of images that are obtained at the interface between exsolution domains.

We have investigated moonstone (USSR) and labradorite crystals, two feldspars liable to exhibit exsolution phenomena. Figs. 33 and 24 show exsolution domain arrangements in labradorite and moonstone. The aspect is rather similar to the observations of Bolton *et al.* (1966) and Laves *et al.* (1965). Particularly striking was that in either case no sharp boundary was observed. In spite of tilting experiments the boundaries remain vague. Special attention was paid to the diffraction patterns where no detectable splitting could be observed. At most a slight elongation of the spots in a direction perpendicular to the boundary planes could be detected. This may be an indication that the unmixing is not complete or gradual at the boundaries and this also explains the unsharpness of the latter. Inside the bands the exsolution is probably complete. This model is furthermore substantiated by the contrast which we believe to be structure factor contrast (Section 4.4.3.3). The difference in shade between the domains is associated with the difference in composition; the contrast behavior of the wedge fringes at the border of the specimens also supports the interpretation as structure factor contrast. The almost discontinuous displacement of the wedge fringes as visible in Figs. 24 and 33 is related to the difference in structure factor and the associated change in extinction distance.

We also want to draw attention to the difference of these observations with those on (K, Na) feldspars by Fleet and Ribbe (1963), Willaime and Gandais (1972) and Lorimer and Champness (1973, and this volume Chapter 4.1). The exsolution patterns there invariably show periodic twinning in the Na-rich phase (see e.g. Fig. 1 of Willaime *et al.*, Chapter 4.9 of this volume). In these cases the lattices in both domains are different so that stress relaxation by twinning is required. Also the boundaries between exsolution domains are usually better defined. The contrast however is complicated by the presence of twins in the albite phase.

6.4 Layered Structures: Sequential and Compositional Stacking Faults

6.4.1 Bastnaesite – Synchisite Series (Van Landuyt and Amelinckx, 1975)

This rare earth mineral series is structurally characterized by long period stacking of layers of the type B and S, representing respectively bastnaesite and synchisite blocks. The structure for bastnaesite was determined and other species of the series were analyzed by Donnay and Donnay (1953).

By means of direct lattice imaging technique in electron microscopy it has been possible to visualize the individual building blocks for these minerals (Van Landuyt and Amelinckx, 1975). Roentgenite is shown as an example in Fig. 34, which is schematically related to Fig. 35. The imaging code is very simple and amounts in this case to relate a wide dark line with a B-layer and a broad white line with an S-layer. The thinner dark line is then related with a single rare earth-fluor layer.

All species of this series could be analyzed by this technique and it has even been possible to discover three new minerals compositionally situated between bastnaesite and synchisite.

It is clear that by the very nature of the layered structure of these minerals, stacking errors, i.e errors in the sequence of successive layers are possible.

As shown above (Section 3.1) *sequential* faults as well as *compositional* or chemical faults are expected. Fig. 36 shows a particular area of a BS_4-part in

Contrast Effects at Planar Interfaces 109

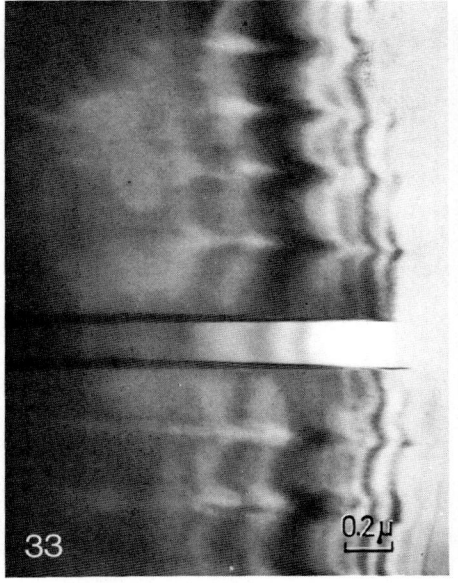
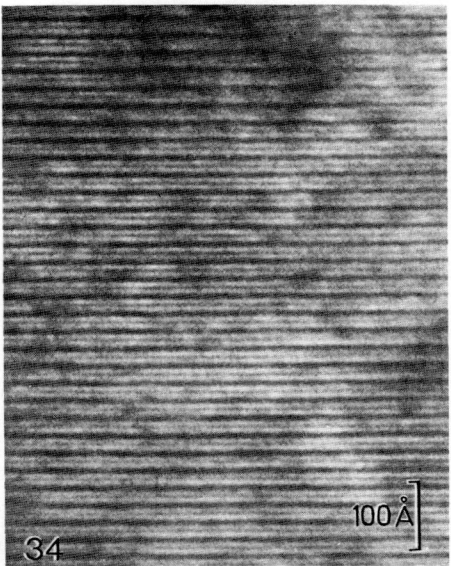
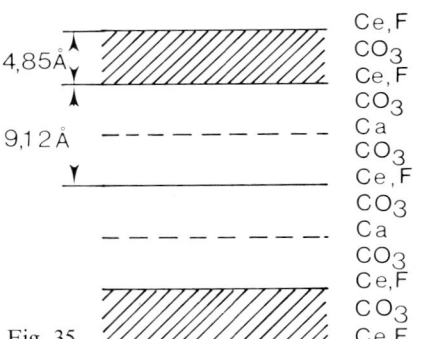

Fig. 33. Exsolution domains in labradorite. The wavy contrast at the exsolution domain boundaries is clearly distinguished from the twin boundaries in the center of the photograph. Notice the shift of the thickness contours

Fig. 34. Lattice imaging of roentgenite

Fig. 35. Structural and compositional stacking model to be related with the lattice image of Fig. 34

syntaxy with a BS_2-part (roentgenite) where both types of faults were observed. Sequential faults (S) simply change the stacking sequence, compositional faults (C) do change locally the composition. Schematically these faults can be represented as:

C_1: $BS_4BS_4BS_2BS_4BS_4$
S_1: $BS_4BS_3BS_5BS_4BS_4$
S_2: $BS_2BS_2BS_3BS\,BS_2BS_2$
C_2: $BS_2BS_4BS_2BS_2$.

6.4.2 Molybdenite (Amelinckx and Delavignette, 1963)

Whereas in the previous case the stacking faults extend usually throughout the crystal fragment and were visualized by observing with the electron beam parallel to the layers, we next describe an example of a layer structure mineral with the

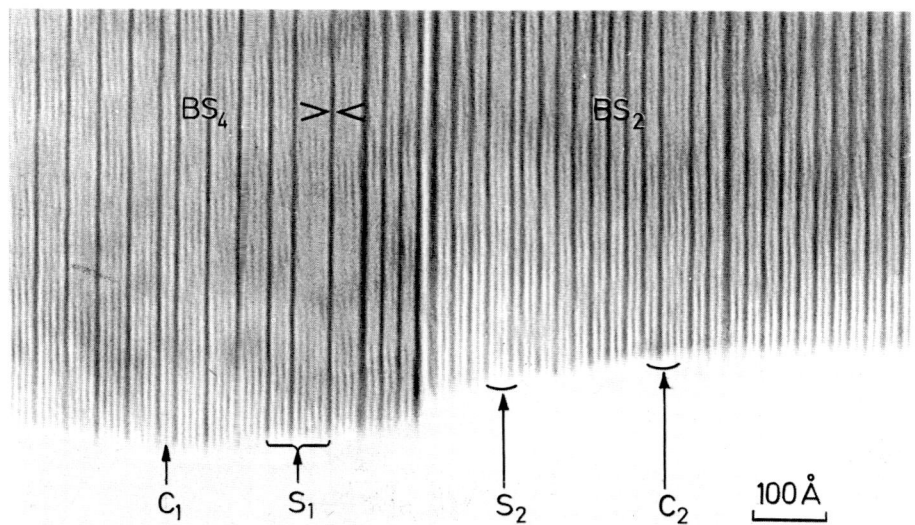

Fig. 36. Sequential (S) and Compositional (C) faults in a roentgenite crystal in syntaxy with "BS$_4$"

MoS$_2$ structure where the defects will be observed with the beam perpendicular to the layers.

The structure of this mineral can conveniently be described by the stacking symbol

$$a\beta a b \alpha b \daleth \beta a b$$

where the Roman letters refer to sulfur atoms and the Greek letters to molybdenum atoms.

Stacking faults are easily generated by glide of partial dislocations between two sulfur layers, e.g. between an a and a b-layer since the bonding is weakest there. Observations as described above yield configurations as shown in Fig. 23.

Ribbons of two partial dislocations delimiting areas of stacking fault can be observed. The presence of the planar defects is now visible as a difference in background intensity because the fault is parallel with the foil surfaces. This difference in intensity with respect to the unfaulted region is due to the phase shift $\alpha = 2\pi g R$ as defined in the introductory Section 4.4.1.4.

References

Amelinckx, S.: The direct observation of dislocations, p. 121. New York: Academic Press 1964.
Amelinckx, S.: In: Modern diffraction and imaging techniques in materials science (eds. S. Amelinckx, R. Gevers, G. Remaut, J. Van Landuyt), p. 257. Amsterdam: North Holland 1970.
Amelinckx, S., Delavignette, P.: In: Electron microscopy and strength of crystals (eds. G. Thomas and J. Washburn), p. 441. New York: J. Wiley and Sons 1963.
Ardell, A.J.: Diffraction contrast at planar interfaces of large coherent precipitates. Phil. Mag. 16, 147–158 (1967).
Art, A., Gevers, R., Amelinckx, S.: The determination of the type of stacking faults in face-centered cubic alloys by means of contrast effects in the electron microscope. Phys. Stat. Sol. 3, 697–711 (1963).

Ashby, M. F., Brown, L. M.: On diffraction contrast from inclusions, Phil. Mag. **8**, 1649–1676 (1963).
Blank, H., Delavignette, P., Gevers, R., Amelinckx, S.: Fault structures in wurtzite. Phys. Stat. Sol. **7**, 747–764 (1964).
Bolton, H. C., Bursill, L. A., McLaren, A. C., Turner, R. G.: On the origin of the colour of labradorite. Phys. Stat. Sol. **18**, 221–230 (1966).
Cahn, R. W.: Twinned crystals. Advan. Phys. **3**, 363–445 (1954).
Chakraverty, B. K.: Verwey ordering on magnetite as a co-operative Jahn-Teller transition. Solid State Comm. **15**, 1271–1275 (1974).
De Ridder, R., Van Landuyt, J., Amelinckx, S.: Diffraction effects associated with shear structures and related structures. Phys. Stat. Sol. (a) **9**, 551–565 (1972).
Donnay, G., Donnay, J. D. H.: The crystallography of bastnaesite, parisite, roentgenite and synchisite. Am. Mineralogist **38**, 932–963 (1953).
Fleet, S. G., Ribbe, P. H.: An electron microscope investigation of a moonstone. Phil. Mag. **8**, 1179–1187 (1963).
Fourdeux, A., Gevers, R., Amelinckx, S.: Electron microscopic contrast effects at the interface between two different polytypes. I. Interface cubic-hexagonal. Phys. Stat. Sol. **24**, 195–206 (1967).
Gevers, R., Art, A., Amelinckx, S.: Electron microscopic images of single and intersecting stacking faults in thick foils. Part I: Simple faults. Phys. Stat. Sol. **3**, 1563–1593 (1963).
Gevers, R., Art, A., Amelinckx, S.: Electron microscopic images of single and intersecting stacking faults in thick foils. Part II: Intersecting faults. Phys. Stat. Sol. **7**, 605–632 (1964a).
Gevers, R., Blank, H., Amelinckx, S.: Extension of the Howie-Whelan equations for electron diffraction to non-centrosymmetrical crystals. Phys. Stat. Sol. **13**, 449–465 (1966).
Gevers, R., Delavignette, P., Blank, H., Amelinckx, S.: Electron microscope transmission images of coherent domain boundaries. Part I: Dynamical theory. Phys. Stat. Sol. **4**, 383–410 (1964b).
Gevers, R., Delavignette, P., Blank, H., Van Landuyt, J., Amelinckx, S.: Electron microscope transmission images of coherent domain boundaries. Part II: Observations. Phys. Stat. Sol. **5**, 595–633 (1964c).
Gevers, R., Van Landuyt, J., Amelinckx, S.: Intensity profiles for fringe patterns due to planar interfaces as observed by electron microscopy. Phys. Stat. Sol. **11**, 689–709 (1965).
Glossop, A., Pashley, D.: The direct observation of anti-phase domain boundaries in ordered copper-gold (CuAu) alloy. Proc. Roy. Soc. (London) Ser. A **250**, 132–146 (1959).
Hayashi, I., Delavignette, P., Amelinckx, S.: Unusual dissociation of dislocations in Ni_3Ta. Phys. Stat. Sol. **42**, 637–644 (1970).
Hirsch, P. B., Howie, A., Nicholson, R., Pashley, D. W., Whelan, M. J.: Electron microscopy of thin crystals, p. 100. London: Butterworth 1965.
Hofmann, W.: Ergebnisse der Strukturbestimmung komplexer Sulfide. Z. Krist. **92**, 161–185 (1935).
Howie, A., Whelan, M. J.: Diffraction contrast of electron microscope images of crystal lattice defects. Part II: The development of a dynamical theory. Proc. Roy. Soc. (London) Ser. A **263**, 217–237 (1961).
Kear, B. H., Giamei, A. F., Laverant, G. R., Oblak, J. M.: On intrinsic/extrinsic stacking-fault pairs in the Ll_2 lattice. Scripta Met. **3**, 123–130 (1969).
Laves, F., Nissen, H., Bollmann, W.: Schiller and submicroscopic lamellae of labradorite. Naturwissenschaften **52**, 427–428 (1965).
Lorimer, G. W., Champness, P. E.: The origin of the phase distribution in two perthitic alkali feldspars. Phil. Mag. **28**, 1391–1403 (1973).
Magnéli, A.: Structures of the ReO_3-type with recurrent dislocations of atoms: homologous series of molybdenum and tungsten oxides. Acta Cryst. **6**, 495–500 (1953).
Marinkovic, V., Amelinckx, S.: Antiphase boundaries and dislocation ribbons in tealite ($PbSnS_2$). Phys. Stat. Sol. **6**, 823–837 (1964).
Meulemans, M., Delavignette, P., Garcia-Gonzales, F., Amelinckx, S.: Regular network of antiphase boundaries in the χ-phase of the systems Fe-Cr-Mo-Ti. Mat. Res. Bull. **5**, 1025–1030 (1970).
Ogawa, S., Watanabe, D., Watanabe, H., Komoda, T.: The direct observations of the long period of the ordered alloy CuAu(II) by means of electron microscope. Acta Cryst. **11**, 872–875 (1958).

Ruedl, E., Delavignette, P., Amelinckx, S.: Electron diffraction and electron microscopic study of long- and short-range order in Ni_4Mo and the substructure resulting from ordering. Phys. Stat. Sol. **28**, 305–328 (1968).

Secco d'Aragona, F., Delavignette, P., Gevers, R., Amelinckx, S.: Electron microscopic contrast effects at the interface between two different polytypes. Part II: Interface 2H — 4H. Phys. Stat. Sol. **31**, 739–755 (1969).

Serneels, R., Snykers, M., Delavignette, P., Gevers, R., Amelinckx, S.: Friedel's law in electron diffraction as applied to the study of domain structures in non-centrosymmetrical crystals. Phys. Stat. Sol. (b) **53**, 277–292 (1973).

Snykers, M., Delavignette, P., Amelinckx, S.: A new concept in the defect structure of alloys: the dissociated antiphase boundary. Phys. Stat. Sol. (b) **48**, K1–K5 (1971).

Snykers, M., Serneels, R., Delavignette, P., Gevers, R., Amelinckx, S.: Inversion domains in the χ-phase alloy $Fe_{36}Cr_{22-x}Ti_x$ with $8 < x < 10$. Crystal Lattice Defects **3**, 99–101 (1972).

Van Landuyt, J.: Determination of the displacement vector at the anti-phase boundaries in rutile by contrast experiments in the electron microscope. Phys. Stat. Sol. **16**, 585–590 (1966).

Van Landuyt, J., Amelinckx, S.: Stacking faults in silicon carbide (6H) as observed by means of transmission electron microscopy. Mat. Res. Bull. **6**, 613–620 (1971).

Van Landuyt, J., Amelinckx, S.: Electron microscope observations of the defect structure of pyrrhotite. Mat. Res. Bull. **7**, 71–80 (1972).

Van Landuyt, J., Amelinckx, S.: Multiple beam direct lattice imaging of new mixed layer compounds of the bastnaesite-synchisite series. Am. Mineralogist **60**, 351–358 (1975).

Van Landuyt, J., Amelinckx, S., Kohn, J.A., Eckart, W.: Direct lattice imaging of the hexagonal ferrites. Mat. Res. Bull. **8**, 339–348 (1973a).

Van Landuyt, J., Amelinckx, S., Kohn, J.A., Eckart, D.W.: Lattice imaging of the hexagonal ferrites — The M_nS series. Mat. Res. Bull. **8**, 1173–1182 (1973b).

Van Landuyt, J., Amelinckx, S., Kohn, J.A., Eckart, D.W.: Multiple beam direct lattice imaging of the hexagonal ferrites. J. Solid State Chem. **9**, 103–119 (1974).

Van Landuyt, J., De Ridder, R., Brabers, V.A.M., Amelinckx, S.: Jahn-Teller domains in $Mn_xFe_{3-x}O_4$ as observed by electron microscopy. Mat. Res. Bull. **7**, 327–338 (1972).

Van Landuyt, J., De Ridder, R., Gevers, R., Amelinckx, S.: Diffraction effects due to shear structures: a new method for determining the shear vector. Mat. Res. Bull. **5**, 353–362 (1970).

Van Landuyt, J., Gevers, R., Amelinckx, S.: Electron microscopic study of twins, antiphase boundaries and dislocations in thin films of rutile. Phys. Stat. Sol. **7**, 307–329 (1964a).

Van Landuyt, J., Gevers, R., Amelinckx, S.: Fringe patterns at antiphase boundaries with $\alpha = \pi$ observed in the electron microscope. Phys. Stat. Sol. **7**, 519–547 (1964b).

Van Landuyt, J., Gevers, R., Amelinckx, S.: On the determination of the nature of stacking faults in fcc metals from the bright field image. Phys. Stat. Sol. **18**, 167–172 (1966).

Van Tendeloo, G., Amelinckx, S.: The domain structure of the δ-phase alloy NiMo. Mat. Res. Bull. **8**, 721–732 (1973).

Van Tendeloo, G., Amelinckx, S.: Group-theoretical considerations concerning domain formation in ordered alloys. Acta Cryst. A**30**, 431–440 (1974a).

Van Tendeloo, G., Amelinckx, S.: Lattice relaxation at non-conservative antiphase boundaries in Ni_3Mo. Phys. Stat. Sol. (a) **22**, 621–629 (1974c).

Van Tendeloo, G., Amelinckx, S.: Lattice relaxation at glide antiphase boundaries and stacking faults in Ni_3Mo and Ni_4Mo. Phys. Stat. Sol., Section A, **27**, 903–908 (1975).

Van Tendeloo, G., Van Landuyt, J., Delavignette, P., Amelinckx, S.: Compositional changes associated with periodic antiphase boundaries in the initial stages of ordering in Ni_3Mo. Part I: Crystallographic analysis. Phys. Stat. Sol. (a) **25**, 697–707 (1974a).

Wadsley, A.D.: Non-stoichiometric compounds (ed. L. Mandelcorn), p. 98. New York: Academic Press 1964.

Willaime, C., Gandais, M.: Study of exsolution in alkali feldspars calculation of elastic stresses inducing periodic twins. Phys. Stat. Sol. (a) **9**, 529–539 (1972).

Wyart, J.: Etude crystallographique d'une leucite artificielle. Structure atomique et symmétrie du minéral. Bull. Français Minéralogie **63**, 5–7 (1940).

Yamada, T., Juzuki, K., Chikamuri, S.: Electron microscopy of orthorhombic phase in magnetite. Appl. Phys. Letters **13**, 172–175 (1968).

Yoshioka, H.: Effect of inelastic waves on electron diffraction. J. Phys. Soc. Japan **12**, 618–622 (1957).

CHAPTER 2.4

Computer Simulation of Dislocation Images in Quartz

J.W. MCCORMICK

1. Introduction

Transmission electron microscopy (TEM) has become one of the most important tools in the characterization of crystal defects. TEM yields images of individual dislocations, making it invaluable in the determination of slip systems. These images are a result of a diffraction process and as such must be interpreted with considerable care. Since the introduction of the two-beam dynamical theory of electron diffraction for imperfect crystals by Howie and Whelan (1961) many calculations of the contrast produced by various crystal defects have been made and they showed good agreement with experiments. From this theory many general rules for the contrast of various defects have been formed. The invisibility criteria for dislocations in an isotropic crystal ($g \cdot b = 0$ and $g \cdot b \times u = 0$) can be derived directly from the equations of the dynamical theory. In general, however, such rules cannot be established for more complicated situations. It has been shown, for example (Head et al., 1967; Head, 1967; Humble, 1968), that the above invisibility criteria are not valid for anisotropic crystals. In such cases numerical solutions of the Howie-Whelan equations may be necessary to interpret the contrast arising from the dislocations. Since most rock-forming minerals have compositions, structures, and properties more complicated than the materials (primarily metals) for which many of these general rules have been developed, it is important that theoretical calculations be made before attempting to apply any of them. It is the purpose of this paper to investigate the nature of dislocation images in quartz using numerical solutions of the Howie-Whelan equations.

2. Theoretical Calculation of Dislocation Images

Computations are presented in the form of "half-tone" images produced by the computer line printer. This technique was introduced by Head (1967) and while more sophisticated techniques, such as photographing the display on a cathode ray tube, have been developed (Bullough et al., 1969) the line printer technique remains the most widely used because of its availability and simplicity. The symbols used for the grey scale in these calculations were selected on the basis of photometric measurements in a manner described by Head et al. (1973). Eleven shades of grey (including blank paper) are obtained by overprinting up to three

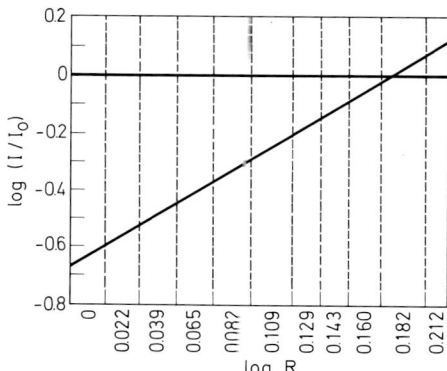

Fig. 1. Graph of log (calculated intensity/background intensity) as a function of log (reflectivity of grey scale symbols). The values along the abscissa are the logs of the relative reflectivities of the different symbol combinations. The dashed lines indicate where one shade of grey is replaced by another

different symbols. The allocation of the grey scale symbols to the calculated intensities is shown in Fig. 1. Details in the image topology which may be overlooked when the calculations are presented as intensity profiles can be observed in these theoretical micrographs.

The computer program for a single dislocation by Head et al. (1973) has been modified for the trigonal symmetry of quartz. For a complete description of the principles of computation the reader is referred to Head et al. (1973) and the review by Skalicky (1973). The problem of computing a simulated micrograph consists of two major portions: (a) calculation of the displacement field of the defect and (b) solution of the diffraction equations. The solution of the displacement field of a dislocation in an anisotropic material has been given by Eshelby et al. (1953) and the results are formulated by Stroh (1958). It should be kept in mind that this solution is valid for an infinite defect in an infinite medium and when it is applied to a dislocation in a thin foil no relaxations of the free surfaces are considered.

The theory of electron diffraction used is the two-beam dynamical theory originally formulated by Howie and Whelan (1961) for centrosymmetric crystals. Quartz is non-centrosymmetric and it would be more appropriate to use the extended formulation of the Howie-Whelan equations for non-centrosymmetric crystals derived by Gevers et al. (1966). However, these authors pointed out that electrons emerging from a crystal in the transmitted beam will have been reflected an equal number of times from both sides of the diffracting planes. They therefore suggested that the effects of enantiomorphism are probably averaged out in bright field images. This is consistent with our experience that theoretical brightfield images of dislocation, calculated with the centrosymmetric assumption, generally match actual micrographs. Fig. 2 shows an example of a good match.

The formation of the Howie-Whelan equations used in the program is:

$$\frac{dT}{dZ} = -\mathcal{N} T + (i - \mathcal{A}) S$$

$$\frac{dS}{dZ} = (i - \mathcal{A}) T + (-\mathcal{N} + 2iw + 2\pi i \beta) S$$

Computer Simulation of Dislocation Images in Quartz

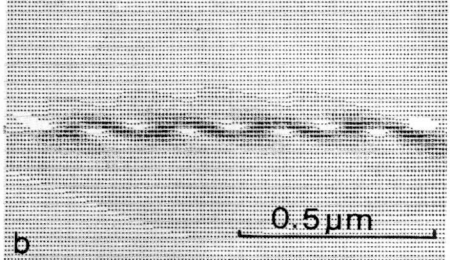
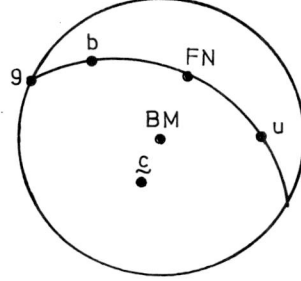

Fig. 2a and b. Comparison of (a) an actual micrograph with (b) a computed micrograph. In both cases $\mathbf{g}=(\bar{1}010)$, $\mathbf{b}=1/3[2\bar{1}\bar{1}0]$, $\mathbf{u}=[11\bar{2}0]$, the beam direction ($\mathbf{BM}$) is $[8\cdot 16\cdot\bar{8}\cdot 13]$, the foil normal ($\mathbf{FN}$) is $[\bar{1}2\bar{1}0]$, $\xi_g/\xi'_g=0.03$, and $w=0$. This discrepancy on the left side is probably due to interference with other dislocations in (a). Bright field images. Dislocations intersect the top surface of the foil on the right

where T and S are the amplitudes of the electron waves in the directions of the incident and diffracted beams, respectively, Z is in the direction of the incident beam and has units ξ_g/π, $\mathcal{N}$ is the normal absorption coefficient (ξ_g/ξ'_0), $\mathscr{A}$ is the anomalous absorption coefficient (ξ_g/ξ'_g), w is the dimensionless deviation parameter, and β is the derivative of the scalar product of the diffracting vector and displacement vector, $\dfrac{d(\mathbf{g}\cdot\mathbf{R})}{dZ}$. The intensities of the transmitted and scattered beams are proportional to $|T|^2$ and $|S|^2$ respectively.

In order to produce a theoretical micrograph about 8000 separate values of intensity are required. One way to obtain these is to integrate the Howie-Whelan equations for 8000 different columns, the length of each being the thickness of the foil. Head (1967) has developed a method by which the number of integrations is effectively reduced to 240. This method takes advantage of two facts: (1) the displacements around a dislocation are independent of the coordinate parallel to the dislocation, and (2) the Howie-Whelan equations are a pair of coupled linear differential equations. Since the displacements in columns in a row parallel to the dislocation (e.g., parallel to plane ABFE in Fig. 3) are simply related it is possible to form a generalized cross-section which contains all of the displacement information in the foil. Fig. 3 shows the geometry of a single dislocation in an untilted foil and the projection used to form the generalized cross-section. Modifications of this geometry are made to handle more complex geometries such as tilted foils, multiple dislocations, and stacking faults (Humble, 1968). By solving the Howie-Whelan equations for columns in this generalized cross section it is possible, by applying appropriate boundary conditions, to solve for all the intensities over the specified region of the foil around the dislocation. Using this method each computed micrograph requires 4 sec to compute on an IBM 360/91.

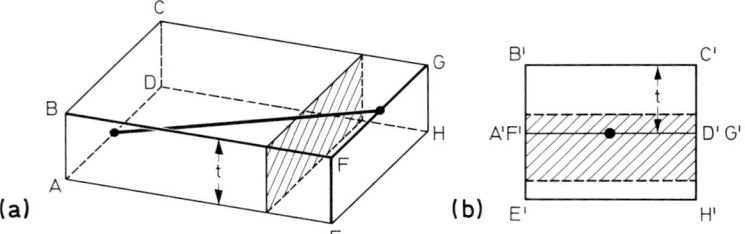

Fig. 3. (a) Geometry and (b) generalized cross-section of a dislocation in an untilted foil. The undashed letters in (a) project parallel to the dislocation to the dashed letters in the generalized (b). An arbitrary profile and its projection in the generalized cross-section are shown shaded. The dislocation in (b) is coming out of the phage but not at right angles to it

3. Computed Electron Micrographs of Dislocations in Quartz

3.1 General Statement

Quartz has trigonal symmetry ($P3_22$, $P3_12$), so six independent elastic constants are required to determine the displacements of a dislocation. To determine the effect of elastic anisotropy on dislocation images in quartz computed micrographs using the full anisotropy of quartz (Simmons and Wang, 1971) and computed micrographs using the isotropic elastic coefficients measured on a quartzite (Birch, 1966) were compared. There is little difference between the two sets of micrographs (Fig. 4). Similar comparisons made for six other orientations and Burgers

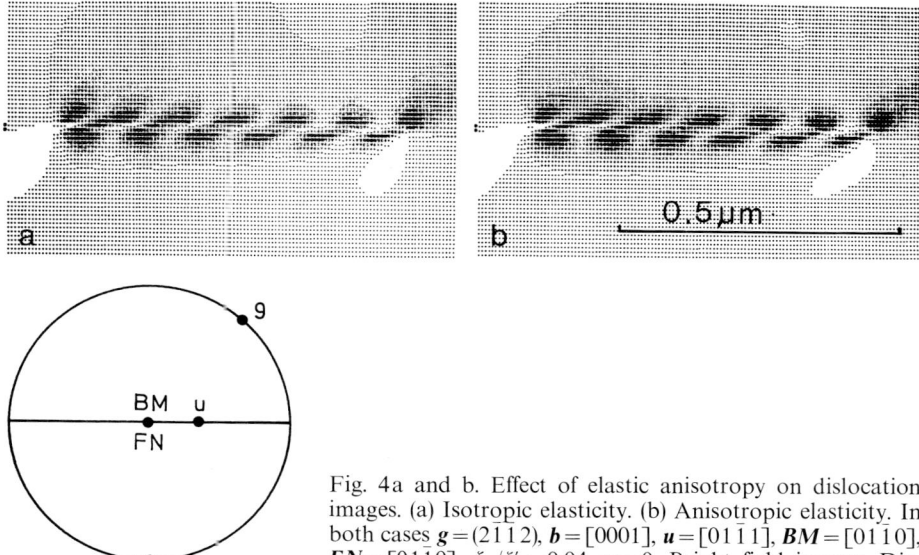

Fig. 4a and b. Effect of elastic anisotropy on dislocation images. (a) Isotropic elasticity. (b) Anisotropic elasticity. In both cases $g = (2\bar{1}\bar{1}2)$, $b = [0001]$, $u = [01\bar{1}1]$, $BM = [01\bar{1}0]$, $FN = [01\bar{1}0]$, $\xi_g/\xi'_g = 0.04$, $w = 0$. Bright field images. Dislocations intersect the top surface of the foil on the right

vectors also show negligible differences due to anisotropy. Of these comparisons, a c-axis screw dislocation showed the greatest effect of anisotropy.

A series of theoretical bright field electron micrographs has been computed in order to determine the effects of various parameters on the dislocation images in quartz. Particular attention has been paid to parameters involved in the determination of Burgers vectors either by the invisibility criteria or by image matching.

3.2 Extinction Distance

Because of the relatively low atomic scattering factor of oxygen, quartz and most other silicates have large extinction distances. Extinction distances for commonly used reflections in quartz have been calculated for 120 kV electrons (Ardell et al., 1974). Fig. 5 demonstrates the effect of the extinction distance on a dislocation image. Fig. 5a was computed with $g=(01\bar{1}3)$ ($\xi_g=2890$ Å) and Fig. 5b was computed with $g=(0\bar{1}13)$ ($\xi_g=7954$ Å). The image in Fig. 5b is much broader and shows less image detail than that in Fig. 5a. This effect can be seen in actual micrographs using these same reflections (Fig. 5, Ardell et al., 1974). Another effect of extinction distance on dislocation images will be discussed in the following section.

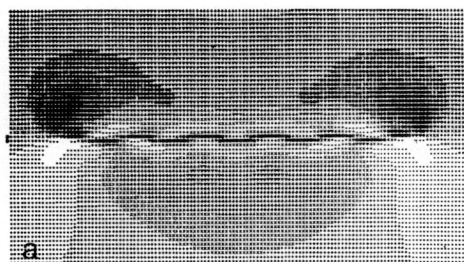
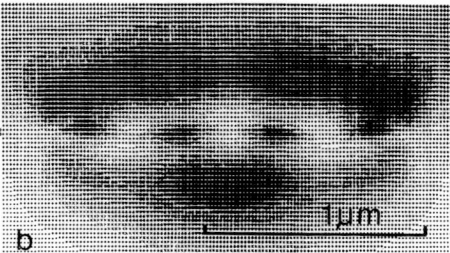

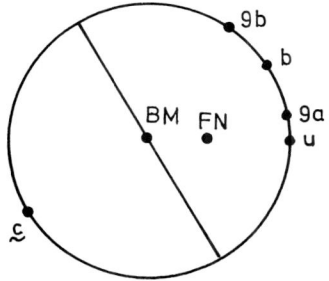

Fig. 5a and b. Effect of extinction distance on dislocation images. (a) $g=(01\bar{1}3)$, $\xi_g=2890$ Å. (b) $g=(0\bar{1}13)$, $\xi_g=7954$ Å. In both cases $b=[0001]$, $u=[01\bar{1}2]$, $BM=[2\bar{1}\bar{1}0]$, $FN=[10\bar{1}1]$, $\xi_g/\xi'_g=0.06$, $w=0$. Bright field images. Dislocations intersect the top surface of the foil on the right

3.3 Deviation Parameter

The deviation parameter is a measure of the deviation from exact Bragg diffracting conditions. The dimensionless deviation parameter (w) used in the Howie-Whelan equations is related to the angular deviation ($\Delta\theta$) from the exact Bragg condition by: $w = \xi_g |g| \sin \Delta\theta$ where $|g|$ is the magnitude of the diffracting vector. Calculated images indicate that a common effect of $w \neq 0$ is the narrowing of dislocation images. In practical microscopy use is made of this effect by taking micrographs with slight positive values of w so as to yield more "pleasing" images. It is found, however, through calculated images and actual microscopy (France and Loretto, 1968; Loretto and France, 1969) that image contrast disappears almost entirely for values of $w > 3$. With the large extinction distances of many of the commonly used reflections in quartz there is loss of image contrast with only very small $\Delta\theta$. Fig. 6 shows the effect of small deviations on a dislocation image using one of the more commonly used reflections in quartz ($g = (0003)$, $\xi_g = 10.171$ Å). The angular deviation in Fig. 6d, in which the dislocation is almost completely out of contrast, is only ~ 0.06 degrees.

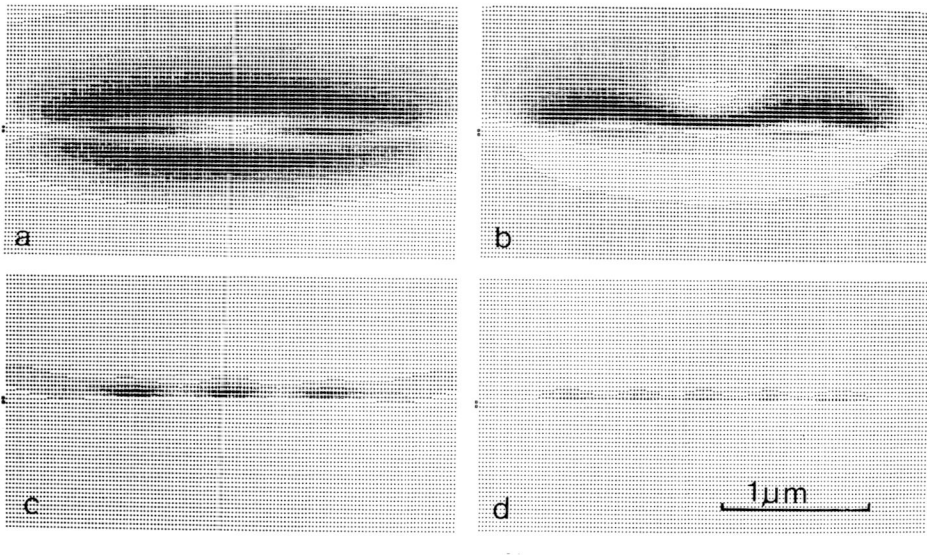

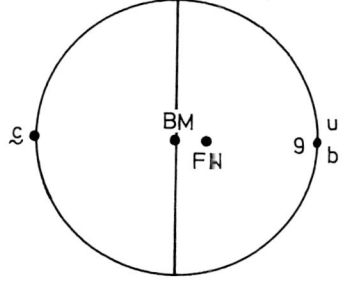

Fig. 6a-d. Effect of deviation from the exact Bragg conditions on dislocation images. (a) $w = 0$, (b) $w = 2$, (c) $w = 4$, (d) $w = 6$. In all cases $g = (0003)$, $b = u = [0001]$, $BM = [2\bar{1}\bar{1}0]$, $FN = [2\bar{1}\bar{1}1]$, $\xi_g/\xi'_g = 0.04$. Bright field images. Dislocations intersect the top surface of the foil on the right

3.4 Anomalous Absorption Length

The anomalous absorption length (ξ'_g) is known to affect the oscillating contrast near the center of the foil (Howie and Whelan, 1962). The sensitivity of the computed image topology to ξ'_g is found to depend on the particular reflection used. Head et al. (1973) have found that the sensitivity of computed images to changes in ξ'_g decreases with increasing foil thickness. Some of the values of ξ'_g used in these calculations were obtained by matching computed and actual micrographs. Other values of ξ'_g were interpolated from these. However, the computed images shown here were found to be relatively insensitive to changes in ξ'_g.

3.5 Foil Orientation and Beam Direction

These two parameters are necessary in setting up the geometry of the foil for the computations. Small variations in these parameters generally have only small effects on the computed image topology. In a few cases, however, small changes in these parameters skewed the computed image topology.

4. Burgers Vector Determination in Quartz

This investigation of the nature of dislocation images in quartz was undertaken in order to examine the feasibility of determining Burgers vectors in quartz. The two basic methods of determining Burgers vectors in TEM are (1) application of the invisibility criteria and (2) image matching. The use of the isotropic invisibility criteria is by far the simplest and most commonly applied method. For an isotropic material a screw dislocation will show no contrast if $\mathbf{g} \cdot \mathbf{b} = 0$. The additional criterion $\mathbf{g} \cdot \mathbf{b} \times \mathbf{u} = 0$ must be fulfilled, in principle, for edge dislocations to be out of contrast.

It has been shown here that the anisotropy of quartz has negligible effect on dislocation images. Thus the above isotropic invisibility criteria should be applicable. Calculated dislocation images in quartz show no or only very slight contrast when these invisibility criteria are met (Fig. 7). The contrast of the c-axis screw dislocation in Fig. 7a is the *maximum* contrast so far observed for diffracting conditions fulfilling the isotropic invisibility criteria. The contrast of the $\mathbf{a}+\mathbf{c}$ screw dislocation shows less contrast even though in this orientation the displacements have the maximum elastic anisotropy (as measured by J_1, the first stress invariant) of any screw dislocation in quartz (Heinish et al., 1975). Computed images with $\mathbf{g} \cdot \mathbf{b} = 0$ and $\mathbf{g} \cdot \mathbf{b} \times \mathbf{u} \neq 0$ indicate that an edge dislocation in quartz will be effectively out of contrast if $\mathbf{g} \cdot \mathbf{b} \times \mathbf{u} < 0.64$ (Figs. 7c and 8). This agrees with the isotropic calculations by Howie and Whelan (1962). Image calculations made for many mixed dislocations also show little or no contrast when $\mathbf{g} \cdot \mathbf{b} = 0$ and $\mathbf{g} \cdot \mathbf{b} \times \mathbf{u} < 0.64$ (Fig. 7d). Ardell et al. (1974) have shown experimentally that the isotropic invisibility criteria appear to be applicable for a number of dislocations in quartz.

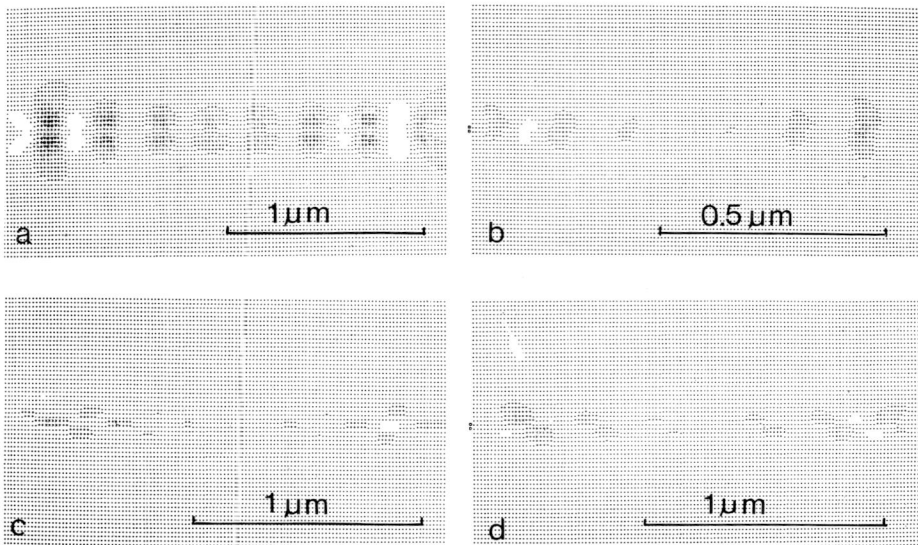

Fig. 7a–d. Dislocations out of contrast. (a) Pure screw dislocation with $g=(2\bar{1}\bar{1}0)$, $b=u=[0001]$, $BM=[0\bar{1}10]$, $FN=[0\bar{1}11]$, $\xi_g/\xi'_g=0.04$, $w=0$. (b) Pure screw dislocation with $g=(0\bar{1}\bar{1}0)$, $b=1/3[2\bar{1}\bar{1}3]$, $u=[2\bar{1}\bar{1}3]$, $BM=FN=[0001]$, $\xi_g/\xi'_g=0.04$, $w=0$. (c) Pure edge dislocation with $g=(1\bar{1}00)$, $b=[0001]$, $u=[2\bar{1}\bar{1}0]$, $BM=FN=[11\bar{2}0]$, $\xi_g/\xi'_g=0.04$, $w=0$, $g \cdot b \times u = -0.64$. (d) Mixed dislocation with $g=(0\bar{1}\bar{1}0)$, $b=1/3[2\bar{1}\bar{1}0]$, $u=[11\bar{2}0]$, $BM=FN=[2\bar{1}\bar{1}0]$, $\xi_g/\xi'_g=0.04$, $w=0$, $g \cdot b \times u=0.0$. In all cases $g \cdot b=0$. Bright field images. Dislocations intersect the top surface of the foil on the right

Calculated images show, however, that considerable care should be taken in applying these invisibility criteria to quartz. Because of the large extinction distances of many of the commonly used reflections, dislocations can easily be put out of contrast because of small angular deviations from exact Bragg conditions even though the invisibility criteria have not been met (Fig. 6). The use of dark field images may help to overcome this sensitivity to deviations as it is easier to determine from the contrast at bend contours where $w \approx 0$. The contrast also tends to be enhanced generally in dark field images (Hirsch et al., 1965). The use of high order reflections of the type (0006) and (0009), which have smaller extinction distances than (0003), has been shown to have certain advantages in Burgers vector determinations (Ardell et al., 1974). In either bright or dark field, great care should be exercised in obtaining a diffraction pattern from the area around the specific dislocation under examination, since w will generally vary over the field of view. The large magnitudes of the Burgers vectors in quartz generally yield values of $g \cdot b \times u$ greater than 0.64. Thus although $g \cdot b = 0$ the dislocation may still be in contrast because the second criterion for invisibility (which is not always considered) has not been met. Thus one needs to determine the approximate value of $g \cdot b \times u$ for each diffraction experiment.

Because of the limitations mentioned above it may not always be possible to obtain suitable diffraction conditions to determine a Burgers vector uniquely and

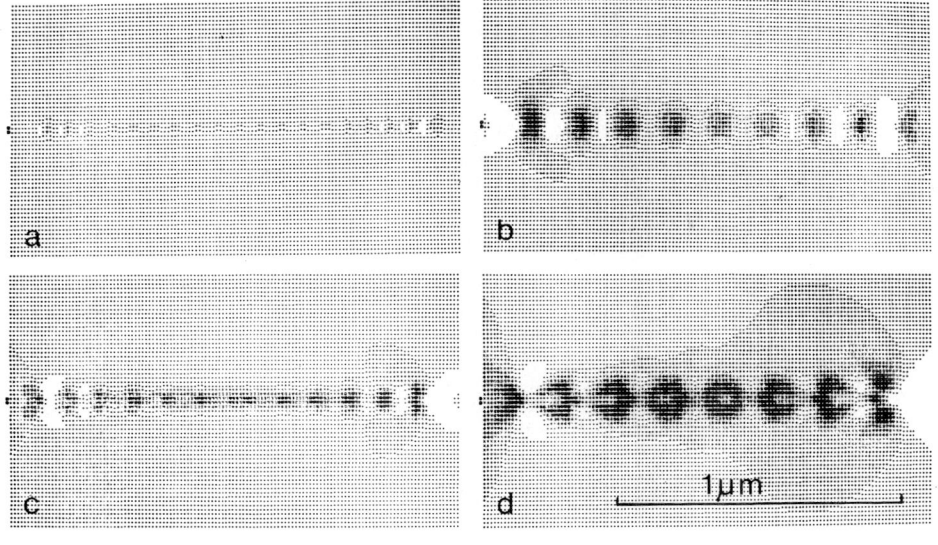

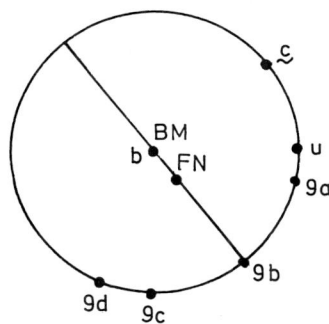

Fig. 8a–d. Effect of $g \cdot b \times u$ on dislocation contrast; $g \cdot b = 0$. (a) $g = (01\bar{1}1)$, $g \cdot b \times u = -0.345$, (b) $g = (01\bar{1}0)$, $g \cdot b \times u = -0.973$, (c) $g = (0\bar{1}11)$, $g \cdot b \times u = 1.470$, (d) $g = (0\bar{1}12)$, $g \cdot b \times u = 2.032$. In all cases $b = 1/3[2\bar{1}\bar{1}0]$, $u = [01\bar{1}2]$, $BM = [2\bar{1}\bar{1}0]$, $FN = [10\bar{1}0]$, $\xi'_g/\xi_g = 0.05$, $w = 0$. Bright field images. Dislocations intersect the top surface of the foil on the right

it may then be necessary to use image matching. Details of this technique are found in Head et al. (1973). While this technique has not yet been extensively applied to quartz, in the present study, image matching with dislocations of known character has shown that the technique, while laborious, may be effective. Problems arise in the determination of the various parameters, such as thickness, ξ'_g, and w, needed to compute a simulated image. Also it is difficult to take the large number of micrographs required for image matching because of the tendency of quartz to damage under the electron beam (McLaren and Phakey, 1965).

References

Ardell, A.J., Christie, J.M., McCormick, J.W.: Dislocation images in quartz and the determination of Burgers vectors. Phil. Mag. **29**, 1399–1411 (1974).
Ardell, A.J., McCormick, J.W., Christie, J.M.: Diffraction contrast experiments on quartz using high-order (000*l*) reflections. Proc. 8th Int. Congress on Electron Microscopy, Canberra, **1**, 486–487 (1974).
Birch, F.: Compressibility; elastic constants. Geol. Soc. Am. Memoir **97**, 167 (1966).

Bullough, R., Maher, D. M., Perrin, R. C.: Accurate computer simulation of electron microscope images. Nature **224**, 364–365 (1969).

Eshelby, J.D., Read, W.T., Shockley, W.: Anisotropic elasticity with applications to dislocation theory. Acta Met. **1**, 251–259 (1953).

France, L. K., Loretto, M. H.: The influence of the size of diffracting vectors on the visibility of dislocations in α-iron. Proc. Roy. Soc. (London) Ser. A **307**, 83–96 (1968).

Gevers, R., Blank, H., Amelinckx, S.: Extension of the Howie-Whelan equations for electron diffraction to non-centrosymmetrical crystals. Phys. Stat. Sol. **13**, 449–465 (1966).

Head, A. K.: The computer generation of electron microscope pictures of dislocations. Austr. J. Phys. **20**, 557–566 (1967).

Head, A. K., Humble, P., Clarebrough, L. M., Morton, A. J., Forwood, C. T.: Computed electron micrographs and defect identification. Amsterdam: North Holland Press 1973.

Head, A. K., Loretto, M. H., Humble, P.: The influence of large elastic anisotropy on the determination of Burgers vectors of dislocation in β-brass by electron microscopy. Phys. Stat. Sol. **20**, 505–519 (1967).

Heinisch, A. K., Jr., Sines, G., Goodman, J. W., Kirby, S. H.: Elastic stresses and self energies of dislocations of arbitrary orientation in anisotropic media: olivine, orthopyroxene, calcite and quartz. J. Geoph. Res. **80**, 1885–1896 (1975).

Hirsch, P. B., Howie, A., Nicholson, R. B., Pashley, D. W., Whelan, M. J.: Electron Microscopy of thin crystals. Washington: Butterworth 1965.

Howie, A., Whelan, M. J.: Diffraction contrast of electron microscope images of crystal lattice defects II. The development of a dynamical theory. Proc. Roy. Soc. (London) Ser. A **263**, 217–236 (1961).

Howie, A., Whelan, M. J.: Diffraction contrast of electron microscope images of crystal lattice defects III. Results and experimental confirmation of the dynamical theory of dislocation image contrast. Proc. Roy. Soc. (London) Ser. A **267**, 206–230 (1962).

Humble, P.: Computed electron micrographs for tilted foils containing dislocations and stacking faults. Austr. J. Phys. **21**, 325–336 (1968).

Loretto, M. H., France L. K.: The influence of the degree of the deviation from the Bragg condition on the visibility of dislocations in copper. Phil. Mag. **19**, 141–154 (1969).

McLaren, A. C., Phakey, P. P.: A transmission electron microscopy study of amethyst and citrine. Austr. J. Phys. **18**, 135–141 (1965).

Simmons, G., Wang, H.: Single crystal elastic constants and calculated aggregate properties. Cambridge, Mass: M.I.T. Press 1971.

Skalicky, P.: Computer-simulated electron micrographs of crystal defects. Phys. Stat. Sol. **20**, 11–52 (1973).

Stroh, A. N.: Dislocations and cracks in anisotropic elasticity. Phil. Mag. **3**, 625–646 (1958).

CHAPTER 2.5

The Direct Imaging of Crystal Structures

J.M. COWLEY and S. IIJIMA

1. Electron Microscope Imaging

The essential stages in the formation of an image in an electron microscope can be described in terms of the Abbe Theory of the formation of the image by the objective lens. Although this is a wave theory, we may illustrate it by reference to the geometric optics diagram, Fig. 1.

For a parallel incident beam the amplitude distribution at the exit face of the object is given by the transmission function $q(xy)$. Ideally, the lens takes the radiation transmitted and scattered from each point of the object and brings it together in the image plane with the correct relative phases to form a wave function

$$\psi(xy) = q\left(-\frac{x}{M}, -\frac{y}{M}\right)$$

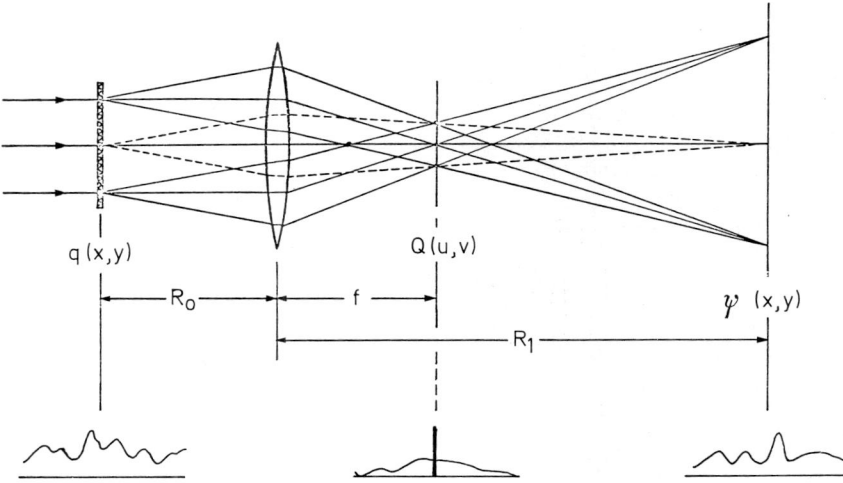

Fig. 1. Geometric optics diagram used to illustrate the principles of the wave-optics, Abbe theory of image formation

i.e. it recreates the transmission function of the object, inverted and magnified by a factor M. Often we refer the image to the scale of the object and write instead, $\psi(xy) = q(xy)$. Actually because of the limitations of the aperture of the lens and its aberrations, this wave function will be smeared out by a spread function $s(xy)$. The smearing action is represented by a convolution integral

$$q(xy)*s(xy) \equiv \iint q(XY)s(x-X, y-Y)\, dX\, dY. \tag{1}$$

The observed intensity is then given by

$$I(xy) = \psi\psi^* = |q(xy)*s(xy)|^2. \tag{2}$$

From Fig. 1 we see that all radiation scattered at a particular angle, φ, from the object is brought together at one point in the back-focal plane of the lens. The amplitude distribution in this back-focal plane is then the Fraunhofer diffraction pattern of the object and is calculated by taking the Fourier transform of $q(xy)$, thus:

$$Q(uv) = Fq(xy) \equiv \iint q(xy) \exp\{2\pi i(ux+vy)\}\, dx\, dy. \tag{3}$$

We may consider that the observation and aperture limitations of the lens act on this back-focal plane and have the effect of multiplying $Q(uv)$ by the transfer function $S(uv)$. Thus the formation of the image is regarded as a further process of Fraunhofer diffraction, described by a Fourier transform

$$F[Q(uv) \cdot S(uv)] = q(xy)*s(xy), \tag{4}$$

where we have used the Multiplication Theorem which states that the Fourier transform of a product of two functions is the convolution of their Fourier transforms.

If an electron microscope is used to image a small non-crystalline particle (without resolution of the atoms) the diffraction pattern in the back-focal plane will consist of a sharp central maximum, where the transmitted incident beam is brought to a focus, plus the scattering due to the particle shape and its variations of thickness or scattering power.

If the particle is crystalline there will be sharp diffraction maxima in addition to the sharp central spot, and around each diffraction spot there will be a scattering distribution due to the size, shape and thickness variations of the portion of the crystal giving that spot.

Thus if an aperture is used to select one particular diffraction spot and its surrounding scattering from the amplitude distribution in the back-focal plane, the image will show intensity only for those regions of the crystal giving that diffraction spot. There will be modulations of the image intensity corresponding to the variation of the diffracted intensity due to variations of thickness, orientation, composition or degree of imperfection of the crystal regions. This is the customary form of dark field or bright field (for the transmitted beam spot) image of a crystalline object.

If the objective aperture transmits two diffraction spots (of which one may be that for the transmitted incident beam) the image of the regions giving both beams will be crossed by a set of fringes. These fringes can be thought of as being produced in the same way as the fringes in the optical diffraction pattern for two coherent point sources, such as the two pin holes or slits in a Young's fringes experiment. The separation of the fringes is inversely proportional to the separation of the two spots in the diffraction pattern and so corresponds to the periodicity of a set of lattice planes in the crystal. The positions of the maxima and minima of the fringes depend on the relative phases of the two diffracted beams and these phases depend on the orientation of the crystal, its thickness and degree of perfection as well as on the relative phases of the structure amplitudes for the diffraction spots and the aberrations and focusing of the objective lens. In the simplest case that the crystal is perfect and parallel-faced and the two diffraction spots are the zero (incident) beam and the only other strong spot for which the Bragg condition is exactly satisfied, and if these two beams are equally inclined to the axis of the objective lens so that they are influenced in the same way by lens aberrations, then the maxima of intensity of the fringes are known to occur at a distance $d/4$ from the planes of maximum atom concentration for the lattice planes of spacing d. If all these conditions are not exactly satisfied, the fringe positions will be elsewhere (see Hashimoto et al., 1961; Cowley, 1959). In general the scattering conditions or the orientation and thickness of the crystal are rarely known with sufficient accuracy to allow any meaningful interpretation of the positions or contrast of the fringes and so it is difficult to make useful deductions from their observation.

It will not be possible to observe these two-beam interference fringes if the magnification of the electron microscope is insufficient or if the image is blurred by chromatic aberration effects due to variations of the incident electron energy or variations in the objective lens current, or by displacements of the image by specimen drift, stray magnetic or electric fields or mechanical vibrations. The fact that fringes having spacings of less than 1 Å have been observed is a tribute to the quality of the engineering of modern electron microscopes.

If more than two diffraction spots are included in the objective aperture the fringe pattern will be more complicated and if the spots are not collinear a two-dimensional pattern of overlapping fringes will be formed. This pattern will in general vary strongly with crystal thickness and orientation.

In order that the image should provide direct information concerning the actual arrangement of atoms in crystals, some very special conditions must apply. Firstly the wave function at the exit face of the crystal should have a direct relationship to the structure of the crystal. Since it is necessarily a two-dimensional function, the relationship should be with a projection of the structure, preferably along a principal axis. Secondly, the combined action of the irreducible lens aberrations, defocus and aperture limitations of the objective lens should be such as to turn the wave function into an image intensity which has some direct relationship to a projection of the structure.

For very thin biological objects, such a favorable combination of circumstances is known to exist. To a first approximation, the intensity distribution for such an object has been shown to be related to a projection of the potential distribution of the object if the image is taken at an optimum defocus and the resolution is limited to a well-defined optimum value (Scherzer, 1949; Cowley, 1974).

In very thin objects the lateral spread of the electron wave due to Fresnel diffraction effects may be neglected. The electron wave may be considered to travel straight through the object, suffering only a phase change proportional to the electrostatic potential it has experienced along a straight-line path. The object is thus a phase object with transmission function

$$q(xy) = \exp\{-i\sigma\varphi(xy)\}, \qquad (5)$$

where σ is the interaction constant ($\sigma = \pi/\lambda E$; λ is the electron wavelength and E is the accelerating potential) and $\varphi(xy)$ is the projection of the three-dimensional potential distribution, $\varphi(r)$, along the beam direction, defined to be the z-axis.

For thin samples of light-atom materials we may assume the phase change $\sigma\varphi$ to be much smaller than unity and write

$$q(xy) \approx 1 - i\sigma\varphi(xy), \qquad (6)$$

$$Q(uv) \approx \delta(uv) - i\sigma\Phi(uv). \qquad (7)$$

Eq. (7) represents the amplitude in the back-focal plane, consisting of a sharp, delta-function peak and a scattered amplitude $\Phi(uv) = F\varphi(xy)$. As in Eq. (4), $Q(u,v)$ is multiplied by a modification function including the perturbations of phase due to defocus and aberrations and an aperture function:

$$Q'(uv) = Q(uv) \cdot \exp\{i\chi(uv)\} \cdot A(uv). \qquad (8)$$

Putting in Eq. (7) and considering only first-order terms in $\sigma\varphi$, it can be shown that only the imaginary part of the modification function need be considered, so that

$$Q'(uv) \approx \delta(uv) + \sigma\Phi(uv) \cdot \sin\chi(uv) \cdot A(uv). \qquad (9)$$

For a particular, optimum defocus value, $\sin\chi$ has a value close to unity for most of the diffraction pattern within an optimum aperture size. Thus for a limited resolution range we have an image intensity given approximately by

$$I(xy) = \psi\psi^* \approx 1 + 2\sigma\varphi(xy), \qquad (10)$$

which is directly interpretable in terms of the structure of the specimen. The optimum defocus condition is

$$\Delta f = -\tfrac{4}{3}(C_s\lambda)^{1/2} \qquad (11)$$

or -900 Å for 100 keV electrons with a spherical aberration coefficient, $C_s = 1.8$ mm. The interpretable resolution is given by $\Delta x = 0.66(C_s\lambda^3)^{1/4}$, and this is 3.5 Å for these same conditions. With higher voltages in electron microscopes now under construction this limit of interpretable resolution may well be reduced to 1.5 to 2.0 Å.

It must be stressed that this resolution limit is that for an image which may be immediately interpreted in terms of the atom arrangement in the sample. There is no reason why, with larger apertures and other amounts of defocus, it should not be possible to observe details in the image on the same scale as the finest observable fringe spacings and limited only by chromatic aberration and instrumental stabilities. The direct interpretation of the detail on this finer scale is not possible. Various schemes employing image analysis or manipulation, using digital or analog techniques, have been proposed for the interpretation of this finer image detail, but so far without any conspicuous practical success.

This favorable case, giving the linear relationship between image contrast and potential projection, Eq. (10), and the possibility of image interpretation, either by direct inspection or by image manipulation, applies only for the very special cases of very thin specimens giving very small phase changes. For almost all cases of the high resolution imaging of crystals it does not apply. It can, at best, be used to supply rough guidelines as to appropriate imaging conditions. For 100 keV electrons and a resolution of 3.5 Å, the phase object approximation, Eq. (5), applies for thicknesses of 100 to 150 Å but the small-phase-change approximation, Eq. (6), may fail for a thickness of 10–20 Å for a crystal containing medium-weight atoms.

There are alternative approximations available which are useful within other limited ranges of conditions. It was shown by Cowley and Moodie (1960), for example, that for sufficiently small amounts of defocus, the intensity variation in the image of a phase object is proportional to the projection of the density distribution of electrical charge in the specimen (including contributions from both the positive charges on the nuclei and the negative charges on the electrons). There is no requirement that the phase change should be small. The range of validity of this "projected-charge" approximation has been shown by Lynch and O'Keefe (1972) and by Anstis *et al.* (1973) to be considerable, but its value is limited for resolutions for which the spherical aberration of the objective lens is important. An alternative approximation suggested by Cowley (1974) includes spherical aberration effects and reasonably large phase changes.

However, for a more reliable and precise interpretation of the images of crystals we must turn to the more sophisticated theory of n-beam dynamical diffraction of electrons and the methods of computation developed on this basis.

2. The Imaging of Crystals

The modifications of the amplitude and phase of an electron wave transmitted through a thin crystal can be calculated with arbitrary accuracy by use of moderate-sized computers. This can be done for the case of no absorption or for any assumed form for the absorption function. Although there remains some uncertainty as to the nature of the absorption processes for most crystals, the results of the calculations do not differ greatly for any reasonable assumptions, especially for thin crystals. Hence we can conclude with confidence that reliable indications can be given theoretically of the intensity distributions to be observed in electron microscope images and electron diffraction patterns for any postulated distribution of atoms. The inverse process of deriving accurate data on the nature and relative positions of atoms directly from observed intensity data has not been proven possible in general.

The computing methods used can be based on any of the several formulations of the n-beam dynamical diffraction theory, but the methods developed by Goodman and Moodie (1974), based on the formulation due to Cowley and Moodie (1957), have been found to be the most convenient and flexible. In this approach the progress of an electron wave is traced through the crystal, with the accumulating effects of the phase-changes due to variation of the electrostatic potential in the crystal and the spread of the wave due to Fresnel diffraction. The wave function at the exit face of the crystal is calculated and used to derive the intensity distribution in an image by applying the phase changes due to defocus and other lens aberrations. The amplitude and intensity in the diffraction pattern may be derived by taking the Fourier transform of the exit wave. Details given by Goodman and Moodie (1974), Lynch and O'Keefe (1972) and Cowley (1975) are sufficient to allow a satisfactory computer program to be written without difficulty.

For a crystal having a reasonably large unit cell containing many atoms, the number of diffracted beams occurring simultaneously in the crystal is usually large. This is especially the case if, in order to obtain a clear and recognizable image of the structure, the incident beam is taken to be parallel to a principal axis of the crystal lattice. For example, for his calculations of images of $Ti_2Nb_{10}O_{29}$ crystals, Fejes (1973) found it necessary to take into account the interactions of a least 450 beams in the crystal. The number of diffracted beams transmitted through the objective aperture and actually contributing to the image intensity was much smaller, usually about 100. The comparison of the image intensity distributions calculated in this way with those observed experimentally is given in Fig. 2. For this special case that the beam is parallel to a principal axis, the amount of computing can often be reduced by taking advantage of the symmetry of the diffraction pattern.

The various images of Fig. 2 are calculated for various amounts of defocus of the objective lens, assuming the use of 100 keV electrons and a lens having a spherical aberration coefficient, $C_s = 1.8$ mm. It is seen that an image having an obvious relationship to a projection of the structure of $Ti_2Nb_{10}O_{29}$ (see Fig. 5 and Iijima, 1971) is obtained by calculation and experiment for a defocus of -950 ± 100 Å. This applied for crystal thicknesses from 0 up to about 150 Å. In a few cases, good, recognizable images of the structure can also be obtained for thicknesses in the range 400–700 Å (Fejes et al., 1973).

It therefore appears that for crystals which are not too thick, as for the weak phase objects discussed above, an image which is directly interpretable in terms of a projection of the potential distribution can be obtained with an optimum defocus of about -950 Å and a resolution limited to about 3.5 Å for 100 keV electrons. The thickness limitation of 100–150 Å corresponds to the approximate range of validity of the phase-object approximation for this resolu-

The Direct Imaging of Crystal Structures 129

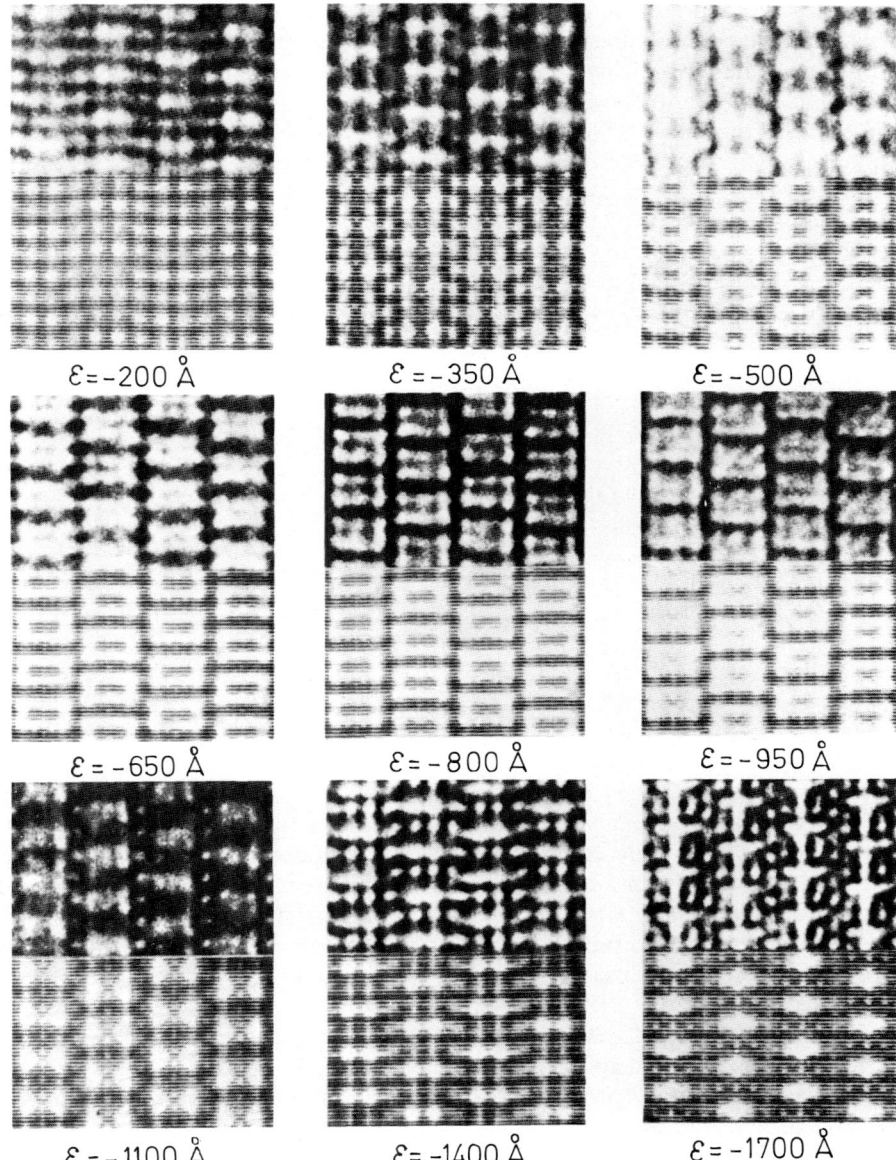

Fig. 2. Comparison of observed (top part of each figure) and calculated intensity distributions in images of the crystal structure of $Ti_2Nb_{10}O_{27}$, obtained for the indicated deviations from the exact Gaussian focus, for 100 keV electrons and a spherical aberration coefficient $C_s = 1.8$ mm. The optimum defocus, giving the best representation of the projected structure of the crystal, is about -900 Å. These observed and calculated images may be compared with the corresponding parts of Fig. 5 (Fejes, 1973)

tion, but the apparent validity of the theory based on the assumption of a small phase change is surprising since, for the type of oxide investigated, there may be a phase change varying by 5–10 radians around the heavy atom positions for a thickness of about 100 Å.

Actually the linear relationship of image intensity to the projected potential value, given by Eq. (10), does not hold except for very small thicknesses. As shown in Fig. 3, the variation of the intensity at the position of a metal atom row is not linear with crystal thickness but does tend almost monotonically towards a constant limiting value. Thus the image may be thought of as representing the projected potential distribution, although with a distorted intensity scale as in the case of an over-exposed photographic negative.

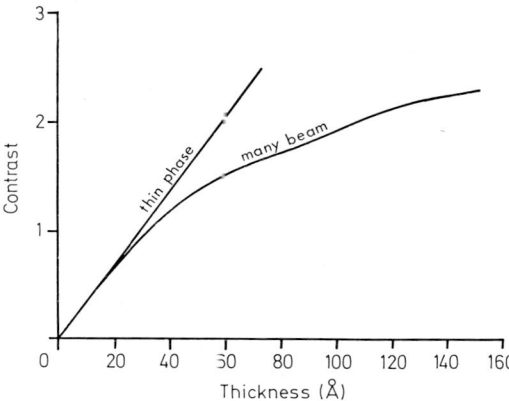

Fig. 3. Theoretical values for the contrast (relative intensity variation) in the image of a row of medium-weight metal atoms (as in a titanium-niobium oxide crystals) as a function of crystal thickness, according to the thin phase-object approximation and as calculated by the full many-beam dynamical diffraction theory (Fejes, 1973)

On the basis of the comparison between calculated and observed image intensities for several types of crystal and on the basis of extensive comparison of images with known structures, a reliable set of rules and prescriptions for obtaining interpretable images of the structures of crystals can be drawn up. For 100 keV electrons and electron microscopes having a limit of interpretable resolution of about 3.5 Å these are as follows:

1. The crystal should be one in which there are atoms or concentrations of atoms at least 3.5 Å apart, and the incident beam direction should be such as to give a projection of the structure which shows these features to good advantage.

2. The thickness of the crystal should be less than 150 Å except in special circumstances. The thickness can often be estimated by observations on the changes of apparent width of planar defects when the crystal is tilted through a known angle.

3. The crystal should be oriented, by use of a goniometer stage, so that the incident beam should be parallel to a principal axis of the lattice with an accuracy approaching 3×10^{-3} radians. This degree of alignment may be achieved by adjusting the crystal orientation until the spots in the diffraction pattern show a close symmetry of intensities about the incident beam spot.

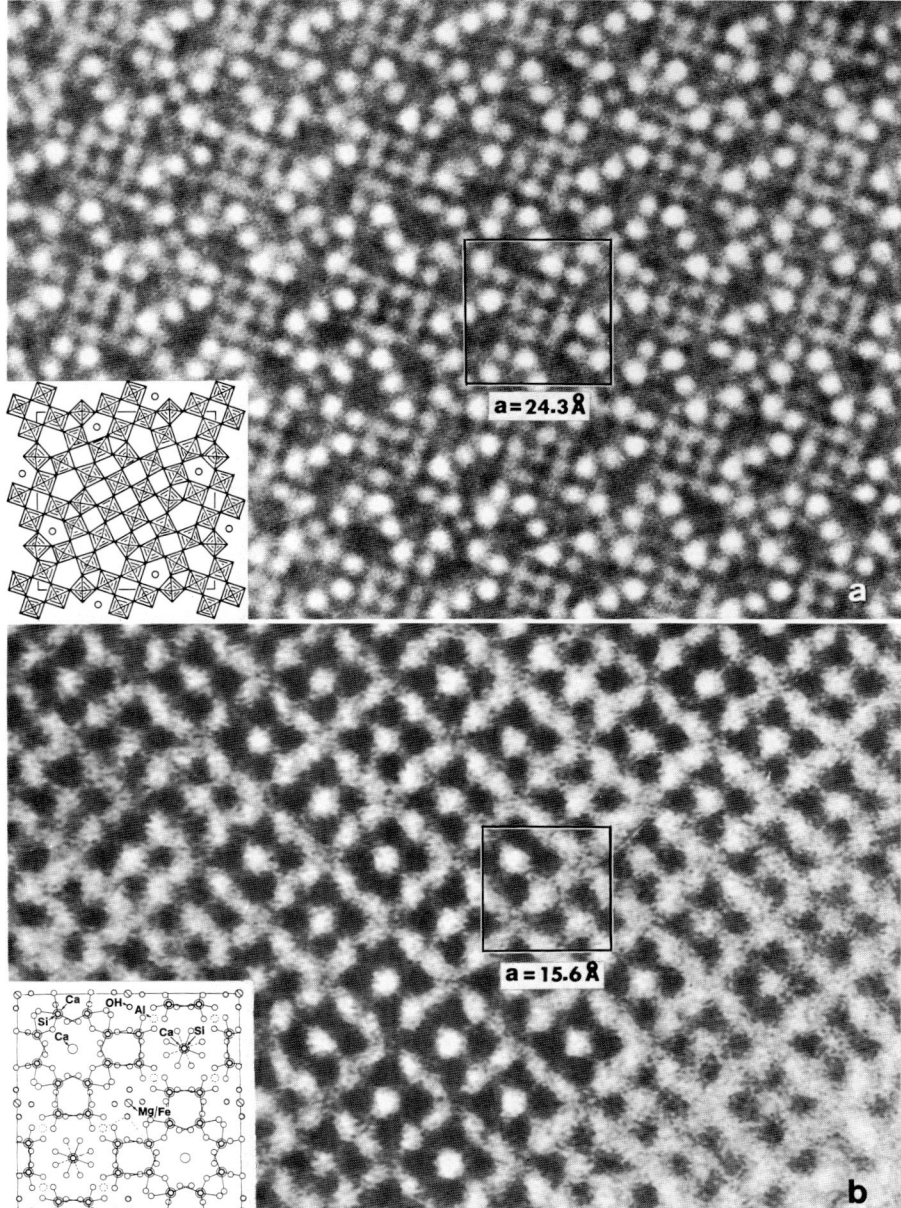

Fig. 4. (a) The determination of the crystal structure of a new phase, $2\,Nb_2O_5 \cdot 7\,WO_3$, by observation of the image of a thin crystal. The diagram represents the structure deduced from the image. Each shaded square represents an octahedron of oxygen atoms containing a metal atom near its center. The small circles in some of the pentagonal tunnels indicate that these tunnels are filled with strings of metal and oxygen atoms and so appear dark in the image (Iijima and Allpress, 1974a). (b) High resolution image showing the distribution of the heavier atom clumps in the structure of idocrase. Taken with a JEM-100B electron microscope operating at 100 keV with optimum defocus. The beam is parallel to the tetragonal z-axis. The correspondence of the image with the projection of the structure in the insert is apparent (Buseck and Iijima, 1974)

4. After tilting the crystal into the correct orientation the objective lens should be stigmated and focused, preferably by observation of the granularity in an adjacent area of amorphous carbon film. The easiest focal position to pick is that of minimum contrast in the image of the carbon film, which corresponds to a defocus of about -400 Å.

5. The objective lens defocus is then changed to the optimum defocus value and the image is recorded. With modern microscopes, and especially with a pointed filament or field emission gun, the image of the crystal structure should be clearly visible on the fluorescent screen at a magnification in excess of $500000 \times$. No through-focus series should in general be necessary.

The specimens of oxides and minerals which have been studied in this way have been prepared by dispersing a finely ground powder of the material in a suitable liquid and then depositing it on a holey carbon film having open holes over a large fraction of its area. Crystal fragments extending over the holes in the film are sought and their edges are examined for suitable thin, parallel-sided or wedge-shaped regions.

It may be possible to use specimens thinned by ion bombardment or chemical etching, although it has not yet been well established that these methods can give good specimens which are sufficiently thin and not excessively damaged on the surfaces.

Examples of images showing direct representations of the projections of crystal structures are shown in Fig. 4a and b.

3. Imaging of Crystal Defects

For crystals which are thin enough to allow a direct interpretation of the image, the image intensity depends on the projection of the structure in the beam direction. This applies for both periodic and non-periodic distributions of atoms. Any defect in a crystal which leaves the alignment of atom rows in the beam direction unchanged, but modifies only the ordering of such rows of atoms in directions perpendicular to the beam, will be imaged with the same clarity as the unmodified crystal structure. This is illustrated in Fig. 5 where there is one-dimensional disorder and in the *bottom* part of Fig. 6 where there is considerable two-dimensional disorder in the arrangement of the rows of heavy metal atoms parallel to the beam but the rows themselves are undistorted. However in the *top* part of Fig. 6 these metal atom rows are distorted or disrupted by defects and the image intensity is strongly modified.

The origin of the very strong contrast in the case of the latter type of defect is not very clear. It has been argued by Cowley and Iijima (1972) that this may be in part a consequence of the non-linearity of the variation of image intensity with projected potential, as in Fig. 3. Two closely-spaced, unresolved rows of atoms will give greater intensity modulation than one row of twice the length. Hence, if one row of metal atoms is distorted so that, in projection, it gives two projected maxima of potential, the corresponding part of the image will appear blacker. However this explanation seems inadequate to explain all the observed effects.

Fig. 5. High-resolution image of thin edge of a disordered crystal of composition near to $Ti_2Nb_{10}O_{29}$. With variation of the metal atom concentrations, the structure alternates irregularly between layers of 4×3 blocks of octahedra between shear planes of spacing 14.3 Å and layers of 3×3 blocks of octahedra, with shear-plane spacing of 10.3 Å (Iijima, 1971)

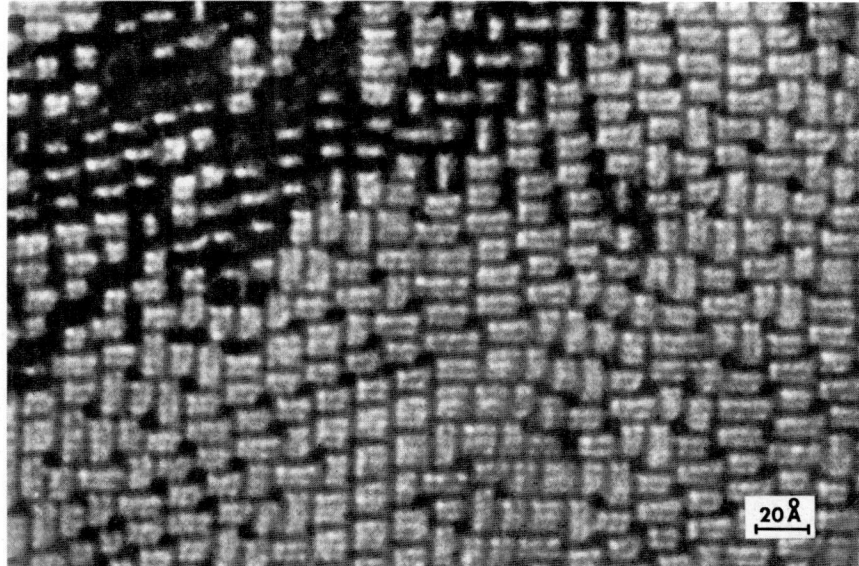

Fig. 6. High resolution electron micrograph of heavily disordered $H-Nb_2O_5$. In the bottom part there is two dimensional disorder with an intergrowth of regions having 4×3, 4×4, 5×3 and other blocks of octahedra (cf. Fig. 5) plus atoms in tetrahedral sites appearing as heavy black spots. In the top part of the picture there is disorder in the beam direction also, giving rise to strong contrast. Most of the black regions probably come from areas where the rows of atoms in the beam direction are stepped sideways to give an increase in the number of rows of metal atoms seen in projection

It is clear that detailed calculations are required for the image intensities for a variety of crystal defects in order to build up experience on the types of contrast to be expected in particular cases. Suitable n-beam dynamical diffraction calculations can be made by modification of the computer programs for perfect crystals, but so far relatively few have been made because of the added complexity involved.

An observation of particular significance has been that of Iijima et al. (1973, 1974) that strong contrast can be produced in images of non-stoichiometric niobium oxides by what appear to be point defects. Chemical evidence suggests that each of the black spots seen involves the presence of interstitial oxygen atoms and displaced metal atoms. It seems clear that the strong contrast observed must result from grouping the displaced metal atoms which array along the incident electron beam direction, forming a chain of the point-defect clusters (Iijima, 1975).

While some degree of uncertainty remains regarding the interpretation of the image contrast in such cases as these, for many cases of interest the interpretation is clear and unambiguous, as in the study of defects and disorders in the niobium- and niobium-titanium oxides (Iijima, 1973; Iijima and Allpress, 1973; Hutchinson, 1974), the observation of shear-plane and rotational faults in the niobium tungsten oxides (Iijima and Allpress, 1974a, b), the study of micro-twinning of ortho- and clinoenstatite (Iijima and Buseck, Chapter 5.4 of this volume) and the study of the ordering of defect clusters in wustite (Iijima, 1974).

4. Dark Field Images

It is well known that very useful information can be obtained regarding the variation of orientation in a distorted crystal or regarding the distribution of different phases within a specimen by using individual diffraction spots to form dark field images. Since for a large unit cell size it is necessary to use an objective aperture of small size to select a diffraction spot, the resolution of the dark field image is necessarily limited. Then it is possible to make the usual assumption that the intensity at any point of the dark field image is proportional to the strength of the diffracted beam from the corresponding region of the specimen.

For high-resolution dark field images the objective aperture must transmit a relatively large proportion of the whole diffraction pattern, including in general a large number of diffracted beams. The relationship of the image intensity to the distribution of atoms in a thin crystal then depends very strongly on the number and nature of the diffracted beams included. The image may show the correct unit cell periodicities but the detailed intensity distribution will in general show no obvious relationship with the atom configuration.

It has been shown by Cowley (1973a) that, for very thin, weakly scattering specimens for which the bright field contrast is given by Eq. (10), the dark field intensity in the ideal case that only the central spot of the diffraction pattern is omitted is given approximately by

$$I_{DF} = \sigma^2 (\varphi - \bar{\varphi})^2, \tag{12}$$

where $\bar{\varphi}$ is the average value of the projected potential. For this ideal case, both positive and negative deviations from the average potential give intensity maxima. For the case of crystals, the phase variations $\sigma(\varphi-\bar{\varphi})$ are usually not small and there are strong deviations from the simple quadratic law of Eq. (12) (Cowley, 1974).

For the more convenient modes of dark field imaging, such as the so-called "high-resolution" mode involving a tilting of the incident beam, the correlation of image intensity with atom distribution is much more complicated. While for any assumed structure of the specimen it is possible to calculate the image intensity for any given dark field imaging mode, it is dangerous to assume *a priori* that the image intensity has any direct relationship to the crystal structure.

The special case of the dark field imaging of disordered structures, using the diffuse scattering in the background of the diffraction pattern to obtain the image, has been treated by Cowley (1973) who suggested that even for the relatively simple case of small deviations from the average potential, when Eq. (12) may be a reasonable approximation, the interpretation of the image must be made with considerable care.

An empirical application of dark field imaging with a high resolution has been attempted by Pierce and Buseck (Chapter 2.6 of this volume).

References

Allpress, J.G., Sanders, J.V.: The direct observation of the structure of real crystals by lattice imaging. J. Appl. Crystallogr. **6**, 165–190 (1973).

Anstis, G.R., Lynch, D.F., Moodie, A.F., O'Keefe, M.: *n*-Beam lattice images III. Upper limits of ionicity in $W_4Nb_{26}O_{77}$. Acta Cryst. A**29**, 138–147 (1973).

Buseck, P.R., Iijima, S.: High resolution electron microscopy of silicates. Am. Mineralogist **59**, 1–21 (1974).

Cowley, J.M.: The electron-optical imaging of crystal lattices. Acta Cryst. **12**, 367–375 (1959).

Cowley, J.M.: High-resolution dark field electron microscopy, I. Useful approximations. Acta Cryst. A**29**, 529–536 (1973a).

Cowley, J.M.: High-resolution dark field electron microscopy, II. Short-range order in crystals. Acta Cryst. A**29**, 537–540 (1973b).

Cowley, J.M.: The principles of high resolution electron microscopy. In: Principles and techniques of electron microscopy (ed. M.A. Hayat). New York: Van Nostrand Reinhold Co. 1974.

Cowley, J.M.: Diffraction physics. Amsterdam: North Holland 1975.

Cowley, J.M., Iijima, S.: Electron microscope image contrast for thin crystals. Z. Naturforsch. **27a**, 445–451 (1972).

Cowley, J.M., Moodie, A.F.: The scattering of electrons by atoms and crystals, I. A new theoretical approach. Acta Cryst. **10**, 609–619 (1957).

Cowley, J.M., Moodie, A.F.: Fourier images IV. Phase gratings. Proc. Phys. Soc. (London) **76**, 378–384 (1960).

Fejes, P.L.: High resolution lattice images. Ph. D. Thesis, Arizona State University (1973).

Fejes, P.L., Iijima, S., Cowley, J.M.: Periodicity in thickness of electron-microscope crystal-lattice images. Acta Cryst. A**29**, 710–714 (1973).

Goodman, P., Moodie, A.F.: Numerical evaluation of *n*-beam wave functions in electron scattering by the multislice method. Acta Cryst. A**30**, 280–290 (1974).

Hashimoto, H., Mannami, M., Naiki, T.: Dynamical theory of electron diffraction for the electron microscope image of crystal lattices. Phil. Trans. Roy. Soc. London, Ser. A**253**, 459–516 (1961).

Hutchinson, J.L.: Direct observation of linear defects in niobium oxides. In: Diffraction studies of real atoms and real crystals, p. 145–146. Proc. of Int. Crystallography Conf., Melbourne (1974).

Iijima, S.: High resolution electron microscopy of crystal lattice of titanium-niobium oxide. J. Appl. Phys. **42**, 5891–5893 (1971).

Iijima, S.: Direct observation of lattice defects in H–Nb_2O_5 by high resolution electron microscopy. Acta Cryst. A **29**, 18–24 (1973).

Iijima, S.: High resolution electron microscopy of wustite crystal. In: Diffraction studies of real atoms and real crystals, p. 217–218. Proc. of Int. Crystallography Conf., Melbourne (1974).

Iijima, S.: Ordering of the Point Defects in Nonstoichiometric Crystals of $Nb_{12}O_{29}$. Acta Cryst. A **31**, 784–790 (1975).

Iijima, S., Allpress, J.G.: High resolution electron microscopy of $TiO_2 \cdot 7 Nb_2O_5$. J. Solid State Chem. **7**, 94–105 (1973).

Iijima, S., Allpress, J.G.: Structural studies by high resolution electron microscopy: Tetragonal tungsten bronze type of structures in the system Nb_2O_5–WO_3. Acta Cryst. A **30**, 22–29 (1974a).

Iijima, S., Allpress, J.G.: Structural studies by electron microscopy: coherent intergrowth of the ReO_3 and tetragonal bronze structure types in the system Nb_2O_5–WO_5. Acta Cryst. A **30**, 29–36 (1974b).

Iijima, S., Kimura, S., Goto, M.: Direct observation of point defects in $Nb_{12}O_{29}$ by high resolution electron microscopy. Acta Cryst. A **29**, 632–636 (1973).

Iijima, S., Kimura, S., Goto, M.: High resolution electron microscopy of non-stoichiometric $Nb_{22}O_{54}$ crystals: point defects and structural defects. Acta Cryst. A **30**, 251–257 (1974).

Lynch, D.F., O'Keefe, M.A.: n-beam lattice images II. Methods of calculation. Acta Cryst. A **28**, 536–548 (1972).

Scherzer, O.: The theoretical resolution limit of the electron microscope. J. Appl. Phys. **20**, 20–29 (1949).

CHAPTER 2.6

A Comparison of Bright Field and Dark Field Imaging of Pyrrhotite Structures

L. PIERCE and P.R. BUSECK

Pyrrhotite, $Fe_{1-x}S$, consists of a number of discrete, but structurally similar compounds (e.g. Morimoto et al., 1970; Amelinckx and Van Landuyt, Chapter 2.3 of this volume; Nakazawa et al., Chapter 5.2 of this volume) related through superstructuring of a hexagonal ($a = 3.45$ Å $\equiv$ A, ~ 5.8 Å $\equiv$ C) subcell. Non-integral as well as integral superstructures apparently result from vacancy ordering (Ovanesyan et al., 1971). High resolution electron microscopy is a logical tool for studying vacancy distributions in pyrrhotite, and has been used to explain c-superstructure spacings (Nakazawa et al., 1974; Pierce and Buseck, 1974).

Cowley and Iijima (Chapter 2.5 of this volume) show that specific instrumental conditions must be used to assure direct interpretation of high resolution electron images. These conditions impose a 3.5 Å limit of resolution. Such high resolution images have been obtained using bright field microscopy (e.g. Cowley and Iijima, 1972; Allpress and Sanders, 1973; Buseck and Iijima, 1974). Theoretical difficulties in the interpretation of dark field images, however, are predicted (Cowley and Iijima, Chapter 2.5 of this volume), and only a few high-resolution dark field photographs of crystals have been published. We have obtained structural information about pyrrhotite from bright and dark field images with resolution greater than 3.5 Å. We base our interpretation on coincidence of the image with the well-known subcell structure of pyrrhotite.

Fig. 1 is a bright field micrograph taken under optimum defocus conditions of pyrrhotite ($Fe_{0.91+0.01}S$ by the X-ray spacing determination of Arnold, 1962) with a c-superstructure of 5C. The objective lens aperture placement is indicated in the accompanying electron diffraction pattern. All fringes are interpretable. The marked 6 Å spacing corresponds to twice a^*, indicating a 2A superstructure. Bright spots correspond to positions of relatively vacancy-rich iron sites, as indicated in the inset drawing. Based on this vacancy distribution, antiphase domains (with an apparent out-of-step vector of $1/2a$) can be recognized. The boundaries of a possible antiphase domain scheme are marked in the figure by dashed lines. Vacancy distributions, and superstructures resulting from such distributions are interpretable, but are not as clear as desired, primarily because the resolution is limited to ~ 3.5 Å.

Experimental indications are that there may be certain instrumental conditions, in addition to those mentioned by Cowley and Iijima (Chapter 2.5 of this volume), for which images can be interpreted in terms of projection of

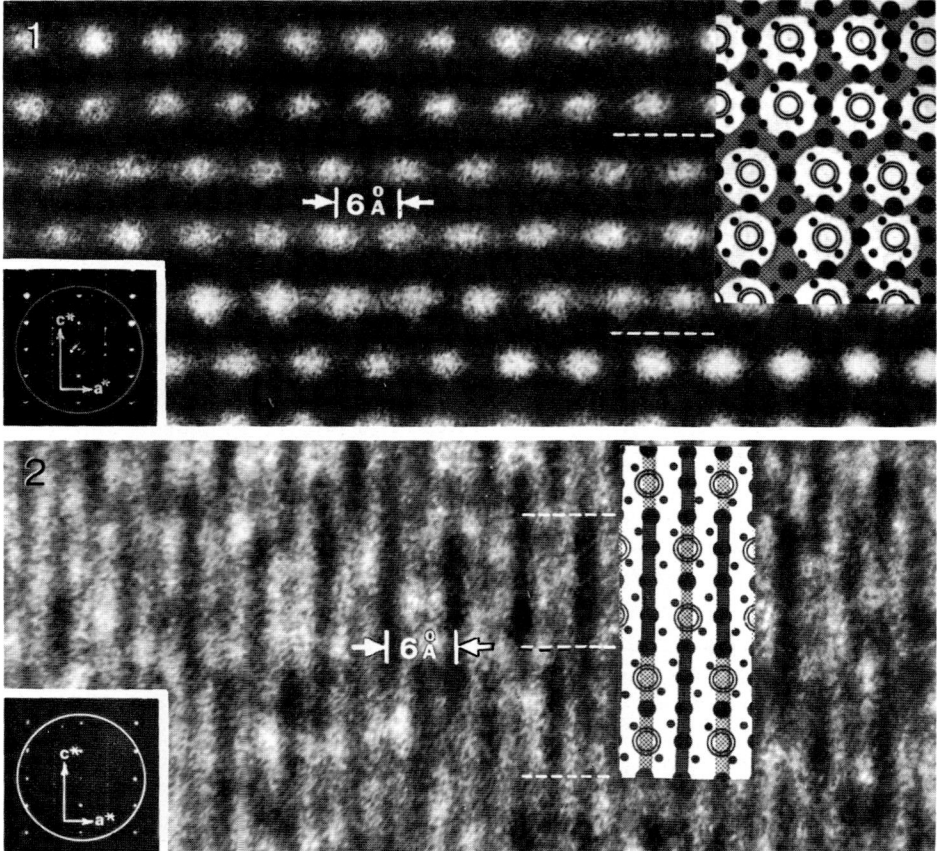

Fig. 1. High-resolution bright field electron micrograph taken at the optimum defocus condition. The inset drawing shows the projected structure of pyrrhotite and the distribution of Fe vacancies, with large solid dots representing projections of nearly filled Fe sites, open circles representing relatively vacancy-rich sites, and small dots representing sulfur atoms. Shaded areas in the drawing represent dark areas in the image. Antiphase boundaries in the image are indicated by dashed lines. Horizontal streaks in the image correspond to either the disordering of vacancies or the superposition of both kinds of ordered domains

Fig. 2. Bright field electron micrograph of pyrrhotite, taken at ~2500 Å underfocus. Vacancy distribution and resulting antiphase boundaries are illustrated by the inset drawing and dashed lines as in Fig. 1

the structure, and for which greater resolutions than 3.5 Å are possible. One indication that favorable circumstances may exist is the imaging of a well-known and identifiable portion of the structure in a recognizable manner. For example, if the subcell structure of pyrrhotite is imaged correctly then superstructures will also probably be imaged correctly. This follows because the fluctuations of projected potential or charge density will be much the same throughout the superlattice cell as in the sublattice cell and so will give much the same image intensity fluctuations for equivalent local structure.

A Comparison of Bright Field and Dark Field Imaging of Pyrrhotite Structures

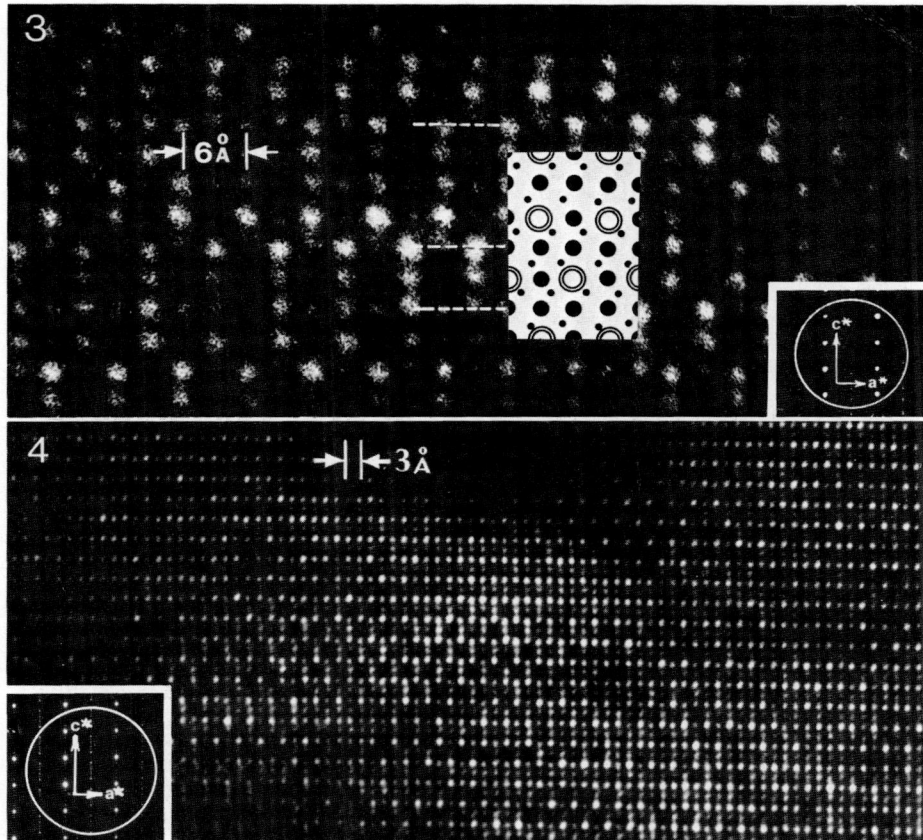

Fig. 3. Dark field electron micrograph, taken at ~ 300 Å underfocus. The vacancy distribution and antiphase boundaries are illustrated as in Fig. 1, except without shading. Compare the relative positions of the objective apertures (white circles) in Figs. 1 and 3

Fig. 4. Dark field electron micrograph of pyrrhotite from Franklin, N.J. The variations in superstructuring, vacancy distribution and antiphase type structure are clearly visible

Fig. 2 shows a bright field electron micrograph of pyrrhotite containing 3 Å fringes, which correspond to the rows of iron atoms along c^* in the subcell structure. The 2A superstructure and the antiphase boundaries perpendicular to c^* (dashed lines) are therefore assumed to be correctly imaged, despite the fact that the photograph was taken at ~ 2500 Å underfocus instead of 900 Å. Under these conditions, 3 Å fringes are visible, but positions of relatively vacancy-rich sites, indicated in the accompanying schematic insert, are not very clear.

For dark field microscopy, the relationship between image intensity and a projection of the structure is not as well established as for bright field microscopy. Calculations for the idealized case of a weak phase object when only the central beam contribution to the image is excluded (Cowley, 1973) indicate that a more complicated relationship than for bright field might be expected. For thin crystals, the experience of Dr. Iijima is that for particular crystals and for particular positions of the objective aperture relative to the diffraction pattern, the dark field image is very close to being the inverse of the bright field image; in general however, this relationship holds only approximately if at all (Cowley, 1974). In practice, as in bright field microscopy, we may rely on the assumption that if a subcell unit is imaged to give a direct representation of the structure then the superlattice will be also.

In this study, dark field aperture placements analogous to that in Fig. 2 produced images that were approximately the inverse of Fig. 2. However, a much clearer representation of structure, shown in Fig. 3, was obtained by an aperture placement (indicated in the accompanying diffraction pattern) that brings four subcell reflections close to the objective lens axis. The bright dots in the image correspond to projections of iron atoms along b, which are separated by 3 Å in the a^*-direction and 2.9 Å in the c^*-direction. Sulfur atoms are not seen in the image. Dim or missing dots correspond to positions of relatively vacancy-rich sites. Dark field images seem interpretable only for much thinner crystals than bright field, as only the extreme edges of crystal images correspond to the structure. Determination of underfocus is difficult in the dark field mode, so the clearest structure representation of a "through-focus-series" is chosen. Fig. 4 is an example of a dark field electron micrograph of pyrrhotite, which shows structural heterogeneity between distances as small as 50 Å. We found that such short range structural variations are common in pyrrhotite, possibly arising from compositional fluctuations within the crystal.

While structural information is obtainable from any of the above examples, the greatest contrast and the clearest representation of vacancy distributions is given by the dark field micrographs. For example, photographs such as Figs. 3 and 4 of pyrrhotite can give details on population densities, stacking sequences in the c^*-direction of filled and non-filled iron layers, and agree with the model that some if not all pyrrhotite superstructures can be attributed to the stacking, ordered and disordered, of antiphase domains as is discussed in detail by Pierce and Buseck (1974). We therefore believe that high-resolution dark field electron microscopy is a fruitful means for the study of superstructuring and vacancy distributions in materials in which the basic structure is known.

References

Allpress, J.G., Sanders, J.V.: The direct observation of the structures of real crystals by lattice imaging. J. Appl. Crystallogr. **6**, 165–190 (1973).
Arnold, R.G.: Equilibrium relations between pyrrhotite and pyrite from 325° to 743 °C. Econ. Geol. **57**, 72–90 (1962).
Buseck, P.R., Iijima, S.: High resolution microscopy of silicates. Am. Mineralogist **59**, 1–21 (1974).
Cowley, J.M.: High resolution dark field electron microscopy I. Useful approximations. Acta Cryst. A **29**, 529–536 (1973).
Cowley, J.M.: Contrast in high resolution bright field and dark field images of thin specimens. In: Electron microscopy and microbeam analysis (eds. B.M. Siegel and D.R. Beaman), p. 3–15. New York: John Wiley Sons 1974.

Cowley, J.M., Iijima, S.: Electron microscope image contrast for thin crystals. Z. Naturforsch. **27**a, 445–451 (1972).
Morimoto, N., Nakazawa, H., Nishiguchi, K., Tokonami, M.: Pyrrhotites: stoichiometric compounds with composition $Fe_{n-1}S_n$ ($n \geq 8$). Science **168**, 964–966 (1970).
Nakazawa, H., Morimoto, N., Watanabi, E.: Direct observation of the non-stoichiometric pyrrhotite. Proc. 8th Int. Congress on Electron Microscopy, Canberra **1**, 498–499 (1974).
Ovanesyan, N.S., Trukhtanov, V.A., Odinets, G.Ym., Novikov, G.V.: Vacancy distribution and magnetic ordering in iron sulfides. Soviet Phys. JETP **33**, 1193–1197 (1971).
Pierce, L.P., Buseck, P.R.: Electron imaging of pyrrhotite superstructures. Science **186**, 1209–1212 (1974).

Section 3 Experimental Techniques

◄ The column of the 3 million volt microscope at the Laboratoire d'Optique Electronique du C.N.R.S. at Toulouse. (Courtesy of G. Dupouy and F. Perrier)

CHAPTER 3

Experimental Techniques

N.J. TIGHE

1. Introduction

The electron microscope and its accessory equipment are being redesigned and developed continuously in order to improve resolution, penetration, and specimen handling facilities. In the early years, considerable effort was expended on lenses, the design of pole pieces, high intensity filament systems, and stable high voltage power supplies. These developments are documented in the conference proceedings of the International and Regional Congresses on Electron Microscopy which have been held since 1948, and will not be discussed in this Chapter. Readers interested in the history of electron microscopy should consult recent reviews by Cosslett (1968), Dupouy (1968, 1974), Gabor (1974), Marton (1968), and Ruska (1974). The present chapter describes techniques for preparing and handling specimens, special microscope stages, and experiments that can be done directly in the transmission electron microscope.

Thin-foil specimens now can be made reproducibly from synthetic and natural rock forming minerals and in such significant quantities that they are used for dynamic *in situ* and *ex situ* experiments. The microstructural characteristics and changes observed in selected specimens can be related to strength and chemical durability of ceramic materials and to the geologic history of rocks and meteorites. Such information is increasingly important as theoretical evaluations based on model structures are applied to real crystal systems and to practical problems.

Stages are essential in electron microscopes to manipulate specimens for diffraction contrast experiments, for mineral identification from series of diffraction patterns, and for *in situ* and *ex situ* heating, cooling, and straining experiments in the microscope vacuum or in special corrosive or protective environments. Stage development was intensive after Hirsch *et al.* (1956) observed dislocations in thin aluminum foils and derived the diffraction-contrast theories necessary to explain their results (Hirsch *et al.*, 1960; Howie and Whelan, 1962).

Metallographic specimen preparation details and stage developments are discussed in the texts of G. Thomas (1962), Amelinckx (1964), Hirsch *et al.* (1965), Kay (1965), Brammer and Dewey (1966), and Glauert (1972), and will not be discussed in the present chapter unless they apply specifically to non-metal specimens. Compared with metals, non-metal specimens were difficult to prepare as electron transparent foils until Paulus and Reverchon (1961, 1962) designed a simple apparatus for thinning specimens by ionic bombardment. The development of ionic bombardment as a standard procedure for thinning ceramic materials (Bach, 1964, 1970; Drum, 1965; Tighe and Hyman, 1968; Barber, 1970; Tighe, 1970; Heuer et al., 1971) coincided with the need to examine lunar rocks by electron microscopy and recent progress has been greatly influenced by the enthusiasm of geologically oriented scientists examining lunar materials (Radcliffe et al., 1970; Barber and Price, 1971; Champness and Lorimer, 1971). The ionic bombardment method is so universally applicable that machines similar to that of Paulus and Reverchon have been made in many laboratories and are available commercially.

2. Preparation of Electron Transparent Specimens

2.1 Introduction

Electron microscopy provides useful information on microstructure and its relationship to physical and chemical properties particularly when defects observed in thin foil specimens can be related directly to microstructural features in bulk and massive samples. Light and electron microscopic examination of the same areas in massive samples and thin foils made from them was shown possible during the development of various thinning techniques for ceramic materials (Barber and Tighe, 1965; Tighe and Christie, 1969).

Most mineralogical and ceramic specimens are brittle, chemically inert insulators which may be polyphase and severely deformed. Ceramic materials are fragile when prepared as thin sections and require careful handling to prevent damage by surface abrasion and fracture. Electron microscopic examination of Al_2O_3 and SiC after abrasion by alumina and diamond particles, and after indentation by Knoop and Vickers indenters showed extensive damage in the form of dislocation arrays, twins, and cracks (Hockey, 1971). Cracks heal spontaneously in many minerals; and electron microscopic examination of such healed cracks in corundum and in SiC showed that the misfit along the healed crack interface is accommodated by a dislocation network (Wiederhorn et al., 1973). Fig. 1 shows the dislocation loops produced by polishing with $^1/_4$ μm alumina particles. Fig. 2 shows misfit dislocations along healed cracks in corundum which propagated at an angle to the surface (Fig. 2a) and laterally from a Knoop indent in the surface (Fig. 2b). These types of defects also occur during rough handling and improper initial preparation of thin foils and preclude the use of the mechanical polishing technique which was developed for making thin foils of glass, and glass ceramics by Doherty and Leombruno (1964).

Electron microscope specimens are made either as disks which fit directly in the specimen holders and are profiled to produce an electron transparent region in the center, or as fracture fragments and cleavage flakes which are electron-transparent near an edge or across the flake. Both types of specimen have their usefulness as seen by the results included in this volume. Disk specimens usually are made to be representative of the bulk microstructure (Paulus and Reverchon, 1962; Tighe, 1964; Drum, 1965; Barber, 1970) whereas fracture fragments are used when fundamental information can be obtained from small areas to identify faults (Amelinckx and Delavignette, 1962; McLaren and Phakey, 1965), for lattice fringe imaging (Allpress et al., 1968; Buseck and Iijima, 1974), and for phase identification (Champness, 1970; Nakahira and Uda, 1967; Evans and Phaal, 1962; Vakdiek, 1966).

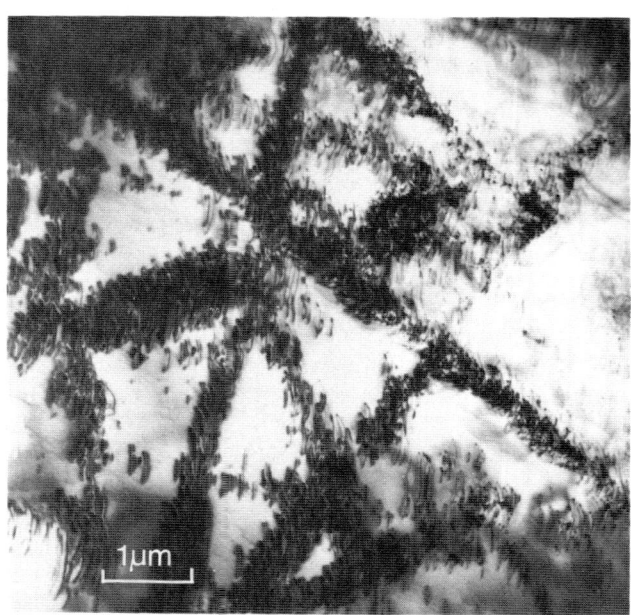

Fig. 1. Dislocations produced in polycrystalline Al_2O_3 by mechanical polishing with $1/4$ μm diamond abrasive. Approximately 2000 Å were removed from the polished surface to reduce the dislocation density sufficiently to resolve the damage tracks (Hockey, 1972)

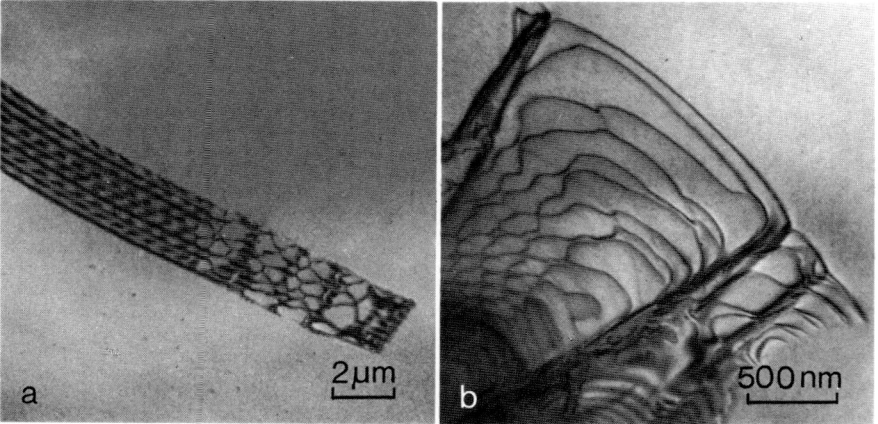

Fig. 2a and b. Spontaneously closed cracks associated with room temperature Knoop indentations in corundum: (a) vertical crack in (0001) surface showing a complex misfit fringe pattern with an interfacial dislocation network in contrast near the crack tip (Wiederhorn et al., 1972); (b) lateral crack in $\{1010\}$ surface with interfacial dislocations $\boldsymbol{b} = 1/3 < 01\bar{1}0 >$ and $1/3 < 0001 >$ forming the network (Hockey and Lawn, 1975)

Disk specimens can be produced so that the electron transparent regions are free of grinding defects and can be prepared from any material by chemical dissolution or ionic bombardment. Fracture fragments can be prepared rapidly from any brittle material but they are more likely to contain defects such as twins and slip bands of dislocations produced by crushing. The details of the methods for preparing these specimens are given in the following sections.

Thinned insulating specimens must be coated with a thin carbon film to prevent charging in the microscope. This carbon layer can be removed by ionic bombardment or by soaking in a solution of 5% $NaOH + H_2O_2$.

2.2 Fragmentation and Cleavage

The simplest, most rapid procedure for preparing electron-transparent specimens is to crush small pieces of a mineral and retrieve the thin flakes. The flakes usually form surfaces parallel to cleavage or parting planes, e.g., micaceous minerals, such as mica, graphite (Amelinckx, 1964), and hematite cleave readily into flakes with surfaces parallel to the basal plane (0001).

McLaren and Phakey (1965) crushed small particles of mineral between glass slides or in a mortar and pestle with a liquid such as alcohol and either (a) evaporated a carbon film into the glass slides, and stripped off the carbon film with the thin flakes adhering to it, or (b) retrieved the crushed fragments on holey carbon grids or onto adhesive coated grids.

Amelinckx (1964), using a slightly different procedure for easily cleaved materials, fixed a thick flake to a glass slide with adhesive or double-sided adhesive tape, and pulled pieces away from the flake by touching it with adhesive tape. This process was repeated until flakes thin enough for electron microscopy adhered to the tape. An adhesive tape was chosen which dissolves readily so that the thinnest flakes were released by soaking the tape in appropriate solvents.

2.3 Thin-section Preparation

Petrographic thin sections are excellent starting samples for cutting disk specimens. However, to reach a thickness of 30–50 µm the initial selection and preparation stage requires some effort and planning. Polishing sections to 30 µm thickness requires considerable skill or semiautomatic polishing machines. The following procedures for preparing electron microscope specimens for ceramic materials have been used for many years in this laboratory.

1. Slabs 100 to 200 µm thick are obtained from massive samples by slicing with a diamond bladed saw, cleaving, or grinding. Diamond wheels, diamond fixed abrasive laps, or silicon carbide abrasive papers can be used for grinding depending on the hardness of the material.

2. The slabs are polished with alumina or diamond abrasives to a thickness of 30–100 µm. Damage produced by abrasive particles is at least as deep beneath the surface as the size of the particles. Specimen surfaces should be polished smoothly before ion thinning.

3. Disks 2.3 or 3 mm in diameter are cut from the polished plate with a sonic drill or a diamond core drill.

4. Disks are polished to 30–50 µm using alumina or diamond abrasives.

5. The disks can be profiled by grinding with a spherical diamond tool or with a sonic tool. This step is recommended when final thinning is done by chemical dissolution but is not necessary for ion thinned samples.

6. Fragile disks are strengthened by cementing them to a supporting ring made from a ceramic tube or from a single hole copper grid. Copper grids with round or oval holes are now readily available from most electron microscope equipment suppliers. These grids can be used also to select specific areas of interest by cementing them to 30–50 µm plates and cutting the disk around the grid.

Strict adherence to a 30 µm thin section can result in material loss or damage through extensive crack propagation. In the case of extremely fragile and porous specimens, it is better to use disks thicker than 30 µm and to accept the longer time required for ion thinning than to damage or lose the specimen. Each new material, particularly heavily deformed rocks, must be evaluated before grinding it too thin. With materials which bend or fracture easily when they are 50 µm thick, disks must be cut from thicker slabs.

Heavily deformed or porous samples of rocks and single crystals usually are strengthened before sectioning by impregnation with an epoxy resin which fills the pores and cracks. Small particles (Barber, 1971; Watson et al., 1960) and fibers (Goodhew, 1971; Shaw, 1971) are compacted and embedded in epoxy resin and the resulting composite samples can be handled like bulk materials. However, the epoxy resin-specimen bond can weaken from heating during the grinding and polishing operations. Barber and Ashworth (1975) found it necessary to embed their deformed meteorite samples several times throughout the mechanical thinning process in order to keep the samples intact.

2.4 Ionic Bombardment

Ionic bombardment or sputter-etching methods are as essential for preparing thin foils from non-conducting materials as electrochemical methods are for preparing thin foils from metals and conducting materials. Paulus and Reverchon (1961, 1962) established the usefulness of ionic bombardment by preparing manganese ferrites for examination in a conventional 100 kV electron microscope; they imaged stacking faults and grain boundaries and discussed the surface roughness characteristic of ion-thinned samples.

At the Bellevue Conference on Ion Bombardment in 1961 (Trillat, 1962a), ion-sputtering machines made by Genty (1962), by Raether (1962), and by Paulus and Reverchon (1962) for preparing thin foils for electron microscopy were described. Trillat (1962) discussed the application of ion sputtering to etching of metals and non-metals and showed how reactive sputtering caused oxidation. Numerous micrographs taken by these authors showed that surface irregularities and radiation damage were undesirable artifacts of the process. Preferential etching of grains and second phases was reduced by having the ion beams impinge at ~15° to the specimen surface and by rotating the specimen throughout the thinning process.

Ionic bombardment produces uniquely etched surfaces which are easily recognized in light and electron micrographs. With stationary specimens, closely spaced grooves and ridges ~800 Å apart are etched parallel to the direction of beam impingment. By slowly rotating the specimen these ridges are smoothed and an undulating orange-peel surface is produced. The severity of etching decreases when the angle of incidence is decreased to near grazing angles but etching is never eliminated. The orange-peel texture or surface bumps are randomly located with respect to grain boundaries, dislocations, and interfaces found in the thin foils. Some etch bumps can be associated with debris on the surface and with grinding damage. There is some evidence from *in situ* experiments with an ion bombardment source in an electron microscope (Castaing and Jouffrey, 1962) that the bumps may be related to point defects and dislocations produced by the ion beam. The depth of ion penetration is ~400 Å at 15° and is decreased to ~100 Å by reducing the angle to grazing incidence (Paulus and Reverchon, 1962). The etch figures also are being analyzed in terms of preferred dissolution of crystal faces (Barber et al., 1973).

The surface etching phenomena and the subsurface ion damage are always present and are accepted, therefore, as conditions for obtaining thin foils. The ion damage may affect the reaction rates in some experiments and can be annealed out or removed from the surface layers by chemical etching when necessary.

2.5 Ionic Bombardment Apparatus

The apparatus designed by Paulus and Reverchon (1962) is shown in Fig. 3a. They provided two ion sources (A), for thinning for both specimen surfaces simultaneously, viewing windows, a tilting, rotating specimen holder, and a probe for measuring ion current. The sputtering chamber is mounted on a vacuum system which maintains a pressure of 10^{-4} torr (1.3 cPa) with the gas load of both sources operating. This pressure is achieved with a 300 to 400 l/sec oil diffusion pump or with a turbo-molecular pump as used in the present Paulus model (Paulus, 1974).

Paulus and Reverchon determined that ion beams from multi-hole cathodes produced a flat surface several millimeters in diameter and equipped their machine with electrodes having 13 holes in a hexagonal array. This design has several disadvantages: the alignment of holes in the anode and cathode is critical for proper operation; the cathode holes enlarge during operation and when they approach the size of the anode holes, or when one hole enlarges more than the others, the discharge becomes unstable; and the cathodes are difficult and expensive to make. In order to alleviate these problems and to increase operational efficiency a single hole electrode assembly was tried and found to work. An electrode assembly similar to that used by Genty (1962) was adapted to the Paulus-type ion source and is used now on most machines (Holland et al., 1971; Gillespie et al., 1971). This single hole source is easy to make and disk specimens thin uniformly without excessive sputtering of the specimen holder.

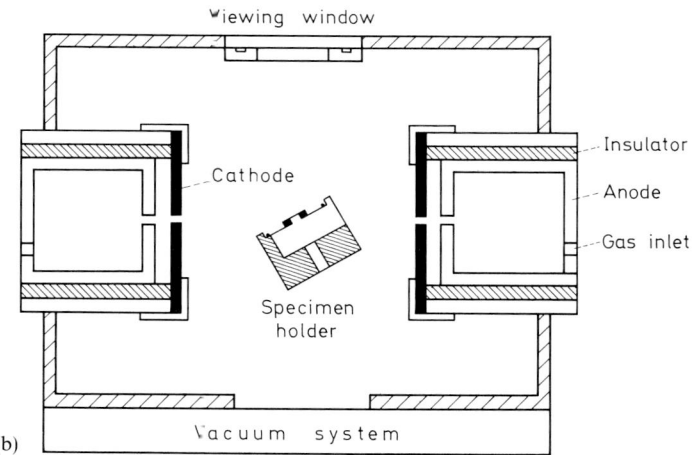

Fig. 3. (a) Ionic bombardment thinning apparatus of Paulus and Reverchon (1962) showing the external arrangement of the ion sources (A), the gas inlets, and the observation windows. (b) Schematic drawing of typical ionic bombardment apparatus showing the arrangement of the ion sources, the electrodes, the gas inlet, and the specimen holder

A modified Paulus-type ionic thinning apparatus is shown schematically in Fig. 3b. The anode and cathode are separated by a Teflon insulator tube which was found necessary because the original glass insulating tubes chipped easily and occasionally fractured during operation. Alumina tubes are used also (Heuer et al., 1971). The ion sources have a hollow anode which provides a gas ballast

Experimental Techniques

chamber for a steady gas supply. The volume of the anode chamber is not critical (Holland et al., 1971).

The gas to be ionized is introduced to the ion source through a tube in the outer case as in Fig. 3a or directly into the anode chamber (Fig. 3b). In the first method the gas leaks around the outside of the insulator and across the insulator-cathode interface. When gas is introduced directly into the anode, vacuum tightness around the insulators is easily achieved, thus improving the ultimate vacuum; and the system works more efficiently. A high-voltage discharge through the gas is eliminated by using ~3 meters of 1–2 mm bore Teflon tubing between the anode and gas leak valve (Holland et al., 1971).

The specimen holder which is placed between the two ion sources is attached to a support which can be tilted and rotated. Disk specimens are held between shims to produce thinned foils which are dished on both sides (Paulus, 1962) or they can be cemented to a shim with one side uncovered as in Fig. 3b to produce thinned foils which have one flat side and one dished side. The adhesive chosen for this purpose must remain soluble after long-time irradiation from the ion beam. Inexpensive glass repair cements are useful. In the holder shown in Fig. 3b the shim is held with a spring clip and can be removed quickly to change specimens.

2.6 Specimen Thinning Procedure

Paulus and Reverchon determined that fastest thinning occurred when the specimen was tilted at an angle of ~15° from the ion beam. Their results are reproduced in Fig. 4 and it is seen that there is only a small variation in thinning rate with crystallographic orientation of the specimen surface; and that the yield increases linearly with accelerating voltage. Bach (1964) obtained similar results.

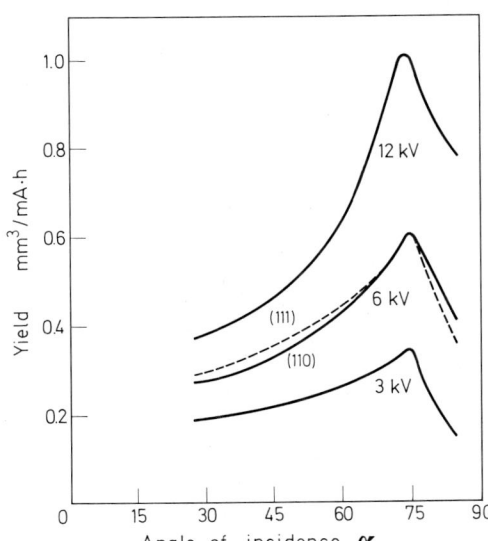

Fig. 4. Thinning rate or yield of $MnFe_2O_4$ versus angle of inclination of the specimen normal to the ion beam for different voltages and crystallographic orientation (Paulus and Reverchon, 1962)

The specimen to be thinned is placed in the holder and the thinning is monitored using a microscope placed over the viewing window. Devices have been designed to cut off the ion source when perforation occurs (Heuer et al., 1971). The sensitivity of these devices can be adjusted to allow their use with transparent and opaque specimens.

For most applications, specimens are thinned simultaneously from both sides. An accelerating voltage of 3–7 kV is applied to the ion sources to ionize the gas which is usually argon. Heuer et al. (1971) adapted their power supply so that the sources were controlled individually, but in most systems both sources have the same control. The ion beam current is adjusted to 50–100 µamp by varying the gas flow and the voltage. For most materials there is a minimum voltage at which little or no thinning occurs and this voltage can be determined readily by trial. The maximum voltage is a function of the power supply and the source configuration. Thinning occurs at higher voltages but surface damage is increased and may be too severe to be acceptable.

Maximum thinning rates are 1–3 µm/h with both sources operating. The thinning rate decreases as the cathode holes enlarge because the chamber pressure increases, and sputtering yield is a function of chamber pressure (Yonts and Harrison, 1960). The thinning rate decreases also when the specimen rotates eccentrically or when the ion sources are not properly aligned with each other.

Specimen disks are thinned from one side in order to examine the near surface layer for damage (Hockey, 1972) or for precipitation (Tighe and Swann, 1974, 1975) and in order to compare light and electron micrographs of the same areas in such specimens. Paulus and Reverchon (1962) tried this technique during development of the ion bombardment apparatus and found that specimens bowed inward when they were thin. It is known now that bowing results from the strain remaining after mechanical polishing and an extreme example is shown in Fig. 5. Here the foil bowed away from a ground surface and then fractured and curled up when perforation occurred. It is necessary to remove a few hundred

Fig. 5. Stereo-pair of Scanning Electron Microscope (SEM) pictures showing how the strain in a ground Si_3N_4 surface caused the thin area of a disk specimen to curl away from the damaged side and fracture during ion thinning from the other side

nanometers from the strained surface before perforation occurs in order to make a useful thin foil; alternatively, the specimen can be annealed to relieve the strain (Hockey, 1972).

When only one ion source is bombarding a specimen, the sputtering debris deposits on the other side (Paulus and Reverchon, 1962) and must be removed because it forms a magnetic film. In machines which have separate power supplies for each ion source (Heuer et al., 1971) one source can be operated at a voltage which will clean but not thin. Otherwise, the contamination can be removed by adjusting both ion sources to a voltage lower than that used for thinning. The unbombarded surface can be protected with a gold or aluminum evaporated coating which is removed after thinning by etching in aqua regia or HF.

Any material can be ion-thinned although special handling techniques are required for small specimens such as fibers and small particles. Particles and coarse fines, 0.1 to 1 mm in diameter, of lunar and meteoric minerals were embedded in a cold-setting epoxy, ion-thinned, and etched chemically in order to examine particle tracks by light and electron microscopy (Barber and Price, 1971). Carbon fibers were either embedded in epoxy (Goodhew, 1971; Shaw, 1971) or stretched across a single hole copper grid and cemented to the grid (Sharp and Burney, 1971; Goodhew, 1973). In order to minimize striations and uneven sides of the individually mounted fibers, the specimen holder was held stationary and the ion beams were directed along the fiber axis (Goodhew, 1973).

Phakey et al. (1974) developed a special technique to thin selected areas of a specimen. Before placing the foil in the ion beam they focus a fine jet of alumina powder on a 100 μm thin section producing a small dimple which thins first.

Bach (1970) prepared transverse sections of a TiO_2 layer on a glass substrate by cementing the interfaces together with epoxy adhesive, cutting out a cylinder and grinding the cylinder to a thin disk for ion thinning. This method of examining interfaces between phases is a fine supplement to sectioning parallel to the interface.

Ion-bombardment thinning was developed to make specimens which were thin enough to be examined at 100 kV. With the advent of HVEM and the resulting increased electron penetration, it is tempting to be satisfied with thicker specimens; however, a good 100 kV foil will be a magnificent 1000 kV foil.

2.7 Chemical Dissolution

Many minerals are soluble enough to be thinned chemically by immersion in a solution or by jetting a solution against a specimen. Chemical dissolution methods were developed before ion thinning was perfected and were adapted from metallurgical practice. The motivation to search for suitable solvents and thinning conditions is diminished now by the availability of ion thinning machines. However, it is advantageous for some problems to use chemical thinning in some stage of electron microscope specimen preparation, for example, to relate etch pits and dislocations (Barber and Tighe, 1965) to prepare a defect-free surface, or to remove the ion-damaged layer from thin foils. Chemical dissolution can produce adverse reactions resulting in etch pits and surface precipitates and the thinning conditions must be chosen to avoid formation of these artifacts. Chemical thinning methods are not suitable for polycrystalline specimens, heavily deformed single-crystal specimens, or multi-phase single crystal specimens because preferential dissolution occurs at grain boundaries, precipitates, and emergent sites of dislocations. The few non-metallic minerals that have been thinned chemically for examination by electron microscopy are listed in Table 1.

Table 1. Chemical etchants used for preparing thin foils from single crystal ceramic materials. Symbols I-immersion method; SFJ-separatory funnel jet; CJ-convection jet; BJ-boiling jet

Material	Etchant		References
Al_2O_3	85% H_3PO_4, 450–500 °C	I, BJ	Tighe, 1964
$BaTiO_3$	H_2SO_4	CJ	Kirkpatrick and Amelinckx, 1962
$CaCO_3$	$H_8O_7C_6$ dil	I	Braillon et al., 1974
CoO	85% H_3PO_4	CJ	Remaut et al., 1964
$LiNbO_3$	KOH, 350–400 °C	I	Wicks and Lewis, 1968
MgO	85% H_3PO_4, 100 °C	I, SFJ	Washburn et al., 1960
$MgAl_2O_4$	85% H_3PO_4, 250–450 °C	I	Lewis, 1966
MnO	$HCl + NO_3$		Barber and Evans, 1970
SiO_2	$NH_4F \cdot HF$, 180–200 °C	I	Tighe (unpubl.)
	HF, 100 °C	I	
TiO_2	NaOH, 550 °C	I	Barber and Farabaugh, 1965
$ZrSiO_4$	$NH_4F \cdot HF + KF$ 1:1, 420–430 °C	I	Tighe (unpubl.)
$Y_3Al_5O_{12}$	85% H_3PO_4, 300 °C	I	Keast, 1967

In the simplest thinning technique, a disk or slab specimen is placed in a holder and immersed in an etching solution. Considerable material is dissolved by this method and it is most useful in the initial thinning stage. The concentration and temperature of etchants are varied to produce constant and predictable dissolution rates. Chemical dissolution can be used to obtain a smooth polished surface or an etched surface with etch pits at the emergent sites of dislocations. In certain instances the same solvents can be used at different temperatures to either etch or polish. For example, corundum is etched in boiling phosphoric acid at a temperature 50 °C lower than that required to produce a polished surface (Tighe, 1964). This behavior is helpful in determining the optimum polishing conditions for different crystallographic orientations.

Immersion-thinning removes material so quickly that it may be difficult to stop the reaction when perforation occurs and for this final thinning a device which squirts a jet of solvent against a specimen is used. Many devices have been made for chemical jet thinning and there is insufficient space to describe them here. The devices and etchants are described in a recent review (Goodhew, 1972) and in the microscopy texts (Hirsch et al., 1965; Thomas, 1962). The jet-thinning devices are of the three types shown in Fig. 6: (a) a directed jet, (b) a convection jet, and (c) a boiling jet. The directed jet is made of glass and has a reservoir for the etchant, a bent delivery tube, a specimen holder, a light source and microscope. Several modifications of this system have been used (Booker and Stickler, 1962; Washburn et al., 1960; Hobbs, 1970). The delivery tube can be wrapped with fine wire to heat the solution (Booker and

Experimental Techniques

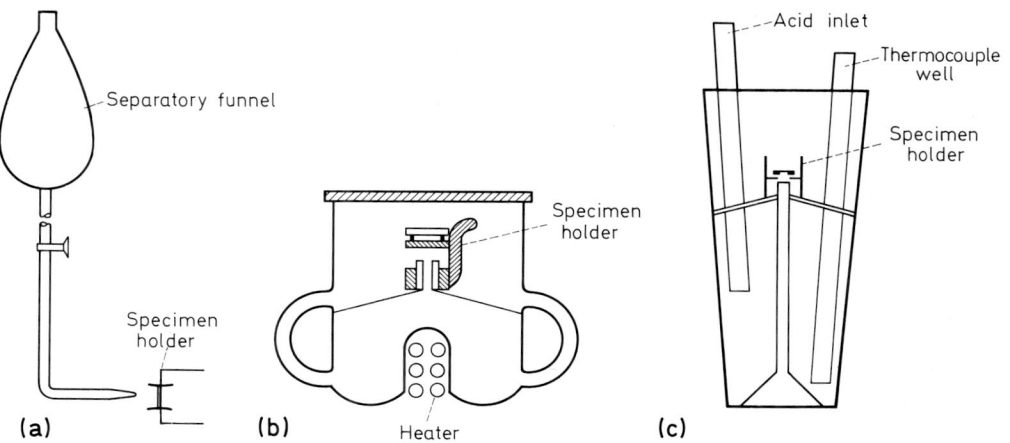

Fig. 6a–c. Jetting devices for final thinning of ceramics by chemical dissolution: (a) directed jet using a separatory funnel with bent delivery tube and having provision for holding disk or plate specimen; (b) convection jet with small heat source, provision for acid recirculation, and specimen holder; (c) boiling jet with provision for adding acid, specimen holder, and thermocouple

Stickler, 1962). Specimen holders for these jet-thinning devices are rotated or moved laterally so that plates as well as disks can be thinned (Washburn *et al.*, 1960).

The convection jet shown in Fig. 6b was designed by Kirkpatrick and Amelinckx (1962) for thinning specimens soluble in solutions heated to 120 °C and has been used successfully for such materials as MgO, $BaTiO_3$, and NbO which are soluble in warm H_3PO_4 or H_2SO_4. The solution is directed against a specimen by a convection jet formed by heating a small area. The specimen is completely immersed and the jetted solution recirculates through the side arms.

The boiling jet device shown in Fig. 6c was designed (Tighe, 1964) for thinning Al_2O_3. It is made of platinum and operates with phosphoric acid heated to ~500 °C. Acid heated in the lower chamber bubbles up through the tube and strikes the specimen. Rapid volatization of the jetted acid requires continuous or intermittent addition of fresh solution.

With all of the jet-thinning devices, provision is made for viewing the specimen continuously and for removing it rapidly when perforation occurs. Additionally, because of the corrosiveness of the solutions, shielding is required to protect the operator and the surroundings.

3. Electron Microscope Stages

The text of Hirsch *et al.* (1965) describes most of the stages which were designed and built by users in order to carry out their experiments, and many design features of those stages are incorporated in commercially available microscopes. The stages described at that time did not meet all the stability requirements for high-resolution imaging. This is no longer the case and many high-angle tilting stages which meet the requirements of freedom from drift, tilting without image shift, and reproducibility of tilting angle are available as standard equipment on 100 kV, 200 kV, and HV electron microscopes. Additionally, comprehensive ranges of special-purpose specimen stages are available for these microscopes and the principle features of specimen stages will be described in this section.

Specimens must be immersed in the field of the objective lens pole piece and can be placed there by means of (a) a top-entry stage whereby the specimen holder fits into the top of the pole piece, or (b) a side-entry stage whereby the specimen holder fits into the side of the pole piece. Both types of stage are used with an air lock so that specimens can be exchanged quickly without affecting the microscope vacuum. The side-entry stage which was standard on Philips microscopes has proved so versatile and simple in design that it is gradually replacing the top-entry stage in the newer electron microscopes.

3.1 Top-entry Stages

The top-entry stage body is fixed to the top of the objective lens and has provisions for the X and Y traverses and for the specimen holder. Specimen tilts and rotation are obtained by moving the specimen holder or the stage body. A wide-bore pole piece is required to provide enough room for high angle tilting and heating stages. Such pole pieces have poorer ultimate resolution capability than small-bore pole pieces used for fixed or small tilt stages. Nevertheless, stages are available which tilt $\pm 25°$ and permit micrographs to be taken with resolution of ~ 2 Å.

3.2 Side-entry Stages

In side-entry stages the specimen is held in a rod and inserted between the upper and lower pole pieces of the objective lens. Specimen rods are supported at one end in a conical socket attached to the stage body and at the other end by an O-ring seal in the rod rotating mechanism attached to the air lock. Side-entry rods have a ball-shaped tip which is machined in the rod (Swann, 1972a) or obtained by inserting a ruby or metal ball bearing in the tip (AEI-EM7; JEOL; Philips).

The microscope vacuum acts along the rod axis to hold it in place and translation can be obtained simply by working against atmospheric pressure (Fisher et al., 1960). In JEOL microscopes the supporting socket is driven in and out to traverse the specimen along the X direction and there is enough movement to have two specimen cups in the stage tip. In Philips and AEI-EM7 microscopes the rod is supported in the socket and the stage body is moved to traverse in the X and Y directions at angles of 30–45° to the rod axis.

Micrometer tilting, straining, and rotating drives can be contained within the specimen rod and the drive motors affixed to one end of the rod (Makin, 1972; Browning, 1974; JEOL; Philips) or these drives can be applied through the rod tip thus leaving the rod solid or free for other facilities and permitting the use of detachable specimen holders as in Fig. 7 (Swann, 1972a, b).

3.3 Tilting Stages

For optimum results samples must be tilted smoothly in any direction over a range of angles with respect to the electron beam in a reproducible manner without image shift. The reproducibility of tilting angle should be $\pm 0.1°$ for a tilting stage to be designated a goniometer stage (Valdre, 1972).

One direction of tilt is provided for side-entry stages by rotating about the specimen rod axis with a motor or hand-operated micrometer drive. Clearly the rotation can be $\pm 180°$ if desired. The tilt angle is read directly from the air-lock mechanism or it can be indicated indirectly with a digital meter coupled to the tilting mechanism (Harrison *et al.*, 1972), or with a fiber optic system (Thalen *et al.*, 1970). A second tilt or rotation is obtained by driving the specimen carriage about a normal to the rod axis. A design for accomplishing this movement is shown in Fig. 7a. In Fig. 7a the specimen tip is clamped rigidly to one end

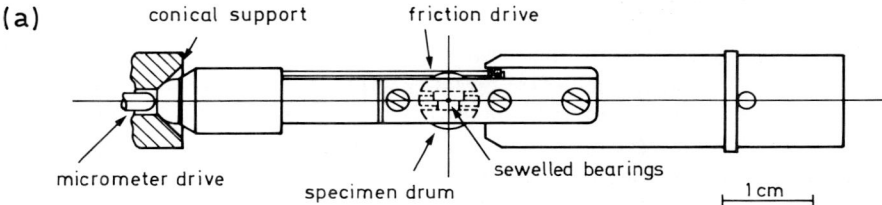

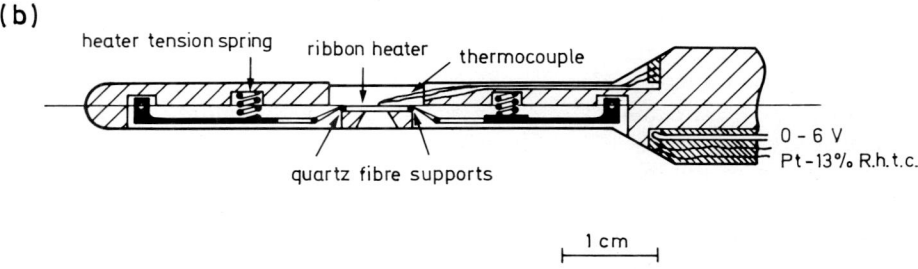

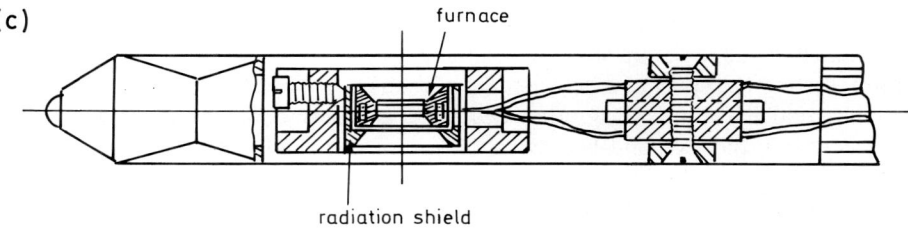

Fig. 7a–c. Side-entry specimen holders: (a) double tilt, detachable type holder which has a friction driven specimen drum set in jewel bearings; conical tip support and micrometer drive are shown (Swann, 1972); (b) single tilt hot stage with ribbon heater (Swann, 1972); (c) single tilt hot stage with furnace (JEOL-200 A)

of an insertion rod, and the tilt is achieved by a friction drive which rotates a specimen drum on its jewelled bearings (Swann, 1972a). In the JEOL stage, tilt or rotation are achieved by a friction drive which touches the sides of the specimen cups. Both cups move identically and are an integral part of the insertion rod.

3.4 Hot Stages

The simplest heater in the electron microscope is the electron beam itself. Beam heating is used for flash heating and is accomplished by removing the condenser aperture and focusing the beam until the desired thermal conditions are reached. Specimens are overheated easily or melted with this technique and considerable care and practice are needed for successful experiments.

The requirements of a heating stage are: a wide temperature range, reproducible heating and cooling rates, good image resolution, and some tilting capability.

Side-entry hot stages are relatively rugged, they have one tilt axis and the power leads and thermocouple leads are readily attached to the heater. Top-entry tilting hot stages are difficult to design (Valdre, 1972) because the tilting mechanisms are complex, and furnace leads and thermocouples must be flexible.

Two types of heater are used in both top-entry and side-entry stages: (a) a strip heater with a grid of holes, and (b) a non-inductively wound furnace (Fig. 7b, c). The strip heaters can achieve temperatures near the melting point of the material used and are made from stainless steel mesh (Fisher et al., 1960; Whelan, 1958), from photo-etched tungsten sheet (McPartland, 1972), and from platinum-rhodium, nichrome, and tantalum sheet with a grid cut by spark machining (Swann, 1972). Furnaces used up to 1000 °C are made with molybdnum, tungsten, or platinum-rhodium windings which are supported on insulated surfaces or embedded in alumina or other insulating cement (Valdre, 1972; Henderson-Brown et al., 1972; JEOL; Philips). Furnaces have a higher thermal mass than strip heaters but may have a more uniform temperature distribution.

Specimens must be clamped firmly against the heater strip or in the furnace to assure good thermal contact and specimen stability. Metal specimens heated at low temperatures can be cemented to the heater (Fisher et al., 1960); however, this system fails at higher temperatures. In a furnace the specimen is clamped between grids with a circlip. In the hot stage shown in Fig. 7b, the heater is a folded ribbon and is stretched across fused quartz fibers by spring-loading the heater leads. Specimen loading is achieved by releasing the spring tension and inserting the specimen between the ribbons.

Thermal drift caused by gradual heating of component parts reduces resolution and is minimized by good heater insulation. Thermal drift characteristics must be related to specific stages; however, the extent of the drift is indicated in Fig. 8 due to Makin (1974). These measurements at 600 °C show that water cooling reduced the stage drift by a factor of 10 to ~ 1 Å/sec.

Heaters are calibrated and used with a thermocouple attached to the heater strip, furnace wall, or specimen. McPartland (1962) used an optical pyrometer to measure temperatures between 1000 °C and 2900 °C. The measured difference

Fig. 8. Measured values of the variation of stage drift with time at 600 °C (Makin, 1974)

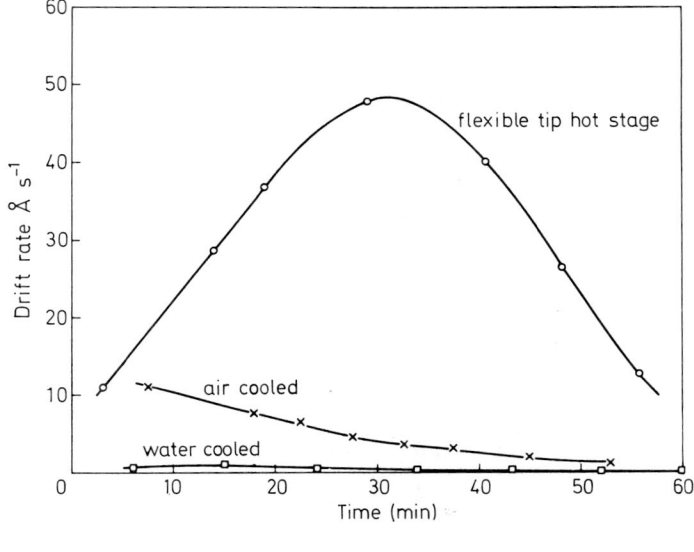

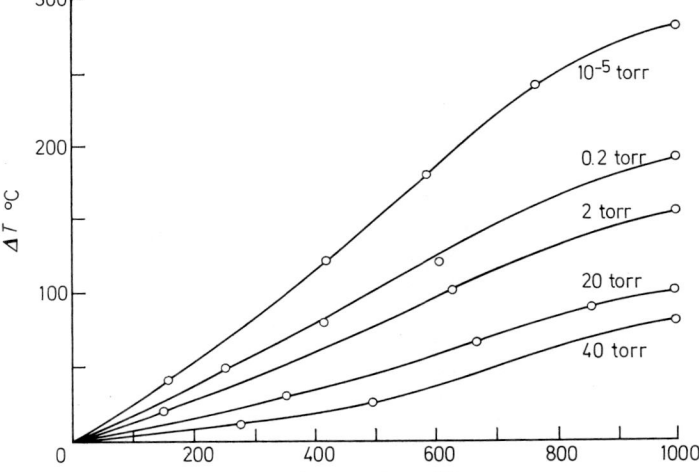

Fig. 9. Temperature difference between a specimen and a strip heater versus heater temperature and air pressure (Swann et al., 1973)

between specimen temperature and heater temperature in a strip heater can be several hundred degrees as shown in Fig. 9 (Swann et al., 1972).

The actual specimen temperature depends on the contact of the specimen and the heater, the heater current, the heat loss through the stage and, in the case of environmental cells, on the pressure and thermal conductivity of the gas around the specimen. Flower (1974) calibrated the platinum grid heater in different atmospheres and his plot of heater temperature versus the current required to drive the heater is shown in Fig. 10. Flower showed that thermal conduction through the gas was important and heat losses with light gases such as helium reduced the maximum temperature of the heater.

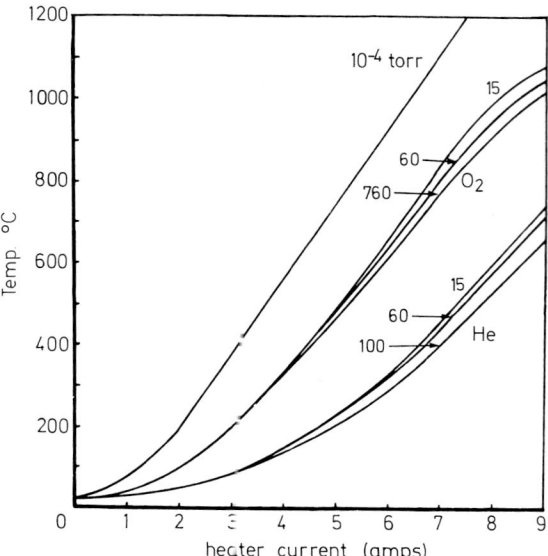

Fig. 10. Calibration chart for a platinum grid heater. Light, highly conducting gases such as helium seriously reduce the maximum attainable temperature (Flower, 1974)

3.5 Cold Stages

Stages cooled by liquid nitrogen or liquid helium are used to obtain temperatures as low as 10 °K in order to observe crystallization and growth of condensed gas crystals (Venables et al., 1968; Price and Venables, 1972), to reduce electron beam damage in alkali halide crystals (Hobbs and Goringe, 1970), to observe low temperature phase transformations (Swann and Swann, 1970), to examine freeze-dried biological specimens (Valdre and Horne, 1972), and to prevent specimen contamination.

Vibration produced by the coolant boiling in the storage dewar or around the cold surface of the stage is the principle factor limiting resolution (Valdre, 1972). Cooling is achieved by circulating the liquid or vapor around a conducting surface surrounding the specimen holder. With vapor designs (Swann and Swann, 1970; Swann and Lloyd, 1974), vibration is minimized, rapid temperature changes can be made during experiments, and the specimen holder is easily removed from the stage by venting the gas and heating the conducting surface of the stage block. In liquid-cooled stages the coolant must be decanted or diverted from the system to let the specimen holder reach ambient temperature for removal.

3.6 Straining Stages

Top-entry and side-entry straining stages have been designed which apply a tensile load to a specimen (Wilsdorf, 1958; Valdre, 1974; Henderson-Brown et al., 1972; Imura et al., 1974; Vesely, 1973; JEOL; Philips). However, the stages are designed to deform metal specimens and the loads are too low to

Experimental Techniques

deform many ceramic materials at room temperature. The stages could be used to propagate cracks; however, the more interesting results would be obtained during high-temperature deformation and such stages are not available.

3.7 Environmental Cells

Environmental cells enable specimens to be surrounded with a protective or reactive gaseous or liquid atmosphere. The test environments should be equivalent to environments produced *ex situ* in order to correlate experiments. The stage should have provision for obtaining diffraction patterns, for tilting specimens, for heating specimens, and should allow good enough resolution to observe specimen changes as they occur. These requirements were not achieved until the development of HVEM stages which utilized fully the increased penetration of the HV electrons to obtain transmission through both the atmosphere and the specimen (Swann and Tighe, 1971). The large HVEM specimen chambers provide space for hot stages and for large gas volumes at pressures $\geq$ one atmosphere. Flower (1973, 1974), and Joy (1973) reviewed environmental stage development and assessed their capabilities; their articles should be consulted for complete references.

Environmental cells for HVEM have: (a) a closed cell design with windows for sealing the atmosphere around the specimens from the microscope vacuum (Allinson, 1970), or (b) an open cell design with apertures which utilize differential pumping between pairs of apertures above and below the specimen to keep the atmosphere from entering the microscope column and to contain it around the specimen (Swann and Tighe, 1971; Ward and Mitchell, 1972; Parsons *et al.*, 1972).

The closed cell design of Allinson (1970) was used to study cement hydration and is shown schematically in Fig. 11. The environmental cell is completely contained in a side-entry rod and is slightly modified (Allinson *et al.*, 1972) from the original. The cell gas volume is ~ 5 ml and the pressure can be varied from vacuum to nearly one atmosphere. Glass or sapphire windows <2000 Å thick are sealed to top- and bottom-removable plugs. The window gap is adjustable from 0.1 to 0.5 mm by screwing the plugs in or out. The specimen holder can

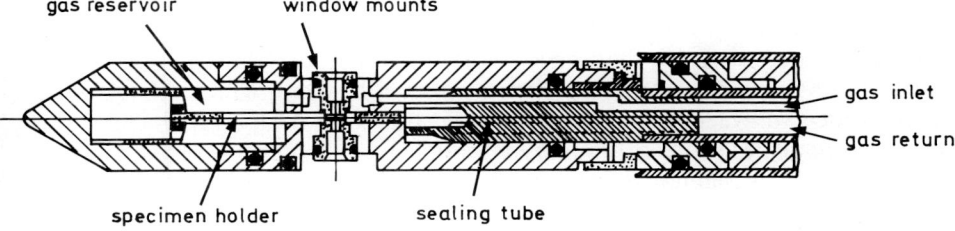

Fig. 11. Side-entry environmental cell MkII designed by Allinson (1972). The cell used windows to keep the environment from entering the microscope column. Evacuation and gas entry lines are enclosed in the side entry rod

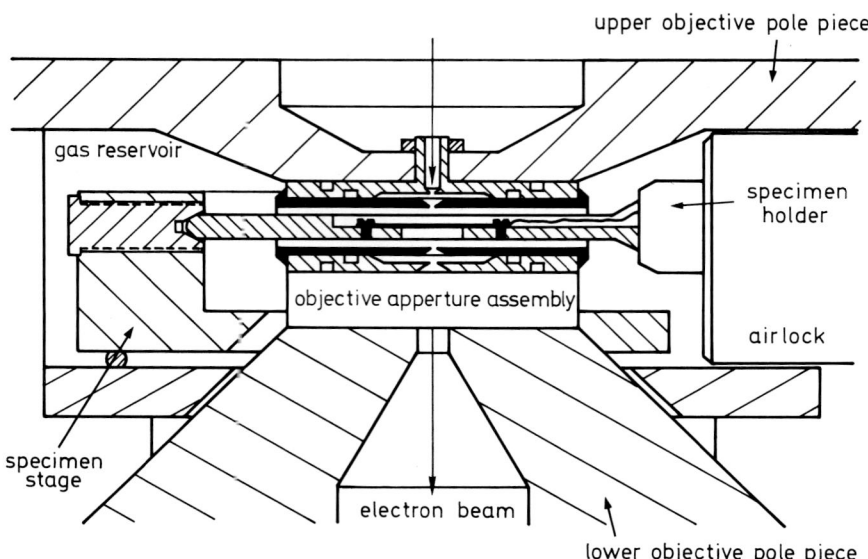

Fig. 12. Differentially pumped environmental cell designed by Swann (Swann and Tighe, 1971). The diagram shows a ±180° tilting hot stage in position in the 5.5 mm gap cell

hold three 3.05 mm grids and is loaded by unscrewing the upper window and removing the "nose cone" from the stage. Some care is required in changing pressure to keep the windows intact. The windows reduce the resolution, and, when water is used inside the cell, the images are very indistinct. Diffraction analysis of the hydrated specimen can be made, however. The Allinson cell can be used interchangeably with other side-entry stages without modifying the specimen chamber of the microscope.

The differentially pumped aperture designs (Swann and Tighe, 1971; Ward and Mitchell, 1972) require opening the electron microscope to insert the cell. In the Swann design, Fig. 12, the environmental cell is attached to the upper pole piece of the objective lens and rests on the bottom pole piece. Vacuum seals around the cell top and bottom are made with O-rings. With the cell in position the remaining volume of the specimen chamber, ~300 ml, is the gas reservoir. The cell has a moveable objective aperture rod installed in a hermetically sealed compartment at the bottom. Gas enters the specimen chamber volume through the port used by the normal objective aperture. The gas pressure is adjusted using a needle valve in line with a capsule dial gauge. A turbo-molecular pump is used to evacuate the space between the pairs of apertures which are 100 μm in diameter. With this pumping arrangement, a pressure of one atmosphere (101 kPa) is maintained around the specimen, while the column vacuum does not exceed 10^{-4} torr.

Calibration of the differentially pumped cell (Swann and Tighe, 1971) showed that acceptable images in stainless steel foil were obtained at pressures of 1 atm. of helium and 160 torr of air with the 5.5 mm gap cell. When this specimen

Experimental Techniques 163

gap was reduced to 1.5 mm, acceptable images were obtained at pressures up to 400 torr of air and one atmosphere of water saturated helium; images obtained in one atmosphere of dry helium were similar to those obtained in vacuum (Swann and Tighe, 1972). Recently the 1.5 mm gap cell has been used up to 1.5 atm. of helium (Swann, 1974).

The 5.5 mm cell has enough space for a tilting hot stage (Fig. 12) and can be used with any gas that is not corrosive to the aluminum cell body. The 1.5 mm cell is large enough to accept a specimen holder with a tilt of $\pm 7°$ and becomes a wet cell when used with water-saturated helium (Tighe et al., 1973).

The Ward and Mitchell (1972) cell also uses differential pumping and was designed as a wet cell for examining biological materials. The test gas is introduced near the specimen and is contained within the space of the cell.

3.8 In situ Experiments

With the availability of HVEM and of the stages just described, it is possible to do experiments which permit direct observation of the microstructural changes that occur during heating and during straining. Observations of the behavior of thin foil specimens during these *in situ* experiments can be correlated with the behavior of massive specimens. When environmental chambers are used, data can be obtained on the oxidation and reduction of minerals in wet and dry gases. A few examples of *in situ* experiments are given below and other experiments are presented elsewhere in this book.

In correlating *in situ* and *ex situ* results, the effects of heating, of radiation damage, and of enhanced carbonaceous contamination from the electron beam must be determined for each dynamic process. Carbonaceous contamination produced during examination of specimens at room temperature enhanced the rate of titanium oxidation (Flower, 1973) and of hematite reduction (Swann, 1972c; Tighe and Swann, 1974) but retarded the rate of tantalum nitridation (Swann et al., 1972). Radiation damage is produced at room temperature and at elevated temperature in most non-metallic materials and, depending on the material, the damage is evidenced by a high density of defect clusters, by precipitates, and by jagged dislocations. Beam heating can be used to advantage in order to control some reactions by confining them to the areas being examined.

Although dislocation movements in metals (Fisher et al.; Whelan, 1958) and in non-metals (Barber and Tighe, 1965) were studied using the early hot stages, the reactions observed were typical only of thin foil specimens. It was not until thicker specimens were examined in HVEM hot stages that the operation of dislocation sources such as grain boundaries (Hale et al., 1972; Henderson-Brown et al., 1972) and precipitates (Imura et al., 1974) were observed and photographed on plates and on video tape. More recently, *in situ* heating experiments were used to study the stability of superstructures in minerals (for example, Chapters 5.2 by Nakazawa et al., on pyrrhotite and 5.9 by Lee on wenkite in this volume). Of particular interest has been the stability of c-APB's in anorthite (Lally et al., 1972; Müller and Wenk, 1973 and the discussion by Heuer and Nord in Chapter 5.1 of this volume).

Growth kinetics for cellular precipitation in an Al-28% Zn alloy were obtained from measurements of photographs taken during carefully controlled experiments in the HVEM (Ramaswamy et al., 1973). Similar growth rates and microstructures were found in both *in situ* and *ex situ* experiments.

The reduction of iron oxides was studied by heating hematite and magnetite in carbon monoxide and in hydrogen atmospheres contained in the environmental cell of the HVEM (Swann, 1972c; Tighe and Swann, 1974). The nucleation and growth of magnetite on hematite surfaces was observed directly and the growth forms were similar to those produced in *ex situ* experiments. Rapid reduction reactions which occurred at temperatures as high as 1200 °C were recorded on video tape.

A potential asset in *in situ* experiments is an ion bombardment source which is mounted in the specimen chamber of the electron microscope. This type of ion source was used to remove surface layers, to thin specimens, to relieve surface strains, and to remove the carbonaceous contamination that formed on the top surface of the foil (Castaing and Jouffrey, 1962; Dupouy and Perrier, 1965).

The wet environmental cell designed by Allinson (1970) was used for some preliminary studies of cement hydration and could be used to study clays. The resolution obtained with environmental wet cells is low because of electron scattering from the water. It is advisable, therefore, to examine material at magnifications of $7-10000 \times$. Experiments with algae and diatoms (Tighe et al., 1973) showed that resolution decreases remarkably when water is introduced and it is doubtful that specimens showing resolution of <200 Å retain any water at all. The dehydration process was observed in algae by using water saturated helium as the medium for controlling the moisture. In a strongly focused electron beam wet objects disintegrated quickly and appeared to be boiling. More successful experiments may be carried out using image intensification systems with low beam intensities.

3.9 Recording of Dynamic in situ Experiments

Although many *in situ* reactions can be recorded on photographic plates, some reactions cannot be slowed sufficiently to record sequential events. These events are recorded by interrupting the reaction at intervals and waiting several minutes for specimen movement to cease before taking photographs. However, a continuous recording system is necessary for observing and recording events without interruption.

Ciné-photography can be used to record images through the front window of the microscope using procedures described by Hirsch et al., 1956; Hirsch et al., 1965, namely a short focal length, wide aperture lens, frame speed of 8–16 frames/sec, and fast film. The ciné-camera also can be mounted beneath the transmission phosphor screen of high-voltage electron microscopes. This ciné-method of recording is less useful than video recording because the area being photographed cannot be seen easily during filming; the film images are of poor quality; there is a long delay between photographing and viewing the film; and the film and its processing are expensive. With video recording, electron images are recorded simultaneously on tape and displayed on a monitor. Video tapes can be erased, reused, and edited easily. Images on the video monitor can be photographed using fast film, HP-4 or Tri-X, at f-5.6 and 1/25 sec when single pictures are required.

Video tape recording systems (VTR) can be internally fitted using a phosphor coated fiber-optir disc (Heerschap and DeCat, 1972; Lehtinen and Roberts, 1973); or the video camera can be used with the above lens system to photograph the transmission phosphor screen or the normal phosphor screen.

4. Stereo Microscopy

Three-dimensional stereoscopic effects result from the parallax between two pictures of identical areas which have the same magnification and contrast but different specimen tilt angles. For crystalline materials, stereo-pairs are obtained by taking the first plate with one set of diffraction planes operating and then tilting along the Kikuchi band for the appropriate angle to obtain similar diffraction conditions; for biological and amorphous materials the specimens simply are tilted over the required angles (Hirsch et al., 1965; Nankivell, 1963).

The optimum stereo-tilt angles are related to the specimen thickness, t, the viewing magnification, M, and the parallax, p, according to the equation:

$$\sin \theta = p/2 M t;$$

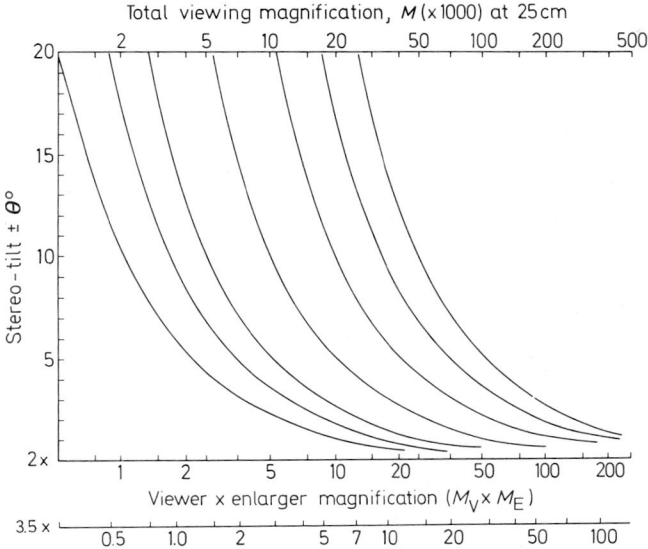

Fig. 13. Optimum stereo-tilt angles for a wide range of microscope magnifications and various specimen thicknesses for viewing with a pocket stereoscope or with a folding mirror microscope. Total viewing magnification $M = M_M \times M_V \times M_E$ (Beeston, 1973)

and this relationship can be plotted as in Fig. 13. The curves in Fig. 13 were plotted by Beeston (1973) from similar curves of Hudson and Makin (1970) so that the optimum tilt angles are read off by selecting the scale for the viewing system to be used, M_V, the microscope magnification, M_M, and the specimen thickness; M_E is the photographic enlargement magnification.

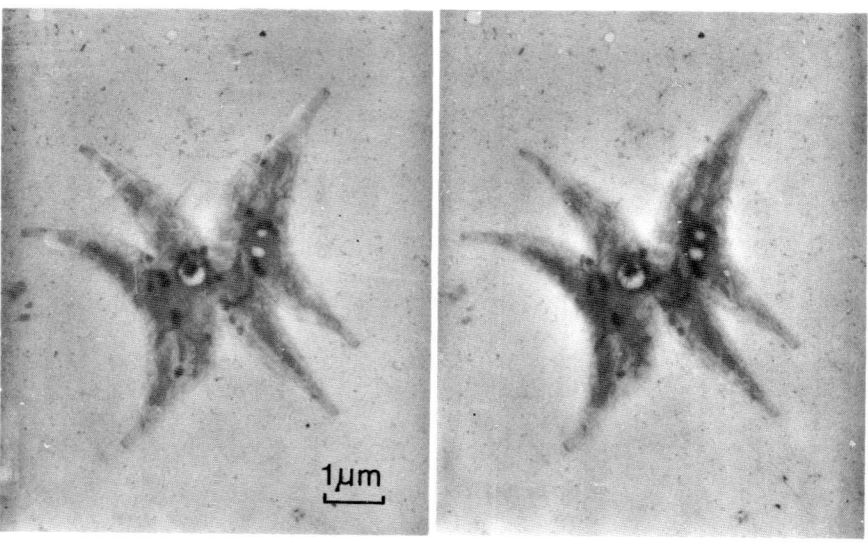

Fig. 14. Stereo pair of pond algae in the 1.5 mm environmental cell taken in one atmosphere of helium saturated with water vapor. The carbon supporting film is covered with a layer of water (Tighe et al., 1973)

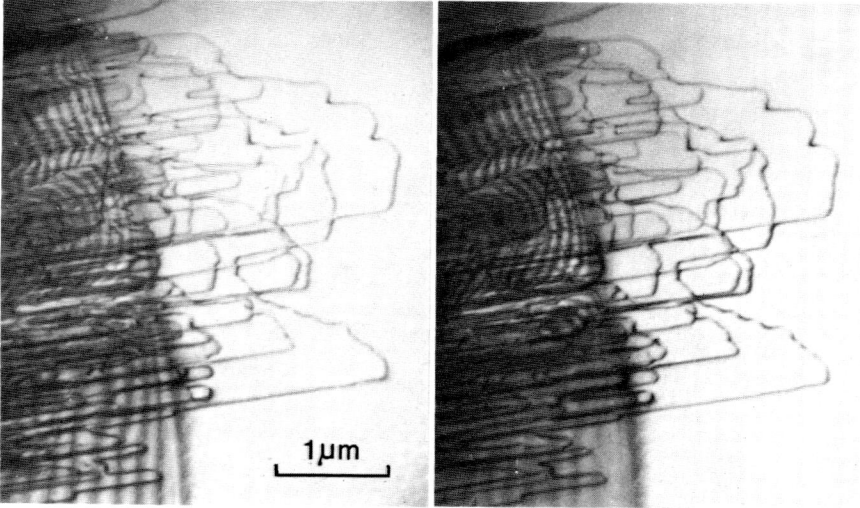

Fig. 15. Stereo pair of a lateral crack in SiC showing the curvature of the crack and the dislocations associated with a Vickers pyramid indenter. (Courtesy of Hockey)

Stereo-pairs of micrographs can be used to obtain quantitative information on precipitate density, foil thickness, and special distribution of defects (L.E. Thomas *et al.*, 1974) or the pictures can be used simply to reveal the three-dimensional microstructure of defects and three-dimensional structure of micro-objects. The examples in Figs. 14 and 15 were obtained using HVEM (Fig. 14) and a 200 kV microscope (Fig. 15).

References

Allinson, D.L.: Environmental cell for use in a high voltage microscope. Proc. 7th Int. Congress on Electron Microscopy, Grenoble, **1**, 169–170.

Allinson, D.L., Gosnold, A.W., Loveday, M.S.: A modified environmental cell for use in a high voltage electron microscope. Proc. 5th Europ. Congress on Electron Microscopy, Manchester, 336–337 (1972).

Allpress, J.G., Sanders, J.V., Wadsley, A.D.: Electron microscopy of high temperature Nb_2O_5 and related phases. Phys. Stat. Sol. **25**, 541–550 (1968).

Amelinckx, S.: The direct observation of dislocations. New York: Academic Press 1964.

Amelinckx, S., Delavignette, P.: Dislocation patterns in graphite. J. Nucl. Mater. **5**, 17–66 (1962).

Bach, H.: Elektronenmikroskopische Durchstrahlungsaufnahmen und Feinbereichselektronenbeugung an Al_2O_3 Keramik. BOSCH Techn. Ber. **1**, 10–13 (1964).

Bach, H.: Application of ion sputtering in preparing glasses and their surface layers for electron microscope investigations. J. Non-Cryst. Sol. **3**, 1–32 (1970).

Barber, D.J.: Thin foils of non-metals made for electron microscopy by sputter-etching. J. Mat. Sci. **5**, 1–8 (1970).

Barber, D.J.: Mounting grains to be examined by high resolution electron microscopy. Am. Mineralogist **56**, 2152–2155 (1971).

Barber, D.J., Ashworth, J.R.: Electron petrography of shock effects in gas-rich enstatite-achondrite. Contrib. Mineral. Petrol. **49**, 149–162 (1975).

Barber, D.J., Evans, R.G.: Dislocations ordering and antiferromagnetic domains in MnO. Proc. EMSA, p. 522–523. Baton Rouge: Claitor 1970.

Barber, D.J., Farabaugh, E.N.: Dislocations and stacking faults in rutile crystals grown by flame fusion methods. J. Appl. Phys. **36**, 2803–2806 (1965).

Barber, D.J., Frank, F.C., Moss, M., Steeds, J.W., Tsong, I.S.T.: Prediction of ion-bombarded surface topographies using Frank's kinematic theory of crystal dissolution. J. Mat. Sci. **8**, 1030–1040 (1973).

Barber, D.J., Price, P.B.: Solar flare particle tracks in lunar and meteoric minerals. In: Electron microscopy and analysis (ed. W.C. Nixon), p. 276–279. London: Institute of Physics 1971.

Barber, D.J., Tighe, N.J.: Observations of dislocations and surface features in corundum crystals by electron transmission microscopy. J. Res. Nat. Bur. Std. **69**A, 271–280 (1965).

Beeston, B.E.P.: High voltage microscopy of biological specimens: some practical considerations. J. Micros. **98**, 402–416 (1973).

Booker, G.R., Stickler, R.: Method of preparing silicon and germanium specimens for examination by transmission electron microscopy. Brit. J. Appl. Phys. **13**, 446–448 (1962).

Braillon, P., Mughier, J., Serughetti, J.: Transmission electron microscope observations of the dislocations in calcite single crystals. Cryst. Lat. Defects **5**, 73–78 (1974).

Brammer, I.S., Dewey, M.A.P.: Specimen preparation for electron metallography. Oxford: Blackwell Sci. Publ. 1966.

Browning, G.: A new axis-centered stage. In: High voltage microscopy (eds. P.R. Swann, C.J. Humphreys, M.J. Goringe), p. 121–128. London: Academic Press 1974.

Bouchard, M., Worthington, W.L., Jurica, D.J., Swann, P.R.: A high-angle tilting stage for the 650 kV Hitachi microscope. Rev. Sci. Instr. **44**, 511–512 (1973).

Buseck, P.R., Jijima, S.: High resolution electron microscopy of silicates. Am. Mineralogist **59**, 1–21 (1974).

Castaing, R., Jouffrey, B.: Effets de bombardements ioniques de courte durée sur des monocristanse metalliques minces. J. Micros. **1**, 201–214 (1962).
Champness, P.E.: Nucleation and growth of iron oxides in olivines (MgFe)$_2$SiO$_4$. Mineral. Mag. **37**, 790–800 (1970).
Champness, P.E., Lorimer, G.W.: Electron microscopic studies of some lunar materials. In: Electron microscopy and analysis (ed. W.C. Nixon), p. 324–327. London: Institute of Physics 1971.
Cosslett, V.E.: The high-voltage electron microscope. Contemp. Phys. **9**, 333–354 (1968).
Doherty, P.E., Leombruno, R.R.: Transmission electron microscopy of glass ceramics. J. Am. Ceram. Soc. **47**, 368–370 (1964).
Drum, C.M.: Electron microscopy of dislocations and other defects in sapphire and in silicon carbide, thinned by sputtering. Phys. Stat. Sol. **9**, 635–642 (1965).
Dupouy, G.: Electron microscopy at very high voltage. Adv. in Opt. and Elec. Microscopy **2**, 168–250 (1968).
Dupouy, G.: Megavolt electron microscopy. In: High voltage electron microscopy (eds. P.R. Swann, C.J. Humphreys, M.J. Goringe), p. 441–457. London: Academic Press 1974.
Dupouy, G., Perrier, F.: Observation en microscopie électronique à haute tension .d'àchantillons métalliques progessivement amincis par bombardement ionique. Compt. Rend. **261**, 4649–4651 (1965).
Evans, T., Phaal, C.: Transmission electron microscopy of diamond. Proc. 5th Int. Congress on Electron Microscopy, Philadelphia, **1**B, 4–5 (1962).
Fisher, R.M., Swann, P.R., Nutting, J.: A new objective pole piece and specimen heating stage for the Elmiskop. Proc. Europ. Reg. Congress on Electron Microscopy, Delft, **1**, 131–133 (1960).
Flower, H.M.: High voltage electron microscopy of environmental reactions. J. Micros. **97**, 171–190 (1973).
Flower, H.M.: Environmental gas reaction cells. I. Review. In: High voltage electron microscopy (eds. P.R. Swann, C.J. Humphreys, M.J. Goringe), p. 383–399. London: Academic Press 1974.
Gabor, D.: The history of the electron microscope from ideas to achievements. Proc. 8th Int. Congress on Electron Microscopy, Canberra, **2**, 6–12 (1974).
Genty, B.: Amincissement par bombardement ionique d'échantillons métalliques en vue de leur examen au microscope électronique. In: Le bombardement ionique. C.N.R.S., p. 95–104 (1962).
Gilliespie, P., McLaren, A.C., Boland, J.N.: Operating characteristics of an ion bombardment apparatus for thinning non-metals for transmission electron microscopy. J. Mat. Sci. **6**, 87–89 (1971).
Glauert, A.M. (ed.): Practical methods in electron microscopy. Amsterdam: North Holland Publ. Co. 1972.
Goodhew, P.J.: Preparation of carbon fibers for transmission electron microscopy. J. Phys. E. Sci. Inst. **4**, 392–394 (1971).
Goodhew, P.J.: Specimen preparation in materials science. In: Practical methods in electron microscopy (ed. A.M. Glauert), p. 3–180. Amsterdam: North Holland Publ. Co. 1972.
Goodhew, P.J.: The sputtering of rough cylinders: applications to the thinning of fibers for transmission electron microscopy. J. Mat. Sci. **8**, 581–589 (1973).
Hale, K.F., Henderson-Brown, M., Ishida, Y.: *In situ* dynamic observations of creep deformation in Al-1% Mg at 1 MV. Proc. 5th Europ. Congress on Electron Microscopy, Manchester, 350–381 (1972).
Harrison, C.G., Leaver, K.D., Swann, P.R.: A versatile specimen chamber for *in situ* experiments on magnetic materials. Proc. 5th Europ. Congress on Electron Microscopy, Manchester, 334–335 (1972).
Heerschap, M., DeCat, R.: Direct filming of transient phenomena with a closed TV circuit as viewfinder. Proc. 5th Europ. Congress on Electron Microscopy, Manchester, 170–171 (1972).
Henderson-Brown, M., Loveday, M.S., Hale, K.F., Gibbons, T.B.: Side entry soft straining and heating stages for *in situ* dynamic experiments in the IMV electron microscope. Proc. 5th Europ. Congress on Electron Microscopy, Manchester, 328–329 (1972).

Heuer, A.H., Firestone, R.F., Snow, J.D., Green, H.W., Howe, R.G., Christie, J.M.: An improved ion-thinning apparatus. Rev. Sci. Instr. **42**, 1177–1184 (1971).
Hirsch, P.B., Horne, R.W., Whelan, M.J.: Direct observations of the arrangement and motion of dislocations in aluminium. Phil. Mag. **1**, 677–684, Pl 23–24 (1956).
Hirsch, P.B., Howie, A., Whelan, M.J.: A kinematical theory of diffraction contrast of electron transmission microscope images of dislocations and other defects. Phil. Trans. Roy. Soc. London, Ser. A **252**, 499–529 (1960).
Hirsch, P.B., Howie, A., Nicholson, R.B., Pashley, D.W., Whelan, M.J.: Electron microscopy of thin crystals. London: Butterworths 1965.
Hobbs, L.W.: Preparation of thin films of moisture sensitive crystals for transmission electron microscopy. J. Phys. E. Sci. Inst. **3**, 85–89 (1970).
Hobbs, L.W., Goringe, M.J.: Electron microscopical observations of *in situ* and external irradiation damage in alkali halide crystals. Proc. 7th Int. Congress on Electron Microscopy, Grenoble, **1**, 239–240 (1970).
Hockey, B.J.: Plastic deformation of aluminum oxide by indentation and abrasion. J. Amer. Ceram. Soc. **54**, 223–231 (1971).
Hockey, B.J.: Observations by transmission electron microscopy on the subsurface damage produced in aluminum oxide by mechanical polishing and grinding. Proc. Brit. Ceram. Soc. **20**, 95–115 (1972).
Hockey, B.J., Lawn, B.R.: Electron microscopy of microcracking about indentations in aluminium oxide and silicon carbide. J. Mat. Sci., **10**, 1275–1284 (1975).
Holland, L., Hurley, R.E., Laurenson, L.: The operation of a glow discharge ion gun used for specimen thinning. J. Phys. E. Sci. Inst. **4**, 198–200 (1971).
Howie, A., Whelan, M.J.: Diffraction contrast of electron microscope images of crystal lattice defects. II. The development of a dynamical theory. Proc. Roy. Soc. (London), Ser. A **267**, 217–237 (1962).
Hudson, B.: The application of stereo-techniques to electron micrographs. J. Micros. **98**, 396–401 (1973).
Hudson, B., Makin, M.J.: The optimum tilt angle for electron stereo microscopy. J. Phys. E. Sci. Inst. **3**, 311 (1970).
Imura, T., Saka, H., Doi, M.: Application of TV-VTR recording system for studying dynamic or transient phenomena by high voltage electron microscopy (HVEM). Proc. 7th Int. Congress on Electron Microscopy, Grenoble, **2**, 331–332 (1970).
Imura, T., Yukawa, N., Saka, H., Nohara, A., Noda, R., Ishikawa, I.: *In situ* dynamic observation of dislocation motion at low and high-temperature by HVEM. In: High voltage electron microscopy (eds. P.R. Swann, C.J. Humphreys, M.J. Goringe), p. 199–205. London: Academic Press 1974.
Joy, T.R.: The electron microscopical observations of aqueous biological specimens. Adv. in Optical and Elec. Microscopy **5**, 297–352 (1973).
Kay, D.: Techniques for electron microscopy. Oxford: Blackwell Sci. Publ. 1965.
Keast, D.J.: A chemical thinning technique for the simultaneous preparation of foils for transmission electron microscopy. Application to yttrium aluminum garnet (YAG). J. Sci. Instr. **44**, 862–863 (1967).
Kirkpatrick, H.B., Amelinckx, S.: Device for chemically thinning crystals for transmission electron microscopy. Rev. Sci. Instr. **33**, 488–489 (1962).
Lally, J.S., Fisher, R.M., Christie, J.M., Griggs, D.T., Heuer, A.H., Nord, G.L., Radcliffe, S.V.: Electron petrography of Apollo 15 and 15 rocks. 3rd Lunar Sci. Conf. Geochim. Cosmochim. Acta, Suppl. 3, **1**, 401–422, MIT Press (1972).
Lehtinen, B., Roberts, W.: *In situ* observations of precipitation and recrystallization with a 1 MeV electron microscope. J. Micros. **97**, 197–208 (1973).
Lewis, M.H.: Defects in spinel crystals grown by the Verneuil process. Phil. Mag. **14**, 1003–1008 (1966).
Makin, M.J.: Time-lapse cine photography on the EM 7 microscope, and its use in radiation damage studies. In: High voltage electron microscopy (eds. P.R. Swann, C.J. Humphreys and M.J. Goringe), p. 365–369. London: Academic Press 1974.
Marton, L.: Early history of the electron microscope. San Francisco: San Francisco Press 1968.

Mazey, D.J., Nelson, R.S., Barnes, R.S.: Observation of ion bombardment damage in silicon. Phil. Mag. **17**, 1146–1161 (1968).
McLaren, A.C., Phakey, P.P.: Transmission electron microscope study of amethyst and citrine. Australian J. Phys. **18**, 135–141 (1965).
McPartland, J.O.: A high temperature stage for transmission electron microscopy. Proc. 5th Int. Congress for Electron Microscopy, Philadelphia, **1**, E3–4 (1962).
Müller, W.F., Wenk, E.-R.: Changes in the domain structure of anorthites induced by heating. Neues Jahrb. Mineral., Monatsh. **1973**, 17–26.
Nakahira, M., Uda, M. Defect structures of clay minerals. Z. Krist. **124**, 6 (1967).
Nankivell, J.F.: The theory of electron stereo microscopy. Optik **2**, 171–197 (1963).
Parsons, D.F., Matricardi, V.R., Subjek, J., Udyess, I., Wray, G.: High voltage electron microscopy of wet whole cancer and normal cells. Biochim. Biophys. Acta **290**, 110–124 (1972).
Paulus, M.: Le bombardement ionique, moyen d'étude de la microstructure des matériaux. Spectra 2000, **2**, 43–53 (1974).
Paulus, M., Reverchon, F.: Dispositif de bombardement ionique pour préparations micrographiques. J. Phys. Radium **22**, 103–107A (1961).
Paulus, M., Reverchon, F.: Étude des paramètres du bombardement ionique des ferrites. In: Le bombardement ionique, C.N.R.S. 223–234 (1962).
Phakey, P.P., Rachinger, W.A., Orams, H.J., Carmichael, G.G.: Preparation of thin section of selected areas. Proc. 8th Int. Congress on Electron Microscopy, Canberra, **1**, 412–413 (1974).
Price, G.L., Venables, J.A.: Apparatus for *in situ* nucleation studies in the T.E.M. Proc. 5th Europ. Congress in Electron Microscopy, Manchester, 338–339 (1972).
Radcliffe, S.V., Heuer, A.H., Fisher, R.M., Christie, J.M., Griggs, D.T.: High voltage transmission electron microscopy of lunar surface material. Science **167**, 638–640 (1970).
Raether, H.: Thin monocrystals produced by cathodic sputtering. In: Le bombardement ionique. C.N.R.S. 129–135 (1962).
Ramaswamy, V., Butler, E.P., Swann, P.R.: Direct observation of discontinuous precipitation in Al.28 at % Zn. J. Micros. **97**, 259–268 (1973).
Remaut, G., Lagasse, A., Amelinckx, S.: Electron microscopic study of the domain structure in anti-ferromagnetic cobalteous oxide. Phys. Stat. Sol. **7**, 497–510 (1964).
Ruska, E.: Zur Vor- und Frühgeschichte des Elektronenmikroskops. Proc. 8th Int. Congress on Electron Microscopy, Canberra, **1**, 1–5 (1974).
Sharp, J.V., Burnay, S.G.: High voltage electron microscopy of fibers. Electron Microscopy and Analysis (ed. W.C. Nixon), p. 28–29. London: Institute of Physics 1971.
Shaw, G.G.: Techniques for transmission electron microscopy of fiber-matrix interfaces in aluminum-alloy-boron composites by ion erosion. Proc. EMSA, p. 204–205. Baton Rouge: Claitor Publ. 1971.
Stringer, R.K., Warble, C.E., Williams, L.S.: Phenomenalogical observations during solid reactions. In: Kinetics of reaction in ionic systems (eds. T.J. Grey, V.D. Frechette). New York: Plenum Press 1969.
Swann, P.R.: A high angle double tilting specimen stage with straining and cooling attachments for the AEI 1 million volt microscope. Proc. 5th Europ. Congress on Electron Microscopy, Manchester, 322–323 (1972a).
Swann, P.R.: Side-entry single tilt specimen holders for heating and stress corrosion cracking of electron microscope specimens. Proc. 5th Europ. Congress on Electron Microscopy, Manchester, 330–331 (1972b).
Swann, P.R.: High voltage microscope studies of environmental reactions. In: Electron microscopy and structure of metals, p. 878–904. Berkeley: University of California Press 1972c.
Swann, P.R., Lloyd, A.E.: A high angle, double tilting cold stage for the AEI EM7. Proc. EMSA, p. 450–451. Baton Rouge: Claitor Publ. 1974.
Swann, P.R., Swann, G.R.: A tilting cold stage for the AEI EM802. Proc. EMSA, p. 372–373. Baton Rouge: Claitor Publ. Co. 1970.
Swann, P.R., Tighe, N.J.: High voltage microscopy of gas oxide reactions. Jerkon Ann. **155**, 497–501 (1971).

Swann, P.R., Thomas, G., Tighe, N.J.: *In situ* observations of the nitriding of tantalum. J. Micros. **97**, 249–258 (1973).
Thalen, J., Spoelstra, J., van Breemen, J.F.L., Mellema, J.E.: A tilting stage for electron microscopy of biological objects. J. Phys. E. Sci. Inst. **3**, 499–500 (1970).
Thomas, G.: Transmission electron microscopy of metals. New York: Wiley and Sons 1962.
Thomas, L.E., Lentz, S., Fisher, R.M.: Stereoscopic methods in the HVEM. In: High voltage microscopy (eds. P.R. Swann, C.J. Humphreys and M.J. Goringe), p. 255–259. London: Academic Press 1974.
Tighe, N.J.: Jet thinning device for preparation of Al_2O_3 electron microscopy specimens. Rev. Sci. Instr. **35**, 520–521 (1964).
Tighe, N.J.: Microstructure of fine-grain ceramics. In: Ultrafine-grain ceramics, p. 249–258. Syracuse: Syracuse Univ. Press 1970.
Tighe, N.J., Christie, J.M.: Deformation structures in quartz rocks. Proc. EMSA, p. 60–61. Baton Rouge: Claitor Publ. 1969.
Tighe, N.J., Hyman, A.: Transmission electron microscopy of alumina ceramics. In: Anisotropy in single-crystal refractory compounds (eds. E.W. Valdiek and S.S. Mersol), p. 121–236. New York: Plenum Press 1968.
Tighe, N.J., Swann, P.R.: Environmental gas reaction cells 2. Direct reduction of haematite. In: High voltage microscopy (eds. P.R. Swann, C.J. Humphreys and M.J. Goringe), p. 391–395. London: Academic Press 1974.
Tighe, N.J., Flower, H.M., Swann, P.R.: An environmental wet cell for high voltage electron microscopy. Proc. EMSA, p. 18–19. Baton Rouge: Claitor Publ. Co. 1973.
Trillat, J.J.: Le bombardement ionique nouvelle méthode d'étude des surfaces. In: Le bombardement ionique. C.N.R.S., p. 113–135, 1962 b.
Trillat, J.J.: Le bombardement ionique; théories et applications. C.N.R.S. 1962a. Ionic bombardment, theory and applications. New York: Gordon and Breach 1964.
Trueb, L.R., Barrett, C.S.: Microstructural investigation of Ballas diamond. Am. Mineralogist **57**, 1664–1680 (1972).
Valdiek, F.W.: Phase transition and imperfections of titanium dioxide and zirconium dioxide under high pressure. Proc. 6th Int. Congress on Electron Microscopy, Kyoto, 443–444 (1966).
Valdrè, U.: General consideration in specimen stages. Proc. 5th Europ. Congress on Electron Microscopy, Manchester, 317–321 (1972).
Valdrè, U.: Side-entry and top-entry straining stages with double tilting facilities. In: High voltage electron microscopy (eds. C.J. Humphrey and M.J. Goringe), p. 124–128. London: Academic Press 1974.
Valdrè, U., Horne, R.W.: A freeze-drying and low-temperature stage for electron microscopy. Proc. 5th Europ. Congress on Electron Microscopy, Manchester, 332–333 (1972).
Venables, J.A., Ball, D.J., Thomas, G.J.: An electron microscope liquid helium stage for use with accessories. J. Phys. E. Sci. Inst. **1**, 121–126 (1968).
Vesely, D.: *In situ* deformation of molybdenum thin foils. J. Micros. **97**, 191–196 (1973).
Ward, P.R., Mitchell, R.F.: A facility for electron microscopy of specimens in controlled environments. J. Phys. E. Sci. Inst. **5**, 160–162 (1972).
Washburn, J., Groves, G.W., Kelly, A., Williamson, G.K.: Electron microscope observations of deformed magnesium oxide. Phil. Mag. **5**, 991–999 (1960).
Watson, J.H.L., Heller, W., Schuster, T.: A study of tactoid forming crystals of βFeOOH. Proc. 2nd. Europ. Congress on Electron Microscopy, Delft, **1**, 229–234 (1960).
Whelan, M.J.: A high temperature stage for the Elmiskop I. Proc. 4th Int. Conf. on Electron Microscopy, Berlin, **1**, 96–100 (1958).
Wicks, B.J., Lewis, M.H.: Direct observations of ferro-electric domains in lithium niobate. Phys. Stat. Sol. **26**, 571–576 (1968).
Wiederhorn, S.M., Hockey, B.J., Roberts, D.E.: Effect of temperature on the fracture of sapphire. Phil. Mag. **28**, 783–796 (1973).
Wilsdorf, H.G.F.: Apparatus for the deformation of foils in an electron microscope. Rev. Sci. Instr. **29**, 323–324 (1958).
Yonts, O.C., Harrison, D.E.: Surface cleaning by cathode sputtering **31**, 1583–1584 (1960).

Section 4 Exsolution

◄ The microstructure of gedrite AV9 showing (010) lamellae of anthophyllite with a complicated morphology and small (010) platelets between them. Some of these platelets have nucleated on planar features parallel to (100). (Photograph by M. Gittos, see article by Gittos *et al.*, Chapter 4.8 in this volume)

CHAPTER 4.1

Exsolution in Silicates

P.E. CHAMPNESS and G.W. LORIMER

1. Introduction

The phase distributions which have been observed in exsolved minerals often bear a marked resemblance, both in scale and morphology, to the phase distributions which have been observed in metallic alloys. At first this similarity appears anomalous for the cooling rates of naturally occurring minerals are 10^4 to 10^6 times slower than those usually encountered in alloys. However, the diffusivities in minerals are typically 10^4 to 10^6 times lower than in metals at equivalent temperatures and hence the size of microstructural features, which is to a first approximation proportional to $\sqrt{Dt}$, where D is the diffusivity and t the time, are similar. This coincidence in the scale of microstructural features is fortunate for, while it is possible to study the development of phase distributions in the laboratory in metallic alloys, it is often not possible for the experimentalist to reproduce the microstructures of naturally exsolved minerals. Care must be taken when comparing simple metallic systems and the structurally and chemically-complex silicates; nevertheless the great wealth of well-documented information on precipitation reactions available in the metallurgical literature can be used to advantage when investigating exsolution in silicates.

In the following review of phase transformation we have drawn extensively from the book by Christian (1965) and the A.S.M. Seminar on Phase Transformations (1970).

2. Precipitate Nucleation and Spinodal Decomposition

The necessary, although not sufficient, condition for exsolution to occur is that there is a decrease in free energy on passing from a single phase to a two-phase system. The mechanism, or route, by which the exsolution reaction proceeds will depend upon both thermodynamics and kinetics.

The two simple schematic phase diagrams shown in Fig. 1 a and Fig. 2 a can be used to illustrate many of the phase transformations which have been reported in silicates. In Fig. 1 a the crystal structures of the two end-members must be the same or very similar; this is reflected in the continuous free energy versus composition curve in Fig. 1 b. In Fig. 2 a C and D have different crystal structures and the free-energy curves of the α and β phases are discrete (Fig. 2 b).

Exsolution in Silicates

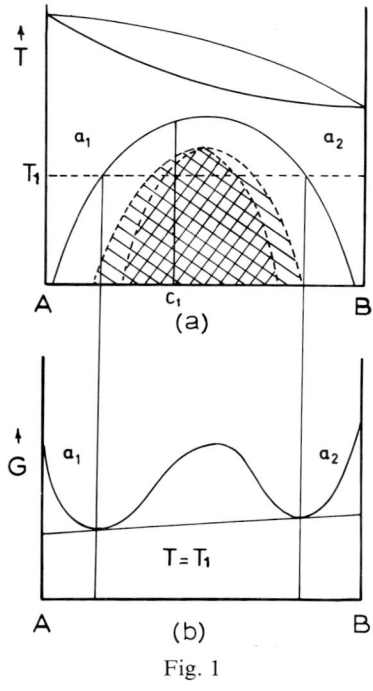
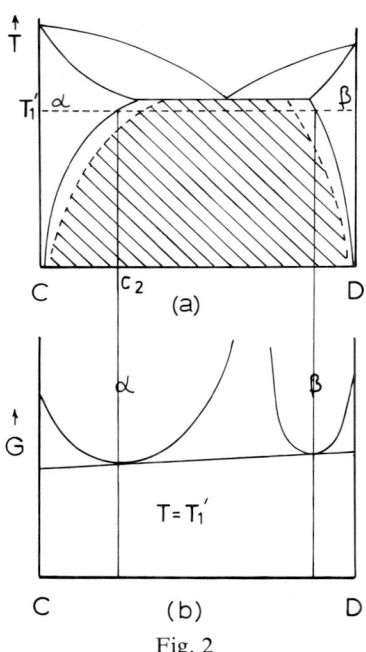

Fig. 1. (a) Schematic phase diagram in which the end members A and B have similar crystal structures. The coherent spinodal and coherent solvus are marked; both are suppressed below the equilibrium solvus due to strain. Spinodal decomposition can occur within the doubly-hatched area, homogeneous nucleation and growth in the singly-hatched area and heterogeneous nucleation and growth beneath the equilibrium solvus (solid line). (b) Schematic variation in free energy as a function of composition at $T=T_1$ showing a continuous free-energy curve between α_1 and α_2 and the composition of the two phases in equilibrium at $T=T_1$

Fig. 2. (a) Schematic phase diagram in which the end members C and D have different crystal structures α and β, respectively. The coherent solvus for nucleation and growth is marked (dashed line). (b) Schematic variation in free energy as a function of composition at $T=T_1'$ showing separate free-energy curves for the α and β phases and the composition of the phases in equilibrium at T_1'

In such a system exsolution can only proceed by a nucleation and growth mechanism, while in the system A–B exsolution can proceed either by a nucleation and growth mechanism or by spinodal decomposition depending on the cooling rate, the original composition and exact juxtaposition of the free energy versus composition curves.

2.1 Homogeneous Nucleation

In the singly-hatched regions of Figs. 1a and 2a exsolution can occur by homogeneous nucleation and growth. The nucleation event involves, in the simplest

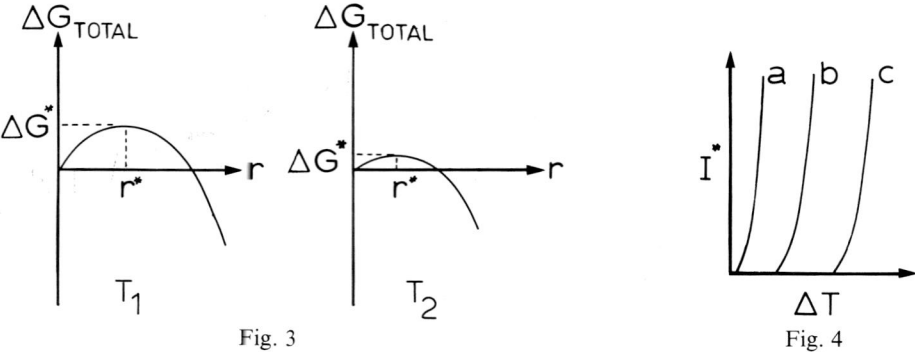

Fig. 3. Schematic variation in free-energy change ΔG as a function of the radius of the nucleus for two values of undercooling, T_1 and T_2, where $T_1 > T_2$. ΔG^* is the activation barrier to nucleation

Fig. 4. Schematic variation in nucleation rate I^* as a function of undercooling ΔT for *a* heterogeneous nucleation at grain boundaries, *b* heterogeneous nucleation at dislocations and *c* homogeneous nucleation

case, the formation of a particle of the exsolved phase of a sufficient size that it can grow with a decrease in free energy. The free-energy change accompanying precipitation, ΔG_{total} is composed of three terms:

$$\Delta G_{total} = \Delta G_{volume} + \Delta G_{surface} + \Delta G_{strain}.$$

ΔG_{volume} is the volume free-energy difference between the parent and product phases; this is the driving force for the reaction. $\Delta G_{surface}$ is the free energy increase accompanying the creation of the interface between the matrix and the precipitate and ΔG_{strain} is the strain energy due to the difference in volume between the precipitate and the matrix. Both $\Delta G_{surface}$ and ΔG_{strain} are positive terms and oppose the nucleation event. The variation in ΔG_{total} with the size of nucleus, r, is shown in Fig. 3. At the critical nucleus size, r^*, growth of the nucleus can proceed with a decrease in free energy; the value of $\Delta G_{total} = \Delta G^*$ is the activation-energy barrier to precipitate nucleation.

The values of r^* and, hence ΔG^*, in any system are sensitive functions of the extent of undercooling below the solvus (Fig. 3). If $\Delta T = 0$, the critical nucleus is infinitely large and the nucleation rate I^* is zero. As the undercooling, ΔT is increased ΔG_{volume} becomes larger and r^* becomes smaller. When r^* is sufficiently small that, by random short-range fluctuations in composition over a sensible period of time it is possible to form a critically-sized nucleus, then the nucleation rate I^* increases rapidly (Fig. 4).

2.2 Heterogeneous Nucleation

If the elastic misfit between the matrix and the precipitate is more than about 2% (Nicholson, 1970) ΔG_{strain} (and hence r^*) will be so large that it will be impossible to form a critically-sized nucleus of the equilibrium phase, even at very large undercoolings, and homogeneous nucleation will not be observed. One way of overcoming this prohibitively high energy-barrier is by heterogeneous nucleation at either planar or linear defects.

An incoherent planar interface — a grain boundary — is a potent nucleation site; ΔG_{strain} can be reduced because the misfit between the precipitate and the matrix can be accommodated in the disordered structure of the boundary and $\Delta G_{\text{surface}}$ can be reduced if the precipitate "wets" the grain boundary. The efficiency of a grain boundary in reducing ΔG_{strain} and $\Delta G_{\text{surface}}$ depends upon the exact atomic structure of the boundary, as well as the orientation of the precipitate nucleus on the boundary with respect to the two grains. In natural minerals grain boundaries are usually irregular features and the nature of the misfit across the boundary varies from point-to-point. This variation in misfit is mirrored in the variability of precipitate density along grain boundaries.

A linear defect — a dislocation — provides a nucleation site whose potency is intermediate between that of the planar grain-boundary and the perfect matrix. The misfit energy associated with the formation of a precipitate nucleus is balanced by the strain energy of the dislocation and ΔG_{strain} for nucleation is reduced. Dislocations can only relieve strain in one dimension, and thus are not as effective nucleation sites as are grain boundaries, at which two-dimensional relaxation is possible, or grain corners.

The relative potency of grain boundaries, dislocations and the matrix as sites for precipitate nucleation are reflected in the amount of undercooling required for a nucleation event to occur at these sites as shown by curves a, b, and c, respectively, in Fig. 4. If a mineral of composition C_2 (Fig. 2a) is cooled below the equilibrium solvus, heterogeneous nucleation will be initiated at grain boundaries at undercoolings of a few °K. Undercoolings of several tens or hundreds of °K may be required for homogeneous nucleation to occur, while nucleation on dislocations will take place at intermediate undercoolings.

2.3 Transitional Phases

A second mechanism by which the large free-energy barrier opposing the nucleation of the equilibrium phase can be circumvented is by the formation of a transitional (intermediate) phase which is structurally more similar to the matrix than is the equilibrium phase. Although the change in ΔG_{volume} during the precipitation of the intermediate phase is less than that which would result if the equilibrium phase had formed (Fig. 5), the activation barrier opposing nucleation will be less, due to the reduction in ΔG_{strain}. The quality of "fit" between the lattice of the matrix and the intermediate precipitate is often reflected in the nucleation site. If the transitional phase is structurally fully coherent with the matrix, i.e. the lattice planes are continuous across the interface as shown schematically in Fig. 6a, then homogeneous nucleation of the intermediate phase may occur. When the misfit can no longer be accompanied by elastic strains dislocations are inserted at the boundary and the precipitate/matrix interface is structurally semi-coherent (Fig. 6b). An incoherent precipitate does not exhibit any con-

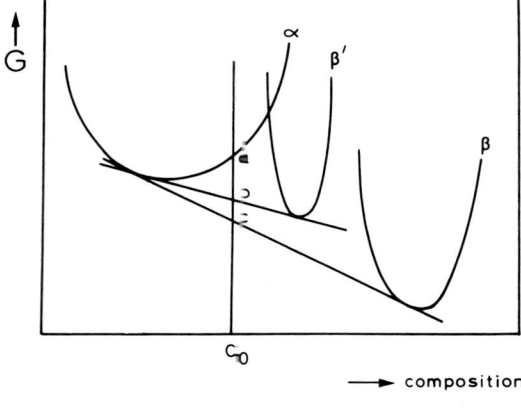

Fig. 5. Variation in free energy G with composition showing matrix α, equilibrium precipitate β and the intermediate phase β'. The decrease in free energy for a composition C_0 which exsolves the equilibrium phase is a–c. Exsolution of the intermediate phase β' results in a decrease in free energy of a–b

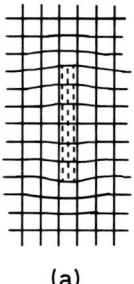
(a)

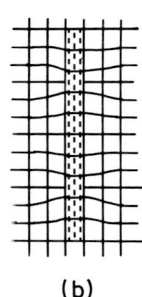
(b)

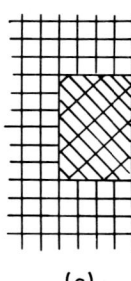
(c)

Fig. 6a–c. Illustration of a precipitate which is (a) coherent, (b) semi-coherent and (c) incoherent with the matrix

tinuity of structure with the matrix (Fig. 6c). Both semi-coherent and incoherent precipitates normally nucleate heterogeneously, the latter invariably at grain boundaries. Coherent precipitates may become first semi-coherent and eventually incoherent during growth.

2.4 Spinodal Decomposition

Detailed accounts of spinodal decomposition and the criteria by which it may be recognized have been given by Cahn (1968) and Hilliard (1970). It involves the evolution of sinusoidal composition waves with progressively increasing wavelength and amplitude in which the overall symmetry reflects that of the parent structure (Cahn, 1961, 1962). The peaks and valleys in these waves eventually become two discrete phases. This behavior is in contrast to the classical nucleation and growth mechanism in which two phases with a distinct interface are present from the very beginning of exsolution. The temperature-composition curve which defines the upper limit of possible spinodal behavior is known as the "coherent spinodal"; it is the "chemical spinodal" (the temperature-composition curve along which the second partial derivative of the free energy of the solid solution with respect to composition is zero) modified to take account of elastic coherency strains. Fig. 1a shows a simple phase diagram in which the coherent solvus and coherent spinodal are indicated.

Spinodal decomposition is associated with uphill diffusion and a specific variation of the amplitude and wavelength of the composition modulations with time. Absolute proof of the operation of a spinodal mechanism involves sophisticated small-angle X-ray scattering experiments which monitor the variation of these parameters during decomposition. Such experiments have been carried out on very few metallic systems (Hilliard, 1970) and on no mineral systems to date. Because of the difficulties of carrying out the crucial X-ray experiments, many materials scientists accept the following microstructural criteria as being indicative of a spinodal reaction.[1] In systems which are elastically isotropic, e.g. glasses and Al–Zn, the amplification of the composition modulations is equally probable in all directions and a highly interconnected structure with a characteristic wavelength is produced (Hilliard, 1970). In elastically anisotropic crystals composition modulations are amplified preferentially in the elastically soft directions. In cubic crystals this is usually $\langle 100 \rangle$ and leads to three sets of perpendicular modulations. In the initial stages there is a diffuse interface between the solute-rich and solute-poor component which becomes progressively sharper once the initial decomposition reaction has occurred and coarsening is initiated. However, at the stage when there is a sharp interface between two distinct phases it is impossible to determine whether the reaction has been nucleated or produced by spinodal decomposition.

In silicates, which have low symmetry, one, two or three composition modulations are observed. In the early stages they appear, from their diffraction contrast in the electron microscope, to have a small amplitude and a diffuse interface and are large in extent (see for instance Figs. 9 and 15b). These microstructural features are not consistent with a nucleated reaction, but are fully consistent with spinodal decomposition.

The diffraction patterns from these fine-scale modulated structures are also consistent with spinodal decomposition. Satellites, streaks or diffuse scattering are convoluted about a *single* reciprocal lattice. A nucleated reaction produces two phases from the beginning and two distinct reciprocal lattices should be detected.

3. Precipitation Growth and Coarsening

3.1 Precipitate Growth

The driving force for precipitate growth is the same as that for precipitate nucleation—the reduction in solute supersaturation accompanying the formation of the exsolved phase. The morphology of the growing particle is determined by the nucleation site, the orientation relationship between the precipitate and the matrix and the nature of the precipitate/matrix interface. A disordered (incoher-

---

[1] D.E. Laughlin and J.W. Cahn [Acta Met. **23**, 329 (1975)] have recently described a definitive experimental method by which spinodal decomposition can be recognized from the microstructure.

ent) boundary, such as that which could be produced by the heterogeneous nucleation of the equilibrium phase at a grain boundary, does not present a structural obstacle to growth and the growth rate is controlled either by the rate of solute migration to the precipitate/matrix interface (volume-diffusion control) or by the rate of transfer of the solute at the interface itself (interface-control).

A fully coherent or semi-coherent interphase boundary presents a barrier to growth, particularly when the phase transformation requires a change in stacking sequence perpendicular to the boundary. The addition of a small "cap" of precipitate to a planar, fully-coherent interphase boundary and the non-conservative migration of a semi-coherent boundary (i.e. perpendicular to the boundary plane) are both energetically expensive processes. The mechanism by which this problem is often overcome is by the introduction of ledges or steps at the interphase boundary. Thickening of the precipitate perpendicular to the coherent or semi-coherent boundary proceeds by the migration of the ledges along the boundary plane. Confirmation of the validity of the ledge mechanism of precipitate growth has been obtained by Aaronson and co-workers during *in situ* experiments with aluminum and iron alloys (Laird and Aaronson, 1969; Aaronson et al., 1970). Ledges have been observed at lamellar precipitate/matrix interfaces in pyroxenes (Champness and Lorimer, 1973, 1974; Copley et al., 1974; Vander Sande and Kohlstedt, 1974; Kohlstedt and Vander Sande, Chapter 4.7 of this volume), amphiboles (Gittos et al., 1974 and Chapter 4.8 of this volume) and plagioclase feldspars (Lorimer and Champness, unpublished), and they must be accepted as an important mechanism of precipitate thickening in silicates.

3.2 Rates of Transformation

The rates of precipitate nucleation and growth in solid-state transformations which are thermally-activated are governed by the same thermodynamic and kinetic factors: the thermodynamic driving force ΔG_{total} and the diffusion coefficient, D. At high temperatures, corresponding to small amounts of undercooling below the equilibrium solvus, ΔG_{volume} is small and D is large. Conversely, at low temperatures ΔG_{volume} is large but D is small. Clearly there is an optimum degree of undercooling at which the transformation rate is a maximum.

Information concerning the rates of transformation can be conveniently illustrated on Time-Temperature-Transformation (TTT) diagrams (see for example Chadwick, 1972) where the times for a given fraction of a reaction to proceed are plotted as a function of temperature. Fig. 7a is a schematic TTT diagram (actually a continuous-cooling-transformation diagram) for a system in which either heterogeneous nucleation and growth or spinodal decomposition can occur, depending on the cooling rate, and might represent the behavior of a mineral of composition C_1 in Fig. 1. Four cooling curves have been superimposed on the TTT diagram. For slow cooling rates, A in Fig. 7a, heterogeneous nucleation of the equilibrium phase—probably at grain boundaries—will occur when the cooling curve crosses the "start" line. Growth of the equilibrium phase will continue as the temperature falls until the solute supersaturation has been elim-

Fig. 7. (a) Schematic time-temperature-transformation diagram for a mineral of composition C_1 in Fig. 1. The curves A, B, C and D correspond to slow, intermediate, fast and very fast (geological) cooling rates. (b) A schematic TTT diagram similar to (a) but which includes curves for homogeneous nucleation and growth of the equilibrium phase

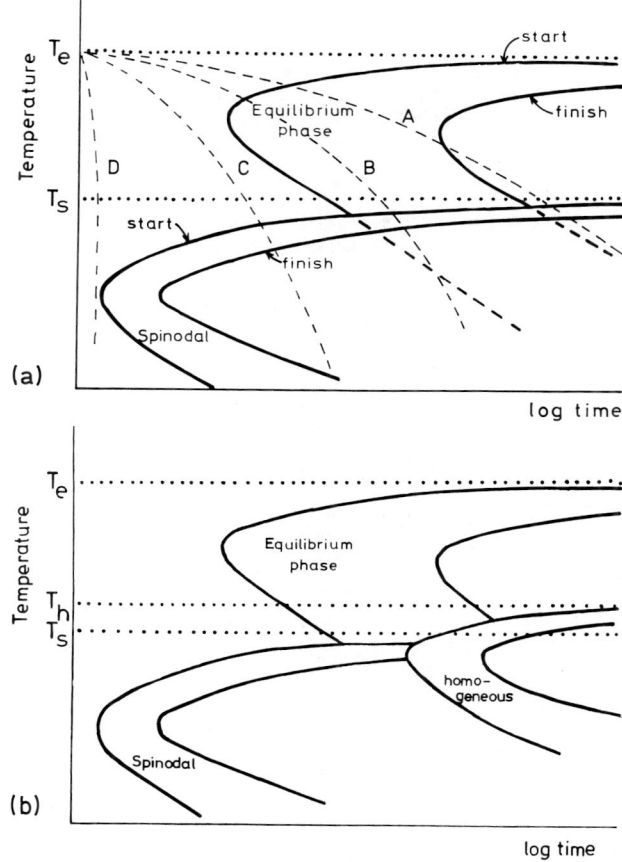

inated — when the cooling curve crosses the "finish" line for the equilibrium phase. At fairly rapid cooling rates, C in Fig. 7a, precipitation of the equilibrium phase will not take place at a detectable rate and phase separation will proceed completely by a spinodal mechanism; at very fast cooling rates, D, no decomposition will occur. At an intermediate cooling rate (B in Fig. 7a) nucleation of the equilibrium phase will occur at a high temperature, but only a proportion of the reaction will have occurred before the "start" of the spinodal reaction is crossed. Those regions of the matrix which still contain sufficient solute supersaturation will then exsolve by the spinodal mechanism.

Fig. 7b is a TTT diagram in which the curves for homogeneous nucleation and growth of the equilibrium phase are also included. For homogeneous nucleation the strain and surface-energy terms, together with the larger diffusion distances involved, leads to a lower rate of transformation compared with spinodal decomposition. This is reflected in the relative positions of the "noses" of the two curves. Consequently, homogeneous nucleation of the equilibrium phase is only expected for a fairly narrow range of intermediate cooling rates.

3.3 Coarsening

When precipitate growth has stopped – the solute supersaturation has been eliminated – the phase distribution does not have the minimum free energy; there is a tendency for the smaller particles to dissolve and for the larger particles to grow. This is the phenomenon of precipitate coarsening, the driving force for which is the reduction in the total interfacial area. The solute concentration in equilibrium with a small particle is larger than that in equilibrium with a large one as shown by the Gibbs-Thompson equation:

$$\ln\left[\frac{C_r}{C_\infty}\right] = \frac{2\gamma\Omega}{kTr},$$

where C_r is the solubility of a particle of radius r, C_∞ the solubility at a planar interface, γ the interfacial energy, Ω the atomic volume of the particle of radius r, k is Boltzmann's constant and T is the absolute temperature. Thus there will be a solute concentration-gradient between the large and the small particle with the net result that solute will be continually drained from the matrix near the small particles to the matrix near the larger ones.

The detailed variation in particle growth rate as a function of the radius of the particle, relative to the average radius, depends upon whether volume diffusion through the matrix or transfer of solute across the precipitate/matrix interface is the rate-controlling step. The exact shape of the particle-size distribution for both interface control and volume-diffusion control can be calculated, as can the variation in average particle size with time at temperature (see, for example, Greenwood, 1969).

The precipitation site can alter the coarsening rate for individual particles due to the effects of short-circuit diffusion, for example at grain boundaries. If the particles have different elastic constants from the matrix, particle alignment can take place during the coarsening reaction (Ardell et al., 1966), leading to the "rafting" of precipitates.

4. Exsolution in Pyroxenes

4.1 Phase Relationships

The pyroxenes in igneous rocks are essentially confined to the quadrilateral of the ternary system shown in Fig. 8a. During the crystallization of a typical tholeiitic magma a calcium-rich and a calcium-poor pyroxene crystallize together; Fig. 8a shows the crystallization trends which result from fractional crystallization of a slowly-cooled intrusion. Superimposed are three tie-lines which join co-precipitating phases.

Augite (the Ca-rich phase) and pigeonite (one of the Ca-poor phases) are both monoclinic. At high temperatures both have the $C2/c$ space group and the structures are almost identical, differing only in the detailed configurations of the Si–O chains and the coordination of one of the cation sites (G.E. Brown *et al.*, 1972). At a lower temperature pigeonite undergoes a polymorphic transition to the $P2_1/c$ space group.

Orthopyroxene, the Ca-poor phase which crystallizes with augite at the beginning of fractionation, is orthorhombic and has a different stacking sequence of the Si–O chains from the monoclinic phases. Pigeonites in igneous intrusions transform to the orthopyroxene structure on cooling.

Hess (1941) first recognized that the oriented lamellae of Ca-poor pyroxene in augite (and *vice versa*) were the result of exsolution during subsolidus cooling and he illustrated the phase relationships with a schematic phase diagram similar to Fig. 8b. Generally, monoclinic exsolution lamellae in a monoclinic matrix are parallel to (001) and monoclinic lamellae in an orthorhombic matrix (and *vice versa*) are parallel to (100). Morimoto and Tokonami (1969) showed that

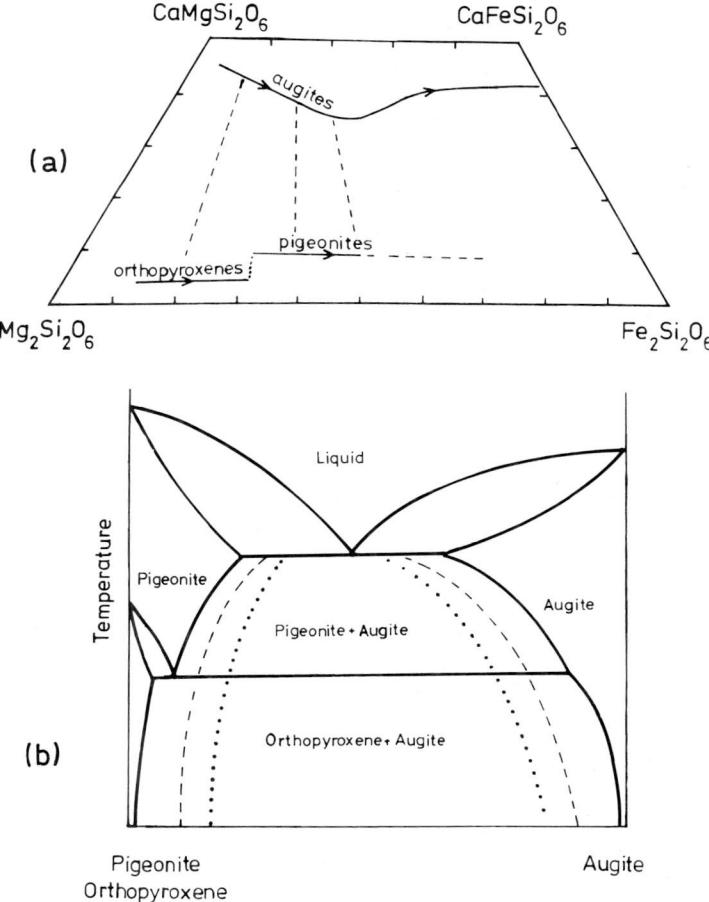

Fig. 8. (a) The pyroxene quadrilateral. The solid lines show the crystallization trends in tholeiitic intrusions and three tie lines joining co-precipitating phases are shown. (b) A schematic diagram for the clino-pyroxenes. The high → low pigeonite transformation is not shown. The dashed line represents the coherent solvus and the dotted line is the coherent spinodal

the elastic strain energy for coherent exsolution of augite in pigeonite was a minimum for lamellae parallel to (001). They did not consider the surface-energy component of the interphase-boundary energy, but this may be unimportant compared with the strain energy (especially for coarsened intergrowths) because, with two such closely similar silicates, only second, or even third, nearest-neighbor atoms may be in "wrong" positions across the interface. No calculations have been made for ortho-clino intergrowths, but the observed orientation is explained qualitatively by the fact that (100) is the only plane whose structure is common to the two phases. Both the strain and surface energies will be a minimum for this orientation.

The subsolidus phase relationships inferred by Hess from careful optical observations have been amply confirmed by X-ray diffraction experiments, particularly of Skaergaard pyroxenes (Bown and Gay, 1960), and by experimental studies of the synthetic system $MgSiO_3$–$CaMgSi_2O_6$ (Atlas, 1952; Boyd and Schairer, 1964).

Recently Ross *et al.* (1973) attempted to define the coherent solvus between augite and pigeonite by heating experiments on exsolved lunar clinopyroxenes. The single crystals were heated at different temperatures and the compositions of the two phases were estimated from measurements of the angle β. Although the composition estimates were only semiquantitative, the results confirm the existence of a solvus which extends across most of the quadrilateral. The coherent solvus was found to be asymmetric towards pigeonite and to be steeper on the pigeonite than on the augite side as shown in Fig. 8b.

4.2 Clinopyroxenes

Although pyroxenes were the first mineral group in which exsolution was recognized and documented, electron microscopic studies are a recent innovation. The phase distributions which have been revealed by the electron microscope are the most diverse of any mineral group and include examples of many of the phenomena outlined in the previous sections.

Augites and pigeonites from lunar and terrestrial basalts were found to have microstructures consistent with exsolution by spinodal decomposition (Christie et al., 1971; Champness and Lorimer, 1971, 1972; Lally et al., 1972). Some of the finest structures consisted of coherent modulations, with diffuse interfaces, parallel to (001) and (100), the former generally having the greater amplitude (Fig. 9a). In most cases the (100) modulation fails to coarsen so that, when the periodicity becomes greater than about 200Å, only the (001) modulation remains (Fig. 9b).

Diffraction patterns from singly-modulated structures indicate the presence of two phases with very similar cell dimensions, while patterns from the finer tweed structures show streaks parallel to a^* and c^* rather than two distinct maxima (Nord et al., Chapter 4.5 of this volume). Champness and Lorimer (1972) found that the type (a) $(h+k=2n)$ reflections from a pigeonite in a Mull andesite, were surrounded by a region of diffuse scattering. The corresponding image showed a "mottled" texture on a scale of about 50 Å which was interpreted as

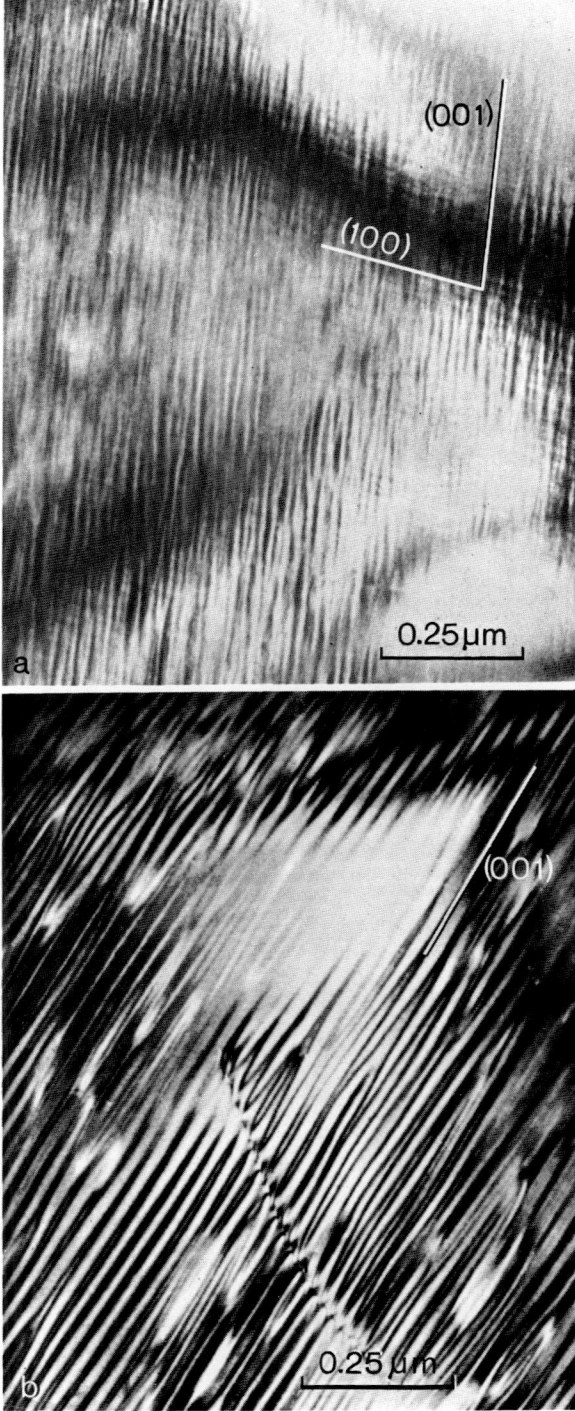

Fig. 9a and b. Modulated structures consistent with spinodal decomposition in Apollo 12 lunar pigeonites showing (a) two sets of modulations parallel to (001) and (100) and (b) coarsened modulations parallel to (001) only representing an early stage of phase separation by spinodal decomposition. The perturbation of the structure was approximately longitudinal with propagation vectors randomly distributed in reciprocal space. In a lunar pigeonite a slightly later stage of decomposition was recognized in a crystal which showed a single reciprocal lattice where each reflection was flanked by two pairs of satellites, one approximately along c^* and the other along a^*; the microstructure exhibits a faint "tweed" texture with a wavelength of about 100 Å.

In contrast, clinopyroxenes from plutonic environments show microstructures which are consistent with heterogeneous nucleation and growth (Copley, 1973; Copley et al., 1974; Champness and Copley, Chapter 4.6 of this volume). When the precipitates have a strictly lamellar morphology their interface with the matrix is coherent, but at an advanced stage of coarsening, when the lamellae begin to break up into a "shish-kebab" morphology, the interface loses coherency.

Examples of the microstructures produced by an intermediate cooling rate have been reported by Nord et al. (1973) in pigeonites from an Apollo 15 Mare basalt in which a tweed structure is present between heterogeneously-nucleated augite lamellae. Similar phase distributions have been found in an augite and a calcium-rich pigeonite (which is thought to have crystallized along a metastable extension of the solidus below the solvus, Dunham et al., 1972) in a dolerite from the Whin Sill, Northern England (Copley, 1973). In Fig. 10, which shows the structure in the augite, the matrix between the pigeonite lamellae contains either a single coherent "modulation" parallel to (100) or, where the texture is finer, two modulations parallel to (100) and (001). The scale of the structure

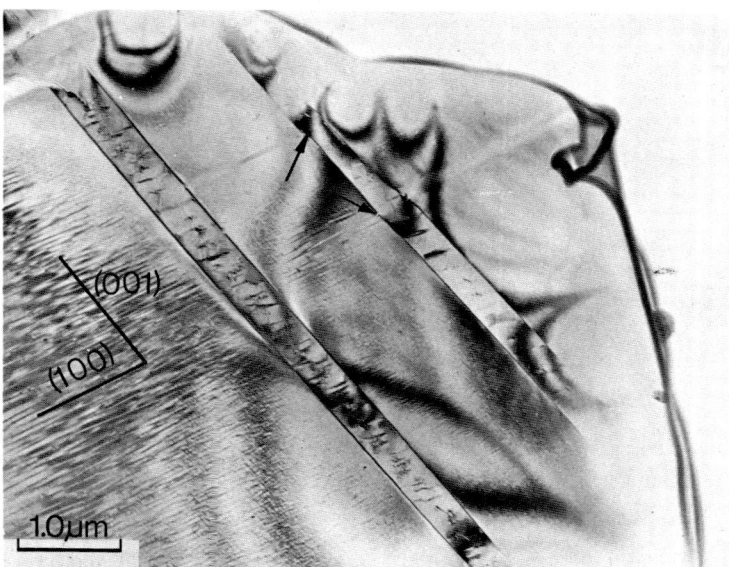

Fig. 10. Phase distribution in an augite from the Whin Sill showing heterogeneously-nucleated (001) pigeonite lamellae with a tweed structure between them. Heterogeneous nucleation of (100) pigeonite lamellae has occurred at ledges on the interface of the (001) lamellae (arrowed)

becomes smaller as a lamella is approached, and is absent immediately adjacent to it. This indicates that the lamellae were formed first, denuding the matrix around them of solute and that, on further cooling, the modulated structure was formed, probably by spinodal decomposition, in areas where the supersaturation was sufficient.

The microstructures observed in clinopyroxenes with different thermal histories can be illustrated with reference to the TTT diagram in Fig. 7; quickly-cooled samples such as the Mull pigeonite and some lunar basalts (curve C) contain only spinodal structures; samples from plutonic rocks only contain nucleated structures (curve A), while samples which have cooled at an intermediate rate, such as the Whin Sill augite, contain both nucleated and spinodal structures (curve B).

Nord et al. (1973) reported another "two-stage" exsolution microstructure in pigeonite from an Apollo 16 basalt which the authors concluded had cooled more slowly than the Apollo 15 rocks. Between the (001) augite lamellae (although not immediately adjacent to them) were thin (~ 40 Å), homogeneously-distributed (001) platelets of augite about 0.2 μm long. Their interfaces with the matrix were considerably sharper than those of precipitates on a comparable scale which have been attributed to spinodal decomposition (cf. Fig. 9b). This fact, and their distinctive morphology suggests that they may have formed by homogeneous nucleation and growth. Similar microstructures have been found in certain areas of a calcic pigeonite from the Whin Sill dolerite (Copley, 1973).

It may be significant that, for fast cooling rates, only modulated structures (consistent with spinodal decomposition) are found. Fig. 7b shows that homogeneous nucleation of the equilibrium phase is not expected under these conditions, but might occur in a narrow range of intermediate cooling rates after some heterogeneous nucleation has already taken place.

In this volume Nord et al. describe an augite from an Apollo 17 basalt which contains two sets of fine pigeonite precipitates between larger (heterogeneously-nucleated) pigeonite lamellae. The larger precipitates are parallel to (001) while the smaller ones are parallel to (100) and, although their scale is similar to the modulated structure in Fig. 9b, their interfaces appear to be sharper and their morphology is less continuous. These precipitates may also have formed by homogeneous nucleation and growth, as suggested by Nord et al., although it is difficult to see how two sets of platelets could have formed by homogeneous nucleation in crystallographically non-equivalent orientations, because the anisotropy of the surface energy is important from the nucleation stage.

4.3 Ortho-clinopyroxenes

The exsolution of clinopyroxene from orthopyroxene (or *vice versa*) does not permit operation of the spinodal mechanism because of the difference in structure of the two phases (Section 2). In their studies of orthopyroxenes from the Stillwater and Bushveld complexes, Champness and Lorimer (1973, 1974) identified two transition phases in addition to the heterogeneously-nucleated equilibrium precipitate, augite. A general view of the microstructure in the Stillwater sample

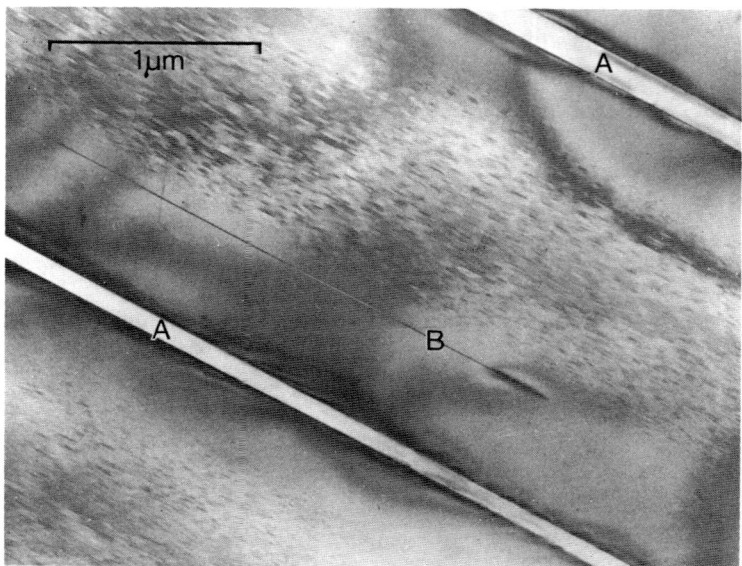

Fig. 11. Microstructure of the Stillwater orthopyroxene showing large (100) augite lamellae *A*, intermediate phase *B* and a fine distribution of homogeneously-nucleated G.P. zones

is shown in Fig. 11. The (100) augite lamellae, A, nucleate at grain boundaries, are semi-coherent with the matrix and thicken by the movement of ledges, from one to four lattice parameters high, along the interface. The intermediate precipitate labelled B in Fig. 11 forms thin, coherent (100) plates and nucleates at sub-grain boundaries (Champness and Lorimer, 1973a, Fig. 7) and other intragranular defects. Using analytical electron microscopy, high-resolution electron microscopy and electron diffraction, Lorimer and Champness (1973a) and Champness and Lorimer (1974) have shown that the phase is richer in calcium than the matrix, has an *a* lattice parameter of approximately 9.6 Å, and is primitive monoclinic.

The finely-dispersed phase which can be seen between the augite lamellae in Fig. 11 is a second intermediate phase which takes the form of small, coherent (100) platelets less than 0.25 μm in diameter and 18 Å (one matrix lattice parameter) thick. It has a structure different from that of the augite and the coherent, lamellar transition-phase (the platelets are imaged under different diffracting conditions). The origin of the precipitate-free zone (PFZ) of the platelets which occurs adjacent to both types of lamella (Fig. 11) was shown to be a lower calcium concentration than obtained in other areas of the matrix (Lorimer and Champness, 1973a). Growth of the augite lamellae and the lamellar transitional phase drained calcium from the matrix and suppressed later nucleation of the Ca-rich platelets. Champness and Lorimer (1973a, 1974) suggested that the platelets are G.P. zones (Guinier, 1938; Preston, 1938) on the grounds of their morphology, homogeneous distribution, coherency and the fact that they are calcium-rich with a structure similar to the matrix.

The sequence of precipitation in the Stillwater and Bushveld orthopyroxenes—heterogeneous nucleation of the equilibrium precipitate at grain boundaries, followed by the nucleation at dislocations of a coherent intermediate phase and, finally, homogeneous nucleation of a second intermediate phase with a structure very similar to that of the matrix—illustrates well the principles of intermediate phases and nucleation site discussed in Sections 2.2 and 2.3.

5. Exsolution in Amphiboles

Although the existence of two or more distinct amphibole phases within the same rock was first recognized over 50 years ago, it is only recently that exsolution textures have been described and investigated by optical microscopy and X-ray diffraction (Ross et al., 1969; Robinson et al., 1971). Electron microscopic studies are even more recent.

Gittos et al. (1974) described the microstructures of the coexisting hornblendes and grunerites from two metamorphic assemblages. In both cases the Ca-rich and Ca-poor phases contained coherent (100) and ($\bar{1}$01) lamellae of the complementary phase. The grunerites contained, in addition, a second generation of precipitates between the large lamellae. These were small coherent (100) platelets less than 1 μm long with a distribution similar to that of the G.P. zones in orthopyroxenes (Fig. 11). In Chapter 4.8 of this volume Gittos et al. report similar phase distributions in a third coexisting hornblende-grunerite pair. In all cases the small platelets are assumed to have nucleated homogeneously after precipitation of the large lamellae, the PFZ's results from calcium being drained into the hornblende lamellae during growth.

Christie and Olsen (1974) investigated the phase distribution of a blue schiller orthoamphibole (60% gedrite) by electron microscopy. It contained a homogeneous distribution of anthophyllite lamellae parallel to (010) with a periodicity of about 0.16 μm. The microstructure is consistent with exsolution having occurred by spinodal decomposition followed by coarsening.

Gittos et al. (Chapter 4.8 of this volume) describe gedrite lamellae parallel to {120} in a specimen containing 34% gedrite (their Fig. 4a). A more complex phase distribution is reported in an orthoamphibole containing 60% gedrite. The reader is referred to the paper for details.

6. Exsolution in Feldspars

6.1 Alkali Feldspars

Alkali feldspars exsolve on a coarser scale than any other silicate. This phenomenon can be mainly attributed to the relatively high diffusivities of potassium and sodium ions (Orville, 1963; Lin and Yund, 1972; Petrović, 1974) and to the fact that (unlike the case of the plagioclases) exsolution does not require diffusion of silicon and aluminum. A consequence of the rapid exsolution in alkali feldspars is that, in comparison with most silicates, the equilibrium sol-

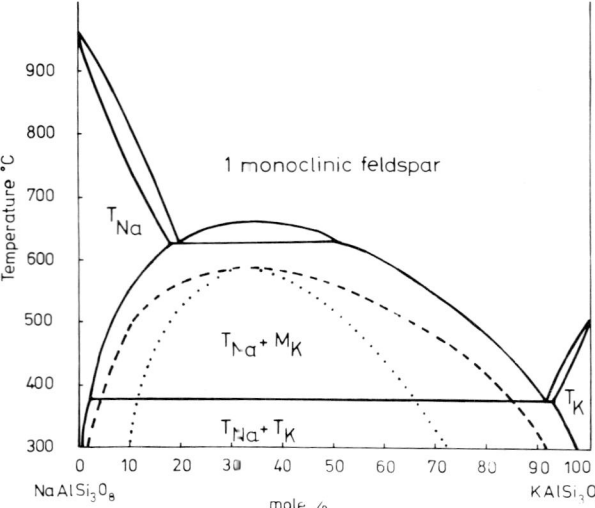

Fig. 12. The subsolidus phase diagram for the alkali feldspars at 1 kbar as calculated by Robin (1974). The dashed line is the coherent solvus and the dotted line is the coherent spinodal. The monoclinic → triclinic inversion temperatures are taken from MacKenzie (1952) and the inversion in orthoclase is from Steiger and Hart (1967) and Wright (1967). The thermodynamic nature of these transitions is not known, but they are shown as first order. T_{Na}=triclinic sodium feldspars, M_K=monoclinic potassium feldspar, T_K=triclinic potassium feldspar

vus has been well defined by synthetic studies (Bowen and Tuttle, 1950; Luth and Tuttle, 1966; Morse, 1970; Luth et al., 1974; Goldsmith and Newton, 1974; Smith and Parsons, 1974). Fig. 12 shows the subsolidus phase diagram for a pressure of 1 kbar. It consists of a miscibility gap with a crest at about 660 °C; the phases in equilibrium at ambient temperatures are almost pure $NaAl-Si_3O_8$ (Ab) and $KAlSi_3O_8$ (Or). During cooling the initially-monoclinic phases both undergo transformations to triclinic symmetry, and twinning results. In the case of the sodic phase the transformation is displacive and rapid, but in $KAlSi_3O_8$ the transformation involves ordering of aluminum and silicon ions and is very sluggish.

Recently Robin (1974) calculated the positions of the coherent solvus and coherent spinodal for the monoclinic phases using the elastic constants and cell dimensions. Surface energy, ordering of Al/Si, and variation of the elastic constants with composition were neglected. The coherent phase diagram was shown to be depressed by 70–85° from the equilibrium solvus (Fig. 12).

Fleet and Ribbe (1963) first examined a moonstone by transmission electron microscopy and showed that it contained exsolution lamellae parallel to ($\bar{6}01$), the schiller plane. They also identified regularly-spaced albite twins in the Na-phase (as predicted by Laves, 1952, from the presence of superlattice reflections parallel to b^* in X-ray diffraction patterns). Laves (1952) suggested that periodic albite-twinning of the triclinic phase would reduce the strain energy of the interface with the monoclinic phase. Willaime and Gandais (1972) explained the variation in spacing of the albite twins with lamellar thickness from calculations of the strain energy.

McConnell (1969) examined a volcanic alkali feldspar (36 mole-% Or) and showed that it contained a modulated structure approximately parallel to (100) with a wavelength of about 100 Å. The diffraction pattern showed a single reciprocal lattice with intense streaks through each reflection normal to b^*. The intensity distribution of the streaks indicated that the lattice distortion was longitudinal and was consistent with a variation in chemical composition. When the specimen was heated at 685 °C the modulated structure disappeared; McConnell concluded that the structure represented "incipient exsolution".

Owen and McConnell (1971, 1974), using previously-homogenized material, reproduced the modulated structure by hydrothermal treatment at 1 kbar (Fig. 13). Early stages of the reaction were characterized by fine-scale textures (~ 90 Å) with a broad spectrum of wavelengths, as evidenced by the streaks in electron diffraction patterns. The average (preferred) wavelength increased from 90 Å for an annealing temperature of 420 °C to 160 Å at 540 °C. As the reaction proceeded the wavelength distribution narrowed; the streaks convoluted about each diffraction spot became a pair of sharp satellites oriented approximately normal to $(\bar{1}0 \cdot 0 \cdot 1)$.

For annealing temperatures between 650 °C and 630 °C Owen and McConnell found only two-phase structures and suggested that these had formed by nucleation and growth. That alkali feldspars can certainly exsolve by nucleation and growth has been shown by Carstens (1967) who identified cellular (discontinuous) precipitates at the grain boundaries of alkali feldspars in an anorthosite gabbro.

Yund et al. (1974) also reproduced modulated structures by heating an initially-homogeneous alkali feldspar in air. Their sample, which contained 33 mole-% Or (the critical composition according to Robin, 1974) was prepared by alkali-ion exchange from a sample containing about 89 mole-% Or. Like Owen and McCon-

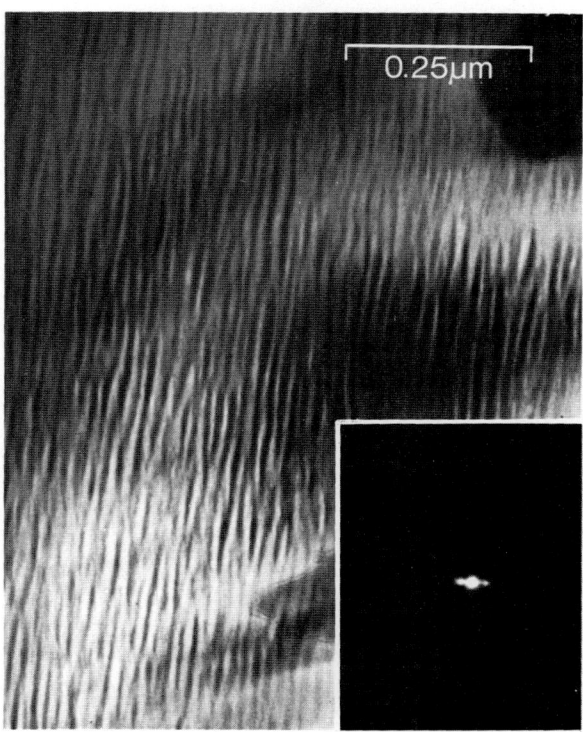

Fig. 13. Modulated structure consistent with spinodal decomposition in an alkali feldspar. The microstructure was produced by heating previously homogenized sample of composition 36 mole-% Or hydrothermally at 540 °C for 48 hrs at 1 kbar. Inset is an enlargement of a diffraction spot which shows satellites. (From Owen and McConnell, 1974)

nell they found that the early stages of the reaction were characterized by modulated structures which produced satellites in the diffraction pattern, but, after the feldspar had been annealed for several days at 600 °C, they found evidence for two phases from splitting of the Bragg reflections. Yund et al. plotted the lamellar spacing against (annealing time)$^{1/3}$ at 600 °C and 700 °C and obtained a straight-line relationship, thus showing that the coarsening was diffusion-controlled. The graph showed a non-zero value of the wavelength at zero time which the authors interpreted as evidence for spinodal decomposition.

Owen and McConnell (1974) also claim to have proved the operation of the spinodal mechanism from heating experiments in air at 525 °C. Their plot of wavelength against time shows an initial transient section followed by a curve which is approximately parallel to the time axis. This, the authors say, shows that one wavelength is specific to the annealing temperature. The kinetic results presented in the two papers are not *proof* of spinodal decomposition. Yund et al. produce no evidence of the kinetics of the early stages of the reaction. An apparent non-zero intercept on the wavelength axis could result from a change in reaction rate. The results of Owen and McConnell indicate that there *is* a two-stage process and the apparently constant wavelength may represent the initial stage of a coarsening reaction. The results of Yund et al. indicate that the change in wavelength with time is slow at 600 °C; the experiments of Owen and McConnell were carried out at 525 °C.

We feel that the strongest evidence for spinodal decomposition in the alkali feldspars is the morphology of the decomposition product and the presence of satellites associated with an (apparently) single reciprocal lattice at the early stages of the reaction. That spinodal decomposition is possible in this system is indicated by Robin's (1974) calculation which shows that the coherent spinodal is depressed by only 70–85° below the equilibrium solvus.

Recent, detailed studies of moonstones by transmission electron-microscopy have revealed complex microstructures. W.L. Brown et al. (1972) found that an Na-rich alkali feldspar in which the K-phase was triclinic and entirely in the diagonal association (Smith and MacKenzie, 1959) contained continuous, zig-zag or lozenge-shaped lamellae with interfaces near to $(\bar{6}61)$ and $(\bar{6}6\bar{1})$. In a later paper Willaime et al. (1973) described another specimen in which the morphology of the Na-phase varied from "zig-zags" to wavy or straight $(\bar{6}01)$ lamellae (see also Chapter 4.9 by Willaime et al. in this volume).

Lorimer and Champness (1973b) described two alkali feldspar specimens with similar compositions. One (Spencer M) contained "wavy" lamellae of Na-feldspar approximately parallel to $(\bar{6}01)$ together with an apparently monoclinic K-phase (Fig. 14a) and the other (Spencer N), whose microstructure was considerably coarser, contained discrete particles of Na-feldspar in the form of rhombic prisms with boundaries close to $(\bar{6}61)$ and $(\bar{6}6\bar{1})$ (Fig. 14b). The Na-rich particles were found to be aligned in two directions, approximately $(\bar{8}61)$ and $(\bar{8}6\bar{1})$, one lineation being more prominent than the other. They concluded that the phase distribution in Spencer N developed in the following way: initial exsolution occurred (probably by spinodal decomposition) in an initially monoclinic phase, producing lamellae parallel to $(\bar{6}01)$. The structure then coarsened extensively (Section 3.3) and the phases adopted interfaces of low energy parallel to $(\bar{6}61)$

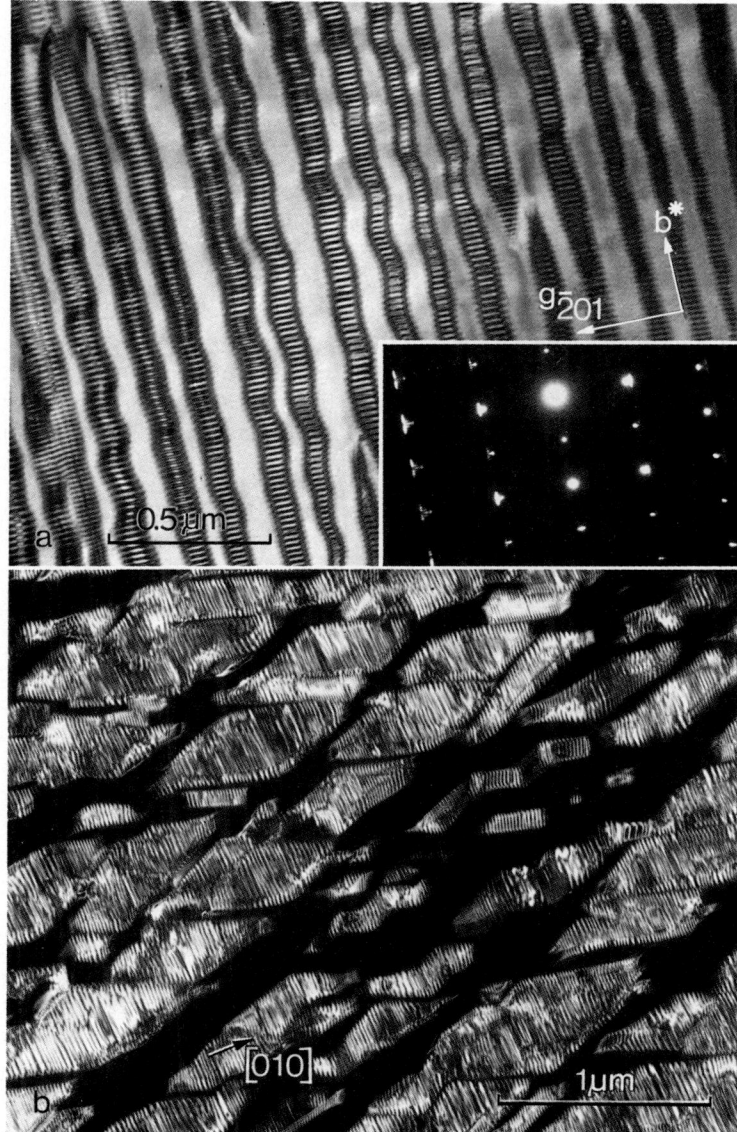

Fig. 14. (a) Microstructure of an alkali feldspar of composition 57.3 wt.-% Or (Spencer M). The Na-phase is albite-twinned. (001) Foil, 1 000 kV. Inset is the diffraction pattern in the correct orientation. (b) Microstructure of an alkali feldspar of composition 53.7 wt.-% Or (Spencer N) showing discrete particles and zig-zag lamellae of Na-feldspar in a K-feldspar matrix. (001) Foil, 1 000 kV

and ($\overline{6}6\overline{1}$), initially by forming zig-zag lamellae, but later by forming discrete rhombic prisms of Na-feldspar (to decrease the surface area). Finally "rafting" of the particles took place due to the interaction of their strain fields and produced the two lineations observed. Coarsening in Spencer M has not proceeded as far as in Spencer N.

The change in the orientation of the interphase boundary from (601) to ($\overline{6}6\overline{1}$) probably occurs as a result of the transformation of the K-feldspar to triclinic symmetry. W.L. Brown and Willaime (1974) calculated the elastic energy for an intergrowth of two coherent alkali-feldspars as a function of the boundary orientation. They showed that, where both phases are monoclinic or where the Na-feldspar is triclinic and intimately twinned, but the K-feldspar is monoclinic, a minimum occurs near ($\overline{6}01$) in agreement with Bollmann and Nissen's (1968) calculation using the coincident lattice (0-lattice) theory. On the other hand, for two triclinic phases they found that the minimum elastic energy occurs for a ($\overline{6}6\overline{1}$) interface. Neither Brown and Willaime's nor Bollmann and Nissen's method of calculation takes into account the surface energy component of the interface-boundary energy. However, as was pointed out in Section 4.1, this may be unimportant for fairly coarse intergrowths in silicates if the structures of the two phases are identical or nearly so.

6.2 Plagioclase Feldspars

Phase relationships in the plagioclases are not well understood. At high temperatures complete solid solution exists between the end-members albite and anorthite, but the presence of two-phase structures in some specimens is now well-established. Lamellar two-phase or modulated structures occur in three composition regions: An_{2-25} (peristerites), An_{40-60} (labradorites) and An_{65-90} (bytownites).

6.2.1 Peristerites

Laves (1954) showed by X-ray diffraction that two phases were present in plagioclases of compositions An_{2-15} which exhibit optical iridescence (schiller). Fleet and Ribbe (1965) observed the two phases in the transmission electron-microscope and confirmed the (approximately) ($0\overline{4}1$)[2, 3] orientation of the lamellae predicted from optical measurements by Bøggild (1924) (Fig. 15a). Bown and Gay (1958) have shown by single-crystal X-ray diffraction that, at least in some cases, the two phases are albite and an intermediate plagioclase with e and f satellites. In other specimens, both phases appear to have the albite structure (S. Carter, personal communication). Measurements of the cell dimensions of peristerites have indicated that the compositions are approximately An_0 and An_{25} (Laves, 1954; Gay and Smith, 1955; W.L. Brown, 1960) and qualitative chemical analysis of a peristerite with red schiller by analytical electron-microscopy has confirmed this (Cliff et al., Chapter 4.10 of this volume).

---

[2] We will index all planes and directions in the plagioclases using the anorthite ($c=14$Å) unit cell.
[3] Olsen (1974) has also found that a second set of lamellae parallel to approximately ($\overline{7}12$) sometimes occurs.

Plagioclases in the composition range An_{15}–An_{25} have been found to exhibit a "tweed" microstructure on a scale of about 100Å (Korekawa et al., 1970; McLaren, 1974); the major and minor modulations are approximately parallel to $(0\bar{4}1)$ and (100) respectively. Korekawa et al. (1970), using X-ray diffraction to investigate the nature of the structures, found that only one reciprocal lattice was present, but that each reflection was convoluted with a pair of satellites normal to $(0\bar{4}1)$. Detailed analysis of the distribution of satellite intensity in reciprocal space showed that the major modulation could be described as a wave function which was of mainly transverse character (with amplitude parallel to a), but with a small longitudinal component (with amplitude parallel to b). The modulation was caused mainly by a periodic variation in lattice parameters (mainly the angles α and γ, which impart the transverse nature, and b). Korekawa et al. did not investigate the nature of the minor (100) modulation as no X-ray scattering could be detected from it. McLaren (1974), using quantitative electron microscopy (the real space equivalent of the technique used by Korekawa et al.) confirmed that the $(0\bar{4}1)$ modulation is predominantly transverse in nature and showed that the (100) modulation was invisible in $(h0\bar{h})$ reflections, but visible in $(0k0)$ reflections.

Lorimer et al. (1974) and McLaren (1974) have described specimens of compositions An_{12} and $An_{2.6}$ respectively, which contain only a single $(0\bar{4}1)$ modulation and a periodicity of about 200Å. Diffraction patterns from the former specimen indicate that two phases with similar cell dimensions are present. It is probable that the two samples represent a later stage in the exsolution process than those containing "tweed" structures and that the peristerites resemble the pyroxenes in producing two initial spinodal modulations, only one of which coarsens.

The absence of coarse exsolution structures in plagioclases in the composition range An_{15-25} has puzzled mineralogists for some time. The discovery of "tweed" textures in some of these specimens suggests that they have undergone exsolution, but the reaction has not progressed beyond a very early stage. There are two possible reasons for this; firstly the solvus may be asymmetric with the temperature maximum occurring near An_5 (as suggested by the work of Crawford, 1966, on the compositions of co-existing metamorphic plagioclases) so that, for plagioclases with an An content greater 15 mole-%, the solvus is reached at a temperature where diffusion is very sluggish and the spinodal exsolution-structure cannot coarsen. A variant on this suggestion is the possibility that a binary loop rather than a solvus occurs at the albite end of the plagioclase phase diagram (Smith, 1972; Orville, 1974). This would mean that the spinodal textures which develop are metastable. The second possible cause of the failure of the tweed textures to coarsen in plagioclases of composition An_{15-25} is the presence of the superlattice typical of intermediate plagioclase in one of the two phases. Once the superlattice has formed it will be energetically very difficult for the phases to coarsen, as alteration of the ordering sequence would be required.

6.2.2 Labradorite

Examination in the electron microscope of labradorites which show intense optical iridescence reveals a periodic, lamellar structure whose orientation and spacing

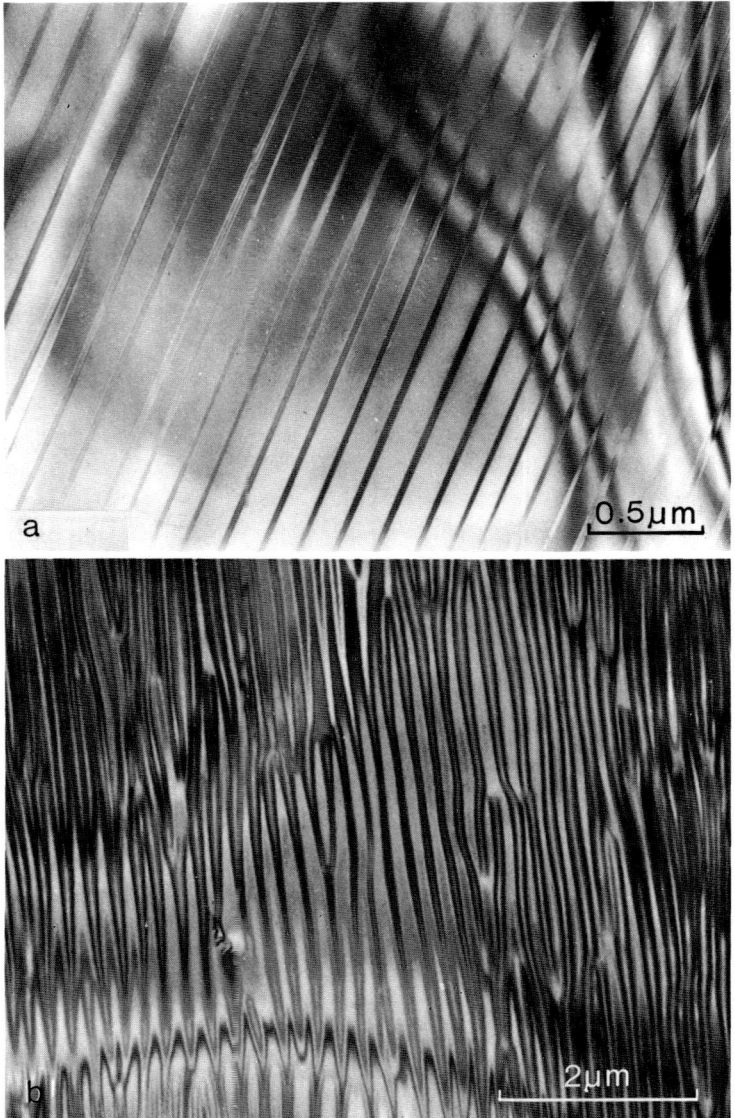

Fig. 15. (a) Microstructure of a red peristerite from Hybla, Ontario showing lamellae approximately parallel to (0$\bar{4}$1). The lamellae have a sharp interface with the matrix. This sample has been analyzed by analytical electron microscopy (Cliff et al., Chapter 4.10, this volume). 1000 kV. (b) Microstructure of a labradorite from Labrador showing lamellae with a diffuse interface with the matrix (courtesy H.-U. Nissen). (c) Microstructure of bytownite, An_{82}, from the Stillwater complex showing modulations parallel to (0$\bar{3}$1) and ($\bar{1}$01). Inset is a type (b) reflection with curved type (e) satellites. (From Heuer et al., 1972 and Nord et al., 1974.) Compare the scale of this microstructure with that of Cliff et al., Fig. 1a)

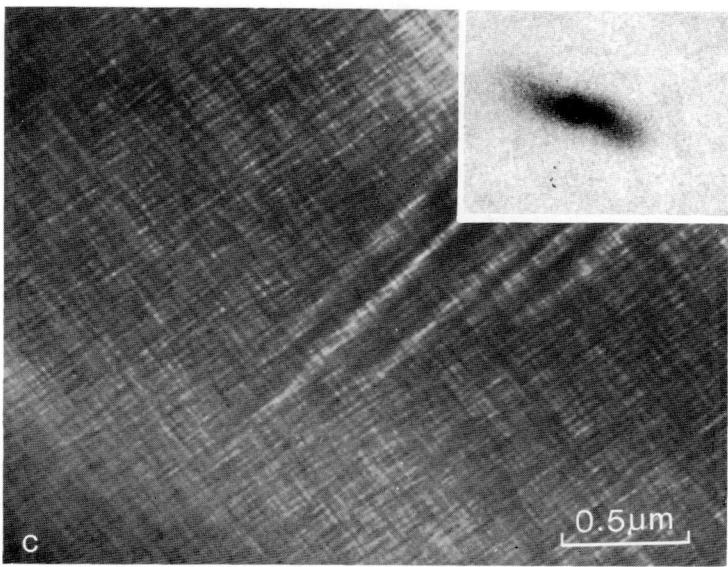

Fig. 15c

is consistent with the optical phenomenon (Baier and Pense, 1957; Laves et al., 1965; Bolton et al., 1966). The lamellae have a diffuse interface and a morphology which is consistent with their having formed by spinodal decomposition without a significant amount of coarsening (Fig. 15b). Although an exsolution origin was first suggested by Laves in 1960, X-ray diffraction has failed to confirm the presence of two phases. However, electron diffraction patterns show split Kikuchi lines (Nissen and Bollmann, 1968) which can be attributed to two phases of very similar lattice parameters.

Direct evidence of a difference in composition between the lamellae has been provided by Nissen et al. (1973) and Cliff et al. (Chapter 4.10 of this volume) who analyzed a labradorite of composition $An_{53.8}$ directly in the electron-microscope. They concluded that the lamellae differed in An content by at least 12%[4].

Korekawa and Jagodzinski (1967), using high-resolution X-ray diffraction, found that the lamellar structure in schiller labradorites produced a pair of closely-spaced "super satellites" about each of the type (a) ($h+k=2n$, $l=2n$) reflections. The nature of the lamellar modulation (longitudinal or transverse) could not be established with certainty, but the absence of satellites about the strong reflections 040 and 060 suggested that it is predominantly transverse [the lamellae are approximately parallel to (010)]. McLaren (1974) also concluded from contrast experiments performed in the electron microscope that the lattice

---

[4] Nissen et al. (1967) pointed out that all schiller labradorites contain at least 1.6 mole-% K-feldspar. As chemical analysis has shown that the difference in composition between the lamellae is predominantly one of anorthite content, we conclude that the potassium may act as a "catalyst" for the exsolution reaction, but does not necessarily segregate to one of the phases.

perturbation had a large component of the propagation vector in the plane of the lamellae. This is in agreement with Nissen and Bollmann's (1968) analysis of split Kikuchi lines which suggested that the lamellae were related by a shear. The lack of a correlation between the satellite intensity and the Bragg angle as reported by Korekawa and Jagodzinski suggests that the modulation is produced by a variation in the scattering as a result of segregation of Ca and Na between the lamellae.

McConnell (1974a), McLaren (1974), McLaren and Marshall (1974) and Hashimoto *et al.* (Chapter 5.6 of this volume) have resolved the (e) superlattice in labradorites. McConnell found that, in a specimen of composition $An_{54.4}$, the superlattice fringes were present in both lamellae, and their spacing and orientation appeared to vary continuously on passing from one to the other. Hashimoto *et al.* found that in a specimen of composition $An_{53.8}$ the superlattice was also present throughout the sample, but its orientation and spacing did not differ significantly in the two types of lamellae. In the sample examined by McLaren and Marshall (composition $An_{52.4}$) the superlattice was only present in the thicker lamellae. Exsolution is apparently initiated in labradorite when it has the $C\bar{1}$ structure and the more calcic of the two lamellae "orders" first (as concluded by Gay and Muir, 1962). In some cases the sodic lamellae retain the high-temperature structure, while in others they become ordered at a lower temperature. The behavior of a particular sample will depend upon its initial composition and cooling history.

6.2.3 Bytownites

Huttenlocher (1942) first attributed optically-visible $(0\bar{3}1)$ lamellae in a bytownite of composition 75% An to exsolution, but was unable to prove the existence of two phases by X-ray diffraction (Jaeger and Huttenlocher, 1955). Nissen (1968) found both (b) and (e) reflections in precession photographs of the same material and concluded that two phases were present, one with the intermediate-plagioclase structure and the other with the anorthite structure. He subsequently confirmed this by selected-area electron diffraction and by resolution of the (e) superlattice in one of the two phases (Nissen, 1974).

Grove (Chapter 4.11 of this volume) reports exsolution lamellae of transitional anorthite parallel to $(\bar{2}01)$ or $(\bar{3}01)$ as well as $(0\bar{3}1)$ in a sillimanite-grade metamorphosed calc-silicate rock from Western Massachusetts. The $(0\bar{3}1)$ lamellae have a lenticular shape, in contrast to Nissen's samples where the interface is strictly planar. Grove also reports secondary exsolution of intermediate plagioclase parallel to $(0\bar{3}1)$ inside the $(\bar{3}01)$ lamellae (but not within the primary $(0\bar{3}1)$ lamellae). This suggests that the $(\bar{3}01)$ lamellae formed first and that at a lower temperature the exsolution plane changed to $(0\bar{3}1)$. The change in the relative interfacial energies of the two planes may have coincided with the change in structure of the matrix phase from primitive anorthite to intermediate plagioclase. Cliff *et al.* (Chapter 4.10 of this volume) also report secondary exsolution inside the calcic lamellae of a bytownite. Both the primary and secondary lamellae are parallel to $(0\bar{3}1)$.

Direct evidence of a difference in anorthite content between the two phases was provided by qualitative scans for Ca and Na in the electron microprobe analyzer (Nissen, 1968, 1974). Cliff *et al.* (Chapter 4.10 of this volume), using analytical electron-microscopy, have shown that the compositions of the two phases in an An_{73} bytownite are $An_{69\pm4}$ and $An_{88\pm4}$.

The coarse exsolution textures described above are rare and confined to high-grade metamorphic rocks. Much finer structures have been found in igneous plagioclases with compositions ranging from An_{74} to An_{94} (Heuer *et al.*, 1972; Wenk *et al.*, 1972; Nord *et al.*, 1974; McLaren, 1974; McConnell, 1974b). These are "tweed" textures similar to those found in peristerites. The more prominent of the two modulations is parallel to $(0\bar{3}1)$ and the second is parallel to $(\bar{1}01)$ (Fig. 15c).

Nord *et al.* (1974) found that in the diffraction pattern from a Stillwater bytownite of composition An_{82} both (a) and (b) reflections had curved satellites (Fig. 15c). The satellites indicate a continuous chemical and crystallographic transition between two "phases" and support the suggestion that exsolution occurred by spinodal decomposition (Nord *et al.*, 1974).

Contrast experiments by Heuer *et al.* (1972), Wenk *et al.* (1973), Nord *et al.* (1973) showed that the $(0\bar{3}1)$ modulation is out of contrast for $1\bar{3}0$ and 020 reflections and the $(\bar{1}01)$ modulation is out of contrast for $\bar{1}01$ and $\bar{2}02$ reflections, suggesting that both modulations are primarily transverse in character. This is consistent with McLaren's (1974) conclusions that the $(0\bar{3}1)$ modulation is the result of the rotation of the lattice by about 3.5 minutes of arc. Consideration of the cell dimensions of plagioclases in the bytownite range shows that those which vary most are c, γ, α and β (Bambauer *et al.*, 1967). On this basis (assuming that variation of scattering power is unimportant in comparison with the cell parameters) a fluctuation of anorthite content parallel to $(0\bar{3}1)$ would be transverse and a fluctuation of anorthite content parallel to $(\bar{1}01)$ would have a large component of transverse character.

Although exsolution apparently takes place very sluggishly in the bytownites McConnell (1974b) and Nord *et al.* (1974) were able to homogenize tweed structures in the laboratory. McConnell heated an An_{76} sample hydrothermally at pressures of 400 and 1000 atmospheres and found that the crystals were homogenized above 960 °C, while Nord *et al.* heated an An_{82} sample in air at 1 atmosphere and found that it homogenized at 1225 °C. However Grove's (Chapter 4.11 of this volume) observation of coarse exsolution in a metamorphosed calc-silicate rock suggests that the solvus is below 700–750 °C. His results also suggest that the composition of the crest of the solvus lies between An_{72} and An_{75}.

6.2.4 The Plagioclase Phase Diagram

The details of the phase diagram for the plagioclases are still a matter of speculation. The evidence published until recently suggested that there are three miscibility gaps or solvi corresponding to the peristerites, labradorites and bytownites. However, Voll (1971, 1972) has reported (although not described in detail) unmixing of a metamorphic plagioclase of composition An_{50-55} into two plagioclases of composition An_{18} and An_{93}. The plagioclase grains exsolved by cellular (discontinuous) precipitation at grain boundaries. Voll estimated that the two-plagioclase assemblage represents a temperature below about 500 °C. It is therefore possible that, below about 500 °C, a solvus stretches from approximately An_{18} to An_{93}.

There are numerous examples of coexisting primary An_{20-30} grains in regionally-metamorphosed rocks (e.g. Evans, 1964; Crawford, 1966) which provide convincing evidence for a solvus corresponding to the "peristerite gap". However, petrographic evidence for two other distinct miscibility gaps is less substantial. The lamellae in labradorites clearly have not coarsened significantly and their compositions probably do not represent stable, equilibrium phases. It is possible that there is one solvus from about An_{20} to An_{90} but coarsening of spinodal structures in this composition range is inhibited by the slow diffusion of Al and Si and by the development of the intermediate-structure superlattice.

7. Summary and Discussion

Microstructures consistent with spinodal decomposition have been observed in clinopyroxenes, orthoamphiboles, alkali feldspars and plagioclase feldspars. Modulations in two directions ("tweed" structures) occur in the pyroxenes, peristerites and bytownites[5] and in all cases are precursors of coarser structures with only one modulation. The presence of two primary modulations in directions which are crystallographically non-equivalent is unexpected, but indicates that the relevant elastic constants are very similar in the high-temperature single phase. It is possible that, when the composition difference between the two incipient phases becomes significant and a distinct interface begins to develop, the anisotropy of the surface energy promotes preferential coarsening of one set of lamellae.

The reasons for the common occurrence of spinodal decomposition in silicates are the similarity of the structures and molar volumes of the exsolved phases and the absence of a high-density of suitable nucleation sites – dislocations and grain boundaries (Aaronson et al., 1974). In all cases, except the plagioclase feldspars (and only rarely in the alkali feldspars), nucleation and growth reactions occur at slow cooling rates (Fig. 8). Nucleation of the equilibrium phase invariably takes place at grain boundaries and growth produces a lamellar morphology (though in alkali feldspars the reaction is often cellular).

In the only silicate system in which spinodal decomposition cannot occur because of the difference in crystal structure of the parent and product phase, the orthopyroxenes, both equilibrium- and transitional-nucleated phases have been observed. The two intermediate phases, one of which nucleates on dislocations while the other nucleates homogeneously in the matrix, have structures which are more similar to the matrix phase than does the equilibrium phase. Similar phase distributions, involving heterogeneously-nucleated lamellae with homogeneously-nucleated platelets between them, have been found in clino- and orthoamphiboles, although the exact nature of the platelets is not known.

---

[5] Both Christie and Olsen (1974) and Slimming (Proc. EMAG-75, Institute of Physics, London, in press 1976) have reported modulations in three directions in labradorites, although the former authors describe two of them as "distortion waves".

It is interesting to reflect that the orthopyroxenes and plagioclases in the Bushveld and Stillwater complexes, two of the largest intrusions known, contain phase distributions which are far from equilibrium. In the pyroxenes the precipitation of intermediate phases reflects the very slow rate of calcium diffusion (the rate-controlling step in the exsolution of augite), while in the plagioclases the failure of the "tweed" structure to coarsen is a result of the even slower rate of Si and Al diffusion and the difficulty of altering the ordering sequence in the intermediate structure.

References

Aaronson, H.I., Laird, C., Kinsman, K.R.: Mechanisms of diffusional growth of precipitate crystals. In: Phase transformations. A.S.M., 313–390 (1970).
Aaronson, H.I., Lorimer, G.W., Champness, P.E., Spooner, E.T.C.: On differences between phase transformations (exsolution) in metals and silicates. Chem. Geol. **14**, 75–80 (1974).
Ardell, A.J., Nicholson, R.B., Eshelby, J.D.: On the modulated structure of aged Ni-Al alloys. Acta Met. **14**, 1295–1309 (1966).
Atlas, L.: The polymorphism of $MgSiO_3$ and solid-state equilibria in the system $MgSiO_3 - CaMgSi_2O_6$. J. Geol. **60**, 125–147 (1952).
Baier, E., Pense, J.: Elektronenmikroskopische Untersuchungen an Labradoren. Naturwissenschaften **44**, 110–111 (1957).
Bambauer, H.U., Eberhard, E., Viswanathan, K.: The lattice constants and related parameters of "plagioclases (low)". Schweiz. Mineral. Petrog. Mitt. **47**, 351–364 (1967).
Bøggild, O.B.: On the labradorization of the feldspars. Koninkl. Danske Vidensk. Selsk. math.-fys. Medd. **6**, 1–79 (1924).
Bollmann, W., Nissen, H.-U.: A study of optimal phase boundaries: the case of exsolved alkali feldspars. Acta Cryst. A **24**, 546–557 (1968).
Bolton, H.C., Bursill, L.A., McLaren, A.C., Turner, R.G.: On the origin of the colour of labradorite. Phys. Stat. Sol. **18**, 221–230 (1966).
Bowen, N.L., Tuttle, O.F.: The system $NaAlSi_3O_8 - KAlSi_3O_8 - H_2O$. J. Geol. **58**, 489–511 (1950).
Bown, M.G., Gay, P.: The reciprocal lattice geometry of the plagioclase feldspar structures. Z. Krist. **111**, 1–14 (1958).
Bown, M.G., Gay, P.: An X-ray study of exsolution phenomena in the Skaergaard pyroxenes. Mineral. Mag. **32**, 379–388 (1960).
Boyd, F.R., Schairer, J.F.: The system $MgSiO_3 - CaMgSi_2O_6$. J. Petrol. **5**, 275–309 (1964).
Brown, G.E., Prewitt, C.T., Papike, J.J., Sueno, S.: A comparison of the structures of low and high pigeonite. J. Geophys. Res. **77**, 5778–5789 (1972).
Brown, W.L.: X-ray studies in the plagioclases. Part 2. The crystallographic and petrologic significance of peristerite unmixing in the acid plagioclases. Z. Krist. **113**, 297–344 (1960).
Brown, W.L., Willaime, C.: An explanation of exsolution orientations and residual strain in cryptoperthites. In: The feldspars (eds. W.S. MacKenzie and J. Zussman) Proc. NATO Adv. Study Inst. p. 440–459. Manchester: University Press 1974.
Brown, W.L., Willaime, C., Guillemin, C.: Exsolution selon l'association diagonale dans une cryptoperthite: étude par microscopie électronique et diffraction des rayons X. Bull. Soc. Franç. Minéral. Crist. **95**, 429–436 (1972).
Cahn, J.W.: On spinodal decomposition. Acta Met. **9**, 795–801 (1961).
Cahn, J.W.: On spinodal decomposition in cubic crystals. Acta Met. **10**, 179–184 (1962).
Cahn, J.W.: Spinodal decomposition. Trans. AIME **242**, 166–180 (1968).
Carstens, H.: Exsolution in ternary feldspars. I. On the formation of antiperthites. Contrib. Mineral. Petrol. **14**, 27–35 (1967).
Chadwick, G.A.: Metallography of phase transformations. London: Butterworths 1972.
Champness, P.E., Lorimer, G.W.: An electron microscopic study of a lunar pyroxene. Contrib. Mineral. Petrol. **33**, 171–183 (1971).

Champness, P.E., Lorimer, G.W.: Electron microscopic studies of some lunar and terrestrial pyroxenes. Proc. 5th Intl. Materials Symp., p. 1245–1255. Berkeley: University of California Press 1972.

Champness, P.E., Lorimer, G.W.: Precipitation (exsolution) in an orthopyroxene. J. Mat. Sci. **8**, 467–474 (1973).

Champness, P.E., Lorimer, G.W.: A direct lattice-resolution study of precipitation (exsolution) in orthopyroxene. Phil. Mag. **30**, 357–365 (1974).

Christian, J.W.: The theory of transformations in metals and alloys. Oxford: Pergamon Press 1965.

Christie, J.M., Lally, J.S., Heuer, A.H., Fisher, R.M., Griggs, D.T., Radcliffe, S.V.: Comparative electron petrography of Apollo 11, Apollo 12 and terrestrial rocks. Proc. 2nd Lunar Sci. Conf. Geochim. Cosmochim. Acta, Suppl. 2, **1**, 69–89 (1971).

Christie, O.H.J., Olsen, A.: Spinodal precipitation in minerals. Review and some new observations. Bull. Soc. Franç. Mineral. Crist. **97**, 386–392 (1974).

Copley, P.A.: Electron microscope studies of phase transformations in pyroxene minerals. Ph.D. Thesis, Manchester University (1973).

Copley, P.A., Champness, P.E., Lorimer, G.W.: Electron petrography of exsolution textures in an iron-rich clinopyroxene. J. Petrol. **15**, 41–57 (1974).

Crawford, M.L.: Composition of plagioclase and associated minerals in some schists from Vermont, U.S.A. and South Westland, New Zealand with interferences about the peristerite solvus. Contrib. Mineral. Petrol. **13**, 269–294 (1966).

Dunham, A.C., Copley, P.A., Strasser-King, V.H.: Submicroscopic exsolution lamellae in pyroxenes in the Whin Sill, Northern England. Contrib. Mineral. Petrol. **37**, 211–220 (1972).

Evans, B.W.: Coexisting albite and oligoclase in some schists from New Zealand. Am. Mineralogist **49**, 173–179 (1964).

Fleet, S.G., Ribbe, P.H: An electron microscope investigation of a moonstone. Phil. Mag. **8**, 1179–1187 (1963).

Fleet, S.G., Ribbe, P.H.: An electron-microscope study of peristerite plagioclases. Mineral. Mag. **35**, 165–176 (1965).

Gay, P., Muir, I.D.: Investigation of the feldspars of the Skaergaard intrusion, eastern Greenland. J. Geol. **70**, 565–581 (1962).

Gay, P., Smith, J.V.: Phase relations in the plagioclase feldspars: composition range An_0 to An_{70}. Acta Cryst. **8**, 64–65 (1955).

Gittos, M.F., Lorimer G.W., Champness, P.E.: An electron microscopic study of precipitation (exsolution) in an amphibole (the hornblende-grunerite system). J. Mat. Sci. **9**, 184–192 (1974).

Goldsmith, J.R., Newton, R.C.: An experimental determination of the alkali feldspar solvus. In: The feldspars (eds. W.S. MacKenzie and J. Zussman). Proc. NATO Adv. Study Inst. p. 297–312. Manchester: University Press 1974.

Greenwood, G.W.: Particle coarsening. In: The mechanism of phase transformations in crystalline solids. Institute of Metals, 103–110 (1969).

Guinier, A.: Structure of age-hardened aluminium-copper alloys. Nature **142**, 569–570 (1938).

Hess, H.H.: Pyroxenes of common mafic magmas. Am. Mineralogist **26**, 515–535, 573–594 (1941).

Heuer, A.H., Lally, J.S., Christie, J.M., Radcliffe, S.V.: Phase transformation and exsolution in lunar and terrestrial calcic plagioclases. Phil. Mag. **26**, 465–482 (1972).

Hilliard, J.E.: Spinodal decomposition. In: Phase transformations. A.S.M., 497–560 (1970).

Huttenlocher, H.: Beiträge zur Petrographie des Gesteinzuges Ivrea Verbano. I. Allgemeines. Die gabbroïden Gesteine von Anzola. Schweiz. Mineral. Petrog. Mitt. **22**, 326–366 (1942).

Jaeger, E., Huttenlocher, H.: Beobachtungen an basischen Plagioklasen der Ivrea-Zone. Schweiz. Mineral. Petrog. Mitt. **35**, 199–203 (1955).

Korekawa, M., Jagodzinski, H.: Die Satellitenreflexe des Labradorites. Schweiz. Mineral. Petrogr. Mitt. **47**, 269–278 (1967).

Korekawa, M., Nissen, H.-U., Philipp, D.: X-ray and electron microscopic studies of a sodium-rich low plagioclase. Z. Krist. **131**, 418–436 (1970).

Laird, C., Aaronson, H.I.: The growth of γ plates in an Al-15% Ag alloy. Acta Met. **17**, 505–519 (1969).

Lally, J.S., Fisher, R.M., Christie, J.M., Griggs, D.T., Heuer, A.H., Nord, G.L., Radcliffe, S.V.: Electron petrography of Apollo 14 and 15 rocks. Proc. 3rd Lunar Sci. Conf. Geochim. Cosmochim. Acta Suppl. **1**, 401–422 (1972).
Laves, F.: The phase relations of the alkali feldspars. II J. Geol. **60**, 549–574 (1952).
Laves, F.: The co-existence of two plagioclases in the oligoclase composition range. J. Geol. **62**, 409–411 (1954).
Laves, F.: Die Feldspäte, ihre polysynthetischen Verzwilligungen und Phasenbeziehungen. Rend. Soc. Miner. Ital. **16**, 37–68 (1960).
Laves, F., Nissen, H.-U., Bollmann, W.: On schiller and submicroscopic lamellae of labradorite, (Na, Ca) $(Si,Al)_4O_8$. Naturwissenschaften **52**, 427–428 (1965).
Lin, T.H., Yund, R.A.: Potassium and sodium self-diffusion in alkali feldspar. Contrib. Mineral. Petrol. **34**, 177–184 (1972).
Lorimer, G.W., Champness, P.E.: Combined electron microscopy and analysis of an orthopyroxene. Am. Mineralogist **58**, 243–248 (1973a).
Lorimer, G.W., Champness, P.E.: The origin of the phase distribution in two perthitic alkali feldspars. Phil. Mag. **28**, 1391–1403 (1973b).
Lorimer, G.W., Nissen, H.-U., Champness, P.E.: High voltage electron microscopy of a deformed plagioclase from an Alpine gneiss. Schweiz. Mineral. Petrog. Mitt. **54**, 707–715 (1974).
Luth, W.C., Tuttle, O.F.: The alkali feldspar solvus in the system $Na_2O-K_2O-Al_2O_3-SiO_2-H_2O$. Am. Mineralogist **51**, 1359–1373 (1966).
Luth, W.C., Martin, R.F., Fenn, P.M.: Peralkaline alkali solvi. In: The feldspars (eds. W.S. MacKenzie and J. Zussman). Proc. NATO Adv. Study Inst., p. 297–312. Manchester: University Press 1974.
MacKenzie, W.S.: The effect of temperature on the symmetry of high-temperature soda-rich feldspars. Am. J. Sci. Bowen Vol. **250**A, 319–342 (1952).
McConnell, J.D.C.: Electron optical study of incipient exsolution and inversion phenomena in the system $NaAlSi_3O_8-KAlSi_3O_8$. Phil. Mag. **19**, 221–229 (1969).
McConnell, J.D.C.: Electron-optical study of the fine structure of a schiller labradorite. In: The feldspars (eds. W.S. MacKenzie and J. Zussman). Proc. NATO Adv. Study Inst., p. 478–490. Manchester: University Press 1974a.
McConnell, J.D.C.: Analysis of the time-temperature-transformation behaviour of the plagioclase feldspars. In: The feldspars (eds. W.S. MacKenzie and J. Zussman). Proc. NATO Adv. Study Inst., p. 460–477. Manchester: University Press 1974b.
McLaren, A.C.: Transmission electron microsopy of the feldspars. In: The feldspars (eds. W.S. MacKenzie and J. Zussman). Proc. NATO Adv. Study Inst., p. 378–424. Manchester: University Press 1974.
McLaren, A.C., Marshall, D.B.: Transmission electron microscope study of the domain structures associated with the b-, c-, d-, e- and f-reflections in plagioclase feldspars. Contrib. Mineral. Petrol. **44**, 237–249 (1974).
Morimoto, M., Tokonami, M.: Oriented exsolution of augite in pigeonite. Am. Mineralogist **54**, 1101–1117 (1969).
Morse, S.A.: Alkali feldspars with water at 5 Kb pressure. J. Petrol. **11**, 221–251 (1970).
Nicholson, R.B.: Nucleation at imperfections. In: Phase transformations. A.S.M., 269–309 (1970).
Nissen, H.-U.: A study of bytownite plagioclases in amphibolites of the Ivreazone (Italian Alps) and in anorthosites: a new unmixing gap in the low plagioclases. Schweiz. Mineral. Petrog. Mitt. **48**, 53–55 (1968).
Nissen, H.-U.: Exsolution in bytownite plagioclases. In: The feldspars (eds. W.S. MacKenzie and J. Zussman). Proc. NATO Adv. Study. Inst., p. 491–521. Manchester: University Press 1974.
Nissen, H.-U., Bollmann, W.: Doubled kikuchi lines as a means of distinguishing quasi-identical phases. Proc. 4th Europ. Congress on Electron Microscopy, Rome, 321–322 (1968).
Nissen, H.-U., Champness, P.E., Cliff, G., Lorimer, G.W.: Chemical evidence for exsolution in a labradorite. Nature Phys. Sci. **245**, 135–137 (1973).
Nissen, H.-U., Eggmann, H., Laves, F.: Schiller and submicroscopic lamellae of labradorite. A preliminary report. Schweiz. Mineral. Petrog. Mitt. **47**, 289–302 (1967).

Nord, G.L., Heuer, A.H., Lally, J.S.: Transmission electron microscopy of substructures of Stillwater bytownites. In: The feldspars (eds. W.S. MacKenzie and J. Zussman). Proc. NATO Adv. Study Inst., p. 522–535. Manchester: University Press 1974.

Nord, G.L., Lally, J.S., Heuer, A.H., Christie, J.M., Radcliffe, S.V., Griggs, D.T., Fisher, R.M.: Petrologic study of igneous and metaigneous rocks from Apollo 15 and 16 using high voltage transmission electron microscopy. Proc. 4th Lunar Sci. Conf. Geochim. Cosmochim. Acta, Suppl. **4**, 1, 953–970 (1973).

Olsen, A.: Schiller effects and exsolution in sodium-rich plagioclases. Contrib. Mineral. Petrol. **47**, 141–152 (1974).

Orville, P.M.: Alkali ion exchange between vapour and feldspar phases. Am. J. Sci. **261**, 283–316 (1963).

Orville, P.M.: The "peristerite gap" as an equilibrium between ordered albite and disordered plagioclase solution. Bull. Soc. franç. Mineral. Crist. **97**, 202–205 (1974).

Owen, D.C., McConnell, J.D.C.: Spinodal behaviour in an alkali feldspar. Nature Phys. Sci. **230**, 118–119 (1971).

Owen, D.C., McConnel, J.D.C.: Spinodal unmixing in an alkali feldspar. In: The feldspars (eds. W.S. MacKenzie and J. Zussman). Proc. NATO Adv. Study Inst., p. 424–439. Manchester: University Press 1974.

Petrović, R.: Diffusion of alkali ions in alkali feldspars. In: The feldspars (eds. W.S. MacKenzie and J. Zussman) Proc. NATO Adv. Study Inst., p. 174–182. Manchester: University Press 1974.

Preston, G.D.: Structure of age-hardened aluminium-copper alloys. Nature **142**, 570 (1938).

Robin, Y.-P.F.: Stress and strain in cryptoperthite lamellae and the coherent solvus of alkali feldspars. Am. Mineralogist **59**, 1299–1318 (1974).

Robinson, P., Ross, M., Jaffe, H.W.: Composition of the anthophyllite-gedrite series, comparisons of gedrite and hornblende and the anthophyllite-gedrite solvus. Am. Mineralogist **56**, 1005–1041 (1971).

Ross, M., Huebner, J.S., Dowty, E.: Delineation of the one atmosphere augite-pigeonite miscibility gap for pyroxenes from lunar basalt 12021. Am. Mineralogist **58**, 619–635 (1973).

Ross, M., Papike, J.J., Shaw, K.W.: Exsolution textures in amphiboles as indicators of subsolidus thermal histories. Mineral. Soc. Am. Spec. Paper **2**, 275–299 (1969).

Smith, J.V.: Critical review of synthesis and occurrence of plagioclase feldspars and a possible phase diagram. J. Geol. **80**, 505–525 (1972).

Smith, J.V., MacKenzie, W.S.: The alkali feldspars. V. The nature of orthoclase and microcline perthites and observations concerning the polymorphism of potassium feldspar. Am. Mineralogist **44**, 1169–1186 (1959).

Smith, P., Parsons, I.: The alkali-feldspar solvus at 1 kilobar water-vapour pressure. Mineral. Mag. **39**, 747–767 (1974).

Steiger, R.H., Hart, S.R.: The microcline-orthoclase transition within a contact aureole. Am. Mineralogist **52**, 87–116 (1967).

Vander Sande, J.B., Kohlstedt, D.L.: A high-resolution electron microscopy study of exsolution lamellae in enstatite. Phil. Mag. **29**, 1041–1049 (1974).

Voll, G.P.L.E.: Entmischung von Plagioklasen mit An_{50-55} zu $An_{18} + An_{93}$ (Abs.) Fortschr. Mineral. **49**, 61–63 (1971).

Voll, G.P.L.E.: Unmixing of An_{50-55} to An_{17+93} and cellular growth of precipitates. (Abs.) NATO Adv. Study Inst. on Feldspars, 16–17 (1972).

Wenk, H.-R., Müller, W.F., Thomas, G.: Antiphase domains in lunar plagioclase. Proc. 4th Lunar Sci. Conf. Geochim. Cosmochim. Acta, Suppl. **4**, **1**, 909–923 (1973).

Wenk, H.-R., Ulbrich, M., Müller, W.F.: Lunar plagioclase: a mineralogical study. Proc. 3rd Lunar Sci. Conf. Geochim. Cosmochim. Acta, Suppl. **3**, **1**, 569–579 (1972).

Willaime, C., Brown W.L., Gandais, M.: An electron-microscopic and X-ray study of complex exsolution textures in a cryptoperthitic alkali feldspar. J. Mat. Sci. **8**, 461–466 (1973).

Willaime, C., Gandais, M.: Study of exsolution in alkali feldspars. Calculation of elastic stresses inducing periodic twins. Phys. Stat. Sol. **9**, 529–539 (1972).

Wright, T.L.: The microcline-orthoclase transformation in the contact aureole of the Eldora stock, Colorado. Am. Mineralogist **52**, 117–136 (1967).

Yund, R.A., McLaren, A.C., Hobbs, B.E.: Coarsening kinetics of the exsolution microstructure in alkali feldspar. Contrib. Mineral. Petrol. **48**, 45–55 (1974).

CHAPTER 4.2

Coarsening in a Spinodally Decomposing System: TiO$_2$–SnO$_2$

M. PARK, T.E. MITCHELL, and A.H. HEUER

1. Introduction

Solid solutions can decompose (i.e. exsolve) *via* two distinct mechanisms – spinodal decomposition or nucleation and growth – and there are several easily defined differences between them, particularly during the early stages of decomposition (Cahn, 1968). Of these differences, probably the most important is that there is no nucleation barrier for spinodal decomposition, the kinetics of decomposition being controlled entirely by diffusion; in contrast, the nucleation barrier in conventional nucleation and growth theories gives rise to the well-known idea of a critical nucleus. During spinodal decomposition, therefore, spontaneous composition fluctuations of various wave numbers form, and those with maximum amplification factors dominate the microstructural evolution. When the amplitude of the composition fluctuations approaches that given by the equilibrium phase diagram, a late stage of decomposition has been reached and further microstructural changes represent a coarsening phenomenon. At this stage of decomposition, there is little conceptual difference in the coarsening behavior between systems that initially decompose spinodally and those that decompose by nucleation and growth, in that coarsening in both is controlled by the diffusion kinetics.

The usual kinetic equation for describing coarsening is

$$\lambda^m \propto Kt$$

where λ represents the size of the average precipitate particle at time t and K and m are constants. In the present work, it will be shown that this equation adequately describes coarsening in TiO$_2$–SnO$_2$ "alloys" with $m=3$. Before describing our results, however, it will be useful to review briefly the relevant literature on coarsening in other systems.

Ardell and Nicholson (1966) were among the first to study coarsening using electron microscopy. They determined the growth kinetics of precipitates in a Ni-Al alloy which had formed *via* a nucleation and growth mechanism, and showed that the kinetic equation accurately described their data at temperatures between 625 and 775 °C with $m=3$, in agreement with the coarsening theories of Lifshitz and Slyozov (1961) and Wagner (1961). In systems that decompose spinodally, however, the earliest study was apparently that performed by Hillert *et al.* (1961) on a Cu-Ni-Fe alloy; they estimated the magnitude of composition

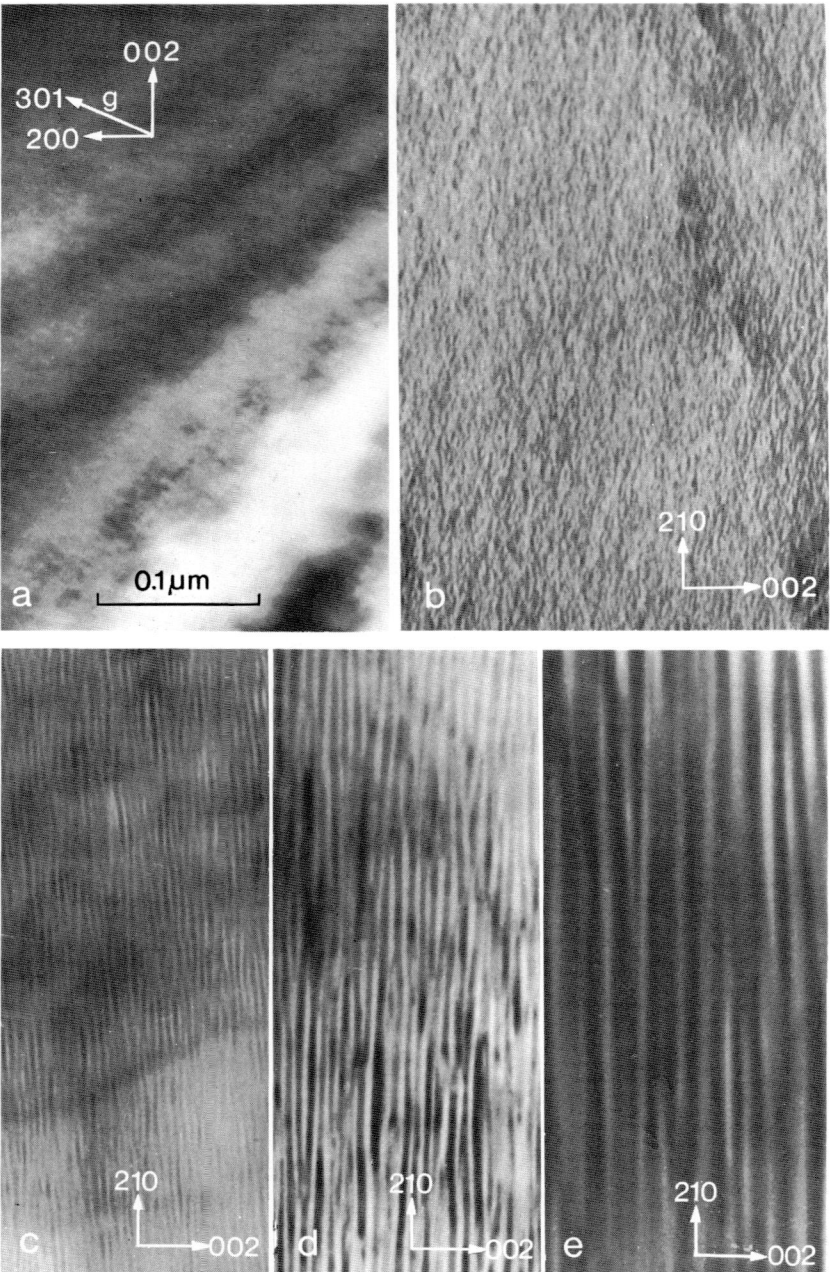

Fig. 1. (a) Single-phase equimolar TiO_2-SnO_2 alloy annealed at 1580 °C and quenched into water. (b) Equimolar TiO_2-SnO_2 specimen annealed at 1500 °C and air-quenched. The background structure is evidence that some decomposition has occurred. (c) Sample of Fig. 2 after annealing for 5 min at 1000 °C; (d) after annealing for 60 min at 1000 °C; (e) after annealing for 15 h at 1000 °C. All micrographs are same magnification

fluctuations from X-ray side band measurements using the equation of Daniel and Lipson (1943) and found $m=4.8$. Subsequently, however, Butler and Thomas (1970) used transmission electron microscopy to study this same alloy system and found $m=3$ (i.e. $t^{1/3}$ behavior); the zero time intercept gave $\lambda \sim 30\text{--}35$ Å, a reasonable estimate for the magnitude of the composition fluctuations in the as-quenched sample. Butler (1971) subsequently extended this study to the initial stages of spinodal decomposition and found departures from a $t^{1/3}$ behavior at the very early stages. Similar departures from a $t^{1/3}$ behavior during the early stages of decomposition were found by Cornie et al. (1973) on a Cu-Ti alloy, and in single phase TiO_2-SnO_2 "alloys" when decomposed at 900 °C (Park, 1974). If sufficient decomposition occurred during quenching of these TiO_2-SnO_2 alloys, however, $t^{1/3}$ growth kinetics were observed during further coarsening, as will now be discussed.

2. Results

Samples of equimolar TiO_2-SnO_2 alloys were hot-pressed at 6000 psi ($=0.41$ kbar) and at 1450 °C, which is just above the critical temperature of 1430 °C in this system (Park et al., 1975), using a graphite die with an alumina liner; they were then annealed for 24 h at 1500–1600 °C to improve homogeneity and quenched. Samples for electron microscopy were prepared by ion thinning and examined in a Hitachi HU-650B electron microscope operating at 650 kV. If quenching was rapid, as in the example of Fig. 1a, a good single phase alloy was obtained. However, Park (1974) has shown that decomposition occurs very rapidly at temperatures above 1000 °C in this system, and some as-quenched samples

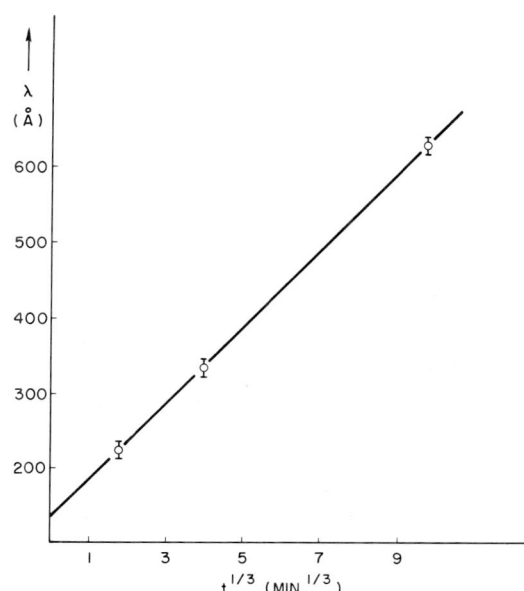

Fig. 2. Average lamellar spacing from Fig. 1c–e versus $t^{1/3}$

(Fig. 1b) show evidence of decomposition. (The elongation of diffraction spots in the electron diffraction pattern of the region shown in Fig. 1b gave further evidence that the particular quenching treatment used for Fig. 1b allowed some decomposition to occur.) Coarsening experiments on the sample used for Fig. 1b are shown in Fig. 1c–e; each sample was isothermally annealed for 5 min, 60 min and 15 h, respectively, at 1000 °C. The foils shown in Fig. 1b–e were of grains in the polycrystalline sample chosen such that $\langle 001 \rangle$ lay in the plane of the foil. Since decomposition in this system occurs *via* composition fluctuation perpendicular to $\langle 001 \rangle$[1], this allowed the lamellar spacing to be determined accurately. The average lamella spacing in these figures is plotted versus $t^{1/3}$ in Fig. 2; it is clear that coarsening in this system is described quite well by the kinetic equation with $m=3$. It is furthermore interesting to note that extrapolation of the curve to zero annealing time yields an estimate of the average lamellar spacing in the as-quenched sample of 135 ± 10 Å while direct measurement of Fig. 1b yields 150 ± 10 Å.

References

Ardell, J., Nicholson, R.B.: On the modulated structure of aged Ni-Al alloys. Acta Met. **14**, 1295–1309 (1966).
Butler, E.P.: The initial stages of spinodal decomposition in a Cu-Ni-Fe alloy studied by electron microscopy. Metal Sci. J. **5**, 8–10 (1971).
Butler, E.P., Thomas, G.: Structure and properties of spinodally decomposed Cu-Ni-Fe alloys. Acta Met. **18**, 347–365 (1970).
Cahn, J.W.: Spinodal decomposition. Trans. AIME **242**, 166–180 (1968).
Cornie, J.A., Datta, A., Soffa, W.A.: An electron microscopy study of precipitation in Cu-Ti sideband alloys. Met. Trans. **4**, 727–733 (1973).
Daniel, V., Lipson, H.: Proc. Roy. Soc. (London) **182**, 378 (1943).
Hillert, M., Cohen, M., Averbach, B.L.: Formation of modulated structures in copper-nickel-iron alloys. Acta Met. **9**, 536–546 (1961).
Lifshitz, I.M., Slyozov, V.V.: The kinetics of precipitation from supersaturated solid solutions. J. Phys. Chem. Solids **19**, 35–50 (1961).
Park, M.W.: Spinodal decomposition in the $TiO_2 - SnO_2$ system. Ph. D. Thesis, Case Western Reserve University (1974).
Park, M.W., Mitchell, T.E., Heuer, A.H.: Subsolidus phase equilibria in the $TiO_2 - SnO_2$ system. J. Am. Ceram. Soc. **58**, 44–47 (1975).
Schultz, A.H., Butler, W.R., Stubican, V.S.: Spinodal decomposition in the tetragonal system. Phys. Stat. Sol. **32**, K117–K119 (1969).
Schultz, A.H., Stubican, V.S.: Modulated structures in the system $TiO_2 - SnO_2$. Phil. Mag. **18**, 929–937 (1968).
Wagner, C.: Theory of the ageing of precipitates by redissolution (Ostwald Maturing). Z. Elektrochem. **65**, 581–591 (1961).
Wu, K., Mendelson, K.S.: Spinodal decomposition in the tetragonal system. J. Chem. Phys. **58**, 2929–2933 (1973).

---

[1] The lamellar nature of the decomposition in this system was first observed by Shultz and Stubican (1968) and was rationalized in terms of elastic anisotropy by Shultz *et al.* (1969). Furthermore Wu and Mendelson (1973) have pointed out that possible anisotropies in diffusion kinetics and in gradient energy can also contribute to the pronounced lamellar decomposition structure in this system.

CHAPTER 4.3

Magnetite Lamellae in Reduced Hematites

N.J. TIGHE and P.R. SWANN

1. Introduction

A lamellar texture in natural and synthetic oxidized magnetites and reduced hematites has been observed for many years. Hematite (Fe_2O_3, $R\bar{3}c$) and magnetite (Fe_3O_4, Fd3m) lamellae were found in sinters and burden pellets made by oxidizing magnetite ores at $T > 1000$ °C (Schwartz, 1929; Cooke and Ban, 1952; Tigerschiöld, 1954; Ponghis et al., 1967). Greig et al. (1935) and Gruner (1926) used light optical and X-ray techniques to identify the $\{111\}_m \| \{0001\}_h$ relationship in oxidized magnetites.

Brill-Edwards et al. (1965) showed that magnetite laths grew at acute angles to some hematite grain surfaces in polycrystalline specimens reduced at $T > 800$ °C and suggested the laths conformed to the $\{111\}_m \| \{0001\}_h$ relationship. This orientation relationship corresponds to the close-packed oxygen planes in the two structures; in hematite, oxygen stacking is hcp with ferric ions in 2/3 of the octahedral interstices; in magnetite, oxygen stacking is fcc with ferric and ferrous ions in some octahedral and tetrahedral interstices (Néel, 1949; Nicholls, 1955; Fasiska, 1967). Baker and Whelan (1968) reduced hematite in the vacuum of the electron microscope, identified the epitaxial relationship $\{111\}_m \| \{0001\}_h$ with $\pm \langle 1\bar{1}0 \rangle_m \| \langle 01\bar{1}0 \rangle_h$ and found planar faults in magnetite on $\{110\}$ and $\{111\}$. Reduction experiments in the high-voltage electron microscope (Swann, 1972; Tighe, 1973; Flower et al., 1974) showed that magnetite (1) nucleates preferentially at grain boundaries, cleavage steps and fracture surfaces, (2) grows radially into the hematite with no preferred orientation to form hemi-spherical colonies at 400–600 °C, and (3) grows with the $\{111\}_m \| \{0001\}_h$ habit to form faceted plates at $T > 700$ °C.

Hematite is thermodynamically stable as long as the oxygen partial pressure ($P(O_2)$) at the gas-solid interface is $\geq$ the dissociation pressure at that temperature. Thus phase equilibria are a function of $P(O_2)$ and precipitation and phase transformations in the iron oxides occur by removing oxygen from or adding oxygen to a given phase. The required $P(O_2)$ for reduction of hematite to magnetite is readily obtained with CO/CO_2 mixtures over the temperature range 400–1000 °C (Richardson and Jeffes, 1948).

2. Experimental Procedure

Magnetite plates were produced by heating natural and synthetic hematites in CO/CO_2 atmospheres at 1000 °C and the reduced specimens were examined by light and electron microscopy. Reduction experiments were carried out *in situ* in the environmental cell designed for the EM-7 HVEM (Swann and Tighe, 1971) and *ex situ* in order to observe the nucleation and growth of the magnetite plates and to compare thin foil behavior with bulk behavior. The reduction process was controlled by varying the CO/CO_2 ratio and the time. Thin foils were prepared from specimens before and after they were reduced by the ion thinning techniques described in Chapter 3 of this volume.

3. Results

In foils and bulk specimens of hematite the magnetite plates formed at 1000 °C were coherent with the matrix during nucleation and early growth but lost coherency as they increased in size. The initial growth forms found in specimens from *in situ* and *ex situ* experiments are shown in Figs. 1 and 2 respectively. The magnetite plates in Fig. 1 have nucleated preferentially along cleavage steps produced during specimen preparation. On smooth surfaces such as those in Fig. 2 an induction period is observed before magnetite precipitates are seen and nucleation occurs randomly over the surface. The *ex situ* specimen of Fig. 2 was thinned from one side and shows rhombic plates and needles parallel to the surface of the hematite. In this specimen the foil surface is parallel to the rhombohedral plane $(10\bar{1}2)$. The initially rhombic plates seen in Fig. 2 grew faster normal to the hematite surface than parallel to it and usually cooperative growth of twin variables occurred to form a polycrystalline nodule.

In more reducing CO/CO_2 atmospheres, magnetite needles started to grow in from the hematite surface to produce lenticular plates approximately 10 times

---

Fig. 1. Magnetite plates grown during *in situ* reduction in 40 torr (53.3 mbar) CO at $T=1000$ °C. Foil surface is $\|(0001)$. Plates nucleated along cleavage steps

Fig. 2. Magnetite plates grown during *ex situ* reduction in $CO/CO_2=0.032$ by heating 1 min at $T=1000$ °C. Foil surface is $\| (1\bar{1}02)$; habit $(1\bar{1}5)_m\|(1\bar{1}02)_h \langle 110\rangle_m\|\langle\bar{1}101\rangle$

Fig. 3. Magnetite plates grown during *ex situ* reduction in $CO/CO_2=0.059$ at $T=1000$ °C. Sample surface $\|\{10\bar{1}2\}$

Fig. 4a and b. Magnetite plates grown during *ex situ* reduction in $CO/CO_2=0.059$ by heating 1 min at $T=986$ °C. (a) Light micrograph of specimen after ion thinning, (b) electron micrograph of area within the box (a). Specimen surfaces are $\|\{10\bar{1}2\}$

Fig. 5. Polished section of an hematite pellet reduced during heating in $CO/CO_2=0.125$ for 10 min at $T=1000$ °C. Dark phase magnetite, light phase hematite

Fig. 6a and b. Electron micrographs of hematite pellet reduced during heating in $CO/CO_2=0.125$ for 3 min at $T=1000$ °C; (a) magnetite-hematite lamellae, (b) grain with ion-thinned surfaces $\| (0001)$ showing magnetite plates, pores and interfacial dislocations

longer and wider than the plates shown in Fig. 2. This anisotropic growth and lenticular shape are seen clearly in the light micrographs of matching surfaces across a fracture interface, Fig. 3. This figure clearly shows that the plate interface is approximately parallel to $(0001)_h$. Obviously growth parallel to this common interface is faster than growth in any other direction or along the surface. The microstructure of the phases can be examined in the electron microscope by thinning from one side of a bulk specimen as was done for Fig. 4. It is apparent from the dislocation substructure in the micrograph that the formation of the

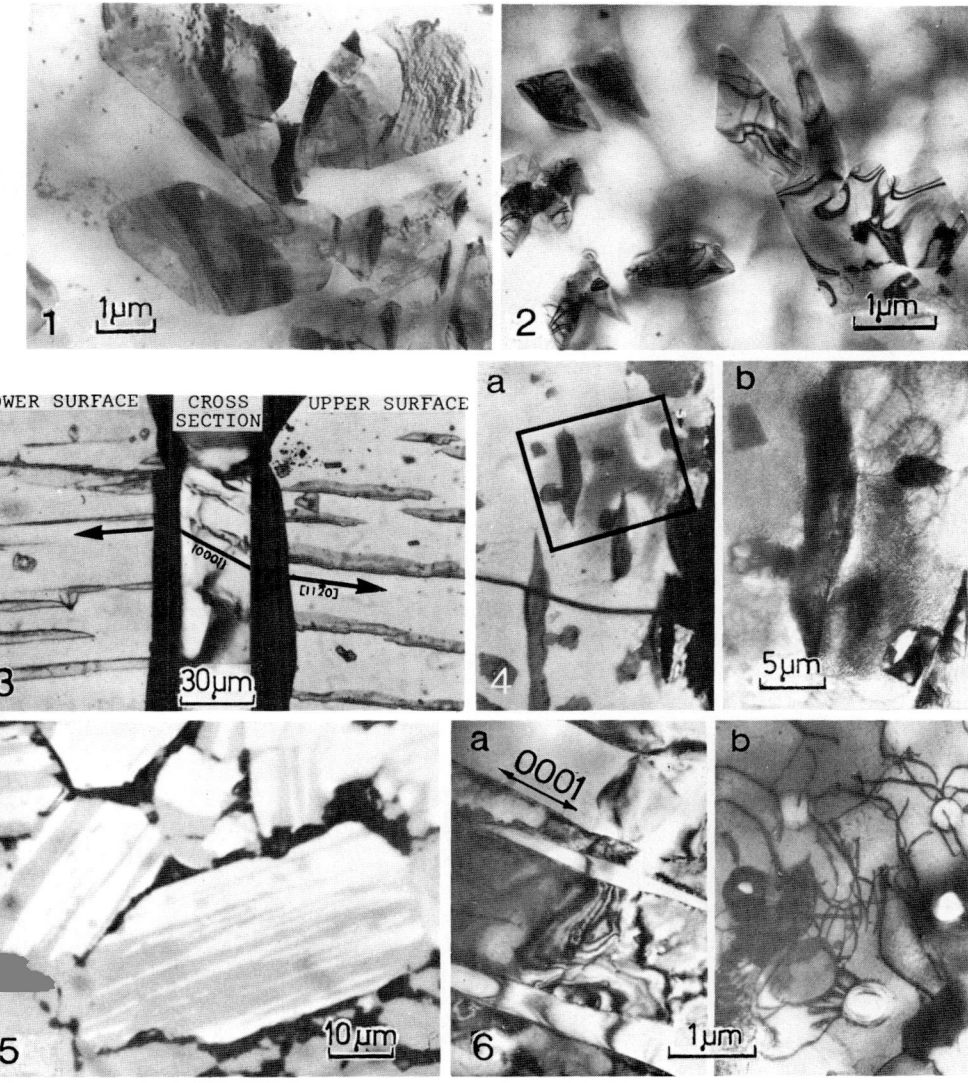

Figs. 1-6

magnetite plates is accompanied by plastic deformation of the magnetite and to a lesser extent of the hematite.

When polycrystalline hematites are reduced and then sectioned, a layer of grains containing hematite-magnetite lamellae is found between the fully reduced magnetite and the unreduced hematite core. This layer is several grain-diameters thick as seen in Fig. 5. Extensive porosity is visible around the two-phase grains. When partially reduced polycrystalline specimens are examined by electron microscopy the relative sizes of the phases vary considerably because of the random orientation of the grain sections. Fig. 6 shows examples of the microstructures in thin foils made from the two-phase layer of bulk specimen. Fig. 6a shows alternate hematite-magnetite plates with $(0001)_h$ at an acute angle to the foil surface. Fig. 6b is of a grain which was sectioned approximately parallel to $(0001)_h$ and shows extensive arrays of dislocations in the hematite matrix produced during the reduction reaction.

4. Discussion

In order for plates to grow at an angle to the surface at a faster rate than along the surface there must be a concentration gradient of oxygen which extends a considerable distance below the hematite surface. Oxygen diffusion from the hematite to the gas-solid interface must be rapid enough at 1000 °C to produce this concentration gradient and to reduce the oxygen concentration to that of the spinel composition.

The plates grow into this oxygen-depleted zone in a direction determined by the crystallographic orientation of the surface and of $(0001)_h$ with respect to the surface. Close correspondence of the oxygen stacking along basal and octahedral planes of hematite and magnetite respectively insures at least a semicoherent interface. Growth into the oxygen-depleted zone involves the operation of a shear mechanism which is accomplished by glide of dislocations lying in the interface and moving with it. In this respect the formation of magnetite plates has certain similarities with the hcp→fcc martensite transformation in cobalt (Christian, 1965). For isothermal martensites growth is constrained by the surrounding matrix and the free-energy change may include negative terms from the chemical driving force and positive terms from the strains needed to accommodate the shape deformation. There is a balance between the driving stress and the opposing stress. Isothermal martensites have lenticular shapes similar to those shown in Figs. 3 and 4 and the shape deformation of such constrained plates leads to plastic deformation in the surrounding matrix at temperatures above 1000 °C. At temperatures below 1000 °C the growth of oriented magnetite plates in hematite occurs less readily presumably because plastic deformation becomes more difficult. At these temperatures a different reduction mechanism predominates and will be discussed in a later publication.

5. Conclusions

At 1000 °C magnetite nucleates on hematite surfaces exposed to a reducing atmosphere and grows as plates with preferred orientation $\{111\}_m \| \{0001\}_h$ with $\langle 1\bar{1}0 \rangle_m \| \langle 01\bar{1}0 \rangle_h$. The plate growth is similar in many respects to isothermal martensite growth, the driving force in this case being dependent upon oxygen partial pressure as well as temperature.

References

Baker, G.S., Whelan, M.J.: The electron microscopy of oxide phases of iron. 4th Europ. Reg. Congress on Electron Microscopy, Rome, p. 449–450 (1968).
Brill-Edwards, H., Daniell, B.L., Samuel, R.L.: Structural changes accompanying the reduction of polycrystalline haematite. J. Iron Steel Inst. London **203**, 361–368 (1965).
Christian, J.W.: Military transformation: An introductory survey. In: Physical properties of martensite and bainite. Iron and Steel Inst., p. 1–19 (1965).
Cooke, S.R.B., Ban, T.E.: Microstructures in iron ore pellets. Trans. AIME **193**, 1053–1058 (1952).
Fasiska, E.J.: Structural aspects of the oxides and hydroxides of iron. Corr. Sci. **7**, 833–839 (1967).
Flower, H.M., Tighe, N.J., Swann, P.R.: Environmental gas reaction cells. In: High voltage electron microscopy (eds. P.R. Swann, C.J. Humphreys and M.J. Goringe), p. 383–395. London: Academic Press 1974.
Greig, J.W., Posnjak, E., Merwin, H.E., Sosman, R.B.: Equilibrium relationships of Fe_3O_4, Fe_2O_3 and oxygen. Am. J. Sci. **230**, 239–316 (1935).
Gruner, J.W.: Magnetite-martite-hematite. Econ. Geol. **21**, 375–393 (1926).
Néel, L.: Essai d'interprétation des propriétés magnétiques du sesquioxyde de fer rhomboédrique. Ann. d. Physiq. **17**, 250–267 (1949).
Nicholls, G.D.: The mineralogy of rock magnetism. Advan. Phys. **4**, 113–190 (1955).
Ponghis, N., Vidal, R., Poos, A.: Degradation of iron rich pellets during reduction. C.N.R.M. **12**, 3–15 (1967).
Richardson, F.D., Jeffes, J.H.E.: The thermodynamics of substances of interest in iron and steel making from 0 °C–2400 °C. B.I.S.R.A. **160**, 261–270 (1948).
Schwartz, G.M.: Iron-ore sinter. Trans. Amer. Inst. Mining Met. Eng. **84**, 39–67 (1929).
Swann, P.R.: High voltage microscope studies of environmental reactions. In: Electron microscopy and structure of materials (eds. G. Thomas, R. Fulrath, R.M. Fisher), p. 878–903. Berkeley: University of California Press 1972.
Swann, P.R., Tighe, N.J.: High voltage microscopy of gas oxide reactions. Jernkont. Ann. **155**, 447–501 (1971).
Tigerschiöld, M.: Aspects on pelletizing of iron ore concentrates. J. Iron Steel Inst. London **177**, 13–24 (1954).
Tighe, N.J.: High voltage electron microscopy of the reduction of iron oxides. Ph.D. Thesis. University of London (1973).

CHAPTER 4.4

Precipitation in the Ilmenite-hematite System

J.S. LALLY, A.H. HEUER, and G.L. NORD, JR.

1. Introduction

Although a miscibility gap in the $FeTiO_3 - Fe_2O_3$ system has been known from optical examination for many years (for a review see Ramdohr, 1969), the fact that the interface between precipitate and matrix is *semicoherent*, i.e. is composed of an array of *misfit dislocations*, was first reported by Christie et al. (1971). Cullen et al. (1973) have since suggested that the Burgers vectors of the misfit dislocations are of the type $1/3\langle 11\bar{2}0\rangle$ and that the dislocations are dissociated into quarter partials, consistent with the model of Kronberg (1957) for corundum. In the present paper we describe further aspects of precipitation in this system, and focus on the precipitate morphology and on additional analysis of the array of misfit dislocations. It is emphasized that our conclusions on the nature of the misfit dislocations differ significantly from those of Cullen et al. (1973).

2. Observations

Fig. 1a shows hematite precipitates in a basal section of an ilmenite from a pegmatite obtained by Dr. Chao at Bancroft, Ontario and illustrates the precipitate morphology normally observed in this system. The three-dimensional shape of two intermediate-size precipitates can be seen from the stereo pair in Fig. 2a (also a basal section) to be flattened lenses, with the lens axis parallel to the c direction of the hexagonal unit cell. This morphology is consistent with the nature of the lattice mismatch between matrix and precipitate, which is greatest in the c direction and least in the basal plane. The interface separating particle and matrix is composed of three sets of misfit dislocations arranged in a hexagonal grid pattern (Fig. 1b), which appears to effectively accommodate the differing lattice parameters of the two phases.

Fig. 6b of the review article by Champness and Lorimer (Chapter 4.1 of this volume) illustrates schematically how edge dislocations can restore the long-range registry of two lattices of different lattice parameter (except of course for a narrow cylinder of distorted crystal near the dislocation core). The simplest type of interface of this kind would be a planar boundary that has mismatch in one direction only. Under this condition, long-range stresses are minimized if an array of parallel, equally spaced dislocations of Burgers vector $|b| \approx d_1$ are

Precipitation in the Ilmenite-hematite System

separated by distance, S, given by the relation

$$S = \frac{d_1 d_2}{d_1 - d_2} \qquad (1)$$

where d_1 and d_2 are the lattice plane spacings in the matrix and precipitate, respectively, parallel to the interface. For small differences in lattice parameter, the boundary energy is given by the following expression due to Brooks (1952)

$$E = \frac{\mu |\boldsymbol{b}|}{4\pi S d_2 (1-v)} \left[1 + \ln \frac{|\boldsymbol{b}|}{2\pi r_0} - \ln \frac{1}{S d_2} \right] \qquad (2)$$

where v is Poisson's ratio, r_0 is the core radius of the dislocation, and μ is the average shear modulus of precipitate and matrix. The energy per unit area of boundary calculated from Eq. (2) is ~ 25 ergs/cm^2 for misfits of 1%. Lattice strain is relieved at the ilmenite-hematite interface by three sets of dislocations (Fig. 1b) and therefore the interface would have a slightly higher surface energy

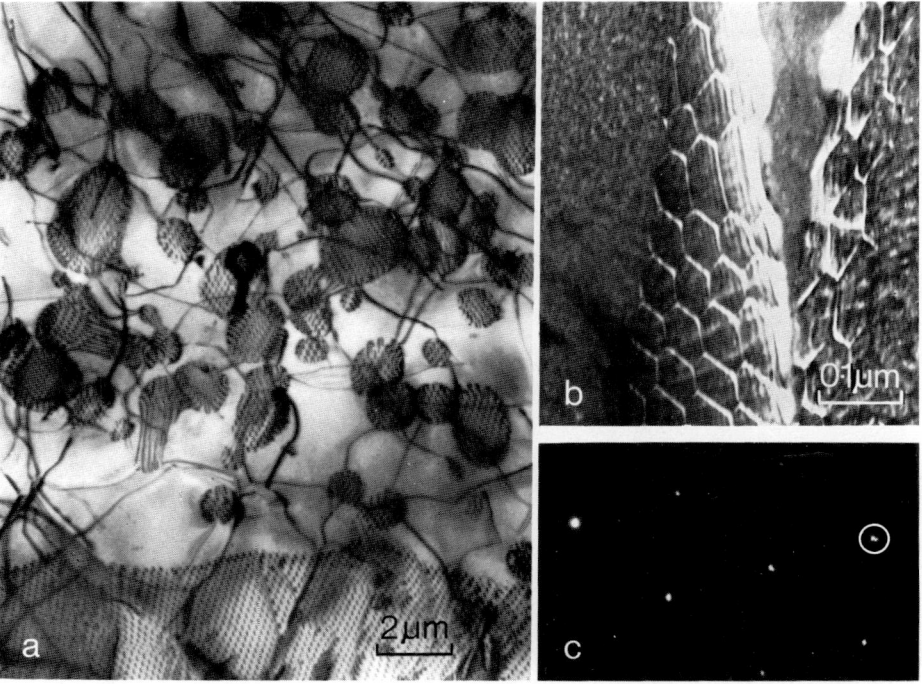

Fig. 1. (a) Bancroft, Ontario ilmenite showing small lenticular hematite particles in ilmenite, as well as a larger particle (bottom of micrograph), which shows the boundary of misfit dislocations separating hematite and ilmenite regions. (b) Dark field micrograph taken at large deviations from the exact Bragg condition illustrating the hexagonal grid of dislocations. The dislocation nodes are not alternately extended and contracted. (c) Diffraction pattern across the host, precipitate boundary in an ilmenite-hematite system. The splitting of the circled reflection is a measure of the mismatch across the boundary

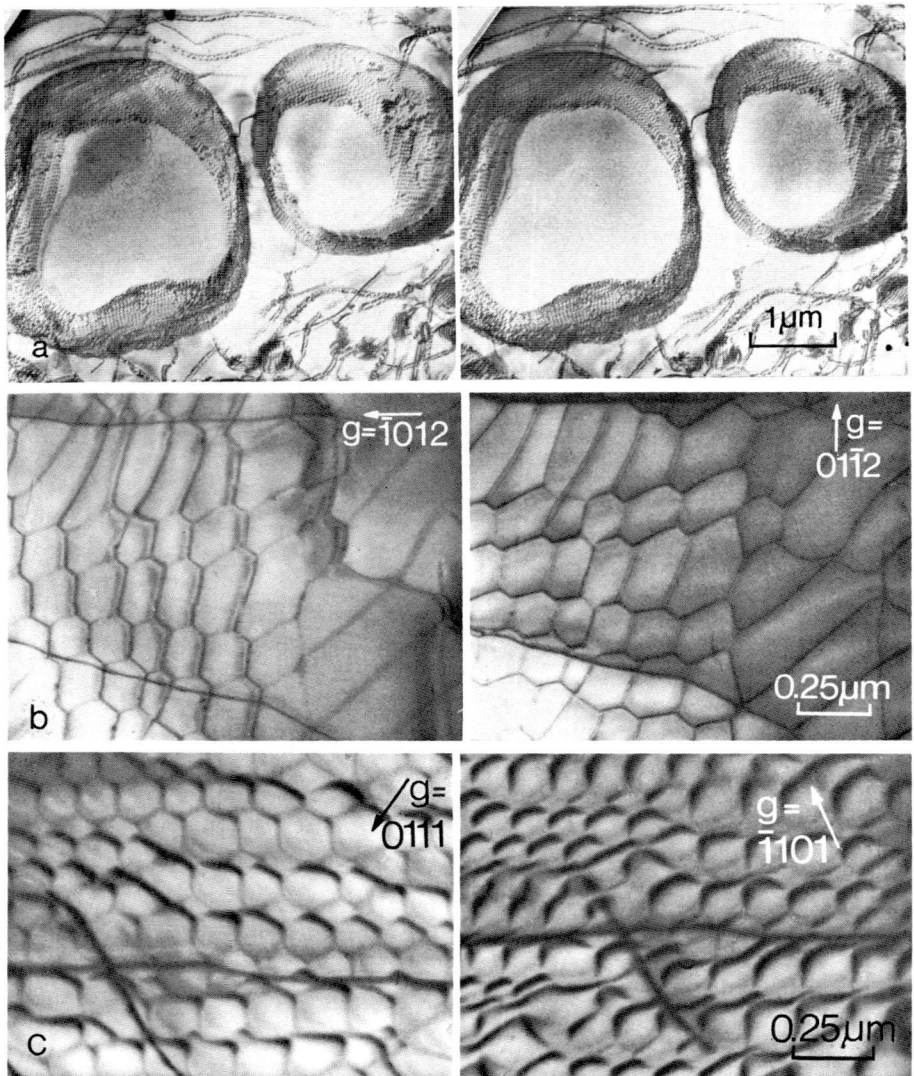

Fig. 2. (a) Stereographic pairs of dislocation networks. Lenticular nature of the precipitates of hematite in ilmenite. The dislocations in the surrounding ilmenite matrix are also decorated with small hematite particles. (b) Dislocation contrast of hexagonal misfit dislocation network using two $10\bar{1}2$-type reflections in the $[\bar{2}201]$ zone axis. (c) Dislocation contrast of a hexagonal misfit dislocation network using two $10\bar{1}1$-type reflections from the $[10\bar{1}1]$ zone axis

than this estimate. Nevertheless this interface has very low energy, much lower than free-surface and grain-boundary energies of oxides.

The misfit between the host phase and precipitate can be measured by obtaining electron diffraction patterns from a two-phase region and measuring the difference in the c spacings. In Fig. 1c, a diffraction pattern that included both the matrix and a precipitate shows splitting of the spots which corresponds to a mismatch of 1.1% in a direction roughly perpendicular to $(10\bar{1}2)$ planes. The lattice parameter measurements of Ishikawa and Akimoto (1958) indicate that this splitting corresponds to compositions of 0.8 and 0.2 mole fraction of $FeTiO_3$ for the matrix and precipitate phase, respectively. Using the solvus of Uyeda (1958) these compositions are in equilibrium at 700 °C, suggesting that this was the temperature at which this sample equilibrated during its cooling history. However this temperature estimate has a large uncertainty because the solvus has not been accurately determined. The 1.1% lattice mismatch can also be used to calculate the dislocation spacing at the precipitate-matrix interface, assuming that each set of dislocations relieves one-third of the total mismatch. Using Eq. (1), a dislocation spacing of 1500 Å is obtained, which compares closely to the 1200 Å measured spacing for this particle.

The dislocations in the interface separating ilmenite and hematite must have Burgers vectors corresponding to a lattice translation in an appropriate unit cell of this system. In Fig. 3, the rhombohedral and hexagonal unit cell are shown to illustrate possible Burgers vectors. The two shortest Burgers vectors, in order of increasing size, are the basal plane translation vectors a_1, a_2, and a_3 and the rhombohedral unit cell edges r_1, r_2, and r_3 (see Snow and Heuer 1973, for a fuller treatment of dislocations in this system). Any vector sum of these unit vectors is also a possibility (e.g. $r_1+r_2+r_3=c$) but under most circumstances these longer Burgers-vector dislocations would not be stable and would spontaneously decompose into their lower-energy components. The types of dislocation that constitute the interface will also depend on the nature of the stresses resulting from lattice strain that the interface dislocations can relieve. The lattice-parameter mismatch between the end components of this system is greatest along the c axis (2.8%) and least along the a axis (0.9%). Thus we would expect any of the Burgers vectors at the interface to have a large component in the c direction.

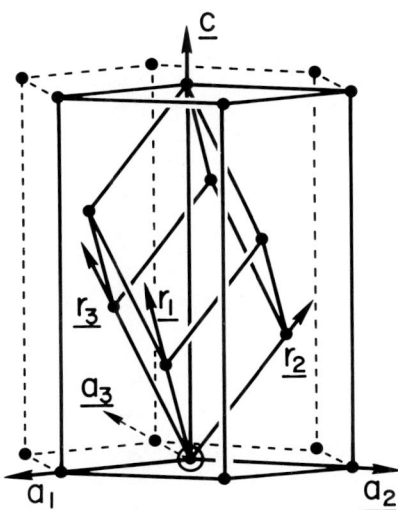

Fig. 3. The rhombohedral and hexagonal unit cells representing the hematite and ilmenite structures

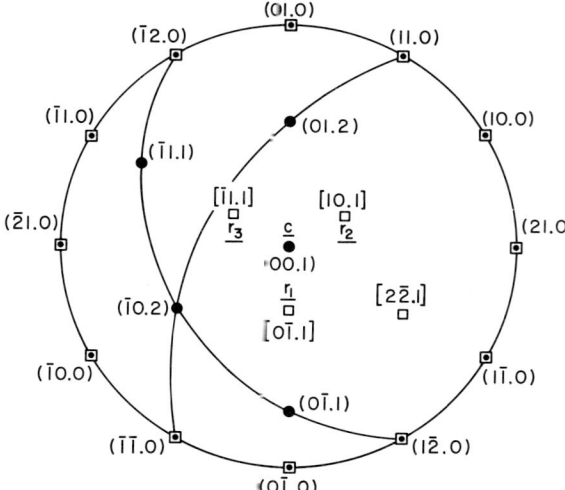

Fig. 4. A hexagonal stereographic projection of ilmenite illustrating the two major zone axes used in the Burgers vector analysis

Standard $g \cdot b$ analyses have been performed to determine the Burgers vectors of the misfit dislocations, using two different foil orientations, corresponding to $[10\bar{1}1]$ and $[2\bar{2}01]$ zone axes, as shown in Fig. 4.

Fig. 2b shows two micrographs taken with two different $10\bar{1}2$-type reflections from the $[2\bar{2}01]$ zone axis. Each $10\bar{1}2$ reflection causes one set of dislocations in the hexagonal grid to disappear. Similarly, Fig. 2c shows two micrographs taken using $10\bar{1}1$-type reflections, from the $[10\bar{1}1]$ zone, and illustrates that under these conditions, two sets of dislocations are out of contrast for each $10\bar{1}1$-type reflection. Table 1 below lists the invisibility conditions ($g \cdot b = 0$) for rhombohedral and basal Burgers vectors for the diffraction conditions of Fig. 2b and c[1].

Table 1

g	b	g·b	g	b	g·b
$01\bar{1}2$	$1/3\,[\bar{2}110]$	0	$\bar{1}101$	$1/3\,[\bar{2}110]$	+1
	$1/3\,[\bar{1}\bar{1}20]$	-1		$1/3\,[11\bar{2}0]$	0
	$1/3\,[\bar{1}2\bar{1}0]$	$+1$		$1/3\,[1\bar{2}10]$	-1
	$1/3\,[10\bar{1}1]$	$+1$		$1/3\,[10\bar{1}1]$	0
	$1/3\,[\bar{1}101]$	$+1$		$1/3\,[\bar{1}101]$	$+1$
	$1/3\,[0\bar{1}11]$	0		$1/3\,[0\bar{1}11]$	0
$\bar{1}012$	$1/3\,[\bar{2}110]$	$+1$	$0\bar{1}11$	$1/3\,[\bar{2}110]$	0
	$1/3\,[\bar{1}\bar{1}20]$	$+1$		$1/3\,[11\bar{2}0]$	-1
	$1/3\,[\bar{1}2\bar{1}0]$	0		$1/3\,[\bar{1}2\bar{1}0]$	-1
	$1/3\,[10\bar{1}1]$	0		$1/3\,[10\bar{1}1]$	0
	$1/3\,[\bar{1}101]$	$+1$		$1/3\,[\bar{1}101]$	0
	$1/3\,[0\bar{1}11]$	$+1$		$1/3\,[0\bar{1}11]$	$+1$

[1] In this paper, planes and directions have been described using the 4-index Miller-Bravais hexagonal notation. Although it is not widely recognized, dot products can be calculated directly using 4-index notation (Frank, 1965).

It is immediately clear that the Burgers vectors of the misfit dislocations are unambiguously identified as r_1, r_2, and r_3. Furthermore, it was observed that the dislocations in the lattice surrounding the particles obeyed the same invisibility criteria as did the interfacial networks, indicating that rhombohedral Burgers vectors are also fundamental slip vectors for the phases in this system.

3. Discussion

It is difficult to reconcile the observations reported here with those of Cullen et al. (1973), who assumed that the misfit dislocations had a basal Burgers vector and that the dislocation nodes were extended due to dissociation into quarter partials separated by stacking faults. For $\langle 10\bar{1}1 \rangle$ type Burgers vectors, there is no common plane on which they can separate into partials, and high-resolution, dark field micrographs of dislocation images (Fig. 1b) show little, if any, evidence of splitting.

The rhombohedral interfacial dislocations observed here occur because the displacements across the precipitate boundary caused by these dislocations closely match the mismatch of the lattices. The rhombohedral vector has a component $c = 0.85|r|$ along the c axis and another $a = 0.53|r|$ along the a axis. However, the lattice mismatch between ilmenite and hematite is too large along the c direction to be relieved by an equally-spaced hexagonal grid of rhombohedral dislocations around a spherical (minimum area) precipitate. Thus, the particles are flattened along the c axis into lenses so that the network of rhombohedral dislocations can effectively cancel long range stresses.

References

Akimoto, S.: Magnetic properties of ferromagnetic oxide minerals as a basis of rock-magnetism. Advan. Phys. **6**, 288–298 (1957).

Brooks, H.: Theory of internal boundaries. Metal Interfaces, p. 20–64. ASM (1952).

Christie, J.M., Lally, J.S., Heuer, A.H., Fisher, R.M., Griggs, D.T., Radcliffe, S.V.: Comparative electron petrography of Apollo 11, Apollo 12 and terrestrial rocks. Proc. 2nd Lunar Sci. Conf. Geochim. Cosmochim. Acta, Suppl. 2, **1**, 69–89. MIT Press (1971).

Cullen, W.H., Marcinkowski, M.J., Das, E.S.P.: A study of the interface boundary between ilmenite and hematite. Surface Sci. **36**, 395–412, (1973).

Frank, F.C.: On Miller-Bravais indices and 4 dimensional vectors. Acta Cryst. **18**, 862–866 (1965).

Ishikawa, H., Akimoto, S.: Magnetic properties and crystal-chemistry of ilmenite and hematite system I., Crystal Chemistry. J. Phys. Soc. Japan **13**, 1110–1118 (1958).

Kronberg, M.L.: Plastic deformation of single crystals of sapphire: Basal Slip and Twinning. Acta Met. **5**, 508–527 (1957).

Ramdohr, P.: Ore minerals and their intergrowths. p. 959. Oxford: Pergamon Press 1969.

Snow, J.D., Heuer, A.H.: Slip systems in Al_2O_3. Am. Ceram. Soc. J**6**, 153–157 (1973).

Uyeda, S.: Thermal remnant magnetization as a medium of paleomagnetism with special reference to reverse thermal magnetism. Japan. J. Geophys. **2**, 1–173 (1958).

CHAPTER 4.5

Pigeonite Exsolution from Augite

G.L. NORD, JR., A.H. HEUER, and J.S. LALLY

1. Introduction

Subsolidus precipitation (exsolution) in rock-forming minerals provides information on the thermal history of natural materials. Several possible mechanisms of precipitation can operate, as reviewed by Champness and Lorimer (Chapter 4.1 of this volume); the particular mechanism operating will in general depend on the rate of cooling. In the case of chain silicates, both the mechanisms of nucleation and growth and of spinodal decomposition followed by coarsening have been suggested.

The continuous[1] precipitation mechanism requiring heterogeneous nucleation as a first step is perhaps the easiest to verify. The nuclei form on pre-existing defects such as stacking faults, low or high-angle boundaries, isolated dislocations, etc.; these thus provide "local" precipitate distributions which are diagnostic. The mechanisms of homogeneous nucleation and spinodal decomposition occur at higher supersaturations and thus at lower temperatures; they give rise to "general" precipitate distributions. It is not a trivial matter to distinguish between these two latter mechanisms by morphological considerations, especially if coarsening has eliminated the "diffuse" matrix-precipitate interface characteristic of spinodal decomposition. In such cases, detailed heating and electron diffraction experiments may be required to determine the particular exsolution mechanism operating.

This paper is concerned with the characterization and distribution of low-calcium clinopyroxene precipitates of pigeonite in a calcium-rich clinopyroxene host of augite. The precipitates in the host are distributed both "locally" and "generally" and indicate that at least two precipitation mechanisms operated during cooling.

2. Observations

The augite crystal studied here is shown in Fig. 1a and was taken from Apollo 17 mare basalt 70017; it was oriented such that [010] was normal to the plane of the thin section. The augites in this basalt have a complex radiating

---

[1] *Continuous* is used here to distinguish the precipitation reactions under discussion from *discontinuous* reactions, in which a duplex product grows into a single phase matrix (Christian, 1965).

subgrain structure, the subgrains being rotated about an axis nearly parallel to [010]. In addition, the subgrain structure is compositionally zoned, both outward from the core area and between growth sectors (sector zoning). A detailed description of the complex zoning and subgrain orientation relationships of similar lunar augites is given by Carter et al. (1970).

The particular area studied by TEM (A in Fig. 1a) is relatively homogeneous, with a composition range from six nearby microprobe spots (plotted in Fig. 1a) of En 47.8–48.8, Wo 35.8–37.5, Fs 14.7–15.5 mole%, Al_2O_3 4.2–4.5 wt% and TiO_2 3.1–3.5 wt%. This composition lies just within the pigeonite-augite solvus of Ross et al. (1973). The area is also characterized optically by subboundaries radiating outward from the center of the crystal.

Fig. 1b–d show pigeonite exsolution lamellae which are associated with low-angle dislocation subboundaries and isolated dislocations. The (001)-sharing lamellae are parallel to the electron beam while the (100)-sharing lamellae are inclined to the beam. Both types of lamellae have nucleated on dislocations. It appears, however, that only certain types of dislocations provide good nucleation sites for a particular lamellar habit. Thus, Fig. 1b and d show that (100) pigeonite lamellae nucleate predominately on subboundary dislocations, although some (001) lamellae are present. Fig. 1c shows an area of abundant isolated dislocations not associated with the subboundaries. Both unit (u) and partial (p) dislocations are present. The partials are straight, with the stacking faults out of contrast, and have fewer pigeonite lamellae associated with them than the unit dislocations. The isolated unit dislocations appear to have been preferred sites for (001) pigeonite nucleation since only a few (100) lamellae are present. The unit dislocations have a "zig-zag" appearance and appear to have rearranged after the precipitation of the (001) pigeonite in order to be along the lamellar interfaces, and thus adopt a lower-energy configuration; similar phenomena are found in metallic systems (Nicholson, 1970). Fig. 1d also shows that distinct "denuded" or precipitate-free zones are associated with the coarse (001) pigeonite lamellae. Similar "denuded" zones are also associated with the coarse (100) lamellae but are not shown.

The background "tweed" texture, in good contrast in Fig. 1d, consists of finer pigeonite lamellae, also on (001) and (100). These finer lamellae are shown more clearly in Fig. 2a in a foil oriented normal to [010]; in this orientation, the lamellar width may be accurately estimated. The (001) lamellae are several times larger than the (100) lamellae and both sets of lamellae have distinct and sharp interfaces with the matrix. $\Delta\beta$, as measured from the $h0l$ diffraction pattern (Fig. 2b), is $\sim 2.6°$, the expected equilibrium value (Papike et al., 1971). The $h0l$ net itself is drawn in Fig. 2c, with reflections from the augite host, the (100)-sharing pigeonite, the (001)-sharing pigeonite, and the magnified insets of Fig. 2b all indicated. The upper inset shows that the host augite reflections are slightly streaked in a direction normal to both sets of pigeonite lamellae; this streaking is due to the narrow lamellar width. The (001)-sharing pigeonite produces sharp reflections with a weak connecting streak with the augite. The (100)-sharing pigeonite reflections, which are weak and streaked in Fig. 2b, are strong and streaked in the lower inset of this figure, which is from the $h0l$ net after tilting $\sim 1°$. The streaked reflections indicate that the (100) pigeonite lamellae are very thin relative to the (001) lamellae, perhaps only 3 or 4 unit cells.

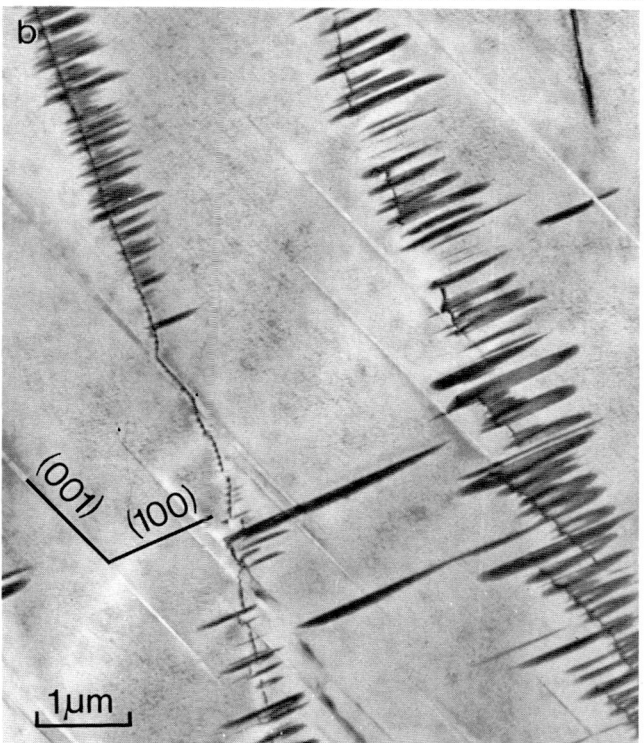

Fig. 1. (a) Is an optical micrograph of a (010) section of augite in Apollo 70017 under partially crossed nicols. (b–d) Are electron micrographs showing both (100) and (001) pigeonite lamellae, the trends of which are indicated. (b and d) Show subboundary dislocations with (100) pigeonite lamellae while (c) shows (001) lamellae heterogeneously nucleated on isolated unit u and partial p dislocations. Micrographs are bright field. (b) 200 kV, (c and d) 1000 kV

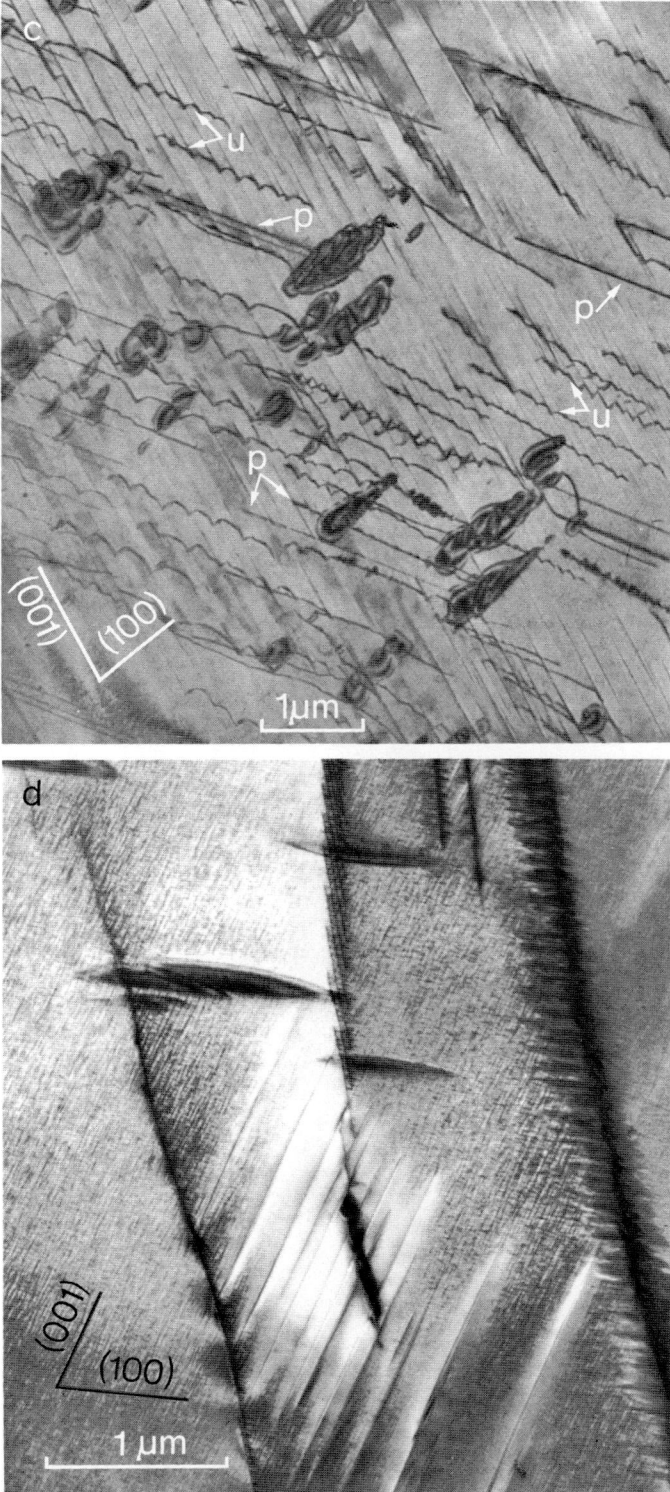

Fig. 1c and d

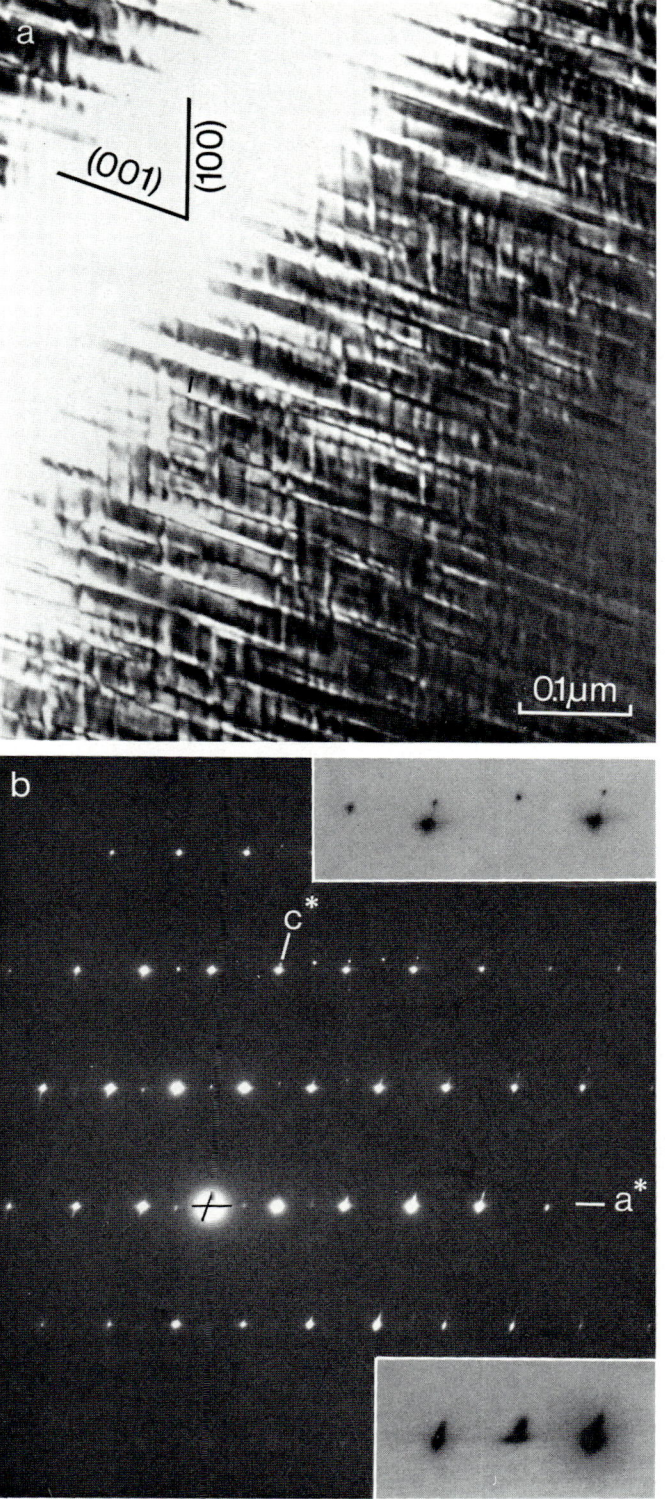

Fig. 2a and b

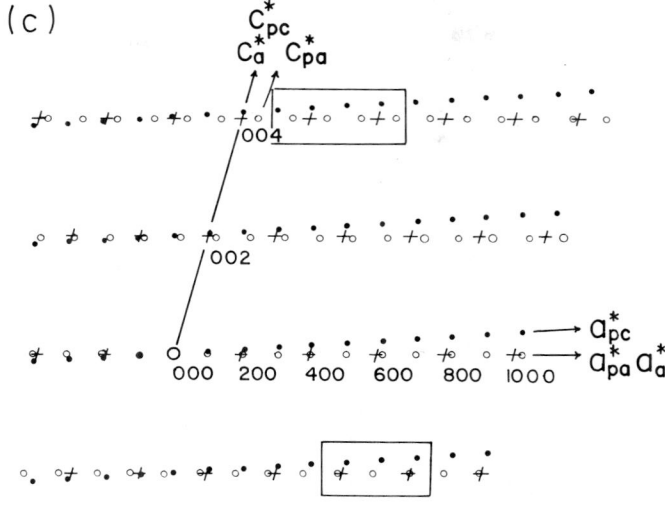

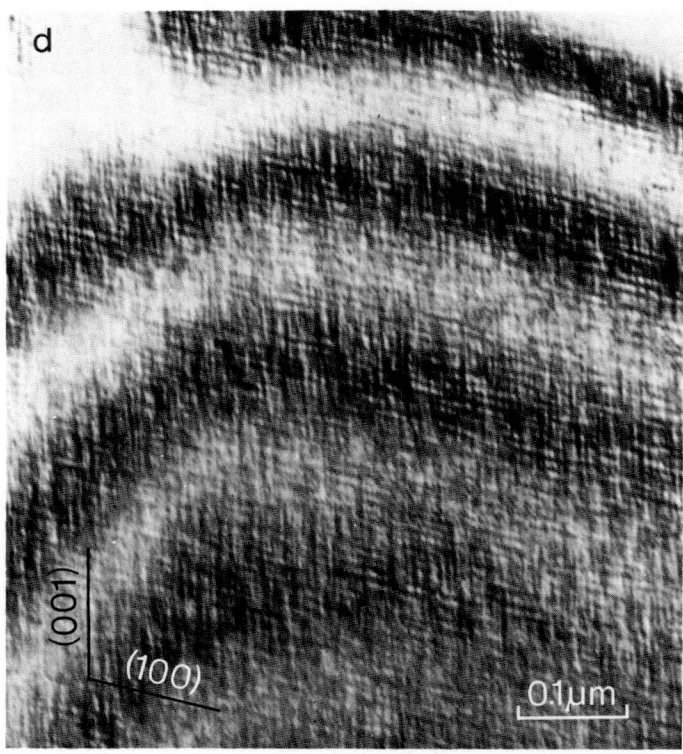

Fig. 2. (a) Is an electron micrograph showing the (001) and (100) fine pigeonite lamellae and (b) is the diffraction pattern from the area shown in (a). (c) Is an $h0l$ net showing the augite matrix reflections as crosses, the (100)-sharing and (001)-sharing pigeonite reflections as open circles and closed circles, respectively. The areas of the insets in (b) are boxed in (c). Micrographs (a) and (d) are bright field, 1000 kV. (d) Shows an (001) and (100) "tweed" texture from an Apollo 12053 augite

For comparison, Fig. 2d shows a similar "tweed" texture at a similar magnification from an Apollo 12053 augite with an identical Ca content ($Wo_{37}En_{40}Fs_{23}$). In contrast to Fig. 2a, the interfaces are not sharp but diffuse and the diffraction pattern shows a continuous strong streak instead of distinct augite and pigeonite maxima; $\Delta\beta$ is 1.6°, significantly less than the equilibrium value (Lally et al., 1975).

3. Discussion and Conclusions

It is clear that pigeonite nucleated heterogeneously on dislocations in 70017. Most simply stated, such heterogeneous nucleation occurs because the presence of the nucleus on the dislocation will lower the dislocation strain energy, thus reducing the nucleation barrier.

In addition, a given dislocation did not appear to be an equally effective nucleation site for both pigeonite habits; (001) occurred mainly on isolated dislocations and (100) occurred mainly on subboundary dislocations. An empirical rule has been developed for precipitation in metallic alloys in which "it is found that the habits which do nucleate are those in which the principal misfit of the precipitate is ~parallel to the Burgers vector of the dislocation" (Nicholson, 1970). In the case of the (001) lamellae, the principle misfit direction, [001]*, lies only 15° from a possible Burgers vector of [001]. Burgers vectors of [001] and [010] have been determined by X-ray topography (Smith and Yu, 1974) in hedenbergite and [001] and [010] for isolated dislocations in orthopyroxenes by TEM (Kohlstedt and Vander Sande, 1973) (also see Kirby, Chapter 6.7 of this volume). The situation with regard to the (100) lamellae nucleating on subboundaries is less clear, since the "preferred" Burgers vector, [100], has a large self-energy. The energy of a dislocation is proportional to the square of the Burgers vector (Cottrell, 1964), i.e. to $(5.25 Å)^2$ for [001] and to $(9.8 Å)^2$ for [100]. However, as stated above [010] was determined as a Burgers vector in both pyroxenes for which the self-energy is proportional to $(9.0 Å)^2$. The absence of significant nucleation on the partial dislocations is almost certainly due to the low-strain energy associated with a partial dislocation compared to its unit dislocation.

In the 70017 augite studied here, nucleation on dislocations predominated until a sufficiently high supersaturation was reached for "general" matrix decomposition to occur. The precipitate-free zones about the coarse heterogeneously-nucleated lamellae developed because of solute depletion during growth of the coarse lamellae. The width of such zones can be used to estimate cooling histories, as discussed by Lally et al. (1975).

The pigeonite lamellae have sharp interfaces and have reached their equilibrium compositions, as shown by the $\Delta\beta$ measurement. While such a texture is *consistent* with homogeneous nucleation and growth, this texture could also arise from spinodal decomposition with later coarsening. Fig. 3 (after Cahn, 1968), shows schematically the evolution of concentration profiles during unmixing by homogeneous nucleation and growth and by spinodal decomposition; it is clear that $\Delta\beta$ should be constant and at the equilibrium value for nucleation

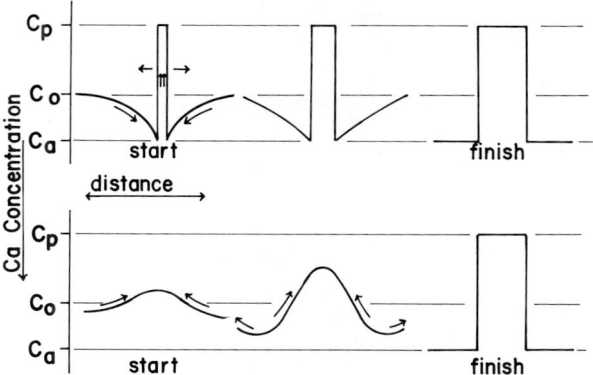

Fig. 3. The evolution of decomposition is shown in a schematic diagram for nucleation and growth (top) and spinodal decomposition and coarsening (bottom). C_p pigeonite, C_o original, C_a augite. (After Cahn, 1968)

and growth but should increase gradually to the equilibrium value for spinodal decomposition. (The final products are of course identical.) Thus, while the small and nonequilibrium $\Delta\beta$'s and diffuse interfaces argue strongly for spinodal decomposition as the operating mechanisms during the low-temperature unmixing of 12053 (Christie et al., 1971 and Lally et al., 1975), no definitive statement can be made for 70017 except that the lower-temperature exsolution was further advanced than for 12053. It is emphasized that both microstructural evidence — diffuse precipitate/matrix interfaces — and diffraction evidence — non-equilibrium $\Delta\beta$'s — are needed before spinodal decomposition can be suggested with rigor as an unmixing mechanism for a particular pyroxene.

References

Cahn, J.W.: Spinodal Decomposition. Trans. AIME **242**, 166–180 (1968).
Carter, N.L., Fernandez, L.A., Ave'Lallemant, H.G., Leung, I.S.: Pyroxenes and olivines in crystalline rocks from the Ocean of Storms. Proc. 2nd Lunar Sci. Conf. Geochim. Cosmochim. Acta, Suppl. 2, **1**, 775–795 (1971).
Christian, J.W.: The theory of transformations in metals and alloys, p. 973. Oxford: Pergamon Press 1965.
Christie, J.M., Lally, J.S., Heuer, A.H., Fisher, R.M., Griggs, D.T., Radcliffe, S.V.: Comparative electron petrography of Apollo 11 and Apollo 12 and terrestrial rocks. Proc. 2nd Lunar Sci. Conf. Geochem. Cosmochim. Acta, Suppl. 2, **1**, 69–90 (1971).
Cottrell, A.H.: Theory of crystal dislocations, p. 91. New York: Gordon and Breach 1964.
Kohlstedt, D.L., Vander Sande, J.B.: Transmission electron microscopy investigation of the defect microstructure of four natural orthopyroxenes. Contrib. Mineral. Petrol. **42**, 169–180 (1973).
Lally, J.S., Heuer, A.H., Nord, Jr., G.L., Christie, J.M.: Subsolidus reactions in lunar pyroxenes: an electron petrographic study. Contrib. Mineral. Petrol. **51**, 263–282 (1975).
Nicholson, R.B.: Nucleation at imperfections. In: Phase transformations. Am. Soc. for Metals, Metals Park, Ohio, p. 625 (1970).
Papike, J.J., Bence, A.E., Brown, G.E., Prewitt, C.T., Wu, C.H.: Apollo 12 clinopyroxenes: exsolution and epitaxy. Earth Planet. Sci. Letters **10**, 307–315 (1971).
Ross, M., Huebner, J.S., Dowty, E.: Delineation of the one atmosphere augite-pigeonite miscibility gap for pyroxenes from lunar basalt 12021. Am. Mineralogist **58**, 619–635 (1973).
Smith, D.K., Yu, S.: Observations on the dislocation substructures in a hedenbergite from Nordmark, Sweden. [abs.] Am. Cryst. Assoc. Abstracts, Berkeley Meeting, March (1974).

CHAPTER 4.6

The Transformation of Pigeonite to Orthopyroxene

P.E. CHAMPNESS and P.A. COPLEY

1. Introduction

Pigeonites in igneous intrusions undergo a structural (polymorphic) transition to orthopyroxene on cooling. Poldervaart and Hess (1951) maintained that the orthopyroxene nearly always retains the b and c axes of the parent phase, but Brown (1957) (from optical examination) and Bown and Gay (1960) (from studies by single-crystal X-ray diffraction) found that, in the inverted pigeonites of the Skaergaard intrusion, the orientation of the orthopyroxene is generally random with respect to the original pigeonite (although Bown and Gay did find one crystal in which orientation had been preserved).

In Fe-rich orthopyroxenes which have inverted from pigeonite in the Biwabik (Bonnichsen, 1969) and Gunflint (Simmons et al., 1974) iron formations it was found that, not only had the original pigeonite crystallographic axes been disregarded during the transformation, but each new orthopyroxene grain had encompassed several previously-existing pigeonite grains. This microstructure has all the features of a massive transformation (Massalski, 1970); i.e. a diffusionless, "civilian" transformation for which nucleation occurs preferentially at grain boundaries and growth proceeds by the rapid propagation of an incoherent interface, disregarding pre-existing grain boundaries. Massive transformations occur in many metallic systems at a cooling rate which is intermediate between that needed to form the equilibrium phase distribution and that needed to produce a martensitic phase. It is possible that, because of the low diffusivity of calcium in pyroxenes (witness the presence in orthopyroxene of metastable phases which form even at the slow cooling rates of intrusions such as the Bushveld complex, Champness and Lorimer, 1974), a massive transformation could occur under the conditions of cooling of metamorphosed iron formations. If the transformation pigeonite→orthopyroxene occurs via a binary loop in the phase diagram (as shown in Fig. 3 of Bonnichsen, 1969) a direct analogy in a metallic system would be the $\beta \to \alpha_m$ transformation in Cu-Ga (Massalski, 1970).

Bown and Gay (1960) and Brown (1972) have stated that the pigeonite which exsolves as lamellae in augite does not invert to orthopyroxene, even in large intrusions, because it is "stabilized" by the augite due to the similarity of their two structures (i.e. by the surface traction across the interface). Binns et al. (1963) and Boyd and Brown (1969) showed that the large exsolved "pigeonite" lamellae in augites from the Bushveld intrusion contain less than 0.8 wt.percent

CaO and concluded that they are clinohypersthene. This paper shows, however, that the large "pigeonite" lamellae in the Bushveld intrusion have inverted and that the narrower lamellae in the Bushveld and Skaergaard intrusions contain faults which are probably the precursors of the orthopyroxene in the transformation sequence.

2. Electron Microscopy

Grains of augite from four rocks of the layered series of the Skaergaard intrusion and two rocks from the main zone and one from the upper zone of the Bushveld intrusion were examined by transmission electron microscopy. Suitably oriented grains were drilled from doubly-polished thin sections and thinned by ion-bombardment. The Skaergaard samples were numbers EG 4369, 4306, 4309, 4316 of Brown (1957). The Bushveld samples were numbers SA 1019 (Boyd and Brown, 1969), 738 and 616 (Atkins, 1969).

Lamellae of pigeonite approximately parallel to (001) and occasional, thin (100) lamellae of orthopyroxene are common to all the samples. All the pigeonite precipitates (except those wider than about 3 μm in the Bushveld samples) contain (100) lamellae exhibiting stacking-fault contrast, and generally the thicker precipitates contain a higher density of faults (Fig. 1a and c). Diffraction patterns from the lamellae show continuous streaks parallel to a^*, but no sign of a second phase (Fig. 1b).

The faults in the upper lamellae of Fig. 1a and c have nucleated at some of the growth ledges at the interphase interfaces while, in the lower lamellae in the two figures, faults have also nucleated at the coherent planar interfaces. The preferential nucleation at ledges occurs because they are the regions of the interphase boundary with the highest strain energy; some of this energy can be used to overcome the activation barrier opposing nucleation of the faults. The increase in the fault density with increasing lamellar thickness can be attributed to the effective dilution of the surface traction force which opposes nucleation. For a lamella of length l, width w and thickness t the total surface-energy barrier to the transformation is $2lw\gamma$ where γ is the effective surface traction (it represents the energy needed to reconstitute the interface). The energy per unit volume is then $2\gamma/t$ and the thicker lamellae will have a lower nucleation barrier.

The two Bushveld samples SA1019 and 738 contain, in addition to the lamellae with (100) faults, large (~ 3 μm wide) lamellae of low birefringence (labeled A in Fig. 1d). They have incoherent, irregular boundaries approximately parallel to (001) and contain (100) lamellae of variable thickness (Fig. 1e). Diffraction evidence shows that the major phase is orthopyroxene and that most of the lamellae are augite, sharing the (100) lattice plane with their host. Diffraction patterns recorded across the incoherent interphase boundary show that the a^* axes of the primary and secondary augite phases are at a small angle to each other (Fig. 1f). The augite lamellae clearly exsolved during or after inversion of the original pigeonite.

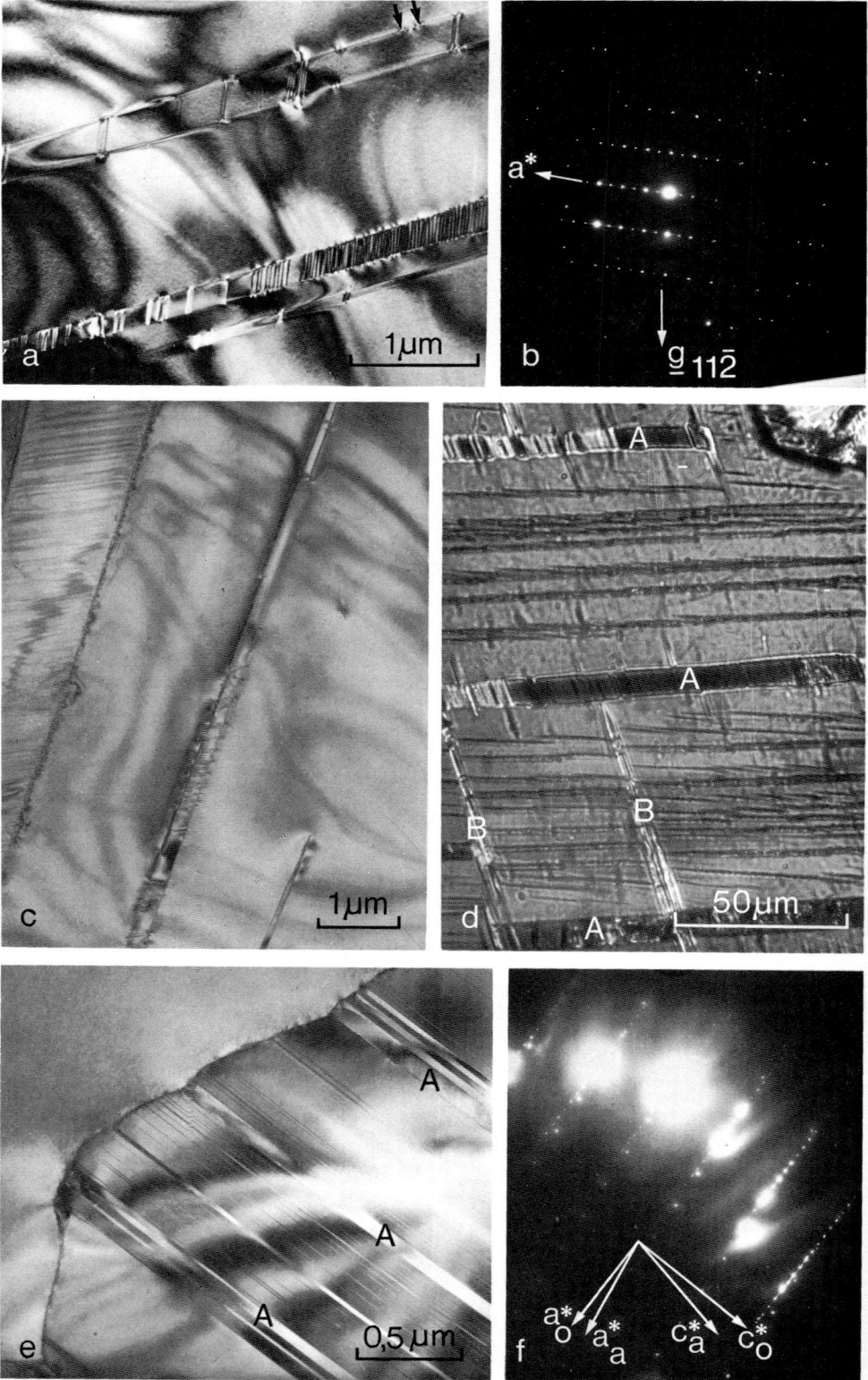

Fig. 1a–f

3. Analysis of the Fault Vector

Contrast experiments were performed on the faults in pigeonite lamellae of specimen SA 1019. The foil was tilted to bring the $(1\bar{1}0)$ reciprocal-lattice section perpendicular to the electron beam and dark field images were recorded with different reflections (Fig. 2). If the fault vector lies in the (100) plane the invisibility of the faults in reflections 220 (Fig. 2b) and 330 indicates that the vector is parallel to [001]. Fig. 2d–f show the faults imaged with a 002, 004 and 006 reflection respectively. They are clearly in contrast in 2d, but the image is weak in e and very weak in f. As the faintness of the images in the 004 and 006 reflections is partly due to the low structure factors for the reflections (F_0 for X-rays is 115, 37 and 16 for 002, 004 and 006 respectively, Morimoto et al., 1960), further experiments were performed. The images obtained with reflections $\bar{2}2\bar{4}$ (Fig. 2c) $\bar{2}2\bar{3}$ (2g) and 113 show the faults clearly in contrast, while the micrograph taken with the $\bar{1}1\bar{6}$ reflection (2h) show only faint "residual" contrast at the faults. The results therefore suggest that the fault vector is 1/6 [001]. However, detailed examination of the fault images show that they do not behave in the way expected from the dynamical theory in which absorption is considered. If the fault vector is 1/6 c, the phase angle for reflections $\bar{2}2\bar{1}$, 002, 004 and $\bar{2}2\bar{4}$ should be $\pi/3$, $2\pi/3$, $4\pi/3$ and $\pi/3$ respectively and the image in dark field should be asymmetrical in each case (Hashimoto et al., 1962). Of the relevant images in Fig. 2, only that taken in reflection $\bar{2}2\bar{4}$ is asymmetrical. Clearly definitive determination of the fault vector requires calculations using the dynamical theory and comparison between calculated and observed images.

4. Discussion and Conclusions

The inversion of pigeonite to orthopyroxene in primary crystals from igneous intrusions usually takes place without an orientation relationship being preserved between the parent and product phases, and may occur under certain conditions

◀

Fig. 1. (a) Microstructure of augite in EG 4369. Lamellae of pigeonite approximately parallel to (001) contain (100) stacking faults which have nucleated at ledges and at planar interphase boundaries. Two ledges which have not nucleated faults are arrowed. 100 kV. (b) (021) diffraction pattern from a pigeonite lamella in the same sample. Notice the streaking along a^*. (c) Microstructure of augite in SA 1019 showing two of the thinner "(001)" pigeonite lamellae. Note the high density of faults in the thicker compared with the thinner precipitate. 1000 kV. (d) Optical micrograph (crossed polars) of augite from SA 1019. Note large coarsened "(001" lamellae, A and narrower "(001)" lamellae between them. Two (100) lamellae labeled B can also be seen. (e) A large lamella in SA 738 augite. The interface with the augite matrix is incoherent. The larger (100) lamellae within the precipitate are augite (some are labeled A) and the matrix is orthopyroxene. 100 kV. (f) Diffraction pattern in the correct orientation relative to (e) taken across the interphase boundary. In the lower left can by seen the (010) reciprocal-lattice section from the augite matrix while in the upper right is the (010) orthopyroxene pattern with a second augite pattern (from the lamellae) superimposed on it. These two latter sections share a common a^* direction, labeled a_0^*, which diverges at an angle of about 4° from a_a^*, the direction of a^* in the matrix. This confirms the lack of coherency across the lamellar boundary

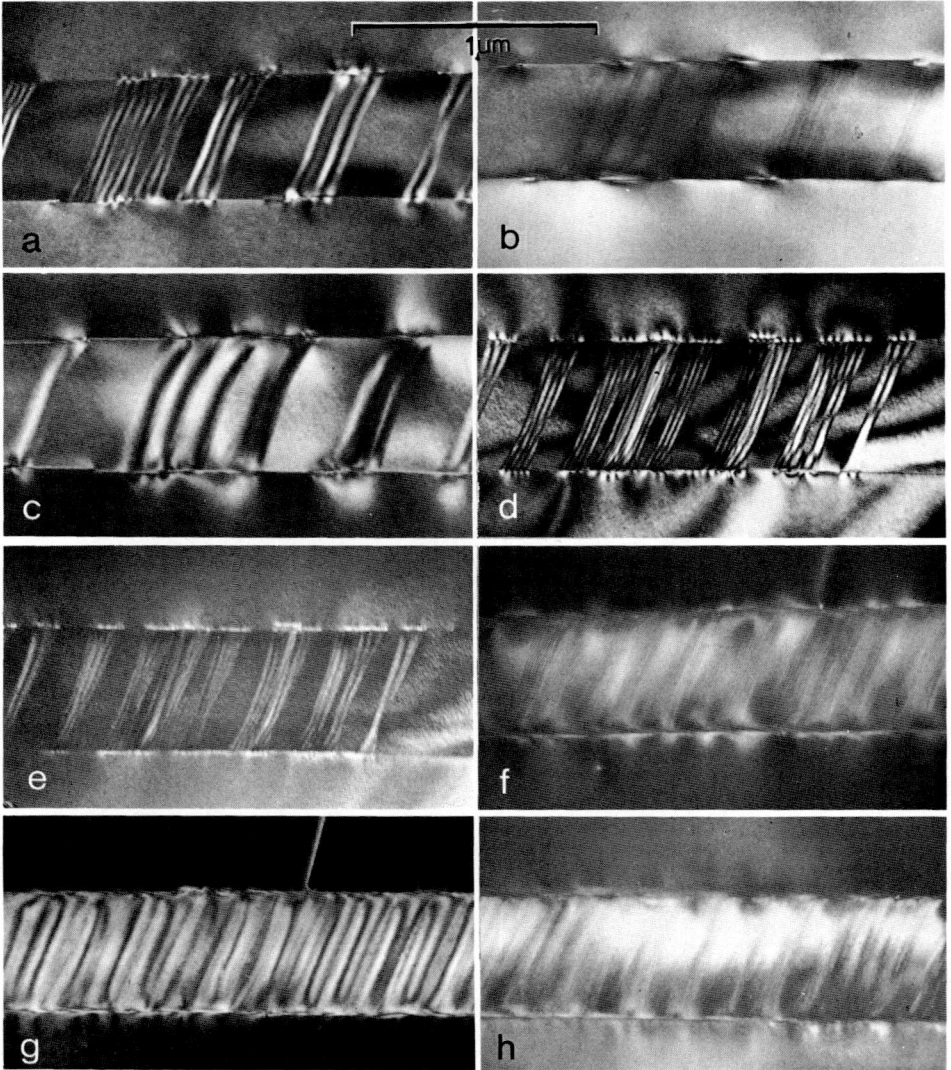

Fig. 2a–h. Dark field micrographs of the lamellae in SA 1019 taken with the following reflections (a) $\bar{2}\bar{2}\bar{1}$, (b) 220, (c) $\bar{2}\bar{2}\bar{4}$, (d) 002, (e) 004, (f) 006, (g) $\bar{2}\bar{2}3$, (h) $\bar{1}\bar{1}6$. 100 kV. Micrographs (a), (b) and (c) are from the same area. (d) and (e) are also from the same area; so are (f) and (g)

by means of a massive transformation. In pigeonites which have exsolved from augites and subsequently inverted to orthopyroxene, the (100) lattice planes of parent and product are parallel.

The existence of (100) stacking faults in the uninverted pigeonite lamellae suggests that they may initiate the inversion reaction, particularly as their frequency for a given lamellar width generally decreases with the iron content of the augite matrix (the temperature of the inversion decreases with Fe-content, Brown, 1957).

The structures of the various pyroxenes differ in the relative positions of the Si−O chains along the c-axis and in the resulting coordination of the cations between them. Morimoto and Koto (1969) showed that the orthoenstatite structure could be "derived" from clinoenstatite by twinning of every unit cell on the (100) plane. Although the structures of $P2_1/c$ pigeonite and clinoenstatite are not identical (they differ for instance in the details of the coordination of the cations), a (100), 1/6 [001] fault is a plausible precursor of orthopyroxene in pigeonite.

McLaren has shown (paper presented at the conference "Deformation mechanisms in rocks" held in Goolwa, South Australia, September, 1974) that the fault involved in the transformation of orthoenstatite to clinoenstatite is 1/6 [001] (100). He also showed that the one structure could be derived from the other by a fault of approximately this magnitude. See also Kirby (Chapter 6.7 of this volume).

References

Atkins, F.B.: Pyroxenes of the Bushveld intrusion, South Africa. J. Petrol. **10**, 222–249 (1969).
Binns, R.A., Long, J.V.P., Reed, S.J.B.: Some naturally occurring members of the clinoenstatite-clinoferrosilite mineral series. Nature **198**, 777–778 (1963).
Bonnichsen, B.: Metamorphic pyroxenes and amphiboles in the Biwabik Iron Formation, Dunka River Area, Minnesota. Mineral. Soc. Spec. Paper **2**, 217–239 (1969).
Bown, M.G., Gay, P.: An X-ray study of exsolution phenomena in the Skaergaard pyroxenes. Mineral. Mag. **32**, 379–388 (1960).
Boyd, F.R., Brown, G.M.: Electron-probe study of pyroxene exsolution. Mineral. Soc. Am. Spec. Paper **2**, 211–216 (1969).
Brown, G.M.: Pyroxenes from the early and middle stages of fractionation of the Skaergaard intrusion, East Greenland. Mineral. Mag. **31**, 511–543 (1957).
Brown, G.M.: Pigeonitic pyroxenes: a review. Geol. Soc. Am. Mem. **132**, 523–534 (1972).
Champness, P.E., Lorimer, G.W.: A direct lattice-resolution study of precipitation (exsolution) in orthopyroxene. Phil. Mag. **30**, 357–365 (1974).
Hashimoto, H., Howie, A., Whelan, M.J.: Anomalous electron absorption effects in metal foils − theory and comparison with experiment. Proc. Roy. Soc. (London), Ser. A **269**, 80–103 (1962).
Massalski, T.B.: Massive transformations. In: Phase transformations, A.S.M., p. 433–486 (1970).
Morimoto, N., Appleman, D.E., Evans, H.T.: The crystal structures of clinoenstatite and pigeonite. Z. Krist. **114**, 120–147 (1960).
Morimoto, N., Koto, K.: The crystal structure of orthoenstatite. Z. Krist. **129**, 65–83 (1969).
Poldervaart, A., Hess, H.H.: Pyroxenes in the crystallisation of basaltic magmas. J. Geol. **59**, 472–489 (1951).
Simmons, E.C., Lindsley, D.H., Papike, J.J.: Phase relations and crystallisation sequence in a contact-metamorphosed rock from the Gunflint iron formation. Minnesota. J. Petrol. **15**, 539–565 (1974).

CHAPTER 4.7

On the Detailed Structure of Ledges in an Augite-enstatite Interface

D.L. KOHLSTEDT and J.B. VANDER SANDE

1. Introduction

Precipitation phenomena in Mg-rich orthopyroxenes have recently been studied by transmission electron microscopy. Three types of precipitates have been observed: the first is a (100) G.P. zone-like precipitate, the second is a monoclinic, Ca-rich lamellar precipitate identified as an intermediate clinopyroxene phase, and the third is the equilibrium phase augite (Champness and Lorimer, 1973, 1974; Lorimer and Champness, 1973; Kohlstedt and Vander Sande, 1973). Discussions of and references to the optical and X-ray data concerning the lamellar precipitates in orthopyroxenes can be found in the papers by Champness and Lorimer.

In the case of the wider ($>0.2\mu$m) augite precipitates, the (100) matrix-lamella interfaces contain misfit dislocations with both [100] and [001] Burgers vector components (Vander Sande and Kohlstedt, 1974). In addition, growth steps or ledges as high as four (100) orthorhombic lattice spacings...$2d_{100}$ (augite) $\cong d_{100}$ (enstatite) $\cong 18.3$ Å...are located at 100 to 1000 Å intervals along the matrix-lamella interfaces.

Although the ledges are undoubtedly important in the growth of the lamellae, little is known about their detailed structure. A complete description of the ledges requires that the matrix, the lamella, and the associated interface be observed as an entity with high spatial resolution. Therefore, in the present paper, high-resolution electron microscopy is used to study the enstatite-augite interface by direct lattice-fringe resolution of both the precipitate and the matrix planes parallel to (100). The nature of the lattice fringes near ledges is examined in detail.

2. Electron Microscopy

Gem-quality fragments of the orthopyroxene enstatite, $(Mg_{1.80}Fe_{0.14}Ca_{0.06})(Si_{1.94}Al_{0.06})O_6$, from near Sultan Hamud, Kenya, were chosen for this investigation. Electron-transparent (at 100 kV) foils were prepared by ion bombardment thinning of 50 μm thick petrographic sections. Ion-thinned samples were coated with approximately 200 Å of carbon prior to observation to eliminate charging effects.

In a high-resolution direct lattice-fringe image of the matrix-lamella interface, a substantial amount of structural detail is apparent, as is observed in Fig. 1. Both matrix-lamella interfaces separating the 0.1 μm-wide augite lamella from the enstatite matrix are seen in this micrograph taken with the electron beam parallel to [010]. The 18.3 Å (100) lattice fringes are imaged in the enstatite. The inclusion of higher-order systematic reflections, e.g., 200, 300, 400..., in the objective aperture results in subfringes in many of these (100) fringes. The nature of the subfringes changes along the length of any given fringe as the diffracting condition changes. In the lamella, the 4.6 Å (200) lattice fringes are imaged; the 100 reciprocal-lattice reflections are forbidden for the space group C2/c and cannot be formed by double diffraction.

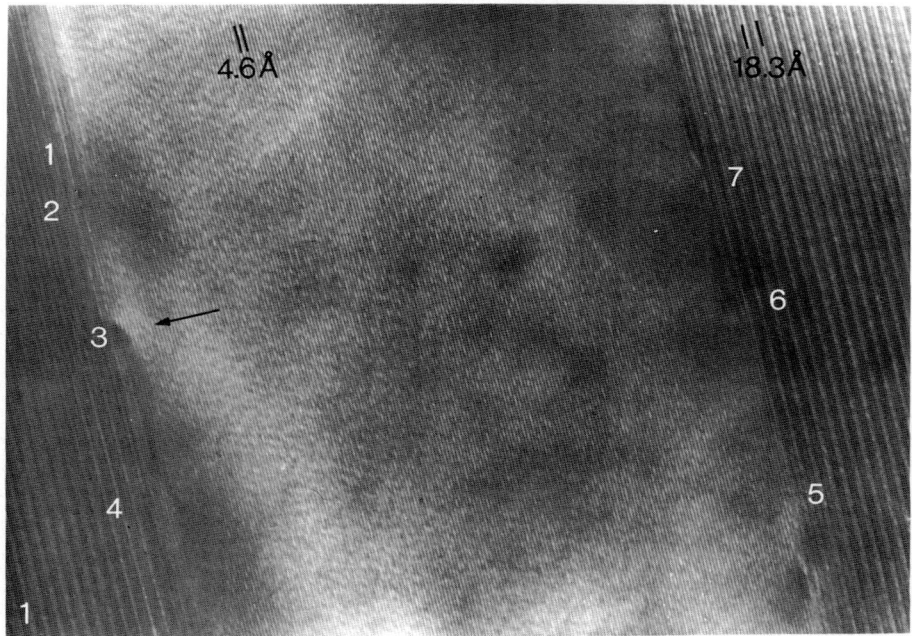

Fig. 1. Direct lattice-fringe image of an augite lamella in an enstatite matrix. The (200) 4.6 Å augite planes and the (100) 18.3 Å enstatite planes are imaged. Seven growth ledges, marked by numbers, are present in the two matrix-lamella interfaces. An intensity cusp due to local elastic strain near ledge #3 is arrowed. The electron beam is parallel to the [010] zone axis

Subtly visible in the interfaces of Fig. 1 are intensity cusps, one of which is arrowed. These intensity cusps are more clearly seen when an interface is viewed in the standard bright field mode with a small objective aperture (cf., Champness and Lorimer, 1974; Vander Sande and Kohlstedt, 1974). The intensity cusps, which appear at each step or ledge in the planar structure of the matrix-lamella interface, evidence the straining of the lattice in the vicinity of the ledge. A total of seven ledges, denoted by arabic numerals, are visible in the two matrix-lamella interfaces.

The structure of the ledges in Fig. 1 is complex. (The detailed nature of a ledge is best viewed by sighting along the lattice fringes while holding the micrograph at a glazing angle.) At ledge #5, for example, two 4.6 Å fringes merge into a single 9.2 Å repeat. Thus the basic unit in the growth of an augite lamella into the enstatite matrix appears to be a ledge only one-half a (100) orthorhombic lattice spacing, rather than a full spacing as had previously been suggested (Champness and Lorimer, 1974; Vander Sande and Kohlstedt, 1974).

Near both #6 and #7, an enlargement and schematic of which are shown in Fig. 2, two 9.2 Å fringes are formed from 4.6 Å fringes. The 9.2 Å fringes then combine to yield a 18.3 Å repeat. Between the two short segments of 18.3 Å spacing exist a 9.2 Å and two 4.6 Å fringes.

Fig. 2. An enlarged view of ledges #6 and #7 in Fig. 1. The inset is a schematic representation of the lattice fringes around these ledges

In contrast to the ledge structure observed in the only other published high-resolution lattice fringe image of an augite-enstatite interface (Vander Sande and Kohlstedt, 1974), no terminating fringes are observed at any of the seven ledges in Fig. 1. In this earlier reported work, terminating lattice fringes were identified (cf. Cockayne et al., 1971) as the [100] components of the misfit dislocations observed along the interface at lower magnification. The factors which determine whether a dislocation will form at a growth ledge are intimately tied to the balance struck between the chemical free energy, the strain energy, and the surface energy. The composition of the matrix and the lamella studied here are nearly identical to those in the earlier investigation. The lamellar widths in the present investigation, however, are smaller by a factor of 2.5. Thus the strain energy created by the (100) lattice-plane mismatch between the augite and the enstatite must become large enough to require misfit dislocations when the lamellar width exceeds 0.1–0.2 μm.

3. Conclusion

Direct lattice-fringe imaging of the (100) interfacial planes between an augite lamella and the enstatite host reveal two important new pieces of information concerning the structure of growth ledges in the interface. First, the fundamental step in the ledges is the addition or depletion of one (100) augite lattice plane, not one (100) enstatite lattice plane. Second, for lamellae less than ~ 0.2 μm in width, no terminating (100) lattice fringes are observed at the ledges, while for wider lamellae, terminating fringes are present at nearly every ledge.

References

Champness, P.E., Lorimer, G.W.: Precipitation (exsolution) in an orthopyroxene. J. Mat. Sci. **8**, 467–474 (1973).
Champness, P.E., Lorimer, G.W.: A direct lattice-resolution study of precipitation (exsolution) in orthopyroxene. Phil. Mag. **30**, 357–366 (1974).
Cockayne, D.J.H., Parsons, J.R., Hoelke, C.W.: A study of the relationship between lattice fringes and lattice planes in electron microscope images of crystals containing defects. Phil. Mag. **24**, 139–153 (1971).
Kohlstedt, D.L., Vander Sande, J.B.: Transmission electron microscopy investigation of the defect microstructure of four natural orthopyroxenes. Contrib. Mineral. Petrol. **42**, 169–180 (1973).
Lorimer, G.W., Champness, P.E.: Combined electron microscopy and analysis of an orthopyroxene. Am. Mineralogist **58**, 243–248 (1973).
Vander Sande, J.B., Kohlstedt, D.L.: A high-resolution electron microscopy study of exsolution in enstatite. Phil. Mag. **29**, 1041–1049 (1974).

CHAPTER 4.8

The Phase Distributions in Some Exsolved Amphiboles

M.F. GITTOS, G.W. LORIMER, and P.E. CHAMPNESS

1. Introduction

Coexisting and intergrown amphiboles have been widely reported in the literature. In 1923 Asklund postulated the existence of a miscibility gap between hornblende and cummingtonite, and since then it has been comprehensively documented. In addition, several other miscibility gaps within the amphibole group have been demonstrated by the discovery of two or more primary amphibole phases in the same rock (see for instance Klein, 1968, 1969; Robinson et al., 1969).

The appearance of lamellae of a second amphibole phase within a grain indicates precipitation from solid solution, referred to in the literature as "unmixing" or "exsolution". A homogeneous amphibole crystallizes during initial metamorphism and undergoes phase separation on cooling below the solvus temperature. The failure of laboratory experiments to reproduce exsolution structures has been attributed to the sluggishness of the reactions, a view supported by the attempts to homogenize exsolved amphiboles (Ross et al., 1969). Exsolution has not been reported in all the systems in which a miscibility gap has been shown to exist, but this may simply be because optical microscopy is inadequate to detect it and, since the unit cells of host and exsolved phases are commonly very similar, the second phase may, in some cases, escape detection by X-ray diffraction. Also a sample must be found which has had the necessary P-T history to produce an exsolved structure.

2. Description of Specimens

The hornblende and grunerite are from the Cairnsmore of Cairsphairn igneous complex, Scotland. Colorless grunerite grains contain irregular patches of brown hornblende. The sample, obtained from S.O. Agrell (no. 59681) has been analyzed by Klein (1968, assemblage 5–22).

Two of the orthoamphiboles are from the Orange area, Massachusetts and New Hampshire. 6A9X and N30X are medium-grained hornblende-anthophyllite amphibolites and are described with analyses in Robinson and Jaffe (1969). AV9 is a medium-grained gedrite-cummingtonite amphibolite from the Ashuelot River near Surry Mountain, New Hampshire about 30 km north of the Orange area and is at approximately the same metamorphic grade as 6A9X and N30X. It was collected by A.B. Thompson.

Samples for electron microscopy were prepared from petrographic thin sections by ion thinning. 100 kV, 1000 kV, high-resolution, analytical and transmission electron-microscopy were used when appropriate.

3. Exsolution in Monoclinic Amphiboles

The chemistry of exsolution in monoclinic amphiboles depends mainly upon the distribution of the relatively large calcium ion. The structure of the monoclinic ferromagnesian amphiboles cummingtonite and grunerite can accommodate little more than 1 wt. % CaO at low temperatures (Klein, 1968, demonstrated that published analyses indicating higher CaO contents may be in error). In contrast tremolites, actinolites, and hornblendes contain nearly 2 calcium ions per formula unit and CaO can constitute over 10 wt. %. Exsolution of calcium-rich from calcium-poor amphiboles and *vice versa* has been detected by optical microscopy and X-ray diffraction (Eskola, 1950; Vernon, 1962; Callegari, 1965; Ross et al., 1969). In an earlier paper (Gittos et al., 1974) we described the microstructures of two examples in this system and Sections 3.1 and 3.2 of this paper describe a further sample.

Although it is clear that a miscibility gap exists between monoclinic and orthorhombic amphiboles, exsolution in such systems is always of the similar amphibole. For example, in a hornblende-anthopyllite assemblage the hornblende will exsolve either cummingtonite or grunerite and the anthophyllite will exsolve gedrite (Ross et al., 1969; Robinson and Jaffe, 1969) (Section 4 of this paper describes the phase distribution in some orthorhombic amphiboles from such assemblages).

3.1 Heterogeneous Nucleation and Growth in Hornblende and Grunerite

In a previous paper (Gittos et al., 1974) we described the phase distributions in the coexisting hornblendes and cummingtonite-grunerites of two metamorphic rocks. While the microstructures of the two hornblendes on the one hand and the two cummingtonites on the other were similar, those of the calcium-rich and calcium-poor amphiboles from the same rock were distinctly different. Here we report our observations on the amphibole microstructures in a third hornblende-grunerite rock.

The phase distribution in the hornblende closely resembles those previously described; it contains coherent lamellae of grunerite, approximately parallel to (100) and ($\bar{1}$01)[1], which nucleated heterogeneously. The grunerite also contains heterogeneously-nucleated lamellae of hornblende approximately parallel to (100) and ($\bar{1}$01): Fig. 1 shows lamellae of hornblende which have nucleated at a (100) twin boundary. The semicoherent interface between intergrown grunerite and hornblende was also found to have acted as a nucleation site.

---
[1] These indices refer to the C2/m unit cell. The corresponding planes in the I2/m unit cell are (100) and (001)

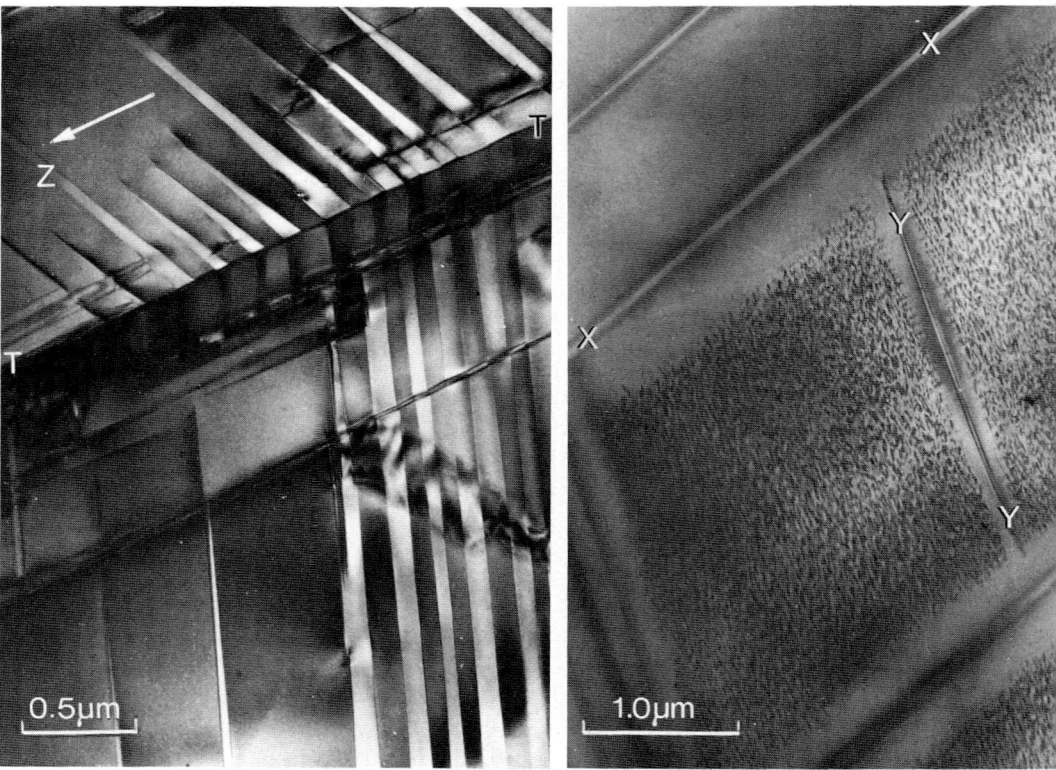

Fig. 1   Fig. 2

Fig. 1. Nucleation of ($\bar{1}01$) hornblende lamellae at a (100) twin boundary T-T in grunerite (sample 59681)

Fig. 2. Homogeneous precipitation of (100) platelets between heterogeneous ($\bar{1}01$), X-X, and (100), Y-Y, lamellae in grunerite (sample 59681)

The motion of small steps or ledges along the lamellar interfaces (cf. Fig. 4 of Gittos et al., 1974) provides the thickening mechanism of the precipitates (Laird and Aaronson, 1969, observed in situ precipitate thickening in Al-Ag alloys). Similar ledges have also been reported in pyroxenes (Champness and Lorimer, 1973; Copley et al., 1974; Vander Sande and Kohlstedt, 1974). If such ledges are arranged in a systematic manner on each side of a lamella they could explain the fact that, although the (100) or ($10\bar{1}$) lattice planes are always nearly parallel in the matrix and precipitate for a lamella approximately parallel to (100) or ($10\bar{1}$), the interface may lie at a small to moderate angle to these planes (Robinson et al., 1971a). This arrangement has been found as a rare phenomenon in pigeonite exsolved from augite by Copley et al. (1974), but has not been observed in amphiboles where ledges occur with less regularity and with lower frequency than in pyroxenes. The rarity of this phenomenon suggests it is an unlikely explanation for the angular relations observed by Robinson et al.

3.2 Homogeneous Nucleation and Growth in Grunerite

The cummingtonite and grunerite previously examined contained a homogeneous distribution of fine (100) platelets between the (100) and (10$\bar{1}$) hornblende lamellae. The microstructure of the grunerite studied here is similar. A precipitate-free zone occurs adjacent to the lamellae (Fig. 2) and indicates that the platelets formed after the lamellae in areas where the calcium supersaturation was high enough to overcome the energy barrier for homogeneous nucleation. The platelets are also absent in areas containing a high density of lamellae, such as at heterogeneous-nucleation sites (Fig. 1) where the calcium has been drained into the calcic phase.

3.3 Summary

In the three Ca-rich and Ca-poor clinoamphiboles which have been examined, precipitates nucleated heterogeneously at phase and twin boundaries and thickened by a ledge mechanism while remaining coherent with the matrix. If the solute supersaturation is high enough in Ca-poor amphiboles, homogeneous nucleation and growth occurs between the heterogeneously-nucleated lamellae. Similar microstructures have been reported in pyroxenes (Champness and Lorimer, 1973; Copley, 1973; Nord et al., 1973; Nord et al., Chapter 4.5 of this volume).

4. Precipitation in Orthoamphiboles

Although the existence of a miscibility gap between ortho- and clinoamphiboles has been known for some time (Rabbitt, 1947) it is only recently that the presence of exsolution in orthoamphiboles has been suggested and the important miscibility gap between anthophyllite and gedrite has been demonstrated (Robinson and Jaffe, 1969; Stout, 1969; Robinson et al., 1971b). Most anthophyllites and gedrites are optically homogeneous, although exsolution may be detected by X-ray diffraction or electron microscopy in many of them.

Gedrites are aluminum-rich with aluminum occupying up to two tetrahedral sites per formula unit.[2] Anthophyllites are correspondingly richer in silicon and magnesium and poorer in aluminum. They also generally contain less sodium than gedrite. The unit-cell dimensions of anthophyllite and gedrite are nearly identical, the only detectable difference being that the b-axis of anthophyllite is slightly larger (by less than 1%) than that of gedrite (Ross et al., 1969; Christie and Olsen, 1974). The plane of 'best fit' between the lattices is thus (010).

Robinson and Jaffe (1969) and Robinson et al. (1971b) reported that X-ray single-crystal diffraction patterns from several anthophyllites and gedrites showed fine splitting of the reflections and concluded that the samples were exsolved. Subsequent optical microscopy revealed exsolution textures in two of the samples.

---

[2] The compositions of orthoamphiboles described in this paper in terms of % gedrite have been calculated assuming that the gedrite end-member contains two tetrahedral aluminums per formula unit.

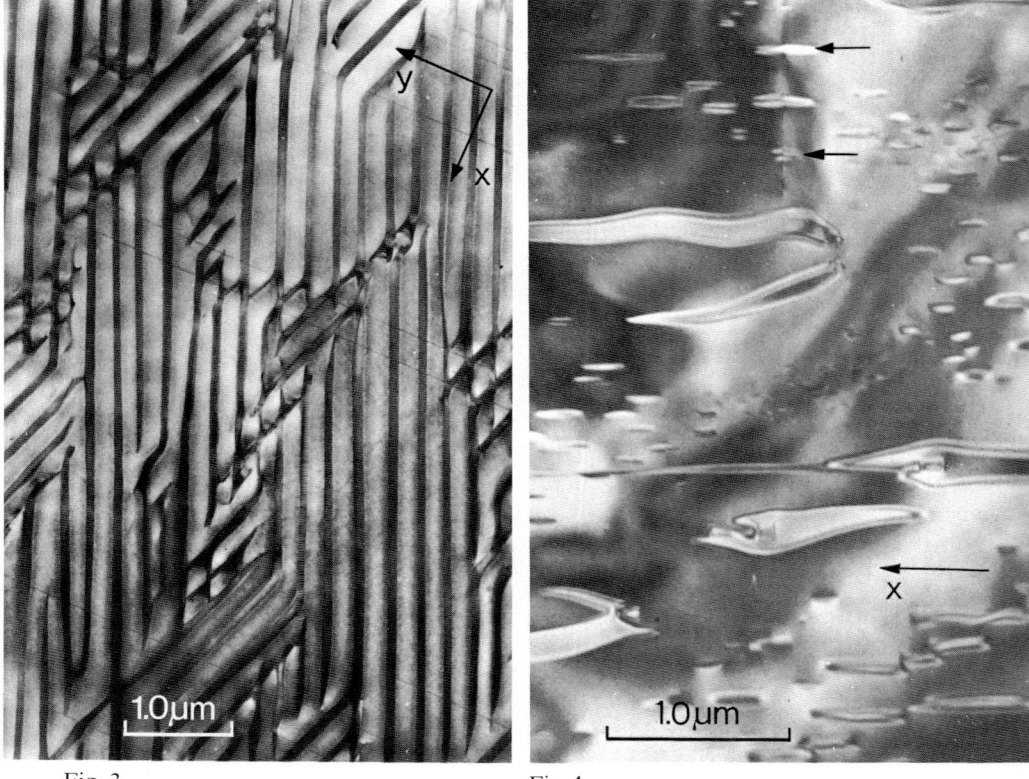

Fig. 3

Fig. 4

Fig. 3. Anthophyllite 5A9X. Interlinked lamellae approximately parallel to {120} which constitute the majority of the grain. Note (100) linear features

Fig. 4. Anthophyllite 6A9X. In localized areas the precipitates are isolated, 'wavy' particles. Between them are smaller (010) platelets, some of which have nucleated on the (100) faults (arrowed)

4.1 Anthophyllite 6A9X (34% Gedrite)

The anthophyllite is optically homogeneous, but electron microscopy reveals that it is exsolved on a fine scale. The bulk of the grain is composed of interlinked lamellae lying on irrational matrix planes close to {120} (Fig. 3). Analytical electron-microscopy has shown that localized areas of the matrix are depleted in aluminum. (For a description of the technique see Lorimer and Cliff, Chapter 7.2 of this volume.) Here the volume fraction of the second phase decreases and the structure degenerates into a series of 'wavy' precipitates straddling the (010) plane or, in regions very low in aluminum, into isolated particles, some of which are surrounded by a second generation of small (010) precipitates (Fig. 4).

A set of fine linear features parallel to (100) has been observed in all the orthoamphiboles which have been studied (Fig. 3). Lattice resolution of the (100)

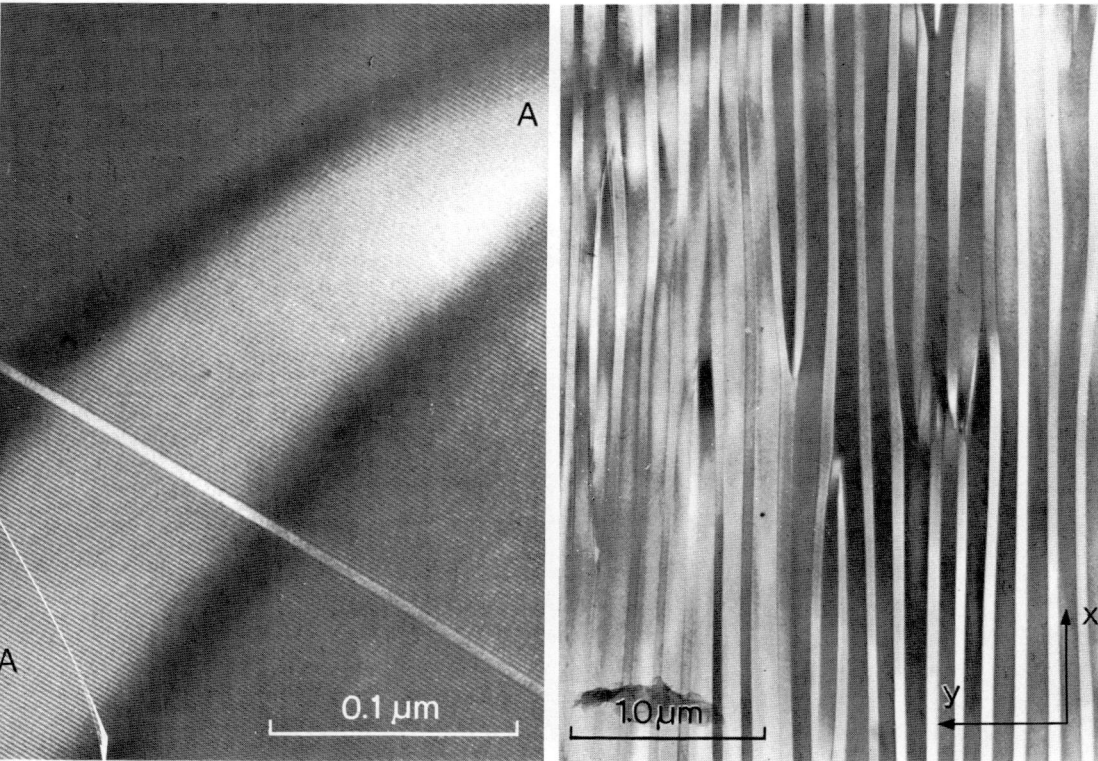

Fig. 5   Fig. 6

Fig. 5. Lattice resolution of the (100) planes of anthophyllite 6A9X (spacing 18.6 Å) shows complete coherency across the interface between the gedrite lamella (*A-A*) and the matrix. A narrow (100) plate 50 Å wide can also be seen

Fig. 6. (010) Lamellae of gedrite in anthophyllite N30X

planes (Fig. 5) shows that the lamellae are only a few unit cells wide. It was not possible to characterize them either by electron diffraction or by analytical electron-microscopy due to their small size and low volume-fraction. They have acted as nucleation sites for some of the (010) platelets (Fig. 4).

4.2 Anthophyllite N30X (41% Gedrite)

In this specimen the gedrite precipitates are periodic, coherent lamellae with interfaces which are not strictly planar, but on average are parallel to (010) (Fig. 6). The microstructure is similar to that observed in an orthoamphibole (60% gedrite) by Christie and Olsen (1974). The (100) 'faults' observed in specimen 6A9X are present in this specimen also.

4.3 Gedrite AV9 (60% Gedrite)

Faint lineations can be seen parallel to (010) when this specimen is examined in the optical microscope (cf. Fig. 12 of Robinson et al., 1971b). In the electron microscope a two-phase microstructure is revealed (see Frontispiece on page 172). Compared with those in N30X, the (010) lamellae in this sample are thicker and more widely spaced. In addition, their morphology is more complex and their distribution is less regular. A second generation of small (010) precipitates has formed between the lamellae (cf. the structures found in grunerites) and, although most of them appear to have nucleated homogeneously, some have nucleated on the thin (100) 'faults'. The anthophyllite lamellae also contain planar features parallel to (010).

4.4 Chemistry

The thickness of the gedrite lamellae in 6A9X and N30X is about $0.1\,\mu m$, which is just within the spatial resolution of the analytical microscope EMMA-4, but the antophyllite lamellae in AV9 are considerably larger (up to $0.3\,\mu m$ wide). The analyses of the matrix and lamellae given in Table 1 (which are averages of four readings) are consistent with the analyses of Stout (1972) of coexisting coarse anthophyllite and gedrite from Norway. (Sodium analyses are not quoted because sodium is near the detection limit of the instrument and it is also possible that it is lost as a result of ionization damage in the electron beam.) As a result of the fine scale of the structures in 6A9X and N30X, it was difficult to position the probe entirely within the minor phase and it is possible that the measured differences in composition between the lamellae and matrix in these samples are slightly greater than is shown in Table 1. However it is clear that aluminum (and possibly calcium) has segregated to the gedrite phase with magnesium and silicon concentrating in the anthophyllite. Iron does not partition between the phases. This pattern of segregation is in agreement with that proposed by Robinson et al. (1971b).

Table 1. Structural formulae of the two phases in the orthoamphiboles as determined in EMMA-4. The analyses are based on a total of 15 cations, excluding sodium. The number of tetrahedral aluminum ions has been calculated assuming $Si + Al^{IV} = 8$ and the percentage of gedrite has been calculated from the Al^{IV} assuming that ideal gedrite has two Al^{IV} per formula unit

		Si	Al	Fe	Mg	Mn	Ca	Al^{IV}	% Gedrite
6A9X	matrix	7.40	0.55	2.36	4.48	0.10	0.11	0.55	27
	lamellae	6.68	2.05	2.45	3.58	0.07	0.17	1.32	66
N30X	matrix	7.22	0.61	1.83	5.21	0.05	0.08	0.61	30
	lamellae	6.69	2.14	1.84	4.15	0.06	0.12	1.31	66
AV9	matrix	6.23	2.57	2.31	3.74	0.09	0.06	1.77	89
	lamellae	7.36	0.68	2.14	4.67	0.07	0.08	0.68	34

4.5 Discussion of Exsolution Mechanisms

No evidence for heterogeneous nucleation of precipitates has been observed in the regions containing a high volume-fraction of precipitate in 6A9X (Fig. 3) or in N30X. In this respect the microstructures are similar to the gedrite described by Christie and Olsen (1974). These authors cite the exsolution of anthophyllite as an example of spinodal decomposition of a homogeneous gedrite. The phase distribution observed in our samples could also have been produced by the coarsening of a spinodal structure.

Both AV9 and 6A9X (in areas such as that shown in Fig. 4) contain (010) lamellae several microns long (in AV9 they are several tens of microns long) with 'wavy' matrix/lamellar interfaces; their distribution is consistent with heterogeneous nucleation. Large, heterogeneously-nucleated lamellae have been observed in other amphiboles, and pyroxenes (Gittos et al., 1974; Copley et al., 1974; Champness and Lorimer, Chapter 4.1 of this volume); however this is the first report of large lamellae which do not have planar (or stepped-planar) interfaces. 'Wavy' interfaces have been observed in feldspars which have probably exsolved by a spinodal mechanism but, when the microstructure has coarsened to a scale comparable to that in AV9, planar interfaces have developed. The 'wavy' interfaces in orthorhombic amphiboles must reflect an almost isotropic matrix/lamellar interfacial energy, a result of the structural similarities of the two amphiboles. The final shape of the precipitates is probably not governed by the criterion of minimum interfacial-energy, but is the result of the impingement of solute diffusion-fields and the original distribution of nucleation sites. The change in habit plane from {120} to {010} over a composition interval of 7% gedrite in samples 6A9X and N30X is possibly also a reflection of the isotropy of interfacial energy.

The origin and identity of the planar faults parallel to (010) inside the large precipitates of AV9 are not known. They appear to affect the growth of the lamellae, for abrupt changes in lamellar shape often occur at their intersection with the lamellar/matrix interface (Frontispiece on page 172).

The planar features parallel to (100), which occur in all three orthoamphiboles, cut through both the matrix and the lamellae and, where they terminate in the field of view, they always do so at a matrix-lamellar boundary (Fig. 3). As they produce a displacement of the exsolution lamellae (Fig. 5), they are probably a deformation feature; they may be thin lamellae of clino-amphibole which form from orthoamphibole by an analogous mechanism to the formation of clinoenstatite from orthoenstatite[3] (Kirby, Chapter 6.7 of this volume). The diffraction contrast exhibited by the faults shows that they occur in two orientations (clinoamphibole, because it has a lower symmetry than orthoamphibole, would necessarily form in two twin-related orientations). The faults must have formed after the large (010) exsolution lamellae (they cut across them) and before the small (010) platelets in AV9 and 6A9X.

The (010) platelets formed between the large lamellae in regions where the solute concentration is sufficiently high for nucleation to occur. The origin of

---

[3] The presence of (100) faults in protoamphibole was inferred by Gibbs (1969) from X-ray diffraction studies. He suggested that they were thin lamellae of orthoamphibole.

the precipitate-free zone is similar to that in the grunerite (Fig. 2) (i.e. a solute concentration-profile adjacent to the large lamellae). The efficiency of the (100) planar faults in reducing the barrier to precipitate nucleation can be clearly seen (Fig. 4 and frontispiece). Heterogeneous nucleation of the (010) platelets has occurred on the faults in regions closer to the large lamellae than the homogeneously-nucleated platelets and they have generally grown to a larger size, i.e. heterogeneous nucleation occurred at a lower solute supersaturation than that required for homogeneous nucleation. The distribution of the fine platelets between the coarse lamellae in AV9 and in some areas of 6A9X infers a three-stage exsolution sequence: heterogeneous nucleation and growth of the large lamellae at a high temperature followed by the heterogeneous nucleation of the platelets on the (100) 'faults' at a lower temperature and, finally, homogeneous nucleation of the platelets at an even lower temperature.

Christie and Olsen (1974) have observed a lamellar microstructure identical to that shown in Fig. 6 in a gedrite of similar composition to AV9. These two distinct phase distributions in gedrites of approximately the same bulk composition, implying different exsolution mechanisms, may reflect differences in their cooling rates during exsolution: the homogeneous distribution and finer scale of the microstructure in Christie and Olsen's sample is consistent with a spinodal mechanism and a faster cooling rate (see Champness and Lorimer, Chapter 4.1 of this volume).

References

Asklund, B.: Petrological studies in the neighbourhood of Stavsjö at Kolmarten. Sveriges Geol. Unders. Arsb. **17**, No. 6 (1923).
Callegari, E.: Osservazioni su alcuni cummingtoniti del massiccio dell'Adamello. Mem. Accad. Patavina, Cl. Sci. Mat.-Nat. **78**, 273–310 (1965).
Champness, P.E., Lorimer, G.W.: Precipitation (exsolution) in an orthopyroxene. J. Mat. Sci. **8**, 467–474 (1973).
Christie, O.H.J., Olsen, A.: Spinodal precipitation in minerals review and some new observations. Bull. Soc. Franç Minéral. Crist. **97**, 386–392 (1974).
Copley, P.A.: Electron microscope studies of phase transformations in pyroxene minerals. Ph. D. Thesis, University of Manchester (1973).
Copley, P.A., Champness, P.E., Lorimer, G.W.: Electron petrography of exsolution textures in an iron-rich clinopyroxene. J. Petrol. **15**, 41–57 (1974).
Eskola, P.: Paragenesis of cummingtonite and hornblende from Muuruvesi, Finland. Am. Mineralogist **35**, 728–734 (1950).
Gibbs. G.V.: Crystal structure of protoamphibole. Mineral. Soc. Am. Spec. Paper **2**, 101–109 (1969).
Gittos, M.F., Lorimer, G.W., Champness, P.E.: Precipitation (exsolution) in an amphibole (the hornblende-grunerite system). J. Mat. Sci. **9**, 184–192 (1974).
Klein, C.: Coexisting amphiboles. J. Petrol. **9**, 281–330 (1968).
Klein, C.: Two amphibole assemblages in the system actinolite-hornblende-glaucophane. Am. Mineralogist **54**, 212–237 (1969).
Laird, C., Aaronson, H.I.: The growth of γ plates in an Al-15° Ag alloy. Acta Met. **17**, 505–519 (1969).
Nord, G.L., Lally, J.S., Heuer, A.H., Christie, J.M., Radcliffe, S.V., Griggs, D.T., Fisher, R.M.: Petrologic study of igneous and metaigneous rocks from Apollo 15 and 16 using high-voltage transmission electron microscopy. Proc. 4th Lunar Sci. Conf. Geochim. Cosmochim. Acta., Suppl. 4, **1**, 953–970 (1973).

Rabbitt, J.C.: Anthophyllite and its occurrence in southwestern Montana. Ph.D. Thesis, Harvard University (1947).
Robinson, P., Jaffe, H.W.: Chemographic exploration of amphibole assemblages from central Massachusetts and southernwestern New Hampshire. Mineral. Soc. Am. Spec. Paper No. 2, 251–274 (1969).
Robinson, P., Jaffe, H.W., Klein, C., Ross, M.: Equilibrium coexistence of three amphiboles. Contrib. Mineral. Petrol. **22**, 248–258 (1969).
Robinson, P., Jaffe, H.W., Ross, M., Klein, C.: Orientation of exsolution lamellae in clinopyroxenes and clinoamphiboles: consideration of optimal phase boundaries. Am. Mineralogist **56**, 909–939 (1971a).
Robinson, P., Ross, M., Jaffe, H.W.: Composition of the anthophyllite-gedrite series, comparisons of gedrite and hornblende and the anthophyllite-gedrite solvus. Am. Mineralogist **56**, 1005–1041 (1971b).
Ross, M., Papike, J.J., Shaw, K.W.: Exsolution textures in amphiboles as indicators of sub-solidus thermal histories. Mineral. Soc. Am. Spec. Paper No. 2, 275–296 (1969).
Stout, J.H.: An electron microprobe study of coexisting orthorhombic amphiboles (abst.). Trans. Am. Geophys. Union **50**, 359 (1969).
Stout, J.H.: Phase petrology and mineral chemistry of coexisting amphiboles from Telemark, Norway. J. Petrol. **13**, 99–145 (1972).
Vander Sande, J.B., Kohlstedt, D.L.: A high-resolution electron microscopy study of exsolution lamellae in enstatite. Phil. Mag. **29**, 1041–1049 (1974).
Vernon, R.H.: Co-existing cummingtonite and hornblende in an amphibolite from Duchess, Queensland, Australia. Am. Mineralogist **47**, 360–370 (1962).

CHAPTER 4.9

Physical Aspects of Exsolution in Natural Alkali Feldspars

C. WILLAIME, W.L. BROWN, and M. GANDAIS

1. Introduction

At high temperatures alkali feldspars form a complete series of solid solutions with monoclinic symmetry between albite and K-feldspar. On cooling, different *intracrystalline* processes may occur and we consider that the evolution of perthitic textures is determined by minimizing strain energy. It is possible to calculate the strain energies involved in these processes:

a) diffusive segregation of potassium and sodium atoms giving exsolution with K-rich and Na-rich lamellae either by spinodal decomposition or by homogeneous or heterogeneous nucleation;

b) symmetry change from $C2/m$ to $C\bar{1}$ and twinning due to a reversible displacement of silicon tetrahedra in sodium-rich feldspars, and/or to an ordering of aluminum and silicon atoms of the framework in either lamella. The reversible displacement is instantaneous, whereas the Al/Si ordering, a diffusive process, is geologically slow.

Different geological conditions, particularly cooling rates, give different final textures in the alkali feldspars. Volcanic explosions are followed by a quenching of the ejected materials; in this case alkali feldspars remain homogeneous as

---

Fig. 1a–f. Exsolution and twinning in alkali feldspar. Sample numbers refering to Table 1 are indicated. (a) Two-sanidine cryptoperthite Sample 1 (McConnell, 1974, personal communication; cf. McConnell, 1969, Fig. 1). (b) (010) photograph of two-sanidine cryptoperthite; the more abundant lamellae are potassic, $\sigma = 117°$. Sample 2 (Willaime et al., 1973, Fig. 2). (c) (001) photograph of a plagioclase-low-sanidine cryptoperthite showing variation of albite twin period ($2l$) with lamellar width (L). Sample 4 (Willaime and Gandais, 1972, Fig. 4). (d) (001) photograph of low-albite-orthoclase cryptoperthite with periodical albite twins. Sample 6 (Brown et al., 1972, Fig. 1). (e) (001) transmission photograph of a low albite-"orthoclase" microperthite (1 million volts) showing lenticular Na-rich domains with periodical albite twins in a matrix of M-twinned microcline (only pericline twins seen in the photo). Edge dislocations occur at the domain boundaries with a period of less than 1 μm. The spacing is much more regular near the outward limits of the lenses as seen in the narrow bands. Sample 13 (cf. Gandais et al., 1974). (f) (001) transmission photograph of a low-albite-microcline cryptoperthite with two textural types (a) diagonal association with zig-zag texture and albite twins (center) and (b) M-type microcline with pericline twinned low albite (twins not seen on (001)). Sample 14 (Willaime et al., 1973, Fig. 5)

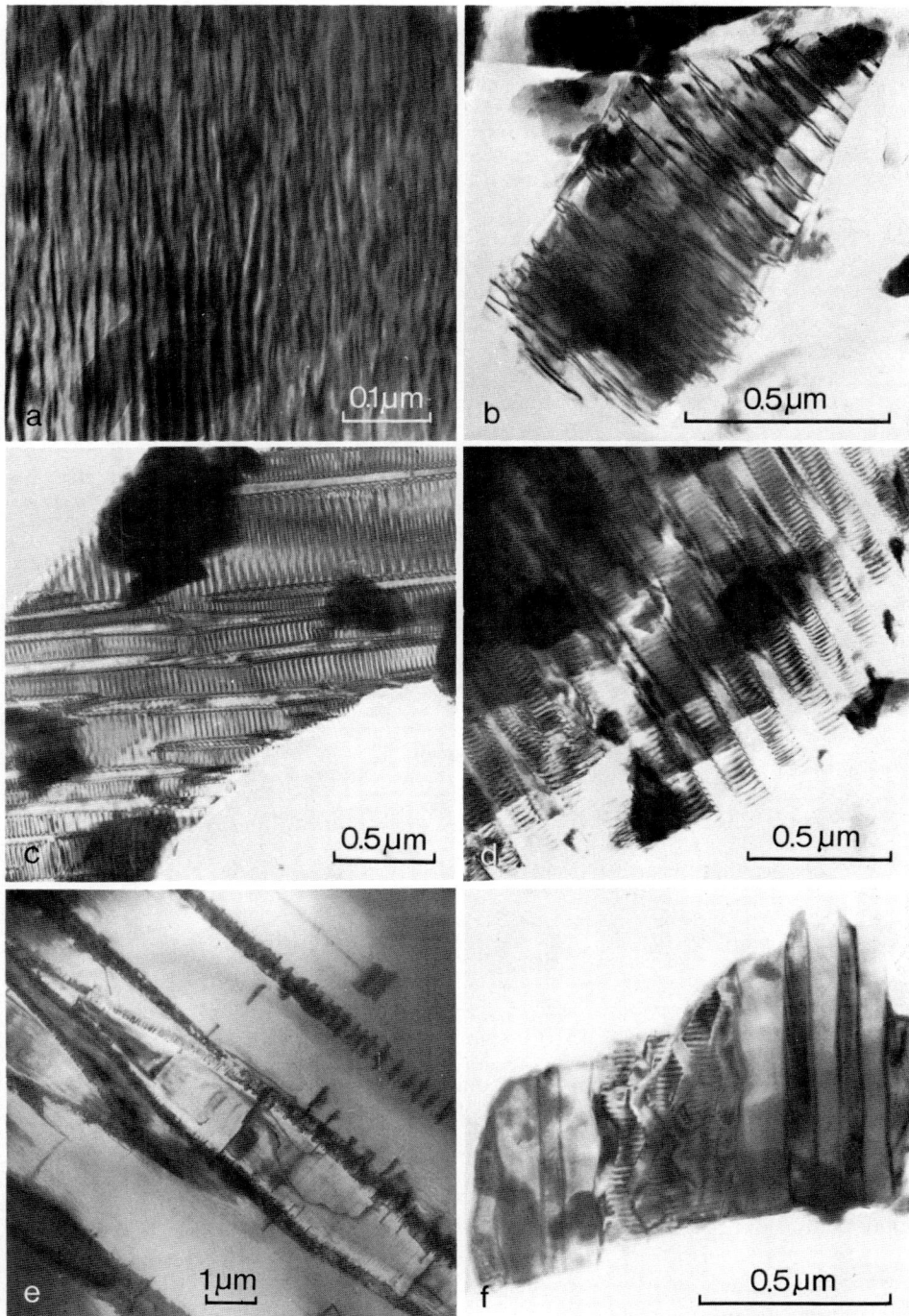

Fig. 1a–f

Table 1. Description of cryptoperthite textures examined by TEM

No.	Sample number	Feldspar and rock type, origin	Bulk composition	Exsolution boundary
1[a] (Fig. 1a)	–	Anorthoclase, Lava Pantelleria, Italy	$Or_{37}Ab_{63}An_0$	$(\bar{6}01)$
2 (Fig. 1b)	F105	Sanidine, Lava, Samothrace, Greece	–	$(\bar{6}01)$
3[a]	Spencer P	Moonstone, Korea	$Or_{42}Ab_{58}An_0$	$(\bar{6}01)$
4 (Fig. 1c)	F99	Anorthoclase, Larvikite, Larvik, Norway	$Or_{19}Ab_{68}An_{13}$	$(\bar{6}01)$
5	Spencer F	Moonstone, Frederiks Värn, Norway	$Or_{44}Ab_{52}An_4$	$(\bar{6}01)$
6 (Fig. 1d)	F97	Moonstone, Syenite, unknown	$Or_{48}Ab_{49}An_3$	$(\bar{6}01)$
7	128	Moonstone, Australia	$Or_{72}Ab_{27}An_1$	$(\bar{6}01)$
8[b]	Spencer M	Moonstone, Ambalangoda, Ceylon	$Or_{57}Ab_{40}An_2$	$(\bar{6}01)$ to $(\bar{6}31)$
9	189	Moonstone, Ceylon	$Or_{61}Ab_{38}An_1$	$(\bar{6}01)$ to $(\bar{6}31)$
10	–	Moonstone, Ceylon	–	$\approx(\bar{6}01)$
11[c]	L29	Anorthoclase, Syenite, Wausau, Wisc.	$Or_{28}Ab_{70}An_2$	$(\bar{6}31)$ to $(\bar{6}61)$
12[b]	Spencer N	Moonstone, Mogok, Burma	$Or_{54}Ab_{45}An_1$	$(\bar{6}31)$ to $(\bar{6}61)$
13 (Fig. 1e)	F91	Moonstone, unknown	$Or_{72}Ab_{26}An_2$	$(\bar{6}01)$ to $(\bar{6}21)$
14 (Fig. 1f)	L31	Anorthoclase, Syenite, Wausau, Wisc.	$Or_{35}Ab_{62}An_3$	$(\bar{8}01)$ $(\bar{8}61)$

[a] Akizuki and Sugawara (1970) described similar materials from Korea with periods of 20–900 Å (Fig. 4B) and 1200–3000 Å (Fig. 4A) respectively. Their Fig. 4A shows pericline twinned high albite like Spencer P and not albite twins as they reported.
[b] See Champness and Lorimer (Chapter 4.1 of this volume, Fig. 14a and b respectively).
[c] See Wenk (Chapter 1 of this volume, Fig. 2).

monoclinic sanidine. In lavas or in small intrusions, the cooling is slower and very fine cryptoperthitic textures are found (Fig. 1a, b and specimens 1 and 2 in Table 1). These textures are consistent with spinodal decomposition without coarsening of the exsolution lamellae. In large intrusions and in metamorphic rocks, the cooling time is much longer and coarser exsolution lamellae and com-

In fact, a K-rich lamella is also strained to adjust its boundary with the neighboring Na-rich triclinic lamella. McLaren (1974) observed that the twin periodicity for a given lamellar width varies inversely with the Or content of the cryptoperthite. This can be explained on taking the deformation of the K-rich lamellae into account. Since the lamellar compositions do not greatly depend on the bulk composition (Brown and Willaime, 1974), the relative volumes of the lamellae must vary with the bulk composition. The relative strains of each kind of lamella then depend on their relative volumes. The greater the volume of a domain, the less must be its deformation. Thus for a given width of the Na-rich lamellae in Or-rich cryptoperthites, the K-rich domains are bigger and so the φ angle used in the previous calculation is larger and this corresponds to a smaller periodicity of the twins. This model is in good agreement with the measurements obtained by McLaren (1974).

In zig-zag or lozenge-shaped Na-rich lamellae (Fig. 1e, f and samples 11, 12, 14) the twin periodicity is constant in a band which varies greatly in width along its length or even wedges out. This observation indicates either that the model does not apply to boundaries other than ($\bar{6}01$), which is nonsense, or that the twins formed when the lamellae had constant width parallel to ($\bar{6}01$) and that the present boundaries arose subsequently by diffusional rearrangement of alkali ions when the potassium feldspar became triclinic. Where the Na-rich zones are lenticular (Fig. 1e and samples 7 and 13) in a K-rich matrix, the twin periodicity varies along the lens, suggesting that twinning occurred after growth of the lenticular zones (see also McLaren, 1974, pp. 414–415).

2.3 Albite and Pericline Twinning

Periodic twinning as explained above occurs in Na-rich triclinic lamellae. In samples 3 and 14 in part, these are pericline twins (Fig. 1f), whereas in others they are albite twins (sometimes with pericline twins in a few lamellae as in sample 6). Albite and pericline twinning correspond to the same geometrical shear (twice the obliquity, $(b \wedge b^*)$, and the difference in the shear energy in both cases is due to the elastic anisotropy of feldspars). The variation of the calculated shear energy in both cases is only due to the variation of a stiffness coefficient when the reference axes are tilted less than $4°$ $(b \wedge b^*$ angle). So the shear energy is very similar for both twins and the production of one or the other must be determined by some other cause not yet determined.

2.4 Coherence Along the Boundary Plane

In cryptoperthites the lamellar compositions are very different, but X-ray measurements show that the lattice parameters of both lamellae are similar in the boundary plane, whereas they differ perpendicularly to this plane. In some cases (Fig. 1d, sample 6), the measured parameters in the boundary plane are nearly identical: the boundary is totally coherent. In other cases, the lattice plane arrangement on the boundary must be interpreted in terms of periodical dislocations (Fig. 1e, sample 13). The dislocations, lying in the boundary plane, occur at the end

Physical Aspects of Exsolution in Natural Alkali Feldspars   253

Taking into account only the elastic energy, we extended Cahn's cubic calculations to any symmetry and applied them to feldspar exsolutions (Williame and Brown, 1972, 1974). Preferred orientations of the exsolution boundary were calculated using as input data the lattice parameters of both crystals and their elastic coefficients. In these calculations, total coherence is assumed between both kinds of lattices. The excellent agreement between calculations and observations justifies the model.

2.1.1 Monoclinic/Monoclinic Boundary

When exsolution first occurs in homogeneous monoclinic alkali feldspars, the two kinds of lamellae retain their monoclinic symmetry. The calculations gave (601) for the preferred orientation of the boundary, as observed in most cryptoperthites (Fig. 1a–d and specimens 1 to 7). This shows that the orientation of the lamellae at the beginning of the exsolution process is determined essentially by the coherent elastic energy of the boundary.

After twinning of the Na-rich lamellae, the orientation remains the same from calculations using the "averaged monoclinic" parameters for periodically albite or pericline twinned Na-rich lamellae and monoclinic K-rich lamellae.

2.1.2 Triclinic/Monoclinic Boundary

In other cases (Fig. 1e, f and samples 11, 12, 14), the K-rich lamellae are triclinic and the Na-rich lamellae are "averaged monoclinic" (periodically twinned triclinic lattice). The calculated preferred orientation varies then from ($\bar{6}31$) to ($\bar{6}61$), depending on the degree of ordering of the potassic feldspar. This agrees with the observed orientations. The average orientation of the zig-zag lamellae is parallel to ($\bar{6}01$) so we assume that this kind of texture is obtained from the previous one (($\bar{6}01$) boundary) by a diffuse rearrangement of the lamellar boundaries, when the K-rich lamellae become triclinic (Fig. 1f). We propose that samples 8 and 9 (and perhaps 10) show an earlier stage in this evolution, before the K-rich domains are noticeably triclinic.

2.2 Periodic Twinning

In the Na-rich lamellae, regularly spaced twins occur and the twin periodicity varies with the square root of the width of the lamella. This variation has been explained by a calculation which minimizes the sum of the elastic energy to produce coherency along the boundary and the twin energy, but where the K-rich lamellae are undeformed.

The minimization of the total energy gives a relation between the periodicity ($2l$) of the twins, the width (L) of the Na-rich lamella, the φ angle ($= b \wedge b^*$), the specific twinning energy (γ_T) and a coefficient (k) depending on the stiffness coefficients (Williame and Gandais, 1972)

$$2l = \sqrt{\frac{\gamma_T L}{2k^3}}.$$

Chapter 4.1 of this volume). It is hazardous to go from the pure geometry of a texture or a series of textures to a genetic explanation without taking into consideration the physical evidence available, of which perhaps the most important is the state of internal strain which can exceed 1% in some cases (Brown and Willaime, 1974).

Exsolution experiments in the laboratory give valuable clues as to the possible behavior of natural alkali feldspars. It is important to realize that most cryptoperthite textures so far studied formed in rocks which cooled at a very much slower rate than the slowest laboratory rate (resulting often in supercooled isothermal coarsening experiments). According to spinodal theory, the wavelength is infinite at the spinodal, therefore fine spinodal textures can only arise on significant undercooling, as in lavas. Relatively coarse cryptoperthitic lamellar textures may arise directly by spinodal decomposition without significant coarsening. From a study of Table 1 and of the micrographs (Fig. 1) several features seem important and need to be explained.

a) The boundary orientation of exsolution lamellae is *either* near (661) *or* near (66̄1) when planar. Wavy boundaries are intermediate.
b) Periodical twinning is observed in the Na-rich triclinic lamellae.
c) In most samples the periodical twinning obeys the albite law; in a few it is pericline twinning or both.
d) In some samples the coherence along the boundary plane seems perfect, whereas in others a dislocation array adjusts the misfit between both lattices. A discussion of the elastic deformation necessary to adjust the different lattices on exsolution, when the symmetry changes, or when twinning occurs, and of the elastic energies involved in each case are presented in the next section.

2. Physical Explanations

2.1 Orientation of the Boundary Plane

Cahn (1962, 1968) developed the theory of spinodal decomposition as applied to cubic crystals, and suggested that the orientation of the boundary of the exsolved lamellae would be determined by the minimum in the elastic energy necessary for coherence between both kinds of lamellae. Other components of the boundary energy, such as the chemical term, could be anisotropic and could influence the orientation of the boundary plane. In alloys, bonding occurs between unlike first neighbors and if exsolution occurs between two kinds of metallic atoms with similar sizes, the anisotropy of the chemical energy may be more important than the anisotropy of the elastic energy. In alkali-feldspar exsolutions, the silicate framework is only slightly changed and bonding between alkali atoms occurs through oxygen and silicon or aluminum atoms; the chemical energy is probably not greatly modified by the orientation of the gradient in the framework. At present, insufficient data exist even to *guess* at the possible effects of chemical gradients on the boundary orientation. In any case they will be small compared to the very large strain and elastic anisotropy.

Physical Aspects of Exsolution in Natural Alkali Feldspars

K-rich domains		Na-rich domains		References
lattice	shape	shape	lattice	
m	$\approx \parallel$	m	$\approx \parallel$	McConnell (1969)
m	$\approx \parallel$	m	$\approx \parallel$	Williame et al. (1973)
m	$\parallel$	t^* p-P	$\parallel$	Brown and Williame (1974); Spencer (1937)
m	lenticular to $\parallel$	t p-A	$\parallel$	Williame and Gandais (1972); Brown and Williame (1974)
m	$\parallel$	t p-A	$\parallel$	Bollmann and Nissen (1968); Spencer (1937)
m	$\parallel$	t p-A	$\parallel$	Brown et al. (1972); Brown and Williame (1974)
m	$\parallel$	t p-A p-P	lenticular	McLaren (1974)
m (?)	wavy	t p-A	wavy	Lorimer and Champness (1973); Spencer (1937)
m (?)	wavy	t p-A	wavy	McLaren (1974)
m	$\parallel$ to wavy	t p-A	$\parallel$ to wavy	Fleet and Ribbe (1963)
t^{**}	$\parallel$	t p-A	lozenge zig-zag or zig-zag	Brown et al. (1972); Brown and Williame (1974)
t^{**} and t^m in part	$\parallel$ zig-zag	t p-A	lozenge	Lorimer and Champness (1973); Spencer (1937)
t^m	$\parallel$	t p-A	lenticular	Gandais et al. (1974)
t^m	$\parallel$	t p-P	$\parallel$	
t^{**}	wavy to zig-zag	t p-A	wavy to zig-zag	Williame et al. (1973)

Symbols: m = monoclinic; t^* = triclinic, high-albite; t = triclinic, ow-albite; t^m = microcline twinning, finely cross-hatched albite and pericline twins; t^{**} = diagonal association, two adjacent K-domains are albite twinned but slightly deformed; $\parallel$ parallel boundary; p-A = periodical albite twins; p-P = periodical pericline twins.

plex lamellar boundaries are found (Fig. 1c–f and specimens 3 to 14). These alkali feldspar textures can be explained by diffusion of alkali atoms in the same crystal; no intercrystalline replacement is needed.

The electron microscope enables determination of the geometry of the exsolution textures and their lattice coherency (Table 1 and Champness and Lorimer,

of a supplementary plane in the Na-rich lamella, and they form a dislocation loop. These dislocations were found to be edge with a Burgers vector lying in the boundary plane and parallel to the b axis. These dislocations are quasi-periodic with a distance varying from 0.2 µm to 1.0 µm. The mean distance between dislocations is in good agreement with the difference in b parameters for both lattices.

3. Conclusions

Mineralogists and geologists are interested in ascertaining the processes which operated during the formation of cryptoperthites. On the basis of the physical explanations given for some of the features in these cryptoperthites, the following scheme is tentatively put forward (Fig. 2):

1. If the bulk composition is intermediate (about Or_{20} to Or_{60}) the Na-rich lamellae are *uniformly distributed* throughout the sample, the boundaries are generally coherent or highly coherent and the composition occurs within the range for a reasonable position for the coherent spinodal. These observations are consistent with exsolution by spinodal decomposition (samples 1–6, 8–12, 14).

a) A homogeneous monoclinic feldspar exsolves by spinodal decomposition into two sets of monoclinic lamellae with a boundary near $(\bar{6}01)$ (Fig. 1a, b, samples 1 and 2), the scale being inversely proportional to the cooling rate.

b) Coarsening of the lamellae may or may not occur.

c) In coarse Na-rich lamellae, the symmetry changes from monoclinic to triclinic and periodical twinning occurs (albite or pericline law) (Fig. 1c, d and samples 3, 4, 5, 6). The boundary remains near $(\bar{6}01)$.

d) K-rich lamellae tend to become triclinic (microcline). They twin and the boundaries are readjusted to $(\bar{6}31)$ to $(\bar{6}61)$ by K and Na diffusion in order to decrease the boundary energy (Fig. 1f and samples 11, 12, 14). Since microcline has a lower obliquity than albite (and is unaffected by temperature changes at a given degree of order) the twin periodicity will be much larger than for the albite lamellae. It is assumed that the wavy textures in samples 8, 9 and 14 (in part) are due to incipient twinning (one-dimensional distortion wave) in the K-rich bands and are an intermediate stage in the formation of the diagonal association. As noted before, the regular periodicity of the twins in samples 11 and 12 suggests that the twinning occurred before the formation of the diamond shaped zones.

2. If the bulk composition is Or-rich (near Or_{70}) the Na-rich lamellae are small, tend to have a lenticular shape and are *irregularly* distributed throughout the crystal (Fig. 1e and samples 7 and 13). This evidence is consistent with exsolution by nucleation.

e) Nucleation and growth of lenticular shaped Na-rich zones with boundary near $(\bar{6}01)$.

f) The symmetry changes from monoclinic to triclinic in Na-rich lenses and twinning occurs (as in c). The thickness of the twins varies with the width of the lens so that twinning occurred after the formation of the lenticular shaped domains.

g) The K-rich regions become triclinic and finely twinned with both albite and pericline law (M-type twinning) in most parts of the crystal (Fig. 1e, sample 13). Sample 14 seems to be a combination of textures seen in samples 11 and 3 but with low albite and M-twinned microcline; its formation is not well understood.

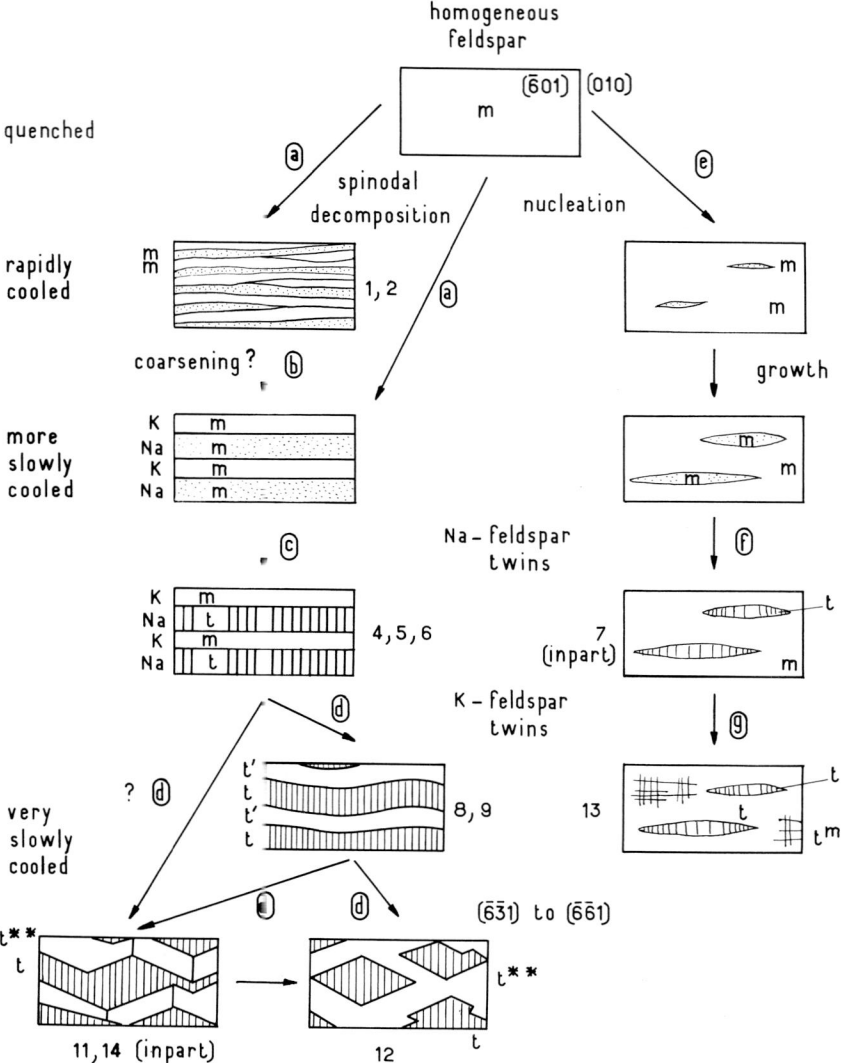

Fig. 2a–g. Proposed scheme for the formation of certain cryptoperthites-for letters (a) to (g) refer to text and for numbers *1–14* refer to Table. Samples *3* and *14* contain Pericline twinned lamellae not shown in this scheme. The left hand side is similar in part to Fig. 6 in Lorimer and Champness (1973); however, the symmetry changes and twinning produced in the potassium feldspar lamellae are considered to be the cause of the textural modifications. Symbols as in Table 1, except *t'* incipient twinning in K-feldspar

Thus, by combining bright field and dark field electron microscopy, electron and X-ray diffraction and elastic energy calculations, we attempted to understand quantitatively the formation of textures in natural cryptoperthites and to relate it to geological conditions.

References

Akizuki, M., Sugawara, H.: The lamellar structure in moonstone and anorthoclase from Korea. Contrib. Mineral. Petrol. **29**, 28–32 (1970).

Bollmann, W., Nissen, H.U.: A study of optimal phase boundaries: the case of exsolved alkali feldspars. Acta Cryst. A **24**, 546–557 (1968).

Brown, W.L., Willaime, C.: An explanation of exsolution orientations and residual strain in cryptoperthites. In: The feldspars (eds. W.S. MacKenzie and J. Zussman). Proc. NATO Adv. Study Inst., p. 440–459. Manchester: University Press 1974.

Brown, W.L., Willaime, C., Guillemin, C.: Exsolution selon l'association diagonale dans une cryptoperthite. Etude par microscopie électronique et diffraction des rayons X. Bull. Soc. Franc. Minéral. Crist. **95**, 429–436 (1972).

Cahn, J.W.: On spinodal decomposition in cubic crystals. Acta Met. **10**, 179–183 (1962).

Cahn, J.W.: Spinodal decomposition. Trans. Met. Soc. AIME **242**, 166–180 (1968).

Fleet, S.G., Ribbe, P.H.: An electron-microscope investigation of a moonstone. Phil. Mag. **8**, 1179–1187 (1963).

Gandais, M., Guillemin, C., Willaime, C.: Study of boundaries in cryptoperthites. 8th Intl. Congress on Electron Microscopy, Canberra, **1**, 508–509 (1974).

Lorimer, G.W., Champness, P.E.: The origin of the phase distribution in two perthitic alkali feldspars. Phil. Mag. **28**, 1391–1403 (1973).

McConnell, J.D.C.: Electron optical study of incipient exsolution and inversion phenomena in the system Na $AlSi_3O_8$-$KAlSi_3O_8$. Phil. Mag. **19**, 221–229 (1969).

McLaren, A.C.: Transmission electron microscopy of the feldspars. In: The feldspars (eds. W.S. MacKenzie and J. Zussman). Proc. NATO Adv. Study Inst., p. 378–423. Manchester: University Press 1974.

Spencer, E.: The potash-soda feldspar. I. Mineral. Mag. **24**, 454–494 (1937).

Willaime, C., Brown, W.L.: Explication de l'orientation des interfaces dans les exsolutions des feldspaths, par un calcul d'énergie élastique. Compt. Rend. **275** D, 627–629 (1972).

Willaime, C., Brown, W.L.: A coherent elastic model for the determination of the orientation of exsolution boundaries: application to the feldspars. Acta Cryst. A **30**, 316–331 (1974).

Willaime, C., Brown, W.L., Gandais, M.: An electron microscopic and X-ray study of complex exsolution textures in a cryptoperthitic alkali feldspar. J. Mat. Sci. **8**, 461–466 (1973).

Willaime, C., Gandais, M.: Study of exsolution in alkali feldspars. Calculation of elastic stresses inducing periodic twins. Phys. Stat. Sol., (a) **9**, 529–539 (1972).

CHAPTER 4.10

Analytical Electron Microscopy of Exsolution Lamellae in Plagioclase Feldspars

G. CLIFF, P.E. CHAMPNESS, H.-U. NISSEN, and G.W. LORIMER

1. Introduction

Lamellar structures occur in plagioclase feldspars representing low-temperature conditions for three composition ranges; An_{2-26} (the peristerites), An_{42-58} (labradorites) and An_{65-95} (bytownites). Diffraction evidence and qualitative scans in the ion- or electron-microprobe have shown that, for the peristerites and bytownites, two phases with different anorthite contents are present (Weber, 1972; Nissen, 1974). Estimates of the compositions of the two phases in peristerites from measured cell dimensions give approximately An_0 and An_{25} (Laves, 1954; Gay and Smith, 1955; Brown, 1960; Ribbe, 1960). In bytownites measurements of the proportions of the two phases as a function of bulk composition lead to estimates of about An_{67} and An_{95} for the compositions of the two phases (Nissen, 1971, 1974).

The case of the labradorites is less clear-cut as only one reciprocal lattice is detected by X-ray diffraction and the lamellae are too fine for analysis by conventional electron probe techniques. However, Nissen and Bollmann (1968) reported that most pairs of Kikuchi lines in electron diffraction patterns were split. Measurement of the proportions of the two lamellae as a function of composition was consistent with a difference in anorthite content of approximately 20 mole-% (Nissen, 1971) and suggested values of An_{35-40} and An_{60-65} for the compositions of the two lamellae.

The present authors recently presented preliminary results on the direct chemical analysis of the lamellae in a labradorite (Nissen et al., 1973) and showed that adjacent lamellae may differ in anorthite content by 12 mole-%. In the present paper the *in situ* chemical analyses of the two phases in an exsolved bytownite are reported, together with a new, extended analysis of the two components in a labradorite and a qualitative analysis of the two phases in a peristerite.

2. Experimental

Ion-thinned foils of the plagioclases were prepared from thin sections cut perpendicular to the lamellae. Analyses were carried out in the analytical electron microscope, EMMA-4, at an accelerating voltage of 100 kV with the electron beam focused down to a probe with a diameter less than $0.1 \mu m$ and a probe current of 1 to 3 n amps. The X-rays were analyzed with an energy-dispersive detector. The analytical procedures and the design of the instrument have been described by Lorimer et al. (1972), Nissen et al. (1973) and in the article by Lorimer and Cliff (Chapter 7.2 of this volume).

2.1 Beam Damage

Some framework silicates suffer ionization damage under the electron beam and light metals such as sodium and potassium are lost. In a previous paper (Nissen et al., 1973) describing the analysis of labradorite using EMMA-4, we reported results which were consistent with the additional loss of calcium when the electron beam was focused to a small probe[1]. We pointed out that the current density when the probe is fully-focused, approximately 20 amp cm^{-2}, is about thirty times higher than in a conventional electron microprobe. Thus, it is to be expected that the ionization damage and local temperature rise would be much more severe under these conditions than in the conventional electron microprobe and could allow the loss of a cation such as calcium. The loss of calcium during analysis has been verified in an experiment in which a homogeneous synthetic anorthite glass[2] was irradiated for runs of one minute duration with the beam spread to a diameter of 10 μm and in others in which it was focused to less than 0.15 μm. The beam current was varied between 30 n amp and 6 n amp. The number of counts recorded for calcium relative to silicon decreased by 25% when the probe was focused from a diameter of 10 to 0.1 μm and a beam current of 30 n amp was used. The ratio of counts for aluminum versus silicon decreased by 12% in the same experiment. For the lower beam currents only the Ca/Si ratios showed significant decreases. A similar set of experiments was carried out on an ion-thinned sample of a homogeneous labradorite from Lake County, Oregon (Stewart et al., 1966) with a composition 67.2 mole-% An. The decrease in the count ratio Ca/Si for a 30 n amp, focused probe compared with that for a 10 μm diameter probe was 5% (3 times the estimated standard deviation) but no significant difference was found in the count ratio for Ca/Si for the other beam fluxes or for Al/Si for any flux.

These results suggest that calcium, and possibly aluminum, are volatilized from plagioclase feldspars and other framework silicates when using probe currents of approximately 20 amp cm^{-2} in thin ($\sim$ 0.1 μm) specimens.

2.2 Analysis

The method by which weight-fraction ratios may be obtained from a thin-film sample using an energy-dispersive detector is described in the paper by Cliff and Lorimer (Chapter 7.2 of this volume). Because of the problems of calcium loss, values of intensity ratio I_{Al}/I_{Si} were used to monitor differences in the

---

[1] Olsen and Lillebø (Proc. EMAG-75, Institute of Physics, London, in press, 1976) have reported similar difficulties when analyzing labradorites in an analytical electron microscope.
[2] Estour (1971) has also reported loss of Ca (together with Ba and Zn) from a television glass during analysis in a conventional electron probe microanalyzer.

anorthite content and the value of the factor $k = c_{Al}/c_{Si} \times I_{Si}/I_{Al}$ (where c_{Al}/c_{Si} is the ratio of the weight fractions of aluminum and silicon) which was used to convert the observed characteristic X-ray intensity ratios into weight fraction ratios was determined from thin specimens of the Lake County labradorite under identical operating conditions to those used to carry out the analysis of the bytownite, labradorite and peristerite.

2.3 Specimens Examined

1. The bytownite, with occasional bluish-white iridescence, occurs in a Lewisian anorthosite from Roneval, Harris, Scotland (No. 45 of Corlett and Eberhard, 1967, also described by Nissen, 1974). The composition was given as 68.7 mole-% An by Corlett and Eberhard, but was found to vary from An_{68-74} by R. Gubser (personal communication). Analyses of material from the same locality performed in the University of Manchester gave a composition of 73 mole-% An. Lamellae approximately parallel to (062) (indexed on the anorthite cell) are visible in the optical microscope. B.J. Wood (personal communication) has calculated the temperature of metamorphism of the anorthosite from the Ca, Mg and Fe contents of the ortho- and clino-pyroxenes of the coexisting gabbro (Wood and Banno, 1973). He estimates that metamorphism took place at 800–850 °C and the rock was later reheated at a temperature less than 800 °C.

2. The labradorite, with bluish schiller, occurs in a coarse anorthosite from Nain, Labrador, Canada. A partial wet-chemical analysis gave $Ab_{43.4}An_{53.8}Or_{2.8}$ (in mole-%). A preliminary EMMA analysis of this specimen was reported by Nissen et al. (1973).

3. The peristerite has schiller colors varying from blue to red and occurs in a pegmatite from Hybla, Ontario, Canada. The thin section was cut as close as possible to the red-schiller area, as the lamellae were expected to be the most widely spaced there. An electron microprobe analysis of the red-schiller area gave a composition of 5.2 mole-% An.

3. Results

3.1 Bytownite

In the electron microscope the bytownite can be seen to be composed of major and minor lamellar phases which are approximately 0.8 μm and 0.4 μm in width respectively (Fig. 1a). The interface between the major and minor phase appears sharp in the electron micrograph. At a high magnification a modulated structure is observed in the minor phase parallel to the interphase interface (inset Fig. 1a). This suggests that the lamellae may have undergone a second stage of exsolution at a lower temperature, and is consistent with the results of B.J. Wood quoted above.

Fig. 1a shows the area of the specimen analyzed; the contamination marks show the actual positions of the analysis probe (the beam width at half-peak-height is estimated to be approximately half the width of the contamination spot; there is a tendency for the carbon to deposit preferentially in the cooler parts of the analysis probe). In Fig. 1b the count ratios which were obtained are plotted and the right-hand scale shows the equivalent anorthite content as calculated using the Lake County labradorite as a standard. The error bars shown are those calculated from the errors in the counts for the two elements ($\sigma = \sqrt{N}$ where N is the number of counts). Clearly the major phase is poorer in anorthite than the minor phase. The horizontal dashed lines represent the average Al/Si count ratios for the two phases and correspond to compositions of 69 ± 4[3] and 88 ± 4 mole-% anorthite for the major and minor phases, respectively.

3.2 Labradorite

The microstructure of a typical labradorite is shown in Fig. 14b of the article by Champness and Lorimer (Chapter 4.1 of this volume). The interface between the two types of lamellae appears diffuse in electron micrographs and there is probably a gradual change in composition and lattice parameters across it (McConnell, 1974). In the labradorite examined here the thinner lamellae were approximately 0.07μm and the thicker lamellae approximately 0.18μm wide. It was therefore impossible to contain the electron probe entirely within the smaller lamellae. It was also very difficult to position the probe accurately because of the low diffraction contrast, especially in the "analysis" mode.

Fig. 2a shows the area of the specimen which was analyzed and Fig. 2b gives the Al/Si count ratios obtained. The right-hand scale shows the equivalent anorthite content as calculated using the Lake County labradorite as a standard.

While the difference in count ratio for each adjacent pair of analyses is almost always significant compared with the error associated with each point, there is a general gradual decrease in the ratio for both types of lamellae with time. No explanation can be given for this trend in the results (no such trend can be detected in the bytownite data).

3.3 Peristerite

The major and minor phases in the peristerite were approximately 0.16μm and 0.04μm in width respectively and the interface between them appeared sharp in electron micrographs (Fig. 3a). It was impossible to contain the electron probe entirely within the minor phase, but with care it was possible to obtain analyses from the major phase without interference from the minor phase. The contamination marks in Fig. 3a show the positions of the analysis probe on the minor phase and the adjacent major phase; the analysis spectra from these two points are shown in Fig. 3b. They confirm that the minor phase is calcium-rich and silicon-poor with respect to the major phase and that there is not a significant amount of calcium in the major phase. In order to gain an estimate of the anorthite content of the minor phase, several electron micrographs were recorded in the high voltage electron microscope (1000 kV) with the lamellar interface

---

[3] $\pm \sigma$ or 67% confidence limit.

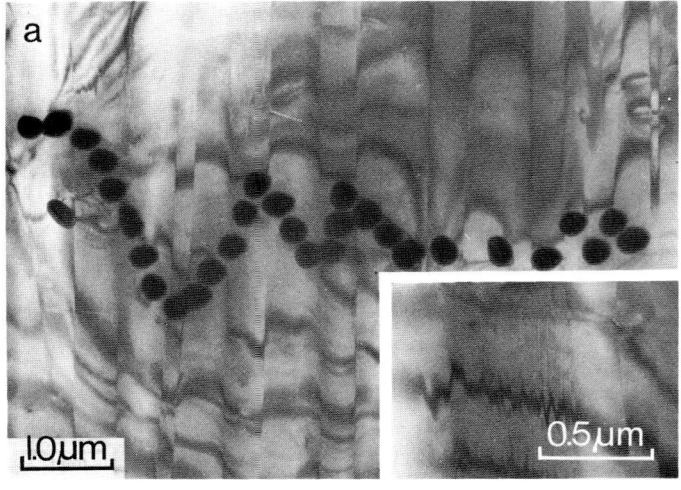

Fig. 1

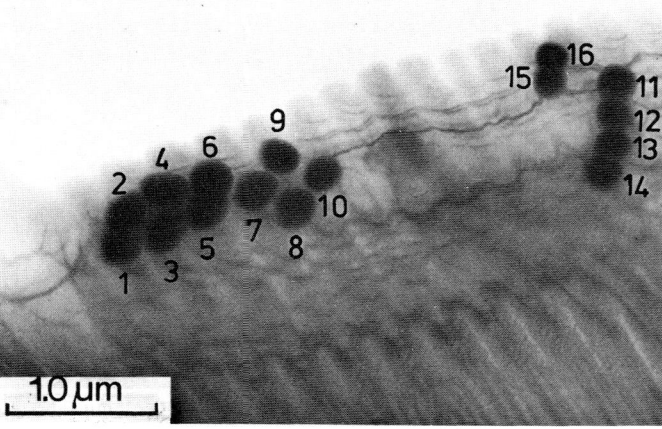

Fig. 2

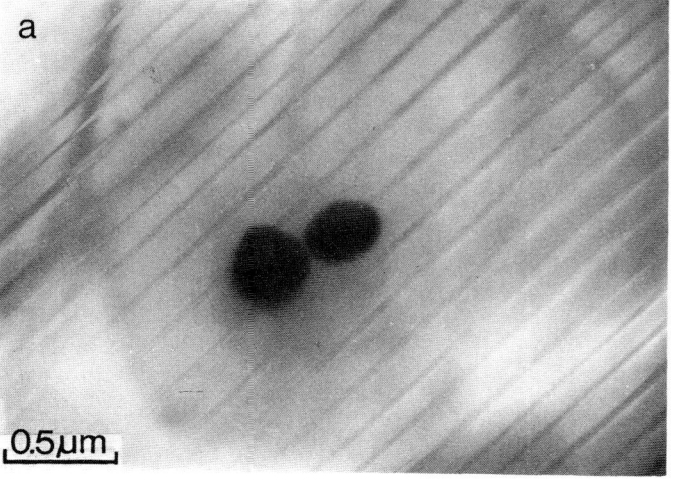

Fig. 3

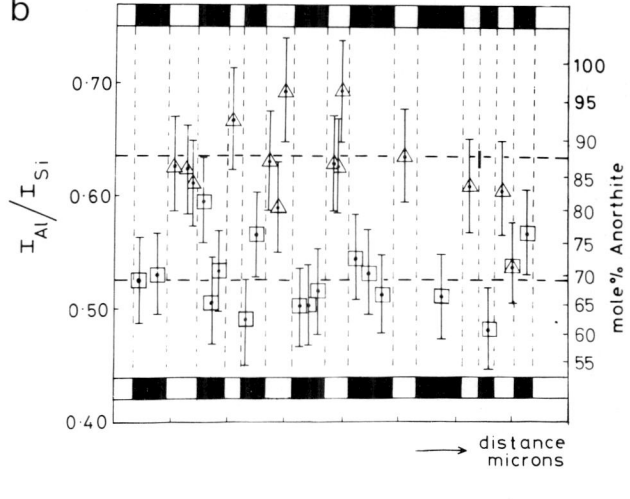

Fig. 1. (a) Electron micrograph of bytownite sample. The contamination marks show the positions of the analysis probe. The inset shows the second-stage exsolution within the minor phase. (b) The results of the analysis of the bytownite. The squares and triangles represent analyses carried out in the major and minor phases respectively. The black bars indicate the positions of the lamellae of the major phase and the horizontal dashed lines give the average values for the two phases

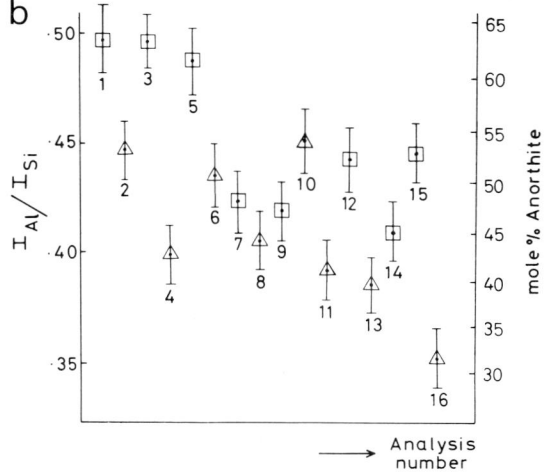

Fig. 2. (a) Electron micrograph of the labradorite sample. The contamination spots are numbered *1–16* in the order in which the analyses were recorded. (b) The results of the labradorite analyses

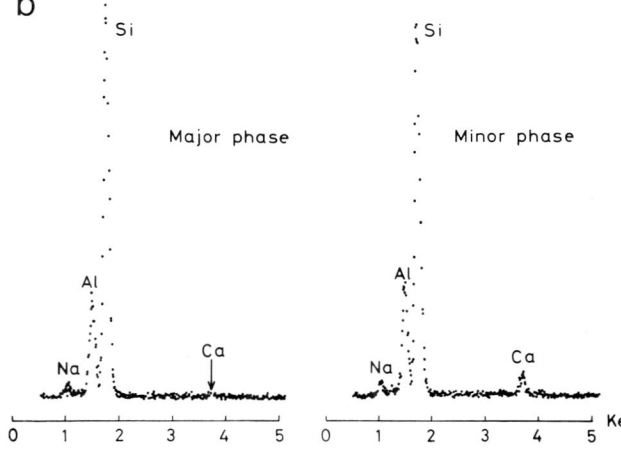

Fig. 3. (a) Electron micrograph of the peristerite sample in which the contamination marks show the two areas analyzed. (b) The analysis spectra recorded from the energy-dispersive detector for the major and minor phases

parallel to the electron beam. Measurement of the proportions of the two phases from the micrographs indicated that the minor phase constituted 17.2% of the sample by volume. Assuming the composition of the major phase to be An_0 and that of the bulk to be $An_{5.2}$ and uniform throughout the sample, gives a composition of the minor phase of An_{30}. If the major phase is assumed to contain 2% anorthite (the estimated maximum content from the analysis results) the corresponding value for the minor phase is 21% An. Thus these qualitative results are in substantial agreement with the estimates of the compositions of the two phases in peristerites by workers using X-ray methods (Laves, 1954; Gay and Smith, 1955; Brown, 1960; Ribbe, 1960).

4. Discussion and Conclusions

The results presented in this paper confirm that the phase distributions which occur in the peristerite, labradorite and bytownite plagioclase-feldspars are the result of exsolution. In the bytownite the spacing and width of the two phases were sufficiently large to be well within the resolution limit of the analysis probe and, assuming that the two phases behaved similarly with respect to the stability of Al and Si in the electron beam, the results indicate that the difference in anorthite content between them is 19 mole-%. If the sample and the Lake County labradorite standard behaved similarly with respect to Al and Si the compositions of the two phases are 69 ± 4 and 88 ± 4 mole-%. These values do not indicate as large a difference in composition as other workers have inferred from the range of bulk compositions of bytownites which are exsolved; Nissen (1974), for instance, has estimated that the solvus ranges from 67 to 95 mole-% An. However, the coarse distribution in the Roneval bytownite may represent a temperature in excess of 800 °C. The second generation exsolution which has taken place inside the Ca-rich phase, is on too fine a scale for the chemical differences to be resolved in EMMA-4.

The scale of the lamellae in the labradorite is too small for accurate quantitative analysis, but the data are consistent with the *difference* in the anorthite content of the two lamellae being at least 12 mole-%, in accordance with our previous report (Nissen *et al.*, 1973). The *absolute values* for the compositions of the lamellae must be viewed with some scepticism, particularly because of the decrease in the count ratio Al/Si with time. The diffuse interface between the lamellae in labradorites suggests that equilibrium has not been achieved. The "compositions" of the two lamellae are therefore likely to vary from one labradorite to another, depending on the initial composition and cooling history of the sample. All specimens with labradorite schiller contain more than 2 mole-% of $KAlSi_3O_8$ and Nissen (1972) has suggested that the sodic lamellae are enriched in potassium. Our attempts to detect a difference in K-content between the two types of lamella were, however, unsuccessful. This is not surprising in view of the low concentration of K in the sample and the small number of X-ray counts produced by the thin specimen.

Although the widths of the minor phase in the peristerite were less than the diameter of the analysis probe, it was possible to contain the electron probe entirely within the major phase and to show that it did not contain a significant amount of calcium. Measurement of the proportions of the two phases leads to the conclusion that their compositions were between An_{0-2} and An_{21-30}.

It is clear that chemical analysis of feldspars and other framework silicates in EMMA-4 (or other instruments with similar beam fluxes) is complicated by the volatilization of cations. Fortunately, our experience shows that other silicates are stable under these conditions (see Lorimer and Cliff, Chapter 7.2 of this volume).

References

Brown, W.L.: X-ray studies in the plagioclases. Part 2. The crystallographic and petrologic significance of peristerite unmixing in the acid plagioclases. Z. Krist. **113**, 297–344 (1960).
Corlett, M., Eberhard, E.: Das Material für chemische und physikalische Untersuchungen an Plagioklasen. Schweiz. Mineral. Petrog. Mitt. **47**, 303–316 (1967).
Estour, H.: Modification in glasses due to the electron beam of a microprobe. Verres. Réfract. **25**, 11–17 (1971).
Gay, P., Smith, J.V.: Phase relations in the plagioclase feldspars: composition range An_0 to An_{70}. Acta Cryst. **8**, 64–65 (1955).
Laves, F.: The coexistence of two plagioclases in the oligoclase composition range. J. Geol. **62**, 409–411 (1954).
Lorimer, G.W., Nasir, M.J., Nicholson, R.B., Nuttall, K., Ward, D.E., Webb, J.R.: The use of an analytical electron microscope (EMMA-4) to investigate solute concentrations in thin metal foils. Proc. 5th Int. Materials Symposium, p. 222–235. Berkeley: University of California Press 1972.
McConnell, J.D.C.: Electron-optical study of the fine structure of a schiller labradorite. In: The feldspars (eds. W.S. MacKenzie and J. Zussman). Proc. NATO Adv. Study Inst., p. 478–490. Manchester: University Press 1974.
Nissen, H.-U.: Exsolution phenomena in bytownite plagioclases. Habilitationsschrift, ETH-Zürich (1971).
Nissen, H.-U.: Electron microscopy of low plagioclases. NATO Advanced Study Institute on Feldspars, Abs. (1972).
Nissen, H.-U.: Exsolution phenomena in bytownite plagioclases. In: The feldspars (eds. W.S. MacKenzie and J. Zussman). Proc. NATO Adv. Study Inst., p. 491–521. Manchester: University Press 1974.
Nissen, H.-U., Bollmann, W.: Doubled Kikuchi lines as a means of distinguishing quasi-identical phases. Proc. 4th Europ. Congress on Electron Microscopy, Rome, 321–322 (1968).
Nissen, H.-U., Champness, P.E., Cliff, G., Lorimer, G.W.: Chemical evidence for exsolution in a labradorite. Nature Phys. Sci. **245**, 135–137 (1973).
Ribbe, P.H.: An X-ray and optical investigation of the peristerite plagioclases. Am. Mineralogist **45**, 626–644 (1960).
Stewart, D.B., Walker, G.W., Wright, T.L., Fahey, J.J.: Physical properties of calcic labradorite from Lake County, Oregon. Am. Mineralogist **51**, 177–197 (1966).
Weber, L.: Das Entmischungsverhalten der Peristerite. Schweiz. Mineral. Petrog. Mitt. **52**, 349–372 (1972).
Wood, B.J., Banno, S.: Garnet-orthopyroxene and orthopyroxene-clinopyroxene relationships in simple and complex systems. Contrib. Mineral. Petrol. **42**, 109–124 (1973).

Chapter 4.11

Exsolution in Metamorphic Bytownite

T. L. Grove

1. Introduction

It is not possible to investigate the exsolution in calcic plagioclases using synthetic materials and standard phase equilibria experiments. The kinetics of the exchange NaSi$\rightleftharpoons$CaAl involved in the unmixing are slow, and cannot be reproduced on a laboratory time scale. Therefore, to characterize the compositional limits and crystallographic relations of coexisting calcic plagioclases, it is necessary to examine natural plagioclases from a variety of environments. To date, TEM studies of exsolution in bytownites and labradorites have been carried out on samples from both igneous and metamorphic rocks. In an attempt to expand our knowledge of the miscibility gap this study of plagioclases formed in a metamorphic environment was undertaken.

2. Materials and Sample Preparation

The plagioclases chosen for this study were formed during sillimanite-grade metamorphism in a calc-silicate rock from Western Massachusetts. Individual plagioclase grains are chemically zoned from a core composition of An_{62} to rim compositions of An_{88} and contain less than 0.9 mole% Or (determined by electron microprobe analysis). Optically-visible intergrowths of two plagioclases are found only in areas having bulk compositions of An_{72-75}. The more sodic cores and calcic rims are devoid of optically-resolvable lamellae. Plagioclases with similar optical microstructures have been observed by Nissen (1974) in bytownites from meta-igneous rocks from the Alps and from Scotland.

Grains containing optically-visible intergrowths were selected from standard thin sections, mounted in 3 mm diameter copper washers and thinned by ion bombardment. The specimens were examined with JEM 7a and Philips EM 300 microscopes operating at 100 kV.

3. Observations and Discussion

Electron diffraction patterns, Fig. 1a, confirm the presence of two plagioclase feldspars. The sodic host is an intermediate plagioclase having type (a) ($h+k=2n$, $l=2n$) reflections and type (e) and (f) satellites (Bown and Gay, 1958). The calcic

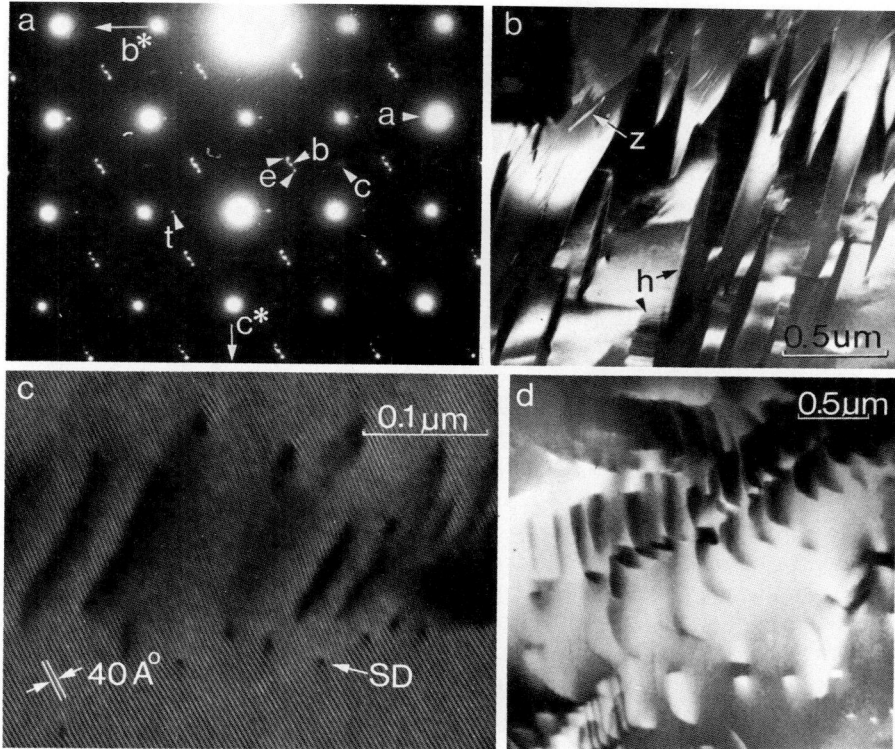

Fig. 1. (a) Electron diffraction pattern from region having two plagioclase intergrowth. Type (a) and (e) reflections from the host and types (a–c) reflections from the lamellae are present. The t indicates an albite twin-related (a) reflection. Streaking is present between (b) reflections and (e) pairs. [100] zone axis. (b) Electron micrograph of $(03\bar{1})$ exsolution lamellae. The lamellae which trend NE-SW are several microns long. Zig-zag APB's are visible in the calcic lamellae (z). Dislocations in superlattice are present at the host-lamellar boundary (h), DF, $g=013$, $s=0$ for the (b) and one (c) reflection. (c) Electron micrograph of sodic host exhibiting superlattice fringes with a periodicity of 40Å. Dislocations in superlattice normal to the foil (SD) have two terminating fringes. Other dislocations in superlattice lie almost in the plane of the foil and trend NE-SW. DF, $g=1\bar{6}3$, $s=0$ for (e) satellite pair. (d) Dislocations in superlattice in intermediate plagioclase. DF, $g=0\bar{5}1$, $s=0$ for one e satellite

lamellae have diffraction geometry consistent with that of transitional anorthite, displaying type (a), type (b) ($h+k=2n+1, l=2n+1$) and diffuse type (c) ($h+k=2n, l=2n+1$) reflections. The calcic lamellae have exsolved along two planes, one parallel to $(03\bar{1})$, Fig. 1b, and another parallel to either the $(30\bar{1})$ or $(20\bar{1})$ plane, Fig. 2b. In contrast to the small scale, 125 to 800 Å, compositional fluctuations observed in plutonic plagioclases (Nord et al., 1974), discrete lamellae, 0.1 to 0.5 μm in width, have developed in the metamorphic plagioclase. The relatively large size of the calcic lamellae resulted from the long-time, low-temperature environment characteristic of regional metamorphism. Exsolution in igneous plagioclases from the Stillwater igneous complex (Nord et al., 1974 and Heuer et al., 1972) and the Duluth gabbro (McConnell, 1974) was limited by the cooling rate of the igneous intrusive, a relatively rapid process when compared to the time scale of regional metamorphism.

Textures of the plagioclase intergrowths viewed near the [100] zone are shown in Fig. 1b. The $(03\bar{1})$ lamellae contain zig-zag antiphase boundaries similar to those observed in plutonic plagioclases by Nord et al. (1974) and Heuer et al. (1972). The antiphase boundaries are parallel to either the $(03\bar{1})$ exsolution direction or $(01\bar{1})$ and are visible in the [100] foil only with type (b) reflections.

Defects found in the host phase, Fig. 1c and d, are observed with (e) and (f) reflections characteristic of the intermediate plagioclase superstructure, and are interpreted as dislocations in the superlattice. Fig. 1c is a dark field image in which the Bragg condition is satisfied for both members of an (e) pair. The image contains superlattice fringes with a periodicity equal in magnitude to the reciprocal of the splitting vector of the (e) pair (McLaren, 1974). The termination of superlattice fringes in pairs, Fig. 1c, is characteristic of defects related to superstructures (Marcinkowski, 1963) and suggests the presence of two dislocations and an antiphase boundary. However, the double termination of fringes does not necessarily imply the presence of superlattice dislocations in the sense of Marcinkowski (1963). The terminations of superlattice fringes at the host-lamellar boundary resemble the dislocations in the superlattice found within the intermediate plagioclase, Fig. 2a. McConnell (1974) has observed similar terminations in a labradorite at the boundary between intermediate plagioclase and calcic lamellae. Determination of the nature of termination of paired fringes awaits further observation.

Although the perturbations of atomic arrangement that lead to the intermediate plagioclase superstructures are not clearly characterized, it has been suggested by several authors that the superperiodicity is related to an alternation of albite-like and anorthite-like domains (Smith and Ribbe, 1969). Therefore, at the host-lamellar boundary the calcic lamellae will be in contact with alternatively albite-like and anorthite-like regions. Dislocations in the superlattice may be present to compensate for the misfit between the albite domains of the superlattice and the anorthite structure of the calcic lamellae.

Planar features distinct from the zig-zag antiphase domains observed on [100] are present in the $(30\bar{1})$ calcic lamellae. Fig. 2c is a DF image taken near the [001] orientation showing the $(03\bar{1})$ and $(30\bar{1})$ calcic lamellae. The planar features within the $(30\bar{1})$ lamellae are visible in dark field with (a) and (e) reflections and are parallel to the $(03\bar{1})$ exsolution direction. When the Bragg condition is satisfied for a pair of (e) reflections of the intermediate plagioclase host, the planar features within the calcic lamellae are in contrast, Fig. 2d, and contain superlattice fringes and superlattice dislocations similar to those of the host. Consequently the $(03\bar{1})$ planar features are best interpreted as exsolution lamellae of intermediate plagioclase within the calcic lamellae. Streaking in electron diffraction patterns between type (b-e) and (a-f) reflection pairs, Fig. 1a, is consistent with the presence of a set of Na-rich lamellae within the calcic phase. The orientation of streaks approximately parallel to $(03\bar{1})$ also suggests the presence of lamellae of intermediate plagioclase within the calcic lamellae.

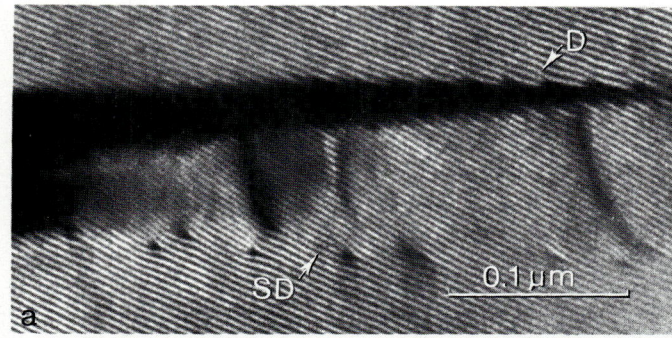
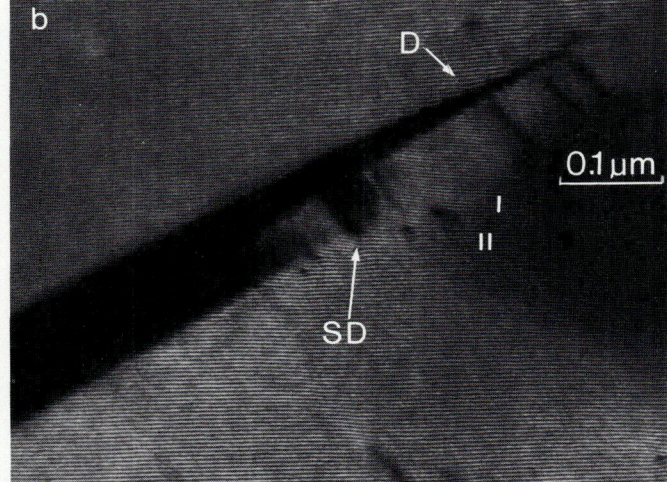
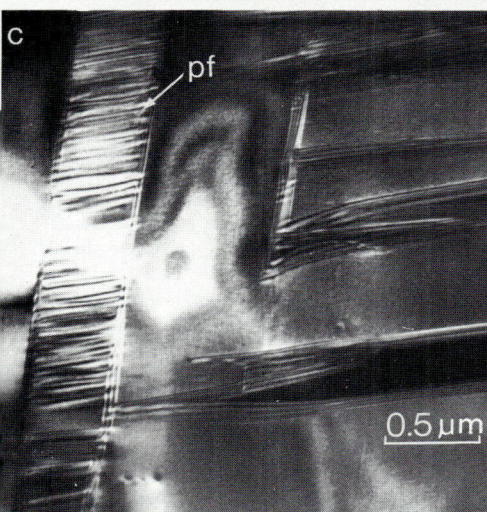
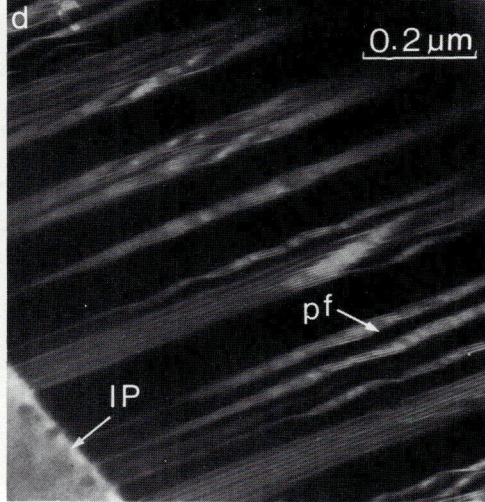

Fig. 2. (a and b) Electron micrograph of sodic host with a single (03$\bar{1}$) calcic lamella. The superlattice fringe periodicity is 40 Å. Dislocations in superlattice normal to the plane of the foil are visible at host-lamellar boundary at (D) and in the host phase at SD. DF, $g = 1\bar{6}3$, $s = 0$ for the (e) satellite pair. (a) is an enlarged part of (b). (c) Electron micrograph of calcic lamellae. The (03$\bar{1}$) calcic lamellae trend E-W and the (30$\bar{1}$) lamellae trend N-S. Sodic lamellae parallel to (03$\bar{1}$) at pf are present in the (30$\bar{1}$) calcic lamellae. DF, $g = \bar{2}20$, $s = 0$ for a reflection. (d) Electron micrograph of a (30$\bar{1}$) calcic exsolution lamella containing sets of sodic lamellae parallel to (03$\bar{1}$) at pf. Note that calcic phase is not in contrast, while the intermediate host phase (IP) does give rise to contrast. DF, $g = 1\bar{6}3$, $s = 0$ for (e) satellite pair

4. Conclusions

The exsolution features created under metamorphic conditions exhibit the results of a long-term, low-temperature process; specifically the development of large discrete calcic lamellae parallel to $(03\bar{1})$ and $(30\bar{1})$ in a host of intermediate plagioclase. Antiphase boundaries in the calcic phase take up low-energy orientations parallel to $(03\bar{1})$ and $(01\bar{1})$. In the $(30\bar{1})$ calcic lamellae exsolution of Na-rich lamellae has occurred along $(03\bar{1})$. Dislocations in the superlattice are present in the intermediate plagioclase host and may also be present at the host-lamellar interface.

The plagioclase intergrowths that developed under sillimanite-grade regional metamorphic conditions are distinctly different from the textures formed in plutonic environments. Two directions of fine-scale exsolution have been observed by Nord et al. (1974) in igneous bytownites, but have orientations of $(03\bar{1})$ and $(10\bar{1})$. $(30\bar{1})$ lamellae have not been observed in other coarsely exsolved plagioclases (Nissen, 1974 and Cliff et al., Chapter 4.10 of this volume). Exsolution of a Na-rich phase within the calcic lamellae has been observed in metaigneous plagioclases (Cliff et al., Chapter 4.10 of this volume, Fig. 2) but it is found within $(03\bar{1})$ lamellae. These textural differences are probably related to differences in the cooling history of the plagioclase.

The recent work of Nissen (1974) and McConnell (1974) suggests that calcic plagioclases have a body-centered structure at high temperatures, but on cooling undergo a phase transformation to primitive or transitional anorthite and intermediate plagioclase. The small-scale compositional fluctuations developed in plutonic plagioclases are consistent with McConnell's hypothesis, and the exsolution features observed in metamorphic plagioclases may be the result of a continuation of the process over longer time periods at lower temperatures.

References

Bown, M.G., Gay, P.: The reciprocal lattice geometry of the plagioclase feldspar structures. Z. Krist. **111**, 1–14 (1958).

Heuer, A.H., Lally, J.S., Christie, J.M., Radcliffe, S.V.: Phase transformations and exsolution in lunar and terrestrial calcic plagioclases. Phil. Mag. **26**, 465–482 (1972).

Marcinkowski, M.J.: Theory and direct observation of antiphase boundaries and dislocations in superlattices. In: Electron microscopy and strength of crystals (eds. G. Thomas and J. Washburn). New York: Interscience Publishers 1963.

McConnell, J.D.C.: Analysis of the time-temperature-transformation behaviour of the plagioclase feldspars. In: The feldspars (eds. W.S. MacKenzie and J. Zussman). Proc. NATO Adv. Study Inst., p. 460–477. Manchester: University Press 1974.

McLaren, A.C.: Transmission electron microscopy of the feldspars. In: The feldspars (eds. W.S. MacKenzie and J. Zussman). Proc. NATO Adv. Study Inst., p. 378–423. Manchester: University Press 1974.

Nissen, H.-U.: Exsolution phenomena in bytownite plagioclases. In: The feldspars (eds. W.S. MacKenzie and J. Zussman). Proc. NATO Adv. Study Inst., p. 491–521. Manchester: University Press 1974.

Nord, G.L., Jr., Heuer, A.H., Lally, J.S.: Transmission electron microscopy of substructures in Stillwater bytownites. In: The feldspars (eds. W.S. MacKenzie and J. Zussman). Proc. NATO Adv. Study Inst., p. 522–535. Manchester: University Press 1974.

Smith, J.V., Ribbe, P.H.: Atomic movements in plagioclase feldspars: Kinetic interpretation. Contrib. Mineral. Petrol. **21**, 157–202 (1969).

Section 5 Polymorphic Phase Transitions

◀ Pyroxene crystal in a diabase (Ironwood, Mich.). Alternate dark and white bands are augite and pigeonite lamellae, respectively. Pigeonite shows coarse antiphase domain boundaries with a displacement vector of $\frac{1}{2}(\boldsymbol{a}+\boldsymbol{b})$. Dark field photograph with an $(h+k)$ odd reflection, 800 kV. (Photograph by J.S. Lally, compare Heuer and Nord, Chapter 5.1 of this volume and Christie et al., 1971.)

CHAPTER 5.1

Polymorphic Phase Transitions in Minerals

A.H. HEUER and G.L. NORD, JR.

(With a section on spinel by O. VAN DER BIEST and G. THOMAS)

Phase transformations in solids can involve changes in composition and/or in symmetry. In recent years, it has become conventional to denote *composition-invariant* transformations as phase *transitions* (Roy, 1973). The title of this review follows this convention; however, *polymorphic* is included to emphasize that we shall be mainly concerned with transitions in which there is a symmetry change but no change in composition. This chapter will be organized as follows: we first present a classification scheme deemed to be useful in mineralogical phase transitions, discuss in general the driving force for phase transitions, then list the types of microstructures resulting from various types of transitions, and finally give a compilation of phase transitions in minerals that have been studied by transmission electron microscopy (TEM).

1. Classification Schemes

Attempts to *understand* and *classify* structural transitions in solids have a long history; classifications have been based on *kinetic* (i.e. sluggish or rapid), *thermodynamic* (i.e. first or second order), or *structural* grounds. A major clarification for mineralogists was made by Buerger (1951, 1972) with his grouping of phase transitions into two types, *reconstructive* and *displacive,* depending on what types of bonds (primary or secondary) were broken during the transition, i.e. whether or not the atoms retained their primary coordination number during the transition.

Roy (1973) has recently attempted to expand Buerger's classification scheme by constructing a two-dimensional matrix in which the thermodynamic order of the transformation—first, second, or mixed—represented one coordinate axis, and the structural aspects of the transition—reconstructive or displacive—represented the other. Roy further suggested that the structural axis be expanded to include a martensitic transition to be placed between the reconstructive and displacive types. The underlying basis for Roy's structural classification is the degree of "relatedness" in the structures of the two phases. In his scheme, reconstructive transitions are those with no relationship between parent and product, martensitic transitions have the parent and product related in two dimensions, and displacive transitions are those in which the parent and product are related in three dimensions. Roy also suggested that his scheme could be extended to include kinetic considerations by adding a third axis to the matrix, but observed that there was sufficient matching of the kinetic and structural scales to obviate the need for the former.

In planning the writing of this review, it was intended to use either Buerger's or Roy's classification scheme as a theoretical framework in which the various transitions in minerals could be discussed; it became obvious, however, that such a classification was not useful for discussing the application of transmission electron microscopy to the study of phase transitions. In particular, there appeared to be insufficient information about any phase transition that had been studied by electron microscopy to find the appropriate "slot" in the Roy matrix and the observations were of no help in this regard. Furthermore, although a very convenient *experimental* criterion exists to decide whether a particular transition is martensitic or not—a *shape change* associated with a *lattice correspondence* but with *no diffusive mixing* of the atoms other than the "shuffles" needed to complete the structural change (Christian, 1969) the same is not necessarily true of the reconstructive and displacive transitions common in rock-forming minerals. To begin with, Buerger and Roy focus on different structural aspects of phase transitions (changes in coordination and orientation relationships, respectively). Secondly, there is an inherent association of kinetic considerations with these structural classifications, i.e. reconstructive = sluggish and displacive = rapid. Thus, while transitions in and between the silica polymorphs lend themselves well to this classification scheme, a diffusion controlled order-disorder reaction with a well defined orientation relationship between parent and product would be difficult to classify—primary bonds are broken but the phases are related in three dimensions, and such reactions may be either fast or slow!

A classification scheme developed by Christian (1965) and based on the nature of the growth processes involved in converting a phase from the parent to the product structure proved to be more useful. This scheme was preferred by Christian because many transformations in solids can be characterized by the way in which atoms are transferred to the growing product. While it neglects other aspects of phase transformations, i.e. considerations involved with nucleation of the product phase, the notion of lattice mode softening (Shirane, 1974), etc., the scheme was found to be quite convenient for the purposes of the present chapter. A modified version is shown in Fig. 1. While Christian's corresponding figure is more general, involving changes in both composition and symmetry and also including other physical processes such as recrystallization and grain growth, Fig. 1 is more extensive in its treatment of polymorphic phase transitions, reflecting the greater diversity of such transitions in mineral systems compared with metallic systems.

The first distinction is whether the transition is *heterogeneous* or *homogeneous*. Heterogeneous transitions are those in which the sample, at an intermediate stage of the reaction, can be divided (at least conceptually) into microscopically distinct regions, of which some have transformed and some have not. It is clear that such transitions must be first order, i.e. that experimental conditions exist wherein the product and parent phases can coexist, and that the products form at discrete (but unspecified) nuclei. Homogeneous transitions are those of second or higher order, where the product phase begins to form everywhere within the parent, and a diffuse interface exists between parent and product. [In this

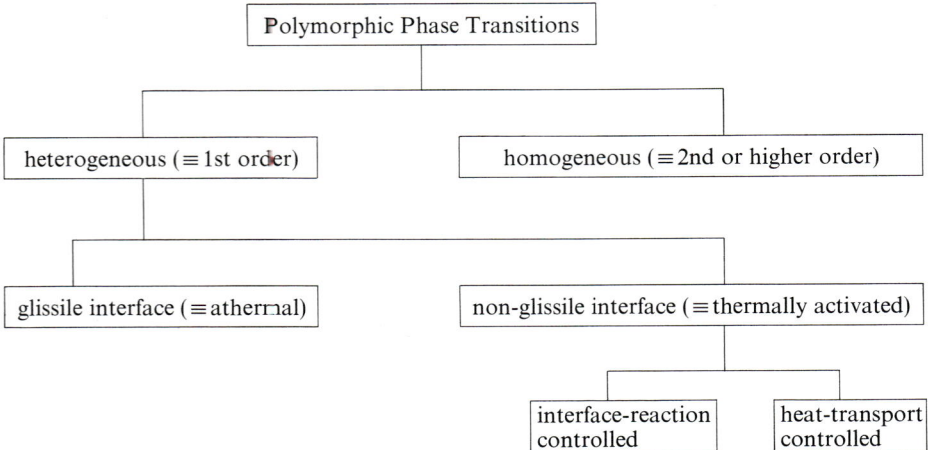

Fig. 1. Modified version of Christian's (1965) classification scheme for phase transformations in solids based on growth processes. This version is appropriate for polymorphic phase transitions, i.e. phase transformations which do not involve a change of composition

context, a diffuse interface is one whose "width" is large compared to interatomic dimensions. Spinodal decomposition (see Champness and Lorimer, Chapter 4.1 of this volume) is the analogous homogeneous second order transformation involving composition changes; it also involves diffuse interfaces between parent and product.]

The main distinction invoked within the class of heterogeneous transitions is whether the interface separating the product or parent is *glissile* or *non-glissile*. This focus on the interface is especially attractive for the present purposes, in that one of the most important contributions that electron microscopy can provide concerns the nature of the interface between product and parent in a partially-transformed material. The notion of converting a parent to a product phase by the migration of a glissile interface is most familiar to metallurgists and it is convenient here to quote from Christian (1965, p. 6): "Boundaries in the solid state may be conveniently regarded as either glissile or non-glissile. A glissile boundary can migrate readily under the action of a suitable driving stress, even at low temperatures, and its movement does not require thermal activation. Mechanical twinning and the growth of martensite plates are examples of the motion of glissile boundaries. It is important to note that the shape of the specimen changes as the boundary is displaced, so that the movement can be regarded as a form of plastic deformation. It follows that suitable external mechanical stress should be able to produce displacement of any glissile interface." It is important to note that in most cases, e.g. the tetragonal→monoclinic inversion in baddelyite (ZrO_2) (Bansal and Heuer, 1972, 1974), the proto→clino transition in enstatite (Smyth, 1974) and possibly the change from the NaCl to the CsCl structure in CsCl (Fraser and Kennedy, 1974), transitions occur in response to temperature (or pressure) changes, while in others, e.g. the ortho→clino inversion in enstatite (Turner et al., 1960; Kirby and Coe, 1974), transitions occur only under the driving force of a shear stress, i.e. if the sample is deformed. Further discussion of the baddelyite and enstatite transitions are given below.

Non-glissile boundaries, on the other hand, can only move by passing through transitory states of higher free energy, i.e. a nucleation barrier exists and their motion is thermally activated. It follows that while such boundaries may be quite mobile at elevated temperatures, they must become immobile at low temperatures. It is further useful to subdivide such thermally-activated transitions into those where the rate of growth of the product phase is determined by atomic processes at the interface, i.e. where the transition is interface-controlled, and those where the growth of the product phase is controlled by heat transport away from the interface. The latter type of transition may be unimportant in metallic systems because of their usually high thermal conductivity.

Specific examples of mineralogical phase transitions believed to proceed by the motion of non-glissile interfaces will be discussed in Section 4. However, it should be noted that virtually all the transitions that Buerger classifies as reconstructive are of this type, as are many of the "disordering" variety he discusses (Buerger, 1972). Furthermore, it is generally impossible to avoid nucleation in transitions between solid phases with different structures, since the structures cannot be continuous and thus the boundary cannot be diffuse; the majority of phase transitions are thus expected to be heterogeneous. Conspicuous exceptions to this generalization are certain ferroelectric transitions, which are known to be second-order, e.g. in colemanite ($CaB_3O_4(OH)_3 \cdot H_2O$) (Wieder, 1959) and certain order-disorder reactions in metals, e.g. β-brass, which are believed to be second-order.

2. Why a Phase Transition?

Why do phase transitions occur with changing temperature or pressure? While the obvious answer is to lower the free energy of the system, the simplest *structural* approach to the thermodynamics involved was given in Buerger's 1951 paper. Buerger emphasized, in considering first order transitions that occur upon heating, that the *internal energy* and *configurational entropy* are lower in the low temperature phase than in the high temperature phase; the *Helmholtz free energy* of the two phases are of course equal at the transition temperature. Thus, a high temperature polymorph of a particular phase is generally of a higher symmetry, i.e. has a higher configurational entropy, than the lower temperature polymorph. Likewise, a "disordered" phase, because of its higher entropy, is the stable high temperature phase.

What is meant by disorder? In the past (Megaw, 1959), *substitutional, positional, and stacking* disorder have been distinguished, although only the first can be rigorously defined. Substitutional disorder occurs when crystallographically-equivalent sites contain atoms of more than one species distributed at random, as in Fig. 2a. (Such disorder may also involve atoms and vacancies, rather than two or more species of atoms.) Fig. 2b shows an ordered arrangement of the crosses and circles of Fig. 2a. Note that a particular ordered array could

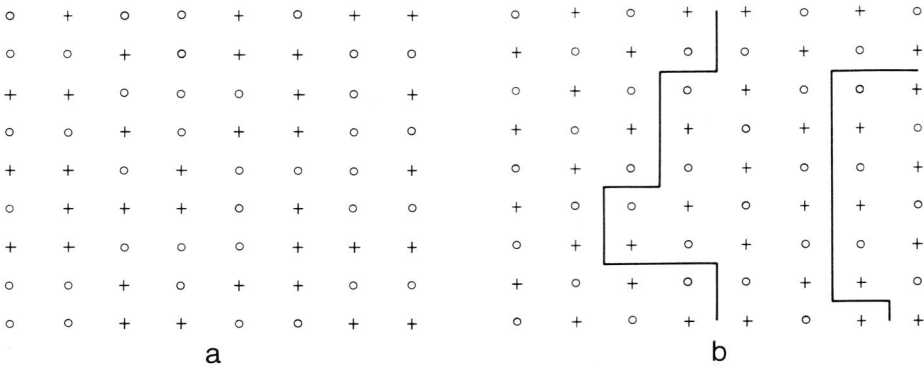

Fig. 2a and b. Schematic diagram showing a substitutionally disordered (a) and ordered (b) array of crosses and circles. The solid lines in (b) delineate antiphase boundaries which originated during the ordering

have grown outward using either a cross or a circle as the nucleation site. An antiphase boundary (as illustrated) will result for arrays, which, upon impingement, are out of register with their neighbors. The low temperature phase of metallic alloys that undergo order-disorder reactions, such as Cu_3Au (Fisher and Marcinkowski, 1961) contain such antiphase domain boundaries, as does the ordered phase of lithium ferrite discussed in Section 4.1.1.[1]

Megaw's ideas of positional and stacking disorder were formulated expressly for feldspars but may be useful for other complex structures. Consider positional disorder first. Feldspars are best described by an (Al, Si)-O tetrahedral framework with large alkali or alkaline earth cations in interstitial positions within the framework. If two (or more) positions of similar energy are available for the large cations, these positions may be occupied at random in the high-temperature (positionally-disordered) phase; anorthite at elevated temperatures may show such disorder (see Section 4.1.4.2). Only one of the positions would be occupied in the ordered structure.

Finally, stacking disorder may be caused by structural sub-units which have a pseudo-translational periodicity. Stacking faults present within polytypical minerals, e.g. wurtzite (ZnS) and carborundum (SiC), are examples of stacking disorder, as are antiphase domain boundaries, and transition-induced twins (see Section 3); note that such disorder occurs *within an ordered phase*. Thus, this last type of terminology is not useful for describing disordered phases. Nevertheless, it is often convenient to refer to a high-temperature phase as a "disordered" polymorph of the low temperature structure. Unless the disorder is clearly of the substitutional type, or the positional disorder can be rigorously specified, quotation marks should be used.

---

[1] Note that a phase transition, i.e. a change in the lattice geometry, must accompany the ordering for such antiphase domains to form. This is not the case for Mg/Fe ordering in olivines, for example, and such ordering is hence not suitable for study by conventional transmission electron microscopy.

3. Types of Interfaces Resulting from a Phase Transition

Evidence that a crystal has undergone a transition can commonly be verified by examining its microstructure in the electron microscope. A transformed crystal often will consist of domains having identical structure but separated by some interface and the distribution, morphology and diffraction contrast from such an interface can provide information on the transition mechanism, the effects of annealing and the interface geometry. The diffraction contrast from and geometry of the interfaces produced by symmetry transitions are reviewed by Amelinckx and Van Landuyt (Chapter 2.3 of this volume) and only a brief description pertinent to polymorphic transitions will be given here.

The domains resulting from a symmetry transition can be geometrically related to each other by a pure translation, as in the case of antiphase domain boundaries (APB's), or by a mirror plane, as in the case of transition-induced twins.

It is often possible to predict what type of interface—APB or twin—will be likely to result from a phase transition by considering the lattice geometry of and orientation relationships between the product and parent phases. The formation of domains generally requires a lowering of the *point group* symmetry; this has been treated rigorously by Van Tendeloo and Amelinckx (1974). If the crystal system does not change, i.e. if the transition is from a multiply-primitive to a primitive lattice, any of the lattice points of the high temperature multiply primitive cell can serve as the origin of the low-temperature primitive cell. Neighboring primitive domains with different origins in the parent phase will then be separated by APB's; these will usually give rise to π fringes. In such cases, it is often useful to think of the primitive cell as being a "superlattice" of the high temperature cell; the new reflections appearing in diffraction patterns are then conventionally referred to as superlattice reflections and can be used to form a dark field image of the domain structures. Order-disorder reactions in metals, oxides (see below) and silicates are often of this type. Similarly, APB's can arise by lowering of the *crystal symmetry*. More commonly, however, transition-induced twinning will result in these cases, due to the fact that a given orientation relationship will often allow more than one equivalent variant. Finally, if the transition is from a non-enantiomorphous to an enantiomorphous space group, as in $\beta \rightarrow \alpha$ cristobalite, a second variety of (parallel-axis) twins can result. Amelinckx and Van Landuyt treat these ideas in more detail in this volume (Chapter 2.3).

4. Specific Examples of Phase Transitions in Minerals

In this Section, we review TEM observations of phase transitions in several minerals and mineral systems. We first consider transitions assumed to be heterogeneous and to involve non-glissile interfaces in spinels (4.1.1), the silica minerals (4.1.2), clinopyroxenes (4.1.3), feldspars (4.1.4), scapolite (4.1.5), and wenkite (4.1.6). We consider the transition in spinel in some length for its pedagogical value; the longest sub-section is devoted to feldspars, the most complex of the mineral systems. We then consider transitions which occur by motion of glissile interfaces in baddelyite (4.2.1) and enstatite (4.2.2), and conclude by discussing briefly transitions in "open" systems (4.3), for which examples will be found elsewhere in this volume.

It will be seen that homogeneous, second-order transitions have been neglected in this survey. While some ferroelectric transitions are known to be of this type, and both the α-β quartz and the $C2/c \rightarrow P2_1/c$ transition in pigeonite has been *suggested* to be of this type, thermodynamic, rather than structural or kinetic data, are usually necessary to prove unambiguously the order of a transition. This has yet to be done satisfactorily for any non-ferroelectric mineralogical phase transition.

4.1 Transitions Involving Non-Glissile Interfaces

4.1.1 Spinels (by O. VAN DER BIEST and G. THOMAS)

We start this survey by considering an order-disorder reaction in spinels in some detail; while this example is not particularly important from a mineralogical viewpoint, the transition does have technological importance. More importantly, the crystallography is relatively simple, and the reaction can thus serve to illustrate many of the important aspects of phase transitions.

Order-disorder reactions were first discovered in I–III spinels ($A^+B_5^{3+}O_8$) by Braun in 1952 using X-ray techniques; although no TEM sudies on such spinel minerals are known to the authors, studies on the synthetic spinel $LiFe_5O_8$ have recently been reported by Lefebvre *et al.* (1974) and, independently, by Van der Biest and Thomas (1975).

Lithium ferrite is an inverse spinel with Fe^{3+} on the tetrahedrally coordinated sites and 3:1 mixture of Fe^{3+} and Li^+ on the octahedral sites. An order-disorder reaction involving the octahedral cations occurs at 750 °C, causing a reduction in the *space group symmetry*. Projections of the configuration of the Li and Fe ions on the octahedral sites in both the left-handed ($P4_332$) and the righthanded ($P4_132$) arrangements are given in Fig. 3a. (The close-packed oxygen sublattice and tetrahedral Fe ions are deleted for clarity.) Within each space group, the ordered structure can be described as an alternation of three iron ions and one lithium ion along $\langle 110 \rangle$ rows. Hence, the set of octahedral sites can be divided into four subsets, one of which contains only lithium ions and the other three only iron ions. When ordering sets in, the lithium ions can occupy any of these four subsets. After ordering, the single crystal is fragmented into domains, in a way similar to that described for the two-dimensional example of Fig. 2. Within each domain, the lithium ions will occupy only one subset; at the boundary between domains, they will be out of phase by a $\frac{1}{2}\langle 110 \rangle$, which is a lattice vector of the disordered structure. Such a translation of the ordered structure does not affect the oxygen ions nor the iron ions on tetrahedral sites, but does transfer the Li ions from one subset to another (for either space group), as required. Thus, there are actually 8 different subsets out of the 16 octahedral sites which the Li ions can occupy and it is possible to have a boundary between

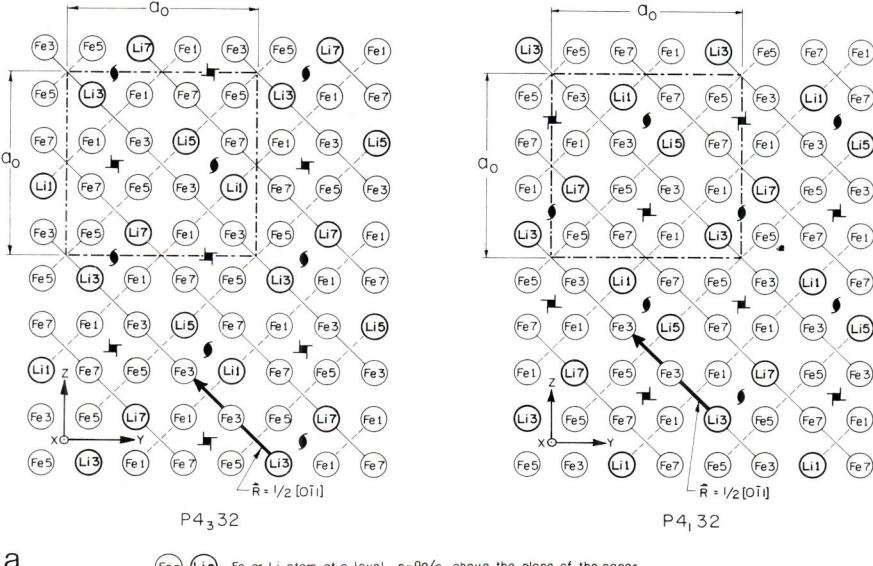

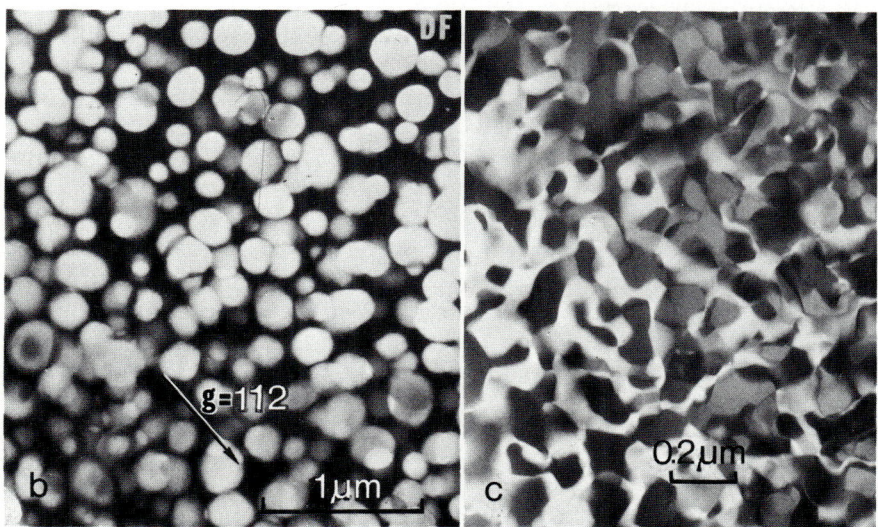

Fig. 3. (a) (100) projection of the octahedral Fe and Li ions in the two enantiomorphs of ordered $LiFe_5O_8$. See text for further discussion. (b and c) are dark-field electron micrographs ($g=112$) showing the domain morphology. (b) is from a specimen quenched from the disordered phase and heat-treated for 10 min at 743 °C, it shows the ordered domain nucleating in the disordered matrix. (c) shows a sample with a similar history but heat-treated for 10 min at 650 °C; it is fully ordered. (Courtesy of Van der Biest and Thomas)

any pair of these, or a total of 28 boundaries between the 8 possible arrangements. However, only 7 boundaries, distinct in the geometrical operations characterizing them, can occur. Among the 7 possible boundaries, 3 involve only translation (translation boundaries), one involves only inversion, and 3 boundaries require both inversion and translation to restore perfect crystal.

Two different methods of analysis can be used to study the domain structure — conventional imaging of the domain boundaries as α fringes, or by taking advantage of the violations of Friedel's law in non-centrosymmetric crystals (Serneels *et al.*, 1973). These different methods are illustrated for $LiFe_5O_8$ by Van der Biest and Thomas (Chapter 2.1 of this volume).

The kinetics involved in this ordering reaction have been studied directly in the microscope making use of an environmental cell (Van der Biest, *et al.*, 1975). It could be shown that nucleation of the ordered phase is homogeneous. Two examples from their work are shown in Fig. 3b and 3c. After annealing for 30 min at 950 °C, samples were water quenched and then annealed for various times at temperatures below the critical ordering temperature (T_c) of 750 °C. Fig. 3b shows a sample after 10 min at 743 °C, in which the ordered domains can be seen to have nucleated within the disordered matrix. At a lower temperature (650 °C), a 10 min anneal leads to a fully ordered structure, albeit with a finer domain size. These different domain morphologies reflect the different nucleation and growth kinetics, their formation depending critically on the amount of supercooling below T_c.

The Verwey electron ordering transition in magnetite (Fe_3O_4) may also be mentioned here (see Bickford, 1953, for a review). This occurs at 119 °K and involves ordering of Fe^{2+} and Fe^{3+} on the octahedral sites and a distortion to orthorhombic symmetry. The low temperature of the transition is consistent with the fact that only "hopping" of electrons is required to convert Fe^{3+} to Fe^{2+}; however, this has also hampered any TEM studies of the transition.

Finally, tetragonal distortions caused by a Jahn-Teller crystal field effect are known in II–III spinels containing $3d^4$ and $3d^9$ high-spin octahedral cations and $3d^3$ and $3d^8$ high-spin tetrahedral cations (Dunitz and Orgel, 1957). The most important mineralogical example is hausmannite (Mn_3O_4) which inverts from a cubic to a tetragonally distorted version of the spinel structure at ~1000 °C and possible substructures resulting from such Jahn-Teller transitions in II–III spinels have been studied by Van Landuyt *et al.* (1972).

4.1.2 Silica Minerals

Much mineralogical thinking has involved transitions between the various polymorphs of silica; the transitions between quartz, tridymite, and cristobalite are the usual examples of reconstructive (and therefore quenchable) transitions, while the α-β transition within each phase are prototypical of displacive reactions. Twin related domains resulting from α-β transitions in cristobalite and quartz have been studied by TEM; similar substructures are presumably present in tridymite but have not yet been studied by TEM.

Consider cristobalite first. The α-β transition involves a change from a high temperature cubic Fd3m structure (stable above ~270 °C) to a low temperature enantiomorphous space group $P4_12_1 - P4_32_1$. Two types of twins result from this transition. One type is due to the fact that any of the 4-fold cube axes of the parent can become the 4-fold tetragonal c axis of the product; this gives rise to twins having a {112} composition plane ($\equiv \{101\}_c$, where the subscript c denotes that the crystal is indexed as pseudo-cubic) (Christie et al., 1971). Secondly, parallel axis twins, which relate enantiomorphous regions of crystals, also occur (Fig. 4a); the boundaries have the appearance of conventional APB's with symmetrical fringes, indicative of π faults. Diffraction contrast analysis (Christie et al., 1971) indicated the displacement vector was $\frac{1}{4}\langle 101 \rangle$ ($\equiv \frac{1}{4}\langle 112 \rangle_c$).

In quartz, two types of twins are found. Brazil twins, which relate enantiomorphic regions of crystals, are produced by growth and deformation (McLaren and Phakey, 1966). Dauphiné (or "electrical") twins, which relate twinned regions of crystal by a 180° rotation around the c-axis, are due (at least in part) to the α-β transition[2]. McLaren and Phakey (1969) were the first to study Dauphiné twins by TEM and elucidated the basic contrast mechanism by which they could be imaged. More recently, Comer (1972) and Kumo and Kato (1974), using TEM and X-ray topography, respectively, studied Dauphiné twins and demonstrated that they did indeed arise from the α-β transition.

4.1.3 Clinopyroxenes

The two primitive clinopyroxenes pigeonite [(Mg, Fe^{2+}, Ca) SiO$_3$] and omphacite [(Na, Ca)(Mg, Fe^{2+}, Al, Fe^{3+}) Si$_2$O$_6$], invert at high temperature to a C-centered structure characteristic of diopside and augite. Morimoto and Tokonami (1969), on the basis of diffuse ($h+k$, odd) X-ray reflections present in certain volcanic pigeonites, first suggested that a C2/c→P2$_1$/c transition (on cooling) caused antiphase domains to form. Structurally, two silicate chains, which are crystallographically equivalent in the high temperature C2/c cell, become crystallographically distinct in the low temperature P2$_1$/c structure. These domains were first imaged using electron microscopy in pigeonites from lunar basalts by Bailey et al. (1970) and Christie et al. (1971), but appear to be ubiquitous to all pigeonites. The frontispiece to this section shows APB's in pigeonite from a terrestrial diabase, imaged with an ($h+k$, odd) reflection. There is general agreement that the fault vector relating the two domains is $\frac{1}{2}$ [110], as originally suggested by Morimoto and Tokonami (1969). The APB's arise because of random nucleation and growth of the low temperature phase in the high temperature matrix, at temperatures ranging from 1000 °C for Mg-rich pigeonites to 500 °C for Fe-rich pigeonites (Prewitt et al., 1971).

---

[2] It is interesting to note that Cooper et al. (private communication), by measuring the strain and temperature on opposite slab surfaces of a quartz crystal as it passed through the α-β phase transition on heating slowly from one surface of the slab, found that the results were consistent with a prediction based on the assumption that the α-β interface was nearly planar and moved through the crystal at a rate controlled by heat transfer.

Furthermore, a TEM study at or near the transition using a heated specimen stage (Van Tendeloo et al., 1975) has shown that the original α phase forms Dauphiné microtwins just below the transition, which decrease in size as T_c is approached. The twin boundaries were seen to oscillate, the frequency increasing near T_c, until the twins were no longer resolvable. The authors thus suggest that the region swept by the oscillating boundaries is the time average of the β-structure and that the β-structure above T_c, is in fact a time averaged Dauphiné micro-twinned α structure.

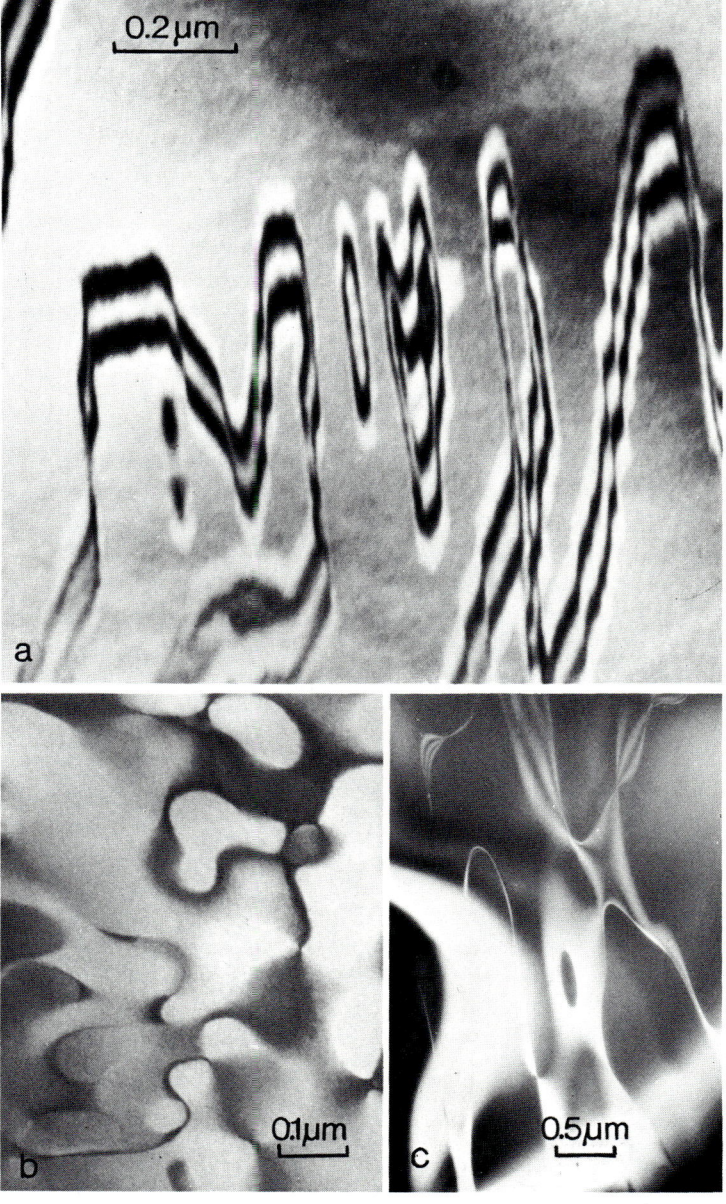

Fig. 4. (a) π fringes on enantiomorphic parallel-axis twins in cristobalite from Apollo 12038 basalt (Christie *et al.*, 1971) ($g=012$, dark field 800 kV). (b) APB's in omphacite, $Jd_{31}Di_{52}Ac_{17}$, in an eclogite from Jenner, California (Champness, 1973) ($g=101$, dark field, 100 kV). (c) APB's in scapolite (Me 37%) from a pegmatite (Phakey and Ghose, 1972) ($g=\bar{2}21$, dark field, 650 kV)

It is possible to use the scale of the domain structure in pigeonite as a measure of geologic cooling rates. However, the necessary calibration is complicated by the fact that many pigeonites contain pre-existing exsolution lamellae of augite, which provide sites for heterogeneous nucleation of the APB's. Such heterogeneous nucleation affects the domain size (Lally et al., 1975) and hence the possible use in estimating geologic cooling history is limited.

Domain structures in omphacite have been studied by Phakey and Ghose (1973) and Champness (1973). As in pigeonite, weak reflections violating a C-centered space group can be used to form dark field images. The example in Fig. 4b shows curved antiphase domain boundaries with a displacement vector of $\frac{1}{2}$ [110]. The structure of high omphacite is not known but is believed to involve a C2/c space group; there is disagreement about the low temperature space group. [Clark and Papike (1968) suggest P2 while Matsumoto and Banno (1970) argue for P2/n.] Phakey and Ghose (1973) and Champness (1973) explained the domain structure as the result of ordering of Mg, Fe^{2+} and Al into the M(1) octahedral chains, with the larger Ca and Na cations occupying the M(2) polyhedral chains, a model developed by Clark and Papike (1968). The omphacites in both studies came from low temperature (300–500 °C), high pressure (> 10 kbar) eclogites and since the proposed transition is diffusion controlled, the thermal environment may be too low for the proposed cation ordering. For this reason, Champness (1973) suggested that the high temperature phase crystallized metastably. Alternatively, the transition that produces the APB's may result from a symmetry change involving the silicate chains, similar to the high→low pigeonite transition, and without the necessity for diffusive ordering. High temperature structural studies for omphacites appear to be necessary to understand this transition fully and thus exploit its potential usefulness for understanding eclogites.

4.1.4 Feldspars

Feldspars are aluminosilicates whose basic three-dimensional framework structure varies as a function of chemical composition and temperature. While the general feldspar structure was determined by Taylor (1933), the structural details of the labradorite, bytownite and anorthite plagioclases are still debated. Electron microscopy has greatly stimulated feldspar research and some new constraints can now be placed on existing models. TEM studies of the transitions in alkali feldspars (sanidine→orthoclase→microcline and monalbite→albite), anorthite ($C\bar{1}\to I\bar{1}\to P\bar{1}$) and intermediate plagioclase will be discussed. For a detailed discussion of the crystal structures, see Smith (1974).

4.1.4.1 Alkali Feldspars

Triclinic potassium feldspar ($KAlSi_3O_8$) has four distinct tetrahedral sites, denoted conventionally T_{1o}, T_{1m}, T_{2o}, T_{2m}, over which Al and Si are distributed. The high temperature phase, sanidine, is monoclinic (C2/m symmetry); the two types of T_1 and T_2 sites are equivalent and are related by a mirror plane, the Al and Si being equally distributed over both sites. Upon cooling, Al is preferentially ordered on the T_1 sites. Before complete ordering in this monoclinic structure occurs, however, the structure inverts to the triclinic symmetry of microcline (C$\bar{1}$ space group[3]), a result of Al occupying T_{1o} site. Since either of the two mirror-plane-related T_1 sites could become the T_{1o} site in the triclinic lattice, twin-related domains are formed, resulting in distinctive "cross-hatched" twinning, called M-twinning (Laves, 1950). Electron microscopy studies of this transition have shown the existence of an intermediate stage (McConnell, 1965; Nissen, 1967), in which the monoclinic K-feldspar contains domains of "incipient" triclinic symmetry, producing a lattice perturbed by two orthogonal "waves", the nodes of which are the incipient triclinic domains. It was proposed by McConnell (1965) that these nodes are the nucleation sites for the later transformation twins. This transition is thus co-operative, with diffusion necessary for ordering and displacement necessary for the twinning (see Willaime et al., Chapter 4.9 of this volume). An analogous transition in Na feldspars (monalbite→albite) exists but has not yet been studied by TEM.

4.1.4.2 Calcic Plagioclases

The plagioclase feldspars (compounds between albite (NaAlSi$_3$O$_8$) and anorthite (CaAl$_2$Si$_2$O$_8$)), once regarded as forming a continuous solid solution series, are now known to include a group of compounds with very complicated structures. In general terms, this structural complexity appears to be due to the nearly equal size of Na and Ca atoms, combined with the completely different Al, Si ordering schemes required by Lowenstein's (1954) "aluminum avoidance" principle for the Al/Si ratios of 1:3 and 2:2 of the end members. Albite itself is triclinic (space group C$\bar{1}$) and has a structure similar to microcline. Anorthite, however, because of the 2:2 ratio of Al/Si, requires a regular alternation of Al and Si in the framework which produces a doubled c-axis compared to albite. The unit cell of anorthite contains four similar but not identical sub-cells, each with two formula units. The subcells are related by C-centering ($\frac{1}{2}$ [110]), $\frac{1}{2}c$ translation ($\frac{1}{2}$ [001]) and body centering ($\frac{1}{2}$ [111]) vectors. The first two vectors relate topologically similar subcells which have a reversed arrangement of Si and Al tetrahedra due to the ordering, thus requiring a change from the C$\bar{1}$ symmetry of albite. A body-centered cell (I$\bar{1}$ symmetry), characterized by (b)-reflections ($h+k$, odd, l, odd) which appear in addition to the basic (a)-reflections ($h+k$, even, l, even) of all the alkali feldspars, is consistent with the ordered Al/Si arrangement.

The subcells related by the body-centering vector ($\frac{1}{2}$ [111]), while topochemically equivalent, have atomic coordinates which differ by some tenths of an Ångstrom. This lowers the symmetry to P$\bar{1}$ and is characterized by (c)-reflections ($h+k$, even, l, odd) and (d)-reflections ($h+k$, odd, l, even) in addition to (a) and (b) reflections. It is important to note for the discussion that follows that the vectors $\frac{1}{2}$ [001] and $\frac{1}{2}$ [110] are equivalent in a truly body-centered cell but not in the P$\bar{1}$ cell. The relationships between the three unit cells, C$\bar{1}$, I$\bar{1}$, and P$\bar{1}$, present in plagioclases are shown in Fig. 5a.

---

[3] All the triclinic feldspar crystal structures must have a primitive unit cell. However, in order to maintain the conventional morphologic continuity, it is traditional to describe the feldspar structures in relation to the monoclinic C-centered lattice of sanidine.

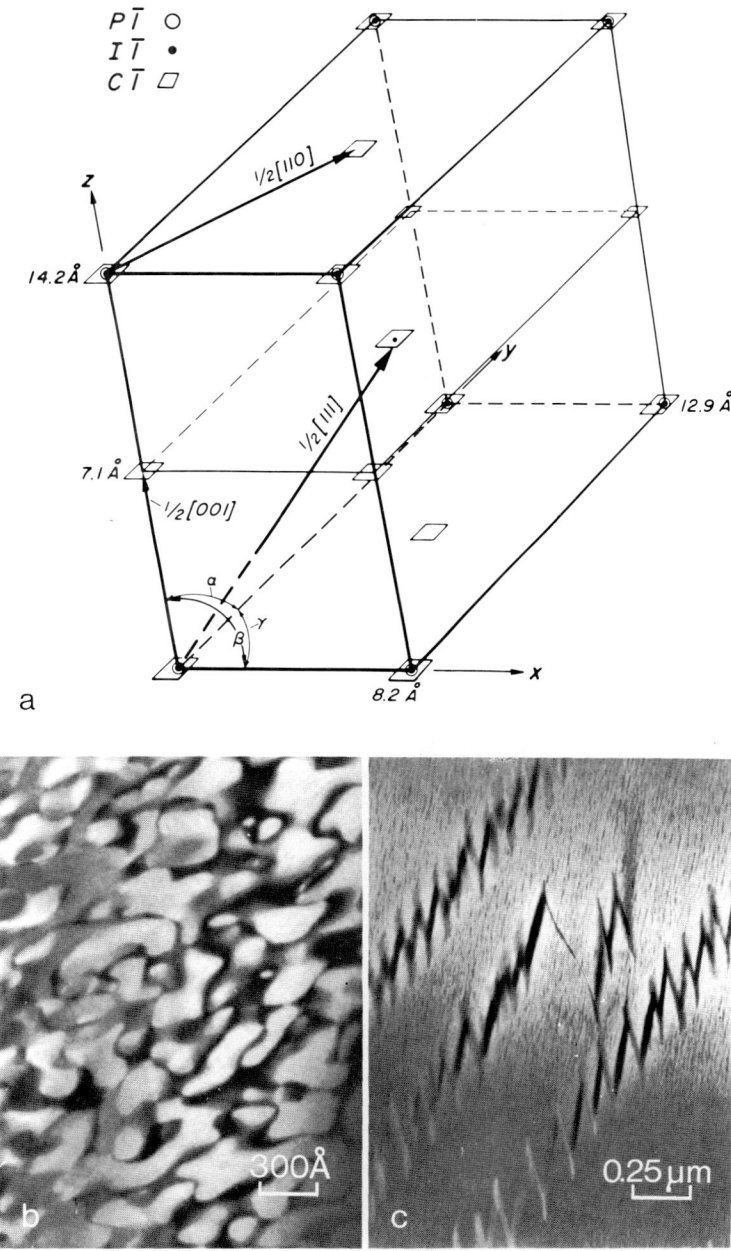

Fig. 5. (a) Schematic diagram showing the unit cell of plagioclase and the several possible displacement vectors. Equivalent lattice points for the $C\bar{1}$, $I\bar{1}$, and $P\bar{1}$ space groups are indicated (after Wenk et al., 1973). (b) 300 Å size APB's in a synthetic bytownite, An_{78-80}, crystallized at 1 100 °C and cooled through the $C\bar{1} \rightarrow I\bar{1}$ transformation at 2 °C/hour ($g=031$, dark field, 1 000 kV). (c) "Zig-zag" APB's in an An_{82} Stillwater bytownite (the background contrast is exsolution on $(03\bar{1})$) (Nord et al., 1974b) ($g=01\bar{1}$, dark field, 800 kV)

In this Section, we discuss transitions in calcic plagioclases with compositions between 70 and 100% anorthite (An_{70}–An_{100}). The existence of antiphase domains associated with both the (b) reflections and the (c) and (d) reflections in calcic plagioclases was predicted by Laves and Goldsmith (1951) because of the diffuseness of these reflections in X-ray photographs. The type (b) domains (Fig. 5b) were first imaged in the electron microscope by Christie et al. (1971) and have since been observed in many calcic plagioclases from igneous rocks. As argued by Christie et al., the (b) domains most likely arose from a $C\bar{1} \to I\bar{1}$ transition, i.e. calcic plagioclases can crystallize with a high-albite like arrangement in which Al and Si are disordered. It is interesting to note that such (b) domains have even been observed in a synthetic An_{100} anorthite (McLaren and Marshall, 1974). For such a transition, the fault vector could be either $\frac{1}{2}$ [001] or $\frac{1}{2}$ [110]; as mentioned above, while these are equivalent in the $I\bar{1}$ cell, and the distinction is meaningless, they are not equivalent in $P\bar{1}$ anorthite and can be distinguished by contrast experiments (Müller et al., 1973). Only a $\frac{1}{2}$ [110] fault vector has been determined unambiguously in the contrast experiments performed to date (Müller et al., 1973; McLaren and Marshall, 1974), the example shown in Fig. 7b and c (below) is consistent with this fault vector.

The size and morphology of (b) domains have proven to be quite variable, being dependent on both composition and thermal history; they are thus potentially useful for determining geologic cooling rates (Heuer et al., 1972). As an example, the 300 Å size APB's in Fig. 5b from a synthetic plagioclase, An_{78-80}, crystallized at 1100 °C by G. Lofgren (private communication) and cooled at 2 °C/hour (Nord et al., 1974a), can be compared with the micron size domains (Fig. 5c) in an An_{82} bytownite from the Stillwater complex, a large slowly cooled igneous body (Nord et al., 1974b). In addition, the curviplanar walls of the synthetic specimen differ from the planar walls of the slowly cooled Stillwater specimen. The preferred crystallographic orientations of the Stillwater APB's reflect an adjustment of the walls to low energy planes during the slow cooling.

The geologic usefulness of domain size is complicated by the fact that at constant cooling rate (as occurs in a single zoned grain in a rock, for example), domain size has been found to vary by an order of magnitude with zoning of ~ 10 percent An over a 30 μm distance (Nord et al., 1973). This variation is no doubt due to a change in the $C\bar{1} \to I\bar{1}$ transition temperature with composition, which causes variation of the degree of supercooling with composition at fixed cooling rate. Information on *crystallization* conditions may be possible in zoned crystals, if the zoning is caused by temperature decreases during crystallization. In this case, the core of a zoned crystal can crystallize with $I\bar{1}$ symmetry (below the transition temperature), while the rim can crystallize with $C\bar{1}$ symmetry (above the transition temperature) if the crystallization path crosses the phase boundary separating the $C\bar{1}$ and $I\bar{1}$ phases (see Nord et al., 1973, for discussion). APB's should then only exist in the rim, and a multi-domain $\to$ single domain change in morphology should be present. This was observed by Nord et al. (1973) in several zoned crystals; an example is shown in Fig. 6a. If the composition of this portion of crystal is known by microprobe analysis, and if the variation of the $C\bar{1} \to I\bar{1}$ transition temperature with composition is available, the temperature of crystallization of the narrow transition region from multi-domain to single domain can be deduced. This use of the presence or absence of (b) domains for estimating geologic histories requires, however, that the $I\bar{1}$ phase not crystallize metastably in igneous rocks (see below).

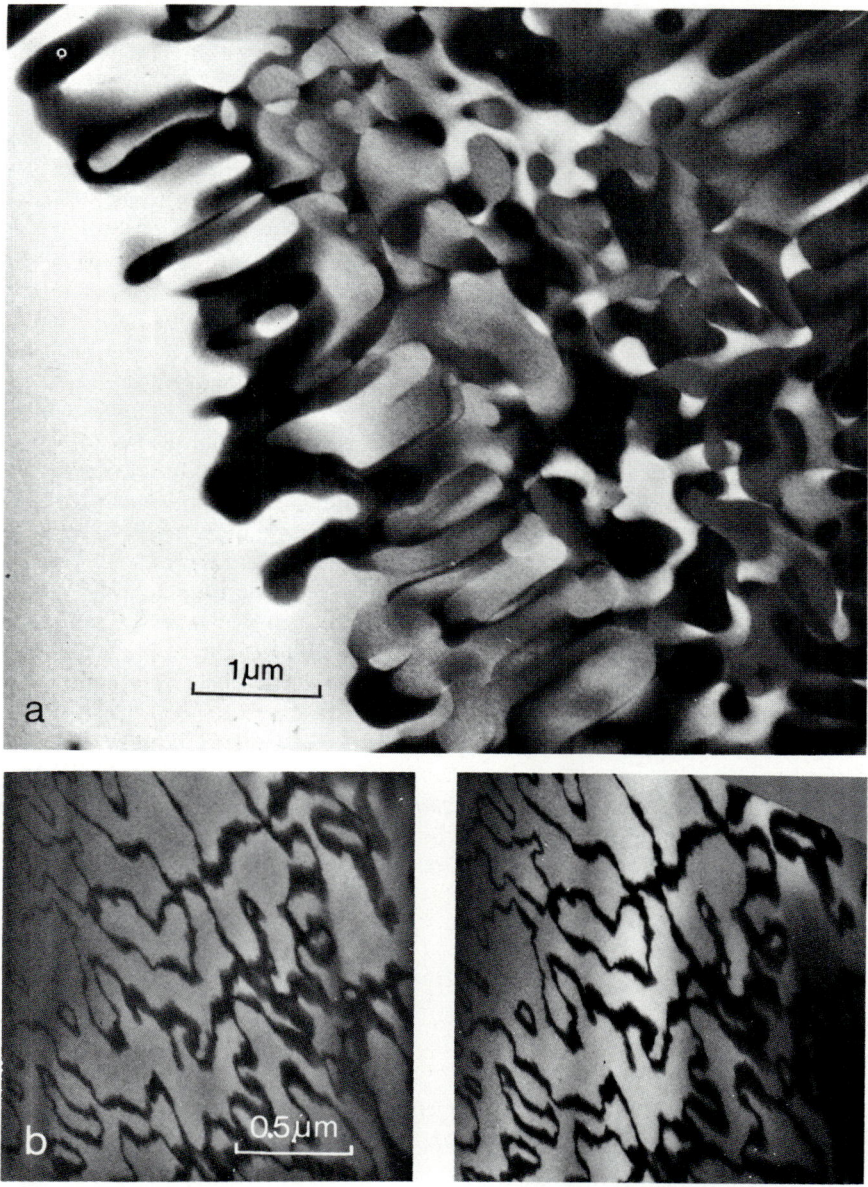

Fig. 6. (a) Transition from single domain (left) to multi-domain morphology in a small groundmass plagioclase from Apollo 60355 (Nord et al., 1973) ($g=21\bar{3}$, dark field, 800 kV). (b) Stereo pair of type (c) domains in Apollo 15415 ($g=02\bar{1}$, dark field, 800 kV)

The first convincing micrographs, e.g. Fig. 6b, of (c) domains were published in 1972 by three different groups — Czank et al. (1972), Heuer et al. (1972), and Müller et al. (1972) — and since then many investigators have imaged them in igneous and metamorphic anorthites. The exact nature of the transformation which gives rise to the type (c) domains has not been completely clarified. Some authors (Müller et al., 1972; and Czank et al., 1973) have interpreted a *fine scale domain structure* as consisting of a mixture of two phases, having $I\bar{1}$ and $P\bar{1}$ symmetry, respectively; direct lattice resolution micrographs by McLaren (1973) (Fig. 7a) and McLaren and Marshall (1974) have been interpreted in this way. On the other hand, there is agreement that the *coarse type (c) domain structures* only occur when the crystal has a uniform primitive structure (space group $P\bar{1}$) (Heuer et al., 1972; Müller et al., 1972, 1973; Czank et al., 1973; McLaren and Marshall, 1974). In this latter case, a simple interpretation is that the domain boundaries formed during a transition from a body-centered ($I\bar{1}$) to a primitive ($P\bar{1}$) lattice due to independent nucleation of the low temperature (superlattice) $P\bar{1}$ phase at 000 or $\frac{1}{2}\frac{1}{2}\frac{1}{2}$ of the parent body-centered $I\bar{1}$ lattice.

The size of the (c) domains have been shown to vary with geological history (Müller et al., 1972); small (70–100 Å) domains have been observed in rapidly cooled An_{95} volcanic anorthite while larger (micron-sized) domains were observed in a metamorphic An_{37} anorthite. The size has also been shown to vary with composition at constant cooling rate (Nord et al., 1973); the domain size decreased in a single zoned crystal from a feldspathic lunar mare basalt from ~ 1500 Å at An_{97} to ~ 100 Å at An_{93}, a compositional change of 4 mol% over $\sim 50 \mu m$.

Most (c) domains to date tend to be columnar, e.g. Fig. 7c. In Fig. 6b, the columnar axis is perpendicular to the plane of the foil, and the domains appear to be relatively equiaxed. X-ray analysis of the direction of streaking of (c) reflections (Ribbe and Colville, 1968) indicated that the columnar axis was $[2\bar{3}\bar{1}]$; electron microscopy results (Heuer et al., 1972 and unpublished results) are consistent with this.

In addition, several groups (Müller et al., 1973; Nord et al., 1973; and McLaren and Marshall, 1974) have been able to image both (b) and (c) domains in the same area of a crystal; an example is shown in Fig. 7b and c. From the above discussion and as suggested by Heuer et al. (1972), the simplest interpretation for such a phenomenon is that the structural evolution follows the sequence

$$\text{melt} \to C\bar{1} \to I\bar{1} \to P\bar{1} \qquad (1)$$

If the crystallization temperature is below the $C\bar{1} \to I\bar{1}$ inversion, then a sequence

$$\text{melt} \to I\bar{1} \to P\bar{1} \qquad (2)$$

would be followed.

In an effort to understand the transition giving rise to the (c) domains more fully, several groups have utilized heating experiments. Czank et al. (1973), for example, heated an $An_{97.6}$ sample using the electron beam and found that the apparent domain size decreased with increasing temperature, a result interpreted in terms of two kinds of domains, one involving the framework and visible only below 230 °C, and a second involving the Ca atoms in either a primitive

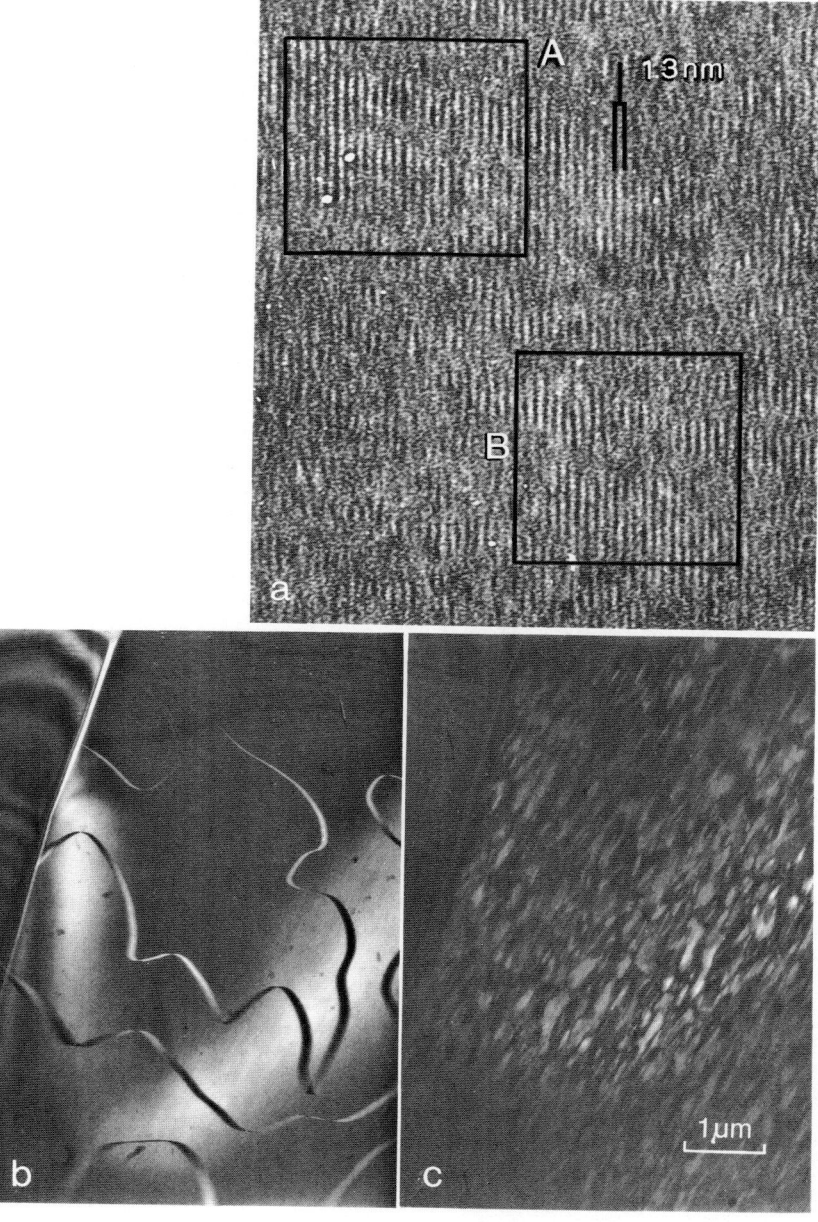

Fig. 7. (a) Direct lattice resolution of (001) planes in anorthite (McLaren, 1973). This micrograph has been interpreted in terms of P$\bar{1}$ domains in an I$\bar{1}$ matrix. Note that P$\bar{1}$ domains are elongated parallel to the trace of (20$\bar{1}$) and that in the area marked A, adjacent domains of P$\bar{1}$ (separated by a domain of I$\bar{1}$) are in register. However, in the area marked B, the adjacent domains of P$\bar{1}$ are displaced by $\frac{1}{2}$[001] (200 kV). (b and c) are the same area in an anorthite from Apollo 68415 showing type (b) and type (c) domains, respectively. [**g**=03$\bar{1}$ for (b) and 041 for (c), dark field, 1 000 kV]

or a body-centered (or "disordered") arrangement. Müller and Wenk (1973) performed more controlled experiments to 575 °C on An_{97} and An_{100} samples using a hot stage within the electron microscope. Although they found that domain contrast was lost at 200–250 °C, the domain structure reappeared unchanged after cooling. However, heat treatment at 1200 °C or above (outside the electron microscope) caused a "disordering" such that (c) domains could no longer be imaged; (c) reflections were weak and diffuse. Lally et al. (1972) also conducted *in situ* heating experiments; they reported reversible loss of the domains on heating to 400 °C but observed that there still was some intensity in the (c) reflections at this temperature; this is in agreement with earlier X-ray diffraction heating experiments (Laves et al., 1970). Heuer et al. (Chapter 5.7 of this volume) have also conducted *in situ* heating experiments, and have attempted to correlate all these observations with a model involving two $P\bar{1}$ phases, one stable below ~ 200 °C and a second stable until a transition to $I\bar{1}$ symmetry occurs; this latter transition occurs at higher temperatures (600–1200 °C) and appears to depend on An content.

Finally, reaction sequences other than (1) or (2) should be considered. McLaren and Marshall (1974) considered the possibility that the (c) domains might have arisen from a $C\bar{1} \rightarrow P\bar{1}$ transition. In such a case, a "foam" structure (Bragg, 1940) might be expected but has not yet been observed. Furthermore, as seen in Fig. 7b and c, many anorthites contain large equiaxed type (b) domains in the same region of crystal where smaller (c) domains are found, suggesting an independent origin for the two types of domains; as already mentioned, both types have been found even for a pure synthetic anorthite crystallized at 1400° C over 8 days and furnace cooled (McLaren and Marshall, 1974).

We have also considered, and rejected, the possibility that the (b) and (c) domains are the result of growth mistakes rather than the result of a phase transition. It is not easy to envisage growth mistakes during crystallization giving rise to the equiaxed (b) domains commonly observed, and the columnar nature of the (c) domains is almost certainly a crystallographic effect. The notion that anorthite may crystallize with $C\bar{1}$ or $I\bar{1}$ symmetry *metastably* is also not appealing. The former case (crystallization as a $C\bar{1}$ phase) would suggest that Al/Si ordering could occur more easily in the solid than in a magma at the crystal-melt interface, an unlikely circumstance, while for the latter case, it is hard to imagine a structural difference favoring the $I\bar{1}$ over the $P\bar{1}$ structure during crystallization but which is still sufficiently small that the $I\bar{1}$ phase cannot be maintained metastably on quenching. However, the presence of (c) domains in amphibolite-grade (~ 700 °C) metamorphic calcsilicate rocks is not easily interpreted in terms of the structural evolution implied by reactions 1 or 2, if the $I\bar{1} \rightarrow P\bar{1}$ transition occurs at a higher temperature, as suggested by Heuer et al. (Chapter 5.7 of this volume).

4.1.4.3 Intermediate Plagioclase

Equally as complicated as the domain structures in anorthite is the superstructure geometry of intermediate plagioclase (i.e. plagioclases between An_{30} and An_{75}). The problems have been reviewed in detail by Smith (1974). This structure is characterized by satellite reflections around the (b)-positions ((e)-reflections) and (a)-positions ((f)-reflections) and arises from a periodic superstructure which changes its orientation and periodicity with composition. This superlattice has been interpreted by Megaw (1960) as a periodic stacking of faults with ordering of Al_2Si_2 and $AlSi_3$ subcells (the misregistry of the Al_2Si_2 units providing the fault) with bridging Si subcells. Korekawa and Jagodzinski (1967) proposed a model involving alternating layers of albite and anorthite-like regions with a periodic antiphase structure of about 8 albite unit cells. Toman and Frueh (1973) similarly suggested an ordering model but involving a simple sinusoidal deformation of the lattice with no antiphase structure.

Prolonged heating of the superlattice phase causes loss of intensity or disappearance of the satellites. McConnell (1974a), using hydrothermal heating, has performed the most extensive experiments. The (e)-superlattice of an An_{37} sample was eliminated at 600 °C, of an An_{53} sample at 800 °C and of an An_{65} sample near 1 000 °C. The satellites of the An_{65} sample were replaced by sharp (b) reflections. Thus, the (e)-superlattice structure appears to involve an ordering phenomenon for which several complex models have been proposed (see Smith, 1974).

As discussed by McLaren (1974), fringe contrast originates from forming an image by combining the 000 or any other (a) reflection with its (f)-satellites or by combining any pair of (e)-satellites and several groups have imaged these fringes (McConnell and Fleet, 1963; McConnell, 1974b; McLaren, 1974, and McLaren and Marshall, 1974). The fringe images, since they appear only in the superlattice phase, have been especially useful for determining the phase relations present in the complex submicroscopic exsolution textures found in the plagioclase series. McLaren and Marshall (1974) also attempted to determine the nature of the superlattice boundaries by direct lattice imaging while simultaneously imaging the superlattice. Since no displacement was observed for various lattice fringes across the superlattice fringes, $\frac{1}{2}[001]$, $\frac{1}{2}[110]$, $\frac{1}{2}[11\bar{1}]$, $\frac{1}{2}[100]$, and $\frac{1}{2}[010]$ were eliminated as possible fault vectors. The authors concluded that "the superlattice is not due to the ordering of planar defects between out-of-step domains" and thus favor the Toman and Frueh model which does not depend on periodic faulting. Hashimoto et al. (Chapter 5.6 of this volume) have also studied the superlattice structure using high resolution microscopy and discuss in detail the relationship of their images with the structures proposed by X-ray crystallographers.

4.1.4.2 Other Feldspars

Müller (1974a and Chapter 5.5 of this volume) has begun studies on Sr and Ba feldspars to gain further insight into ordering processes in alumosilicates. Type (b) APB's (as in anorthite) have been observed in a synthetic $SrAl_2Si_2O_8$ sample crystallized from the melt and containing weak but sharp (b) reflections. No domains have yet been observed in either natural or synthetic celsian (the Ba feldspar) (Müller, 1974a). However, "hexacelsian", a layer alumosilicate with the same chemical formula as celsian, occurs as both an hexagonal and an orthorhombic phase and Müller (Chapter 5.8 of this volume) has observed domains in the orthorhombic phase.

4.1.5 Scapolite

Scapolite is a complex framework mineral which can be chemically represented as an incomplete solid solution between $NaCl \cdot NaAlSi_3O_8$ and $CaCO_3 \cdot CaAl_2Si_2O_8$. The sodic end member, marialite, has space group I4/m. Intermediate compositions, and the calcic end member, show weak extra reflections resulting in a lowering of the symmetry to P42/n (Papike and Stephenson, 1966; Ulbrich, 1973). The loss of symmetry has been suggested to be due to ordering of Al/Si and Cl^-/CO_3^{-2}, or to "positional" disorder. An electron microscopy study (Phakey and Ghose, 1972) of a scapolite with a primitive space group showed the presence of smooth antiphase domain boundaries (Fig. 4c), imageable only in type (b) ($h+k+l,odd$) reflections, i.e. those violating the body centering symmetry. Contrast experiments suggested a displacement vector of $\frac{1}{2}[111]$ and an interpretation of ordering of Cl^- and CO_3^{-2} was given.

4.1.6 Wenkite

Wenkite is a zeolite-type alumosilicate mineral with a formula $Ba_4(Ca_{9.1}Na_{0.1})_6(Al_{0.4}Si_{0.6})_{20}O_{39}(OH)_2(SO_4)_3$ whose structure was determined by Wenk (1973) as trigonal (space group P31m). The original X-ray study (Wenk, 1973) suggested the possibility of domain structures, which was subsequently confirmed by Lee (Chapter 5.9 of this volume), where the reader is referred to for more details.

4.2 Transitions Involving Glissile Interfaces — Martensitic Reactions

Because the martensitic reaction in Fe-C alloys is crucial to controlling the properties of steel, martensitic reactions in general have been extensively studied in metallic alloys (see Christian, 1965, 1969, for review); concern has been with both *kinetic* and *crystallographic* aspects of such transitions. Although the martensite formed in Fe-C alloys on rapid cooling is not an equilibrium product, the term is used for a particular type of phase transition and many examples are known where an equilibrium phase forms martensitically[4]. The following characteristics of martensitic transitions are usually cited:

1. A *shape deformation* occurs during the transition. This implies that a definite and constant *orientation relationship* exists between the product and parent phases, and furthermore that a definite *habit plane* exists, which is common to both parent and product lattices. The macroscopic shape deformation is usually taken to be a shear parallel to the habit plane plus a simple (unaxial) tensile or compressive strain perpendicular to the habit plane. Such an *invariant plane strain* is the most general that can occur while still maintaining the invariance of the habit plane, and indicates the transition occurs by co-ordinated atom movements. As has been mentioned in Section 1, a shape change is the *experimental* observation necessary to confirm that a reaction is martensitic.

---

[4] In practice, the equilibrium phases formed martensitically often contain imperfections such as twins or dislocations.

2. The transition is *athermal*, i.e. the extent of the reaction is characteristic of temperature but does not increase with time. This is related (in part) to the strain energy associated with the shape change, which opposes the progress of the reaction when it is still only partially completed. Hence, a greater driving force, e.g. a larger undercooling, is required to advance the reaction further.

3. The *diffusionless* nature of the reaction is the reason why the strain energy cannot be relieved by atomic migration, and arises because the co-operative nature of the atom movements means that most atoms have the same neighbors (differently arranged) in the product and parent phases.

A phenomenological but mathematically elegant theory exists to calculate the orientation of the habit plane (Wechsler et al., 1953; Bowles and Mackenzie, 1954). The theory is conventionally considered in terms of three basic steps: a Bain (or lattice) deformation which transforms the parent into the product lattice, a simple lattice invariant shear (slip or twinning) and a rigid body rotation. In matrix algebraic notation

S = RBP

where **S, R, B,** and **P** designate matrices describing the total shape deformation, a rigid body rotation, a lattice deformation, and a simple shear, respectively. To the authors' knowledge, the only mineralogical system where the theory has been applied and compared with experiment is in baddeleyite (ZrO_2).

4.2.1 Baddeleyite (ZrO_2)

At temperatures near its melting point ($\sim 2500°$ C) ZrO_2 has the cubic fluorite structure. On cooling below $2285 \pm 50°$ C (Smith and Cline, 1962), the structure inverts to a tetragonally distorted version of the fluorite structure (space group $P4_2/nmc$) and a further inversion occurs at $\sim 1000°$ C to a phase with monoclinic symmetry ($P2_1/c$). While the cubic phase can be "stabilized", i.e. retained on cooling to room temperature by solid solution "alloying" with Ca, Mg, Y etc. (see Garvie, 1970), all mineral specimens of baddeleyite are monoclinic. Because of its technological significance, much work has been devoted to the tetragonal →monoclinic inversion.

The structure of the monoclinic and tetragonal unit cells are shown as a projection on the *a-c* plane in Fig. 8a (Wolten, 1964). As can be seen, only small atom movements are required to convert one phase into the other; this projection also suggests that the following orientation relationships should be observed: $[010]_m \| [010]_{fct}$ and $[001]_m \| [001]_{fct}$. These also imply that $(001)_m \| (100)_{fct}$ and $(010)_m \| (010)_{fct}$. In these relations, *m* and *fct* refer to the monocline and face-centered tetragonal unit cell, respectively (the latter is used, rather than the conventional, and smaller body-centered cell, to facilitate comparison with the monoclinic cell). These orientation relationships were first observed by Bailey (1964) in an *in situ* electron microscopic study of the transformation, and were confirmed by Bansal and Heuer (1974) in their study of materials that had transformed in bulk. While two other orientation relationships were also observed by Bansal and Heuer, they concluded that only the cited relationships were important in the martensitic reaction. (These workers used synthetic single crystals that had been grown from a fluxed melt at temperatures below the tetragonal→monoclinic reaction. On first heating into the tetragonal phase, a different orientation relationship, $(001)_m || \{100\}_{fct}$ and $[100]_m || <100>_{fct}$, was observed. Also prior to the martensitic reaction, some material transformed back according to $(100)_m || \{100\}_{fct}$ and $[001]_m || <100>_{fct}$, an orientation relationship previously found by Patil and Subbarao (1970) in an X-ray study.)

The first step in calculating the orientation of the habit plane is determining the magnitudes and direction of the principal strains (η_i) of the lattice deformation **B**. This is given by the eigenvalues and eigenvectors of the equation

$$(\mathbf{TC'M})(\mathbf{M^*GM})(\mathbf{MCT}) - \eta_i^2 (\mathbf{T^*GT}) = 0$$

where (**MCT**) is the correspondence matrix between product and parent lattice $\begin{pmatrix} 1 & 0 & 0 \\ 0 & 1 & 0 \\ 0 & 0 & 1 \end{pmatrix}$ for the present orientation relationship, (**TC'M**) its transpose, and (**M*GM**) and (**T*GT**) the metrics for the product and parent phases. (A metric is a matrix which relates a direct lattice (**M,T**) to its reciprocal lattice (**M*,T***).) The metrics for the tetragonal and monoclinic phases are

$$\begin{pmatrix} a_t^2 & 0 & 0 \\ 0 & b_t^2 & 0 \\ 0 & 0 & c_t^2 \end{pmatrix} \text{ and } \begin{pmatrix} a_m^2 & 0 & a_m c_m \cos \beta \\ 0 & b_m^2 & 0 \\ a_m b_m \cos \beta & 0 & c_m^2 \end{pmatrix}$$

respectively, where a_t, b_t, c_t, and a_m, b_m, and c_m are the lattice parameters of the coexisting tetragonal and monoclinic phases and β is the monoclinic angles. The components of **B** are thus found to be

$$|\eta_1| = 1.08289 \qquad |\eta_2| = 1.01444 \qquad |\eta_3| = 0.92702$$

$$\eta_1 = 0.68485\,\boldsymbol{i} + 0.72863\,\boldsymbol{k} \qquad \eta_2 = \boldsymbol{j} \qquad \eta_3 = 0.72863\,\boldsymbol{i} - 0.68485\,\boldsymbol{k}$$

where $\boldsymbol{i}, \boldsymbol{j}$, and $\boldsymbol{k}$ and η_1, η_2, and η_3 are unit vectors in the tetragonal basis and along the principal axes of the Bain deformation, respectively. The magnitudes

---

Fig. 8. (a) Atom positions projected onto the a-c plane for monoclinic (left) and tetragonal (right) ZrO_2 (solid circles are Zr and open circles are O). Four unit cells of the monoclinic phase are shown. A fifth, indicated by the dashed line and differing from the others only by the choice of origin, is shown to indicate the atom movements that occur during the transition (after Wolten, 1964). (b) Type B martensite plate in synthetic baddeleyite, internally twinned on (100) (Bansal and Heuer, 1972, 1974) ($\boldsymbol{g}=002$, dark field, 650 kV) (c) (100) lattice fringes in experimentally deformed Bamble enstatite showing both the 9.1 Å and 18.2 Å spacings of clinoenstatite and orthoenstatite, respectively (Müller, 1974b) (100 kV). (d) A schematic diagram showing the relationship between the orthoenstatite unit cell (heavy line) and twinned clinoenstatite unit cells (after Smith, 1969)

Polymorphic Phase Transitions in Minerals

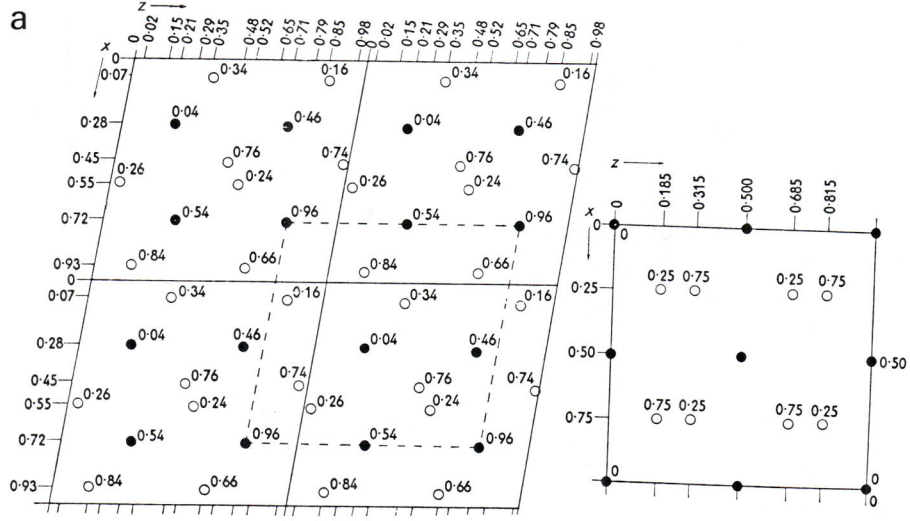

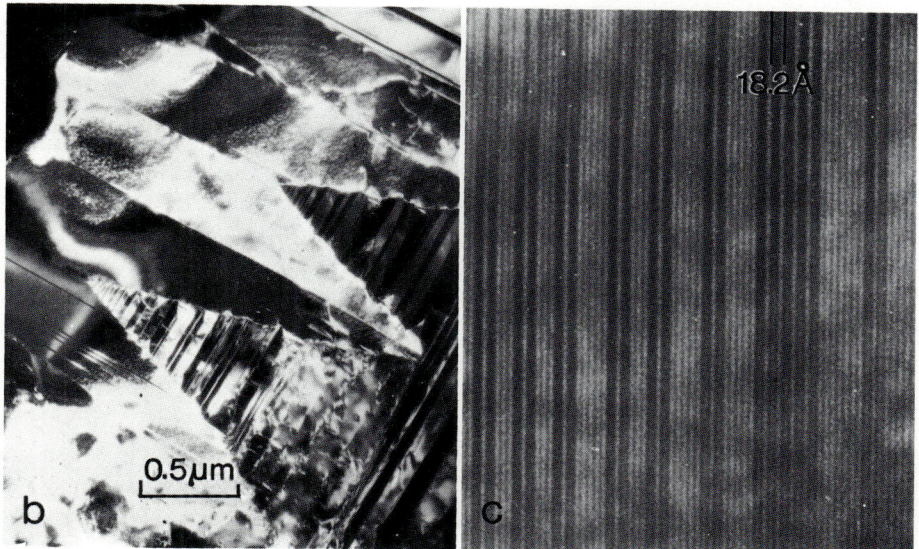

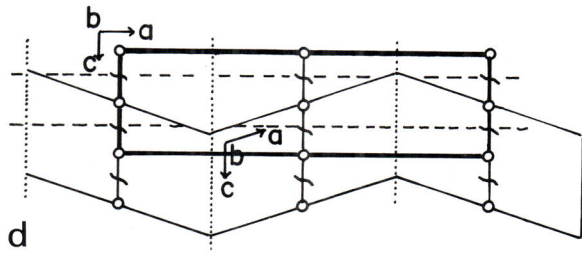

Fig. 8a–d

of the principal strains satisfy the requirement for an invariant plane strain, i.e. two are greater than unity and one is less, in other words, two of the principal strains have the opposite sign to the third.

It next remains to consider the lattice invariant deformation (a simple shear – either slip in the parent phase or deformation twinning in the product phase), which in combination with the Bain deformation, produces an undistorted habit plane. In the baddeleyite case, it is convenient to treat the simple shear as occurring in the tetragonal phase because of its orthogonal basis; it can easily be shown that a given habit plane can arise from shear in either the parent or product lattice. A graphical technique (Lieberman, 1958; Bansal and Heuer, 1974) can be used to predict which lattice invariant shear modes will yield real (as opposed to imaginary) solutions for the habit plane. In the present case, considering slip on low index (100, 110, and 111) planes and directions, $(1\bar{1}0)$ [001], $(1\bar{1}0)$ [110], $(10\bar{1})$ [010], and (111) $[1\bar{1}0]$ yielded real solutions. These slip systems define the **P** matrix, which in combination with the Bain deformation **B**, can be used to calculate the orientation of the habit plane. (Actually, each slip system can give rise to four different habit planes.) The predictions are then compared to experiment to decide which lattice invariant shear actually operated.

Two distinct habit planes (termed types A and B), have been found for the martensite product in the baddeleyite transformation, and each exhibits two variants; an example of the type B product is shown in Fig. 8b. The orientations of the two type A habit plane variants are $\sim$(671) and $\sim$(761), and was shown to arise from $(1\bar{1}0)$ [001] slip; the type B habit plane orientations are (100) and $\sim$(010) and arose from $(1\bar{1}0)$ [110] slip. Note that the latter are sometimes internally twinned on (100) (Fig. 8b).

The final step in the application of the crystallographic theory is to use the calculated habit plane normals to determine the rotation matrix **R** needed to bring the parent and product lattice into coincidence, and then to calculate **S**, the total shape deformation, which finally must be compared with experiment. Although Bansal and Heuer have calculated **S**, the high temperature of the transition has thus far prevented experimental determination of the direction and magnitude of the shape deformation; this last comparison is necessary to completely assure the correctness of the theory.

4.2.2 Orthopyroxene

The polymorphism of low calcium pyroxenes is quite complex (see Smith, 1969, for a review). Four polymorphs exist in the magnesium-rich enstatites ((Mg, Fe) SiO_3) – protoenstatite (Pbcn), stable at high temperature, high clinoenstatite (C2/c), low clinoenstatite ($P2_1/c$) and orthoenstatite (Pbca), the stable room temperature form. The ortho to clino transition can be accomplished by heating (Smyth, 1969) or by shear stress. The transition on heating always leads to the C2/c form, which invariably transforms to the $P2_1/c$ structure on cooling. The shear-induced transition always gives a $P2_1/c$ product and apparently occurs by the motion of a glissile interface; it was first investigated by Turner *et al.* (1960). Kirby and Coe (1974) have reported that a shape change does occur during the transition, confirming its martensitic nature[5]. The orthorhombic unit cell can be considered as essentially two monoclinic unit cells in a twinned relationship (Fig. 8d). Thus, by "untwinning" the subcells of the orthorhombic unit cell, two untwinned monoclinic unit cells will result. Brown *et al.* (1961) gave a detailed description of the atomic movements involved and suggested that (100) [001] shear was necessary for the transition. Their model involves no Si-O bond disruption and only half of the Mg-O bonds are broken. Recently, partially transformed enstatite has been studied by conventional TEM by Coe and Müller (1973) and by lattice imaging by Müller (1974b), Boland (1974) and by Iijima and Buseck (Chapter 5.4 of this volume). All the lattice imaging studies have been similar in that the (100) lattice fringes are imaged, and show the 9.1 Å repeat distance of clinoenstatite and the 18.2 Å repeat distance of orthoenstatite. An example of these fringes is seen in Fig. 8c from Müller (1974b) in an experimentally deformed specimen. There are always an even number of clinoenstatite fringes, consistent with the untwinning mechanism. Boland's study of orthoenstatites from ultramafics indicated that some of the specimens consisted only of orthoenstatite, others contained orthoenstatite with a nonperiodic distribution of the clinoenstatite fringes, while still others contained orthoenstatite with periodic clinoenstatite fringes, the latter regions giving rise to "superlattice" reflections corresponding to a 73 Å repeat distance. As with the previous investigators, the origin of the clinoenstatite was attributed to deformation; a model was proposed involving the propagation of a dislocation in orthoenstatite with Burgers vector $b = 0.83$ [001] which would produce 2 unit cells of clinoenstatite with only minor atomic shuffles. (See also Kirby, Chapter 6.7 of this volume.) A similar model has been suggested for a transition in wollastonite ($CaSiO_3$) by Wenk *et al.* (Chapter 5.5 of this volume).

4.3 Transitions in "Open" Systems

We finally consider transitions which occur in "open" systems, i.e. non-stoichiometric solids (typically oxides or sulphides) where the cation/anion ratio is variable. The changes in structure which can accompany changes in composition in such systems are not strictly classified as polymorphic phase transitions; they are included in this chapter for convenience and hence discussion will be brief.

For cations of variable valency, e.g. Fe, the simplest way to imagine a crystal accommodating a metal-deficient type of non-stoichiometry is by the creation of cation vacancies. This appears to be the case in wüstite ($Fe_{1-x}O$) (although vacancy clusters, rather than isolated vacancies appear to form, Koch and Cohen, 1969, and in pyrrhotite ($Fe_{1-x}S$). The metal vacancies in pyrrhotite form an ordered arrangement and lead to several different superstructures. The literature is further reviewed and TEM studies are presented by Nakazawa *et al.* (Chapter 5.2 of this volume).

---

[5] Smyth (1974) has suggested that the transition from protoenstatite to clinoenstatite is also martensitic; this transition should thus also be studied by transmission electron microscopy.

The other major category of non-stoichiometric oxides of interest to electron microscopists is typified by rutile (TiO_2); these types of solids form "shear structures" to accommodate metal deficiencies. These structures usually appear in transition metal oxides which structurally can be described as consisting of metal/oxygen octahedra exhibiting corner sharing. As the metal/oxygen ratio is lowered, some edge-sharing of octahedra occurs along certain well defined "crystallographic-shear" planes. These ideas are developed in more detail, and the literature dealing with electron microscopy studies on rutile is reviewed by Hyde (Chapter 5.3 of this volume).

5. Concluding Remarks

Transmission electron microscopy is the tool *par excellence* for studying substructures (i.e. domains) resulting from phase transitions in minerals. Aside from the great potential the scale of these substructures possess in elucidating geologic cooling rates, the presence of various combinations of domains permit identification of the sequence of *structural evolution*; this can be especially valuable in minerals with complex crystal structures.

References

Bailey, J.C., Champness, P.E., Dunham, A.C., Esson, J., Fyfe, W.S., MacKenzie, W.S., Stumpfl, E.F., Zussman, J.: Mineralogy and petrology of Apollo 11 lunar samples. Proc. Apollo 11 Lunar Sci. Conf. Geochim. Cosmochim. Acta, Suppl. 1, **1**, 169–194 (1970).

Bailey, J.E.: The monoclinic tetragonal transformation and associated twinning in thin films of zirconia. Proc. Roy. Soc. (London) **279**, Ser. A 395–412 (1964).

Bansal, G.K., Heuer, A.H.: On a martensitic phase transformation in zirconia (ZrO_2)- I. Metallographic evidence. Acta Met. **20**, 1281 (1972).

Bansal, G.K., Heuer, A.H.: On a martensitic phase transformation in zirconia (ZrO_2)- II. Crystallographic aspects. Acta Met. **22**, 409–417 (1974).

Bickford, L.R.: The low-temperature transformation in ferrites. Rev. Mod. Phys. **25**, 75–79 (1953).

Boland, J.N.: Lamellar structures in low-calcium orthopyroxenes. Contrib. Mineral. Petrol. **47**, 215–222 (1974).

Bowles, J.S., Mackenzie, J.K.: The crystallography of martensite transformations. Acta Met. **2**, 129–137 (1954).

Bragg, W.L.: The structure of a cold-worked metal. Proc. Phys. Soc. (London) **52**, 105 (1940).

Braun, P.B.: A superstructure in spinels. Nature **170**, 1123 (1952).

Brown, W.L., Morimoto, N., Smith, J.V.: A structural explanation of the polymorphism and transitions of $MgSiO_3$. J. Geol. **69**, 609–616 (1961).

Buerger, M.J.: Crystallographic aspects of phase transformation. From: Phase transformations in solids (eds. R. Smoluchowski, J.E. Mayer, W.A. Weyl), p. 183–209. New York: John Wiley 1951.

Buerger, M.J.: Phase transformations. Soviet Phys.-Cryst. **16**, 959–968 (1972).

Champness, P.E.: Speculation on an order-disorder transformation in omphacite. Am. Mineralogist **58**, 540–542 (1973).

Christian, J.W.: The theory of transformations in metals and alloys. p. 975. Oxford: Pergamon Press 1965.

Christian, J.W.: Martensitic transformations: a current assessment. From: The mechanisms of phase transformations in crystalline solids, p. 129–142. London: Institute of Metals 1969.

Christie, J.M., Lally, J.S., Heuer, A.H., Fisher, R.M., Griggs, D.T., Radcliffe, S.V.: Comparative electron petrography of Apollo 11, Apollo 12, and terrestrial rocks. Proc. 2nd Lunar Science Conf. Geochim. Cosmochim. Acta, Suppl. 2, **1** (1971).

Clark, J.R., Papike, J.J.: Crystal-chemical characterization of omphacite. Am. Mineralogist **53**, 840–868 (1968).

Coe, R.S., Müller, W.F.: Crystallographic orientation of clinoenstatite produced by deformation of orthoenstatite. Science **180**, 64–66 (1973).

Comer, J.J.: Electron microscope study of Dauphiné microtwins formed in synthetic quartz. J. Crystal. Growth **15**, 179–187 (1972).

Czank, M., Schulz, H., Laves, F.: Investigation of domains in anorthite by electron microscopy. Naturwiss. **59**, 77–78 (1972).

Czank, M., Van Landuyt, J., Schulz, H., Laves, F., Amelinckx, S.: Electron microscopic study of the structural changes as a function of temperature in anorthite. Z. Krist. **138**, 403–418 (1973).

Delavignette, P., Amelinckx, S.: Precipitation phenomena in MnO containing excess O_2. Mat. Res. Bull. **5**, 1009–1014 (1970).

Dunitz, J.D., Orgel, L.E.: Electronic properties of transition metal oxides. I. Distortions from cubic symmetry. J. Phys. Chem. Solids **3**, 20–29 (1957).

Fisher, R.M., Marcinkowski, S.: Direct observation of antiphase boundaries in the $AuCu_3$ superlattice. Phil. Mag. **6**, 1385–1405 (1961).

Fraser, W.L., Kennedy, S.W.: The crystal-structure transformation NaCl-type→CsCl-type. Acta Cryst. A**30**, 13–22 (1974).

Garvie, R.C.: Zirconium dioxide and some of its binary systems. In: High temperature oxides, part II (ed. A.M. Alper), Chapter 4, p. 118–166. New York: Academic Press 1970.

Heuer, A.H., Lally, J.S., Christie, J.M., Radcliffe, S.V.: Phase transformations and exsolution in lunar and terrestrial calcic plagioclases. Phil. Mag. **26**, 2, 465–482 (1972).

Kirby, S.H., Coe, R.S.: The role of crystal defects in the enstatite inversion. EOS (abstract). Trans. Am. Geoph. Union **55**, 419 (1974).

Koch, F.B., Cohen, J.B.: The defect structure of $Fe_{1-x}O$. Acta Cryst. B**25**, 275–287 (1969).

Korekawa, M., Jagodzinski, H.: Die Satellitenreflexe des Labradorites. Schweiz. Mineral. Petrog. Mitt. **47**, 269–278 (1967).

Kumo, S., Kato, N.: A furnace for high-temperature X-ray diffraction topography. J. Appl. Cryst. **7**, 427–429 (1974).

Lally, J.S., Fisher, R.M., Christie, J.M., Griggs, D.T., Heuer, A.H., Nord, Jr., G.L., Radcliffe, S.V.: Electron petrography of Apollo 14 and 15 rocks. Proc. Third Lunar Sci. Conf. Geochim. Cosmochim. Acta, Suppl. 3, **1**, 401–422 (1972).

Lally, J.S., Heuer, A.H., Nord, Jr., G.L., Christie, J.M.: Subsolidus reactions in lunar pyroxenes: an electron petrographic study. Contrib. Mineral. Petrol. **51**, 263–281 (1975).

Laves, F.: The lattice and twinning of microcline and other potash feldspars. J. Geol. **58**, 548 (1950).

Laves, F., Czank, M., Schulz, H.: The temperature dependence of the reflection intensities of anorthite ($CaAl_2Si_2O_8$) and the corresponding formation of domains. Schweiz. Mineral. Petrog. Mitt. **50**, 519–525 (1970).

Laves, F., Goldsmith, J.R.: Short-range order in anorthite. (Abs. p. 10). Am. Cryst. Assoc. Meeting, Chicago, Oct. 1951.

Lefebvre, S., Portier, R., Fayard, M.: Structure en domaines de l'oxyde non centrosymmétrique $LiFe_5O_8$. Phys. Stat. Sol. (a) **24**, 79–89 (1974).

Lieberman, D.S.: Martensitic transformation and determination of the inhomogeneous deformation. Acta Met. **6**, 680–693 (1958).

Lowenstein, W.: The distribution of aluminum in the tetrahedra of silicates and aluminates. Am. Mineralogist **39**, 92–96 (1954).

Matsumoto, T., Banno, S.: A natural pyroxene with the space group C_{2h}^4-P2/n. Proc. Japan. Acad. **46**, 173–175 (1970).

McConnell, J.D.C.: Electron optical study of effects associated with partial inversion in a silicate phase. Phil. Mag. **11**, 1289–1301 (1965).

McConnell, J.D.C.: Analysis of the time-temperature-transformation behaviour of the plagioclase feldspars. In: The feldspars (eds. W.S. MacKenzie and J. Zussman). Proc. NATO Adv. Study Inst., p. 460–477. Manchester: University Press 1974a.

McConnell, J.D.C.: Electron-optical study of the fine structure of a schiller labradorite. In: The feldspars (eds. W.S. MacKenzie and J. Zussman). Proc. NATO Adv. Study Inst. p. 478–490. Manchester: University Press 1974b.

McConnell, J.D.C., Fleet, S.G.: Direct electron-optical resolution of antiphase domains in a silicate. Nature 199, 586 (1963).

McLaren, A.C.: The domain structure of a transitional anorthite; a study by direct-lattice resolution electron microscopy. Contrib. Mineral. Petrol. 41, 47–52 (1973).

McLaren, A.C.: Transmission electron microscopy of the feldspars. In: The feldspars (eds. W.S. MacKenzie and J. Zussman). Proc. NATO Adv. Study Inst., p. 378–423. Manchester: University Press 1974.

McLaren, A.C., Marshall, D.B.: Transmission electron microscope study of the domain structure associated with the b-, c-, d-, e-, and f-reflections in plagioclase feldspars. Contrib. Mineral. Petrol. 44, 237–249 (1974).

McLaren, A.C., Phakey, P.P.: Electron microscope study of Brazil twin boundaries in amethyst quartz. Phys. Stat. Sol. 13, 413–422 (1966).

McLaren, A.C., Phakey, P.P.: Diffraction contrast from Dauphiné twin boundaries in quartz. Phys. Stat. Sol. 31, 723–737 (1969).

Megaw, H.D.: Order and disorder in the feldspars. Mineral. Mag. 32, 226–241 (1959).

Megaw, H.D.: Order and disorder. Proc. Roy. Soc. (London), Ser. A 259, 59–78, 159–183, 184–202 (1960).

Morimoto, N., Tokonami, M.: Domain structure of pigeonite and clinoenstatite. Am. Mineralogist 54, 725–740 (1969).

Müller, W.F.: Antiphase domains in $CaAl_2Si_2O_8$ (anorthite), $SrAl_2Si_2O_8$, and $BaAl_2Si_2O_8$. 8th Int. Congress on Electron Microscopy, Canberra 1, 472–473 (1974a).

Müller, W.F.: One-dimensional lattice imaging of a deformation-induced lamellar intergrowth of orthoenstatite and clinoenstatite [$(Mg,Fe) SiO_3$]. Neues Jahrb. Mineral., Monatsh. 2, 83–88 (1974b).

Müller, W.F., Wenk, H.-R.: Changes in the domain structure of anorthites induced by heating. Neues Jahrb. Mineral., Monatsh. 1, 17–26 (1973).

Müller, W.F., Wenk, H.-R., Bell, W.L., Thomas, G.: Analysis of the displacement vectors of antiphase domain boundaries in anorthites ($CaAl_2Si_2O_8$). Contrib. Mineral. Petrol. 40, 63–74 (1973).

Müller, W.F., Wenk, H.-R., Thomas, G.: Structural variations in anorthites. Contrib. Mineral. Petrol. 34, 304–314 (1972).

Nissen, H.-U.: Direct electron-microscope proof of domain texture in orthoclase. Contrib. Mineral. Petrol. 16, 354 (1967).

Nord, Jr., G.L., Heuer, A.H., Lally, J.S.: Transmission electron microscopy of substructures in Stillwater bytownites. In: The feldspars (eds. W.S. MacKenzie and J. Zussman). Proc. NATO Adv. Study Inst., p. 522–535. Manchester: University Press 1974b.

Nord, Jr., G.L., Lally, J.S., Heuer, A.H., Christie, J.M., Radcliffe, S.V., Fisher, R.M., Griggs, D.T.: A mineralogical study of rock 70017, an ilmenite-rich basalt, by high voltage electron microscopy. Lunar Sci. V, Houston, Lunar Sci. Inst. p. 556, 1974a.

Nord, Jr., G.L., Lally, J.S., Heuer, A.H., Christie, J.M., Radcliffe, S.V., Griggs, D.T., Fisher, R.M.: Petrologic study of igneous and metaigneous rocks from Apollo 15 and 16 using high-voltage transmission electron microscopy. Proc. 4th Lunar Sci. Conf. Geochim. Cosmochim. Acta, Suppl. 4, 1, 953–970 (1973).

Papike, J., Stephenson, N.C.: The crystal structure of mizzonite, a Ca- and carbonate-rich scapolite. Am. Mineralogist 51, 1014–1027 (1966).

Patil, R.N., Subbarao, E.C.: Monoclinic tetragonal phase transition in zirconia: mechanism, pretransformation, and coexistence. Acta Cryst. A26, 535–542 (1970).

Phakey, P.P., Ghose, S.: Scapolite: observation of anti-phase domain structure. Nature, Physical Sci. 38, 78–80 (1972).

Phakey, P.P., Ghose, S.: Direction observation of anti-phase domain structure in omphacite. Contrib. Mineral. Petrol. 39, 239–245 (1973).

Prewitt, C.T., Brown, G.E., Papike, J.J.: Apollo 12 clinopyroxenes: high temperature X-ray diffraction studies. Proc. 2nd Lunar Sci. Conf. Geochim. Cosmochim. Acta, Suppl. 2, **1**, 59–68 (1971).

Ribbe, P.H., Colville, A.A.: Orientation of the boundaries of out-of-step domains in anorthite. Mineral. Mag. **36**, 814–819 (1968).

Roy, R.: A syncretist classification of phase transitions. From: Phase transitions (eds. Henisch, H.K., Roy, R., Cross, L.E.), p. 13–28. New York: Pergamon Press 1973.

Serneels, R., Snykers, M., Delavignette, P., Gevers, R., Amelinckx, S.: Friedel's law in electron diffraction as applied to the study of domain structures in non-centrosymmetrical crystals. Phys. Stat. Sol. **58** (b), 277–292 (1973).

Shirane, G.: Neutron scattering studies of structural phase transitions at Brookhaven. Rev. Mod. Phys. **46**, 437–450 (1974).

Smith, D.K., Cline, C.F.: Verification of existence of cubic zirconia at high temperature. J. Am. Ceram. Soc. **6**, 156 (1962).

Smith, J.V.: Crystal structure and stability of the $MgSiO_3$ polymorphs; physical properties and phase relations of Mg,Fe pyroxenes. Mineral. Soc. Am. Spec. Papers **2**, 3–29 (1969).

Smith, J.V.: Feldspar minerals, vol. 1, Crystal structure and physical properties. Berlin-Heidelberg-New York: Springer 1974.

Smyth, J.R.: Orthopyroxene-high-low clinopyroxene inversions. Earth and Planet. Sci. Letters **6**, 406–407 (1969).

Smyth, J.R.: Experimental study on the polymorphism of enstatite. Am. Mineralogist **59**, 345–352 (1974).

Taylor, W.H.: The structure of sanidine and other feldspars. Z. Krist. **85**, 425–442 (1933).

Toman, K., Frueh, A.J.: Patterson function of plagioclase satellites. Acta Cryst. **A29**, 127–133 (1973).

Turner, F.J., Heard, H., Griggs, D.T.: Experimental deformation of enstatite and accompanying inversion to clinoenstatite. Int. Geol. Congr., 21 session, Copenhagen, Part 18, p. 399–408 (1960)

Ulbrich, H.H.: Crystallographic data and refractive indices of scapolites. Am. Mineralogist **58**, 81–92 (1973).

Van der Biest, O., Thomas, G.: Identification of Enantiomorphism in Crystals by Electron Microscopy. Acta Cryst. **A31**, 1, 70–76 (1975).

Van der Biest, O., Butler, E.P., Thomas, G.: Hot stage HVEM dynamic studies of order-disorder transitions in lithium ferrite. Proc. EMSA, 26–27 (1975).

Van Landuyt, J., DeRidder, R., Brabers, V.A.M., Amelinckx, S.: Jahn-Teller domains in $Mn_x Fe_{3-x} O_4$ as observed by electron microscopy. Mat. Res. Bull. **5**, 353–362 (1970).

Van Tendeloo, G., Amelinckx, S.: Group-theoretical considerations concerning domain formation in ordered alloys. Acta Cryst. **A30**, 431–440 (1974).

Van Tendeloo, G., Van Landuyt, J., and Amelinckx, S.: Electron microscopic observations of the domain structure and the diffuse electron scattering in quartz and in aluminium phosphate in the vicinity of the $\alpha \rightleftharpoons \beta$ transition: Phys. Stat. Sol., (A) **30**, pp. K11–K15, July 16, 1975.

Wechsler, M.S., Lieberman, D.S., Read, T.A.: On the theory of the formation of martensite. Trans. AIME **197**, 1503–1515 (1953).

Wenk, H.-R.: The structure of wenkite. Z. Krist. **137**, 113–126 (1973).

Wenk, H.-R., Müller, W.F., Thomas, G.: Antiphase domains in lunar plagioclase. Proc. 4th Lunar Sci. Conf. Geochim. Cosmochim. Acta, Suppl. 4, **1**, 909–923 (1973).

Wieder, H.H.: Ferroelectric properties of colemanite. J. Appl. Phys. **30**, 1010–1018 (1959).

Wolten, G.M.: Direct high-temperature single-crystal observation of orientation relationship in zirconia phase transformation. Acta Cryst. **17**, 763 (1964).

CHAPTER 5.2

Direct Observation of Iron Vacancies in Polytypes of Pyrrhotite

H. NAKAZAWA, N. MORIMOTO, and E. WATANABE

1. Introduction

The existence of a variety of closely related superstructures in certain sulphide minerals is usually attributed to ordered and partially ordered arrangements of metal vacancies; pyrrhotite, $Fe_{1-x}S$ being a particularly good example. Three different types of pyrrhotite have been found in nature (Carpenter et al., 1964; Morimoto et al., 1975 a and b), all of them having superstructures of the NiAs type structure ($A \sim 3.45$ Å and $C \sim 5.8$ Å). They are monoclinic pyrrhotite, 4C, intermediate pyrrhotite, nC and hexagonal troilite, 2C. The 4C and 2C types are stoichiometric with compositions Fe_7S_8 and FeS, respectively, while the nC type is non-stoichiometric with a composition from approximately Fe_9S_{10} to $Fe_{11}S_{12}$ (Morimoto et al., 1970; Nakazawa and Morimoto, 1971; Morimoto et al., 1975b). A wider composition range of nC is observed above 100°C. The iron-deficient nC transforms to a high-temperature phase, NA, above 220°C (Nakazawa and Morimoto, 1971).

The crystal structures of the 4C and 2C types of pyrrhotite have been determined by X-ray single-crystal methods (Bertaut, 1953; Evans, 1970; Tokonami et al., 1972). The nC type pyrrhotite is characterized by its long c period, a non-integral multiple of the original C-length. The NA type is also characterized by superstructures of $c = 3C$ and $a = NA$ ($N = 30 \sim 90$) (Nakazawa and Morimoto, 1971; Morimoto et al., 1975b).

Although the average structure of the nC type has already been worked out (Koto et al., 1974), the details of the *real* structure are as yet unknown.

Recent developments in high resolution electron microscopy permit the observation of two-dimensional lattice images of crystals that can be directly correlated with structural details (Pierce and Buseck, Chapter 2.6 of this volume) the interpretation of the image contrast is not straightforward unless the structure is well known because of the dynamical effect of electron beams in crystals. Thus the X-ray diffraction and lattice image techniques are complementary.

In this investigation, the lattice imaging technique has been applied to the 4C type to elucidate details of the pyrrhotite lattice, the structure of which is well known. Based on the results from the 4C type, this technique has been extended to the study of the nC type, the average structure of which is known, and of its transformation to the NA type.

2. Experimental

The crystals of pyrrhotite were crushed into fine fragments. They were mounted on grids using the usual procedures of sample preparation for high-resolution electron microscopy (Allpress and Sanders, 1973). Lattice images were obtained at several values of underfocusing and appropriate orientations in a JEM 100B electron microscope, operated at 100 kV, with a high-resolution goniometer stage. An aperture of 40 µm diameter gave the best results. Optimum images for structural details were obtained at the range of underfocusing of approximately 1250 ± 50 Å relative to the Gaussian image plane.

Structural change and decomposition of the pyrrhotite samples were often observed after exposure to the electron beam. However, the thermal effects of the electron irradiation on the samples were different from grain to grain, due to the size and thickness of the grains and the relation with the perforated support film of the grid.

3. Results

3.1 The 4C Type (Fe_7S_8)

Because the ordered arrangement of iron vacancies in the 4C type had been determined in detail (Bertaut, 1953; Tokonami et al., 1972), the lattice image technique was first applied to this type (Nakazawa et al., 1974a). Fig. 1 shows a lattice image of a portion of the 4C type from the Chichibu mine, Japan. A unit cell ($a = 11.9$, $c = 22.8$ Å) is outlined. The contrast in the lattice image is explained by the structure of the 4C type. The white spots correspond to the low atomic density in the projection and represent the projection of Fe sites of which half are vacant (squares in the inset diagrams), while the dark regions correspond to high atomic density and include the projection of Fe sites, all of which are completely filled (black circles), and sulfur atoms (small open circles).

Two different ways of stacking the rows of the white spots result in a fine band structure parallel to (001). The alternating bands with the thickness of one to ten stacking layers correspond to two projections along [110] (upper inset diagram in Fig. 1) and [010] (lower inset diagram in Fig. 1) of the 4C type structure. They are in twin relation of 60° or 120° rotation around the c^* axis. The unit cells in the different bands are outlined in the high-resolution micrograph (Fig. 1).

There are a number of short diffuse streaks in the rows of the white spots in the central part of Fig. 1. On both sides of the streaks, the stacking sequence of the rows change by a translation of $\frac{1}{4}c = C$. Thus the diffuse streaks are explained by superposition of two structures with C phase difference.

From the lattice image of the 4C type, we can conclude (in agreement with Pierce and Buseck, 1974, and Chapter 2.6 of this volume) that the disordered as well as ordered arrangements of iron vacancies of pyrrhotites can be observed at appropriate conditions by high-resolution electron microscopy.

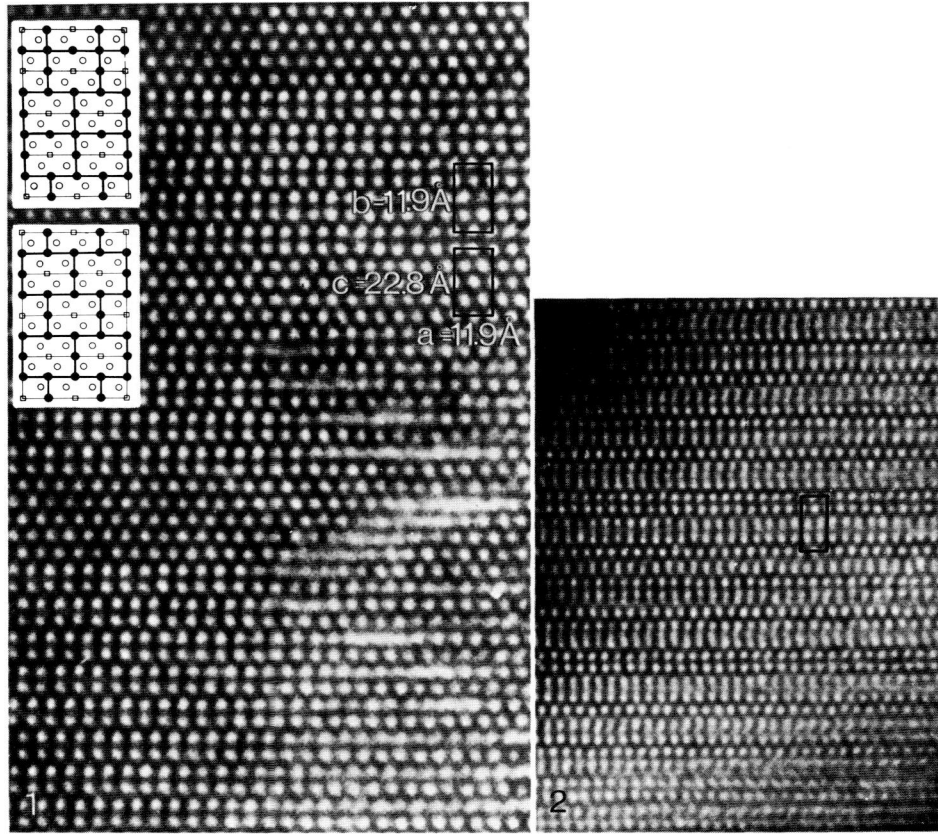

Fig. 1. High magnification lattice image of the 4C type from the Chichibu mine, Japan. Two unit cells with the dimensions of 11.9 Å and 22.8 Å are indicated. Inset diagrams are projected structures of the 4C type along [110] and [010], respectively. Filled circles and open squares indicate the projection of Fe sites completely filled and half filled, respectively. Open small circles correspond to S sites. The two different projections correspond to the unit cells indicated in the lattice image

Fig. 2. High magnification lattice image of the thin portion of the nC type ($n=5.0$) from the Yanahara mine, Japan. The distribution of vacancies in the nC type is apparently random in comparison with that of the 4C type (Fig. 1). The unit cell with $n=5$ is indicated

3.2 The nC Type

Fig. 2 and Fig. 3a show lattice images of the nC type ($n \sim 5$) from the Yanahara mine, Japan. In the thicker portion of the specimen in Fig. 3a (lower and right side in the Figure), contrast changes of somewhat regular intervals take place along the c axis. The unit cell corresponding to 5C ($b=11.9$ Å, $c=28.7$ Å) is outlined. The thicker portion seems to be outside the range for direct interpretation of the image.

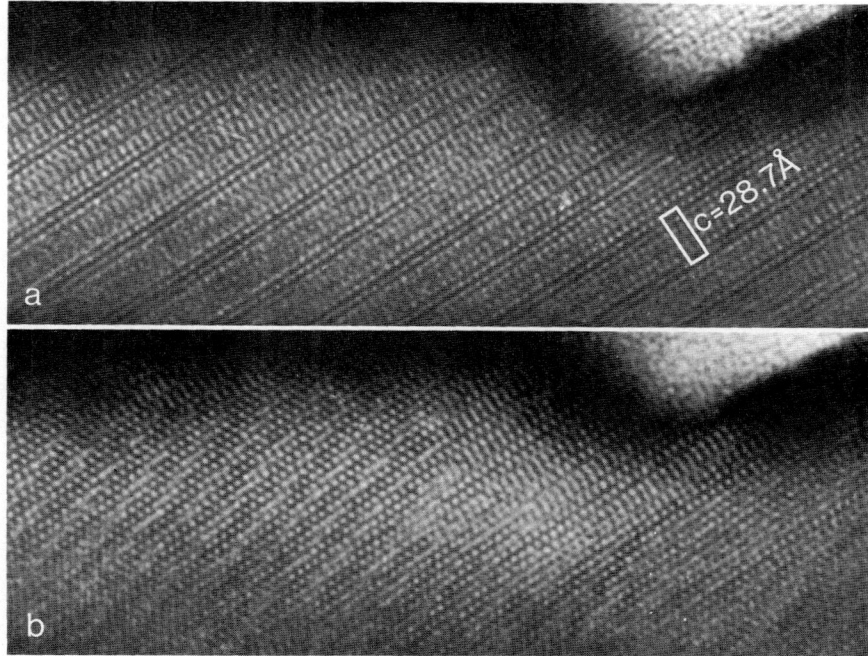

Fig. 3. (a) Lattice image of the nC type from the Yanahara mine, Japan. The intervals along the c axis are nearly $5C$, but slightly variable. The unit cell with $n=5$ is indicated. (b) Lattice image of the NA type taken from the same portion of the specimen as Fig. 3a after exposure to the electron beam. Because this image was taken at the transition from the nC type to NA type, some portion of the sample (the upper central part) are still of the nC type

The regular contrast intervals disappear in the portion near the thin edge of the specimen (in the upper central part), and white short lines are observed together with white spots along the c axis. They are considered to correspond to the statistical distribution of vacancies in two or more successive layers. A high magnification lattice image from a thin portion or a different grain is reproduced in Fig. 2. The distribution of vacancies is apparently random in this image.

The lattice image of the nC type, therefore, indicates that this type has an average structure consisting of a statistical distribution of iron vacancies along the c axis, rather than that of domains with different superstructures (Nakazawa et al., 1974b, 1975).

3.3 The NA Type

During electron irradiation, the nC type occasionally transforms into the NA type, which is stable above 220° C (Nakazawa et al., 1974b). Based on the X-ray diffraction pattern of the NA type, a structure was presented which consists of three domains related by $(1/3)c$ glide along the a axis (Nakazawa, 1968).

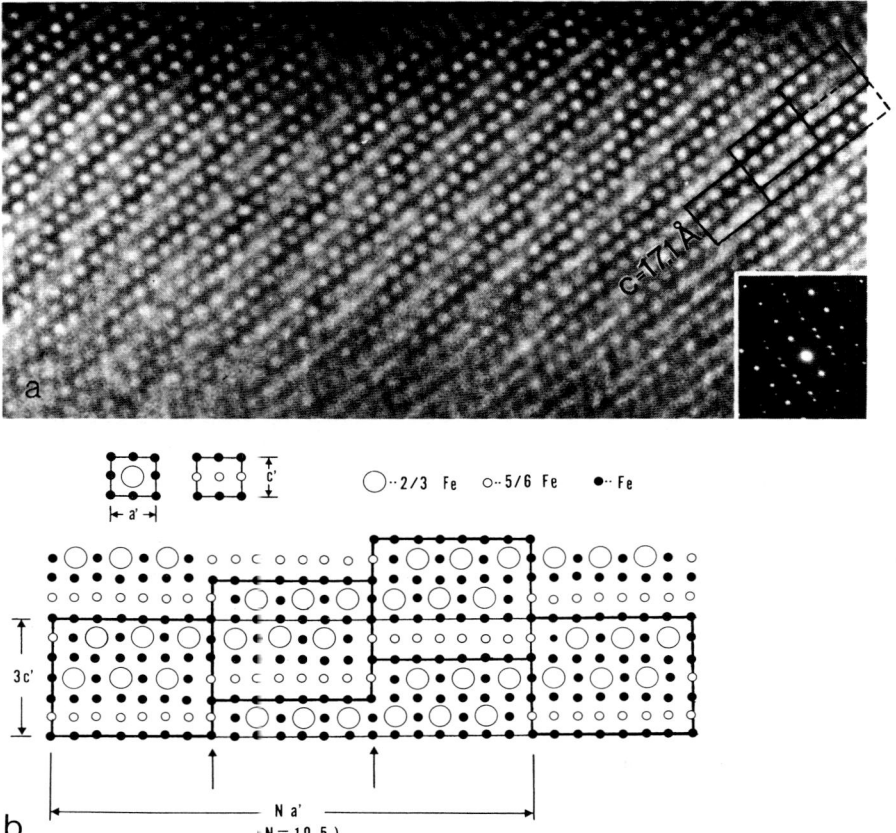

Fig. 4. (a) High magnification lattice image of the *NA* type. A unit with dimensions of a (82.6 Å) and c (17.1 Å) is indicated. An electron diffraction pattern is shown. (b) A structure model of the *NA* type. Filled circles indicate Fe sites that are completely filled. Large and small open circles are the Fe sites of 1/3 and 1/6 vacancies, respectively. Sulfur atoms are omitted for simplicity

It was, however, not possible to determine the arrangement of the iron vacancies in domains, precisely because there are too many parameters to determine, compared with the number of observed reflections.

A lattice image of the *NA* type is reproduced in Fig. 4a. The electron diffraction pattern is also shown. Three domains with $(1/3)c$ glide along the a axis are indicated. From the lattice image, a structure model has been constructed (Fig. 4b) which is consistent with the three dimensional X-ray diffraction data (Nakazawa et al., 1975).

3.4 Transition of *nC* into *NA*

The two lattice images in Fig. 3 are obtained from the same portion of the specimen before and after the transition from *nC* and *NA* at about 220° C caused by exposure to the electron beam. In Fig. 3b, the *NA* structure is observed in most of the figure, while the *nC* structure still remains at the thinner portion (upper central part). This result shows that the transition occurs first in the thick portion of the specimen and then extends from the thicker to the thinner portion. Because a random arrangement of vacancies is observed at the transition front in Fig. 3b, the vacancies of the *nC* structure appear to move randomly at the transition and settle down at new sites resulting in the *NA* structure (Nakazawa et al., 1976).

References

Allpress, J.G., Sanders, J.V.: The direct observation of the structures of real crystals by lattice imaging. J. Appl. Cryst. **6**, 165–190 (1973).
Bertaut, E.F.: Contribution à l'étude des structures lacunaires: la pyrrhotine. Acta Cryst. **6**, 557–561 (1958).
Buseck, P.R., Iijima, S.: High-resolution electron microscopy of silicates. Am. Mineralogist **59**, 1–21 (1973).
Carpenter, R.H., Desborough, G.A.: Range in solid solution and structure of naturally occurring troilite and pyrrhotite. Am. Mineralogist **49**, 1350–1365 (1964).
Evans, H.T., Jr.: Lunar troilite: Crystallography. Science **167**, 621–623 (1970).
Iijima, S., Allpress, J.G.: High resolution electron microscopy of $TiO_2 \cdot 7Nb_2O_5$. J. Solid State Chem. **7**, 94–105 (1973).
Koto, K., Morimoto, N., Gyobu, A.: The crystal structures of the $6C$ ($Fe_{11}S_{12}$) and the $4C$ (Fe_7S_8) type pyrrhotites. Proc. Int. Cryst. Conference on Diffraction Studies on Real Atoms and Real Crystals, Melbourne, 161–162 (1974).
Morimoto, N., Gyobu, A., Mukaiyama, H., Izawa, E.: Crystallography and stability of pyrrhotite. Econ. Geol. (1975a).
Morimoto, N., Gyobu, A., Tsukuma, K., Koto, K.: Superstructure and nonstoichiometry of the intermediate pyrrhotite. Am. Mineralogist **60**, 240–248 (1975b).
Morimoto, N., Nakazawa, H., Nishiguchi, K., Tokonami, M.: Pyrrhotites: stoichiometric compounds with composition $Fe_{n-1}S_n$ ($n \geq 8$). Science **168**, 964–966 (1970).
Nakazawa, H.: Pyrrhotite: new types and their structures with long periods. Dissertation Osaka University, Japan (1968).
Nakazawa, H., Morimoto, N.: Phase relations and superstructures of pyrrhotite, $Fe_{1-x}S$. Mat. Res. Bull. **6**, 345–358 (1971).
Nakazawa, N., Morimoto, N., Watanabe, E.: Direct observation of pyrrhotite (Fe_7S_8) by electron microscopy. Proc. Int. Cryst. Conference on Diffraction Studies on Real Atoms and Real Crystals, Melbourne, 223–224 (1974a).
Nakazawa, H., Morimoto, N., Watanabe, E.: Direct observation of the nonstoichiometric pyrrhotite. Proc. 8th Int. Congress on Electron Microscopy, Canberra **1**, 498–499 (1974b).
Nakazawa, H., Morimoto, N., Watanabe, E.: Direct observation of metal vacancies by high-resolution electron microscopy: $4C$ type pyrrhotite (Fe_7S_8). Am. Mineralogist **60**, 359–366 (1975).
Nakazawa, H., Morimoto, N., Watanabe, E.: Direct observation by high-resolution electron microscopy: the *NA* pyrrhotites. In preparation (1976).
Pierce, L.P., Buseck, P.R.: Electron imaging of pyrrhotite superstructures. Science **186**, 1209–1212 (1974).
Tokonami, M., Nishiguchi, K., Morimoto, N.: Crystal structure of a monoclinic pyrrhotite (Fe_7S_8). Am. Mineralogist **57**, 1066–1081 (1972).

CHAPTER 5.3

Rutile: Planar Defects and Derived Structures

B.G. HYDE

1. Introduction

Over the last 25 years there has been a considerable change in the approach to solid-state crystal chemistry, mainly derived from thorough studies of a few transition metal oxide systems among them rutile — reduced and/or admixed with other oxides of lower valent metals, i.e. $M_2O_3 + TiO_2$, where M = Ti or another trivalent cation. Structural progress was considerably accelerated by the application of the electron microscope and we will summarize some of this work. We demonstrate thereby the inherent weakness of X-ray diffraction and especially of powder techniques to study these systems.

All the work described has utilized synthetic specimens, but is nevertheless relevant also to natural rutiles.

2. "Reduced Rutile"? TiO_{2-}

Twenty years ago, phase analysis by X-ray powder diffraction indicated that, at higher oxygen contents, there were two homogeneous, non-stoichiometric phases: approximately $TiO_{1.65}$–$TiO_{1.80}$ and $TiO_{1.90}$–$TiO_{2.00}$. The latter in particular was the subject of a great deal of research and speculation as to whether the oxygen deficiency = titanium excess was accommodated by random distributions of anion vacancies or interstitial cations in the parent rutile structure. A more careful X-ray study, using a more powerful type of powder camera (focusing, Guinier-Hägg) suggested that these classical ideas had to be abandoned. Apart from the ordered phases Ti_2O_3 (corundum type) and Ti_3O_5, a sequence of low symmetry, fully-ordered structures (with no apparent composition range) occupied the interval $TiO_{1.75}$ to $TiO_{1.89}$. They were separated by diphasic regions, and were closely related in structure — to each other and to rutile. Stoichiometries were Ti_4O_7, Ti_5O_9, Ti_6O_{11}, Ti_7O_{13}, Ti_8O_{15}, and Ti_9O_{17}, i.e. Ti_nO_{2n-1} with $n = 4, 5, 6, 7, 8, 9$. The structure of one member of this *homologous series*, Ti_5O_9, was determined by single crystal X-ray diffraction and yielded the structural principle common to all members (Andersson and Jahnberg, 1963, R).[1] Each structure consists of infinite slabs of rutile type, n octahedra wide and joined

---

[1] R indicates a review type reference substituting for, and containing a number of original references.

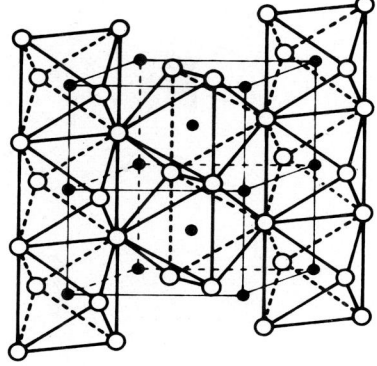

Fig. 1. Clinographic projection of the rutile structure. Two unit cells are lightly outlined. The heavily outlined [TiO$_6$] octahedra are connected by edges to give ribbons which are, in turn, connected by corners. ● = Ti, ○ = O

to adjacent parallel slabs at oxygen-deficient boundaries — described as being corundum type. They are most easily described in terms of idealized rutile-related layers.

Rutile has a simple tetragonal structure composed of strings of edge-shared [TiO$_6$] octahedra (parallel to c) in two mutually perpendicular orientations and joined by corner-sharing (Fig. 1). It is commonly and conveniently idealized by a small distortion [mainly rotation of the strings by $\pm \tan^{-1}(4u-1) = \pm 12.5°$] which puts the anions into perfect hexagonal close-packing (Fig. 2). This idealized structure is orthorhombic with $c_{h.c.p.} = a_r.$[2] All the Ti$_n$O$_{2n-1}$ structures of Andersson are describable in terms of single $(100)_r$ layers of this idealized type. In rutile itself, alternate rows of octahedral interstices (parallel to c_r) are occupied by titanium. In the reduced rutiles these cation rows are interrupted at the boundaries of the rutile slabs, across which the filled and empty rows of octahedral sites are interchanged (Fig. 3). The trace of the boundary on $(100)_r$ is always $[0\bar{1}2]_r$. In each case the complete structure is obtained by stacking identical layers along $[1\bar{1}1]_r$, and the boundary plane is $(121)_r$. It is a simple α-boundary with an ideal displacement vector $\boldsymbol{R} = \frac{1}{2}[0\bar{1}1]_r$, corresponding to a collapse of the structure across an eliminated oxygen plane. Differing values of n in Ti$_n$O$_{2n-1}$ simply mean different spacings between the regularly arrayed boundaries (Terasaki and Watanabe, 1971). Such boundaries have been termed *crystallographic shear* (*CS*) planes. (Some idealization is essential: real rutile slabs can only be joined in this way after some distortion; $\frac{1}{2}[0\bar{1}1]_r$ is approximately, but not exactly, an O–O vector.)

The situation at higher oxygen contents, TiO$_{>1.89}$, could not be resolved by X-ray diffraction, and seemed an ideal subject for study by electron microscopy/diffraction. The technique had already been used to examine the effect of deforming rutile single crystals at high temperature. As well as perfect dislocations, antiphase boundaries (APB's), $\{011\}\frac{1}{2}\langle 0\bar{1}1\rangle$ had also been observed (Ashbee and Smallman, 1963). These differ from *CS* planes by being stoichiometric. Note that the displacement vector is the same for both faults, but that in the APB it is parallel to the fault plane: no material need be added to or subtracted from the boundary for the operation to be carried out. Again the empty and

---

[2] Subscript r denotes axes/indices referred to the rutile cell or sub-cell.

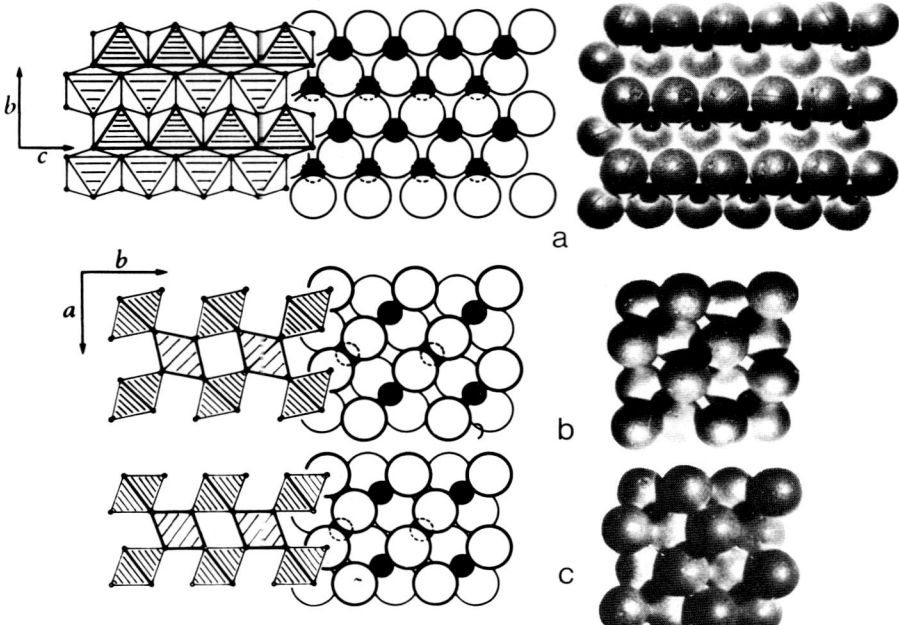

Fig. 2a–c. The rutile structure projected along a in (a) and c in (b); (c) shows the same projection as (b) but with the (100) oxygen planes flattened (by rotating the octahedral strings by $\pm 12.5°$ about c) and the octahedra made regular so that the anion arrangement is perfect hcp. – the idealization utilized in this chapter. In each case we show: on the left, [TiO$_6$] octahedra; in the center, the atoms; on the right, photographs of corresponding ball models

filled rows of octahedral interstices are interchanged across the APB. Eikum and Smallman (1965) observed another type of planar boundary, parallel to $\{132\}_r$ but of unknown $\boldsymbol{R}$.

More thorough studies followed (Bursill and Hyde, 1972, R): $\{132\}$ boundaries were confirmed to occur in slightly-reduced rutile (TiO$_{>1.99}$), and a whole family of obviously CS structures with ordered (132) boundaries was found to inhabit the composition range TiO$_{\sim 1.93}$–TiO$_{>1.97}$. Single crystals large enough for structure determination by X-ray diffraction could not be prepared. The only other route for elucidating the structures was to determine the displacement vector at the (132) CS plane. By analogy with Andersson's (121) CS structures it had been guessed that the ideal displacement vector would be the same, $\boldsymbol{R} = \frac{1}{2}[0\bar{1}1]$. But it was essential that this be confirmed experimentally: from a number of observation of the diffraction contrast at isolated $\{132\}$ faults a vector $\boldsymbol{R} = \frac{1}{6}\langle 211 \rangle$ had been deduced (Van Landuyt, 1966), which did not make crystallochemical sense. Contrast experiments (Bursill and Hyde, 1970), comparing observed fringe intensity profiles with those computed for various crystal thicknesses and values of $\boldsymbol{g} \cdot \boldsymbol{R}$, ruled out the proposed $\frac{1}{6}\langle 211 \rangle$. They showed that the simple criteria, $\boldsymbol{g} \cdot \boldsymbol{R} \approx 0$ for extinction, and $\boldsymbol{g} \cdot \boldsymbol{R} \approx \frac{1}{2}$ for π-contrast, were inade-

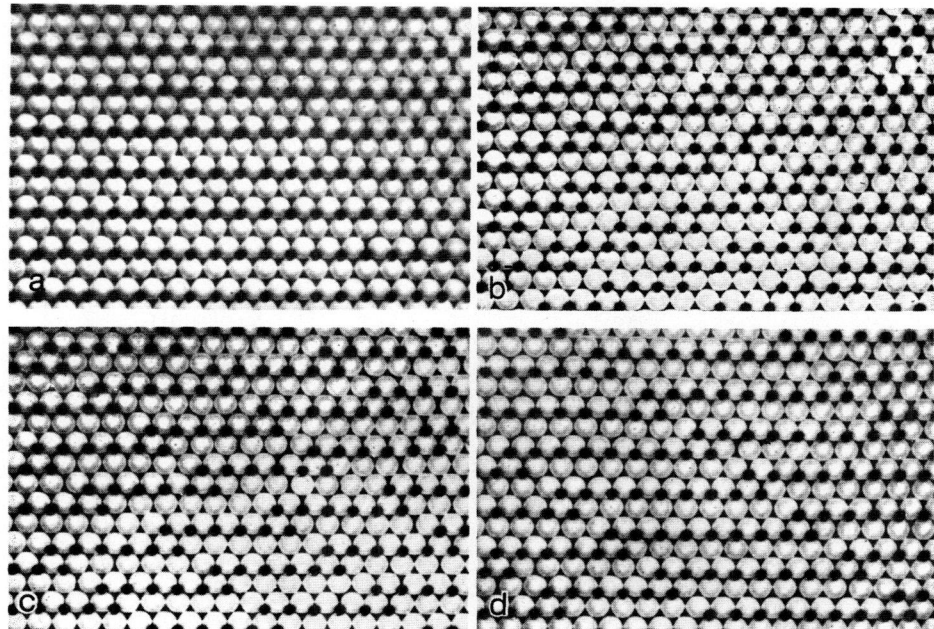

Fig. 3a–d. Idealized $(100)_r$ layers of (a) rutile, TiO_2, (b–d) Ti_nO_{2n-1} with $n=4, 6, 9$ and (121) type CS planes

quate for thin crystals: and that the degree of contrast varied markedly and periodically with t, being a maximum for $t/\xi_g = N + \frac{1}{2}$, especially for small $t (< 6\xi_g)$. (t is the crystal thickness, ξ_g is the extinction distance, and N is an integer). This explained the misinterpretation of the earlier results, and all observations were seen to be consistent with the guessed, ideal $\boldsymbol{R} = \frac{1}{2} \langle 0\bar{1}1 \rangle$, being in fact a non-ideal $\boldsymbol{R} = \frac{1}{2}\langle 0, -0.90, 0.90 \rangle$. A similar departure from ideality had been noted by Andersson for the (121) CS structures, and ascribed to cation-cation repulsion between face-sharing $[TiO_6]$ octahedra at the CS plane. (This is additional to the distortion required to fit rutile slabs together, mentioned above.) As one might expect, the departure is somewhat less in the case of (132) CS since there is a lower density of face-shared pairs at these CS planes. Similar contrast experiments with lamellae of closely-spaced (132) CS planes showed that the same displacement vector operates at these too. This is confirmed by the observations of Van Landuyt et al. (1970), using a quite different method. From the diffraction pattern of rutile coherent with a $(132)_r$ CS structure they deduced $\boldsymbol{R} = \frac{1}{10}[155]_r \approx \frac{1}{2}[011]_r$. This procedure assumes that the rutile cell is identical in size and orientation with the rutile subcell in the CS derivative. This cannot be perfectly true, as we have already seen. But, by the same token, the $\boldsymbol{R}$ values for isolated and closely-spaced CS planes will also be different. Hence neither of these methods is a substitute for the standard, single-crystal methods of X-ray or neutron diffraction. However, in the present instance there is no reason to suspect the electron microscopy results.

An idealized layer of one such (132) *CS* structure is shown in Fig. 4: $n=20$ (132), $Ti_{20}O_{39} = TiO_{1.950}$. The full structure is again produced by stacking identical layers along $[1\bar{1}1]_r$. All the (132) *CS* structures correspond, as before, to the general formula Ti_nO_{2n-1}, but probably only with even n. (It is difficult to index single-crystal electron-diffraction patterns for high values of n.) The range is $12 \lesssim n \lesssim 40$, and the lower limit is temperature-dependent.

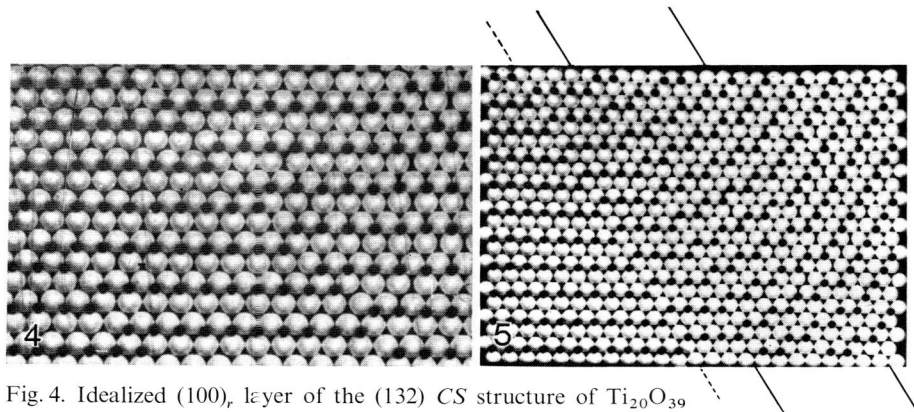

Fig. 4. Idealized $(100)_r$ layer of the (132) *CS* structure of $Ti_{20}O_{39}$

Fig. 5. From left to right is shown an idealized (100) layer of rutile with, first, a single (011) $\frac{1}{2}[0\bar{1}1]$ APB; then the same operation repeated on alternate (011) oxygen planes to give the $TiO_2(II)$ structure type; and then repeated on every (011) plane to give rutile in a twin orientation

Inspection of Fig. 4 shows that the rutile slabs are now $n/2$ octahedra wide, and that the steps in the cation rows parallel to c_r are now of two types, thus doubling the repeat distance along the *CS* plane (as compared with $\{121\}$ $\frac{1}{2}\langle 0\bar{1}1\rangle$). One, which we will call *C*, is the sort that occurs along (121) $\frac{1}{2}[0\bar{1}1]$ *CS* planes; the other, which we will call *A*, occurs along the (011) $\frac{1}{2}[0\bar{1}1]$ APB plane. In (121) structures only *C* steps occur; in (132) structures *C* and *A* alternate along the *CS* plane. An example of an ordered structure with only *A* steps is the high-pressure polymorph of rutile, $TiO_2(II)$ which has the α-PbO_2 type of structure (Fig. 5).

Between the lowest (132) structure and the highest (most oxidized) (121) structure, i.e. in the (temperature-dependent) composition range $TiO_{\sim 1.93}$ to $TiO_{\sim 1.88}$, a new phenomenon is observed: the ordered *CS* planes gradually change their orientation from (132) to (121) through high index planes, (253), (374), (495), etc. These correspond to different proportions of *C* to *A* steps along the *CS* plane, i e. intergrowths of different proportions of the elements (121) $\frac{1}{2}[0\bar{1}1]$ and (011) $\frac{1}{2}[0\bar{1}1]$. The general *CS* plane is

$$(hkl) = p \cdot (121) + q \cdot (011) = (p, 2p+q, p+q),$$

where p and q are both integers. As the reduction of TiO_2 proceeds, the ratio p/q increases from 1 in the (132) region to ∞ in the (121) region. In the intermediate (swinging) region there is, at least in principle, a *continuous* sequence of ordered structures: any stoichiometry, Ti_nO_{2n-p}, can be accommodated by a perfectly ordered (hkl) CS structure. This phenomenon is most strikingly evident when Ti^{3+} is replaced by Cr^{3+}.

3. Chromium(III)-doped Rutile, $Cr_2O_3 + TiO_2$

X-ray powder diffraction studies by Andersson *et al.* (Andersson and Jahnberg, 1963) revealed $(121)_r$ CS structures at relatively high chromia contents, $Cr_2Ti_{n-2}O_{2n-1}$ with $n = 6, 7, 8, 9$ (i.e. 22.2 to 33.3 mole percent $CrO_{1.5}$). As in the reduced rutile system, these were separated by diphasic regions, but the situation at higher oxygen (lower chromia) contents could not be resolved. Later workers (Flörke and Lee, 1970) confirmed these results, but interpreted their own observations at $T \leq 1450$ °C as showing that the (121) series continued to $n = 17$ (11.8% $CrO_{1.5}$). A diphasic gap then separated the CS region from a homogeneous solid solution of up to ~ 8 mol% $CrO_{1.5}$ in rutile (Flörke and Lee, 1970). Careful electron microscope/diffraction studies largely confirm the earlier X-ray work but reveal a different situation at lower chromia contents (Philp and Bursill, 1974). The highest value of n is 8 rather than 9 or 17. Then, as the chromia content is reduced, the CS planes swing away from (121) towards (132) – exactly as in the binary system. But they never get there: the minimum value of $p/q \approx 2$, i.e. $(hkl) \approx (253)$, in the range $1.93 \lesssim O/(Cr + Ti) \lesssim 1.96$, or chromia contents of ~ 14 to ~ 8 mol% $CrO_{1.5}$. Electron diffraction patterns are very sharp (Fig. 6), and lattice images apparently reveal very well ordered structures. At least 50 different structures were observed in the 80 or so crystals examined: it appears that every crystal has its own structure (although similarities are so great it is not always possible to distinguish them). In all cases it was absolutely essential to orient the crystals accurately in the $[1\bar{1}1]$ zone in order to identify the CS plane with confidence.

It is hardly necessary to add that chromian variants are well known in natural rutiles. It seems likely that some unanswered problems remaining from the work described above might be solved by a study of these minerals.

At still higher chromia contents the so-called E-phase appears at ~ 46–50 mole% $CrO_{1.5}$, i.e. approx. $Cr_2Ti_2O_7$ (Flörke and Lee, 1970). At high temperatures there is a range of "homogeneity", but again it is certain that a continuous sequence of ordered structures obtains. Since it is difficult to resolve experimentally the close spacings of the CS planes (and consequent large reciprocal spacings in the "superlattice rows" in the diffraction patterns) it is difficult to recognize the structural relations. Some progress has been made with the related ternary and quaternary systems $(Fe,Cr,Ti)O_x$ and $(Fe,Cr,Ti,Zr)O_x$ in this composition range (Grey *et al.*, 1973, R). The structures may be described as CS derivatives of the α-PbO_2 structure type or, equally well, as high index CS derivatives of

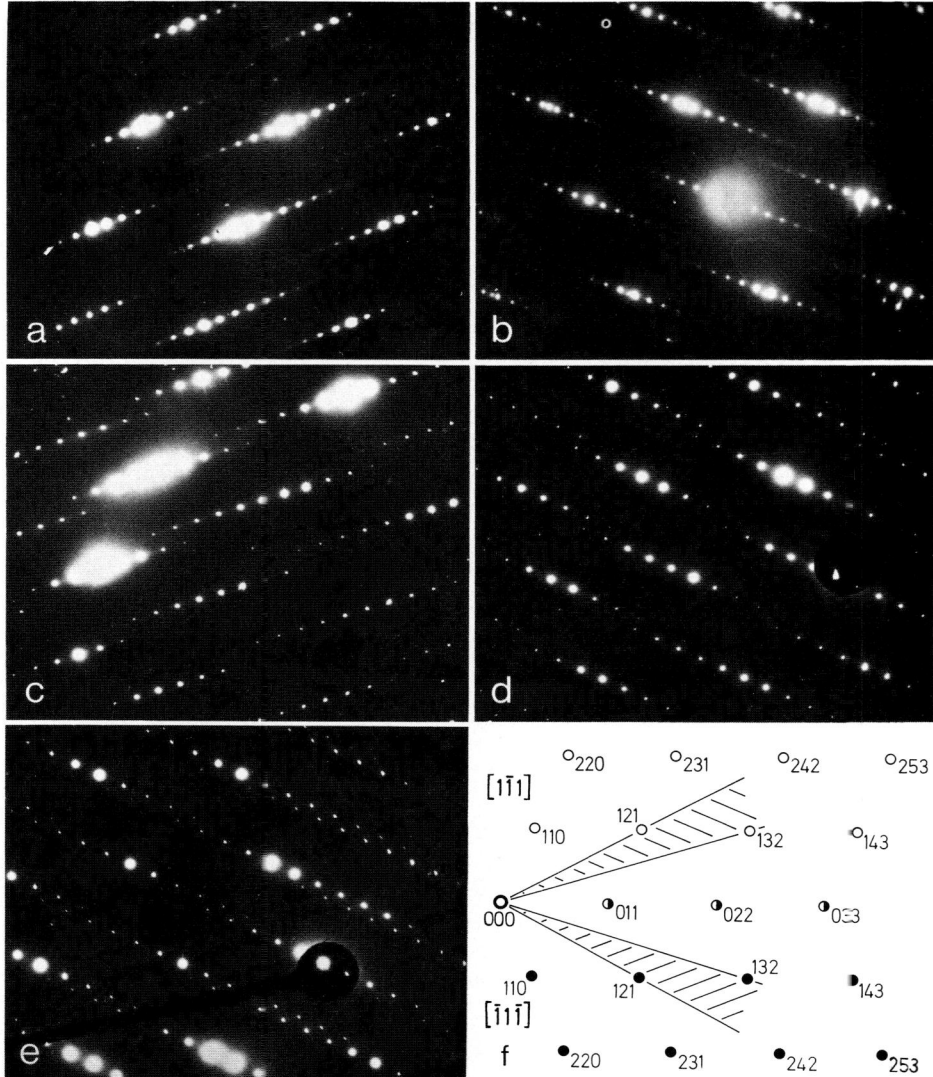

Fig. 6a–f. Electron diffraction patterns of chromia-doped rutile: p/q values are (a) 1.87, (b) 2.00, (c) 3.32, (d) 5.1, (e) 7.0. (f) shows the reciprocal lattice nets for $1\bar{1}1$ and $\bar{1}1\bar{1}$ zones of the rutile sub-cell, the shaded zones being the limits of the observed reciprocal lattice directions for $1 < p/q < \infty$. All SAD's are in the same orientation as (f): a and c being $1\bar{1}1$ zones, b, d and e being $\bar{1}1\bar{1}$ zones. Rows of superlattice spots pass through each node of the rutile reciprocal lattice in the direction hkl of the CS planes

the rutile type, e.g. $(253)_r$, $(385)_r$, with very closely-spaced CS planes, i.e. low n values. $CrFeTi_2O_7$ is $n=8$ $(253)_r$; $(Cr,Fe)_2Ti_3O_9$ is $n=5$ $(132)_r$ (Grey et al., 1973; Hyde et al., 1974). In general, and in contrast to the swinging CS region at higher oxygen contents, it appears that q/p is now often >1.

4. Iron-doped Rutile, $(Fe,Ti)O_x$

This is another system of mineralogical interest: iron is a common impurity in natural rutile. Bursill (1974) has carried out the most thorough study of this very complex system. Samples prepared in the temperature range 1300–1500 °C, and then quenched, are CS structures from $(253)_r$ up to $(121)_r$ in the composition range $1.970 \gtrsim O/(Fe+Ti) \gtrsim 1.930$ (cf. $Cr_2O_3 + TiO_2$). The precise indices depend on the oxygen/metal ratio, the Fe^{2+}/Fe^{3+} ratio and the temperature. At lower temperatures, these structures decompose eutectoidally to rutile and pseudobrookite. The rutile-phase field is always narrow, and the oxygen deficit accommodated by widely-spaced CS planes. At still lower temperatures there is also a phase $Fe_2Ti_3O_9$, which is $n=5$ $(132)_r$ (cf. the last section).

Very recently, some new features of this system have been discovered (Bursill et al., 1974). Above 1450 °C the CS structures of the type already discussed give way to another set, in the same composition range but in a different zone, i.e. $[100]_r$. They are all of the type

$$(0kl) = p \cdot (010) + q \cdot (011) = (0, p+q, q).$$

The APB component (011) and the displacement vector $R = \frac{1}{2}[0\bar{1}1]$ are the same as before, but the CS component is different, (010) $\frac{1}{2}[0\bar{1}1]$. Again, variation of composition and temperature result in a continuous swinging of the CS plane orientation from (031), $p/q=2$, through (041), (051), etc. to (010), $p/q=\infty$.

This is a radical departure: in all other structures (down to the very highly oxygen-deficient corundum type) no $[MO_6]$ octahedron shares more than one face with another such octahedron, and it was thought that the *raison d'être* of CS was that it avoided $[MO_6]$ octahedra sharing two opposite faces with similar octahedra — a high energy situation since relaxation by off-center displacement of the central cations is then strictly circumscribed. (Note that this situation must arise if an interstitial cation is inserted into any empty octahedral site in rutile.) However, this "unfavorable" situation will occur in $(0kl)$ $\frac{1}{2}[0\bar{1}1]$ CS planes, unless some of the cations (probably Fe) are tetrahedrally-coordinated.

5. Gallium(III)-doped Rutile, $Ga_2O_3 + TiO_2$

An as yet unsolved problem appears in this system (Gibb and Anderson, 1972). Disordered and ordered structures are observed, quite analogous to the CS structures described earlier, but the fault plane is now $(210)_r$. The displacement vector, a new one, appears to be $R = \frac{1}{2}[010]_r$, but no successful model has yet been proposed for such a boundary. (Unlike $\frac{1}{2}[0\bar{1}1]$ it is not even approximately an O—O vector.) It seems likely (Hyde et al., 1974) that some new factor operates here; that the conventional idealization of the rutile structure (to hcp. oxygens) is no longer appropriate. This is in accord with the structure of β-Ga_2O_3 being different from the corundum type of α-Fe_2O_3 and Cr_2O_3, in that its anions are ccp.

References

Andersson, S., Jahnberg, L.: Crystal structure studies on the homologous series Ti_nO_{2n-1}, V_nO_{2n-1} and $Ti_{n-2}Cr_2O_{2n-1}$. Arkiv Kemi **21**, 413–426 (1963)

Ashbee, K.H.G., Smallman, R.E.: The plastic deformation of titanium dioxide single crystals. Proc. Roy. Soc. (London), Ser. A **274**, 195–205 (1963)

Bursill, L.A.: An electron microscope study of the $FeO-Fe_2O_3-TiO_2$ system and of the nature of iron-doped rutile. J. Solid State Chem. **10**, 72–94 (1974)

Bursill, L.A., Grey, I.E., Lloyd, D.J.: High-temperature disordered crystallographic shear structures in the $Fe_2O_3-TiO_2$ system. In: Diffraction studies of real atoms and real crystals, p. 265. Melbourne: Austr. Academy of Science 1974

Bursill, L.A., Hyde, B.G.: The displacement vectors of $\{1\bar{3}2\}$ and $\{101\}$ faults in rutile. Proc. Roy. Soc. (London), Ser. A **320**, 147–160 (1970)

Bursill, L.A., Hyde, B.G.: Crystallographic shear in the higher titanium oxides: structure, texture, mechanisms and thermodynamics. In: Progress in solid-state chemistry, vol. 7 (eds. H. Reiss and J.O. McCaldin), p. 177–253. Oxford: Pergamon Press 1972

Eikum, A., Smallman, R.E.: A Note on the transformation in non-stoichiometric rutile. Phil. Mag. **11**, 627–632 (1965)

Flörke, O.W., Lee, C.W.: Andersson-Phasen, dichteste Packung und Wadsley-Defekte im System $Ti-Cr-O$. J. Solid State Chem. **1**, 445–453 (1970)

Gibb, R.M., Anderson, J.S.: Electron microscopy of solid solutions and crystallographic shear structures II. $Fe_2O_3-TiO_2$ and $Ga_2O_3-TiO_2$ systems. J. Solid State Chem. **5**, 212–225 (1972)

Grey, I.E., Reid, A.F., Allpress, J.G.: Compounds in the system $Cr_2O_3-Fe_2O_3-TiO_2-ZrO_2$, based on intergrowth of the α-PbO_2 and V_3O_5 structural types. J. Solid State Chem. **8**, 86–99 (1973)

Hyde, B.G., Bagshaw, A.N., Andersson, S., O'Keeffe, M.: Some defect structures in crystalline solids. Ann. Rev. Mat. Sci. **4**, 43–92 (1974)

Philp, D.K., Bursill, L.A.: Continuous structure variation and rotating reciprocal lattices in the titanium-chromium oxides. Acta Cryst. A **30**, 265–272 (1974)

Terasaki, O., Watanabe, D.: Electron microscopic study on the structure of Ti_nO_{2n-1} ($4 \leq n \leq 10$) phases. Japan. J. Appl. Phys. **10**, 292–303 (1971)

Van Landuyt, J.: Determination of the displacement vector at the anti-phase boundaries in rutile by contrast experiments in the electron microscope. Phys. Stat. Sol. **16**, 585–590 (1966)

Van Landuyt, J., Gevers, R., Amelinckx, S.: Electron microscope study of twins, anti-phase boundaries, and dislocations in thin films of rutile. Phys. Stat. Sol. **7**, 307–329 (1964)

Van Landuyt, J., Ridder, R. de, Gevers, R., Amelinckx, S.: Diffraction effects due to shear structures: a new method for determining the shear vector. Mat. Res. Bull. **5**, 353–362 (1970)

CHAPTER 5.4

High-resolution Electron Microscopy of Unit Cell Twinning in Enstatite

S. IIJIMA and P.R. BUSECK

Unit-cell level twinning is prominent in many solids (Andersson and Hyde, 1974). The problem of detecting such twinning, if of limited extent, is difficult if not impossible using X-ray techniques. High resolution electron microscopy, on the other hand, is well suited for this problem (Iijima, 1973; Buseck and Iijima, 1974).

In this paper we describe procedures used to detect and confirm unit-cell twinning in monoclinic enstatite (CLEN). Such procedures can demonstrate twinning, even if the concentration of twins is too low to be detected from electron or X-ray diffraction patterns. Dark field techniques are commonly used to image twins. These techniques are limited or precluded if the reciprocal lattice spacings are sufficiently small so that the objective aperture cannot be restricted to the diffracted beam (operating vector) of interest, as is the case for most pyroxenes. The twinning features under discussion will be illustrated using enstatite from Bamble, Norway[1]; other pyroxenes would serve equally well.

Our methods of detecting and investigating twinning utilize bright field microscopy. Two procedures were used. The first entails slight tilting of the sample relative to the electron beam in order to develop differential contrast between the twin and matrix images. The second method requires tilting of the beam with the sample stationary; the latter results in higher resolution and thus permits the imaging of closely spaced (<3Å) fringes. As the twinning of CLEN occurs with (100) as the composition plane (Iijima and Buseck, 1975), a–c sections show the twinning well.

Fig. 1 is an example of electron images of a region of CLEN. This sample was originally orthorhombic, but was heated to 1485 °C for 72 h and then quenched, thereby converting it to CLEN. Fig. 1a is a selected area diffraction pattern showing two sets of spots. The set indicated by the black arrows can be explained in terms of twinning on (100).

The bright field images of Fig. 1b–d were obtained by imaging the reflections enclosed by the white circle in Fig. 1a. If the crystal is oriented with its b^* axis exactly coincident with the electron beam (i.e., with the optical axis of the microscope), the resulting image displays relatively uniform contrast (Fig. 1b). If, on the other hand, the crystal is tilted slightly, perhaps 2.5°, around an axis inclined to a^*, contrast develops between adjacent regions. The particular direc-

---
[1] As Bamble pyroxene contains 10.5 weight percent FeO, it is properly considered bronzite. For brevity we refer to it as enstatite.

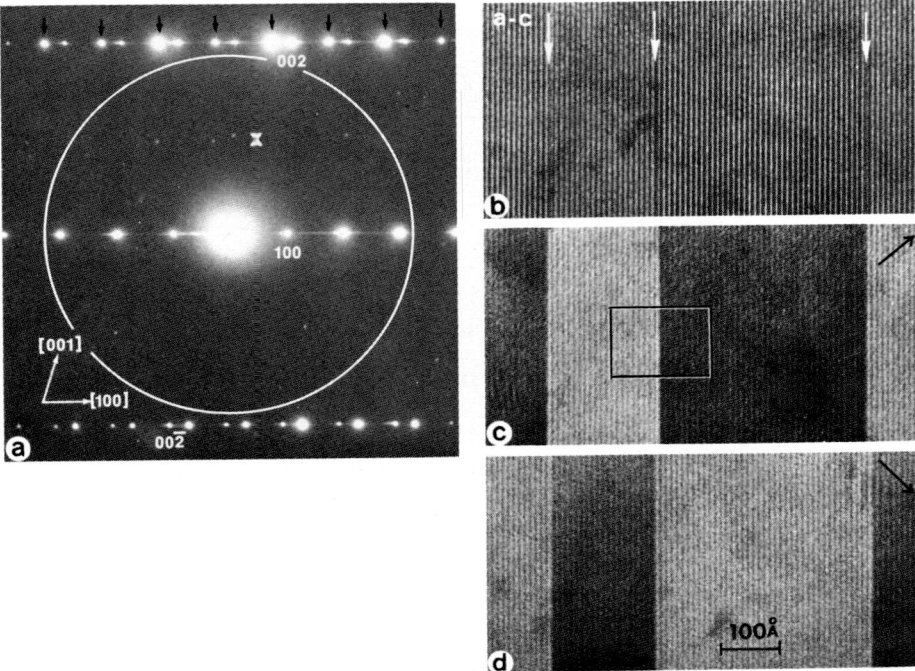

Fig. 1. (a) Electron diffraction pattern of twinned Bamble enstatite that was heated and quenched: the arrows point to twin reflections in the $h02$ layer line. Streaking, resulting from microtwinning, is prominent. (b) $a-c$ section showing 100 fringes, corresponding to the diffraction pattern in (a). The [010] zone axis is exactly coincident with the electron beam, resulting in uniform contrast across the image. The arrows point to twin planes. (c and d) The crystal has been tilted slightly around the directions indicated in the upper right corners. Adjacent twins show differing contrast, and the sense of contrast is reversed when the direction of tilt is changed

tions around which the sample was tilted are shown by the black arrows in the upper right corners of Fig. 1c and d. Note that there is a reversal in contrast between Fig. 1c and d.

Based on the electron diffraction pattern (Fig. 1a) it is reasonable to assume that the bands of differing contrast in Fig. 1c and d are in a twin relationship. Nonetheless, in light of (100) being the composition plane for CLEN, it is surprising that tilting produces a change in contrast when only the $h00$ spots are used to obtain an image. We believe that the explanation lies in the effects on $h00$ of multiple diffraction from the $h0l$ and $h0\bar{l}$ reflections from the twin and matrix portions of the crystal. Although the $h0l$ and $h0\bar{l}$ beams were excluded by the objective aperture from forming the image, they nevertheless may have contributed to the $h00$ intensities.

Confirmation that the areas of differing contrast are truly in a twin relationship is provided by observing the offset of interference fringes at the twin plane. Komoda (1968) used such offsets to detect twinning in microcrystalline particles

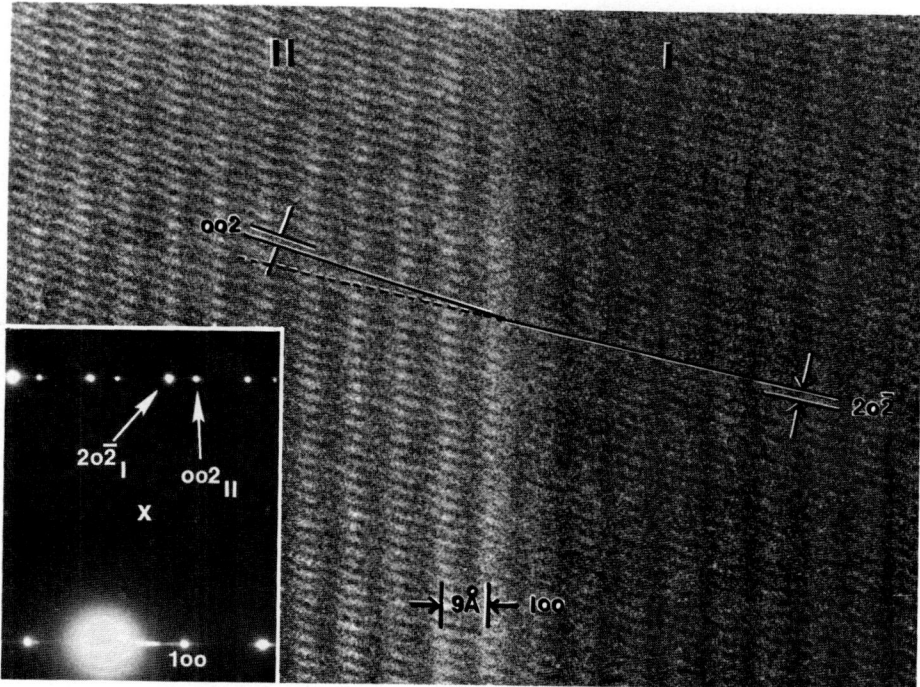

Fig. 2. High magnification (a–c) section/image of the area in the box in Fig. 1c. The image was obtained by beam tilting, and shows one twin boundary, located along the discontinuity between the light and dark regions. Note the respective terminations of the (002) and (202) fringes at the twin boundary; these reflections are shown in the electron diffraction pattern in the lower left insert. The 9Å spacing corresponds to the a unit cell repeat of CLEN

of metals. Fig. 2 is a high-resolution image of the region of the boundary shown in the enclosure in Fig. 1c. Fringes produced by interaction of the incident beam and 002 or, $h0\bar{2}$ of the twin, were observed by using the tilting method of Dowell (1963). In this technique the electron beam is tilted so that it lies midway between the optical axis of the microscope and the particular spots of interest in the first reciprocal lattice rows of the diffraction pattern (insert in Fig. 2). The discontinuity of both sets of fringes at the boundary between the light and dark portions of the image is prominent, and confirms this interface as the twin boundary.

Having demonstrated that the alternating light and dark zones in CLEN are twins, it is of interest to consider a region of highly disordered CLEN, *i.e.* a crystal containing many narrow bands and a highly streaked diffraction pattern. Experiments similar to those described for Fig. 1 were performed. Fig. 3a shows nearly uniform contrast while 3b and c, in which the crystal has been tilted slightly, show adjacent light and dark bands. The contrast is reversed between Fig. 3b and c. The effect is analogous to that in Fig. 1, where we have shown that the bands result from twinning. We thus conclude that the narrow zones in Fig. 3 also are produced by twinning, the dark areas belonging to one

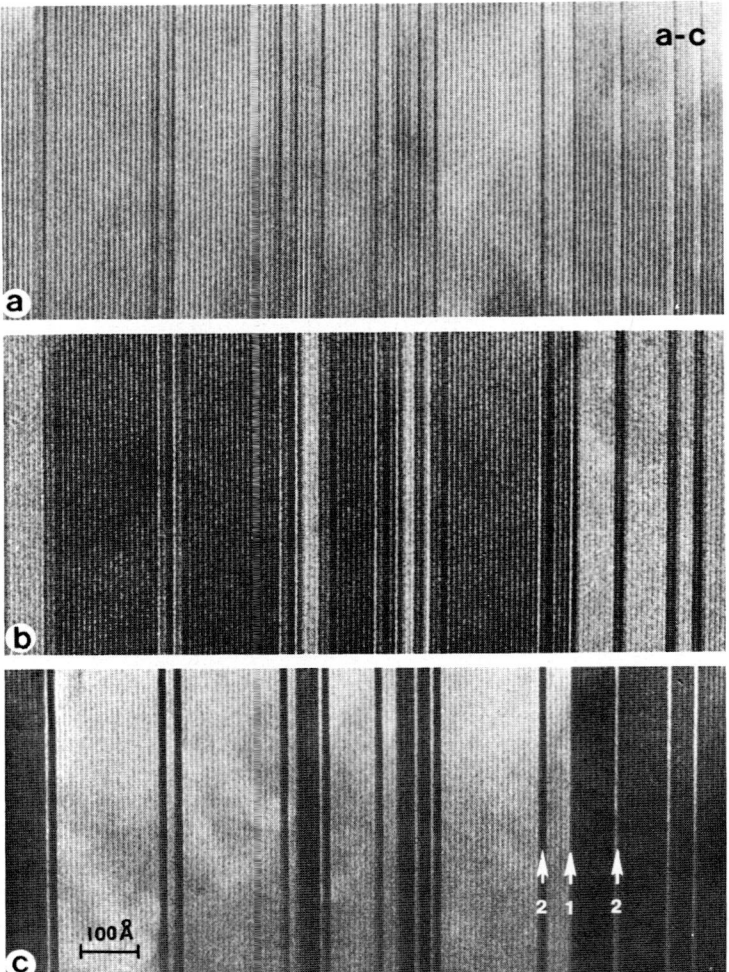

Fig. 3 (a–c). Section of enstatite showing numerous twins. The images were obtained in an analogous fashion to those in Fig. 1 b–d and show similar changes in contrast upon tilting. In this case individual and paired twin planes are shown, arrows #1 and #2 respectively. The latter corresponds to a "plate" of OREN that is one unit cell wide

set of twins and the light ones to the complimentary set. This is an example of polysynthetic twinning, with twinned portions of the crystal only one or a few unit cells wide. As shown by Iijima and Buseck (1975), when such twin planes are 9Å apart a "plate" of orthorhombic enstatite (OREN) results. A few such regions of OREN within CLEN are shown in Fig. 3.

The implications of such twinning are important for understanding polymorphism, polytypism and for the interpretation of geologic history from electron micrographs of pyroxene (Iijima and Buseck, 1975; Buseck and Iijima, 1975).

References

Andersson, S., Hyde, B.G.: Twinning on the unit cell level as a structure-building operation in the solid state. Solid. State Chem. **9**, 92–101 (1974).
Buseck, P.R., Iijima, S.: High resolution electron microscopy of silicates. Am. Mineralogist **59**, 1–21 (1974).
Buseck, P.R., Iijima, S.: High-resolution electron microscopy of enstatite II: Geological applications. Am. Mineralogist **60**, 771–780 (1975).
Dowell, W.C.T.: Das electronenmikroskopische Bild von Netzebenenscharen und sein Kontrast. Optik **20**, 533–566 (1963).
Iijima, S.: Direct observation of lattice defects in $H-Nb_2O_5$ by high resolution electron microscopy. Acta Cryst. A **29**, 18–24 (1973).
Iijima, S., Buseck, P.R.: High-resolution electron microscopy of enstatite I: Twinning, polymorphism and polytypism. Am. Mineralogist **60**, 758–770 (1975).
Komoda, T.: Study on the structure of evaporated gold particles by means of a high resolution electron microscope. Japan. J. Appl. Phys. **7**, 27–30 (1968).

CHAPTER 5.5

Polytypism in Wollastonite

H.-R. WENK, W.F. MÜLLER, N.A. LIDDELL, and P.P. PHAKEY

1. Introduction

The structure of wollastonite $CaSiO_3$ is characterized by chains of $[SiO_4]$-tetrahedra which extend parallel to the y-axis and have a three-repeat (b = 7.32 Å). The chains are held together by interstitial Ca which is in a distorted six fold co-ordination. Fig. 1a shows a z-projection of the structure of triclinic wollastonite as determined by Mamedov and Belov (1956) from two-dimensional film data. Ito (1950) recognized the pseudosymmetry which is evident both in lattice constants ($\cos\gamma \approx b/4a$) and by comparing intensities on X-ray diffraction patterns (see, e.g. Wenk, 1969). He pointed out a pseudomonoclinic cell (dashed area in Fig. 1a) but this setting does not do justice to the symmetry centers in the $P\bar{1}$ structure.

Pseudosymmetry is the underlying principle governing polytypism of wollastonite, various aspects of which have been discussed by Peacock (1935), Ito (1950), Jeffery (1953), Dornberger et al. (1955), Prewitt and Buerger (1963), Trojer (1968), Wenk (1969), and Jefferson and Bown (1973). Jeffery attributed streaking parallel to a^* of all $k = 2n+1$ reflections to a $b/2$-translation and Wenk suggested that wollastonite polytypism could best be explained by periodic (100)-stacking faults with a $b/2$-displacement vector. He observed that streaking was particularly common in naturally and experimentally deformed wollastonite and concluded that stacking faults may be produced by application of shearing stress. Using the principle of simple translation of unit cells of the triclinic wollastonite a large number of polytypes appears possible. If A denotes the basic unit cell, and B the translated cell, ordered stacking sequences such as $ABAB...$, $AABB$-$AABB...$ (Fig. 1b), $AABAAB...$, $AAABBB...$, are all recognizable by specific systematic extinctions in the diffraction patterns. A disordered stacking sequence $ABBBAABAAAA...$ gives rise to continuous diffuse streaks parallel to a^* for $k = 2n+1$ as has been observed in some strongly deformed specimens. In these crystals $k = 2n$ reflections are still sharp (cf. e.g. Fig. 2e in Wenk, 1969). In most wollastonites which have been analyzed so far an intermediate state with sharp reflections but superposed diffuse streaks is present, indicating only partial order of the stacking faults. The diffuse streaking complicates the quantitative interpretation of X-ray data and the electron microscope is the obvious instrument to study polytypism and to investigate the local variations in stacking order.

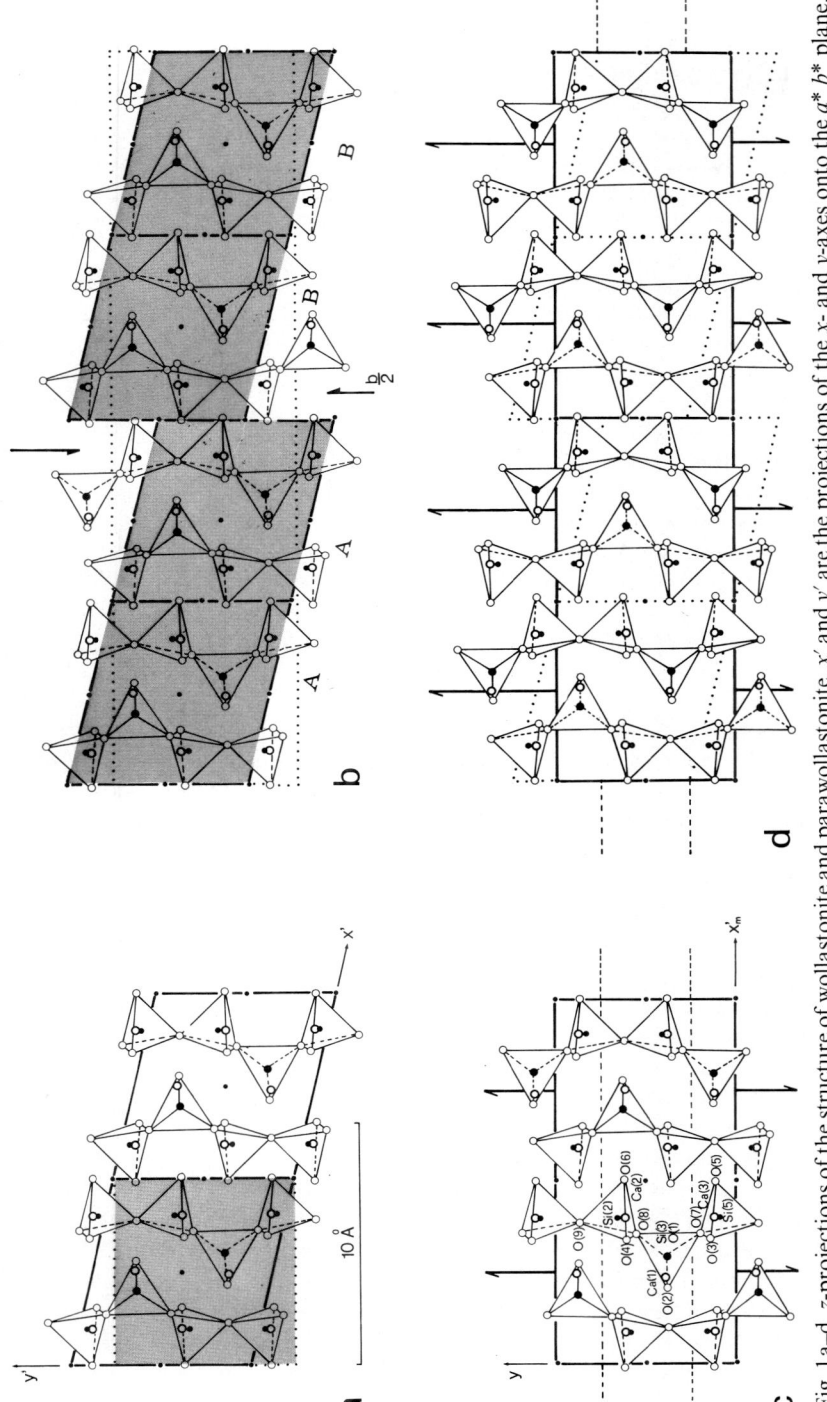

Fig. 1a–d. z-projections of the structure of wollastonite and parawollastonite. x' and y' are the projections of the x- and y-axes onto the $a^* b^*$ plane. (a) Two unit-cells of basic triclinic wollastonite. The pseudomonoclinic cell of Ito (1950) is shaded. (b) Proposed stacking in the translation model for 15.3 Å wollastonite: The unit-cell corresponding to Trojer's parawollastonite is indicated by dotted lines. The displacement on (100) is $\frac{1}{2}\boldsymbol{b}$. (c) Structure of monoclinic parawollastonite according to Trojer (1968), x'_m is the projection of the monoclinic x-axis. (d) Two unit-cells of parawollastonite (Trojer). The corresponding unit-cells of basic triclinic wollastonite in the translation model are outlined by dotted lines. Notice the subtle differences with the translation model (b)

2. Material and Techniques

Three specimens are described in this paper. One, Riv. 22, is from Claro, Ticino, in the center of the regionally metamorphic Lepontine complex in the Alps (corresponding to sample Os 99 in Wenk, 1969). The second one is WSC 56, from Death Valley, California, a massive wollastonite rock at the contact between mesozoic granites and carbonate rocks. WSC 56 is characterized by unusually large (up to 10 cm) single crystals. The third is specimen 8364 from Mungana, North Queensland.

TEM specimens of Riv. 22 and WSC 56 were prepared by crushing wollastonite grains about 0.5 mm in size between two glass slides, adding a drop of ethyl alcohol or acetone and dipping a copper grid coated with a formvar film into the suspension. A JEM 100 B at the University of Frankfurt equipped with a side entry goniometer was used for the study. The high angle tilt capability of the goniometer was used in obtaining the desired orientation of the cleavage fragments, namely the z-axis parallel to the electron beam. Foils of specimen 8364 were prepared by ion beam thinning from a standard petrographic thin section cut approximately normal to the z-axis. This specimen was examined in a JEM 200A at Monash University equipped with a 30° tilting rotating stage which enabled the contrast at a variety of diffracting conditions to be studied.

3. Results

Bright field and dark field images of foils oriented with the z-axis parallel to the beam reveal that samples Riv. 22 and WSC 56 both have a lamellar texture parallel to (100) on a submicroscopic scale. Selected area diffraction patterns taken on a small area (1 μm) resemble X-ray photographs. Characteristically they display for $k=2n+1$ diffraction maxima connected by streaks parallel to a^* (Fig. 2b, d). Diffraction maxima allow the determination of the predominant polytype while streaks indicate considerable stacking disorder which can be seen directly in one-dimensional lattice images. The fringes observed in Fig. 2a have often a spacing of 15 Å, corresponding to twice the $d_{100}=7.7$ Å spacing in the basic wollastonite structure, although the periodicity is frequently interrupted. A small region with a 23 Å-periodicity is visible at the right side of the photograph. Occasionally there appear to be fringes only 7.7 Å apart which are not resolved in our photograph and show up as thick bright lines. In the diffraction pattern (Fig. 2b) maxima corresponding to parawollastonite are strong. In a second sample (Fig. 2c, d) the electron diffraction pattern indicates that the basic 7.7 Å wollastonite dominates but also here the sequence is often interrupted. The dark field image in Fig. 2c taken with the reflection $\bar{2}10$ (including the streak) shows frequent fringes parallel to (100) which can be interpreted as faults interrupting the regular stacking in basic wollastonite. In some places these faults are again periodically spaced for a short distance. The a^*–b^* diffraction pattern of specimen 8364 (Fig. 3c) resembles Fig. 2b in that the diffraction maxima correspond to parawollastonite. However, in this case, there is little streaking parallel to a^*, indicating that this is a fairly well ordered specimen of parawollastonite. Bright

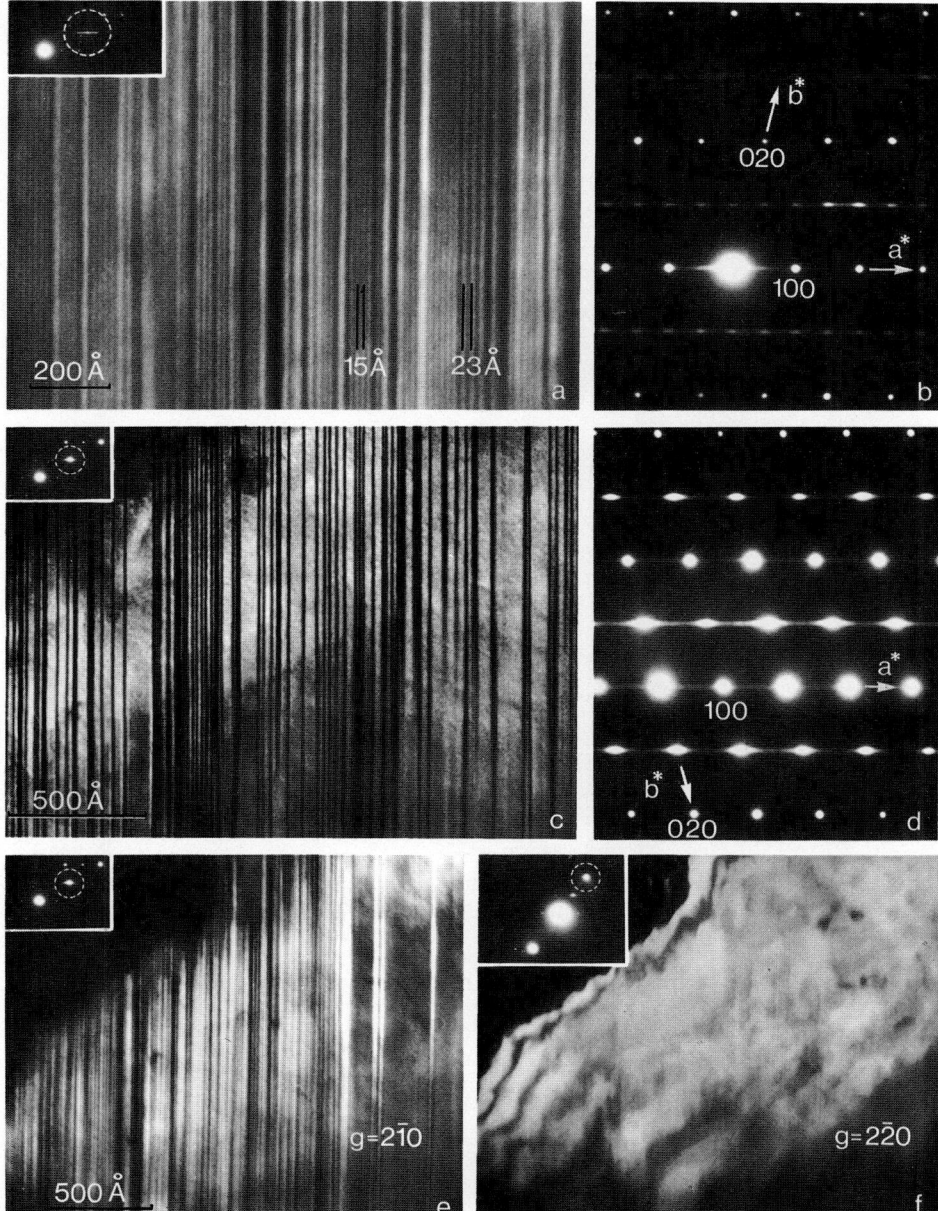

Fig. 2. (a) Lattice image of wollastonite WSC 56 in dark field mode using the reflections encircled (inset). The sequence of fringes parallel to (100) with 15.3 Å spacing is frequently interrupted by faults. (b) Diffraction pattern of wollastonite shown in (a) with diffraction maxima and extinctions characteristic for parawollastonite. (c) Dark field image of wollastonite Riv. 22 using the reflection encircled (diffraction maximum $g=\bar{2}10$). Numerous faults parallel to (100) are seen. (d) Diffraction pattern of wollastonite shown in c) with diffraction maxima and extinctions characteristic for basic wollastonite. (e, f) Pair of dark field images of the same grain as in (c). The faults parallel to (100) are in contrast with $g=2\bar{1}0$ (e) and out of contrast for $g=2\bar{2}0$ (f)

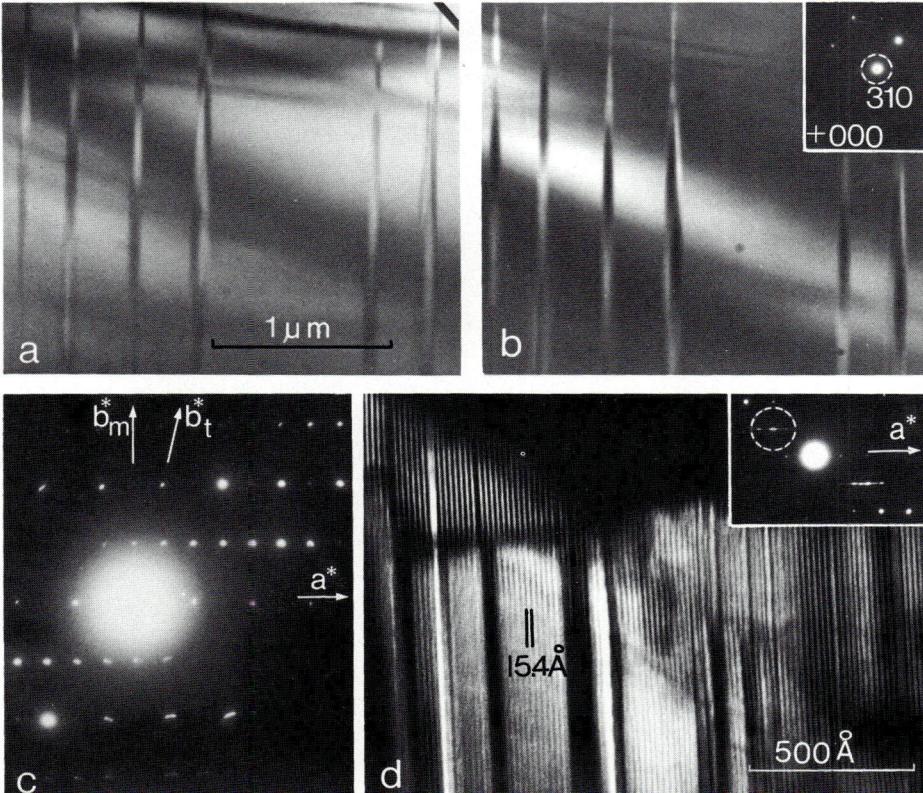

Fig. 3a–d. Parawollastonite 8364 from Mungana (a–c taken with JEM 200A, d with JEM 100B). (a) Bright field image showing typical density of (100) faults. (b) Dark field image of area shown in (a) ($g=310$). In the region of the bend contour, where $s=0$, the fringes have the characteristics of π-fringes. Indices of the inserted diffraction pattern are for parawollastonite unit cell. (c) Diffraction pattern with maxima and extinctions characteristic for parawollastonite. The asterism of spots is due to extreme buckling of the foil. (d) High-resolution dark field micrograph showing the 15 Å-fringes and faults interrupting the periodic stacking

field and dark field images not resolving the 15Å faults (Fig. 3a, b) again show many fringes interrupting the regular stacking in parawollastonite. The density of faults is much lower than in the previous samples (as expected from the diffraction patterns) and only the 15Å periodicity has been observed in high resolution micrographs (Fig. 3d). Comparing observations on the ion-thinned sample (Fig. 3a, b) with those on the crushed grain mount (Fig. 3d) we notice a difference in fault density which can be attributed to heterogeneities in the crystal. It is possible that some planar defects were produced during crushing.

The displacement vector R of the faults has been determined by two beam contrast experiments on specimen 8364. The stacking fault fringes were visible for a variety of k odd reflections ($h10$, $h11$, $h31$ and $h32$ type reflections were used) but were out of contrast for k even and $k=0$ reflections (h00, h20, h40, h21 and h22 type reflections were used). The diffraction contrast observations show that the fault vector is $\mathbf{R}=\frac{1}{2}\mathbf{b}$(all other fault vectors which satisfy the observations are structurally equivalent to this vector). Contrast experiments on the faults in the basic wollastonite sample, Riv. 22, yield the same fault vector (Fig. 2e, f), which corresponds to the displacement vector suggested by Jeffery (1953). This fault vector means that when the faults are in contrast, π-fringes should be observed (since $\alpha=2\pi\mathbf{g}\cdot\mathbf{R}=\pi$ when k is odd). At the exact Bragg condition ($s=0$), π-fringes are characterized by complementary BF and DF images, which are symmetrical with respect to the foil center, the center fringe being bright in bright field and dark in dark field. Away from $s=0$ the behavior is complex (Van Landuyt et al., 1964). Fig. 3a, b illustrates this behavior for $\mathbf{g}=310$. The other micrographs are all consistent with π-fringes. For two sufficiently close, overlapping π-faults fringe contrast will be absent.

4. Discussion

The TEM observations reported here support Wenk's (1969) suggestion that polytypism of wollastonite can be explained by periodic (100) faults with a displacement vector $\frac{1}{2}\mathbf{b}$ which separate blocks consisting of one or several units of basic triclinic wollastonite ($d_{100}=7.7$ Å). An infinity of stacking sequences can be visualized but our observations indicate that besides fault-free sequences only the 15 Å spacing is often observed and extends over a large range. It is interesting to note that in two of the samples illustrated here the same polytype as was found with electron diffraction on a very small volume, has been determined on X-ray precession photographs (Fig. 2b corresponds to Fig. 2g in Wenk, 1969, and Fig. 2d to his Fig. 2b). This indicates that despite local heterogeneities the polytype seems to be characteristic of an occurrence, i.e. of the geological conditions during formation and is not accidental.

A structural model for the 15 Å-wollastonite is shown in Fig. 1b. Every two basic unit cells a fault displaces the next two units, thereby creating a new triclinic unit cell with a four times larger a lattice constant. The $\frac{1}{2}\mathbf{b}$ translation effectively averages top and bottom halves of the basic unit cell integrated over four cells along X and this gives rise to systematic extinctions $H \neq 4n$ for $k=2n$ (lower case indices characterize the basic $a=7.7$ Å structure, upper case H is a running index for the $a=32$ Å superstructure with H=0, 1, 2, 3). The new arrangement is a C-facecentered superlattice (Fig. 4a) and the faults separating the subunits correspond to the 15 Å-fringes observed in Fig. 2a and can be interpreted as periodic stacking faults. Due to this geometry there is a further constraint limiting the presence of reflections to $H+k=2n$. The reciprocal lattice which corresponds to Fig. 4a including only allowed reflections is shown in Fig. 4b. It compares well with the observed patterns in Fig. 2b, e. Thris triclinic but strongly pseudomonoclinic structure with space group $C\bar{1}$ with $a=32$ Å (Fig. 1b) is very similar to the monoclinic $P2_1/a$ $a=15$ Å model for parawollastonite (Fig. 1d) which was suggested by Trojer based on a refinement from three-

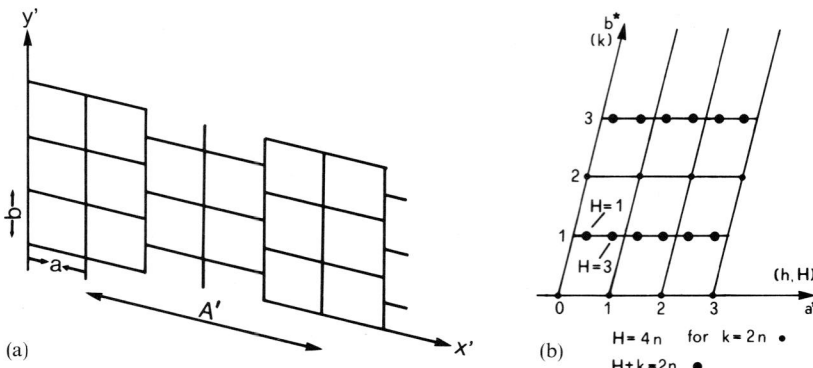

Fig. 4. (a) Stacking scheme in the **b**/2 translation model for parawollastonite, displaying the face-centered pattern, z-projection onto the $a^* b^*$ plane (x' and y' are projections of the x and y axes, a, A, b are projections of the corresponding lattice constants in basic cell and the superstructure cell). (b) Corresponding reciprocal space indicating systematic extinctions due to lattice translation and due to face-centering. (h, k are indices of the basic structure, H a running index for the superstructure)

dimensional X-ray data. Although differences in atomic coordinates are very minor (Table 1), topologically the two models are quite different. Particularly, systematic extinctions in Trojer's model are restricted in $hk0$ planes to $h=2n+1$ for $h00$ reflections and $k=2n+1$ for $0k0$ reflections (based on the monoclinic $a=15.4$ Å parawollastonite unit cell). Trojer emphasizes that for $2h+k=4n+2$ which ought to be absent according to our translation model he observes very weak reflections. This is not the case for several 15 Å wollastonites which we have analyzed so far. We did not find any violations of extinctions even in strongly exposed X-ray diffraction patterns. Some streaking between $k=2n$ reflections and weak maxima in the SAD shown in Figs. 2d and 3c are attributed to double diffraction. Therefore, even if Trojer's model exists, it may be a rather

Table 1. Comparison of Si-coordinates in the $P2_1/a$ (Trojer, 1968) and the $C\bar{1}$ model for parawollastonite. Monoclinic parawollastonite unit cell (compare Fig. 1a)

	Origin (Trojer, 1968)	Symmetry-related atoms Si'		
		$P2_1/a$	$C\bar{1}$	Δ
Si(1) x	0.4076	0.0924	0.0925	0.0001
y	0.0907	0.5907	0.5902	0.0005
z	0.2313	0.7687	0.7687	0.0000
Si(2) x	0.4075	0.0925	0.0924	0.0001
y	0.6598	0.1598	0.1593	0.0005
z	0.2313	0.7687	0.7687	0.0000
Si(3) x	0.3016	0.1984	0.1984	0.0000
y	0.3761	0.8761	0.8739	0.0022
z	0.4432	0.5568	0.5568	0.0000

rare type of parawollastonite. Judging from TEM observations on the heterogeneity of the 15 Å structure we expect that the refinement of parawollastonite with least squares methods may be rather difficult and will only yield an average structure. Furthermore in a superstructure with such a pronounced pseudo-symmetry a low R-factor does not necessarily indicate the correct structure.

Wollastonite is thus another mineral for which TEM observations were instrumental in interpreting X-ray data. We conclude that polytypism of wollastonite can be explained by introducing periodic $\frac{1}{2}$ b (100)-faults in the stacking of basic wollastonite unit-cells. In the case of parawollastonite a C-face centered pattern is created by periodic faults with a 15 Å-spacing. A related type of polytypism exists in enstatite (compare the articles by Iijima and Buseck, Chapter 5.4, and Kirby, Chapter 6., in this volume).

A pure translation can be achieved by mechanical shear with a glide movement on (100) along [010] without breaking any Si-O bonds. A high density of faults has been observed in deformed crystals and disordered sequences of stacking faults have actually been produced in deformation experiments from parawollastonite (Wenk, 1969). Phakey and Liddell (1975) have observed partial dislocations with $b=\frac{1}{2}$ [010] which occasionally terminate a fault within the crystal and support the model that stacking faults in wollastonite can be produced by a movement of dislocations.

Jefferson and Bown (1973) pointed out that translation may be combined with twinning. Due to the pseudosymmetry, twinned parawollastonite would produce a diffraction pattern resembling the "4T" pattern with a stacking sequence $AAABAAAB$... This polymorph which was postulated by Wenk (1969) has not been observed in TEM micrographs. Instead lamellar twins on (100) were observed on a very fine scale. Twinning, combined with a $\frac{1}{4}$ b translation, would also produce a structurally stable arrangement and its role in polytypism has to be further investigated.

References

Dornberger, K., Liebau, E., Thilo, E.: Zur Struktur des β-Wollastonites, des Maddrellschen Salzes und des Natriumpolyarsenats. Acta Cryst. **8**, 752–754 (1955).
Ito, T.: X-ray studies on polymorphism. Tokyo: Maruzen 1950.
Jefferson, D.A., Bown, M.G.: Polytypism and stacking disorder in wollastonite. Nature Phys. Sci. **245**, 43–44 (1973).
Jeffery, J.W.: Unusual X-ray diffraction effects from a crystal of wollastonite. Acta Cryst. **6**, 821–825 (1953).
Mamedov, K.S., Belov, N.V.: Crystal structure of wollastonite. Dokl. Akad. Nauk. SSSR **107**, 463–466 (1956).
Peacock, M.A.: On wollastonite and parawollastonite. Am. J. Sci. **30**, 495–529 (1935).
Phakey, P.P., Liddell, N.A.: Partial dislocations in wollastonite. In preparation (1975).
Prewitt, C.T., Buerger, M.S.: Comparison of the crystal structure of wollastonite and pectolite. Mineral. Soc. Am. Spec. Paper **1**, 293–302 (1963).
Trojer, F.J.: The crystal structure of parawollastonite. Z. Krist. **127**, 291–308 (1968).
Van Landuyt, J., Gevers, R., Amelinckx, S.: Fringe Patterns at anti-phase boundaries with $\alpha=\pi$ observed in the electron microscope. Phys. Stat. Sol. **7**, 519–546 (1964).
Wenk, H.-R.: Polymorphism of wollastonite. Contrib. Mineral. Petrol. **22**, 238–247 (1969).

CHAPTER 5.6

High-resolution Electron Microscopy of Labradorite Feldspar

H. Hashimoto, H.-U. Nissen, A. Ono, A. Kumao, H. Endoh, and C.F. Woensdregt

1. Introduction

The "intermediate plagioclase structure" (Cole et al., 1951), the object of this study, occurs in single-phase low plagioclases between approximately 25 and 68 mole percent anorthite and, in exsolved specimens, extends from at least 8 to 84 mole percent anorthite. It is defined (Bown and Gay, 1958) as lacking the (b)-, (c)-, and (d)-type reflections present in diffraction patterns of anorthites. Besides the (a)-type reflections observed in all feldspars, it contains rows of (e)-satellites in symmetric positions to the (b)-type reflection positions for plagioclases having the $1\bar{1}$ structure as well as rows of (f)-satellites (parallel to those of the (e)-type satellites) in symmetric positions to the (a)-reflections.

Specimens with compositions between 42 and 58 mole percent anorthite (and mostly containing 2–4 mole percent potassium feldspar) are complicated by inhomogeneities on three different scales. First they contain a lamellar structure on the 1000 Å scale which causes iridescence (or schiller) and satellites in close proximity to (a)-reflections (Korekawa and Jagodzinski, 1967). These lamellae are associated with exsolution of two chemically different plagioclases (cf. Nissen, 1974 and Cliff et al., Chapter 4.10 of this volume, for discussion and references). Within the lamellar structure is a domain texture on the scale of several hundred Å which is particularly clear in dark field micrographs and was described by McLaren and Marshall (1974, Fig. 7) for a plagioclase with 32 mole % anorthite. Crossing these features in dark field micrographs are superlattice fringes with distances of 30–50 Å associated with the (e)-type and (f)-type satellite spots due to the superlattice of the intermediate plagioclase structure.

For clarity, the term lamellae is used throughout this paper for the texture in the 1000 Å scale. (Minor and major lamellae are terms used to describe the two components of the lamellar texture contributing minor and major proportions of the total volume.) The term domain is used for the intermediate-scale feature and the term satellite fringes for the fringe patterns associated with (e)- or (f)-satellites. This paper discusses two aspects of new electron microscopic observations on a specimen of labradorite with schiller:

The first deals with the relation between lamellae and fringe patterns and (e)- or (f)-satellites caused by the superlattice of the "intermediate structure". In dark field micrographs of different labradorites using a pair of (e)-satellites Nissen (1972) and McLaren (1974), as well as McLaren and Marshall (1974) found that the corresponding fringe pattern could only be resolved in the major lamellae and concluded that the structure factors for the (e)-satellites were different for major and minor lamellae and might even approach zero for the minor lamellae. In the latter case the minor lamellae would have the high plagioclase or albite structure lacking (e)- and (f)-satellites (C $\bar{1}$) and characterized by relatively little Al/Si order. McConnell (1972, 1974) observed small areas showing (e)-fringes also in the minor lamellae of another labradorite and concluded that spacing and orientation of the (e)-frings indicated differences in anorthite content between the major and minor lamellae. Chemical differences had been suggested on grounds of changing proportions between major and minor lamellae with bulk composition by Nissen (1971) and are also indicated by recent X-ray microanalysis (Cliff et al., Chapter 4.10 of this volume). New dark field electron micrographs are described in order to clarify the question whether a difference in the superlattice associated with the (e)- and (f)-satellites in minor and major lamellae of labradorite exists in the case of the specimen selected and in the intermediate plagioclases containing lamellae in general. The complete description of geometrical relations between the lamellae and the (e)-or (f)-fringe patterns necessitates the inclusion of the geometrical relations between these two features and the domains.

The second aspect involves the nature of the superlattice of the intermediate structure which has been a matter of controversy for nearly twenty-five years. The method of high resolution electron microscopy (n-beam structure images; cf. Buseck and Iijima, 1974; Cowley and Iijima, Chapter 2.5 of this volume) is applied to this problem. In order to demonstrate the possibility of resolving details about cation occupancies in a single superlattice unit of labradorite, examples of n-beam structure images of labradorite in several orientations are given showing details of the atomic positions in the feldspar structure in general. Subsequently this method is applied to a case of an n-beam structure image involving (e)-satellite intensities, and an interpretation in terms of a model for the superlattice in labradorite is proposed. The structure images may thus contribute to a crystallographic problem which has not yet been solved by X-ray methods alone.

2. Experimental

The specimen, No. 1513c, was a labradorite of low structural state with greyish-blue to red-violet schiller color in very coarse-grained labradorite anorthosite from Tabor Island near Nain, Labrador. A partial wet-chemical analysis gave Ab 43.5, An 53.6, Or 2.9 mole percent.[1] The same specimen has been used by Cliff et al. (Chapter 4.10 of this volume). For electron microscopy of the

---

[1] Mr. J. Sommerauer kindly contributed a new electron microprobe analysis of 15 spots of part of the material used for this paper. This showed the material to be very homogeneous and gave (in weight percent): $Na_2O=4.97$, $CaO=11.2$, $K_2O=0.39$, $SiO_2=54.4$, $Al_2O_3=28.9$ (values corrected after Bence-Albee, 15 kV, 50 namp, beam diameter 0.5 μm. This corresponds to Ab 45.6, An 53.6, Or 0.8 (mole percent). The difference in k may be due to antiperthite.

textures, oriented petrographic thin sections were further thinned by argon ion bombardment, while for very high resolution work small amounts of the material were ground in an agate mortar, and the powder was transferred to a holey plastic film. Thin cleavage flakes protruding over the margin of a hole and the ion-thinned specimens were studied in a JEM 100 C electron microscope operating at 100 kV. In order to achieve high resolution, a specially designed objective lens pole piece was used for taking the electron microscopic images (Yanaka et al., 1974). This lens has a spherical aberration coefficient $C_s = 0.7$ mm, a chromatic aberration coefficient $C_c = 1.05$ mm and focal length $f_0 = 1.2$ mm.

3. Results on the Relation between Lamellae, Domains and Superlattice

In order to obtain micrographs showing all three features described in the introduction, the ion-thinned sample was prepared in such a way that the normal to the boundaries of the lamellae as well as the direction connecting the pairs of (e)- and (f)-type satellites were parallel to the plane of the section. The axis [411] is therefore close to the direction of the electron beam. Fig. 1a shows a dark field micrograph made by selecting an (a)-type reflection with the adjacent pair of (f)-type satellites. These three reflections are marked by the outlines of the selective aperture in the inserted corresponding diffractogram, Fig. 1b. Fig. 1c is a diffraction pattern made from the same preparation and in the same orientation. Fig 1a shows two minor lamellae with the adjacent major lamellae running vertically over the micrograph. No definite boundary between major and minor lamellae can be recognized, and in this and similar dark field micrographs the more prominent contrast is that of the domains referred to in the introduction. They have elongate outlines and diffuse boundaries. The elongation direction is nearly parallel to the diagonal of the micrograph. The average size of the domains is smaller within the minor than within the major lamellae. This can be seen more clearly in Fig. 2 which shows part of Fig. 1 at higher magnification.

The contrast pattern in Figs. 1 and 2 is complicated and lacks sharp boundaries, but some regularities are nevertheless evident in this and many similar micrographs. A Bragg contour with diffuse outlines is seen in Fig. 1 running diagonally across the micrograph. Outside this contour the minor lamellae marked X are surrounded by a zone of relatively dark contrast, marked Y and, due to the numerous small domains described above which appear bright, are generally slightly brighter in contrast than the major lamellae which are marked Z. Inside the area of the Bragg contour the contrast is reversed: the minor lamellae marked X' are full of relatively small elongate domains appearing dark and are surrounded by a relatively bright zone marked Y'. The major lamellae are here generally brighter and are marked Z'. The contrast change in the major lamellae adjacent to the minor lamellae may be attributed to strain between the material of the major and that of the minor lamellae. Lattice resolution images resolving various spacings all show complete continuity of lattice fringes over the lamellar boundary indicating that the boundaries between major and minor lamellae are coherent.

Fig. 2 shows two minor lamellae which are approximately in reversed contrast. If we consider only the right half of the micrograph, we recognize three different degrees of brightness, examples of which are marked as A, B and C: First, very dark areas A showing well-developed (f)-fringes. These areas occur near the minor lamellae, in the zone marked Y in Fig. 1. Secondly,

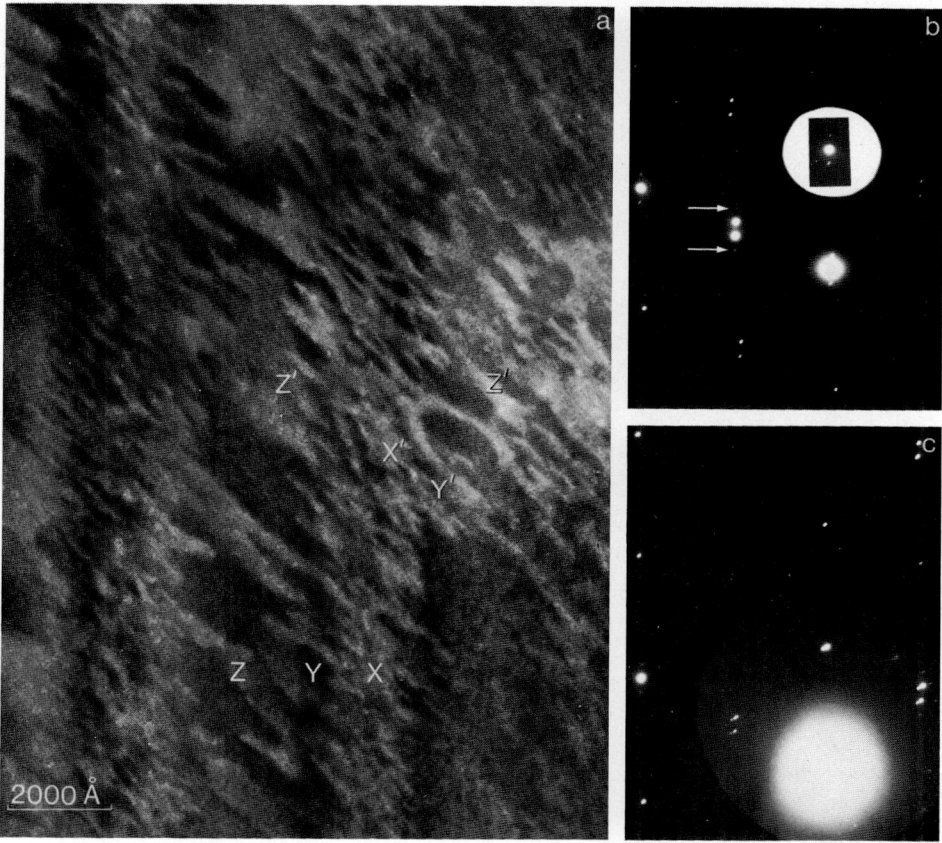

Fig. 1. (a) Dark field micrograph of specimen 1513c taken with an (a)-type reflection and the adjacent pair of (f)-reflections. Vertically two minor lamellae can be recognized, separating major lamellae. The (f)-fringe contrast is vertical, the domain contrast is oriented diagonally. The orientation of the micrograph is approximately normal to the major and minor lamellae and normal to the lamellar units causing the (f)-fringes. Different areas mentioned in the text are marked as X, Y, Z and X', Y', Z'. (b) Electron diffraction pattern corresponding to Fig. 1a and in correct orientation to the micrograph. The white circle denotes objective aperture including the triplet of reflections selected for the dark field micrographs (Figs. 1 and 2). Inserted is that triplet taken from another print of the same negative (this was necessary because of overexposure through the aperture). Note third order (e)-type satellites indicated by arrows. (c) Same orientation as 1b, showing spots of types (a, e and f)

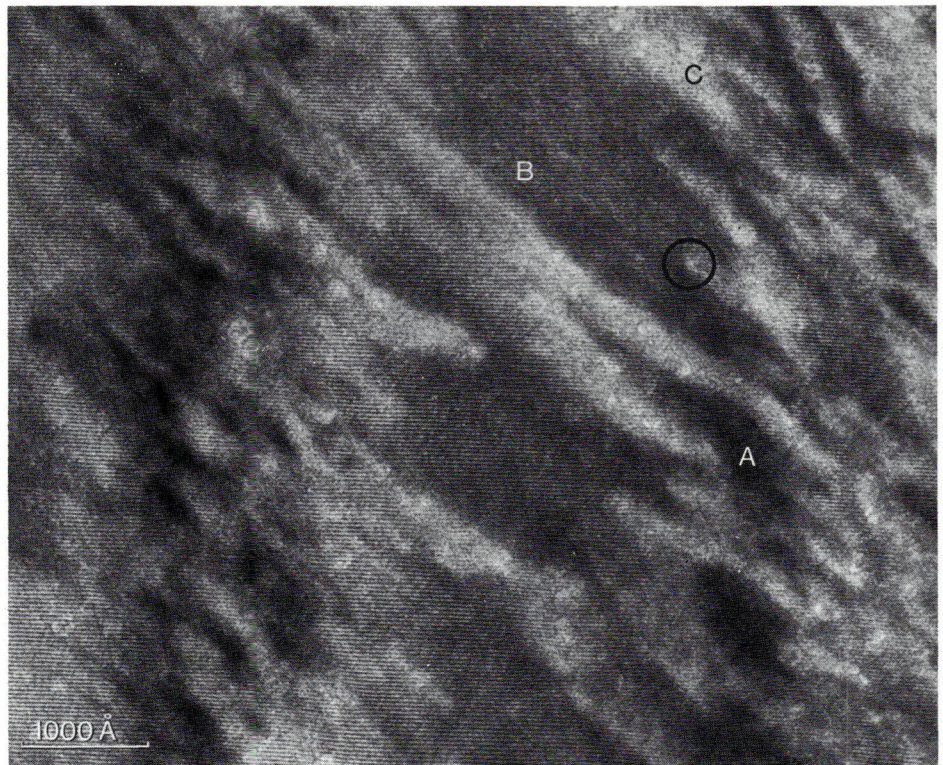

Fig. 2. Portion of Fig. 1 in higher magnification. Black ring denotes double "dislocation" in (f)-fringe pattern. Three different shades of contrast marked by A, B and C (for explanation, see text)

areas of intermediate contrast B containing well developed (f)-fringe contrast. These areas occur in the major lamellae, where they form a smaller portion and occur as domains of smaller size. There is a gradual transition of areas of type A into those of type B. Thirdly, marked C, there are elongated domains appearing brightest, in which the (f)-fringe contrast is only weakly developed or missing.

In the left half of Fig. 2 areas corresponding to those marked A, B and C can also be recognized but with reversed contrast. In the domains marked C it is not quite clear whether the (f)-fringe pattern is completely absent or only weakly developed because the domain boundaries appear to be frequently inclined to the plane of the micrograph which will produce an overlap with domain material containing the superlattice.

The superlattice causing the (e)- and (f)-satellites is only observed in the low plagioclases. Since these low plagioclases are generally believed to have a relatively high degree of Al/Si ordering and since also the superlattice itself can be associated with ordering of Al/Si and Ca/Na (e.g. Korekawa, 1967) it can be concluded that the domains which do not or only very weakly show the (f)-fringes (C in Fig. 2) are areas having a low degree of Al/Si ordering. This hypothesis is corroborated by the fact that the fringe contrast in those domains does not reappear in micrographs of the same area made after slight tilt in arbitrary directions. Reappearance would be expected if the domains represented slight differences in the orientation of one identical lattice. McLaren and Marshall (1974) have given the same explanation, "a random distribution of domains with the C$\bar{1}$ structure" for the domains lacking the (e)-fringe contrast in a specimen with 32 mole percent anorthite. These domains are less elongated than those in Figs. 1 and 2 and occur in a specimen lacking the lamellae causing schiller. The domains described above for specimen 1513c have been observed in the whole compositional range of the intermediate plagioclases, i.e. between approximately 30 and 67 mole percent anorthite.

Darkfield images of groups of four a-reflections each resolve several different lattice fringes and show that the boundaries of the lamellae as well as the boundaries of the domains are completely coherent for these indices.

The most important observation to be made in Figs. 1 and 2 regards the distribution of the (f)-type fringes. In all earlier micrographs showing the relation between (e)- or (f)-fringes and "schiller lamellae" (Nissen, 1972; McLaren, 1974; McLaren and Marshall, 1974), the fringe contrast was clearly visible only in the major lamellae while in the minor lamellae it was either completely invisible or only resolved in a few small regions (McConnell, 1972, 1974). Fig. 1 and 2 on the other hand show that both spacing and orientation of the (f)-fringes are identical within error limits in the major and minor lamellae of specimen 1513c. (The angle between the (f)-fringes and the lamellae in this section is approximately 86°.) Since two other labradorite specimens with more sodic bulk composition also show identical orientations and spacings of (e)- or (f)-fringes, it may be assumed that this is not an exception but the general case. The microanalytical results (Cliff *et al.*, Chapter 4.10 of this volume) clearly indicate a compositional difference between major and minor lamellae; but, as Nissen (1974) has pointed out, the fact that spacing and orientation of (e)- or (f)-fringes are the same in minor and major lamellae implies that these fringes cannot be used as an indication of composition of the minor and major lamellae in labradorite as has been suggested (McConnell, 1974). The mean spacings and orientations of the (e)- and (f)-type satellites and hence of the corresponding fringes are, however, a function of the bulk anorthite composition (Bown and Gay, 1958). The (e)- and (f)-fringes in labradorite 1513c are usually very straight. Only near double dislocations in the fringe pattern, in which two dark fringes always end, do slight deviations in the orientation of the fringes occur. Frequently these dislocations lie within the "poorly-ordered" domains. One of the dislocations is marked by a circle in Fig. 2.

The concept of the major and minor lamellae representing two separate, compositionally different phases appears to contradict the fact that the "well-ordered" domains and the (f)-fringes within them, in many places (e.g. in Fig. 2) cross the boundaries between major and minor lamellae without interruption and that the (f)-fringes have the same orientation and spacing in minor and major lamellae. It may therefore be assumed that two compositions, one lower and one higher than An 50 in anorthite content by the same amount, can have a superstructure (associated with satellites) with the same spacing and orientation (see Wenk et al., 1975). The second structure can be obtained from the first one by exchanging the Na with the Ca positions and exchanging the corresponding Al/Si configurations as in antiphase domains. Spacing and orientation of the superstructure is thus determined not by chemical causes (anorthite content) but by physical causes (stresses between the two structures, cooling history, etc.). The superstructure may have originated *after* the compositional lamellae.

4. Structure Images of Labradorite

Cowley and Moodie (1960) and Allpress et al. (1972) have shown that the slightly defocused image of a very thin crystal gives the projected charge-density distribution of the crystal (see also Cowley and Iijima, Chapter 2.5 of this volume). The powdered labradorite specimen gave thin fragments with wedge-shaped margins of an estimated thickness around 400 Å. High resolution images were taken in orientations with the electron beam parallel to the axes [010], [110], [1$\bar{1}$0] and [5$\bar{1}$1], with 900 Å underfocus at an electron-optical magnification of 250000 times. The orientation of the specimen was determined from the electron diffraction pattern. Optical diffraction patterns were made of areas in the micrographs approximately 100 Å in diameter in order to check which beams actually contributed to the image in these particular areas. The electron diffractograms corresponding to the high-resolution micrographs showed that approximately 80 excited beams were included in the objective aperture.

Fig. 3 shows a micrograph of a fragment oriented with the axis [010] parallel to the beam, together with an inserted projection of the structure of anorthite, $CaAl_2Si_2O_8$, along the axis [010] onto the $(a'b')$-plane. One unit cell is outlined. This projection as well as those inserted in Figs. 4 and 5 were made using the data in Kempster et al. (1962), since atomic positions for "mean" labradorite were not yet available (compare however Toman and Frueh, 1972).

In the inserted projection (Fig. 3b) the oxygen positions have been neglected; this rough approximation could be made because of the relatively small scattering factors of oxygen. The projection of the structure along the axis [010] onto a plane normal to this axis can be regarded as an alternation of rows of "channels" arranged parallel to [010]. There are two types of "channels": one type is composed of two non-tetrahedral cation positions and twelve (Al, Si) positions (marked A in Fig. 3b) the other of four nontetrahedral cation positions and eight (Al, Si) positions (marked B in Fig. 3b). Rows of type A "channels" alternate with rows of type B "channels" in the c-direction. These "channels" can be compared in Fig. 3a with rounded regions of bright contrast arranged in rows in the case of the type B "channels" and as more or less continuous band of bright contrast for the rows of type A "channels".

Fig. 3c shows an area of the micrograph used for Fig. 3a at smaller magnification. Irregularly bounded domains can be recognized within which the regions of bright contrast visible in Fig. 3a show slightly varying outlines and positions along the a'-axis when the micrograph is viewed at small angles along lattice

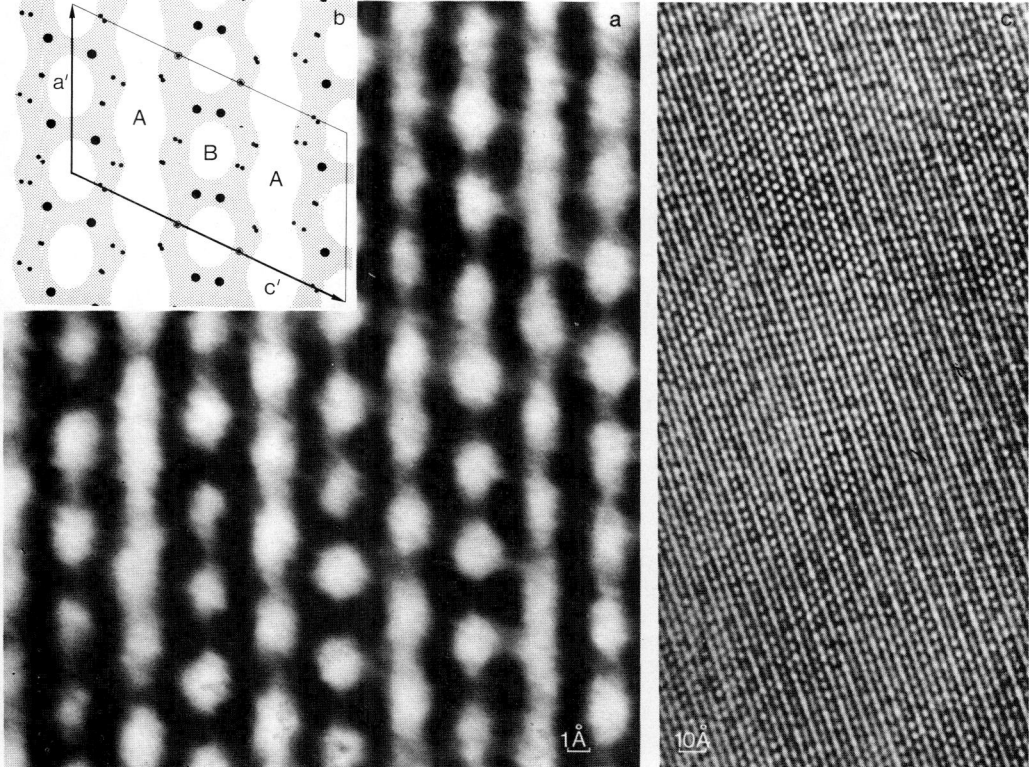

Fig. 3. (a) High resolution electron micrograph of a particle of labradorite (54 mole percent anorthite) oriented with its axis [010] parallel to the electron beam. (b) (Insert in 3a). Portion of the projection of the anorthite structure along the axis [010] onto the (a^*c^*)-plane, in the same scale as Fig. 3a. One unit cell is outlined; a' and c' are the projections of the axes a and c onto the (a^*c^*)-plane. Areas of high electron density in the projection are shaded. Oxygen positions and bands are omitted. Large dots indicate nontetrahedral cation positions, small dots (Al, Si) positions. Small dot with circle indicates two (Al, Si) positions superposed. (Note that a' and c' indicate different directions in Figs. 4 and 5.) (c) Same as Fig. 3a, at lower magnification. The micrograph is rotated 18° anticlockwise against Fig. 3a

rows. This indicates considerable irregular shifts of the atoms contributing to the contrast in areas which have approximately the size of the "poorly-ordered domains" discussed in the previous section.

Fig. 4 shows a small section of a micrograph of a fragment oriented with the axis [1$\bar{1}$0] parallel to the beam, together with a projection of the anorthite structure along the axis [1$\bar{1}$0] onto the plane normal to that axis. In the micrograph dark spots which may be interpreted as the sites of the non-tetrahedral atoms (8 Ca, Na per cell) merge, with dark spots possibly indicating the positions of the tetrahedral atoms (Al, Si, which have comparable electron scattering amplitudes), into ring-shaped areas surrounding "channels" in the structure.

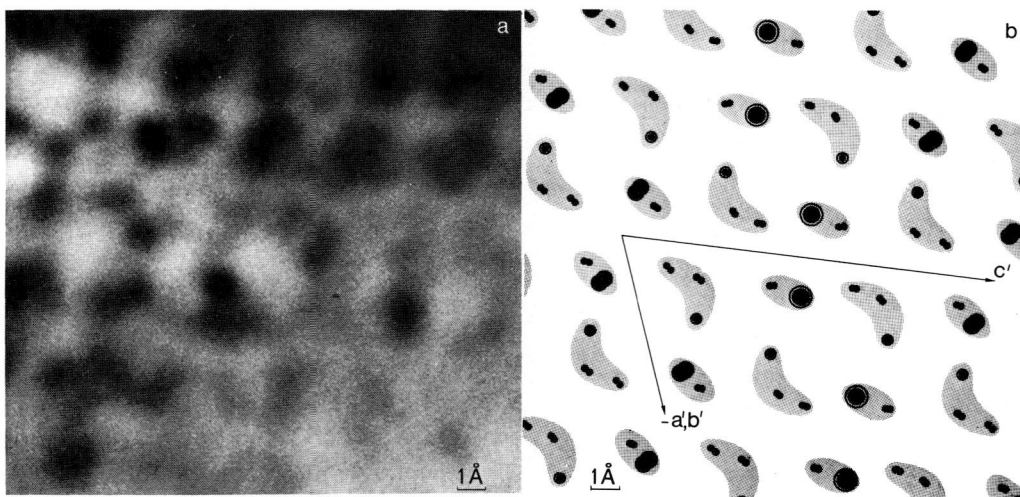

Fig. 4. (a) High resolution electron micrograph of a particle of labradorite (54 mol% anorthite) with its axis [1$\bar{1}$0] parallel to the electron beam. (b) Portion of the projection of the anorthite structure along the axis [1$\bar{1}$0] onto the plane normal to this axis. The projection is arranged in continuous orientation and translational position with regard to the atomic detail indicated by the contrast in Fig. 4a. The key in the projection is the same as for Fig. 3b. The projections of the axes $-a$, b and c onto the plane normal to the axis [1$\bar{1}$0] are marked as $-a'$, b' and c'. (Note that a' and c' are not identical to a' and c' in Fig. 3b)

It is thus possible to obtain information on structural details within the unit cell of feldspar, and it is feasible therefore to resolve details in the superlattice of the "intermediate plagioclase structure" provided that enough satellite reflections are included in the objective aperture. Fig. 5 shows a structure image of a fragment oriented with the axis [010] parallel to the electron beam. The corresponding electron diffraction pattern with indices in correct orientation is inserted. A total of 71 excited beams were included in the objective aperture of which 22 were (a)-type reflections, 39 (e)-type reflections (satellites) and 10 (f)-type reflections (satellites). In order to ensure precise orientation, this micrograph, contributed by E. Watanabe, was made with a tilting goniometer stage for which an objective lens with $C_s = 3.5$ mm and $f_0 = 2.5$ mm was used.

When Fig. 5 is viewed at a small angle to its diagonal (top left to bottom right), dark bands parallel to the [101]-direction approximately 8 Å apart can be recognized which are periodically shifted approximately every 30 Å by one half of their separation along planes which are nearly parallel to the trace of the [301]-direction.

The dark bands may indicate layers in the structure in which the non-tetrahedral positions are occupied by Ca, which has a higher scattering factor than Na, while in the layers alternating with those indicated by the dark the non-tetrahedral atom positions may be occupied by Na. The shifts in the dark bands thus define a superstructure with a period of approximately 30 Å. Along

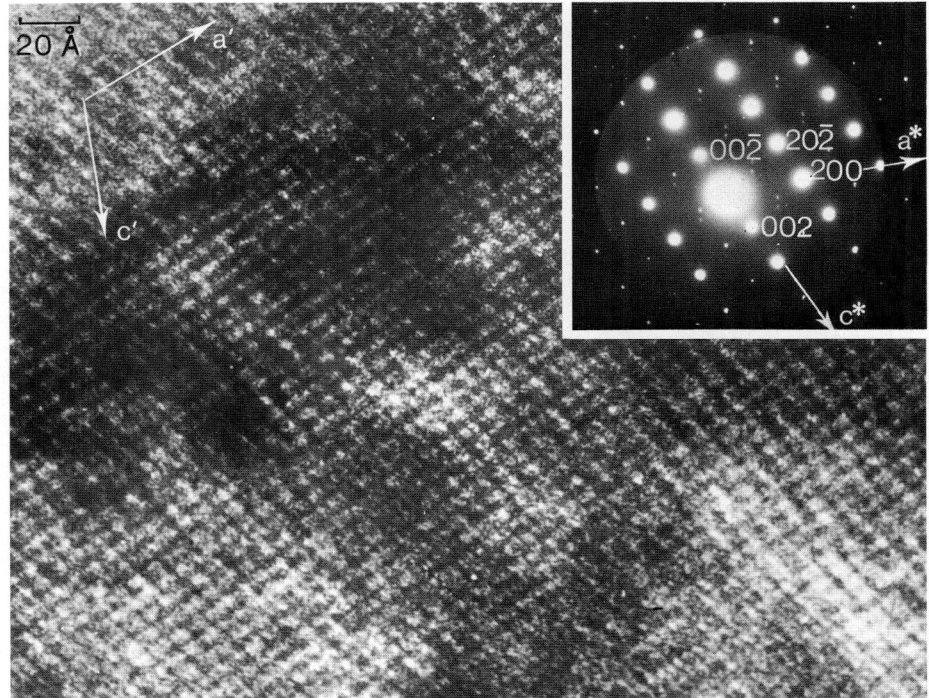

Fig. 5. Labradorite An 54 oriented with [010] parallel to the electron beam. Domain boundaries approximately parallel to (11$\bar{4}$), are horizontal. a' and c' are projections of axes. The domain structure can be seen best by viewing from the lower right to the upper left

Fig. 6. Same as Fig. 5, but [5$\bar{1}$1] parallel to electron beam. Domain boundaries approximately parallel to (11$\bar{4}$) (horizontal)

the boundaries of the superlattice units Ca-positions are replaced by Na-positions as in an antiphase boundary.

Since so far only two Al/Si ordering schemes, i.e. that of low albite and that of anorthite, have been observed in plagioclases, it may be assumed that, in order to achieve charge balance in the labradorite structure, the Ca atoms are surrounded by Al and Si tetrahedra in a configuration as in anorthite and the Na atoms in a configuration as in low albite.

The superlattice indicated in Fig. 5 and also resolved in a powder fragment oriented with [5$\bar{1}$1] parallel to the electron beam (Fig. 6) is consistent with the superlattice indicated by the (e)- and (f)-type satellites in the diffraction patterns of this material and with a mean spacing of 33 Å of the (f)-fringes measured in Figs. 1 and 2, corresponding to a superlattice with a periodicity of 66 Å.

Korekawa (1967) and Korekawa and Jagodzinski (1967) have qualitatively interpreted the satellite orientations and extinctions of a labradorite with 52 mole percent anorthite. They assumed that Ca and Na atoms are periodically separated in the non-tetrahedral cation positions and form alternating double rows of Ca and Na atoms parallel to the *a*-axis which are offset every seven unit cells by half their distance along planes parallel to (11$\bar{4}$) such that the material has a superlattice consisting of antiphase domains in the form of layers parallel to (11$\bar{4}$) with a periodicity around 56 Å (Korekawa, 1967, Fig. 35).

Recently, Korekawa and Horst (1974) refined the model by including an interpretation of the asymmetries in intensity of the pairs of satellites and assumed that the superlattice units are separated by a boundary zone 7 Å wide consisting of alternating albite-like and anorthite-like regions.

Fig. 5 may be directly compared with an (*ac*)-section of the structure because the (*ac*)-plane is inclined by only a few degrees with respect to the plane normal to the axis [010]. The geometry of the superlattice and the size of the identical units are the same in Fig. 5 and the model of Korekawa and Horst; the bands indicating non-tetrahedral positions occupied by Ca atoms run along the [101]-direction in Fig. 5. A model showing the Ca- and Na-positions in the labradorite structure which can be deduced from Fig. 5 is in preparation.

Since Fig. 5 represents only one projection of the structure, further structure images in other projections should be taken into account before a complete comparison with the model of Korekawa and Horst can be made.

Fig. 5 demonstrates that structure images can be used successfully to resolve atomic details of superlattice units in crystals with superstructures; this is similarly shown for pyrrhotite (Nakazawa et al., Chapter 5.2 of this volume). It may be concluded from the interpretation of Fig. 5 that the superlattice of the intermediate plagioclase structure is mainly due to the Ca/Na occupancy. Since the position of the steps in the Ca chains varies irregularly, the thickness of a superstructure unit may also vary in places. This can explain the fact that dislocations in the (e)- or (f)-fringe pattern always involve the appearance of *two* dark fringes at each dislocation. It may be assumed that in these dislocations the width of one unit of the superstructure gradually decreases and the unit ends in a wedge-shape.

It may be assumed that the cation-distribution and hence the bulk anorthite content is the same for the "well-ordered domains" in the major and the minor lamellae of the labradorite investigated since the spacing and orientation of the fringe pattern is the same. From the contrast pattern in Fig. 5a attributed to the non-tetrahedral cations, the bulk composition of this material is roughly 50 mole percent anorthite, i.e. the total volume of dark and light bands parallel to a' is equal. However, in order to account for the bulk composition of the specimen (54 mole percent anorthite) slightly more Ca than Na must be present, e.g. in the boundary zones. Alternatively, the major and minor lamellae differ in their average composition in spite of having the same spacing and orientation of their superstructure, as discussed in the preceeding section.

For a specimen thickness of about 50–100 Å, as estimated from tilting experiments with ion-etched foils, the number of superposed Ca/Na atoms along the normal to the crystal surface is 5 to 10. For this specimen thickness the dynamic effect of electron diffraction cannot be disregarded: The image contrast changes with the crystal thickness (Hashimoto et al., 1961). The comparison between structure projections of plagioclase feldspar and the labradorite micrographs are therefore regarded as only a rough first approximation to structure images. The application of the many-beam theory of electron diffraction which takes into account the phase shift due to spherical aberration and defocussing allows a comparison of calculated and observed contrast and thus refines the information which can be deduced from structure images such as Figs. 3 and 4 (Hashimoto et al., 1975).

It may be concluded from the high resolution micrographs of labradorites, that structure images with an estimated point resolution of 3 Å can be obtained in mineral structures in which the heaviest atom is a medium-heavy element such as Ca. In combination with calculated images this opens a wide field for the study of crystal structures, especially those with superstructures caused by positional disorder or differences in site occupancy.

References

Allpress, J.G., Hewart, E.A., Moodie, A.F., Sanders, J.V.: N-beam lattice images. I. Experimental and computed images from $W_4Nb_{26}O_{77}$. Acta Cryst. A**28**, 528–536 (1972).

Bown, M.G., Gay, P.: The reciprocal lattice geometry of the plagioclase feldspar structures: Z. Krist. **111**, 1–14 (1958).

Buseck, P.R., Iijima, S.: High-resolution electron microscopy of silicates, Am. Mineralogist **59**, 1–21 (1974).

Cole, W.F., Sörum, H., Taylor, W.H.: The structure of the plagioclase feldspars I. Acta Cryst. **4**, 20–29 (1951).

Cowley, J.M., Moodie, A.F.: Fourier images IV: The phase grating. Proc. Phys. Soc. (London) **76**, 378–384 (1960).

Hashimoto, H., Kumao, A., Endoh, H., Nissen, H.-U., Ono, A., Watanabe, E.: Lattice image contrast by many beam dynamical theory and structure determination of labradorite feldspar. Proc. EMAG 75 conference, Bristol (1975).

Hashimoto, H., Mannami, M., Naiki, T.: Dynamical theory of electron diffraction for the electron microscopic images of crystal lattices, Phil. Trans. Roy. Soc. London **253**, 459–516 (1961).

Kempster, C.J.E., Megaw, H.D., Radoslovitch, E.W.: The structure of anorthite, $CaAl_2Si_2O_8$. Structure analysis. Acta Cryst. **15**, 1005–1017 (1962).

Korekawa, M.: Theorie der Satellitenreflexe. Habilitationsschrift. Universität München (1967).

Korekawa, M., Horst, W.: Überstruktur des Labradorits An_{52}. Fortschr. Mineral. **52**, Beih. **2**, 37–39 (1974).

Korekawa, M., Jagodzinski, H.: Die Satellitenreflexe des Labradorits. Schweiz. Mineral. Petrog. Mitt. **47**, 269–278 (1967).

McConnell, J.D.C.: Electron-optical study of the fine structure of a schiller labradorite. Abstr. NATO Adv. Study Inst. on Feldspars, p. 14. Manchester: 1972.

McConnell, J.D.C.: Electron-optical study of the fine structure of a schiller labradorite. In: The feldspars (eds. W.S. MacKenzie and J. Zussman). Proc. NATO Adv. Study Inst., p. 478–490. Manchester: University Press 1974.

McLaren, A.C.: Transmission electron microscopy of the feldspars. In: The feldspars (eds. W.S. MacKenzie and J. Zussman). Proc. NATO Adv. Study Inst., p. 378–423. Manchester: University Press 1974.

McLaren, A.C., Marshall, D.B.: Transmission electron microscope study of the domain structure associated with b-, c-, d-, e- and f-reflections in plagioclase feldspar. Contrib. Mineral. Petrol. **44**, 237–249 (1974).

Nissen, H.-U.: End-member compositions of the labradorite exsolution: Naturwiss. **58**, 454 (1971).

Nissen, H.-U.: Electron microscopy of low plagioclases. Abstr. NATO Adv. Study Inst. on Feldspars, 29–30. Manchester: University Press 1972.

Nissen, H.-U.: Exsolution lamellae in plagioclase feldspars: electron microscopy and X-ray microanalysis. Proc. 8th Int. Congress on Electron Microscopy, Canberra, **1**, 468–469 (1974).

Toman, K., Frueh, A.J.: Intensity averages of plagioclase satellites: distribution in reciprocal space. Acta Cryst. B**28**, 1657–1662 (1972).

Wenk, E., Wenk, H.-R., Glauser, A., Schwander, H.: Intergrowth of andesine and labradorite in marbles of the Central Alps. Contrib. Mineral. Petrol. **53**, 311–326 (1975).

Yanaka, T., Shirota, K., Yonezawa, A., Arai, Y.: Capability of high resolution objective lens with miniaturized lower pole piece top. I. Proc. 8th Int. Congress on Electron Microscopy, Canberra, **1**, 128–129 (1974).

CHAPTER 5.7

Origin of the (c) Domains of Anorthite

A.H. HEUER, G.L. NORD, JR., J.S. LALLY, and J.M. CHRISTIE

The structural changes in calcic plagioclases giving rise to the (c) domains are not yet well understood (see Heuer and Nord, Chapter 5.1 of this volume). In this paper, we first illustrate the nature of the problem, particularly the effect of temperature on domain morphology, and then propose a phenomenological model which attempts to integrate these observations with earlier electron microscopical and X-ray diffraction results.

Two conflicting views have appeared in the literature as to the effect of heating to modest temperatures ($\sim 500°$ C) on the size of (c) domains in anorthite: that held by Lally et al. (1972) and Müller and Wenk (1973) — that such heat treatments do not effect domain size, and that held by Czank et al. (1973) — who suggested that domain size decreases with increasing temperature. It is important to note that Czank et al. in their study of this phenomenon used the electron beam itself to heat the samples. However, both Lally et al. and Müller and Wenk, using stages incorporating a heating coil, reported the occurrence of electron radiation damage during *in situ* heating experiments; we therefore were concerned that such radiation damage may have caused the apparent decrease in domain size observed by Czank et al. (1973). The experiment shown in Fig. 1 confirmed that radiation damage cannot be discounted. The sample, an An_{97} anorthite fragment in lunar breccia 67075, a cataclastic anorthosite, contained a domain structure (Fig. 1a) typical of slowly cooled lunar anorthites, $\sim$ micron size elongated domains visible using $(h+k+l, odd)$ reflections. In anorthites containing such a coarse domain structure, there is general agreement that the structure is "fully" primitive (space group $P\bar{1}$), and that the fault vector is $\frac{1}{2}$ [111]. After one minute of viewing, however, considerable radiation damage producing "black spots" had occurred (Fig. 1b), which had the result of reducing the *apparent* domain size. After a further viewing time of 5 min (Fig. 1c), the domain structure was barely recognizable. Fig. 1d, a dark field image of a bend contour using a $(h+k+l, even)$ reflection from the same area as Figs. 1a to c, indicates the damage is only visible when a $(h+k+l, odd)$ reflection is used. Along with the degradation of the domain image, the (c) reflections in electron diffraction patterns became streaked. *In situ* "annealing" at 370° C for 10 min had no effect on the damaged (c) domain images or reflections. Based on this experiment, we believe that Fig. 6b of Czank et al. (1973) may be interpreted as due to radiation damage rather than as a decrease in domain size, i.e. their Fig. 6b is analogous to Fig. 1c.

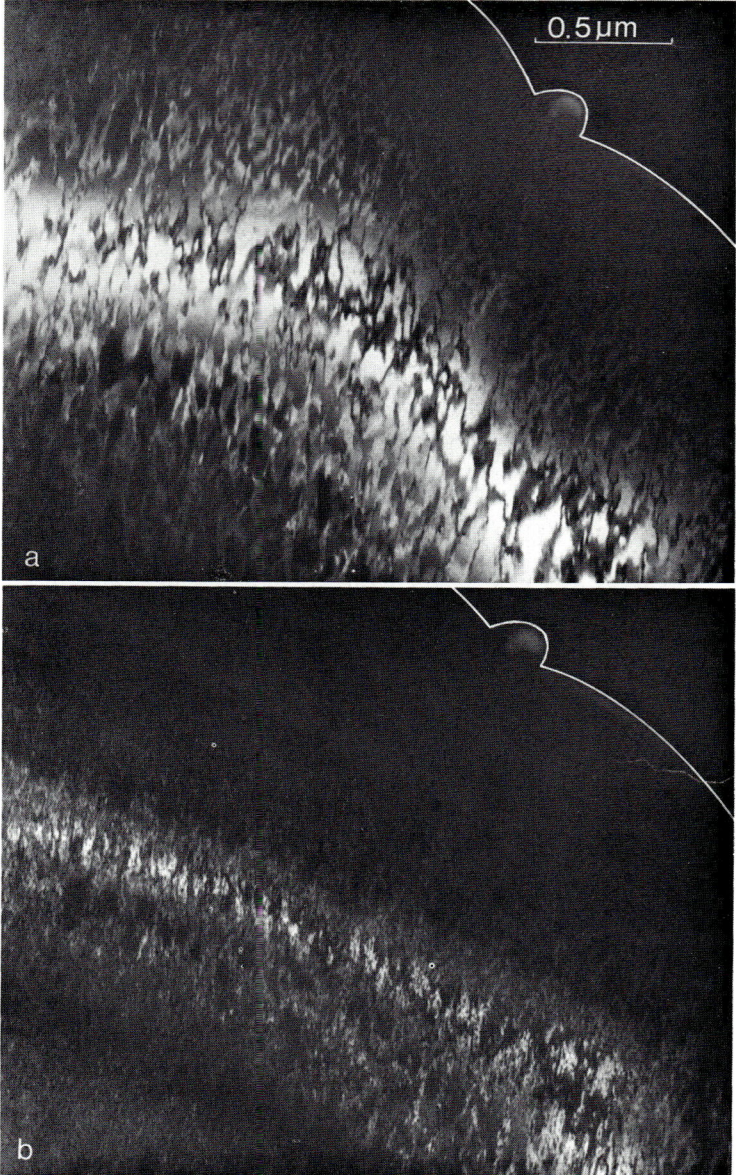

Fig. 1 a–d. Electron radiation damage experiment in 67075. The sequence shown in (a, b, c) corresponds to the start of the experiment, after 1 min and after 5 min viewing time, respectively; (d) was the last micrograph taken. (The white lines denote the foil edge and serve as a fiducial mark.) (a–c) are dark field micrographs using a type (c) reflections while (d) was taken with a type (a) reflection. 1 000 kV

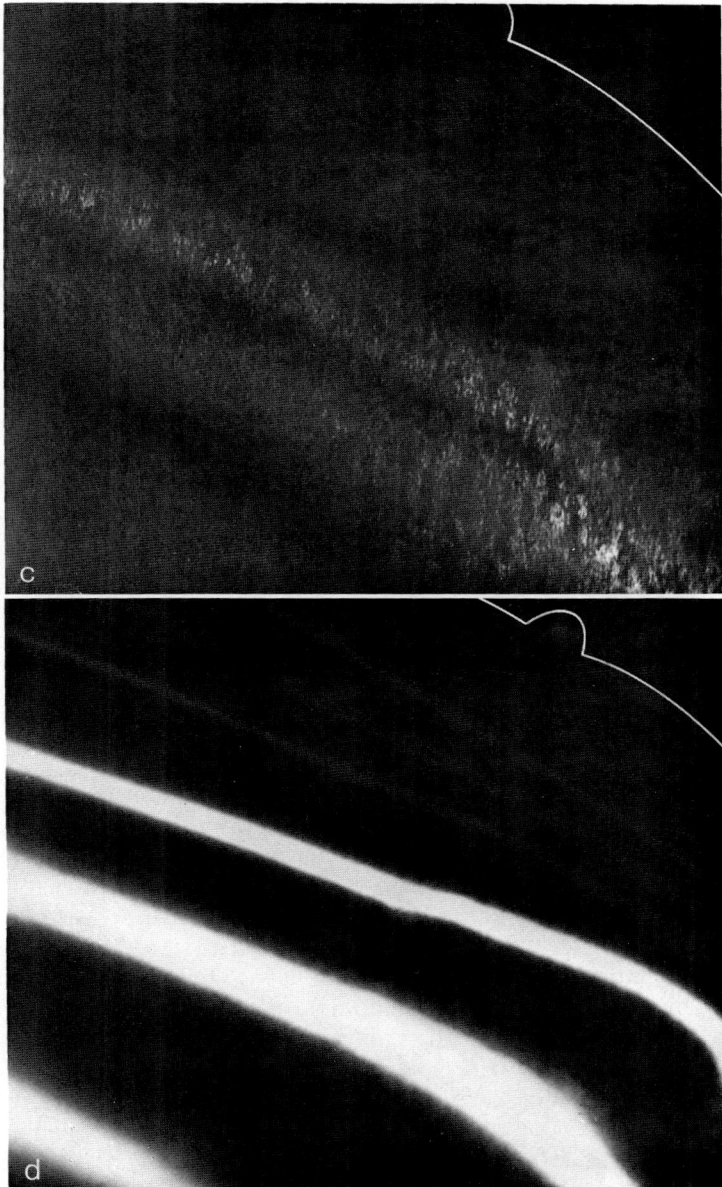

Fig. 1c and d

In further support of this interpretation, we have performed *in situ* heating experiments on An_{97} anorthite from lunar rock 15415 with results which agree totally with those of Müller and Wenk; domain contrast is lost at some quite low temperature (160° C for 15415 compared to 200–250° C for the An_{100} Pasmeda anorthite of Müller and Wenk) but the domain structure is unchanged on cooling back to room temperature. This perfect reversibility is maintained to at least 575° C for Pasmeda (Müller and Wenk, 1973) and 400° C for 15415 (Lally et al., 1972 and unpublished results). (The original report of Lally et al., 1972, that "the domain structure after cooling is similar in scale to the initial structure but the domains are not identical in shape" is now known to have been caused by foil movement during the *in situ* heating experiments.)

The changes in intensity and the shape of the (c) reflections that occur during controlled *in situ* heating experiments are also very informative but have not previously been reported. Consider the series of diffraction patterns taken during heating of a foil of 15415 shown in Fig. 2. At room temperature (Fig. 2a), the pattern consists of only (a) and (c) reflections and the (c) reflections are quite sharp. At 262° C, however, the (c) reflections are streaked out, as shown in Fig. 2b, and at 357° C, there is no visible intensity left in these reflections (Fig. 2c). The specimen was heated to a maximum temperature of 623° C, at which temperature, the (c) reflections had apparent zero intensity. On cooling back to room temperature, the (c) reflections were seen to be streaked, as shown in Fig. 2d. (Note that Fig. 2d has a slightly different orientation from Figs. 2a to c and contains sharp (a) and (b) reflections, diffuse (c)'s and no (d) reflections. This different orientation was caused by the need to remove the sample from the hot stage after the 623° C heating, due to a sticking problem; the specimen was tilted to a close-by zone axis on reinsertion into the electron microscope.) Along with the streaked (c) reflections, the domain size was very much smaller, as shown by comparison of Fig. 2e, the initial microstructure corresponding to the diffraction pattern of Fig. 2a and f, the microstructure after heating to 623° C and cooling back to room temperature, which corresponds to Fig. 2d. (The small orientation difference (22°) between Fig. 2e and f is not enough to account for the difference in apparent domain size.) It thus appears that heating to a temperature between 400° and 623° C for the An_{97} anorthite from 15415 causes irreversible changes in domain size and shape; because of the importance of this observation to the model to be discussed below, this experiment was repeated twice, in each case confirming the temperature at which changes occurred as ~600° C. These changes are believed to be a result of a "disordering" reaction on heating, followed by an "ordering" reaction on cooling during which the (c) domains formed but with a smaller size than in the original sample. No

---

Fig. 2a–f. *In situ* heating experiment in 15415. (a) shows a [512] zone axis pattern at the start of the experiment, while (b and c) show the same pattern at 262° C and 357° C, respectively. (The $\bar{1}31$ reflection is circled in these diffraction-patterns.) This crystal was heated to 623° C; (d) shows a nearby zone axis ([211]) after cooling back to room temperature. ($\bar{1}11$ is circled). (e and f) show the dark field images ($g=02\bar{1}$ and $\bar{1}11$, respectively) corresponding to (a) and (d). See text for further discussion

Origin of the (c) Domains of Anorthite

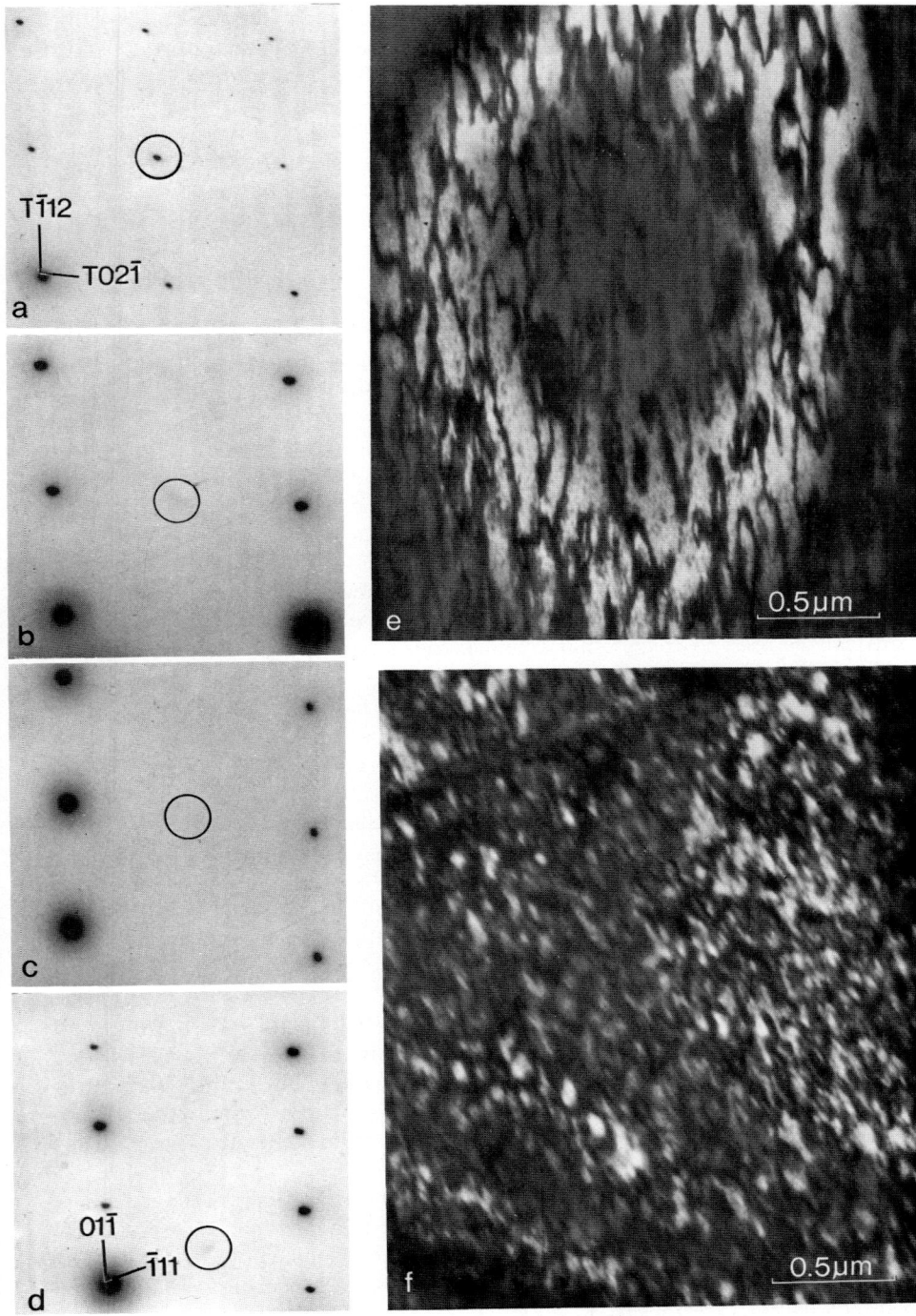

Fig. 2a–f

coarsening of the domains occurred during subsequent short term *in situ* annealing experiments (e.g. 10 min at 375° C). The corresponding temperature at which irreversible changes were noted for the An_{100} Pasmeda anorthite is between 575° and 1200° C (Müller and Wenk, 1973).

The phenomenological model proposed here, which attempts to explain the electron microscopy results from natural and heat-treated anorthite, and to integrate these with X-ray and other physical property data, is that two different $P\bar{1}$ phases exist for anorthite, one stable below 160–250° C, and a second stable from this temperature to a temperature at which an inversion to a "disordered" phase occurs. For convenience, the two primitive phases will be referred to as $P\bar{1}(1)$ and $P\bar{1}(h)$, respectively. Although the structural evolution of calcic plagioclases are still not completely understood (see Heuer and Nord, Chapter 5.1 of this volume), this model can be most readily comprehended if it is assumed (as will be justified below) that the "disordered" phase is body-centered anorthite (symmetry $I\bar{1}$); (c) domains thus arise from an $I\bar{1} \rightarrow P\bar{1}(h)$ transition, and have a fault vector of $\frac{1}{2}[111]$. If it were possible to examine the structure at a temperature immediately below the "ordering" temperature, the domains would not be visible because of the weak and streaked nature of the (c) reflections in the $P\bar{1}(h)$ phase (Fig. 2b and c). The results quoted above on the *reversible* change in domain contrast at $\sim 200°$ C suggest that this streaking is due to anisotropic thermal vibrations, most likely of Ca atoms. On cooling below the $P\bar{1}(h) \rightarrow P\bar{1}(1)$ transition temperature, the (c) domains become visible because the now-stronger (c) reflections can be used to form a dark field (superlattice) image. Note that in this view, the domain morphology depends only on the $I\bar{1} \rightarrow P\bar{1}(h)$ transition; the change from the $P\bar{1}(h)$ to the $P\bar{1}(1)$ structure has no effect on the domain morphology. In addition to the loss in domain contrast at $\sim 200°$ C, support for the existence of a low temperature transition is provided by the X-ray results of Laves *et al.* (1970), who found that the temperature dependence of the intensity of some (a), (b), and (c) reflections of anorthite changed sign at $\sim 230°$ C, and the change in the NMR spectra at 230–250° C (eight non-equivalent → four non-equivalent Al sites) reported by Brinkman and Stähli (1968). (As suggested by Smith and Ribbe, 1969, a $P\bar{1}$ structure above 250° C is consistent with the NMR data if Ca atoms "bounce" more rapidly than 10^{-7}/sec but less rapidly than $10^{-11} - 10^{-15}$/sec, i.e. although the framework and hence the basic structure is primitive, the vibrations of the Ca atoms impart a pseudo-body-centeredness.) That this transition is from one $P\bar{1}$ phase to another (rather than from a $P\bar{1}$ to an $I\bar{1}$ phase, as interpreted by Smith (1974) in his review of the feldspar literature to 1973) is suggested by the reversibility of domain contrast on heating to 400–575° C; in addition, Foit and Peacor (1973), in their structure determinations of an An_{98} Miyake specimen, satisfactorily refined the structure as being $P\bar{1}$ at room temperature, 410°, and 830° C.

The evidence concerning the transition to $I\bar{1}$ symmetry is contradictory. Smith (1974) assumed that the alumosilicate structure was at least *statistically* (his italics) body-centered above about 300° C. On the other hand, Laves *et al.* (1970) found that the residual intensity ($\sim 3\%$ of the room temperature value) present in (c) reflections above 230° C persisted up to the melting point, suggesting a $P\bar{1}$ structure at all temperatures; this is in contrast to Foit and Peacor (1967), who had estimated that the (c) reflections reached zero intensity at ~320° C. Support for a high-temperature inversion (henceforth assumed to be to $I\bar{1}$ symmetry) is found in Czank and Schulz's (1971) measurements of the cell dimensions of a Vesuvius anorthite An_{98} up to 1500° C, where anamolous expansions near 200° C *and* between 650° C and 1000° C were found for the *b* and *c* unit cell dimensions, in the optical anamoly found by Bloss (1964) near 800° C in a Vesuvius An_{97} sample, and by the differential thermal analysis study by Köhler and Wieden (1954), who found an endothermic peak, also near 800° C, in a Pasmeda An_{100} sample. Lastly, Czank (1973) has refined an An_{97} Vesuvius anorthite at room temperature, 240°, and 1430° C. (Smith, 1974, has presented Czank's data in the form of atomic coordinates for body-centered symmetry.) Only type (a) and (b) reflections could be collected at 240° and 1430° C; furthermore, at the higher temperatures, the Ca atoms occupied "split-positions" (Czank *et al.*, 1973). At 240° C, the occupation frequency of these part atoms deviated from one half, but at 1430° C, these split positions were occupied with equal probability.

It appears that the inversion temperature to the $I\bar{1}$ phase varies markedly with An content. For example, while a temperature of ~600° C for An_{97} was found in this work, Foit and Peacor (1967) found reversibility (for the X-ray intensity) on heating to 1000° C for an An_{98} sample, in agreement with their $P\bar{1}$ structure determination at 830° C, and suggesting an inversion temperature $>1000°$ C; this is in agreement with Müller and Wenk's experiments of the An_{100} Pasmeda anorthite, which suggest an inversion between 575° and 1200° C. Bloss also found irreversible changes in room temperature optical extinction on heating above 800° C. Likewise, Laves and Goldsmith's (1954) early results — that an An_{100} synthetic anorthite had sharp (c) reflections when quenched from 1100° but diffuse (c)'s when quenched from 1500° C, can be interpreted as indicating an inversion temperature $>1100°$ C.

To summarize this portion of the paper, abundant evidence exists to support a $P\bar{1} \to I\bar{1}$ transition at elevated temperatures, although the transition temperature appears to vary with An content. The existence of such a transition permits ready explanation for the (c) domain structures found in anorthites of igneous origin.

Returning to the electron microscopy studies, the nature of fine-scale domain structures, such as Fig. 2f, the An_{95} anorthite in the Miyake volcanic anorthite of Müller *et al.* (1972), and the heat-treated and quenched Grass Valley anorthite of Müller and Wenk (1973), require further comment. Two different interpretations are possible. Consistent with the foregoing, it is suggested that for samples rapidly cooled through the $I\bar{1} \to P\bar{1}(h)$ transition, the fast cooling causes a fine domain size (this transition, while probably displacive, involves nucleation and growth of domains and thus the domain morphology depends on the thermal conditions during the transition.) On the other hand, direct lattice resolution studies (McLaren, 1973) have suggested that the fine-scale domain images may in fact be due to $P\bar{1}$ "domains" in an $I\bar{1}$ matrix; this model had previously been suggested by Czank *et al.* (1973). This would suggest that the $I\bar{1} \to P\bar{1}(h)$ transformation is partly quenchable, and that on rapid cooling, a reaction

$$I\bar{1} \to P\bar{1}(l) + I\bar{1}$$

can occur. While this reaction scheme provides an explanation for the weak and diffuse (c) reflections found on fast cooling, the structural and thermodynamic implications of such a reaction are not clear. We thus favor the fine domain size explanation.

Structurally, the $I\bar{1} \to P\bar{1}(h)$ transition should be thought of in terms of framework displacements. The $P\bar{1}(h) \to P\bar{1}(1)$ transition, on the other hand, involves only "quenching" of the "bouncing" motion of the Ca ions suggested by Smith and Ribbe. Some structural aspects of an $I\bar{1} \to P\bar{1}$ transition have previously been considered by Czank et al. (1973) but their point of view differs in several substantial aspects from that espoused here.

The effects of electron radiation must be carefully considered in studying anorthite by TEM. The diffraction patterns of radiation-damaged regions showed weak and streaked reflections, which rendered imaging of the (c) domains difficult. It is suggested that this damage takes the form of short range displacements of the Ca atoms; since Ca makes the largest single contribution to most (c) reflections (Foit and Peacor, 1973), this results in the streaking of the (c) reflections. In other words, the damage is not believed to affect the domain structure, which involves the entire framework, but only the ability to image the domains.

In summary, an attempt has been made to provide a model which integrates the numerous X-ray and electron microscopical evidence on (c) domains in anorthite. The key elements of the model are a structural evolution on slow cooling

$$I\bar{1} \to P\bar{1}(h) \to P\bar{1}(1).$$

The $I\bar{1} \to P\bar{1}$ (h) transition appears to depend on An content and occurs at 600–1200° C; the $P\bar{1}(h) \to P\bar{1}(1)$ transition occurs at $\sim 200°$ C and may also vary with An content. However, it is clear that further work is needed to substantiate this model as well as other aspects of the anorthite "problem".

References

Bloss, F.D.: Optical extinction of anorthite at high temperatures. Am. Mineralogist **49**, 1125–1131 (1964).
Brinkmann, D., Stähli, J.L.: Magnetische Kernresonanz von ^{27}Al im Anorthit, $CaAl_2Si_2O_8$. Helv. Phys. Acta **41**, 274–281 (1968).
Czank, M.: Strukturen des Anorthits bei höheren Temperaturen. Ph.D. Thesis, E.T.H., Zürich (1973).
Czank, M., Schulz, H.: Thermal expansion of anorthite. Naturwiss. **58**, 2, 94 (1971).
Czank, M., Van Landuyt, J., Schulz, H., Laves, F., Amelinckx, S.: Electron microscopic study of the structural changes as a function of temperature in anorthite. Z. Krist. **138**, 403–418 (1973).
Foit, F.F., Peacor, D.R.: High temperature diffraction data on selected reflections of an andesine. Z. Krist. **125**, 1–6 (1967).
Foit, F.F., Peacor, D.R.: The anorthite crystal structure at 410 and 830° C. Am. Mineralogist **58**, 665–675 (1973).
Köhler, A., Wieden, P.: Vorläufige Versuche in der Feldspatgruppe mittels der DTA. Neues Jahrb. Mineral., Monatsh. **12**, 249–252 (1954).

Lally, J.S., Fisher, R.M., Christie, J.M., Griggs, D.T., Heuer, A.H., Nord, G.L., Radcliffe, S.V.: Electron petrography of Apollo 14 and 15 rocks. Proc. 3rd Lunar Sci. Conf. Suppl. 3, Geochim. Cosmochim. Acta **1**, 401–422 (1972).
Laves, F., Czank, M., Schulz, H.: The temperature dependence of the reflection intensities of anorthite ($CaAl_2Si_2O_8$) and the corresponding formation of domains. Schweiz. Mineral. Petrog. Mitt. **50**, 519–525 (1970).
Laves, F., Goldsmith, J.R.: Long-range-short-range order in calcic plagioclases as a continuous and reversible function of temperature. Acta Cryst. **7**, 465–472 (1954).
McLaren, A.C.: The domain structure of a transitional anorthite; a study by direct lattice-resolution electron microscopy. Contr. Mineral. and Petrol. **41**, 47–52 (1973).
Müller, W.F., Wenk, H.R.: Changes in the domain structure of anorthites induced by heating. Neues Jahrb. Mineral., Monatsh. 17–26 (1973).
Müller, W.F., Wenk, H.R., Thomas, G.: Structural variations in anorthites. Contrib. Mineral. Petrol. **34**, 304–314 (1973).
Smith, J.V.: Feldspar minerals. vol. 1, 627 p. Berlin-Heidelberg-New York: Springer 1974.
Smith, J.V., Ribbe, P.H.: Atomic movements in plagioclase feldspars: kinetic interpretation. Contrib. Mineral. Petrol. **21**, 157–202 (1969).

CHAPTER 5.8

On Polymorphism of BaAl$_2$Si$_2$O$_8$

W.F. MÜLLER

1. Introduction

The polymorphs of BaAl$_2$Si$_2$O$_8$ known so far are celsian, paracelsian and hexacelsian. Celsian is a monoclinic feldspar mineral with a 14 Å c-axis and a body-centered lattice (Newnham and Megaw, 1960). Paracelsian, which also occurs as a mineral, has a framework structure similar to the feldspar structure (Smith, 1953; Bakakin and Belov, 1961; Craig et al., 1973). Hexacelsian, which is only known as a synthetic product is a layer alumosilicate built up by double sheets of Al$_2$Si$_2$O$_8$ (Ito, 1950). Takéuchi (1958) described a hexagonal and an orthorhombic polytype. According to Lin and Foster (1968), celsian is stable at atmospheric pressure below 1590° C. From 1590° C up to the melting point ($\sim$1760° C) hexacelsian, is the stable polymorph. It persists metastably when cooled below 1590° C. Paracelsian appears to be a metastable phase.

The work presented here continues previous studies on anorthites and Sr-feldspar (see Müller, 1974), which were undertaken in order to gain more insight into the polymorphism and order-disorder phenomena of feldspars and other alumosilicate structures. In this chapter, TEM observations made on a synthetic BaAl$_2$Si$_2$O$_8$ sample containing the two polytypes of hexacelsian and a polytype of paracelsian will be reported.

2. Experimental

Synthetic BaAl$_2$Si$_2$O$_8$ (about 1500 g) was prepared by melting SiO$_2$, Al$_2$O$_3$, and BaCO$_3$ at 1800° C in a 70% Pt $-$30% Rh crucible 100 mm in diameter and 150 mm high. The crucible was cooled down to 1400° C within 2 min, kept at this temperature for about 2 hrs, cooled down to 600° C in air and finally water-quenched to room temperature. The BaAl$_2$Si$_2$O$_8$ sample was removed from the crucible as fragments of 20 to 40 g. A fragment of about 30 g was heated at 1300° C for 10 days and air-quenched. Suitably thin specimens for transmission electron microscopy were prepared from the original and the heat-treated samples by ion-thinning (see Tighe, Chapter 3 of this volume) and by crushing. Observations were made with a JEOL JEM 100 B electron microscope equipped with a side-entry goniometer.

3. Results

3.1 Hexacelsian

Takéuchi (1958) found that there are two polytypes of hexacelsian: α-hexacelsian consists of pseudo-hexagonal $Al_2Si_2O_8$ sheets which are trigonally distorted, with the Ba^{2+} ions between the layers. Due to the distortion, the true structure contains two such layers in a body-centered orthorhombic cell (Fig. 1). The lattice parameters of α-hexacelsian are $a=5.3$, $b=9.2$, $c=15.6$ Å. At temperatures above 300° C, hexacelsian is believed to have an exactly hexagonal structure caused by a reversion of the distorted structure of the $Al_2Si_2O_8$-sheets to the ideal hexagonal structure. The lattice parameters of this β-hexacelsian are $\alpha=5.3$ and $c=7.8$ Å. TEM analysis showed that the prevailing phase of our $BaAl_2Si_2O_8$ sample is hexacelsian. Two polytypes with lattice constants corresponding to those described by Takéuchi (1958) were observed. In disagreement to this results however, the electron diffraction patterns of the polytype with $c=15.6$ Å contain reflections of the type $h+k+l$ odd, i.e. the structure is primitive. In spite of this it is likely that the polytypes discussed here are identical to α- and β-hexacelsian and thus Takéuchi's nomenclature is used in this paper. α- and β-hexacelsian usually occur as oriented intergrowths with $c_1||c_2$ and $a_1||a_2$ (Fig. 2). The most common composition plane is (001) which is the plane of the $Al_2Si_2O_8$ sheets. α-hexacelsian displays planar lattice defects predominantly oriented parallel to (001); they were in contrast with reflections of the type $h+k+l=$odd operating (Fig. 3) and out of contrast with reflections of the type $h+k+l=$even. These faults are interpreted as antiphase domain boundaries (APB's) and according to contrast experiments their displacement vector is

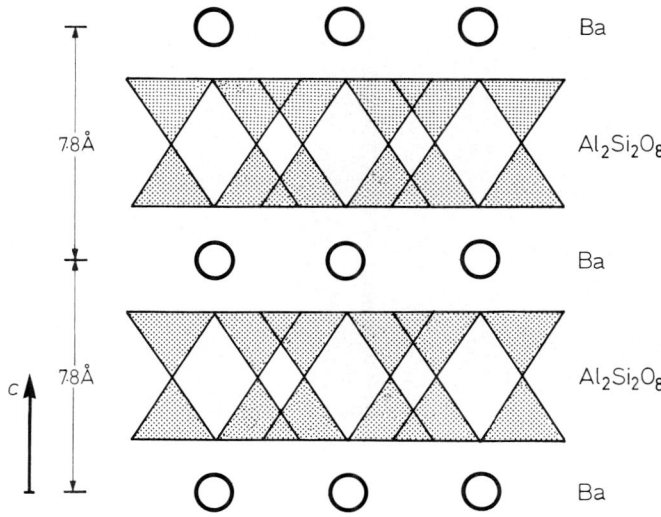

Fig. 1. Schematic drawing of idealized structural units of hexacelsian viewed along the a-axis. β-hexacelsian has $c=7.8$ Å, α-hexacelsian has $c=15.6$ Å

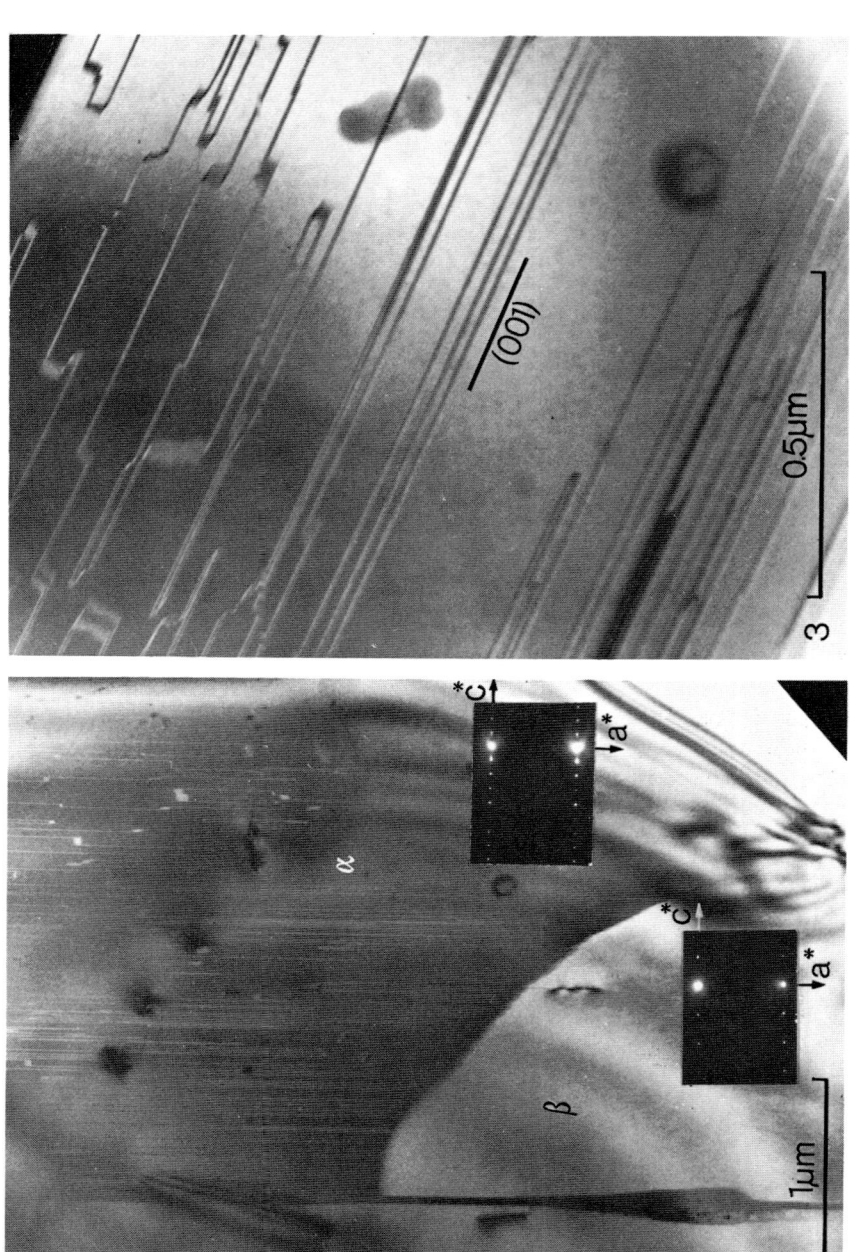

Fig. 2. Oriented intergrowth of α- and β-hexacelsian. Note the APB's in α-hexacelsian. Bright field. This and the following micrographs are taken at 100 kV

Fig. 3 APB's in α-hexacelsian. Dark field. **g** = 207

$1/2\ (\boldsymbol{a}+\boldsymbol{b}+\boldsymbol{c})$. This vector suggests that the antiphase domains form as a result of the transformation of a body-centered high-temperature phase to a primitive structure.

APB's can be imaged and analyzed by means of diffraction contrast. They can also be directly visualized by the transmission electron microscopic method of lattice imaging (cf. Allpress and Sanders, 1973; Cowley and Iijima, Chapter 2.5 of this volume). Fig. 4 shows a one-dimensional lattice image of α-hexacelsian obtained by using the systematic reflections $(00l)$ with $|l| \leq 2$. The regular sequence of $(00l)$ lattice fringes with 15.6 Å spacing is interrupted by APB's. It can be seen that in this orientation the domain designated (a) is displaced relative to domain (b) across the APB by $1/2\ \boldsymbol{c}$. Another image of $(00l)$ lattice fringes of α-hexacelsian (also showing APB's) is displayed in Fig. 5. As the crystal thickness decreases, additional fringes at half the 15.6 Å spacing become increasingly pronounced. The fringes parallel to $(00l)$ with a spacing of 7.8 Å have the periodicity of the structural units, namely of the $Al_2Si_2O_8$ sheets with Ba^{2+} ions in between (see Fig. 1). Since the scattering potential of the Ba^{2+} ions is dominant in hexacelsian, it is not unlikely that the dark lines with the spacing of 7.8 Å in the thinner areas of the crystal (Fig. 5) correspond to the position of the Ba^{2+} ions (cf. Cowley and Iijima, Chapter 2.5 of this volume).

3.2 Paracelsian

Smith (1953) used natural single crystals from the Benallt Mine, Wales, for the crystal structure determination of paracelsian, and found it to be monoclinic but strongly pseudo-orthorhombic with $a=9.08$, $b=9.58$, $c=8.58$ Å, $\beta \approx 90°$.

The $BaAl_2Si_2O_8$ sample as obtained from the melt consists of α- and β-hexacelsian and of glass. Celsian, the Ba-feldspar, was not observed. Examination of the sample heated at 1300° C for 10 days showed that the glassy areas had become crystalline. The electron diffraction patterns of these crystals can be indexed with the paracelsian lattice parameters given by Smith (1953), but additional reflections at $1/2\ a^*$ $(d=18.2$ Å$)$ demand a doubling of the a-axis. The unit cell appears to be orthorhombic and preliminary data suggest that the lattice parameters are $a=18.2$, $b=9.6$, $c=8.6$ Å. The present electron diffraction data are insufficient to determine the space group. It may be mentioned that no systematic absences were observed for reflections of the type $h01$, $h00$ and $00l$. No reflections of the type $0k0$, $0kl$, and hkl with k odd have been found so far. In some electron diffraction patterns the intensity of $h00$ reflections is systematically stronger for h even than for h odd. One-dimensional lattice images did not reveal lamellar intergrowths on a unit cell scale. Because of the similarities of the lattice parameters with those of paracelsian reported by Smith (1953), it is assumed that the crystals described here may be regarded as a new polytype of paracelsian and will be called in the following paracelsian (II).

Paracelsian (II) and hexacelsian often display oriented intergrowths. The a-axis of paracelsian (II) is parallel to the c-axis of hexacelsian; the predominant composition plane is (100) or (001), respectively, as shown in Fig. 6.

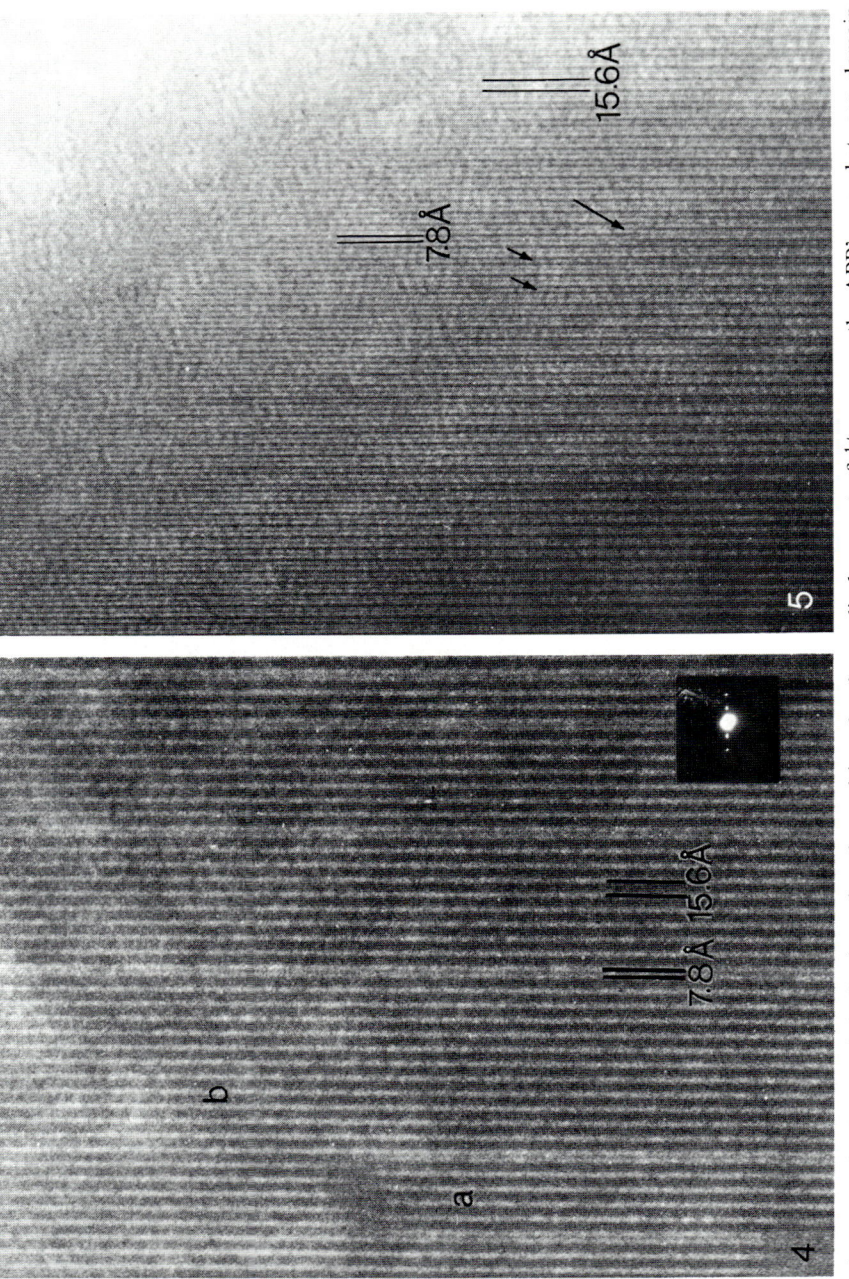

Fig. 4. Lattice image of (001) planes in α-hexacelsian showing a displacement of $1/2\,c$ across the APB's, e.g. between domain a and b

Fig. 5. Lattice image of (001) planes in α-hexacelsian with APB's (arrows). See text

On Polymorphism of BaAl$_2$Si$_2$O$_8$ 359

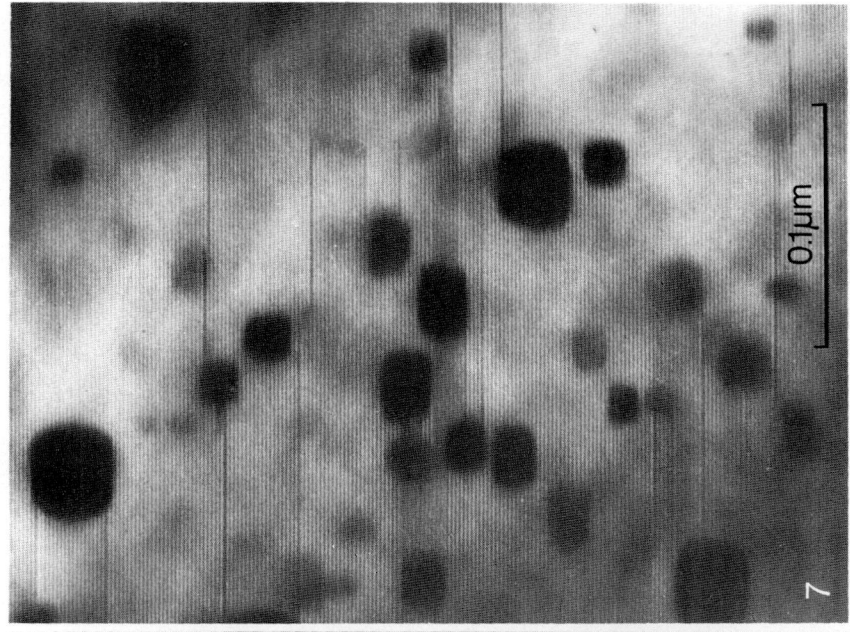

Fig. 6. Oriented intergrowth of paracelsian (II) and hexacelsian
Fig. 7. Precipitates of paracelsian (dark areas) in α-hexacelsian with APB's. Lattice image of (001) planes of α-hexacelsian

Precipitates of paracelsian (2) frequently occur in α-hexacelsian in both the original and heat-treated samples but were not observed in β-hexacelsian. The reason for this preference is probably that nucleation is easier in α-hexacelsian because it contains APB's. Actually, it is observed that the precipitates occur preferentially at APB's and at those places where the APB does not lie parallel to (001) (Fig. 7), i.e. a good example of heterogeneous nucleation on APB's.

Lin and Foster (1968) concluded that paracelsian is "a completely metastable form, which changes monotropically through metastable hexacelsian to the stable celsian". The observations presented here rather suggest that (at temperatures below 1590° C) hexacelsian changes to paracelsian. Celsian, the stable modification below 1590° C, has not been observed in specimens from the $BaAl_2Si_2O_8$ fragment heated at 1300° C for 10 days, but was found in a specimen from a $BaAl_2Si_2O_8$ sample ground to a grain size smaller than 200 μm and heated at 1300° C for 5 days. This is in agreement with an investigation by Bahat (1970) who reported that the hexacelsian-celsian phase transformation is very sluggish in hexacelsian-specimens about 0.6 cm in size but much more rapid in hexacelsian powder.

References

Allpress, J.G., Sanders, J.V.: The direct observation of the structure of real crystals by lattice imaging. J. Appl. Cryst. **6**, 165–190 (1973).
Bahat, D.: Kinetic study on the hexacelsian-celsian phase transformation. J. Mat. Sci. **5**, 805–810 (1970).
Bakakin, V.V., Below, N V.: Crystal structure of paracelsian. Soviet. Phys.-Cryst. **5**, 826–829 (1961).
Craig, J.R., Louisnathan, S.J., Gibbs, G.V.: Al/Si order in paracelsian. EOS, Transactions Austr. Geoph. Union **54**, 497 (1973).
Ito, T.: X-ray studies on polymorphism. Tokyo: Maruzen 1950.
Lin, H.C., Foster, W.R.: Studies in the system $BaO-Al_2O_3-SiO_2$ I. The polymorphism of celsian. Am. Mineralogist **53**, 134–144 (1968).
Müller, W.F.: Antiphase domains in $CaAl_2Si_2O_8$ (anorthite), $SrAl_2Si_2O_8$ and $BaAl_2Si_2O_8$. Proc. 8th Int. Congress on Electron Microscopy, Canberra, **1**, 472–473 (1974).
Newnham, R.E., Megaw, H.D.: The crystal structure of celsian (barium feldspar). Acta Cryst. **13**, 303–312 (1960).
Smith, J.V.: The crystal structure of paracelsian, $BaAl_2Si_2O_8$. Acta Cryst. **6**, 613–620 (1953).
Takéuchi, Y.: A detailed investigation of the structure of hexagonal $BaAl_2Si_2O_8$ with reference to its α-β inversion. Mineral. J. (Sapporo) **2**, 311–332 (1958).

CHAPTER 5.9

The Submicroscopic Structure of Wenkite

F. LEE

1. Introduction

Wenkite (after Prof. E. Wenk) is an alumosilicate mineral which was named and found by Papageorgakis (1962) in highly metamorphosed calcitic marbles of the Ivrea zone. The structure determined by H.-R. Wenk (1973) yielded the formula $Ba_4(Ca_{9.1},Na_{0.9})_6(Al_{0.4}Si_{0.6})_{20}O_{39}(OH)_2(SO_4)_3$. Wenkite has a true trigonal symmetry of P31m and a pseudohexagonal symmetry of $P\bar{6}2m$ ($a_0 = 13.515$ Å, $c_0 = 7.465$ Å). It has an interrupted $[SiO_4]-[AlO_4]$ framework with hexagonal building units similar to that of the zeolites cancrinite and gmelinite. The structure refinement revealed partial occupancy of certain lattice sites in the mineral, suggesting that wenkite may possess a domain structure due to growth twins, submicroscopic twins or antiphase domains.

The purpose of this study was to consider the usefulness of transmission electron microscopy (TEM) in the structural investigation of a mineral where the X-ray evidence suggests domain structures (Wenk, 1973). The loose framework is very susceptible to radiation damage and poses serious experimental problems which were partly overcome by using HVEM, specifically a 650 kV Hitachi TEM. In the first part of this paper, some models are suggested as the source of domains in wenkite. The balance of the paper discusses the experimental evidence and its relationship to some of the suggested models. (For a more detailed description, see Lee, 1973.)

2. Possible Origin of Domains

Domains are parts of a crystal that have structural differences relative to the matrix. The relationship between domain and matrix can be defined by a simple displacement vector **R** which describes the geometry of stacking faults, twins, or antiphase boundaries.

Wenkite is the first reported silicate mineral (Wenk and Lee, 1972) with an interrupted $[SiO_4]-[AlO_4]$ framework (Fig. 1). The interrupting element is an extra TO_4 tetrahedron (T equals Si or Al) in the site T(2) which has the flexibility to point either $+z$ or $-z$ in a particular unit cell and links the cancrinite building units. The structure refinement suggests that this $+z$ or $-z$ disposition occurs in an equal number of unit cells in the whole crystal because the whole

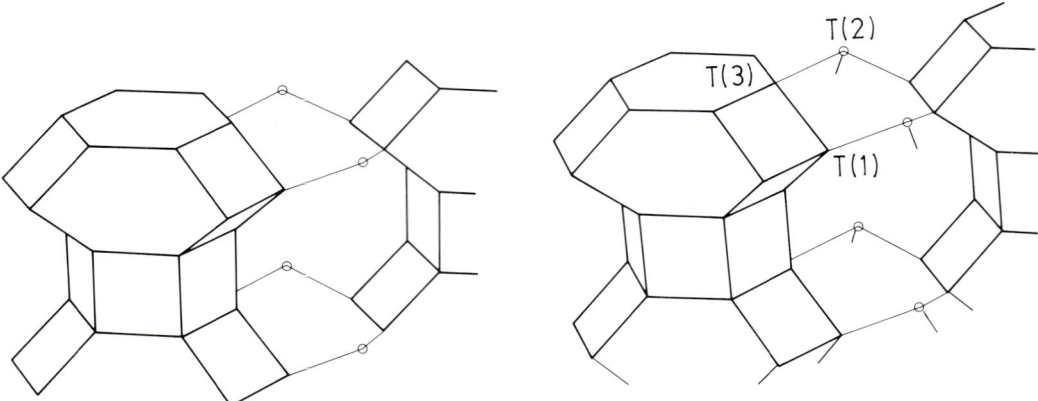

Fig. 1. The framework of wenkite represented as a T-polyhedron. The hexagonal building unit is also found in cancrinite. Circles indicate the T(2) tetrahedron which either point upwards ($+z$) or downwards ($-z$). (From Wenk, 1973)

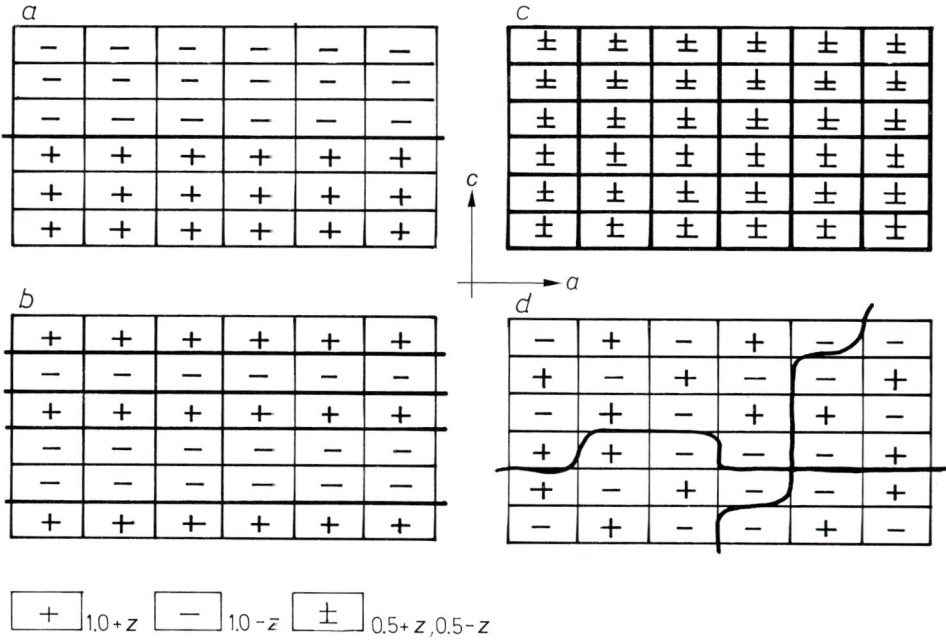

Fig. 2a–d. Models for domains in wenkite: (a) Normal microscopic twinning, (b) submicroscopic twin domains, (c) positional disorder, (d) antiphase domains

tetrahedral site can be expressed as partially occupied by a $+z$ tetrahedron as well as by a $-z$ tetrahedron (Wenk, 1973).

Domains in wenkite have several possible sources. These are (1) ion exchange through the silicate ring channels, (2) partial ordering of the Al-Si among the tetrahedral sites, and (3) the $+z$ or $-z$ disposition of the SO_4 or the linking TO_4 tetrahedra. The first two possibilities are not considered in this discussion because the structure determination showed that channel atoms Ba and Ca are ordered, and that the Al-Si arrangement is disordered. The SO_4 tetrahedron makes a relatively small contribution to domain structure as compared to the T(2) tetrahedron. Hence it is the T(2) tetrahedron that will be considered as the major variable giving rise to domains. Four models are proposed in Fig. 2 to explain the $P\bar{6}2m$ symmetry and the half occupancies of T(2).

The simplest explanation would be a growth twin of two P31m single crystals of identical size with (001) as the twin plane (Fig. 2a). These twins would not be visible in the diffraction pattern because of the coincidence of twin and host in the hexagonal system. No twins were visible at the optical scale. This model alone would require equal sizing of twin and host in order to produce $P\bar{6}2m$ diffraction symmetry which has been observed in all the crystals that have been analyzed. This chance equality is unlikely.

A modification of this first model includes submicroscopic basal twin domains which occur with equal frequency (Fig. 2b). Based on energy considerations, submicroscopic twin lamellae with straight and parallel boundaries are expected. Such twins would again be invisible in the diffraction pattern but should show contrast in bright field and dark field electron images.

Fig. 2c shows a model of positional disorder on a unit-cell level such that cells with $+z$ and $-z$ pointing T(2) tetrahedra are randomly scattered, and statistically occurring 50% $+z$ and 50% $-z$. Related to Fig. 2c is an ordered superstructure with $+z$ and $-z$ pointing tetrahedra regularly alternating (Fig. 2d), thereby doubling a_0 and c_0. This model is suggested from weak extra reflections on single crystal precession X-ray photographs of very long exposure (Fig. 3). Such a superstructure could be uniform throughout the crystal. On the other hand, it is often the result of an ordering process with nucleation centers which can be out of phase. If this phase difference is one-half period (one-half superlattice cell edge) as illustrated, then the boundary is an antiphase domain or π boundary. Several of the structures illustrated in Fig. 2 may occur simultaneously in the crystal.

3. Models for the Superstructure

The evidence for a superstructure of dimensions $2a_0 \times 2b_0 \times 2c_0$ ($a_0 = b_0$) in wenkite was first given by TEM experiments, and subsequently confirmed by single crystal X-ray precession photographs. The calculation of a model for the superstructure based solely on a periodic variation in orientation of the T(2) tetrahedron is now described. It should be kept in mind that only one of the possible variables in the structure is being considered.

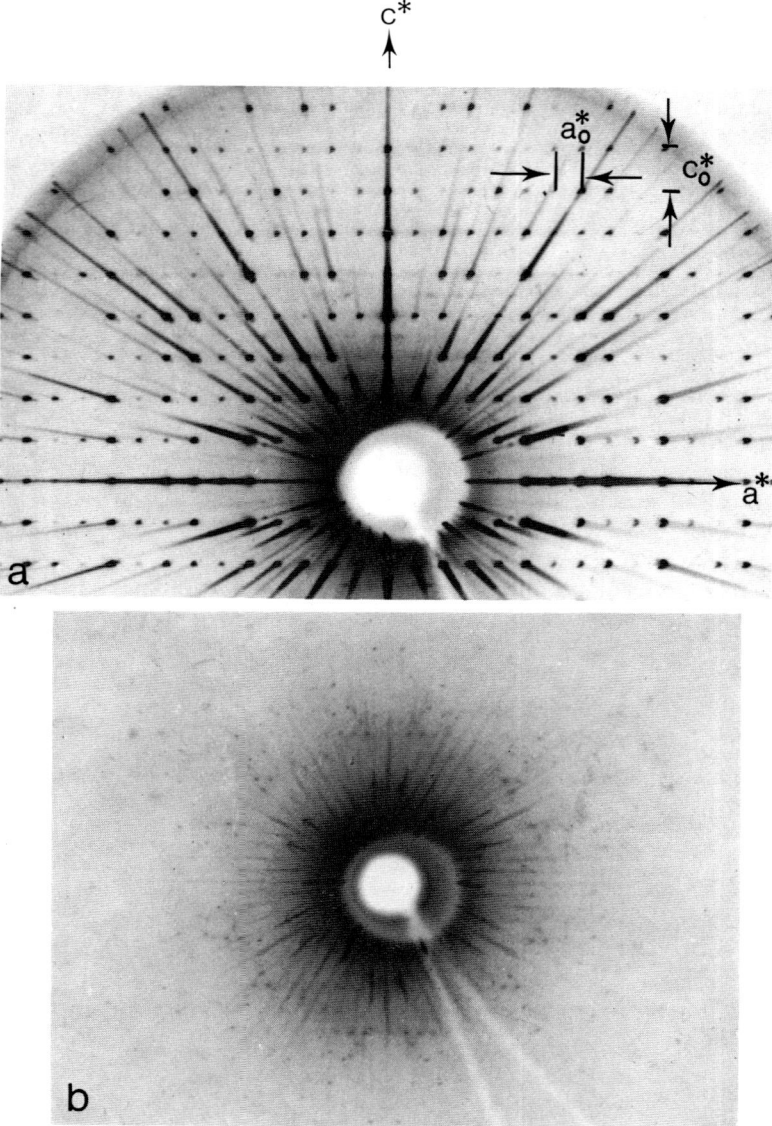

Fig. 3a and b. Single crystal X-ray precession photographs of the superstructure in wenkite: (a) Reciprocal lattice plane $(h0l)$ with weak super-reflections (one week exposure), (b) (hkl) plane recorded for $l=1$ in the $2a_0 \times 2c_0$ supercell (two weeks exposure). Notice hexagonal symmetry

In each basic unit cell $(a_0 \times b_0 \times c_0)$, there are two whole T(2) tetrahedra. For the calculations, both of these tetrahedra were assigned as $+z$ or $-z$. Eight of these basic unit cells (characterized $+z$ or $-z$) make up a supercell

$2a_0 \times 2b_0 \times 2c_0$. There are 20 stacking schemes for eight such cells that are unrelated by simple translation.

The program NUCLS6 (Busing et al., 1962, modified by Ibers and Raymond) was used to compute the structure factors based on these 20 supercell models for 56 reflections. These super-reflections were chosen from the strongest ones visible on X-ray photographs and from ones expected if there were a supercell. All but the coordinates of the T cation (computed as T = Si), and the OH (which occupies the apex of the T(2) tetrahedron) were held constant. The Si and OH are the only members of the T(2) tetrahedron that would contribute to superlattice intensities.

Those super-reflections which had a computed intensity of more than 14 square electron units (based on $I_{000} = 1000$ square electron units) were counted for each of the 20 models and compared with the number of such super-reflections observed. Four models gave high ratios S = (number of observed super-reflections)/(number of calculated super-reflections). They are shown in Fig. 4. Because the superlattice reflections in the X-ray photographs (Fig. 3b) display at least three-fold symmetry, this trigonality must be preserved in the computed models. This is only true for model (c). Model (c) has a high ratio S with 12 observed reflections out of 25 predicted by calculations. Hence model (c) appears as a likely candidate for a model of a superstructure in wenkite. The new unit cell $2a_0 \times 2b_0 \times 2c_0$ is outlined in Fig. 4c with the origin at the three-fold axis at the junction of the three B asymmetric units.

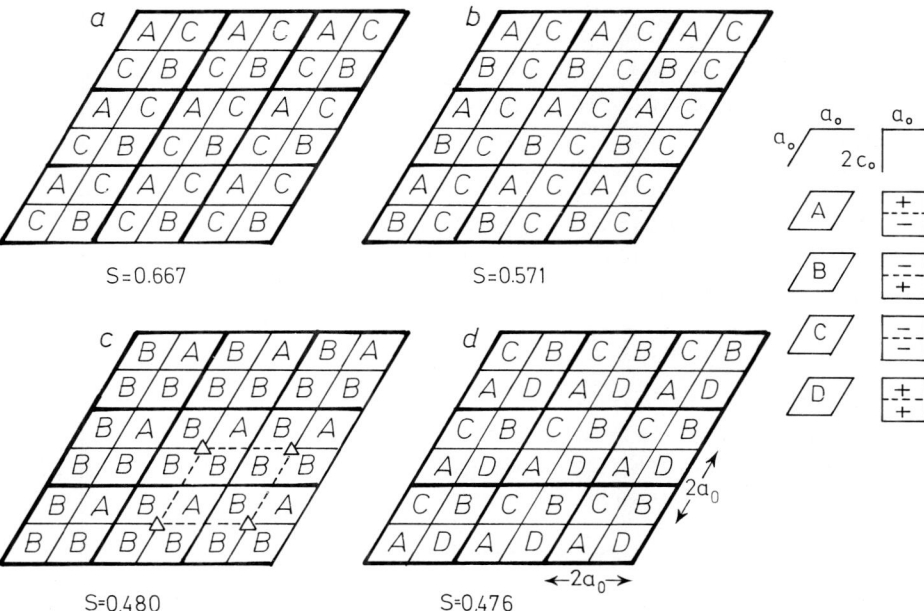

Fig. 4a–d. Models for a superstructure in wenkite. (c) Fits the crystal reflection data best with a supercell defined by the dotted lines. S indicates the ratio of observed to calculated super-reflections (see text)

4. Transmission Electron Microscopy

The TEM experiments (see Figs. 5 and 6) were conducted to look for the sub-microscopic domains that had been suggested from the X-ray diffraction evidence. The wenkite specimens from the type locality Cava Mergozzoni, Candoglia, Italy were furnished by Prof. H.-R. Wenk. They were prepared from crushed grains and petrographic thin sections as described in the literature (McLaren and Phakey, 1965; Radcliffe *et al.*, 1970). Several aspects of these micrographs will now be discussed.

4.1 Wenkite Superstructure

The indexed electron diffraction patterns (SAD) revealed extra reflections at one-half the distances calculated for the basic $a_0 \times b_0 \times c_0$ unit cell (Fig. 5b, d). Since this superstructure is not stable under irradiation by electrons, the lattice constants of the $a_0 \times b_0 \times c_0$ basic unit cell serve best to describe the mineral wenkite.

4.2 Superstructure Instability

The SAD pattern in Fig. 5b shows that all reflections are initially of comparable intensity. After less than one minute of exposure to a 100 kV electron beam, the superlattice reflections (hkl, h, k, or l odd) become progressively less intense (Fig. 5d), split up, and (as has been seen in other specimen areas) eventually become extinguished. If the specimen has not buckled because of beam heating, the initial strength of these reflections may mean that all the atoms are originally in special sites in a $2a_0 \times 2b_0 \times 2c_0$ supperlattice such as those in which the Si and OH have been located. Exposure to any high energy and high frequency radiation (such as that of an electron beam) produces displacement damage, an elastic effect which ejects atoms from the crystal and progressively destroys the order, with the Si and OH of the T(2) tetrahedron becoming disordered last in the process (Fig. 5c, d). It is possible that high-energy X-rays (W-radiation) produce similar damage since they cause coloration in the crystals.

4.3 Lamellar Structures

The imaging of lamellar structures in Fig. 6 does not depend on superlattice reflections. If they are antiphase domain boundaries (APB's) then the APB vector **R** is not one of the simple values $\frac{1}{2}[001]$, $\frac{1}{2}[100]$, $\frac{1}{2}[101]$, $\frac{1}{2}[110]$, or $\frac{1}{2}[111]$

Fig. 5a–e. Image and diffraction effects on sinuous boundaries caused by beam damage. (a) and (b) are before prolonged exposure, and (c) and (d) are after exposure. (a) and (c) are bright field images. The SAD's are indexed on the $2a_0 \times 2c_0$ supercell. (Crushed grains; 100 kV). (e) Unfringed sinuous boundaries in wenkite (dark field). (Thinned foil; 650 kV)

The Submicroscopic Structure of Wenkite

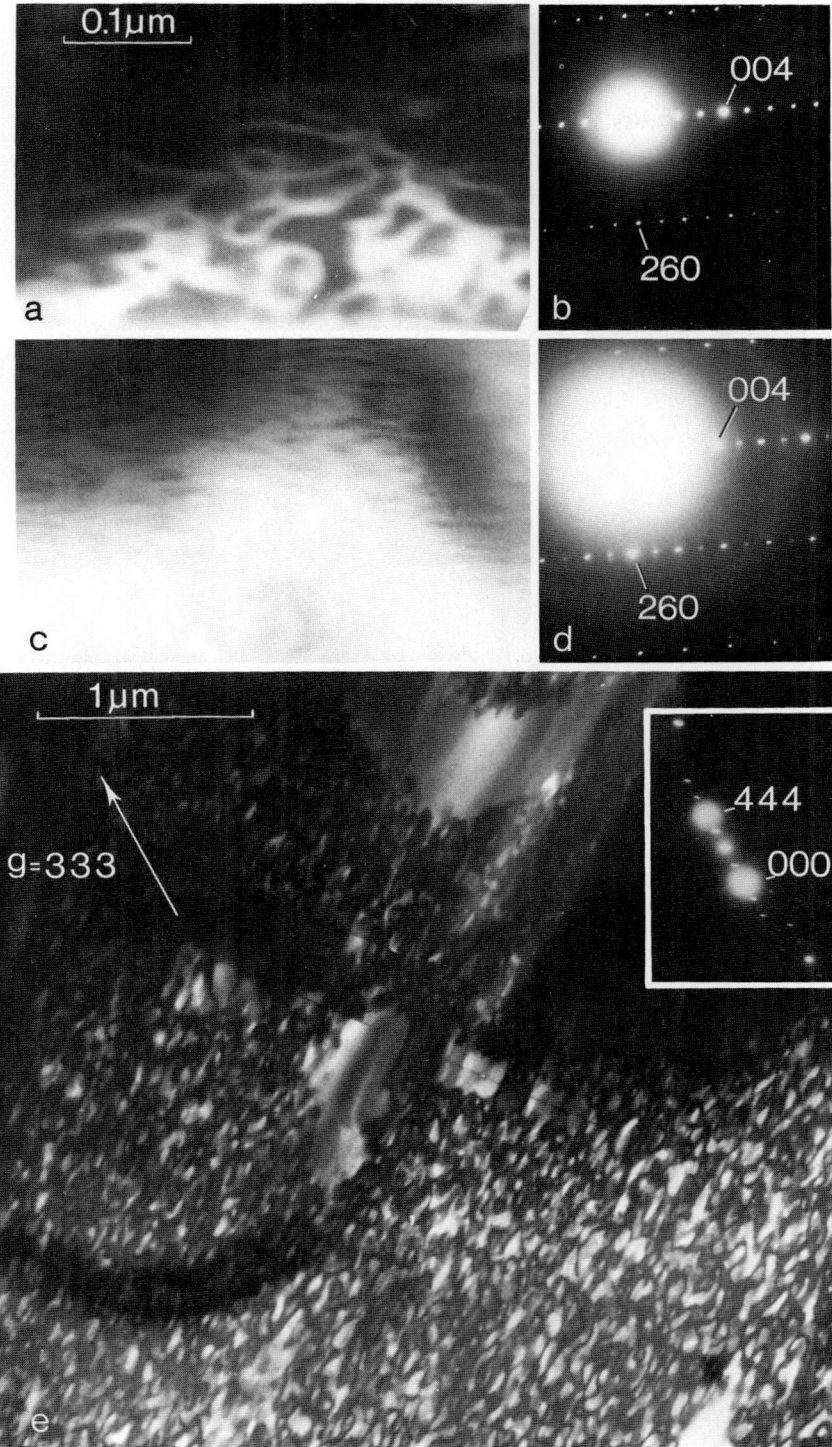

Fig. 5a–e

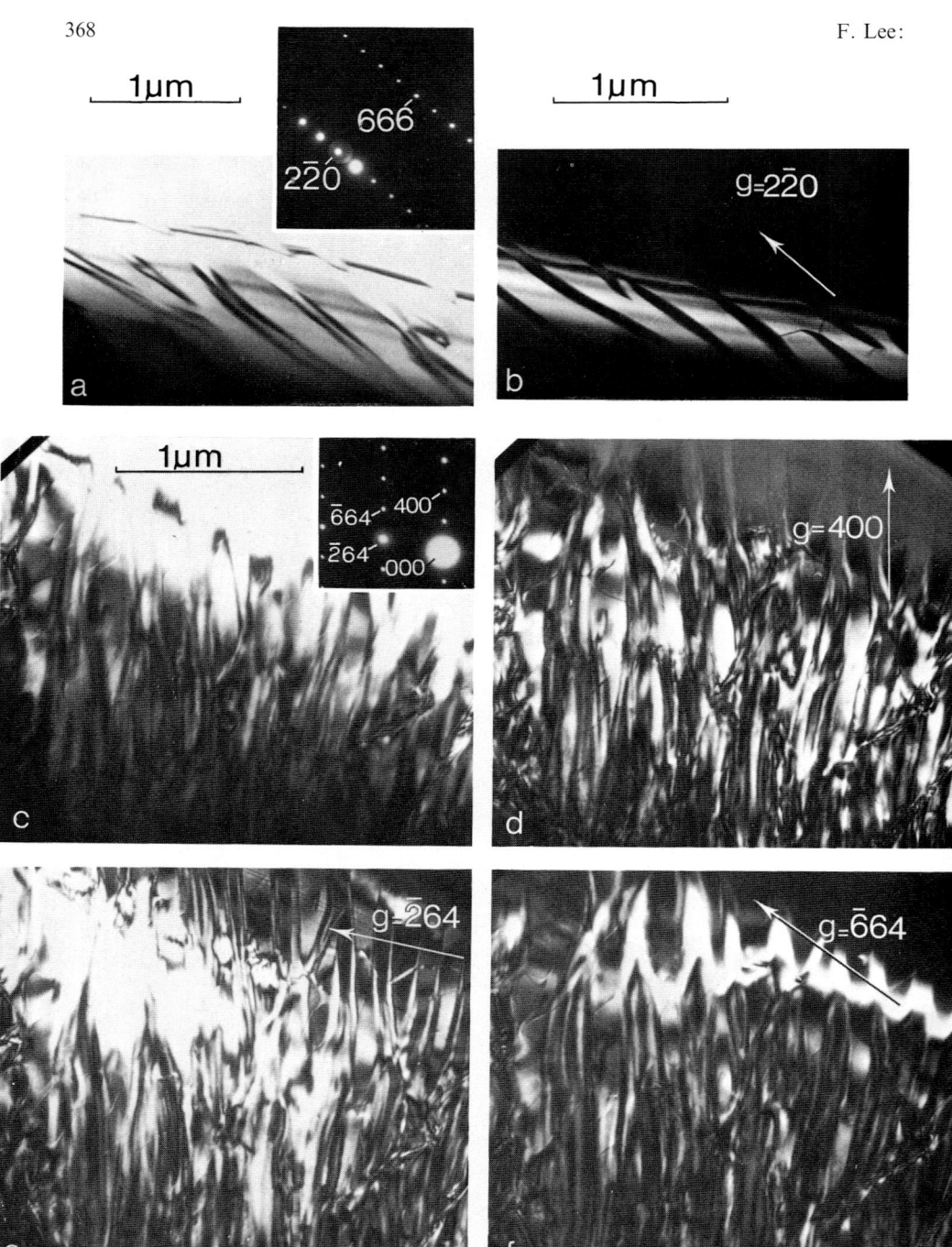

Fig. 6a and b. Fringed lamellar features in wenkite in (a) bright field and (b) dark field. (Crushed grains; 100 kV.) (c–f) Fringed sinuous boundaries superimposed on lamellar features. (c) Bright field and SAD. (d–f) Dark field with various reflections operating. (Thinned foil; 650 kV)

since $g \cdot R$ for any of these R's and the imaged g's is integral, i.e., $\alpha = 0$ (Table 1). The fringe symmetry between bright field and dark field suggests that these lamellar structures may be δ boundaries, i.e., twin boundaries resulting from polysynthetic twinning on a submicroscopic scale (Fig. 2b). No extra reflections need appear because of the twin-host coincidence that is characteristic of the hexagonal lattice.

4.4 Sinuous Structures

Of the two sets of sinuous structures, those enclosing areas less than 500 Å wide show no fringes to characterize the boundaries (Fig. 5a, e) but can only be imaged with superlattice reflections. Since APB's are typical of superstructures such as occur in wenkite, the labeling of these domains as such appears reasonable. A large extinction distance can account for the absence of fringes here.

The second set of sinuous structures (Fig. 6c–f) is fringed and is in contrast throughout the bright and dark field images, though no superlattice reflections are apparent. They cross the lamellar features without change in curvature. These boundaries enclose much larger areas than do those that are not fringed. The extinction distance is shorter than for these others since fringes are present. These are possible APB's with a displacement vector that is a smaller fraction of the ones mentioned before, or that is more complex.

4.5 $\alpha = 2\pi\, g \cdot R$ Calculations

The results from the micrographs are pooled for evaluation. In Table 1, the $\alpha = 2\pi g \cdot R$ values are listed in the first five columns for simple displacement vectors. The last three columns are a tally of the dark field images that are in contrast for particular reflections.

The vectors $\frac{1}{2}[001]$, $\frac{1}{2}[100]$, and $\frac{1}{2}[111]$ identically suggest either π type contrast (APB's) or no contrast at all, depending on the imaging reflection. The reflections for which $\alpha = \pi$ coincide with the reflections for which sinuous unfringed APB's are in contrast. $\frac{1}{2}[001]$, $\frac{1}{2}[100]$, and $\frac{1}{2}[111]$ are all identical lattice vectors for the model shown in Fig. 4c and are evidently responsible for the unfringed APB's (Fig. 5).

Table 1. $\alpha = 2\pi\, g \cdot R$ for the cases of simple displacement vectors R

$R=$	$\frac{1}{2}[001]$	$\frac{1}{2}[100]$	$\frac{1}{2}[101]$	$\frac{1}{2}[110]$	$\frac{1}{2}[111]$	Feature in contrast?		
g						Sinuous unfringed	Sinuous fringed	Lamellar
333	π	π	0	0	π	yes	no	no
2$\bar{2}$0	0	0	0	0	0	no	no	yes
400	0	0	0	0	0	no	yes	yes
$\bar{2}$64	0	0	0	0	0	no	yes	yes
$\bar{6}$64	0	0	0	0	0	no	yes	yes
111	π	π	0	0	π	yes	no	no

The unfringed APB's and the lamellar structures (twins?) do not owe their contrast to any of the simple **R** vectors suggested here. They are probably the result of less simple lattice displacement, or depend on the other possible variables (the SO_4 tetrahedron, the Al-Si positional disorder, and ion exchange) not considered in the superstructure model used. In their texture they resemble exsolution lamellae, but on crystallographic grounds this is unlikely.

References

Busing, W.R., Martin, K.O., Levy, H.A.: A FORTRAN crystallographic function and error program. ORNL-TM-305 (1962).

Lee, F.: The submicroscopic structure of wenkite, a transmission electron microscopy study. M.S. Thesis, University of California, Berkeley (1973, unpubl.).

McLaren, A.C., Phakey, P.P.: A transmission electron microscope study of amethyst and citrine. Austr. J. Phys. **18**, 135–141 (1965).

Papageorgakis, J.: Wenkit, ein neues Mineral von Candoglia. Schweiz. Mineral. Petrog. Mitt. **42**, 269–274 (1962).

Radcliffe, S.V., Heuer, A.H., Fisher, R.M., Christie, J.M., Griggs, D.T.: High-voltage (800 kV) electron petrography of the type B rock from Apollo 11. Proc. Apollo 11 Lunar Sci. Conf. Geochim. Cosmochim. Acta, Suppl. 1, **1**, 731–748 (1970).

Wenk, H.-R.: The structure of wenkite. Z. Krist. **137**, 113–126 (1973).

Wenk, H.-R., Lee, F.: Wenkite, an interrupted framework silicate with domain structure (abs.). Geol. Soc. Am. Abst. with Progr. **4**, 7, 703–704 (1972).

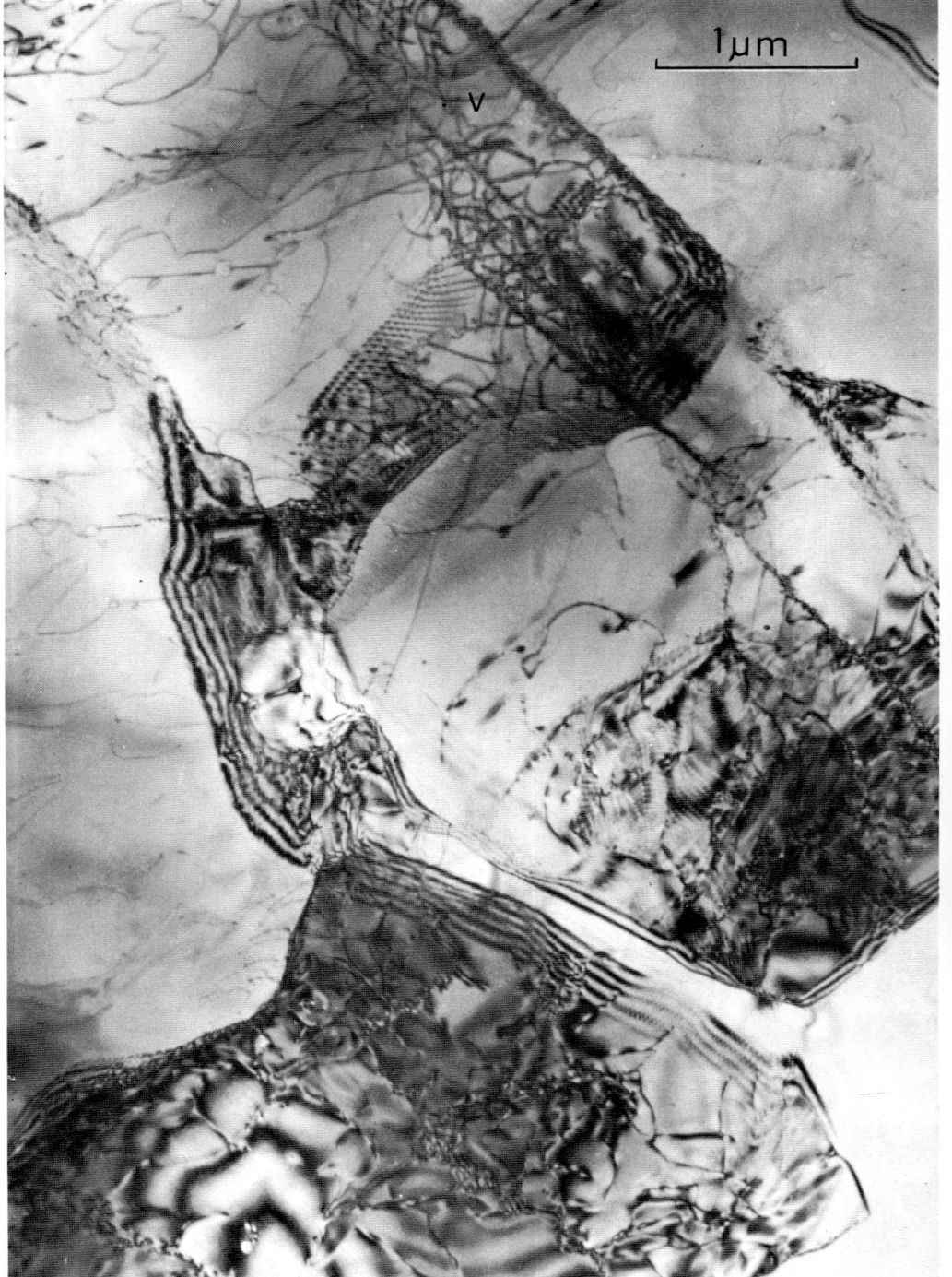

Section 6 Deformation Defects

◄ Dislocation substructures and bubbles (probably filled with H_2O) in a naturally deformed quartzite from the Arltunga Nappe Complex, Central Australia. Notice recovery structures such as networks, dipoles and low angle boundaries. (Photograph by J. Boland)

CHAPTER 6.1

Deformation Structures in Minerals

J.M. CHRISTIE and A.J. ARDELL

1. Introduction

Although it is four decades since it was proposed (Orowan, 1934; Polanyi, 1934; Taylor, 1934) that linear defects or dislocations in crystals play a major role in the plastic deformation of crystalline solids, it is little more than two decades since the existence of dislocations in crystals was established and their motions clearly associated with deformation. Direct observation of dislocations and associated deformation structures in very thin specimens by transmission electron microscopy (TEM), initially achieved by Hirsch et al. (1956) and Bollman (1956), has proved to be the most fruitful method of study. It is interesting to note that mineral crystals figured significantly in these early developments; for example, the first observations of surface growth steps at the emergence of screw dislocations on crystal surfaces were made in natural beryl crystals (Griffin, 1950); and some of the classic early TEM studies of dislocations, partial dislocations and stacking faults in layer structures were made on graphite and talc, muscovite and chlorite (Amelinckx and Delavignette, 1960a, b, 1961), in specimens produced by cleavage. However, largely because of difficulties in producing suitably thin "foils" of minerals with poorly-developed cleavage (Tighe, Chapter 3 of this volume) TEM studies of minerals proceeded slowly until late in the 1960's.

The minerals in which deformation structures have been most intensively studied by TEM are quartz, the pyroxenes and olivine. Fortunately, these minerals, or solid-solution series of minerals, are representative of three of the major groups in the structural classification of silicates (Bragg, 1937) (that is, the framework silicates[1], the chain silicates and the orthosilicates, respectively). Among the other quantitatively important mineral groups, such as the phyllosilicates, oxides, and the rhombohedral carbonates, there have been relatively few TEM studies of deformation structures; exceptions are the early studies of Amelinckx and Delavignette, noted above, on some of the layer silicates, the initial studies of deformed calcite and limestone by Barber and Wenk (1973 and Chapter 6.5 of this volume) and HVEM studies of deformation structures in the lunar breccias (e.g. Christie et al., 1973).

---

[1] Quartz and other silica polymorphs are included among the framework silicates because of the similarity of structure.

Following a brief discussion of the range of conditions under which minerals and rocks may be deformed in the earth and of the TEM techniques used in the examination of deformation structures, we shall review these applications and attempt to make some generalizations regarding the submicroscopic aspects of mineral deformation.

2. Conditions of Mineral Deformation

The deformations of geological and geophysical interest in the earth range over at least twenty orders of magnitude in strain-rate, from shock deformation associated with meteoritic impact, at strain-rates of 10^7 to 10^9 sec^{-1} and stresses in the Megabar range, to the slow "tectonic" deformations in the earth's crust and upper mantle responsible for mountain-building and the motions of lithospheric plates over the asthenosphere; these deformations commonly take place at moderate to high temperatures and pressures (depending mainly on depth) and are believed to occur at rates ranging from 10^{-10} to 10^{-15} sec^{-1}, under stresses less than a kilobar. It is still a moot question whether there is a lower limit for flow rates in solid rocks, but there is some evidence that rocks may possess a "fundamental strength", or limiting stress below which no flow will occur (Griggs, 1936, 1974).

Field and petrographic studies of the structures resulting from these deformations have stimulated experimental work on rock and mineral deformation, aimed at determining the mechanical properties of rocks and elucidating the mechanisms of flow of mineral crystals and aggregates over the experimentally accessible range of conditions (see, for example, Griggs and Handin, 1960; Heard et al., 1972). Rapid loading can be simulated experimentally to some extent by shock deformation experiments on rocks (Müller and Hornemann, 1969; Hörz and Ahrens, 1969; Stöffler, 1972, 1974) and the shocked products may in some instances be recovered for study and comparison with deformed rocks from impact sites on the terrestrial and lunar surfaces. Slower deformations are studied within the range of engineering capability in "static" tests. These tests may be conducted at temperatures up to the melting points of silicates and at "confining pressures" up to several tens of kilobars; the high pressures are employed partly to simulate increased overburden pressure but are often essential to suppress fracture and increase the ductility of brittle silicates. Stress differences up to ~ 50 kbar may be applied to specimens and strain-rates as slow as 10^{-9} sec^{-1} may be measured reliably in creep tests in some types of apparatus. The flow rates of the slower natural deformations are, of course, inaccessible and a major objective of the experimental studies is to determine the flow laws and deformation mechanisms of rocks over a sufficiently wide range of temperature, pressure, stress and strain-rate to permit extrapolation to the conditions and time-scales of geological deformations.

Such extrapolations require the assumption that the deformation mechanisms do not change over the range of extrapolation. The identification of the flow mechanisms, before the advent of TEM techniques, depended on the results of optical (petrographic and metallographic) studies of the microstructures and textures resulting from deformation, and the justification for extrapolation from laboratory data to natural deformations rested on comparison of the microstructures in naturally deformed rocks with those in laboratory specimens. It is now clear that the deformation of minerals, like metals and other crystalline materials, is controlled by the generation and motion of dislocations which are easily imaged by TEM. Thus TEM provides a convenient means of analyzing the submicroscopic structures of deformed rocks and determining the fundamental mechanisms of flow. It also provides a much more sensitive tool for comparing the deformation structures of experimentally deformed minerals and natural tectonites so that similarities or differences in the deformation mechanisms may be established with more confidence.

3. The Role of Defects in Deformation

From optical studies of deformed minerals, the microscopic structures resulting from deformation are well known. They include *slip bands* and other evidence of crystallographically-controlled slip, such as kink bands and bending (undulatory extinction); *mechanical twinning* and *shear* (or *martensitic*) *transformations*. High-temperature deformation commonly results in development of *sub-grain structures* or in *recrystallization* of aggregates. It is now recognized that the plasticity of crystals depends entirely on crystal defects. The most important are the linear imperfections or *dislocations* whose motions through a crystal produce small displacements and strains. The migration of point defects (diffusion) is also significant at high temperatures ($T > 0.6 T_m$), either in association with dislocations and contributing to their motion (dislocation climb) or independently, giving rise to changes in crystal shape, as in the Coble (grain-boundary diffusion) and Nabarro-Herring (volume diffusion) mechanisms of creep (Ashby, 1972).

For a detailed treatment of dislocations and their physical properties and behavior, the reader is referred to the standard works on dislocation theory (e.g. Read, 1953; Cottrell, 1953; Friedel, 1964; Hirth and Lothe, 1968). Here we shall discuss the elementary characteristics of dislocations only in sufficient detail to show how they are imaged and characterized in the electron microscope.

The structure of dislocations is illustrated in Fig. 1. The material above the crystal plane *P* (Fig. 1a) and inside the loop $s\,e\,s'\,e'$ is uniformly displaced with respect to the material below, by an amount specified by the vector **b**. The line between the displaced and fixed regions of the plane is a dislocation line or dislocation loop and the displacement vector **b** is the Burgers vector of the dislocation. Because of the periodic force field in a crystal, the atoms slip from one position of equilibrium to another, so that the displacement **b** generally corresponds to a full lattice vector in the crystal structure. The configurations of the lattice planes perpendicular to **b** at *e* and *s* are illustrated in Fig. 1b and c, respectively. Where the Burgers vector is perpendicular to the dislocation line (at *e* and *e'*) the structure corresponds to the insertion of a "half-plane" of atoms (of thickness $|b|$) above and below the slip-plane *P*, respectively, at *e* and *e'*; this is an *edge* dislocation. Where the Burgers vector and

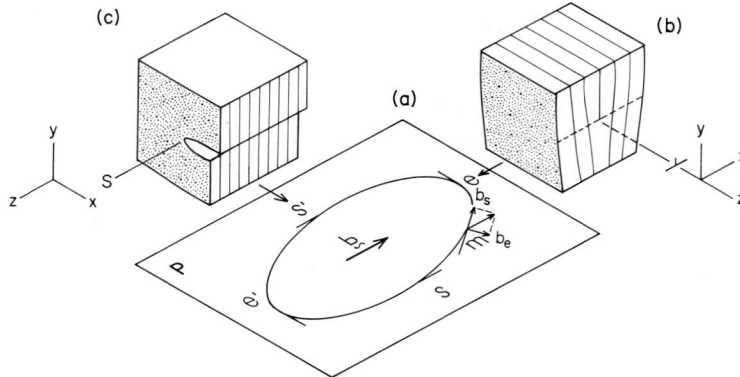

Fig. 1. (a) Dislocation loop with Burgers vector b in the plane P. (b) Structure of an edge dislocation. (c) Structure of a screw dislocation. For fuller explanation see text

dislocation line are parallel (at s and s') the lattice planes perpendicular to b are transformed into a continuous helical surface around the dislocation line; these are *screw* dislocations, of opposite sign at s and s'. Elsewhere round the loop (as at m) the dislocation is *mixed*, with both edge and screw components, and the distortions are correspondingly more complex.

Except in a small region near the dislocation line (the *core*) the properties of a dislocation in an elastic solid are completely specified by the orientation of the dislocation line, the Burgers vector b and the elastic coefficients of the solid. Hence, with suitable specification of the boundary conditions, the displacement field, strain and stress fields may be determined. (In a continuum, the dislocation line is a singularity, where the stresses become infinite, and the region near the line cannot be treated by elastic theory; in crystals the core region, considered to be a small multiple of $|b|$ in dimensions, is a region of uncertain structure and properties.)

It can be shown that in an elastically isotropic medium the displacement field of an edge dislocation has the form $R=[r_x, r_y, 0]$ in the coordinate system of Fig. 1b and that of a screw dislocation is of the form $R=[0, 0, r_z]$ in the coordinate system of Fig. 1c. Thus the atomic displacements in a hypothetical elastically isotropic crystal are such that the atom planes perpendicular to the dislocation line (z) remain planar for an edge dislocation and all other atom planes are bent; for a screw dislocation, all atom planes parallel to the dislocation line (z) are planar. For a mixed dislocation the displacements are again more complex; there are non-zero displacements parallel to all three coordinates, though the component in the plane P perpendicular to b is small. In this case (Fig. 1a) the components of b parallel and perpendicular to the dislocation line are referred to as the screw and edge components of the dislocation, respectively. These summary features are important for understanding the mechanism of imaging of dislocations by diffraction contrast in an electron beam, discussed below, and for the interpretation of *diffraction contrast experiments*, which permit us to identify the Burgers vector of individual dislocations.

The strain field of edge dislocations is strongly dilational, with compressional regions corresponding to the extra half-planes and symmetrical dilational regions on the opposite side of the slip plane. This accounts for the interaction of edge (and mixed) dislocations with the fields of point defects, which may diffuse to or from the edge of the half-plane, causing *climb* of the dislocations out of their slip planes. The strain field of screw dislocations, by contrast, is non-dilational (in the isotropic case), characterized by pure shear, so that they do not interact with point defects (unless the latter produce non-spherical distortions). Hence pure screw dislocations do not climb. Because of the symmetry of their strain fields however, they are not strongly tied to specific slip-planes in a crystal and may *cross-slip* from one slip-plane to another which contains the Burgers vector. Cross-slip and climb are characteristic of dislocation motion at high temperatures, in contrast to the *conservative* motion of dislocations in their slip planes that occurs at low temperatures. The elastic interactions between the stress fields of different dislocations result in attractive and repulsive forces between them. During deformation or annealing at high temperatures, the free energy of a deformed crystal may be reduced by elimination of dislocations (whose strain energy, proportional to $|b|^2$, contributes to the free energy of the crystal) or rearrangement into low-energy configurations (sub-grain boundaries and networks). This process of *recovery* is achieved by attraction and annihilation of dislocations of like Burgers vectors and opposite sign by climb.

The metallurgical term *cold-working* refers to plastic deformation at temperatures appreciably below that at which recrystallization occurs: recovery is slight or absent. The deformation is generally characterized by *work-hardening*: successive increments of strain require higher applied stresses. During work-hardening there is often a marked increase in dislocation density; several processes may be responsible for the restricted motion of dislocations associated with the hardening, such as "pinning" of the dislocations at grain boundaries or precipitates, or restriction of their motion by obstacles such as other dislocations, precipitates and boundaries. At temperatures near or above that at which recrystallization occurs, deformation is referred to in metals as *hot-working*. Hardening is less significant or absent and deformation is accompanied by extensive recovery and recrystallization. Recovery gives rise to sub-grain structures, with dislocation networks forming low-angle ($<1°$) sub-boundaries. Recrystallization occurs by nucleation of strain-free grains in regions of crystals with the highest dislocation densities and migration of the new (high-angle) grain boundaries, under the influence of differences in strain energy, to consume the deformed matrix. The term *creep* is commonly used to denote slow deformation under small stresses, generally under hot-working conditions, but should refer strictly to deformation *under constant stress*. Three stages may be recognized in the strain-time curve in creep tests: *primary* or *transient* creep, characterized by diminishing strain-rate; *secondary* or *steady-state* creep, in which the strain-rate (as well as stress) is invariant with time; and sometimes (in tensile tests) *tertiary* or *accelerating* creep, which generally terminates in creep fracture.

From the large body of observations and theory on metals and other crystals, the characteristic configurations of dislocations resulting from these processes are known. Hence it is possible to infer the operation of specific processes, to some extent, from TEM observations of the density, nature and geometry of dislocations. Such inferences, if they are based solely on analogy with structures in metal crystals, are of course suspect. However, the increasing volume of TEM data on experimentally deformed or annealed mineral crystals and rocks is providing a more direct basis for the interpretation of the mineral sub-structures. These data show that the fundamental processes of deformation, recovery, and recrystallization are at least qualitatively similar in all types of crystalline materials.

Where the Burgers vector *b* is a lattice vector, the dislocation is a *unit* dislocation and its slip over a crystal plane restores the identity of the structure, leaving no trace of its passage. However, in complex structures, especially those with large unit cells and superstructures, dislocations with Burgers vectors equal to a fraction of a lattice vector may be favored (since the energy of a dislocation is proportional to $|b|^2$). These *partial dislocations* cannot exist in isolation in a perfect crystal and it is clear that the motion of a partial through a perfect crystal must leave a *stacking fault* on its slip plane. The motion of similar partial dislocations in parallel slip planes through each unit cell in a structure may change the structure entirely. It is now generally believed that such a process is responsible for mechanical twinning (*twinning partials*) and at least some shear transformations, although the details of the process are still controversial. Examples of these structures are illustrated below and in other papers in this Volume. Partial dislocations are similar in their properties to unit dislocations, with the exception that they are invariably associated with stacking faults, and they need not be treated separately here.

4. Contrast at Dislocations

Dislocations and other deformation defects are generally imaged in TEM by diffraction contrast methods. The theory is outlined in earlier papers in this volume by Van der Biest and Thomas (Chapter 2.1 of this volume) and by Amelinckx and Van Landuyt (Chapter 2.3 of this volume) (see also Hirsch *et al.*, 1965; Amelinckx *et al.*, 1970). High-resolution microscopy (Cowley and Iijima, Chapter 2.5 of this volume) has not so far been employed extensively to study deformation defects in minerals. In distorted regions of a crystal, such as around a dislocation, the displacements of atoms from their normal crystallographic positions results in bending of the diffracting planes. Hence the intensity of the direct and diffracted electron waves in such regions differs from that in a perfect crystal. It is the displacement field of the defects, localized at the fault plane in the case of a stacking fault but long-range in the case of a dislocation, that is recorded by diffraction contrast imaging.

The dynamical theory of diffraction contrast was developed for the "two-beam" case, where one set of crystal planes is diffracting (giving rise to the direct beam and a single strongly diffracted beam in the diffraction pattern), by Hirsch *et al.* (1960) for a perfect crystal and adapted to a distorted crystal by Howie and Whelan (1962). From the formal theory some simple general rules have been derived for the contrast near dislocations in an elastically isotropic crystal for two-beam Bragg diffraction; of particular interest are the special cases where no contrast is observed: these are the *isotropic invisibility criteria*. For a pure screw dislocation, there will be no contrast when $g \cdot b = 0$, where g is the reciprocal lattice vector normal to the diffracting planes and b is the Burgers vector of the dislocation. Contrast will result when $g \cdot b \neq 0$. For invisibility of a pure edge dislocation the condition $g \cdot b \times u = 0$ (where u is a unit vector parallel to the dislocation line) must be fulfilled in addition to $g \cdot b = 0$. As Head (1969) has emphasized, these invisibility conditions correspond to the cases where the dislocations are imaged with "flat planes". This is because displacements within the diffracting plane do not affect the diffracted intensities. Invisibility results when any plane parallel to a screw dislocation is diffracting, but only when the unique flat plane perpendicular to an edge dislocation is diffracting. For a mixed dislocation, strictly speaking, there are no flat planes but, to a sufficiently good approximation, the conditions for an edge dislocation also apply to a mixed dislocation. For a mixed, as for a pure edge dislocation the conditions $g \cdot b = 0$ and $g \cdot b \times u = 0$ can only be simultaneously satisfied for one value of g.

The $g \cdot b \times u$ criterion is sometimes neglected in discussions of contrast because for most low-order reflections in metals the contrast associated with this term in the wave equations is small. However, we note that for many minerals, because of systematic extinctions of some low-order planes ("forbidden reflections"), it may be necessary to employ high-order diffraction vectors and $g \cdot b \times u$ may be large enough to give rise to significant contrast when $g \cdot b = 0$, as discussed by McCormick (Chapter 2.4 of this volume).

These invisibility criteria form the basis for the experimental determination of Burgers vectors of individual dislocations by means of tilting experiments in the electron microscope, if it is assumed that the effects of elastic anisotropy are negligible. The procedure consists of imaging the crystal successively under two-beam conditions using different diffraction vectors g to determine which vector(s) give(s) no contrast at the dislocation (or minimum contrast for edge or mixed types). According to the $g \cdot b = 0$ criterion, which is the one usually employed, at least two values of g for which the dislocation is invisible must be found to identify b uniquely. Since the value of $g \cdot b \times u$ can be sufficiently large for some reflections to result in significant contrast at dislocations with a large edge component, true invisibility can be obtained for only one value of g; indeed, for many combinations of foil orientation, Burgers vector and dislocation line direction, it may be impossible to achieve perfect invisibility. Thus in practice the application of the invisibility criteria involves some subjectivity: when weak contrast is observed, is it due to the fact that $g \cdot b = 0$ but $g \cdot b \times u \neq 0$, or is there another reason for the contrast that we see in such weak or "residual" images?

We illustrate the procedure with the example of an array of straight dislocations in a quartz crystal experimentally deformed by compression perpendicular to $(10\bar{1}1)$ at 750 °C (Ardell *et al.*, 1974a). The dislocations are imaged under two-beam conditions for five different diffraction vectors in Fig. 2. The inset (Fig. 2) shows the traces of three sets of dislocations (AA, BB and CC) in Fig. 2a–c, taken with the zone-axis [$2\bar{4}23$] parallel to the electron beam; the zone-axis for Fig. 2d, e was [$1\bar{2}10$]. The dislocations parallel to AA are in strong contrast for $g = 1\bar{1}02$, $\bar{1}010$ and $10\bar{1}\bar{1}$ (Fig. 2a, c, d) but in very weak contrast for $g = 0\bar{1}12$

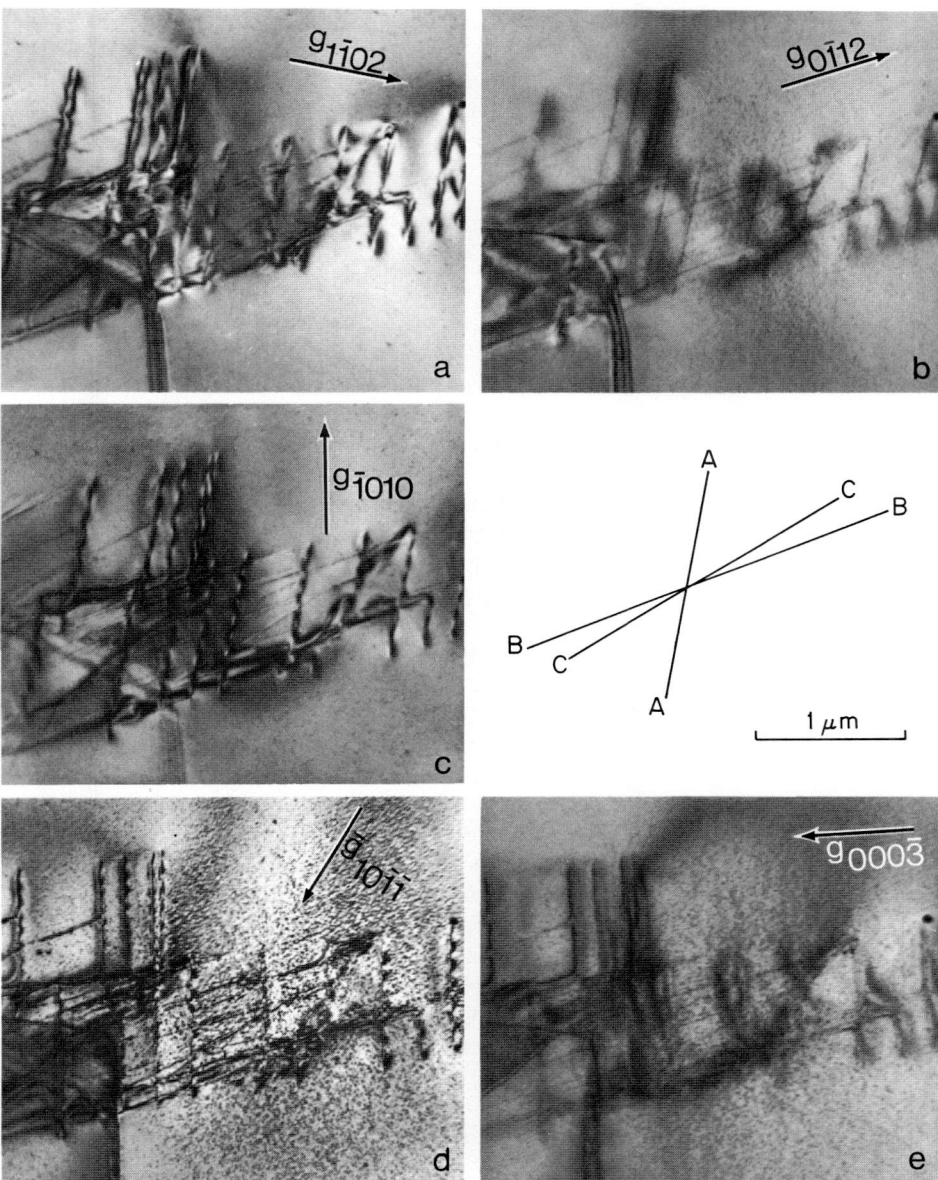

Fig. 2a–e. An array of dislocations in a dry natural quartz crystal, deformed (3%) by compression perpendicular to $(10\bar{1}1)$ at 750° C, $\dot{\varepsilon} = 10^{-6}$ sec^{-1}. Micrographs (a–e) illustrate the contrast obtained for two-beam conditions with five different diffraction vectors **g**, as indicated. AA, BB and CC are the projections of three main orientations of dislocations in a, b and c. See text for explanation. (Reproduced from Ardell *et al.*, 1974a, with permission)

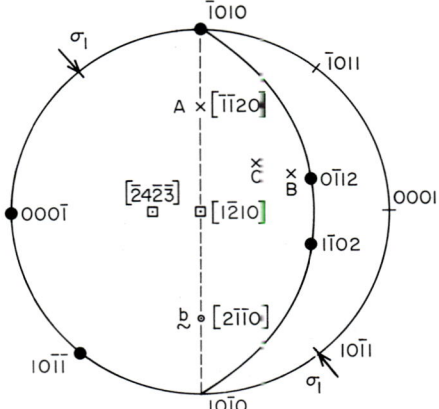

Fig. 3. Stereogram showing the zone axes (squares) that were parallel to the electron beam for the micrographs in Fig. 2. Crosses at A, B and C indicate the orientations of the dislocation groups AA, BB and CC, respectively. Dots represent the orientations of the five diffraction vectors (*g*) employed to obtain the micrographs in Fig. 2. σ_1 is the compression axis and ***b*** is the Burgers vector obtained for the dislocations parallel to A. Note that there is an ambiguity in the orientations of the dislocations, as the top and bottom of the foil are not distinguished. Indexing of crystal directions is for the upper hemisphere of the projection

and $000\bar{3}$ (Fig. 2b, e). The stereogram in Fig. 3 shows the zone-axes $[\bar{2}4\bar{2}3]$ and $[\bar{1}2\bar{1}0]$ employed in the experiment and the orientations of the five diffraction vectors ***g*** in Fig. 2, along with the orientations A $[\bar{1}\bar{1}20]$, B $[0\bar{1}12]$ and C $[0\bar{1}11]$ of the dislocation lines, obtained by trace analysis; the orientation of the compression axis (σ_1) is also shown. If the dislocations are considered to be out of contrast for ***g***$=0\bar{1}12$ (Fig. 2b) and ***g***$=000\bar{3}$ (Fig. 2e), the only Burgers vector that simultaneously satisfies ***g***·***b***$=0$ for both values of ***g*** is parallel to $[2\bar{1}\bar{1}0]$. Since they are presumably unit dislocations, ***b***$=\pm\frac{1}{3}[2\bar{1}\bar{1}0]$. Since the dislocations are parallel to A $[11\bar{2}0]$, they are of mixed type, with the line inclined at 60° to the Burgers vector.

Contrast experiments on olivine (Boland et al., 1971; Phakey et al., 1972; Green and Radcliffe, 1972a), quartz (Ardell et al., 1974a) and pyroxenes (Kirby, Chapter 6.7 of this volume) suggest that the isotropic criteria may be employed to determine Burgers vectors in these minerals but there are difficulties (e.g. Ardell et al., 1974b) and the results may be controversial. The effects of (a) elastic anisotropy and (b) the long extinction distances ξ_g of diffracting planes in many minerals should be examined by computer simulation of dislocation images (Head et al., 1973; McCormick, Chapter 2.5 of this volume) to ascertain the extent to which the invisibility criteria are applicable to specific minerals.

5. Deformation Structures of Minerals

5.1 Quartz

The deformation substructures of quartz have been studied by TEM in naturally and experimentally shocked rocks, in quartzites and natural and synthetic crystals experimentally deformed at elevated temperatures and strain rates in the range 10^{-4} to 10^{-7} sec^{-1} and in naturally deformed quartzites.

The structure of quartz is trigonal and enantiomorphic ($P3_12$, $P3_22$). The SiO_4 tetrahedra all share each oxygen atom with a neighboring tetrahedron, so that the structure consists of a three-dimensional framework of tetrahedra connected by only Si-O bonds. Thus the structure is strongly and relatively uniformly bonded in all directions. This is reflected in the very high strength of most natural quartz crystals, which are also extremely brittle. In spite of repeated attempts to induce plastic deformation in quartz over a period of many decades, large plastic deformations were first obtained quite recently, at relatively high temperatures and confining pressures (Carter et al., 1964; Christie et al., 1964). It was soon demonstrated that minor amounts of water dissolved in the quartz structure have a very marked effect on both the strength and the slip and recovery mechanisms (Griggs and Blacic, 1965; Griggs, 1967). Subsequent work on synthetic quartz crystals (Baeta and Ashbee, 1967, 1969a, b) showed that these could be deformed at atmospheric pressure and high temperatures.

The slip systems in quartz have been identified in several studies (Christie and Green, 1964; Baeta and Ashbee, 1969a, b). Slip takes place in the a, c and $\langle c+a \rangle$ crystal directions on numerous low-index planes containing these directions, but the dominant slip planes under most conditions are the basal (0001) and $\{10\bar{1}0\}$ prism planes. Parallel-axis twins of the Brazil and Dauphiné laws are common in many varieties of quartz and were observed by TEM in amethyst and citrine (McLaren and Phakey, 1965a) and later characterized by diffraction-contrast experiments (McLaren and Phakey, 1966b, 1969). It is well-known that the displacive changes involved in Dauphiné twinning can be produced mechanically (Zinserling and Shubnikov, 1933; Tullis and Tullis, 1972) but it was only recently demonstrated by TEM (McLaren et al., 1967) that Brazil twinning, which is reconstructive, occurs during deformation of dry quartz crystals at high stress and moderate temperatures (500–700 °C).

Dry natural crystals deformed at temperatures of 500–700 °C exhibit strength close to the theoretical value; they contain optical deformation lamellae associated with zones of bending or kinking. TEM of these crystals (McLaren et al., 1967) revealed thin basal Brazil twins formed by shear, either alone or associated with straight dislocations, or in some cases only straight dislocations (as in Fig. 2). It was not clear whether the lamellae were associated with the Brazil twins or arrays of dislocations, since these observations were made on fractured fragments, but later work on ion-thinned foils (Christie and Ardell, 1974) has shown that the lamellae are thin zones of glass, locally associated with dense arrays of dislocations parallel to low-index directions (Fig. 4a, b). The straightness of the dislocations indicates that the Peierls force on the dislocations is strong, causing them to follow low-energy structural configurations.

By contrast, "wet" synthetic quartz containing in extreme cases 10^3–10^4 H/10^6 Si atoms (<0.1 wt. percent) shows quite different mechanical behavior (Griggs, 1967) and the dislocation substructures are also radically different (McLaren and Retchford, 1969). Above a critical temperature (T_c) that decreases with increasing water content, the quartz is anomalously weak. In a crystal containing 5×10^3 H/10^6Si ($T_c \sim 400$ °C) deformation at 300 °C requires high stress and results in high densities ($\geq 10^{12}$ cm^{-2}) of tangled dislocations, even at small strains; this indicates rapid multiplication and tangling of dislocations, consistent with the stress-strain curves, which show marked strain-hardening (McLaren and Retchford, 1969). Annealing of such crystals above T_c resulted

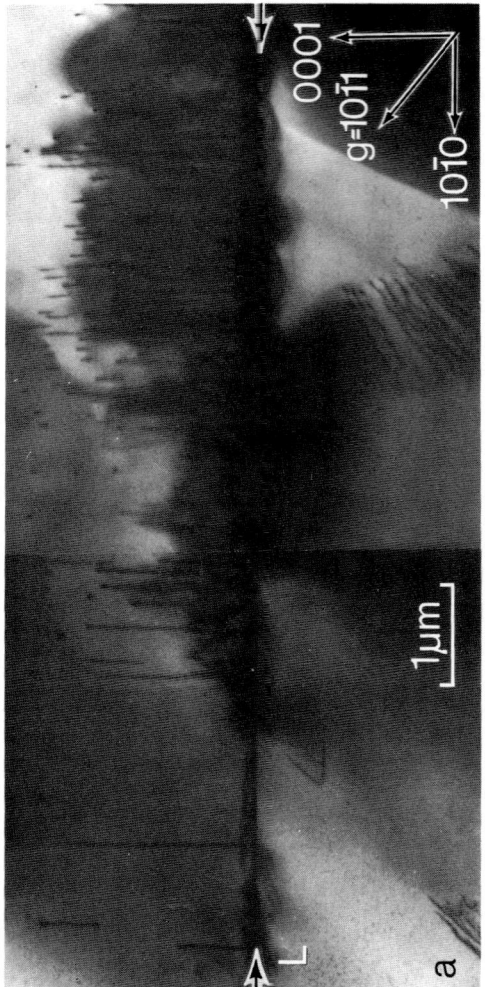

Fig. 4a–e. Deformation structures in quartz. (a) Basal deformation lamella (L) in a dry natural crystal compressed 3% at 750 °C, 10^{-6} sec^{-1}, showing narrow zone corresponding to lamella, with arrays of straight dislocations. (b) Group of closely spaced basal lamellae with a zone of glass (gl) and associated dislocations (d) in a dry natural crystal compressed 1% at 750 °C, 10^{-6} sec^{-1}. (c) Wet synthetic crystal compressed 1.6% at 500 °C and a strain rate of 7×10^{-6} sec^{-1}, showing well-recovered substructure with curved dislocations, small loops (l) and numerous bubbles (b); the latter have strain contrast around them. (d) Sub-grain boundary networks (N) and free dislocations in a quartzite completely recrystallized during compression (65–70%) at 900 °C and 10^{-7} sec^{-1}. (e) Intersecting glassy lamellae (gl) separating crystalline fragments (cr), showing elastic distortion in quartz grain from shocked gneiss from the Ries meteoritic impact structure (Phakey and Christie, unpublished)

in rapid reduction of the dislocation density; prolonged annealing (12h at 425 °C) resulted in a reduction of dislocation density to 10^8 cm^{-2}. Recrystallization occurred after 2 h at 700 °C and was complete in a few minutes at 900 °C. Recovered and recrystallized samples contain low-angle sub-grain boundaries, consisting of generally hexagonal networks of dislocations (as in Fig. 4d). Similar structures are formed during deformation at temperatures above T_c, indicating recovery during deformation. Submicroscopic bubbles or voids resembling those seen in natural milky quartz (McLaren and Phakey, 1965b) are also formed during deformation above T_c and annealing; they are commonly along dislocations or at triple nodes (Fig. 4c). To account for the water-weakening, Griggs and Blacic (1965) postulated that the role of water is to hydrolyze the strong Si-O-Si bonds at the dislocation lines to silanol groups (Si-OH·HO-Si), which

Deformation Structures in Minerals

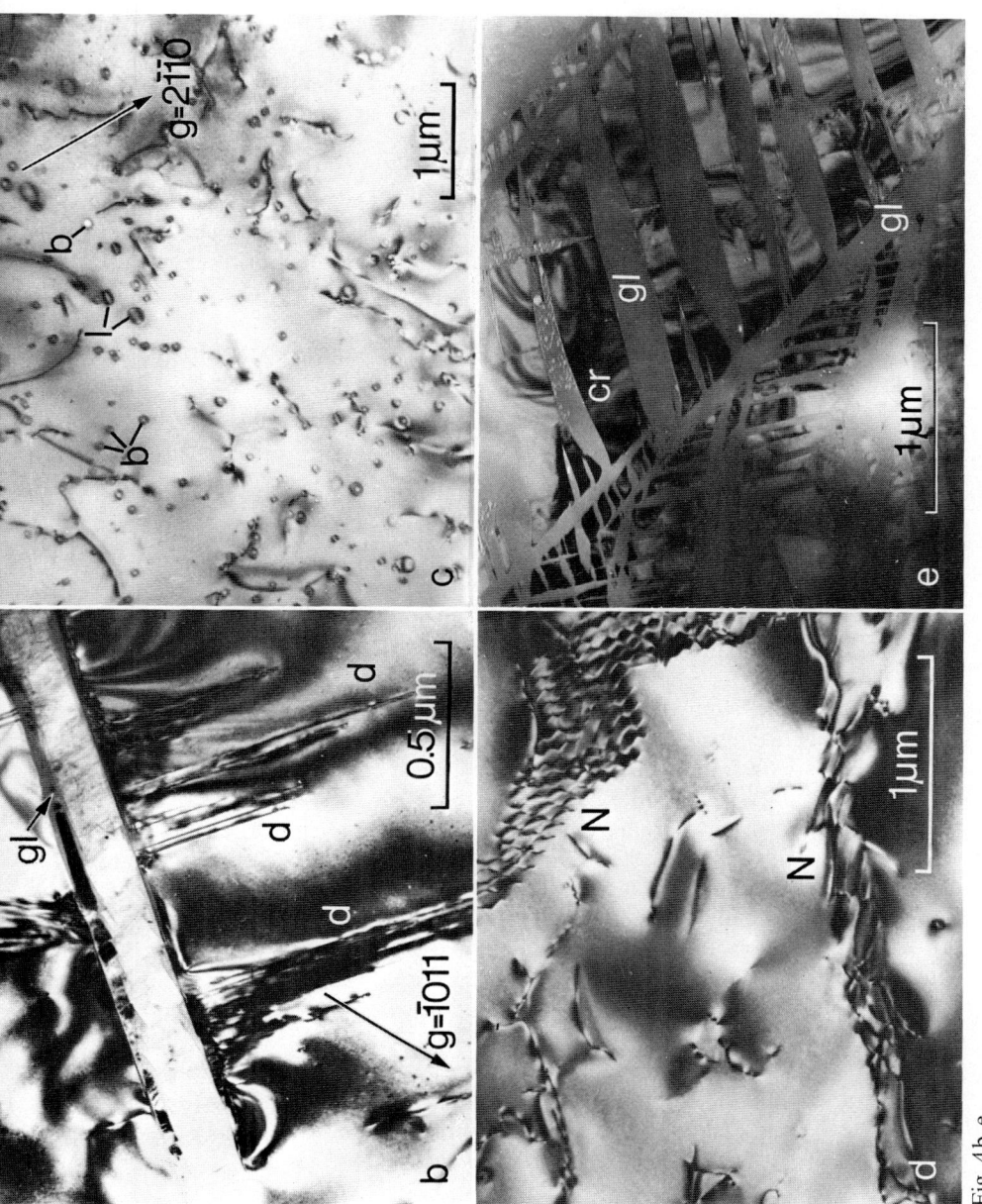

Fig. 4b–e

are more easily exchanged than Si-O bonds, permitting migration of the dislocations. It is clear from the dependence of climb on H_2O content that it also plays an important part in the diffusion of material to and from the dislocations.

Several other experimental studies employing TEM have explored the hydrolytic weakening phenomenon in synthetic crystals of various water contents

(Hobbs, 1968; Hobbs et al., 1972; Baeta and Ashbee, 1970, 1973; Balderman, 1974; Griggs, 1974; Morrison-Smith, 1974; Chapter 6.3 of this volume). Baeta and Ashbee (1970) and Hobbs et al. (1972) identified complex yield-point behavior, and the latter demonstrated that the variation in yield-point characteristics depends on temperature, strain rate and water content of the crystals. They noted the similarity of the stress-strain curves to those observed in materials of diamond structure and discussed the behavior in terms of the microdynamical theory of Alexander and Haasen (1968) for yielding and flow in such materials. This theory accounts for the yield point behavior in terms of the rate of multiplication of dislocations, the dependence of their velocity on stress and a strain-hardening term that is due to elastic interaction between dislocations. The observations of Hobbs et al. (1972) appeared qualitatively consistent with the theory, although the role of H_2O in the process was uncertain. Morrison-Smith's studies indicate extreme variability in dislocation densities within crystals, related to heterogeneity in the distribution of water, and document the development of the dislocation substructure.

On the basis of these studies Griggs (1974) modified the microdynamical theory to incorporate the effect of water concentration on the dislocation velocity and recovery, through specific physical models. By suitable adjustment of parameters, numerical calculations based on Griggs' model provide a remarkably close fit to all of the stress-strain curves available. Clearly a comparison of observed dislocation densities at various stages of strain in crystals deformed under various conditions with those predicted by the model should permit checks on the accuracy of the model and its parameters, but the TEM data are not yet sufficiently numerous to provide such checks.

Quartzites experimentally deformed to large strains at temperatures from 600° to 900 °C at strain rates from 10^{-5} to 10^{-7} sec^{-1} (Tullis et al., 1973) show a wide variety of submicroscopic structures (Ardell et al., 1973). At low temperatures and the faster strain rates, the deformation is very heterogeneous (consistent with optical evidence) and the quartz develops extremely high densities ($>10^{12}$ cm^{-2}) of tangled dislocations. At temperatures above 800 °C at 10^{-5} sec^{-1} and 600 °C at 10^{-7} sec^{-1}, recrystallization occurs in the vicinity of the original grain boundaries; it is important to sample the interiors and boundary regions of the original grains in such specimens, since the substructures are different (Fig. 5a, b). In samples deformed at 900 °C and $10^{-7} sec^{-1}$, recrystallization is virtually complete at high strains and the recrystallized grains have a well-recovered substructure with dislocation densities $\sim 10^8$ cm^{-2} (Fig. 4d; also Ardell et al., 1973, Figs. 9 and 10). The submicroscopic structures of quartzites deformed at the highest temperatures and slowest strain rates in these experiments resemble those of natural quartz tectonites.

It is well known from petrographic work that natural quartz tectonites have generally been partially or completely recrystallized under metamorphic conditions. The degree of internal strain in the grains, revealed by undulatory extinction (bending), deformation bands and deformation lamellae, varies greatly from one rock to another. This is to be expected from the wide range of temperatures and strain rates in tectonic deformations, and the high probability of complexity in their histories, such as annealing after deformation or low-temperature deformation following syntectonic recrystallization at high temperatures. A sedimentary quartzite with few optical deformation structures (Ardell et al., 1973, Fig. 6) showed consistent microstructure in several grains, with widely spaced sub-boundaries and isolated dislocations with a density $\sim 10^8$ cm^{-2}. Metamorphic quartzites have been more extensively studied by TEM to determine the general nature of the substructure and to characterize deformation lamellae and other optical features (White et al., 1971; White, 1971, 1973; McLaren and Hobbs, 1972; Liddell et al., 1974; Chapter 6.4 of this volume). In rocks with no obvious evidence of post-crystalline strain, the grain-boundaries are smooth, meeting at triple junctions at angles close to the equilibrium configuration of 120°, and low-angle sub-boundaries consisting of dislocation networks; dislocation densities are typically in the range 10^8–10^9 cm^{-2}. Quartzites with some optical evidence of strain show higher densities of free dislocations (Fig. 5c), while quartzites with a large amount of undulatory extinction and deformation lamellae may contain very high densities of tangled dislocations (Fig. 5d). A remarkable feature of some partially recrystallized mylonitic quartzites, with very highly strained original grains, is that they show relatively low dislocation densities and considerable evidence of recovery (McLaren and Hobbs, 1972; Liddell et al., Chapter 6.4 of this volume). A comparison of the dislocation structures in natural quartzites with those deformed experimentally (Ardell et al., 1973) reveals that the natural quartzites generally show lower dislocation densities and more extensive recovery, even though they were deformed at lower temperatures (as indicated by the metamorphic facies of the rocks). This is consistent with much slower strain rates (and lower stresses) in the natural deformations. High activity of H_2O during regional metamorphism presumably influences the mechanical behavior and static recovery may be significant in modifying the dislocation structure if temperatures remained high after deformation.

The structure of deformation lamellae in quartz has been a controversial problem. Following the initial suggestion (Christie et al., 1964) that the optical effects result from the long-range stress fields of dislocation arrays approximately parallel to slip planes, TEM studies have revealed that lamellae have very varied substructures. Weak prismatic lamellae in synthetic quartz are zones of tangled dislocations or subgrain boundaries and the optical effects are ascribed to changes of optical path associated with submicroscopic phase objects (McLaren et al., 1970). Some natural lamellae are subgrain boundaries or narrow subgrains (White, 1973), while others are recovered zones, often containing zones of bubbles (Fig. 5d, e). This last type and the experimentally produced lamellae described above (Fig. 4a) clearly have long-range stress fields associated with the dislocation configurations (Christie and Ardell, 1974).

Quartz deformed by shock in meteoritic impact events is characterized optically by reduction of refractive indices and birefringence and frequently contains sets of thin planar features parallel to crystallographic planes; it may be converted completely to "thetomorphic" glass. The electron micrograph (Fig. 4e) of naturally shocked quartz shows parallel sets of submicroscopic glassy SiO_2 layers and lenses separating fragments of crystalline quartz. Similar structures have been observed by TEM in experimentally shocked quartz (Müller, 1969). The total absence of dislocations in the crystal fragments suggests that dislocations

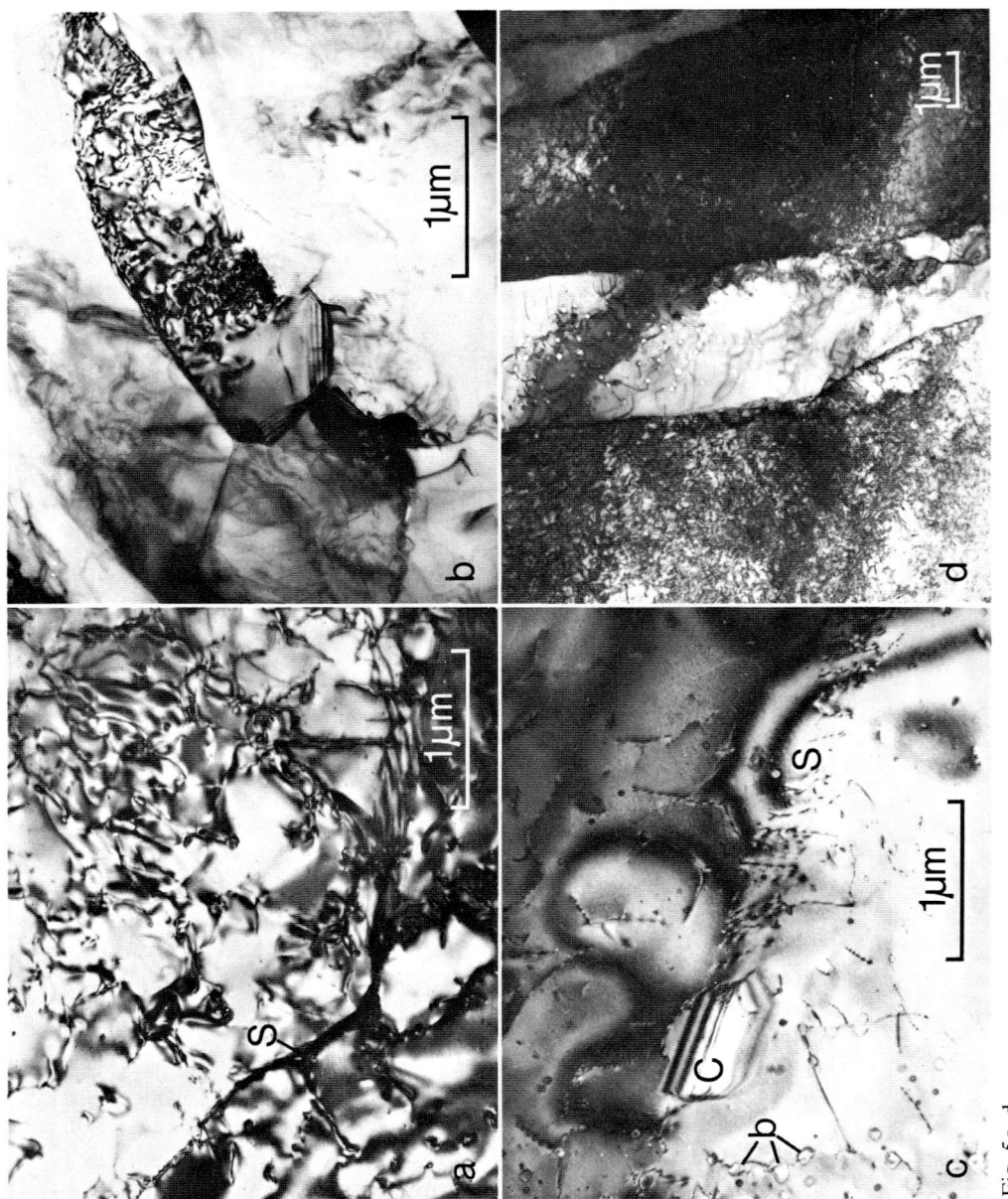

do not play a role in shock deformation. These glass lamellae are also commonly found in shocked feldspars from terrestrial impact sites and in the lunar breccias (Christie et al., 1973; Heuer et al., 1974), so that shock vitrification appears to be characteristic of the framework silicates.

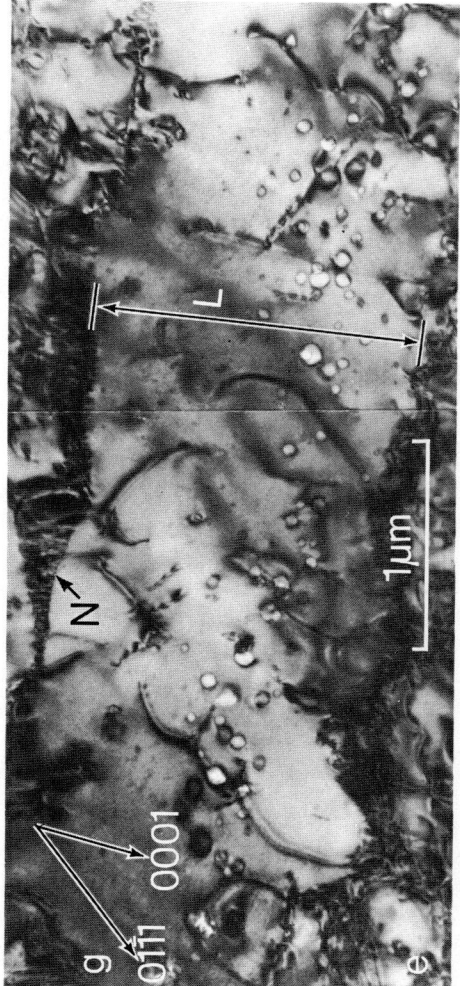

Fig. 5a–e. Deformation structures in quartz. (a) Dislocations and subgrain boundary (S) in one of the deformed original grains in a partially recrystallized quartzite, compressed 45% at 800 °C, 10^{-6} sec^{-1}. (b) Recrystallized region at the boundary between original grains in same sample as (a); note the characteristic fringes on the grain boundaries. (c) Crystal from a metamorphic quartzite (Orocopia schist), showing a subgrain boundary (S), with a large cavity (C) having crystallographic boundaries; smaller bubbles (b) are present, commonly associated with dislocations. (d) Highly deformed Alpine quartzite (Specimen Sci 241, provided by H.-R. Wenk) with deformation lamellae, showing high dislocation density and recovered zones (center, vertical) corresponding to lamellae. (e) Detail of a deformation lamella (L) of the recovered type, with networks (N) at boundaries; note the narrow zone of bubbles within the 1.5 µm thick lamella. Same specimen as (d)

5.2 Pyroxenes

In the more complex structures of the pyroxenes, characterized by layers of parallel Si_2O_6 chains bonded laterally by the metal cations, polymorphism is well-known (Brown et al., 1961). In both orthorhombic and monoclinic pyroxenes the dominant slip system is (100) [001], from optical studies (Griggs et al., 1960; Raleigh, 1965). In addition to slip, a shear transformation occurs during experimental and natural deformation of orthorhombic enstatite, giving rise to lamellae of clinoenstatite parallel to (001) planes and resulting in total conversion to the monoclinic phase at small shear strains ($<15°$) (Griggs et al., 1960; Turner et al., 1960; Trommsdorff and Wenk, 1968; Coe, 1970). At temperatures above 1300 °C at a strain rate of 10^{-3} sec^{-1} and 1000 °C at 10^{-7} sec^{-1}, orthoenstatite deforms by slip without transforming to clinoenstatite (Raleigh *et al.*, 1971). The $C2/c$ clinopyroxenes, such as diopside and augite, deform by slip on the same system and also by mechanical twinning on (100) and 001) composition planes, the twinning elements being $\eta_1 = [001]$, $K_2 = (001)$ and $\eta_1 = [100]$, $K_2 = (100)$, respectively. These are therefore *reciprocal* twinning mechanisms. The (100) twins are more commonly observed in naturally deformed pyroxenes, both are observed in experimentally deformed crystals and rocks, while the (001) twins are commonest in shocked rocks (Hornemann and Müller, 1971).

Submicroscopic exsolution lamellae are present in orthopyroxenes and clinopyroxenes of most compositions (see Champness and Lorimer, Chapter 4.1 of this volume). The habit planes of these lamellae are (100) in the orthopyroxenes and (100) and (001) in clinopyroxenes. Hence these growth- or diffusion-induced structures, with their boundary faults and terminal or boundary partial dislocations, may be superficially similar to the shear-induced (100) clinopyroxene lamellae in orthoenstatite and the mechanical twins on (100) and (001) in clinopyroxenes, especially if they are on a scale too fine for microanalysis. In natural pyroxenes, both exsolution and deformation structures may be present, hence the latter are more reliably distinguished in experimentally deformed crystals that were initially relatively free from exsolution.

The ortho-clinoenstatite transformation has been shown to take place by a shear of 13.3° on the (100) plane parallel to the [001] axis, resulting in a "detwinning" of the orthoenstatite structure, which can be viewed as clinoenstatite twinned on a unit cell scale (Coe and Müller, 1973; Kirby and Coe, 1974). This is achieved by motion of partial dislocations with b parallel to [001], leaving stacking faults on the (100) planes (Coe and Müller, 1973; Kirby and Coe, 1974); these structures in slightly strained Bamble bronzite are illustrated in Fig. 6a. The character of the stacking-fault fringes and the differences in contrast between dislocations at opposite extremities of stacking faults (A and B in Fig. 6a) has been analyzed by Kirby (Chapter 6.7 of this volume), who concludes that the displacement vector on the faults is $R = 0.83$ [001] and that they result from motion of partials with $b = 0.83$ [001], formed by dissociation of unit dislocations with $b = [001]$. Deformation (at 800 °C) beyond the theoretical angle of shear for the transformation (13.3°) results in total conversion to clinobronzite, which continues to deform by slip of unit dislocations with $b = [001]$, as in Fig. 6b. The micrograph shows isolated, straight dislocations (parallel to $\langle 012 \rangle$ directions) and densely populated slip-bands parallel to (100). Annealing for 1 h at temperatures above 1100 °C results in reversion to orthobronzite, which retains a moderate density of unit dislocations, that is reduced by recovery on prolonged annealing (Coe and Kirby, 1975).

A similar shear transformation has been proposed for the related chain silicate wollastonite, in which the SiO_3 chains are parallel to b (Wenk, 1969; Wenk *et al.*, Chapter 5.5 of this volume). Movement of partial dislocations with $b = \frac{1}{2}[010]$ on (100) planes produces stacking faults with displacement vector $R = \frac{1}{2}[010]$, whose periodic or aperiodic occurrence gives rise to superstructures.

An analysis of the atomic "shuffles" necessary to restore the clinopyroxene structure after homogeneous shear of the atoms corresponding to the twinning shear ($s = 0.56$) on (100) and (001) showed that the shuffles for (100) twinning

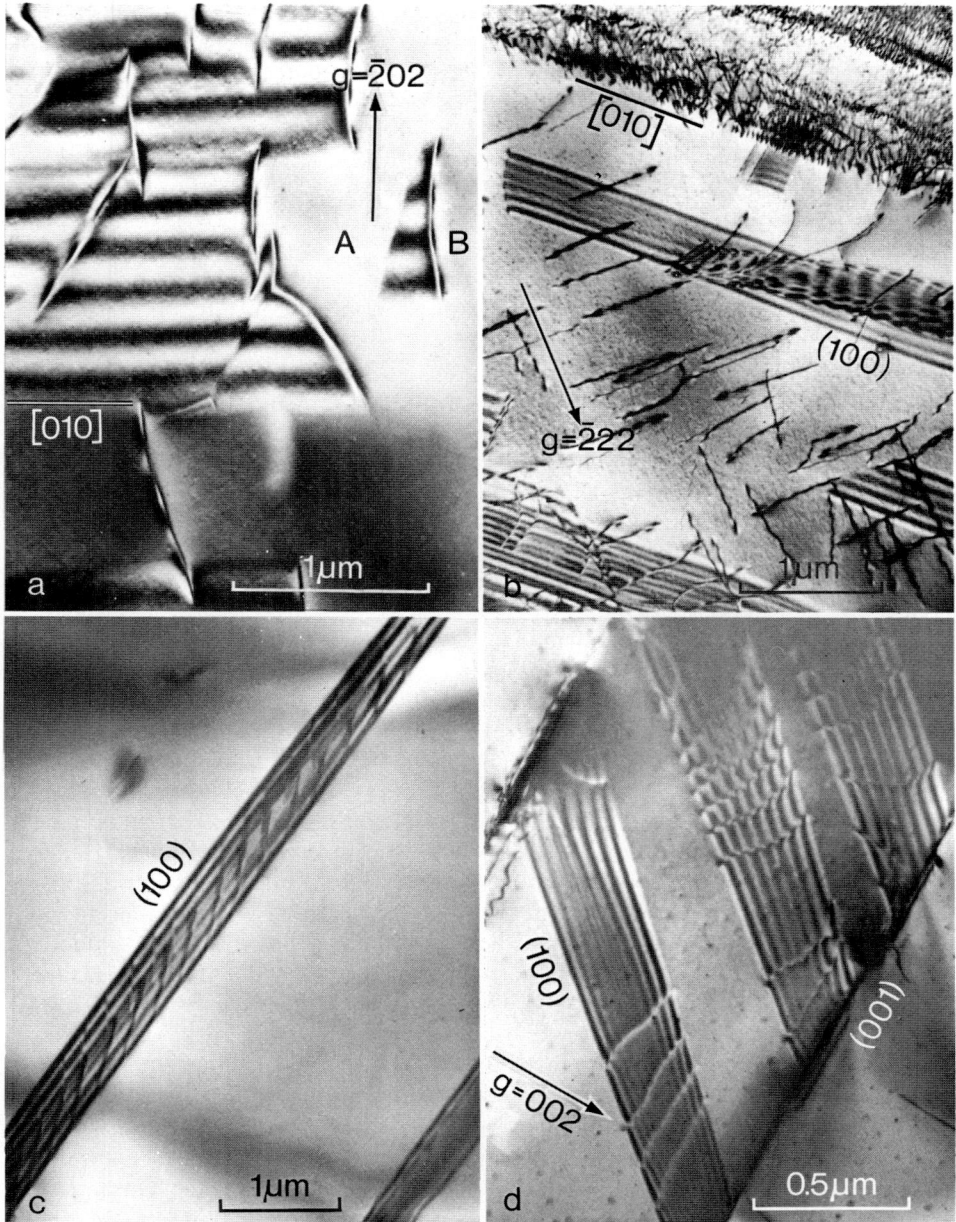

Fig. 6a–d. Electron micrographs of deformation structures in pyroxenes. (a) Stacking faults on (100) planes bounded by partial dislocations with different contrast (at A and B) in slightly strained Bamble orthobronzite. (Courtesy of S.H. Kirby.) (b) Bamble bronzite sheared through an angle of 32° on (100), is transformed to clino-bronzite with slip dislocations, both isolated and in slip bands (top); the crystal retains a few residual (100) stacking faults. (Courtesy of S.H. Kirby.) (c) Thin (100) deformation twins with twinning partial dislocations in the boundaries; experimentally deformed diopside. (Kirby and Christie, unpubl.) (d) Thick (001) twin containing thin (100) twins, the latter with partial dislocations in their boundaries; experimentally deformed diopside (Kirby and Christie, unpubl.)

are small, whereas those for (001) twinning are very large (Kirby and Christie, 1972). This suggests that (001) twinning does not occur by the conventional mechanism of slip of twinning partials on $K_1 = (001)$ with Burgers vectors parallel to $\eta_1 = [100]$. TEM of experimentally deformed diopside crystals with both types of twins showed that the (100) twins are invariably thin and commonly contain numerous partial dislocations in their boundaries (Fig. 6c). By contrast, the thicker (001) twins often contain thin (100) twins, with the same type of partials in the (100) boundaries (Fig. 6d). The (001) twin boundaries are invariably non-coherent, containing such high dislocation densities that individual dislocations are rarely resolved. The Burgers vector of the partials is parallel to [001] but its magnitude has not yet been determined. These observations (Kirby and Christie, unpublished data) suggest that the (001) twins are formed by motion of dislocations with $b = X[001]$ in (100) planes, as in the (100) twins. Such a mechanism, involving motion of twinning dislocations with Burgers vector parallel to η_2 on the plane K_2, is analogous to kinking during slip, and twins produced by this mechanism have been called *incoherent* twins (Friedel, 1964, p. 175). It is interesting to speculate whether other reciprocal twinning mechanisms in low-symmetry crystals may be related in this way.

Shock deformation of clinopyroxenes at peak pressures from 50–390 kbars results in extensive mechanical twinning, dominantly on (001) (Hornemann and Müller, 1971), and such polysynthetic twins are also common in lunar breccias and some meteorites, where they are presumably also caused by shock. HVEM studies of lunar breccias and experimentally shocked lunar soil (Christie et al., 1973) showed, in addition to twinning, extensive conversion of clinopyroxenes to glass at shock pressures of 50 and 100 kbars (Fig. 7b). Shock deformation of Bamble bronzite to 226 kbars resulted in extensive plastic deformation, involving dislocations and stacking faults (Fig. 7a) and local production of glass (Heuer et al., 1974). These studies illustrate that the pyroxenes, under shock conditions, deform plastically by twinning and slip, as well as by shock vitrification. It should be noted here that the mode of deformation and the structures produced by shock are strongly dependent on the porosity of the pre-shock material (Heuer et al., 1974).

All of the deformation mechanisms so far identified by optical or electron microscopy appear to maintain the integrity of the Si_2O_6 chains (parallel to c) and, moreover, appear to involve partial or unit dislocations with Burgers vectors parallel to c. Thus the deformation probably takes place by translation of the layers of chains, involving breaking only of metal-oxygen bonds between the chains, rather than the Si-O bonds in the chains. Mechanical twinning and the ortho-clino transformation involve restacking on the (100) planes, indicating that the stacking fault energy on these planes is very low.

Green and Radcliffe (1972a, b) and Kohlstedt and Vander Sande (1973) have illustrated and discussed the deformation substructures in orthopyroxenes from several natural peridotites. The structures observed by TEM in the natural rocks are similar to those in experimentally deformed specimens (Green and Radcliffe, 1972a; Kirby, Chapter 6.7 of this volume). Annealing of these rocks at high temperatures showed no perceptible change in the defect densities, indicating very slow rates of recovery in orthopyroxene – several orders of magnitude

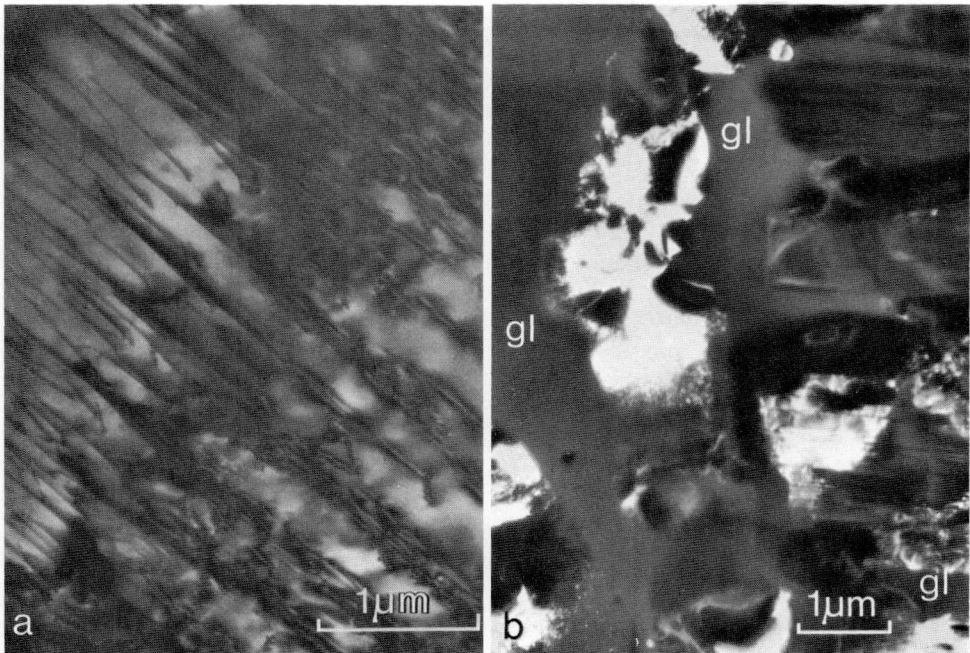

Fig. 7a and b. Deformation structures in pyroxenes deformed by shock (Heuer *et al.*, 1974). (a) Bamble bronzite crystal shocked experimentally to a pressure of ~ 226 kbar showing high density of dislocations. 800 kV. (b) Clinopyroxene crystal in lunar soil shocked experimentally to 50 kbar showing conversion to glass (gl) in veins transecting the crystal. 800 kV, dark field

slower than in coexisting olivine (Kohlstedt and Vander Sande, 1973). Since creep processes depend on climb of dislocations (recovery) these observations suggest that creep rates in pyroxenites may be very slow, compared with other ferromagnesian rocks.

5.3 Olivine

In the orthorhombic (Pbnm) structure of olivine $(Mg, Fe)_2SiO_4$, independent SiO_4 tetrahedra are bonded by the metal ions, which are in octahedral coordination. The mechanical properties of olivine have been the object of many experimental and theoretical studies, since olivine is the major mineral constituent of the upper mantle and presumably dominates the flow of the asthenosphere, which is of great geophysical and geological interest. Since TEM studies of the deformation substructures in peridotites are reviewed extensively elsewhere in this volume by Green (Chapter 6.6), we shall discuss only the deformation mechanisms in olivine crystals and the initial attempts that have been made to correlate quantitatively the dislocation densities and other observable substructural parameters with the flow laws. All of the studies to date appear to have been made with crystals and rocks in the common forsteritic composition range (Fo 85–95).

Optical studies of experimentally deformed olivine (e.g. Raleigh, 1968) indicate that the slip mechanisms vary with temperature and, less sensitively, with strain-rate: at temperatures below 1000 °C ($\dot{\varepsilon} \sim 10^{-1}$ to 10^{-6} sec^{-1}) the mechanisms $\{110\}$ [001], (100)[001] and (100)[010] predominate, whereas at higher temperatures slip parallel to [100] on $\{0kl\}$ planes is predominant. No mechanical twinning has been observed. Electron microscopy (TEM) of experimentally deformed crystals and aggregates has confirmed that the Burgers vectors of dislocations are predominantly [001], [010] and [100], the first being commonest at lower temperatures and the last at temperatures ≥ 1000 °C (Phakey et al., 1972; Blacic and Christie, 1973; Goetze and Kohlstedt, 1973; Kohlstedt and Goetze, 1974). It is remarkable that the dislocations with $b=[010]$ are not dissociated into partials, since the large value of $|b|$ (10.21 Å) should favor dissociation (Phakey et al., 1972). In natural peridotites, dislocations with the same Burgers vectors have also been identified, those with $b=[100]$ being commonest (Green and Radcliffe, 1972a; Goetze and Kohlstedt, 1973). Dislocations with $b=\langle 101 \rangle$ are present in sub-boundaries in natural peridotites, but they never occur as "free" (isolated) dislocations and are therefore believed to originate by combination of dislocations with $b=[100]$ and [001] in the sub-boundaries (Goetze and Kohlstedt, 1973). Tentative identification of dislocations with $b=\langle 112 \rangle$ (Boland et al., 1971) has not been confirmed (J.N. Boland, personal communication). Microfracturing processes have been identified by TEM in peridotite specimens deformed in creep at moderate temperatures (800–950 °C) and stresses (0.5–5.0 kbars) (Boland and Hobbs, 1973) but plastic flow without fracture takes place at higher temperatures.

In olivine crystals deformed at strain rates of 10^{-4} and 10^{-5} sec^{-1}, the dislocation configurations vary considerably with temperature (Phakey et al., 1972). At 600 °C (Fig. 8a) they are straight and strictly parallel to low-index directions (indicating high Peierls forces) and screws predominate over edge segments, suggesting that the edge segments of dislocation loops travel with much higher velocities than screw segments. At 800 °C, jogs in the dislocations indicate cross-slip of the screw dislocations; at 1000 °C, cross-slip is extensive, as indicated by dipoles left by screw dislocations (d in Fig. 8b) and these dipoles in some cases (l in Fig. 8b) have collapsed into rows of small prismatic loops, indicating the onset of climb. The dislocation lines at 1000 °C are more curved (Fig. 8c), which is also an indication of climb. Edge and screw components are more or less equally represented, demonstrating that the velocities of edge and screw dislocations are approximately equal. Loops, networks and helical dislocations are common in crystals deformed above 1000 °C (Fig. 8d) proving that climb is extensive.

The optically visible deformation lamellae in olivine have been shown to be simple arrays of parallel edge dislocations more or less parallel to the active slip-planes (pile-ups) (Green and Radcliffe, 1972b; Phakey et al., 1972). These should not be confused with kink boundaries and tilt-boundaries in subgrain structures, which are almost ubiquitous in deformed olivine. The former are associated with dense tangled zones of dislocations (Blacic and Christie, 1973)

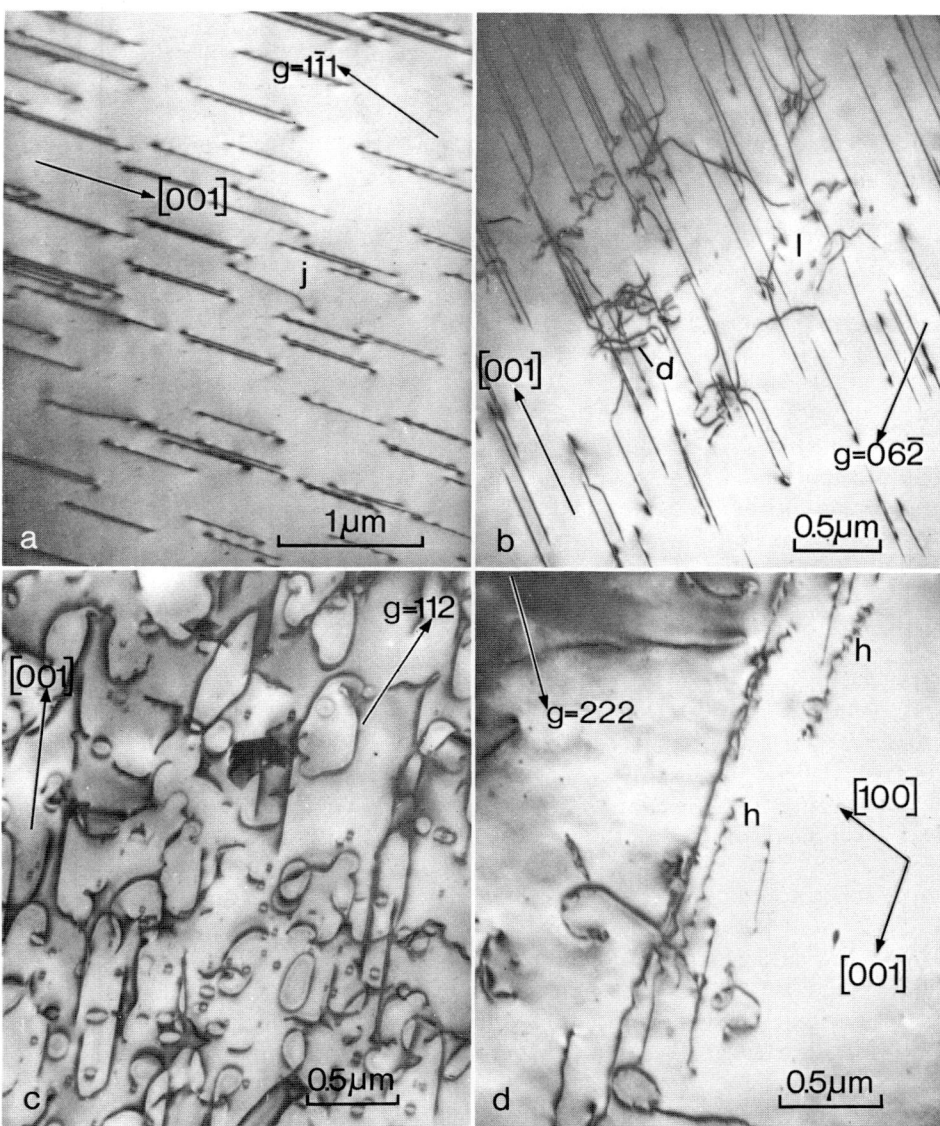

Fig. 8a–d. Dislocation structures in experimentally deformed olivine (Phakey et al., 1972). (a) Crystal deformed at 800 °C, containing generally straight screw dislocations parallel to [001]; note the jog in the dislocation at j. (b) Dislocations in a crystal compressed 8.5% at 1000 °C. They are dominantly screw dislocations parallel to [001] but they show dipoles, (as at d), probably due to cross-slip; the dipoles collapse into loops (e.g. at l), indicating climb. (c) Numerous dislocation loops in a crystal compressed 12.5% at 1000 °C. Dislocations have $b = [001]$. (d) Loops and helical dislocations (h) in a crystal compressed 10% at 1250 °C

and the latter are regular walls of parallel edges, perpendicular to the slip planes, or more general networks, as illustrated by Boland et al. (1973) and several other investigators.

Small amounts of water have a significant weakening effect on olivine, though the effect is much less marked than in quartz (Blacic, 1972; Post, 1973). Dunite samples deformed in hydrous confining media are invariably weaker than vacuum-dried samples deformed in anhydrous media; this complicates the results of experimental studies carried out to determine the flow laws of olivine-bearing rocks. The first studies, made on dunite and peridotite at temperatures ~ 1000–$1200\,°C$, high confining pressure (10–15 kbars) and relatively high differential stresses, indicated a steady-state flow law of the type:

$$\dot{\varepsilon} = A\sigma^n \exp(-Q/RT) \qquad (1)$$

where $\dot{\varepsilon}$ is the axial strain rate, σ is the differential stress $(\sigma_1-\sigma_3)$; R is the gas constant and T is temperature (°K). A and the stress exponent n are constants and Q is an activation energy for creep. Values of A, n and Q obtained were, respectively, 1.2×10^{10}, 4.8 and 119 ± 16 kcal/mole (Carter and Avé Lallemant, 1970) and 10^8, 5, and approximately 100 kcal/mole (Raleigh and Kirby, 1970), where σ is in kbars. Both sets of data were for "dry" rocks. More extensive work by Post (1973) on "wet" and "dry" dunite over the temperature range 750–1350 °C and by Kirby and Raleigh (1973) on dry peridotite is consistent with a more general law of the type

$$\dot{\varepsilon} = B [\sinh(\sigma/\sigma_0)]^n \exp(-Q/RT) \qquad (2)$$

where B, σ_0 and n are constants. At low stresses (<2 kbars) this is well approximated by the simple power law (1) with the parameters:

$$\dot{\varepsilon} = 4 \times 10^8\, \sigma^3 \exp(-Q/RT)$$

Here $\dot{\varepsilon}$ is a natural strain rate and stress is again in kbars. Post showed that Q had a value of 93 ± 2.5 kcal/mole for "wet" experiments and a higher value of approximately 130 kcal/mole for "dry" experiments. Kohlstedt and Goetze (1974) have extended the range of creep studies to higher temperatures (1400–1650 °C) and lower stresses (50 bars–1.5 kbars) with experiments on dry single crystals at atmospheric pressure and controlled oxygen fugacity; they obtained an activation energy $Q = 125 \pm 5$ kcal/mole. They claimed that their data were inconsistent with a simple power law with constant n over the full range of stress in the experimental studies (50 bars–50 kbars) but scatter of their data (due to variable orientation of crystals) is such that this cannot be established. These determinations of the flow law show some degree of consistency but uncertainties in the experimental parameters are still large enough to yield great uncertainties when extrapolations are made to the temperatures and strain rates (or stresses) in the earth's mantle.

Theoretical models of steady-state flow (e.g. Hirth and Lothe, 1968; Weertman, 1968, 1970; Stocker and Ashby, 1973), based on idealized mechanisms involving glide and climb of dislocations and collectively termed *dislocation creep* (Ashby, 1972), also predict behavior of the power law type, but the constant A and exponent n are very dependent on the details of the model. From theoretical considerations it is likely that dislocation creep processes dominate in both the high-temperature experiments (>1 000 °C) and in the upper mantle (Kirby and Raleigh, 1973; Stocker and Ashby, 1973). In order to extrapolate the empirically determined flow laws to slow tectonic deformations it is clearly necessary to establish that the same processes operated. Optical (Raleigh and Kirby, 1970) and TEM observations of microstructures in naturally and experimentally deformed peridotites (Green, Chapter 6.6 of this volume) are at least qualitatively similar and are also generally consistent with the operation of dislocation creep processes. This consistency should be more carefully and rigorously checked, however, by further TEM work.

A significant and potentially useful application of TEM studies has been suggested by Kohlstedt and Goetze (1974), who noted a functional relationship between the differential stress σ and the density ϱ of free dislocations in their olivine crystals deformed by creep at high temperatures:

$$\sigma = C\mu b \varrho^{\frac{1}{2}} \tag{3}$$

where μ is the shear modulus, b is the average magnitude of the Burgers vectors and C is a constant of order unity. This relationship, which is known in some metals (Fe, Fe—3% Si) and ionic crystals (MgO, NaCl), is independent of temperature and strain rate (except through the flow laws) and implies that a steady-state dislocation configuration has developed in equilibrium with the applied stress. We note that such a relationship was not observed in olivine crystals (Phakey *et al.*, 1972) or quartzites (Ardell *et al.*, 1973) deformed at lower temperatures, but in most cases these samples had not attained a steady state, as was the case in the experiments of Kohlstedt and Goetze. If a relationship of this kind is confirmed, in principle it will permit direct estimates of the deviatoric stresses in tectonic deformations from the dislocation densities observed in natural tectonites, subject to limitations arising from complexities in the deformational and thermal histories of the rocks. Goetze and Kohlstedt (1973), for example, estimated that the flow stress in a peridotite xenolith ($\varrho = 10^6$ cm^{-2}), brought up from the mantle in basaltic magma, was approximately 40 bars. In this study, Goetze and Kohlstedt examined the kinetics of the recovery process in olivine by TEM observations of the rate of change in the density of free dislocations and the collapse rate of sessile dislocation loops due to annealing.

Olivine crystals deformed experimentally by shock compression at peak pressures up to 430 kbars (Müller and Hornemann, 1969) contain numerous sets of planar deformation structures parallel to low-index crystal planes (100), (010), (001), {130} and {hkl}. These structures, which have not yet been characterized by TEM, are also found in meteorites and are sufficiently distinctive to be used as indicators of shock deformation. HVEM studies of the lunar rocks indicate that olivine is relatively susceptible to plastic deformation by shock, compared with other silicates. In lunar basalts in which plagioclase and pyroxenes show few or no slip dislocations, moderate densities of straight dislocations, like those produced in static tests at temperatures ~600 °C, have been observed. Moreover, in lunar breccias (which are formed only by shock deformation due to meteoritic impacts) and experimentally shocked lunar soil samples (Lally *et al.,* 1972; Christie *et al.,* 1973) there is no evidence of thetomorphic glass formed within olivine crystals, even when there is extensive shock vitrification of plagioclase and pyroxenes. This suggests either that (a) olivine does not vitrify by shock or (b) thetomorphic glass, if formed, rapidly recovers its crystallinity. TEM examination of experimentally shocked olivine with optically visible planar structures (Müller and Hornemann, 1969), which by analogy with quartz and feldspars would be expected to contain glass, should resolve this problem.

6. Generalizations on Deformation of Silicates

On the basis of the TEM observations that we have reviewed on the substructures of deformed minerals we feel justified in making some general comments regarding their deformation processes. The accompanying speculations are less justifiable but we include them as a stimulus to discussion and further work on some important basic aspects of mineral deformation.

1. In experimental deformation at the strain rates of static tests and in slow natural deformations, minerals behave qualitatively like metals and ionic crystals in that deformation is achieved by slip, mechanical twinning and shear transformations controlled by dislocation motion. At low temperatures, brittle and ductile fracture may occur and at high temperatures, flow is accompanied by recovery and recrystallization, governed by dislocation climb processes, as in the hot-working of metals.

2. At the very high stresses and strain rates of shock deformations, the deformation mechanisms of minerals differ from those observed in metals, at least in the framework and chain silicates. Shock-deformed metals (e.g. Leslie *et al.,* 1965) undergo extensive plastic deformation by twinning, slip and phase transformations. Dislocation motion does not appear to play a role in the shock deformation of the framework silicates; the most obvious mechanism of deformation is vitrification along crystallographically controlled planes. These glass layers may be nucleated on dislocations and other submicroscopic defects that have stress concentrations associated with them. In the chain silicates, if the pyroxenes are typical, vitrification and plastic deformation by dislocation motion both occur. In the orthosilicates (olivine), no evidence of shock vitrification has yet been observed but there is clear evidence of dislocation generation and motion, as in metals. We speculate that this variation in mechanical behavior may be related to the Si-O bonding in the structure: in quartz and the other framework silicates, any permanent distortion involves breaking of Si-O bonds between SiO_4 tetrahedra; in the chain silicates, such bonds must be broken if the silicate chains are displaced, but displacements parallel to the chains (c-axis) may be achieved by breakage and exchange of the weaker metal-oxygen bonds between chains; in the orthosilicates the SiO_4 tetrahedra do not share oxygens, so that general strains can be achieved without disruption of the Si-O bonds. Thus it appears that shock vitrification is associated with the breakage of Si-O bonds

and dislocation motion may accommodate the strain in very rapid deformations if only the exchange of metal-oxygen bonds is required.

3. The effect of dissolved water (H_2O, H^+ or OH^-) on the mechanical behavior of silicates is another distinctive and revealing phenomenon. Of the minerals so far investigated, quartz shows the most extreme effects while in olivine they are quite modest, even at high temperatures and low stresses. Tests on hypersthene also revealed water-weakening (Griggs, 1967; J.D. Blacic, unpublished data) in the chain silicates, though the process has not been investigated quantitatively. These observations are consistent with the hydrolytic weakening model of Griggs and Blacic (1965). Large strains resulting from this process clearly require extensive diffusion of H_2O, H^+ or OH^-. In the chain and orthosilicates, dislocation glide motion may be achieved by breakage of metal-oxygen bonds, but in the framework silicates exchange of Si-O bonds is essential for any slip or twinning process. It is tempting to speculate that bond hydrolysis may be essential for glide processes that involve breaking Si-O bonds, except under stresses at the theoretical strength, but this is not yet established.

4. Under the conditions of experimental and natural deformations, the quartz and olivine structures behave like metals with high stacking-fault energy, whereas the stacking-fault energy on certain planes (notably (100)) in the pyroxenes is very low; partial dislocations and reordering of the chain sequences play a major role in the deformation of the pyroxenes and probably all other chain silicates.

5. At temperatures below about 1000° C, rapid experimental deformation *under dry conditions* of quartz, olivine and pyroxenes produces dislocations that are straight, with strong crystallographic control on their orientations. This indicates strong Peierls (structural) forces. Since little is yet known of the basic mechanism of yielding and flow in minerals under these conditions, we speculate that this may indicate that Peierls stress is controlling the generation and motion of dislocations. In water-weakened minerals and in dry minerals at high temperatures, the dislocations are curved, with frequent nodes and other evidence of interaction due to climb. Here it appears likely that other mechanisms control the initial stages of deformation and that dislocation climb is the rate-limiting process in steady-state flow. There is little evidence that diffusional flow, without dislocations, operated under the experimental conditions so far investigated – or in the naturally deformed rocks so far examined by TEM.

References

Alexander, H., Haasen, P.: Dislocations and plastic flow in the diamond structure. Solid State Phys. **22**, 27–158 (1968).
Amelinckx, S., Delavignette, P.: Observation of dislocations in non-metallic layer structures. Nature **185**, 603–604 (1960a).
Amelinckx, S., Delavignette, P.: Direct evidence for the presence of quarter-dislocations in talc monocrystals. Phil. Mag. **5**, 533–535 (1960b).
Amelinckx, S., Delavignette, P.: Electron microscope observation of dislocations in talc. J. Appl. Phys. **32**, 341–351 (1961).
Amelinckx, S., Gevers, R., Remaut, G., Van Landuyt, J. (eds.): Modern diffraction and imaging techniques in materials science, 745 p. Amsterdam: North Holland Publishing Co. and Elsevier 1970.

Ardell, A.J., Christie, J.M., McCormick, J.W.: Dislocation images in quartz and the determination of Burgers vectors. Phil. Mag. **29**, 1399–1411 (1974a).
Ardell, A.J., Christie, J.M., Tullis, J.A.: Dislocation substructures in deformed quartz rocks. Cryst. Lattice Defects **4**, 275–285 (1973).
Ardell, A.J., McCormick, J.W., Christie, J.M.: Diffraction contrast experiments on quartz using high-order (000l) reflections. Proc. 8th Int. Congress on Electron Microscopy, Canberra, **1**, 486–487 (1974b).
Ashby, M.F.: A first report on deformation-mechanism maps. Acta Met. **20**, 887–897 (1972).
Baeta, R.D., Ashbee, K.H.G.: Plastic deformation and fracture of quartz at atmospheric pressure. Phil. Mag. **15**, 931–938 (1967).
Baeta, R.D., Ashbee, K.H.G.: Slip systems in quartz: I. Experiments. Am. Mineralogist **54**, 1551–1573 (1969a).
Baeta, R.D., Ashbee, K.H.G.: Slip systems in quartz: II. Interpretation. Am. Mineralogist **54**, 1574–1582 (1969b).
Baeta, R.D., Ashbee, K.H.G.: Mechanical deformation of quartz, Parts 1 and 2. Phil. Mag. **22**, 624–635 (1970).
Baeta, R.D., Ashbee, K.H.G.: Transmission electron microscopy studies of plastically deformed quartz. Phys. Stat. Sol. (a) **18**, 155–170 (1973).
Balderman, M.A.: The effect of strain rate and temperature on the yield point of hydrolytically weakened synthetic quartz. J. Geophys. Res. **79**, 1647–1652 (1974).
Barber, D.J., Wenk, H.-R.: The microstructure of experimentally deformed limestones. J. Mater. Sci. **8**, 500–508 (1973).
Blacic, J.D.: Effect of water on the experimental deformation of olivine. In: Flow and fracture of rocks. Geophys. Monogr. **16**, 109–115 (1972).
Blacic, J.D., Christie, J.M.: Dislocation substructure of experimentally deformed olivine. Contrib. Mineral. Petrol. **42**, 141–146 (1973).
Boland, J.N., Hobbs, B.E.: Microfracturing processes in experimentally deformed peridotite. Int. J. Rock Mech. Min. Sci. and Geomech. Abstr. **10**, 623–626 (1973).
Boland, J.N., McLaren, A.C., Hobbs, B.E.: Dislocations associated with optical features in naturally-deformed quartz. Contrib. Mineral. Petrol. **30**, 53–63 (1971).
Bollman, W.: Interference effects in electron microscopy of thin crystal foils. Phys. Rev. **103**, 1588–1589 (1956).
Bragg, W.L.: Atomic structure of minerals, 292 p. Ithaca, N.Y.: Cornell Univ. Press 1937.
Brown, W.L., Morimoto, N., Smith, J.V.: A structural explanation of the polymorphism and transitions in $MgSiO_3$. J. Geol. **69**, 609–616 (1961).
Carter, N.L., Avé Lallemant, H.G.: High temperature flow of dunite and peridotite. Geol. Soc. Am. Bull. **81**, 2181–2202 (1970).
Carter, N.L., Christie, J.M., Griggs, D.T.: Experimental deformation and recrystallization of quartz. J. Geol. **72**, 687–733 (1964).
Christie, J.M., Ardell, A.J.: Substructures of deformation lamellae in quartz. Geology **2**, 405–408 (1974).
Christie, J.M., Green, H.W.: Several new slip mechanisms in quartz. Trans. Am. Geophys. Union **45**, 103 (1964).
Christie, J.M., Griggs, D.T., Carter, N.L.: Experimental evidence of basal slip in quartz. J. Geol. **72**, 734–756 (1964).
Christie, J.M., Griggs, D.T., Heuer, A.H., Nord, G.L., Radcliffe, S.V., Lally, J.S., Fisher, R.M.: Electron petrography of Apollo 14 and 15 breccias and shock-produced analogs. Proc. 4th Lunar Sci. Conf. Geochim. Cosmochim. Acta, Suppl. 4, **1**, 365–382 (1973).
Coe, R.S.: The thermodynamic effect of shear stress on the ortho-clino inversion in enstatite. Contrib. Mineral. Petrol. **26**, 247–264 (1970).
Coe, R.S., Kirby, S.H.: The orthoenstatite to clinoenstatite transformation by shearing and reversion by annealing: mechanism and potential applications. Contrib. Mineral. Petrol. **52**, 29–56 (1975).
Coe, R.S., Müller, W.F.: Crystallographic orientation of clinoenstatite produced by deformation of orthoenstatite. Science **180**, 64–66 (1973).
Cottrell, A.H.: Dislocations and plastic flow in crystals. 223 p. Fair Lawn, New Jersey: Oxford Univ. Press 1953.
Friedel, Jacques: Dislocations, 491 p. Oxford: Pergamon Press 1964.

Goetze, C., Kohlstedt, D.L.: Laboratory study of dislocation climb and diffusion in olivine. J. Geophys. Res. **78**, 5961–5971 (1973).
Green, H.W., Radcliffe, S.V.: Deformation processes in the upper mantle. In: Flow and fracture of rocks. Geophys. Monogr. **16**, 139–156 (1972a).
Green, H.W., Radcliffe, S.V.: The nature of deformation lamellae in silicates. Geol. Soc. Am. Bull. **83**, 847–852 (1972b).
Griffin, L.J.: Observation of unimolecular growth steps on crystal surfaces. Phil. Mag. **41**, 196–199 (1950).
Griggs, D.T.: Deformation of rocks under high confining pressures. J. Geol. **44**, 541–577 (1936).
Griggs, D.T.: Hydrolytic weakening of quartz and other silicates. Geophys. J. Roy. Astron. Soc. **14**, 19–31 (1967).
Griggs, D.T.: A model of hydrolytic weakening in quartz. J. Geophys. Res. **79**, 1653–1661 (1974).
Griggs, D.T., Blacic, J.D.: Quartz: Anomalous weakness of synthetic crystals. Science **147**, 292–295 (1965).
Griggs, D.T., Handin, J. (eds.): Rock deformation. Geol. Soc. Am. Mem. **79**, 382 p. (1960).
Griggs, D.T., Turner, F.J., Heard, H.: Deformation of rocks at 500° C to 800° C. In: Rock deformation (eds. D.T. Griggs, J. Handin). Geol. Soc. Am. Mem. **79**, 39–104 (1960).
Head, A.K.: The invisibility of dislocations. In: Physics of strength and plasticity (ed. A.S. Argon), p. 65–73. Cambridge, Mass.: M.I.T. Press 1969.
Head, A.K., Humble, P., Clarebrough, L.M., Morton, A.J., Forwood, C.T.: Computed electron micrographs and defect identification, 400 p. Amsterdam: North-Holland Pub. Co. 1973.
Heard, H.C., Borg, I.Y., Carter, N.L., Raleigh, C.B.: Flow and fracture of rocks. Geophys. Monogr. **16**, 352 p. (1972).
Heuer, A.H., Christie, J.M., Lally, J.S., Nord, G.L.: Electron petrographic study of some Apollo 17 breccias. Proc. 5th Lunar Sci. Conf. Geochim. Cosmochim. Acta, Suppl. 5, **1**, 275–286 (1974).
Hirsch, P.B., Horne, R.W., Whelan, M.J.: Direct observations of the arrangement and motion of dislocations in aluminium. Phil. Mag. **1**, 677–688 (1956).
Hirsch, P.B., Howie, A., Nicholson, R.B., Pashley, D.W., Whelan, M.J.: Electron microscopy of thin crystals, 548 p. Washington: Butterworths 1965.
Hirsch, P.B., Howie, A., Whelan, M.J.: A kinematical theory of diffraction contrast of electron transmission microscope images of dislocations and other defects. Phil. Trans. Roy. Soc. A**252**, 499–529 (1960).
Hirth, J.P., Lothe, J.: Theory of dislocations. New York: McGraw-Hill Book Co. 1968.
Hobbs, B.E.: Recrystallization of single crystals of quartz. Tectonophysics **6**, 353–401 (1968).
Hobbs, B.E., McLaren, A.C., Paterson, M.S.: Plasticity of single crystals of synthetic quartz. In: Flow and fracture of rocks. Geophys. Monogr. **16**, 29–53 (1972).
Hörz, F., Ahrens, T.J.: Deformation of experimentally shocked biotite. Am. J. Sci. **267**, 1213–1229 (1969).
Horneman, U., Müller, W.F.: Shock-induced deformation twins in clinopyroxene. Neues Jahrb. Mineral. Monatsh. **6**, 247–256 (1971).
Howie, A., Whelan, M.J.: Diffraction contrast of electron microscope images of crystal lattice defects. III. Proc. Roy. Soc. (London), Ser. A**267**, 206–230 (1962).
Kirby, S.H., Christie, J.M.: A comparative study of two modes of deformation twinning in diopside. Trans. Am. Geophys. Union **53**, 727 (1972).
Kirby, S.H., Coe, R.S.: The role of crystal defects in the enstatite inversion. Trans. Am. Geophys. Union **55**, 419 (1974).
Kirby, S.H., Raleigh, C.B.: Mechanisms of high-temperature, solid-state flow in minerals and ceramics. Tectonophysics **19**, 165–194 (1973).
Kohlstedt, D.L., Goetze, C.: Low-stress, high-temperature creep in olivine single crystals. J. Geophys. Res. **79**, 2045–2051 (1974).
Kohlstedt, D.L., Vander Sande, J.B.: Transmission electron microscopy investigation of the defect structure in four natural orthopyroxenes. Contrib. Mineral. Petrol. **42**, 169–180 (1973).

Lally, J.S., Fisher, R.M., Christie, J.M., Griggs, D.T., Heuer, A.H., Nord, G.L., Radcliffe, S.V.: Electron petrography of Apollo 14 and 15 rocks. Proc. 3rd Lunar Sci. Conf. Geochim. Cosmochim. Acta, Suppl. 3, **1**, 401–422 (1972).

Leslie, W.C., Stevens, D.W., Cohen, M.: Deformation and transformation structures in shock-loaded iron-base alloys. In: High-strength materials (ed. V.F. Zackay), p. 382–345. New York: J. Wiley and Sons 1965.

Liddell, N.A., Phakey, P.P., Wenk, H.-R.: TEM – investigation of some quartzites from the Bergell Alps. Proc. 8th Int. Congress on Electron Microscopy, Canberra **1**, 476–477 (1974).

McLaren, A.C., Hobbs, B.E.: Transmission electron microscope investigation of some naturally deformed quartz tes. In: Flow and fracture of rocks (eds. H.C. Heard, I.Y. Borg, N.L. Carter and C.B Raleigh). Geophys. Monogr. **16**, 55–66 (1972).

McLaren, A.C., Phakey, P.P.: A transmission electron microscope study of amethyst and citrine. Austr. J. Phys. **18**, 135–141 (1965a).

McLaren, A.C., Phakey, P.P.: Dislocations in quartz observed by transmission electron microscopy. J. Appl. Phys. **36**, 3244–3246 (1965b).

McLaren, A.C., Phakey, P.P.: Transmission electron microscope study of bubbles and dislocations in amethyst and citrine quartz. Austr. J. Phys. **19**, 19–24 (1966a).

McLaren, A.C., Phakey, P.P.: Electron microscope study of Brazil twin boundaries in amethyst quartz. Phys. Stat. Sol. **13**, 413–422 (1966b).

McLaren, A.C., Phakey, P.P.: Diffraction contrast from Dauphiné twin boundaries in quartz. Phys. Stat. Sol. **31**, 723–737 (1969).

McLaren, A.C., Retchford, J.A.: Transmission electron microscope study of the dislocations in plastically deformed synthetic quartz. Phys. Stat. Sol. **33**, 657–668 (1969).

McLaren, A.C., Retchford, J.A., Griggs, D.T., Christie, J.M.: Transmission electron microscope study of Brazil twins and dislocations experimentally produced in natural quartz. Phys. Stat. Sol. **19**, 631–644 (1967).

McLaren, A.C., Turner, R.G., Boland, J.N., Hobbs, B.E.: Dislocation structure of the deformation lamellae in synthetic quartz. Contrib. Mineral. Petrol. **29**, 104–115 (1970).

Morrison-Smith, D.J.: A mechanical and microstructural investigation of the deformation of synthetic quartz crystals. Ph.D. Thesis, Austr. Nat. Univ., 222 p. (1974).

Müller, W.F.: Elektronenmikroskopischer Nachweis amorpher Bereiche in stoßwellenbeanspruchtem Quarz. Naturwiss. **56**, 279–280 (1969).

Müller, W.F., Hornemann, U.: Shock-induced planar deformation structures in experimentally shock-loaded olivines and in olivines from chondritic meteorites. Earth Planet. Sci. Letters **7**, 251–264 (1969).

Orowan, E.: Zur Kristallplastizität I, II, III. Z. Physik **89**, 605–659 (1934).

Phakey, P., Dollinger, G., Christie, J.: Transmission electron microscopy of experimentally deformed olivine single crystals. Geophys. Monogr. **16**, 117–138 (1972).

Polanyi, M.: Über eine Art Gitterstörung, die einen Kristall plastisch machen konnte. Z. Physik. **89**, 660–664 (1934).

Post, R.L.: The flow laws of Mt. Burnett dunite. Ph.D. dissertation, Univ. of Calif. Los Angeles, 272 p. (1973).

Raleigh, C.B.: Glide mechanisms of experimentally deformed minerals. Science **150**, 739–741 (1965).

Raleigh, C.B.: Mechanisms of plastic deformation of olivine. J. Geophys. Res. **73**, 5391–5406 (1968).

Raleigh, C.B., Kirby, S.H.: Creep in the upper mantle. Mineral. Soc. Am. Spec. Paper **3**, 113–121 (1970).

Raleigh, C.B., Kirby, S.H., Carter, N.L., Avé Lallemant, H.G.: Slip and clinoenstatite transformation as competing rate processes in enstatite. J. Geophys. Res. **76**, 4011–4022 (1971).

Read, W.T.: Dislocations in crystals, 228 p. New York: McGraw-Hill Book Co. 1953.

Stocker, R.L., Ashby, M.F.: On the rheology of the upper mantle. Rev. Geophys. **11**, 391–426 (1973).

Stöffler, D.: Deformation and transformation of rock-forming minerals by natural and experimental shock processes, I. Fortschr. Mineral. **49**, 50–113 (1972).

Stöffler, D.: II. Fortschr. Mineral. **51**, 256–289 (1974).
Taylor, G.I.: The mechanism of plastic deformation of crystals. Proc. Roy. Soc. (London), Ser. A**145**, 362–387 (1934).
Trommsdorff, V., Wenk, H.-R.: Terrestrial metamorphic clinoenstatite in kinks of bronzite crystals. Contrib. Mineral. Petrol. **19**, 158–168 (1968).
Tullis, J., Christie, J.M., Griggs, D.T.: Microstructures and preferred orientations of experimentally deformed quartzites. Geol. Soc. Am. Bull. **84**, 297–314 (1973).
Tullis, J., Tullis, T.: Preferred orientation of quartz produced by mechanical Dauphiné twinning. Geophys. Monogr. **16**, 67–82 (1972).
Turner, F.J., Heard, H., Griggs, D.T.: Experimental deformation of enstatite and accompanying inversion to clino-enstatite. Rep. Int. Geol. Congress 21st Session, Pt. **17**, 399–408 (1960).
Weertman, J.: Dislocation climb theory of steady-state creep. Am. Soc. Met. Trans. **61**, 681–694 (1968).
Weertman, J.: The creep strength of the earth's mantle. Rev. Geophys. **8**, 145–168 (1970).
Wenk, H.-R.: Polymorphism of wollastonite. Contrib. Mineral. Petrol. **22**, 238–247 (1969).
White, S.: Natural creep deformation of quartzites. Nature Phys. Sci. **234**, 175–177 (1971).
White, S.: The dislocation structures responsible for the optical effects in some naturally-deformed quartzes. J. Mat. Sci. **8**, 490–499 (1973).
White, S., Cosby, A., Evans, P.E.: Dislocations in naturally deformed quartzite. Nature Phys. Sci. **231**, 85–86 (1971).
Zinserling, K., Shubnikov, A.: Über die Plastizität des Quarzes. Z. Krist. **85**, 454–461 (1933).

CHAPTER 6.2

Work Hardening and Creep Deformation of Corundum Single Crystals

B.J. PLETKA, T.E. MITCHELL, and A.H. HEUER

1. Introduction

Work hardening was the earliest method used by man to strengthen metals; the phenomena has been understood for many years as being due to elastic interactions between individual mobile dislocations and groups of dislocations. The details of such interactions have been incorporated into a number of work-hardening theories, which have attempted to correlate various experimental observations such as slip-line lengths, slip-step heights, dislocation arrangements and dislocation densities. However, the dislocation substructures are very complicated and many simplifying assumptions are required in order to calculate work-hardening rates quantitatively.

Oxides and silicates also undergo work hardening; however, they have received much less attention than metals both experimentally and theoretically. Our own studies of the deformation of corundum single crystals (Pletka et al., 1974) have shown that work hardening in this material is fairly straight-forward, in as much as glide deformation occurs mostly by single slip on the basal plane. However, there is an additional complicating factor in that dislocation climb occurs at the high temperatures necessary for plastic deformation. Climb plays an important role in the evolution of the dislocation substructure and this aspect of the deformation of corundum will be described first before discussing work hardening. Details of the experimental procedures involved in deforming corundum in compression, preparing thin-foil specimens and their examination by transmission electron microscopy are available elsewhere (Pletka, 1973; Pletka et al., 1974).

2. Dislocation Structures

Fig. 1 illustrates typical stress-strain curves for Czochralski-grown synthetic corundum single crystals deformed in compression at temperatures from 1400 to 1700° C with an initial strain-rate of 1.33×10^{-4} sec^{-1}. On exceeding the elastic limit, distinct upper and lower yield points are observed, which are believed to be due to a dislocation multiplication mechanism (Johnston and Gilman, 1960; Johnston, 1962; Firestone and Heuer, 1973). At strains beyond the yield point, the crystals begin to work harden at a rate which first increases, reaches a maximum and then decreases until the work-hardening rate becomes essentially

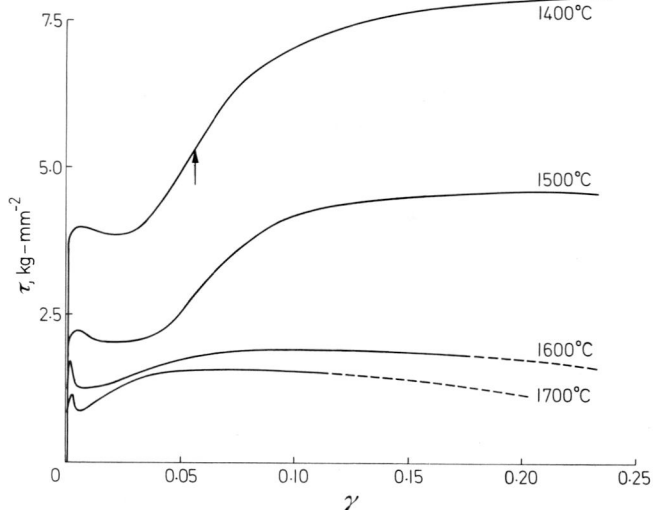

Fig. 1. Shear stress-shear strain curves for corundum single crystals deformed in basal slip at temperatures from 1400–1700° C at an initial strain rate of 1.33×10^{-4} sec^{-1}

zero. The maximum work-hardening rate at 1400° C is $\sim \mu/300$ (where μ is the shear modulus) and is somewhat less at higher temperatures. The shear strain at which the work-hardening rate becomes zero is $\sim 20\%$ at 1400° C and is also less at higher temperatures.

The development of the dislocation structures at different strains along the flow stress curve has been described previously (Pletka et al., 1974) and will be reviewed here only briefly. The structure found at very low plastic strains (between the upper and lower yield point) consists of dipole and multipole configurations linked by glide dislocations. The dipoles lie predominantly along the [1$\bar{1}$00] direction; since their Burgers vector is $\frac{1}{3}$[11$\bar{2}$0], they are primarily edge in character.

The structure at strains just greater than the lower yield point, i.e. where an appreciable work-hardening rate is encountered, is similar to that just described for lower strains, except for an increase in dislocation density. Fig. 2a illustrates the dislocation structure found at this strain (corresponding to the arrow in the 1400° C sample in Fig. 1), and can be seen to be composed of dipole, multipole and linking glide dislocations. In addition, a number of small loops are also present in the background. The dipoles form because dislocations of opposite sign gliding on nearby parallel slip planes interact elastically and become "trapped"; predominantly edge dipoles are left because of the ability of the screw portions of these trapped dislocations to cross-slip and annihilate. This process is known as "edge trapping" and was described in detail by Mitchell and Hirsch (1968) and Hirsch and Lally (1965). The dipoles break up by climb into the observed small loops to reduce their line energy. It can be seen also in Fig. 2a that three different types of elongated loops have formed. One type has the same orientation as the original edge dipoles and has the same $\frac{1}{3}$[11$\bar{2}$0] Burgers vector. The other two types lie at $\pm 30°$ to the [1$\bar{1}$00] direction and

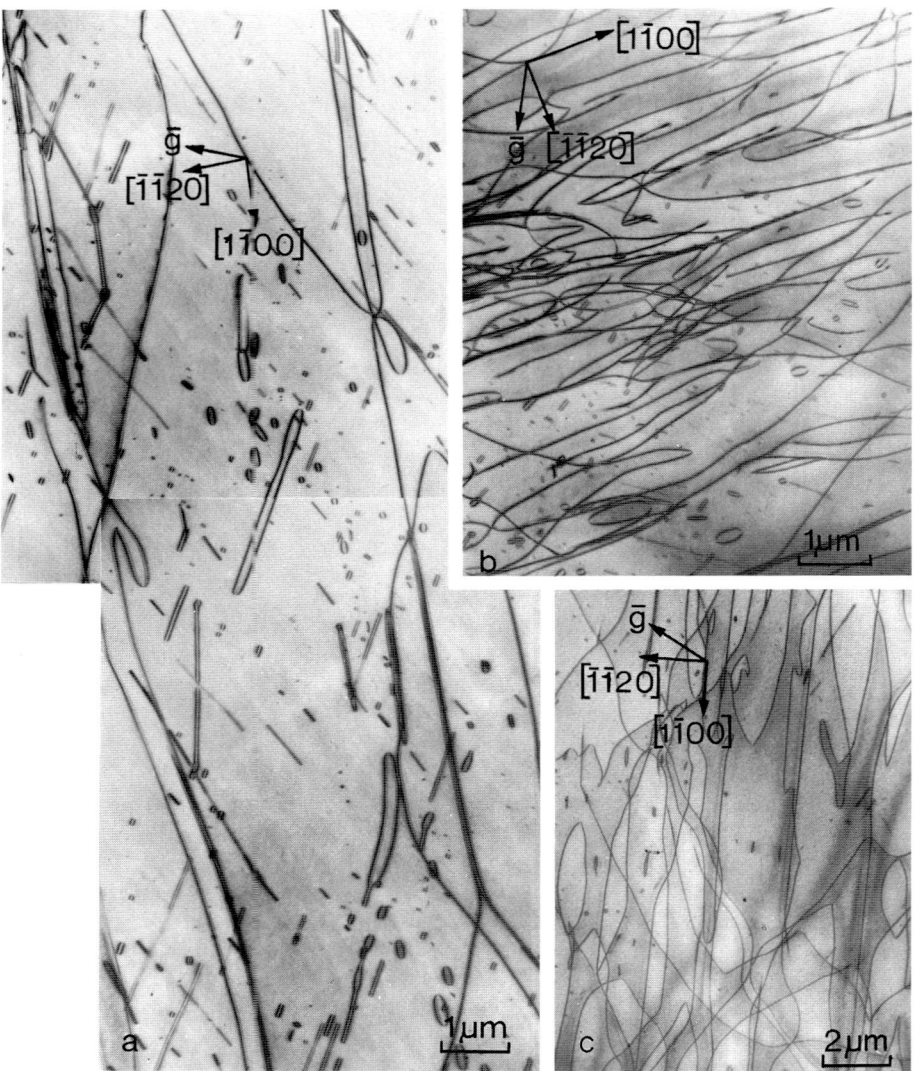

Fig. 2a–c. Dislocation structures in deformed corundum: (a) Shows glide dislocations, dipoles, and dislocation loops in basal foil deformed to 5.5% shear strain at 1400° C (see arrow on Fig. 1); (b) shows a basal foil of a specimen deformed to the region of zero work hardening (22.5% shear strain at 1400° C); (c) shows a basal foil from a creep specimen deformed to 10% shear strain (steady-state creep region) at 1400° C under a stress of 1.5 kg/mm^2

were found (from the usual $\mathbf{g} \cdot \mathbf{b}$ analysis) to have Burgers vector of the type $\frac{1}{3}\langle 10\bar{1}0 \rangle$. This latter type of loop is believed to represent an intermediate stage in the annihilation of dipoles by climb, but the details of the process are still under study.

With increasing strain in the region of finite work-hardening rate, the dislocation density continues to increase. This increase occurs mainly in the density of loops and linking glide dislocations, but the qualitative nature of the substructure remains unchanged. This is observed to be true even for specimens deformed (as in Fig. 2b) into the region of zero work-hardening rate; however, at still larger strains, the dislocation density and substructure are essentially unchanged.

3. Work-hardening Model

The various regions of the flow stress curve can now be correlated with the observed dislocation structures. Where significant work hardening is taking place, the structure is dominated by dipoles, multipoles and linking glide dislocations. Due to their internal stress fields, the dipoles are relatively immobile; if the dipoles pinch off (by climb or cross-slip), there is even greater resistance to movement since the loop ends are sessile jogs. Thus the dipoles and multipoles serve as obstacles to the glide dislocations and result in the observed work hardening. As deformation proceeds, however, recovery processes become important. Dipoles can be eliminated by two processes: (1) narrow dipoles can break up by climb into smaller loops to reduce the line energy; (2) dipoles in excess of the stable width can break up by glide since the stable dipole width is inversely proportional to the flow stress (Hirsch and Lally, 1965). The net effect is that the obstacles to the glide dislocations are continuously forming and breaking up, so that when the rate of accumulation of dipoles is equal to their rate of annihilation, a zero work-hardening rate is reached.

To extend the ideas developed above, dislocation densities have been measured as a function of flow stress at temperatures in the range 1400–1700° C. The densities were measured using a variation of Ham's method (1961), in which the intersection of dislocations with randomly centered circles are counted on a net placed over the micrographs. The thickness of each area (0.5–1.5 μm) was determined by stereomicroscopy.

Preliminary results of the dislocation density analysis are shown in Fig. 3, where total dislocation density is plotted against flow stress at four temperatures

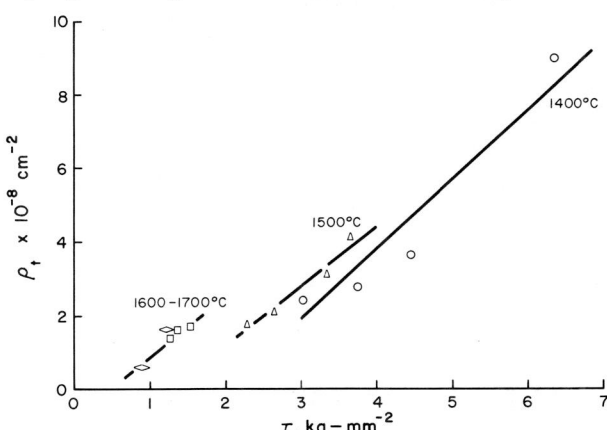

Fig. 3. Total dislocation density versus flow stress for temperatures from 1400–1700° C

corresponding to the stress-strain curves of Fig. 1. The total dislocation density includes contributions from the loops, dipoles and glide dislocations. Fig. 3 shows that the dislocation density increases with increasing flow stress. The rapid increase in dislocation density occurs after yielding at each of the temperatures indicated. Higher dislocation densities are reached at the lower temperatures, corresponding to the higher plateau flow stresses in Fig. 1. The distribution of the total dislocation density amongst dipoles, loops and glide dislocations has also been investigated. It was found that most of the increase of dislocation density during work hardening is due to the linking glide dislocations and loops, while the dipole density remains relatively constant. This is consistent with the idea that the dipoles are constantly forming and breaking up into loops during deformation. The loops, as well as the dipoles, act as effective obstacles to glide dislocations. During work hardening, the buildup in density of loops from the breakup of dipoles causes an increase in density of glide dislocations. When the plateau stress is reached (zero work hardening), the rate of breakup of dipoles into loops and the further annihilation of loops by diffusion is equal to the rate of formation of dipoles by edge-trapping. Furthermore, at higher deformation temperatures, the rate of annihilation of the dipoles and loops should increase so that the plateau stress should be reached at a lower dislocation density. This is in fact observed, as can be seen in Fig. 3, where the highest dislocation density plotted at each temperature corresponds to the plateau stress.

This model should also be applicable to steady-state or secondary creep which is conceptually equivalent to deformation under conditions of zero work hardening. Limited results obtained to date on corundum crystals tested in creep at 1400° C are as follows:

Creep stress (kg/mm^2)	Steady state creep rate (sec^{-1})	Dislocation density (cm^{-2})
1.10	2.05×10^{-7}	0.9×10^8
1.46	1.55×10^{-6}	1.9×10^8

Dislocation densities were obtained as before by transmission electron microscopy, and the same mixture of dipoles, loops and glide dislocations was observed as in constant strain-rate tests (see Fig. 2c). Comparing the above Table with Fig. 3, it is apparent that the dislocation densities in creep at 1400° C correspond to the constant strain-rate data at 1600 to 1700° C because the lower strain rates (and stresses) in creep allow more time for the dipole and loop annihilation processes to proceed. Thus, a lower dislocation density occurs than under constant strain-rate tests at 1400° C. More data are presently being obtained in order to develop a more quantitative theory.

References

Firestone, R.F., Heuer, A.H.: Yield point of sapphire. J. Am. Ceram. Soc. **56**(3), 136–139 (1973).

Ham, R.K.: Determination of dislocation densities in thin films. Phil. Mag. **6**(69), 1183–1184 (1961).

Hirsch, P.B., Lally, J.S.: Deformation of magnesium single crystals. Phil. Mag. **12** (117), 595–648 (1965).
Johnston, W.G.: Yield points and delay times in single crystals. J. Appl. Phys. **33** (9), 2716–2730 (1962).
Johnston, W.G., Gilman, J.J.: Dislocation multiplication in lithium fluoride crystals. J. Appl. Phys. **31**, (4), 632–643 (1960).
Mitchell, T.E., Hirsch, P.B.: In: Work hardening (of metals). (eds. J.P. Hirth and J. Weertman), p. 65–91. New York: Gordon and Breach, Science Publishers, Inc. 1968.
Pletka, B.J.: M.S. Thesis, Case Western Reserve University, Cleveland, Ohio (1973).
Pletka, B.J., Mitchell, T.E., Heuer, A.H.: Dislocation structures in sapphire deformed by basal slip. J. Am. Ceram. Soc. **57** (9) 388–393 (1974).

CHAPTER 6.3

Dislocation Structures in Synthetic Quartz

D.J. MORRISON-SMITH

1. Introduction

There now exists a large amount of data on the dislocation substructures in synthetic quartz as observed by TEM, and although a great deal of useful information on the mechanisms of deformation and dislocation dynamics has been obtained (for instance Morrison-Smith, 1974; Morrison-Smith et al., 1975; Morrison-Smith and Boland, 1975; Hobbs et al., 1972), no real attempt has been made to describe the details of the stress-strain curve in terms of dislocation behavior. This has been precluded by the problem of inhomogeneous deformation in these synthetic crystals. The relationship between the (OH) distribution and the distribution of optical deformation features has been realized for some time (e.g. Blacic, 1971) while the marked effects of variations in (OH) content on the mechanical behavior of specimens is now firmly established (Morrison-Smith et al., 1975). This paper attempts to analyze the available data in such a way that a reasonable description of the ideal, homogeneous stress-strain curve in terms of dislocation structures can be obtained.

2. Experimental Data

Many of the observations utilized in this work are described in detail in Morrison-Smith (1974). The stress-strain curves presented in earlier papers represent the average behavior of each specimen and it has been shown that the response of each specimen depends on its (OH) content and distribution. For instance the stress-strain curves for cores from specimen W2, deformed normal to $r\{10\bar{1}1\}$ show a distinct separation into two groups (Fig. 1a, b) which correlate directly with the cores' positions within the crystal: the soft specimens (Fig. 1a) corresponding to cores taken close to the crystal seed while the harder specimens (Fig. 1b) correspond to cores taken nearer the surface of the crystal. Optical microscopy reveals that the hard specimens contain broad basal deformation bands (~ 1 mm wide) with little other evidence of deformation, while the softer specimens exhibit narrow (~ 0.1 mm wide) basal deformation bands plus a considerable amount of undulatory extinction. In addition it is found that this distribution of deformation corresponds very closely to the detailed distribution of (OH) as measured by infra-red absorption spectroscopy and summarized in Fig. 2. The (OH) content for the whole crystal W2 is 1600 ppm H/Si atomic ratio.

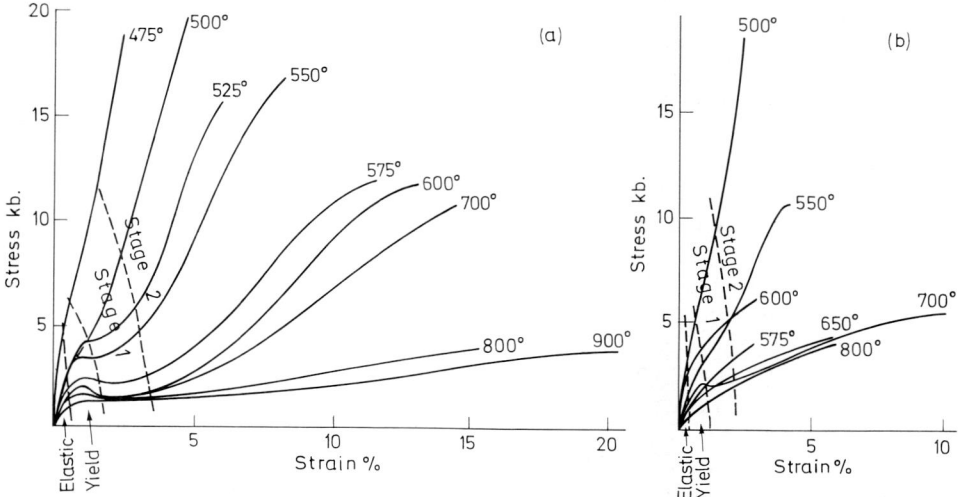

Fig. 1a and b. Stress-strain curves for a series of specimens from crystal W2, compiled from load-displacement data measured for each complete core. Since these stress-strain curves represent the average behavior within the specimen the various stages marked in the figures illustrates the characteristic features of the ideal stress strain curves, but may not describe the dislocation structures within a particular specimen at a given stress. (a) Stress-strain curves for cores taken near to seed which exhibit narrow basal deformation bands. (b) Stress-strain curves for cores taken near surface of specimen which exhibit broad basal deformation bands. Most of these runs ended in the apparent failure of the specimen by fracturing

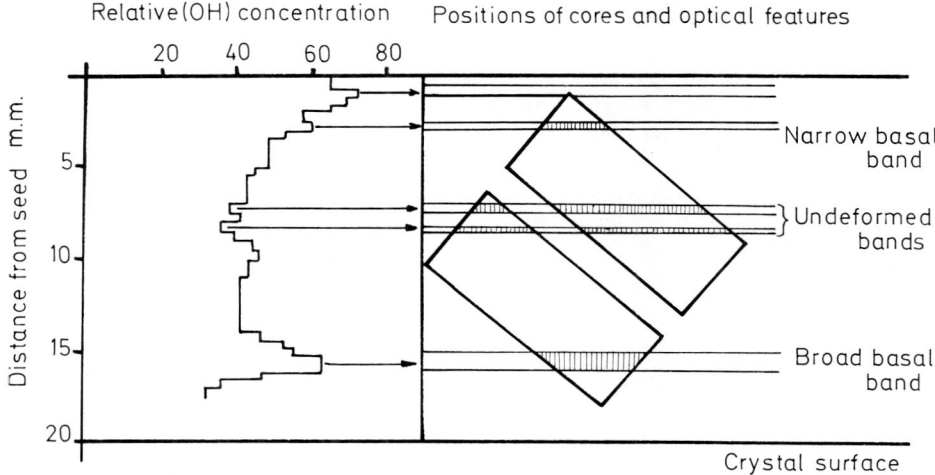

Fig. 2. Schematic section through crystal W2 showing relative positions of the two layers of cores and their relationship to the relative (OH) concentration, measured by infra-red absorption spectroscopy

The initial defect concentration in this material was found to consist of 10^3 cm^{-2} dislocation density and a variable density of small inclusions. The maximum inclusion density was found to be 10^{14} cm^{-3} with sizes ranging from 200 to 2000 Å in diameter (Fig. 3). These features are too small to be analyzed by electron microprobe or diffraction techniques, but one manufacturer of synthetic quartz (Sawyer Research Products Inc., Eastlake, Ohio) has indicated that they commonly find and reject synthetic crystals containing optically observable inclusions of acmite ($NaFeSi_2O_6$). A comparison of the volume fraction of inclusions observed in these specimens of W2 and the measured sodium content (157 ppm Na/Si) is consistent with these features also being composed of acmite.

These features appear to be the major sources of dislocations generated during the early stages of deformation by their interaction with grown-in dislocations (Humphreys and Hirsch, 1970; Brown and Stobbs, 1971) or through the multiplication of interfacial (misfit) dislocations around semi-coherent inclusions. These misfit dislocations may be generated during growth of the inclusions, by prismatic punching or through local stress concentrations due to differences in thermal expansion coefficients or elastic constants (Barnes and Mazey, 1963; Weatherly, 1968; Weatherly and Nicholson, 1968; Brown and Woolhouse, 1970; Hirsch, 1972). Several of these processes appear to have operated in these specimens, producing clusters of irregular dislocation loops as shown in Fig. 4, although processes involving interaction of grown-in dislocations and inclusions are considered to represent only a very small proportion of the observations due to the low densities and velocities of grown-in dislocations.

Further dislocation multiplication leads to growth and coalescence of clusters and interaction of their component dislocations. Frequently, the growth of these loops is anisotropic with marked elongation of loops in one direction which often leads to the development of parallel bands of dislocations. The growth and morphology of these clusters indicate that their component loops are lying in slip planes and that the majority are parallel to $\{01\bar{1}1\}$ planes, while contrast observations indicate that their Burgers vectors are of the type $\langle 2\bar{1}\bar{1}0 \rangle$. There are, however, observations of additional slip on $(0001) \langle 2\bar{1}\bar{1}0 \rangle$, $\{10\bar{1}0\} \langle \bar{1}2\bar{1}3 \rangle$, $\{10\bar{1}0\} \langle \bar{1}2\bar{1}0 \rangle$ and $\{2\bar{1}\bar{1}0\}[0001]$. These systems become increasingly important at higher temperatures, for example $\langle \bar{1}2\bar{1}3 \rangle$ Burgers vectors are only observed in specimens deformed at temperatures greater than 550° C.

In many areas, dislocation multiplication is not achieved solely by the growth of clusters; frequently bands of regular straight-sided loops lying in $\{10\bar{1}0\}$ planes are observed (Fig. 5), apparently comprized chiefly of dislocations on $\{10\bar{1}0\} \langle 2\bar{1}\bar{1}0 \rangle$ slip systems. The arrangement of these bands strongly suggests that they develop by the mechanism suggested by Mendelson (1972) which involves the enhanced generation of loops on neighboring slip planes by double cross slip and the subsequent operation of a Frank-Read source. Cross slip is enhanced by long-range stress fields due to existing loops in the band.

With increasing stress, the growth of loops within clusters leads to the interaction of dislocations from neighboring sources and eventually large scale tangling of dislocations from different sources and different slip systems. These features develop at smaller strains with increasing temperatures due to increases in dislocation velocities and climb rates. At temperatures above about 600° C, climb pro-

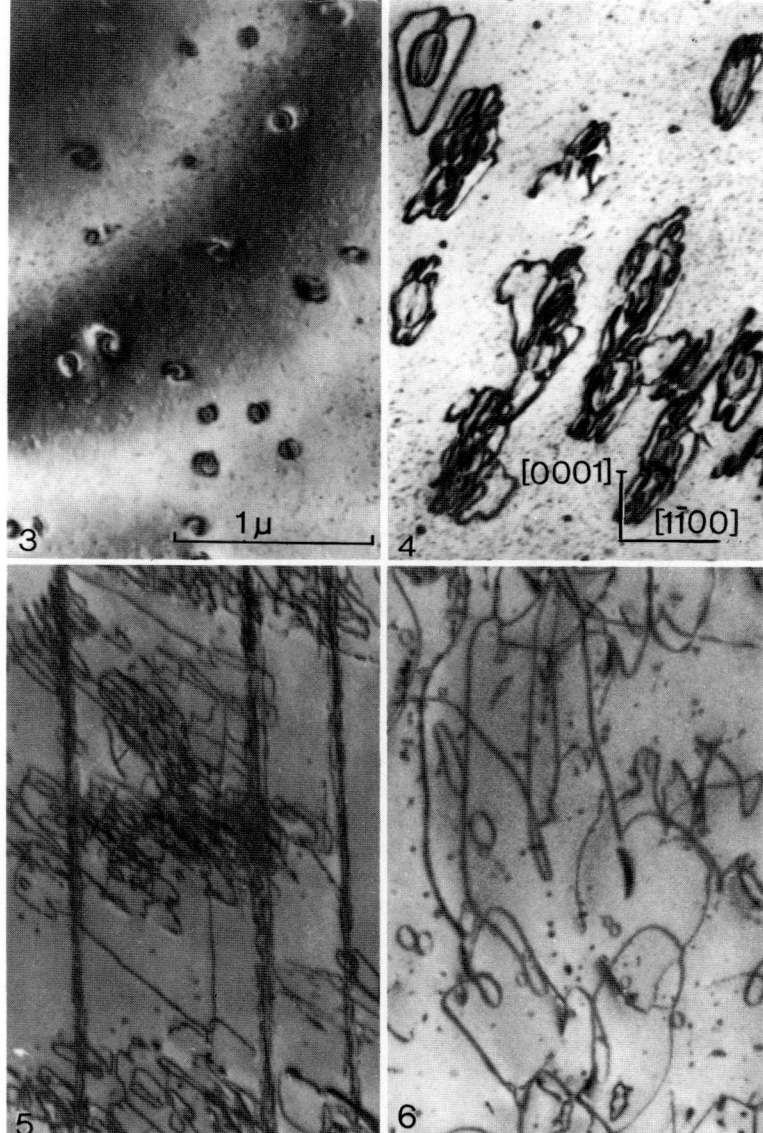

Fig. 3. Images produced by inclusions of acmite in an undeformed specimen, using 1̄101 reflection near exact Bragg condition

Fig. 4. Small dislocation clusters formed during the early stage of deformation in a specimen compressed 0.3% at 475° C

Fig. 5. Areas of clusters intersected by narrow bands of dislocations lying in (1̄100) planes, in a specimen deformed 2% at 550° C

Fig. 6. Typical dislocation structure for a specimen deformed at higher temperatures (600° C) showing mostly curved and tangled dislocations, with a high density of debris loops and dipoles

cesses become significant, resulting in less well-defined clusters, lower dislocation densities and the formation of dipoles and debris loops, as shown in Fig. 6.

3. Discussion

In order to take account of inhomogeneous deformation an attempt has been made to break down each specimen into individual slabs of approximately constant (OH) content, each of which should behave uniformly. The specimen is then reconstructed from these slabs using the infra-red spectroscopy data/optical microscopy correlations (Fig. 2). A simple example of this type of analysis considers all deformation in hard specimens containing broad deformation bands to be restricted to these bands, the remainder of the specimen thus deforming elastically, Using such approximations a reasonable demonstration of the behavior of an homogeneous specimen can be made. There is no attempt to duplicate experimental stress-strain curves using a combination of Griggs' microdynamical theory (Griggs,1974) and this slab model, because it is considered that the results would not justify the amount of computation involved.

It should be noted here that where the basic relationships described below differ significantly from previous work, it is generally because those previous workers were relating their observations to average stress-strain curves.

3.1 Elastic Stage

Initial dislocation densities have been generally observed to be $\sim 10^3$ cm^{-2}, while inclusions have been observed up to $\sim 10^{14}$ cm^{-3}. Little electron microscopy has been performed on other undeformed synthetic crystals, but since our sample W2 has a low Na content compared with other crystals (Table 1 of Hobbs et al., 1972) it seems reasonable that the other crystals contained a similar or higher density of inclusions. During the elastic stage, stress fields would be built up round these inclusions so that in the late stages they become strong enough for dislocations to be generated at the matrix-inclusion interface through one of the mechanisms cited earlier.

3.2 Yield Region

The yield region is the most difficult to deal with using the slab model; hence most of the discussion here will be rather generalized. In the most homogeneous specimens so far deformed it appears that $\sigma_{vy}/\sigma_{ly} = 1.5$, and present deductions indicate that 2.0 is not an unreasonable figure in a homogeneous specimen. (σ_{vy} and σ_{ly} are as defined in Fig. 5 of Hobbs et al., 1972). The precise shape of the yield point in an homogeneous specimen is difficult to define: as Johnston (1962) points out, the shape depends on the elastic properties of the testing machine. For each slab being considered, the remainder of the specimen becomes effectively part of the machine, making the problem more complex. However, it does appear that the yield point is rounded similar to that observed in LiF (Johnston, 1962).

The first departures of the stress-strain curve from linearity appear to be a result of the onset of significant multiplication of the stress-generated dislocations around inclusions. As loops grow and more are generated around the inclusions, resulting in the development of clusters, the dislocation multiplication rate becomes sufficiently high for the stress required to maintain the imposed strain rate falls. Hence the observed yield drops in a manner similar to that described for LiF (Johnston, 1962) and diamond structure materials (Alexander and Haasen, 1968). Johnston (1962) has noted in this case that if it were not for the processes of work hardening, the applied stress would continue to fall with increasing strain, and it is only when the stress required to overcome the resistance to dislocation motion becomes significant compared to the applied stress, that the stress will increase again.

3.3 Stage I

Stage I commences essentially where the stress-strain curve begins to curve up again following the yield drop, and it appears that this increasing stress is required to overcome a backstress component introduced by interactions between dislocations of the same slip system. It seems that even at this stage of deformation (especially in the $\perp r$ orientation) dislocations of more than one slip system are active. However, during Stage I there appears to be no interaction between dislocations on these different systems. At low temperatures the major contribution to work hardening arises because the low dislocation velocity (Morrison-Smith et al., 1975) rapidly saturates sources around inclusions (i.e. the dislocations cannot move away from the source fast enough to allow generation of new loops and consequently a significantly higher stress is required either to activate new, less favorable, sources, or to increase the velocity of existing dislocations). Under these conditions, it is unlikely that the required number of dislocations can be generated to create a yield drop: hence the absence of such a feature in many low-temperature specimens. As deformation temperatures increase, dislocation velocities increase rapidly so that the above mechanisms become less important and yield drops are normally observed. At these higher temperatures ($>550°$ C) the major contribution to work hardening arises through blocking and trapping of the leading dislocations from different sources by such mechanisms as those suggested by Weertman (1957), where dislocations are trapped in each others' stress fields. This causes subsequent dislocations to pile up behind them, and the stress required for the dislocations to pass each other is increased significantly. There is also a stress component favoring climb and subsequent annihilation of dislocations of opposite sign: the rate of work hardening in this case is significantly lower than that in the low-temperature situation. Here the role of "water weakening" becomes increasingly apparent: the dislocation velocity is considered to be controlled by the rate of (OH) diffusion through the lattice, and to be dependent on the concentration of (OH) in the velocity calculations of Griggs (1974). Thus the role of hydrolytic weakening may be conveniently explained in terms of the effect of (OH) diffusion on dislocation glide and climb as they relate to the processes of work hardening.

One consequence of the increasing mobility of dislocations with increasing temperature is the reduction in extent of Stage I with increasing temperature. This is contrary to the findings of Hobbs et al. (1972), but it must be remembered that they were dealing with an average system in which different parts of the specimen were behaving differently. In addition, they were subdividing their stress-strain curves on the basis of the shape of the curve, whereas here the stages of the curve are defined on the basis of dislocation substructures. At temperatures below about 500° C, it appears that specimens may exhibit Stage I up to high stresses although the total strain obtained is small. However, at higher temperatures increased mobility leads to interaction with dislocations on other slip systems at low stresses, and the major portion of the strain is obtained in a multiple slip regime (Stage II).

3.4 Stage II

Stage II is normally associated with a significant increase in the rate of work hardening. However, in these inhomogeneously deforming specimens, the transition from Stage I to Stage II is not well defined. This is not surprising as in many examples, parts of the specimen may be well into Stage II while other parts are deforming in Stage I or even elastically. When the slabs are treated individually, however, the increasing rate of work hardening with strain becomes clearer and the transition from Stage I to Stage II is easier to determine.

The higher rate of work hardening is essentially due to interaction between dislocations on different slip systems and it appears that this produces a considerable amount of tangling and blocking of dislocations. Increased stresses are then required to break dislocations free from these tangles. With increasing temperature, recovery processes become important and some relief is obtained through dislocations cross-slipping and climbing out of tangles, or by screw dislocations dragging out edge dipoles as described by Low and Turkalo (1962). These processes generally lead to a decreasing rate of work hardening in Stage II with both increasing temperature and increasing (OH) content, through their effects on diffusion rates and dislocation mobilities.

The above observations indicate that the processes of Stage II hardening in synthetic quartz are essentially analogous to those in metals (see for example Friedel and Saada, 1958) and there is no direct evidence for the form of work hardening predicted by Griggs' model which could be conveniently described as water-starvation hardening.

An interesting correlation has been established through this analysis between work-hardening processes and the occurrence of features related to optical lamellae. Morrison-Smith and Boland (1975) observed that features corresponding to optical lamellae in these specimens consist of arrays of tangled dislocations with a noticeably higher density than the surrounding areas, both areas being composed of dislocations on at least two slip systems, on the basis of contrast analysis. These observations and the present analysis indicate that these structures are essentially a feature of Stage II behavior in these specimens. It appears that they result from the interaction of dislocations on different slip systems: the alignment possibly corresponding to the edges of Stage I dislocation bands where the dislocations on different systems first interact, and which are therefore the sites of the initial Stage II tangling. The density contrast at these boundaries could then be maintained well into Stage II by continued interactions. However, at some later stage this would become submerged by increasing dislocation densities in the surrounding tangles: behavior which is frequently observed in the lamellae reported by Morrison-Smith and Boland (1975). Unfortunately, there have been insufficient observations of the early stages of lamellae formation to be able to present more justifiable mechanisms.

3.5 Stage III

There have been no definite observations of a final stage of decreasing work hardening in these series of experiments. In those examples where the whole specimen stress-strain curve shows a final decrease, this can normally be associated with a shear failure of the specimen, which frequently runs along the boundary of one of the main deformation (high OH) bands.

4. Conclusions

The work-hardening behavior of synthetic quartz is shown by this discussion to be adequately characterized by a basically simple analysis of the available data. Ideally, of course, experiments should be performed on homogeneous crystals, but so far it has seemed impossible to produce such crystals synthetically. Suitable natural crystals are apparently available, but the impact of (OH) and Na on the deformation characteristics of synthetic quartz implies that this material would behave quite differently from natural quartz, to which these impurities are not intrinsic. In fact, synthetic and natural quartz are quite distinct materials as far as deformation properties are concerned—a point to remember when interpolating synthetic results into real (geological) situations.

References

Alexander, H., Haasen, P.: Dislocations and plastic flow in the diamond structure. Solid State Phys. **22**, 25–158 (1968).
Barnes, R.S., Mazey, D.J.: Stress-generated prismatic dislocation loops in quenched copper. Acta Met. **11**, 281–286 (1963).
Blacic, J.D.: Hydrolytic weakening of quartz and olivine. Ph.D. Thesis. University of California, Los Angeles (1971).
Brown, L.M., Stobbs, W.M.: The work hardening of copper silica: II. The Role of plastic relaxation. Phil. Mag. **23**, 1201–1233 (1971).
Brown, L.M., Woolhouse, G.R.: The loss of coherency of precipitates and the generation of dislocations. Phil. Mag. **21**, 329–345 (1970).
Friedel, J., Saada, G.: Introductory remarks on the nature of strain hardening in single crystals. In: Work hardening, Metallurgical Society Conference no 46 (eds. J.P. Hirth and J. Weertman), p. 1–22. New York: Metallurgical Soc. of Am. Inst. of Mining Engineers 1968.

Griggs, D.T.: A model of hydrolytic weakening in quartz. J. Geophys. Res. **79**, 1653–1661 (1974).

Hirsch, P.B.: Some recent trends in theory of and application of transmission and scanning microscopy of crystalline materials. In: Electron microscopy and structure of materials (eds. G. Thomas, R M. Fulrath and R.M. Fisher), p. 1–22. Berkeley: University of California Press 1972.

Hobbs, B.E., McLaren, A.C., Paterson, M.S.: Plasticity of single crystals of synthetic quartz. In: Flow and fracture of rocks (eds. H.C. Heard, I.Y. Borg, N.L. Carter and C.B. Raleigh), p. 29–53. Richmond: William Byrd Press 1972.

Humphreys, F.J., Hirsch, P.B.: The deformation of single crystals of copper and copper-zinc alloys containing alumina particles: II. Microstructures and dislocation-particle interactions. Proc. Roy. Soc. (London), Ser. A **318**, 73–92 (1970).

Johnston, W.G.: Yield points and delay times in single crystals. J.Appl. Phys. **33**, 2716–2730 (1962).

Low, J.R., Turkalo, A.M.: Slip-band structure and dislocation multiplication in silicon-iron crystals. Acta Met. **10**, 215–227 (1962).

Mendelson, S.: Glide-band formation in silicon. J. Appl. Phys. **43**, 2113–2122 (1972).

Morrison-Smith, D.J.: A mechanical and microstructural investigation of the deformation of synthetic quartz crystals. Ph.D. Thesis, Austr. National University (1974).

Morrison-Smith, D.J., Boland, J.N.: Dislocation structures associated with optical lamellae in experimentally deformed synthetic quartz. In preparation (1975).

Morrison-Smith, D.J., Paterson, M.S., Hobbs, B.E.: The mechanisms of deformation in single crystals of synthetic quartz. Tectonophysics (submitted, 1975).

Weatherly, G.C.: Loss of coherency of growing particles by the prismatic punching of dislocation loops. Ph.l. Mag. **17**, 791–799 (1968).

Weatherly, G.C., Nicholson, R.B.: An electron microscope investigation of the interfacial structure of semi-coherent precipitates. Phil. Mag. **17**, 801–851 (1968).

Weertman, J.: Steady-state creep through dislocation climb. J. Appl. Phys. **28**, 362–364 (1957).

CHAPTER 6.4

The Microstructure of Some Naturally Deformed Quartzites

N.A. LIDDELL, P.P. PHAKEY, and H.-R. WENK

1. Introduction

Deformation processes leave imprints in the submicroscopic structure of minerals which are indicative of the geological history of the rocks. Although we do not yet understand quantitatively the influences of such parameters as temperature, strain, strain rate and even have difficulty in identifying deformation mechanisms in the most common rock-forming minerals, we are able to describe and

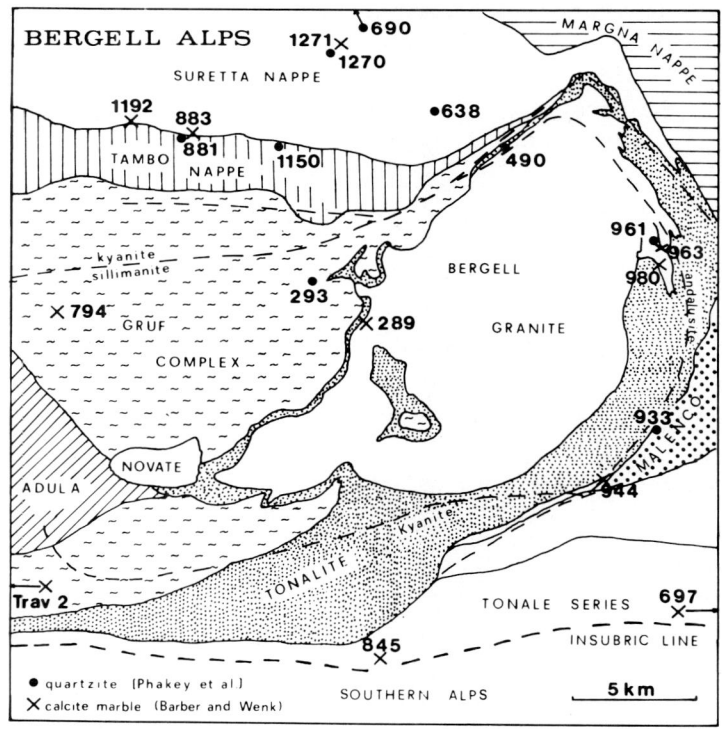

Fig. 1. Tectonic sketch of the Bergell Alps indicating metamorphic zonation and specimen locations

classify observations and try to interpret changes in the microstructure as the result of varying physical conditions using analogies to observations in metals.

It is essential to limit the number of variables and to restrict them to reasonably well-defined parameters. In the case of natural rocks we prefer young rocks with a short history and rocks from a single geological sequence which were subjected to similar deformation processes. A geologically well-defined variable is usually temperature. Quartzites and calcite marbles from the Bergell Alps are such materials (Fig. 1). They were deposited as sedimentary beds in Triassic times, then deformed on thrust planes of the Pennine nappes in early Tertiary (30–50 million years) during the folding of the Alpine belt (Wenk, 1973). In the sample profile under consideration here, metamorphic facies varied from low greenschist to high amphibolite. The evolution of metamorphism can be derived from mineral assemblages (Wenk et al., 1974). Sample locations and their descriptions are given in Table 1. Deformation and subsequent annealing in a series of rocks with a metamorphic gradient produced a varation in microstructure and in this paper we are presenting the transmission electron microscope observations which document processes of recovery and recrystallization in quartz. In a comparative study the microstructure of calcite in adjacent limestones and marbles is analyzed (Barber and Wenk, Chapter 6.5 of this volume). The geological goal is to determine whether metamorphism preceded or outlasted deformation and we hope to get some information from a study of recovery and recrystallization phenomena.

Table 1. Specimen location and description

Sample	Locality	Tectonic unit	Metamorphic index minerals	TEM classification
Sci 690	Avers	Suretta Nappe	Chlorite	Cold work, some recovery
Sci 1270	Gletscherhorn	Suretta Nappe	Chloritoid	Cold work, some recovery
Sci 638	P. Cam	Suretta Nappe	Chloritoid	Cold work, recrystallization beginning
Sci 881	Lago Dentro	Tambo Nappe	Chloritoid	Cold work, some recovery
Sci 1150	Turbine	Tambo Nappe	Chloritoid	Cold work, recrystallization beginning
Sci 933	Preda Rossa	Granite Contact	Sillimanite-andalusite	Recovered
Sci 961	Vazzeda	Granite Contact	Andalusite	Recovered
Sci 490	Albigna	Gruf Complex	Sillimanite-andalusite	Recrystallization
Sci 293	Teggiola	Gruf Complex	Sillimanite	Recrystallization

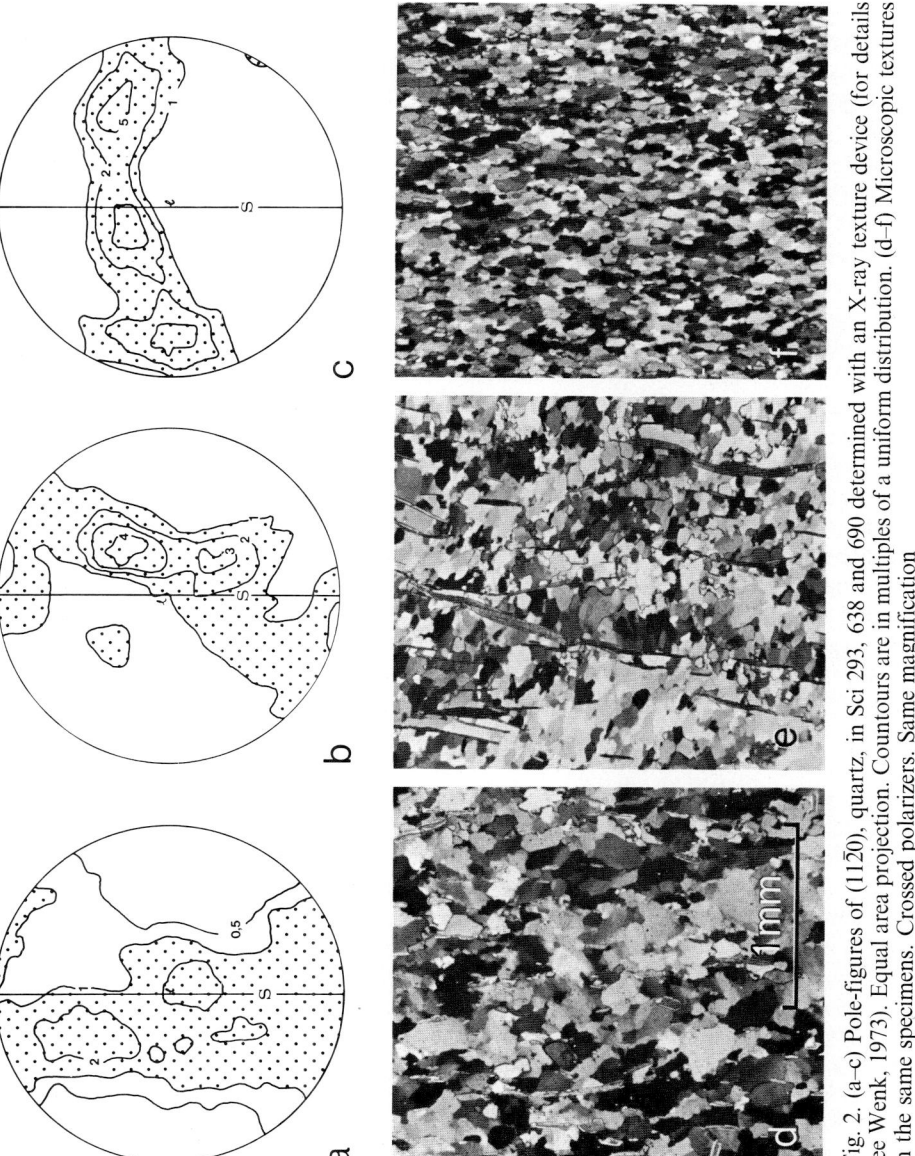

Fig. 2. (a–c) Pole-figures of (11$\bar{2}$0), quartz, in Sci 293, 638 and 690 determined with an X-ray texture device (for details see Wenk, 1973). Equal area projection. Countours are in multiples of a uniform distribution. (d–f) Microscopic textures in the same specimens. Crossed polarizers. Same magnification

2. Observations

The quartzites studied contain small amounts of muscovite and very little feldspar. All samples show strong preferred orientation of quartz grains and pole-figures for Sci 690, 638 and 293 are shown in Fig. 2a–c. (A detailed discussion of preferred orientation in these samples will be given in a separate paper – Bunge and Wenk, in preparation, 1976.) At low metamorphic grade (e.g. Sci 690 and 638) the *c*-axes maximum is roughly perpendicular to the foliation with *a*-axes rather randomly distributed in the plane of the foliation. At high metamorphic grade (e.g. Sci 293) *c*-axes lie in the plane of the foliation normal to the lineation with strong preferred orientation of the *a*-axes. Microscopic textures vary from large deformed crystals with undulatory extinction and occasional deformation lamellae in the greenschist environment (Sci 690, Fig. 2d) to fine-grained recrystallized fabrics in quartzite mylonites and ultramylonites of the high grade zone (Sci 293, Fig. 2h, cf. also Fig. 12 in Wenk, 1973).

Transmission Electron Microscope Observations. Suitable areas were located in thin sections by optical microscopy and ultrathin sections were prepared by ion-beam thinning using a procedure described earlier for fine-grained rocks such as slates (Phakey *et al.*, 1972). These ultra-thin sections were then examined in a JEM 200 electron microscope. The large number of observations can be classified into the following four groups, each of them represented by 2–3 samples from different localities.

1. At lowest metamorphic grade (represented by samples Sci 690, 881 and 1270), the microstructure consists of dense tangles of dislocations forming tightly packed cell walls (Fig. 3a). The distribution of dislocations is heterogeneous and their density is high ($>10^{10}$ cm^{-2}) as is typical of a cold worked material, but does not resemble the extreme dislocation density and tangles observed by Ardell *et al.* (1973) in their coldest experimentally deformed quartzite (600 °C, 10^{-5} sec^{-1}). Sci 1270 which occurs in a similar location to Sci 690, has a lower dislocation density ($<10^9$ cm^{-2}) indicating heterogeneity not only within a specimen but also between related samples. In all specimens of this group some areas contain dislocation loops and evidence of rearrangement of dislocations into low energy configurations (Fig. 3a, b) documenting early stages of recovery.

2. At slightly higher grade, in the stability field of chloritoid, several samples (Sci 1150, 638) are also very heavily deformed with dislocation densities approaching 10^{12} cm^{-2}. Elongated subgrains and tangled walls of dislocations often tending to be parallel to certain crystallographic planes such as $(10\bar{1}0)$ and $(10\bar{1}1)$ are common (Fig. 3c, d). Slight recovery, indicated by dislocation interaction, loops and networks, is apparent in less deformed areas (Fig. 3c) but absent in heavily deformed regions where there are signs of the beginnings of recrystallization (Fig. 3d). Heterogeneity within a specimen is further indicated by sample Sci 638 in which some thin sections reveal only the above features whereas other sections contain many small (0.25–2 µm) recrystallized grains along grain

Fig. 3a–f. Microstructure of greenschist facies quartzites. (a) Heavy deformation in sample Sci 690. (b) Sub-boundaries (S), dislocation interaction (D) and loops (L) are indicative of partial recovery in sample Sci 881. (c) Sample Sci 1150. Tangled walls of dislocations approximately parallel to $(10\bar{1}0)$ and $(10\bar{1}1)$ (traces AA and BB respectively) with subcells tending to be elongated along [0001] and streaking of diffraction spots perpendicular to $\{10\bar{1}1\}$. Note dislocation interaction D and network N. (d) Heavily deformed quartz in sample Sci 1150, with dislocation density $>10^{12}$ cm^{-2}. Note recrystallization at R; M is part of a large mica grain. (e, f) Small recrystallized grains adjoining old deformed grains in sample Sci 638. The diffraction pattern in (f), taken from a collection of crystallites, contains rings A and B due to $\{10\bar{1}0\}$ and $\{11\bar{2}0\}$ respectively, but ring C due to $\{10\bar{1}1\}$ is almost absent

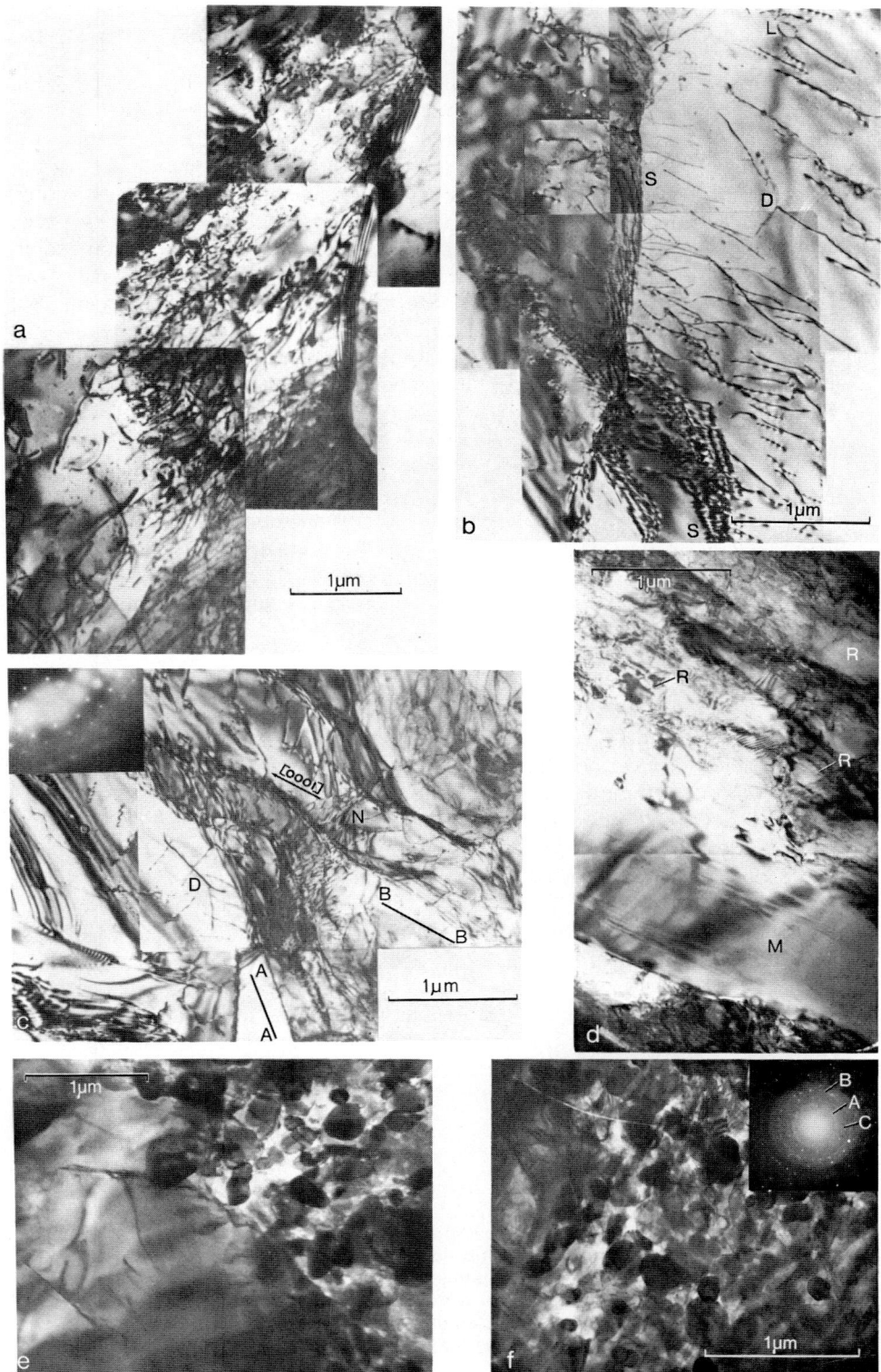

Fig. 3a–f

boundaries or in heavily strained regions (Fig. 3e, f). The diffraction patterns from collections of these grains reveal many orientations. In the diffraction patterns shown in Fig. 3f the rings due to prism reflections are strong but the 3.34 Å ring corresponding to the unit rhomb ($10\bar{1}1$), which should be strong in a randomly oriented polycrystalline quartz sample, contains very few spots. These observations indicate a rather strong preferred orientation for the new crystallites with their c-axes perpendicular to the plane of foliation thus coinciding with the orientation of quartz in the high grade samples (Fig. 2c). However at this stage the correlation between the orientation of the new recrystallized grains and the specimen is poorly established and needs further observations.

3. In the sillimanite zone close to the granite contact (Sci 961, 933) we observe not only a change in preferred orientation (Fig. 2c) but also in the microstructure. In these samples *recovery is very advanced*. A high temperature environment is clearly indicated by low dislocation densities (in some regions about 4×10^8 cm^{-2}, other grains resemble undeformed quartz), loops, bubbles and ample dislocation interaction often forming well-developed low-angle boundaries (Fig. 4a, b). Submicroscopic evidence for recrystallization is absent, although these coarse-grained rocks from the granite contact must have undergone complete recrystallization during their history as suggested by the microscopic textures.

4. Along the northern border of the Bergell granite and in the Gruf complex there are zones of heavily deformed rocks with layers of ultramylonites. Some of them are mono-mineralic quartzites with ribbon texture and extremely high preferred orientation (c-axis maximum in Sci 293 is 25–30 times uniform distribution; see Bunge and Wenk, in preparation, 1976). Smooth-grain boundaries often meeting in triple junctions at angles approaching the 120° equilibrium configuration with low dislocation density (10^8–10^9 cm^{-2}) within the grains characterize the microfabric of the quartz mylonites (Sci 293, 490) as *recrystallized* (Fig. 4c). Long well-developed hexagonal networks of dislocations are common even in the recrystallized grains (Fig. 4c, d) indicating that deformation and recovery continued during grain growth. The samples appear to be *hot worked*.

3. Discussion

Even in the case of these quartzites with a short and simple geological history, the number of variables (such as temperature, pressure, strain rate and annealing time) is large and it is no surprise that the microstructure is very complex. Other factors which limit the interpretation of these features are the heterogeneity

---

▶

Fig. 4a–d. Microstructure of amphibolite facies quartzites. (a,b) Loops (L), dislocation interaction (D), sub-boundaries (S,A,B) and low dislocation density ($<5 \times 10^8$ cm^{-2}) documenting recovery in samples Sci 961 and 933. Sub-boundaries S and A consist of simple arrays of parallel dislocations; the dislocations in B are out of contrast. (a) Sci 961 (b) Sci 933. (c) An example of recrystallized grains in sample Sci 490. Note the nearly equilibrated grain boundaries and the well-developed network within a recrystallized grain. (d) A well-developed hexagonal network of dislocations indicating recovery in sample Sci 293. The dislocation density within subgrains is very low ($<4 \times 10^8$ cm^{-2})

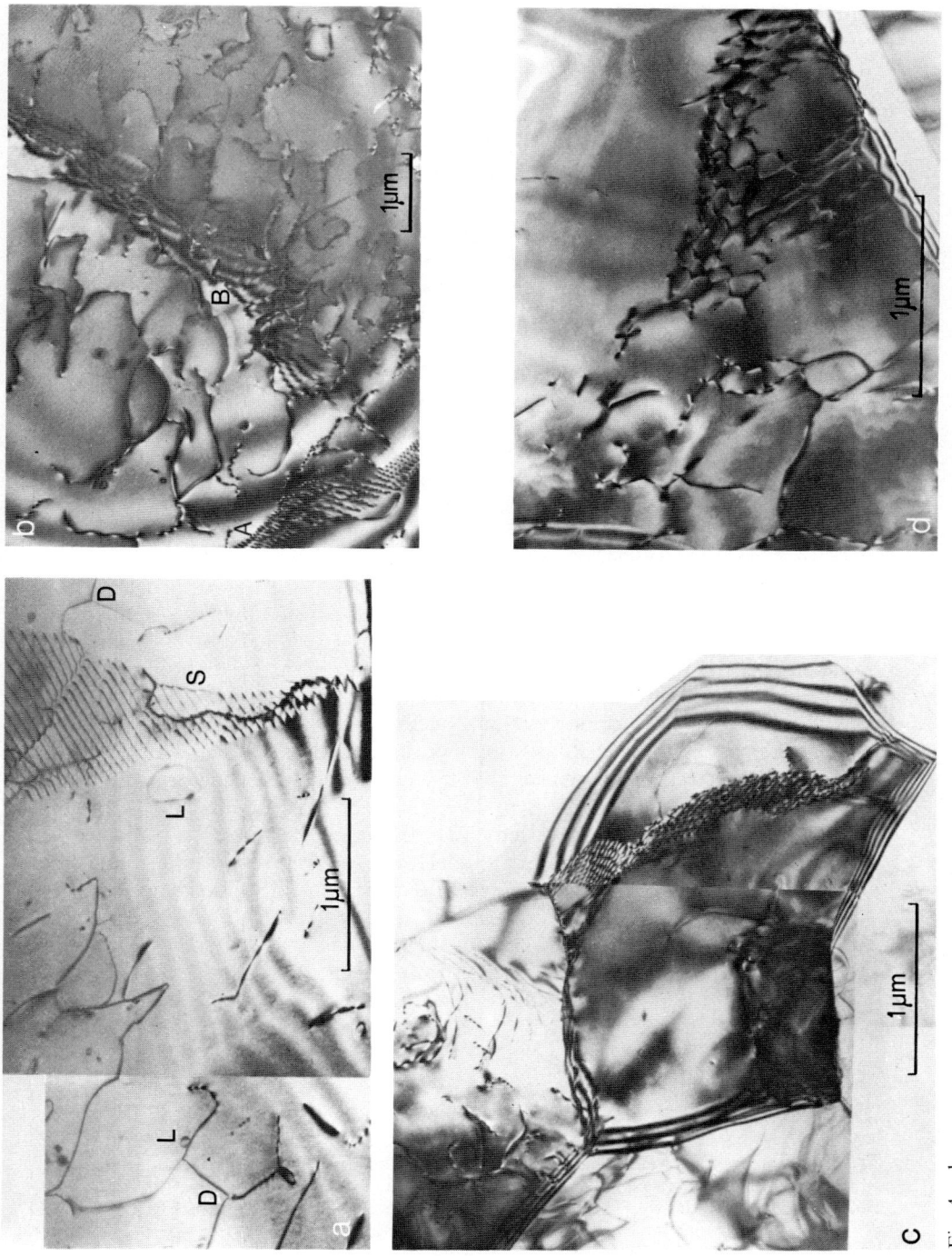

Fig. 4a–d

of processes which control deformation and recrystallization in rocks on a submicroscopic scale and also the ease with which older features are erased in the course of the geological history. Being aware of this and exercising caution, the observations permit us nevertheless to draw some conclusions and illustrate that qualitative microscopy can even at this stage provide significant geological information.

All nine samples from the studied profile show some recovery. High metamorphic grade samples (Sci 490, 293, 961, 933) are definitely more recovered than the lower metamorphic grade samples (Sci 881, 690, 1270, 1150, 638), indicating that the metamorphic gradient outlasted deformation although temperatures were likely lower than during the climax of metamorphism and main Alpine recrystallization in this area. However, as is shown by a study of mineral parageneses (Wenk et al., 1974) dislocations in recrystallized grains (Fig. 4c) point to the fact that rocks were still deformed in the Bergell Alps after the second recrystallization started. On the basis of the microstructure of quartz it is possible to distinguish between amphibolite and greenschist facies regime. Within the greenschist facies due to variables other than temperature the microstructure is too heterogeneous to see a gradient. Recrystallization occurred during and after deformation at moderate to high temperature but seems to be restricted to strongly strained material. An interesting observation is the presence of water bubbles in many of these natural quartzite specimens. Since quartz crystals have reached the limit of H_2O solubility they must have been well within the regime of hydrolytic weakening when they were deformed (Griggs and Blacic, 1965). We emphasize that the two mylonites studied here demonstrate mylonitization as a process of plastic deformation with recovery and recrystallization but without any sign of cataclasis, which supports the conclusions of Carter et al. (1964). These rocks, which look extremely deformed in hand specimens and have therefore been mistaken in the past as products of post-tectonic shear zones and faulting (Schmutz, 1974), have a very clear microstructure proving that they were deformed during metamorphism.

Similar microstructures in naturally deformed quartzites from Central Australia have been described by McLaren and Hobbs (1972). The results presented in this paper compare well with features found in experimentally deformed quartzites (Ardell et al., 1973) but in our case it has not been possible to define an isotherm above which recrystallization starts. The onset of recrystallization appears to be strongly controlled by the strain energy of the crystals.

References

Ardell, A.J., Christie, J.M., Tullis, J.A.: Dislocation substructures in deformed quartz rocks. Crystal Lattice Defects **4**, 275–285 (1973).

Bunge, H.J., Wenk, H.-R.: Three-dimensional texture analysis of three quartzites (trigonal crystal and triclinic specimen symmetry). In preparation (1976).

Carter, N.L., Christie, J.M., Griggs, D.T.: Experimental deformation and recrystallization of quartz. J. Geol. **72**, 587–733 (1964).

Griggs, D.T., Blacic, J.B.: Quartz: Anomalous weakness of synthetic crystals. Science **147**, 292–295 (1965).
McLaren, A.C., Hobbs, B.E.: Transmission electron microscope investigation of some naturally deformed quartzites. In: Flow and fracture of rocks (eds. H.C. Heard, I.Y. Borg, N.L. Carter and C.B. Raleigh). Geophys. Monogr. **16**, 55–66 (1972).
Phakey, P.P., Curtis, C.D., Oertel, G.: Transmission electron microscopy of fine-grained phyllosilicates in ultra-thin rock sections. Clays Clay Minerals **20**, 193–197 (1972).
Schmutz, H.: Geologie des Ophiolitzuges von Chiavenna. Ph. D. thesis, ETH, Zürich (1974).
Wenk, H.-R.: The structure of the Bergell Alps. Eclogae Geol. Helv. **66**, 255–291 (1973).
Wenk, H.-R., Wenk, E., Wallace, H.H.: Metamorphic mineral assemblages in pelitic rocks of the Bergell Alps. Schweiz. Mineral. Petrog. Mitt. **54**, 507–554 (1974).

CHAPTER 6.5

Defects in Deformed Calcite and Carbonate Rocks

D.J. BARBER and H.-R. WENK

1. Introduction

Deformation of calcite crystals and rocks was the subject of an intensive investigation between 1950 and 1960. Through U-stage analysis of experimentally deformed material, deformation mechanisms were established and their variations were studied over a wide range of temperatures (for example Turner et al., 1954; Griggs et al., 1960). From this work it was concluded that at low temperatures, deformation occurs largely by twin gliding on $e=\{01\bar{1}2\}$ [1] with $\langle 01\bar{1}\bar{1}\rangle$ as glide direction and at higher temperature translation gliding on the unit rhomb $r=\{10\bar{1}1\}$ with $\langle \bar{1}012\rangle$ as glide direction dominates. Other mechanisms such as translation on the a-prism (Paterson and Turner, 1970) or on $f=\{02\bar{2}1\}$ (Keith and Gilman, 1960) can also operate but because critical shear stresses are much higher, these systems are more rarely observed. Preferred orientation in deformed aggregates is in agreement with deformation predominantly by r-translation and e-twinning (for example Wenk et al., 1973).

Our aim was to study deformed calcite with the electron microscope in order to determine how microscopical features present themselves on the submicroscopical scale. In this program which we started in 1971, we proceeded along various routes.

1.1 An Empirical Study of Experimentally Deformed Limestone (Barber and Wenk, 1973)

Specimens of fine-grained limestone deformed in uniaxial compression over a wide range of temperature and pressure conditions (Wenk et al., 1973) provided material for a first general survey of microstructures in carbonates. We observed at low temperature (200 °C) heavily "cold-worked" structures, with cracks reducing the material to mosaic blocks containing largely unresolvable defects. At higher temperature (300–600 °C), the microstructure is less confused: some grains

---

[1] Calcite has rhombohedral symmetry, $R\bar{3}c$, but is usually indexed with hexagonal axes. The hexagonal structural unit cell $a=4.990$ Å and $c=17.061$ Å is used in X-ray studies. For morphological work a z-axis of one quarter the length is chosen such that the cleavage rhombohedron r has four-axis hexagonal indices $\{10\bar{1}1\}$. This usage which has been generally adopted in the deformation studies is applied in this article.

contain only tangles of dislocations while others exhibit ragged bands of deformation. Submicroscopic twinning on e is abundant. Above 800 °C twinning still occurs but is less profuse; dislocation densities are generally lower. Dislocations form loose networks and there are many dipoles and loops, indicative of climb.

Quantitative analysis of dislocation contrast was difficult in samples not subject to climb because the large overall strain (30–50%) gave rise to very high dislocation densities.

1.2 An Investigation of Naturally Deformed Carbonate Rocks

Concurrent with the investigation of experimentally deformed material for which conditions are well defined, we studied natural carbonate rocks to explore if similar defects occur and whether the microstructure of calcite contains any useful geological information. We were specifically interested in recovery and recrystallization phenomena, hoping that they might give information on conditions which prevailed during deformation. We collected samples of carbonate rocks along thrust faults in the Central Alps, mainly in the Bergell Alps (cf. Fig. 1 in Liddell et al., Chapter 6.4 of this volume). Most of these rocks were deposited in Triassic in the Alpine geosyncline, then deformed during the folding of the Alpine belt in early Tertiary. Superposed on the deformation was the regional Lepontine metamorphism which ranged in grade from greenschist to high amphibolite facies and in addition in the Bergell Alps an intrusion of granite (Wenk, 1973; Wenk et al., 1974). The time relation between metamorphism, granite intrusion and deformation is not entirely resolved and we expected to get additional information from this study of microstructures. The samples were chosen in order to represent a variety of metamorphic grades. Some were collected along the dominant normal fault in the Alps, the Insubric line, which separates the Tertiary metamorphic Central Alps from old rocks of the Southern Alps and thus definitely postdates the Alpine recrystallization (e.g. Gansser, 1968).

1.3 Experimentally Deformed Calcite Single Crystals

Dislocations introduced by deformation are being studied in experimentally deformed Iceland-spar from Chihuahua (Mexico). Experiments have so far been done with a Griggs-type piston-cylinder apparatus using talc as confining pressure medium (Griggs, 1967). They will be complemented by more controlled experiments with a gas apparatus. Preliminary results from this study, which is still in progress, will be included briefly at the end of the paper.

At this stage we prefer to be descriptive rather than deductive. We would like to make the reader aware of the many open problems, rather than rely heavily on conclusions based on analogies to metals which have such different structural properties from calcite. Thus the following pages are mainly used to present a selection of observations.

2. Specimen Preparation and Experimental Procedure

Specimens thin enough for TEM examination were prepared from optical thin sections by ion-thinning with 5 keV Ar^+ ions. Optical examination of the ion-thinned samples (foils) was found to be particularly valuable in comparison with the TEM results and especially useful in identifying features which possessed overlapping images when examined optically in the standard thin sections.

Calcite suffers more severely from ion and electron damage than do many minerals, the damaged surface layers becoming largely amorphous. In the thinnest region of the foils, these layers appear to strain the calcite which they envelop. Since we have observed that dislocations are sometimes able to move in the thin foils, it is possible that the dislocation structures which can be easily seen with 100 keV electrons are somewhat unrepresentative of the bulk. Higher electron-accelerating voltages must be used, even though the displacement damage caused by the electron beam is increased. Moreover, although the susceptibility

Table 1. Carbonate rocks from the Alps (mainly Bergell region)

Specimen	Origin	Metamorphic facies
Hv 4	S. Grand Chavalard, Valais (thrust plane at base of Morcles nappe, Helvetic nappes, Triassic)	Unmetamorphic
Trav 2	Val Traversagna, Grisons (Lepontine, rootzone)	Amphibolite but deformed later (mylonite)
Sci 697	Triangia-Sondrio, Valtellina (N. of Insubric Line, Tonale series)	Greenschist but deformed later
Sci 845	Serone, Valtellina (N. of Insubric Line, Tonale series)	Amphibolite but deformed later
Sci 1271	Gletscherhorn (Suretta nappe, Triassic)	Greenschist (chloritoid)
Sci 1192	Acqua Fraggia (Tambo nappe, Triassic)	Greenschist
Sci 883	Passo Turbine (Tambo nappe, Triassic)	Greenschist
Sci 289	Pizzo Porcellizzo, Val Codera (thrust contact between Bergell granite and Gruf migmatites)	High-amphibolite
Sci 944	Preda Rossa (Val Masino)	Amphibolite (wollastonite)
Sci 980	Val Sissone-Vazzeda, upper Val Malenco (contact zone of Bergell granite)	Granite contact (tremolite)
Sci 963	Cma. di Vazzeda N crest (contact zone of Bergell granite)	Granite contact (tremolite)
Sci 794	Val Schiesone, Valle della Mera (regionally metamorphic Gruf complex)	Granulite

Defects in Deformed Calcite and Carbonate Rocks   431

to electron-beam damage varies from sample to sample, it can seldom be avoided completely and it leads to somewhat "fuzzy" dislocation images. The observations reported here were obtained with an EM 7 electron microscope operating at 1 MV with the exception of Fig. 5e, which was taken on a JEM 200 kV instrument.

3. Experimental Observations

3.1 Deformed Limestones and Carbonate Rocks

In discussing the TEM-results we have ordered the rocks with increasing metamorphic grade and divided them into three groups. A summary of the specimen description, including locality, geological setting and microstructure is presented in Table 1 (see also Fig. 1 in Liddell *et al.*, Chapter 6.4 of this volume).

Table 1 (continued)

Dislocation density (cm^{-2})	Microstructural characteristics
Variable $5 \times 10^8 \to 5 \times 10^9$	Dislocations forming networks, subcells, some grain boundaries opening, dislocation sources. Recovery.
Very variable $10^8 \to 5 \times 10^9$	Tangles of elongated dislocations, some areas clearing, some small strain-free volumes.
$\sim 3 \times 10^9$	Heavily deformed, no climb or recovery, some cracks, microtwins, shear across "old" walls.
$\sim 2 \times 10^9$	Dislocation density uniform, tangles of elongated dislocations, some signs of recovery.
$> 10^{10}$	Very heavily cold-worked, microcracks, twinning common, no recovery.
Variable $10^7 \to 5 \times 10^8$	Recrystallized followed by slight deformation (cracks, shear across twins etc.), partly recovered.
$\sim 10^9$	Broad bent twins, dislocation density uniform, "creep-like" configurations, recovery subordinate.
$< 2 \times 10^8$	Considerable twinning, majority of dislocation structures are recovered, but not uniformly so.
$10^7 \to 10^8$	Dislocations mostly arrayed in tilt sub-boundaries. Pores, healed cracks in evidence. Recrystallized.
2×10^8 (10^9 occ.)	Mostly highly recovered structures in previously recrystallized fabric. Subsequent slight creep?
$\sim 10^7$	Recrystallized, followed by light deformation and recovery. A few broad twins.
$10^8 \to 7 \times 10^9$	Heavily deformed at lower end of creep regime? Some microtwinning and cracking. Heterogeneous.

3.1.1 Low Grade

In the first group of limestones, on geological evidence, temperature during and after deformation has never exceeded 50–200 °C. Either they are unmetamorphosed sedimentary rocks (Hv 4) or they are medium- to high-grade marbles which have been cold-worked later on, probably in connection with shear movements in the vicinity of the Insubric Line (Trav 2, Sci 697 and Sci 845).

Hv 4 is a strongly deformed very fine grained Triassic limestone. The pole-figure of [0001], determined with X-rays (Fig. 1a) shows a single c-axes maximum perpendicular to the foliation plane, but elongated towards the lineation which appears to be the direction of thrust. The foils have a fairly high but variable dislocation density ($\sim 10^9$ cm^{-2}), with many of the dislocations occurring in the walls of ill-defined sub-cells and others forming loose irregular three-dimensional networks. The occurrence of dislocation sources with dislocations tending to align along crystallographic directions (Fig. 2a) suggests that climb was operative. The absence of the extensive microcracking found in samples experimentally deformed at low temperatures indicates that the deformation rate was never excessively high and that recovery was competitive with the deformation processes so that stresses sufficiently large to initiate fracture did not build up. Twinning is also largely absent. Fig. 2b is another micrograph of Hv 4, illustrating twist and sub-cell boundaries.

The next sample, **Trav 2** is an extremely fine-grained calcite mylonite, which appears in the field like an unmetamorphosed limestone in the midst of the regionally metamorphic Lepontine area which is typical for annealed post-tectonic marble fabrics (Gansser and Dal Vesco, 1962). Inclusions of broken calc-silicate minerals point to the high temperature origin but we attribute the mylonite fabric to deformation in one of the post-metamorphic shear zones. Preferred orientation

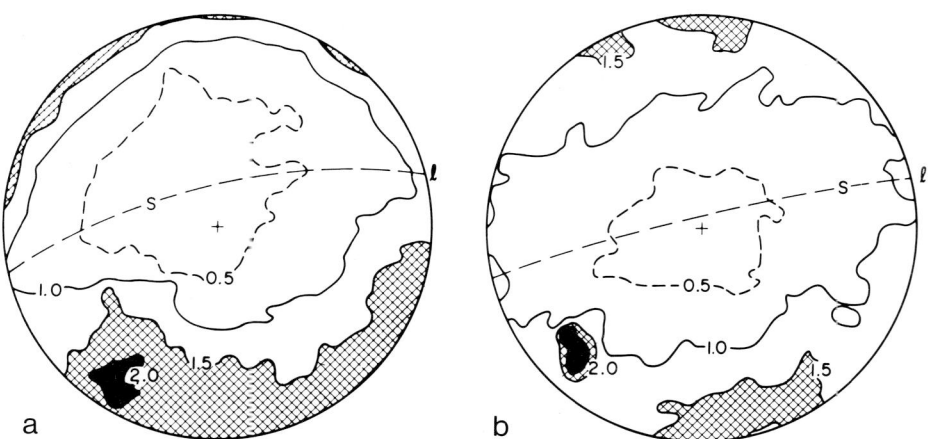

Fig. 1a and b. Pole-figures for 0006 determined with an X-ray texture device. (a) Hv 4, (b) Trav 2. Notice the elongated c-axes maximum perpendicular to the foliation plane s; l is the direction of the lineation. Contours are in multiples of a uniform distribution. Equal area projection

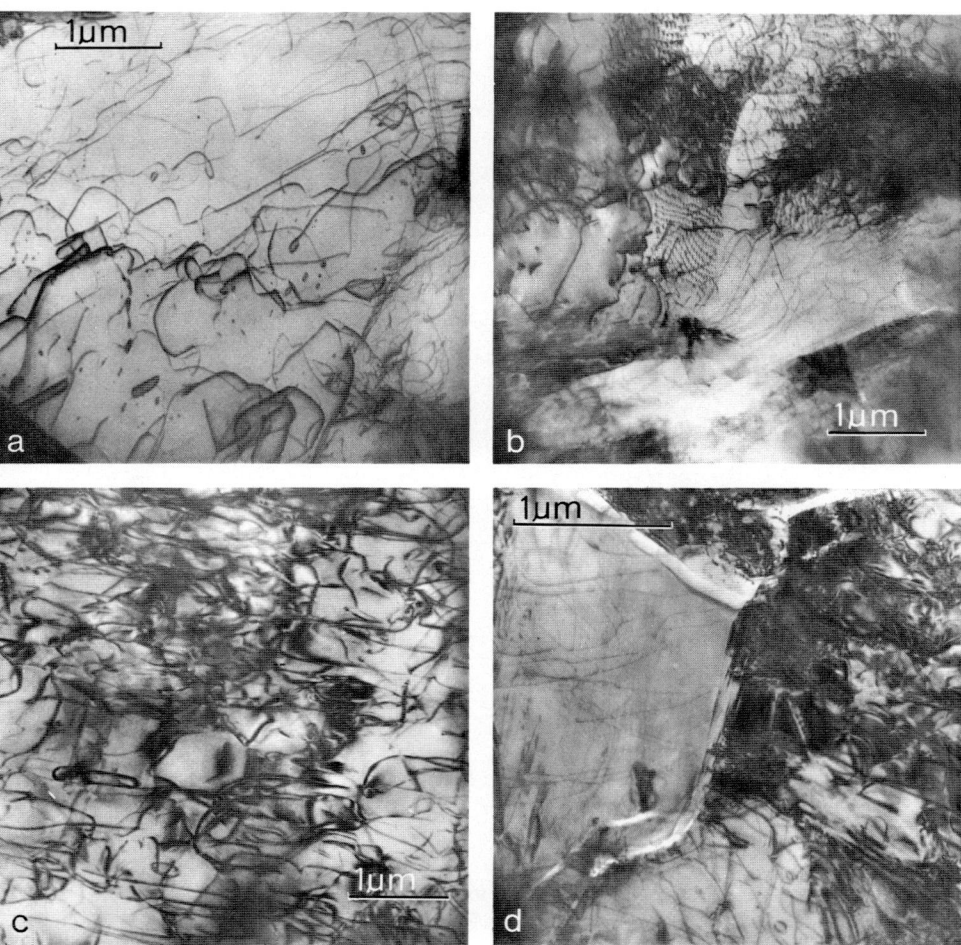

Fig. 2a–d. Microstructures of low metamorphic limestones. (a) Glide dislocations climbing within sub-cell in Hv 4. (b) Twist networks and other sub-cell boundaries in Hv 4. (c) Elongated tangles of dislocations in calcite mylonite Trav 2. (d) Numerous dislocations and some fracture in Sci 697, which is largely unrecovered

is similar to Hv 4 (Fig. 1b). In HVEM the dislocation density is again found to be extremely variable and small grains are present with low dislocation densities, which are apparently the result of recrystallization rather than being relict grains. Tangles of elongated dislocations commonly occur, sometimes crossing what appear to be old grain boundaries. A region with a fairly uniform distribution of dislocations is shown in Fig. 2c. Deformation on well-defined crystal planes and twinning are absent; it appears that we are again witnessing the result of deformation at low strain rates, with continuous concurrent recovery, so that old features are largely obscured. But clearly this rock has been subject of a more complicated history than Hv 4.

In **Sci 697** HVEM reveals that the most recent event has been a fairly heavy deformation without significant recovery. The numerous microtwins also suggest that the temperature of deformation was low. Long dislocations cross relict features such as the locations of old grain boundaries and cracks. It appears that the rock has recrystallized prior to the most recent deformation. As shown in Fig. 2d, cohesion along the grain boundaries is poor; some open up during ion thinning (an effect which can, for example, result from impurity concentrations) but the small crystalline fragments seen along some of the "open" boundaries are evidence of some cataclasis in the last deformation, again pointing to low ambient temperatures and higher strain rates than for Hv 4 and Trav 2. It seems reasonable to conclude that the deformation seen in Sci 697 was indeed associated with vertical movements along the young Insubric fault, even though these rocks probably have a long history of deformation.

Sci 845 is in its geological setting comparable to Trav 2 and Sci 697. The dislocation density in this specimen is uniformly fairly high, about 2×10^9 cm^{-2}, as in Sci 697. But the dislocation configurations show more signs of recovery than in Sci 697, so that some volumes of the rock have similarities when viewed by HVEM to sample Hv 4.

3.1.2 Greenschist Facies

The next group of rock samples comprises deformed Mesozoic rocks of greenschist facies metamorphism in the northern part of the Bergell Alps. The group consists of samples Sci 1271, 1192 and 883 (they should be compared with the quartzites samples Sci 1270 and Sci 881 described by Liddell *et al.*, Chapter 6.4 of this volume). Together with samples of the last group (3.1.3), they are used to discuss the influence of metamorphic grade on the microstructure.

Sci 1271 is the only one of the Bergell limestones so far examined to show features produced by a very intense (i.e. high strain rate) deformation. The dislocation density is very high ($>10^{10}$ cm^{-2}) and cracking is prevalent. As shown in Fig. 3a, the crystallites are fragmented and sheared into elongated blocks, a phenomenon which is very reminiscent of rocks which have been experimentally deformed at low temperatures and/or high strain rates. There is a good correlation, for example, with VW 140 (200 °C, 10^{-4} sec^{-1} strain rate; Barber and Wenk, 1973) and with a shatter-cone limestone from Wells Creek Basin, U.S.A. (Dietz, 1959) which we have examined. The latter is fractured into very many misoriented blocks, i.e. it exhibits the characteristic mosaicism often shown by shocked minerals. A few small blocks of the rock are undeformed, but in most of the blocks the dislocations are so numerous as to be scarcely resolvable, as illustrated by Fig. 3b. Misorientations cause some regions to be out of contrast, but there is a uniformly high density of dislocations. Sci 1271 is also extensively twinned, with the production of the chevron-like structures resulting from the activation of primary and secondary (crossing) *e*-twin planes. Such structures are familiar to us from single crystal HVEM work (see next Section).

It is interesting to note at this point that the corresponding quartz specimen Sci 1270 also is somewhat exceptional in showing very strong "cold-work" deformation (Chapter 6.4 of this volume).

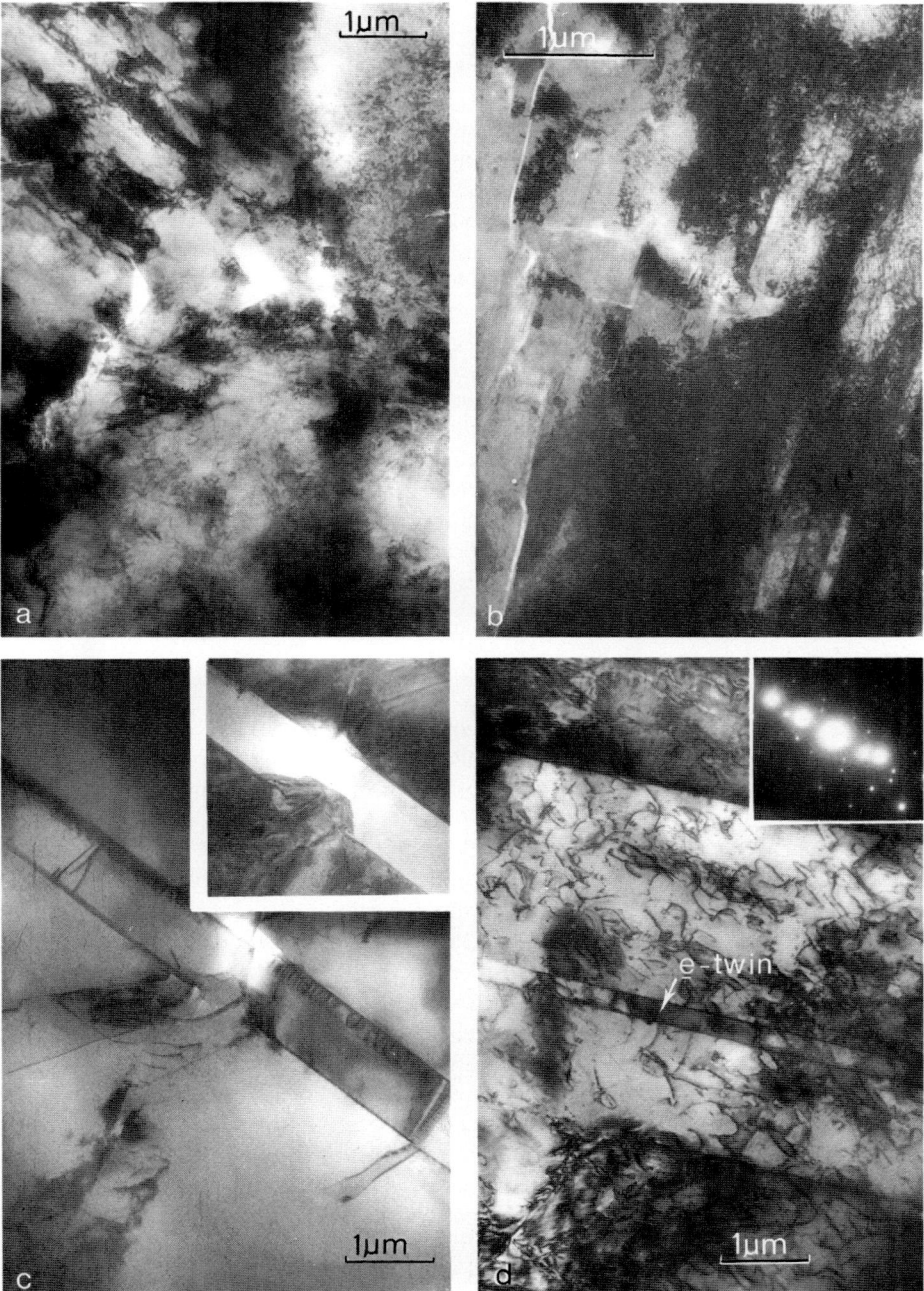

Fig. 3a–d. Microstructure of greenschist facies marbles. (a) High dislocation densities, fracture and the creation of mosaic blocks by cold-working in Sci 1271. (b) High dislocation densities and mosaicism caused by fracture and shear in a shatter-cone limestone from an impact crater (Wells Creek Basin, U.S.A.). (c) Evidence of shear across a twin lamella in marble Sci 1192 (inset: tilted to show intrusion into twin). (d) Twin lamellae in Sci 883; the twins traces correspond to the *e*-twinning law

The second rock in the group, **Sci 1192**, differs markedly from Sci 1271. Its dislocation density is generally low, though variable (10^7 to 5×10^8 cm^{-2}) and it has clearly recrystallized into a marble. There is evidence of healed cracks, there are pores and a few twins, probably incorporated during recrystallization. But there is also some slight deformation and recovery subsequent to recrystallization. Shear traces cross "old" features such as twins and some dislocations have been introduced into the edges of grains. Fig. 3c shows dislocations remaining in a region where a deformation band has intersected a relict twin; microcracks remain in the trace of the band.

A profusion of broad, curved twins and shear zones characterize **Sci 883** Clearly, like Sci 1271, the sample has been heavily deformed. But although the dislocation density reflects this ($\sim 10^9$ cm^{-2}), the fragmented misoriented blocks found in Sci 1271 are absent and cracking is rare. One must conclude that the deformation/temperature regime was less severe for Sci 883 and that some degree of recovery has occurred, probably concurrent with deformation rather than subsequent to it. Fig. 3d shows dislocations in moderate numbers within the twin bands, but not aligned in crystallographic directions as is typical when grains are twinned at low temperatures. The grain size is large, as we would expect for a recrystallized fabric.

3.1.3 High Metamorphic Grade, in Part Rocks from the Granite Contact

The last group of rocks, Sci 289, 944, 980 and 963 have been deformed in the vicinity of the Bergell granite, while Sci 794 is from the Gruf complex, representing material subjected to regional metamorphism of high temperature and high pressure.

Sci 289 is a very pure calcite marble with an exceedingly large grain size. The dislocation density is low but crystallites are commonly twinned. Most of the twins are broad and parallel; here the associated dislocation structures are mostly recovered as shown in Fig. 4a. But bent twin lamellae also occur, sometimes with cracking along the twin boundaries. Nearby, elongated configurations of dislocations have been found which are not tightly tangled and show evidence of climb. From the origin of the sample, we know that it must have undergone severe deformation. The recrystallized and recovered microstructures tell us, however, that the thermal influence has outlasted the main deformation period, although possibly at a lower grade than the maximum attained.

The most completely annealed of all the samples examined is **Sci 944**, a wollastonite marble from the meta-sedimentary mafic-pelitic-carbonate zone surrounding the Bergell granite. It must have been deformed during the granite emplacement but its microstructures show that subsequent annealing at moderate pressure and high temperature has completely erased deformation features. The dislocation density is very low ($\sim 10^7$ cm^{-2}) and most of the dislocations are contained in grain boundaries, partially-healed cracks, around pores etc. Many of the grain boundaries contain polygonal well-equilibrated pores, illustrated in Fig. 4b; elsewhere they contain small precipitates.

Samples **Sci 980** and **963** both belong to a unit of roof pendants in the Eastern contact zone of the Bergell granite. The dislocation densities are similar

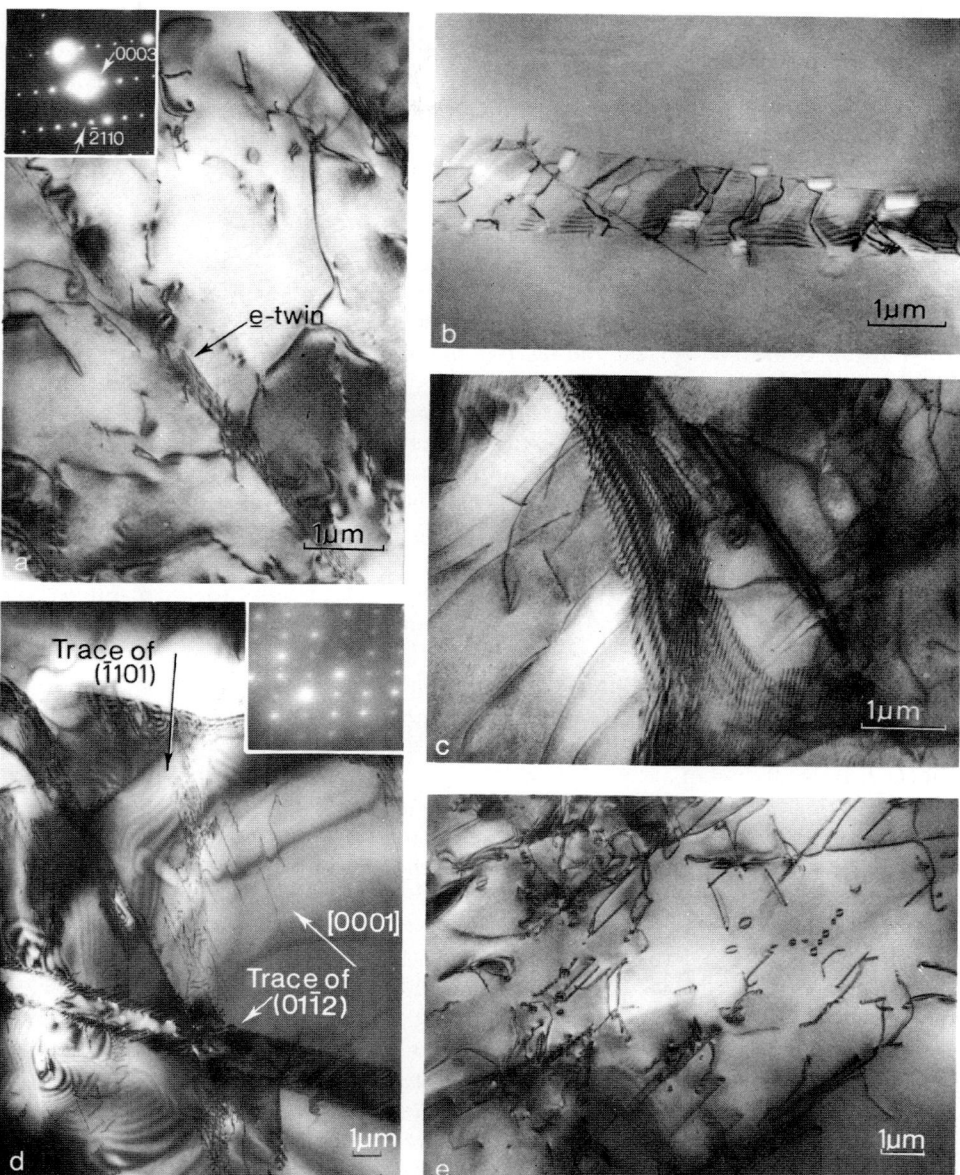

Fig. 4a–e. Microstructure of amphibolite facies marbles. (a) Twins and recovered dislocations in Sci 289. (b) Dislocations and pores in a low angle grain boundary in marble Sci 944. (c) Subgrain boundaries in Sci 980; the background spottiness is incipient radiation damage. (d) Slip dislocations emanating from a sub-grain boundary and crossing an *e*-twin in Sci 963. (e) Dislocations, dipoles and trails of loops in Sci 794

(2×10^8 cm^{-2}) and in both cases there has been some deformation outlasting or postdating the recrystallization of the marble fabric. In Sci 980 the structures are more recovered than in either Sci 1192 or Sci 963, but the operation of dislocation sources still occurs in Sci 980, even though most dislocations are incorporated in or linked to sub-boundaries (Fig. 4c). In Sci 963 there are signs of a late, very gentle plastic deformation which has produced a low density of dislocations (Fig. 4d) in what appear to be ragged slip bands which cut across subgrain boundaries. Serrated microcracks have also appeared, indicating that the deformation occurred at low temperatures and confirming that it was late.

Sci 794 is from a calcsilicate zone in the central part of the Gruf migmatites. Wenk *et al.* (1974) have shown that this tectonic unit has crystallized at 800–1000 °C and 8–9 kbars corresponding to a deep-seated crustal environment. Anorthite from the same calc-silicate sequence has widely spaced (c)-type antiphase boundaries which Müller *et al.* (1972) attributed to slow cooling under regional metamorphic conditions. The dislocation structures in this rock vary somewhat from area to area. There are grains containing broad twins and low numbers of creep-like dislocations, pulled out into dipole forms and generating trails of dislocations loops. This is illustrated in Fig. 4e. Microtwins also occur and the surrounding grains exhibit some cracking, much residual strain and fairly high densities of unrecovered dislocations. The rock appears to have suffered heterogeneous deformation, but the possibility of variations in composition also influencing the degree of recovery cannot be excluded.

In conclusion we would like to point out that there is a good correspondence between geological prediction and observed microstructure. Initially the specimens were classified according to their microstructure and the microscope operator (D.J. Barber) did not at that time have any information about their geological history. In this way it was hoped to avoid the introduction of any bias in interpreting the origins of the microstructures observed and to avoid looking for features which we expected. Yet both classifications agree very closely. We notice a good correlation between the extent of recovery and metamorphic grade and can easily distinguish amphibolite facies from greenschist facies samples on the basis of their microstructure. These results are similar to those observed in corresponding quartzite specimens (Liddell *et al.*, Chapter 6.4 of this volume). Particularly low dislocation densities are typical of calcite in the vicinity of the Bergell granite which indicates rather high temperatures during the final emplacement of the granite. It is striking that recovery – although to a lesser extent – is also present in samples deformed at lowest metamorphic grade (e.g. along the Insubric line or in unmetamorphic nappes) presumably at slow strain rates. Plastic flow with ample movement of dislocations is observed in calcite at much lower temperature than in quartz. This limits the value of the microstructure of calcite as an indicator for conditions during the main tectonic event: it is easily affected by late shearing stresses such as occur on post-tectonic shear zones and local stress heterogeneities in the fabric during cooling. At this stage TEM observations may support a geological model but they cannot yet be used as a decisive argument for a theory. Nevertheless the similarity in microstructure of naturally and experimentally deformed calcite suggest that deformation mechanisms were probably similar even though conditions (especially strain rates) were different.

3.2 Experimentally Deformed Crystals of Iceland Spar

Single crystals of Iceland spar were deformed in a piston cylinder apparatus. All of the results presented here relate to one particular sample, VW 132, which was compressed at 25 °C to a total strain of 10% at a strain rate of 10^{-4} sec^{-1} under 10 kbars confining pressure. This and other crystals which were compressed in an orientation favorable for e-twinning show lamellae in agreement with expectations. There is good correlation between HVEM results and optical observations. In most regions of the test cylinder there is primary twinning on $\{01\bar{1}2\}$ and where secondary twinning also occurs a herringbone or chevron type of structure is created. This is shown in the optical micrograph of an ion-thinned foil (Fig. 5a). An electron micrograph of a region containing only primary e-twins is presented in Fig. 5b. Here the foil has been tilted so that both the matrix and twins are diffracting strongly. Thus the reflections from both lattices are apparent in the selected area diffraction pattern; under these conditions the dislocations images tend to be broad and some are doubled. The secondary e-twin lamellae are always thin ($\gtrsim 1$ μm) but are otherwise unremarkable. We have observed submicroscopic e-twins in many samples. As has been shown by Hauser and Wenk (in press, 1976), these sub-microscopic twins give rise to optical biaxiality, which is common in calcite deformed at low temperature.

A very interesting feature of VW 132 is the occurrence of short deformation features which trace analysis shows to be the activation of $\{10\bar{1}1\}=r$-planes. But the mode of deformation is not simple glide, as might be anticipated for the r-planes but twinning on r, as previously reported to occur during extension below 30° C by Paterson and Turner (1970). VW 132 exhibits these r-twins to a considerable extent away from the ends of the cylinder. Some of the r-twin lamellae are broad enough to be seen in standard thin section and these occur where there are closely spaced, parallel e-twins and cross them. But the r-twins which have dislocation structures simple enough to merit study by TEM are extremely narrow lamellae and can only be resolved as such optically by looking at ion-thinned sections (~ 1 μm thick), as shown in Fig. 5c. For comparison, Fig. 5e shows r-twins, labeled AA' and BB', in a foil (A3) from VW 132 which was cut at approximately 21° off a $\{\bar{2}110\}$ plane. Closely spaced dislocations are visible in the $\{10\bar{1}1\}$ twin boundaries. XX' and similar lamellae are the more common primary e-twins. At Y, close to a region where several deformation mechanisms are operative, there is the extremity of a kink band. The features labeled CC' and DD' are thought to be especially significant, although their definition suffers from the fact that the foil was beginning to show visible electron-beam damage. They are short, slightly wavy faults, exhibiting weak fringes and apparently terminating on dislocations, with traces parallel to the trace of $\{10\bar{1}1\}$ in the foil. It is very unlikely that they are thin r-twins because these are always straight and they usually begin and terminate at e-twin boundaries.

The nature of $\{10\bar{1}1\}=r$ twins will be discussed in detail elsewhere but it is our belief that r-translation gliding, in the sense used by previous researchers working by means of optical microscopy, probably embraces more than one type of dislocation mechanism. We have found that calcite oriented for r-translation and compressed at temperatures $\lesssim 200$ °C (e.g. VW 133) is extremely disrupt-

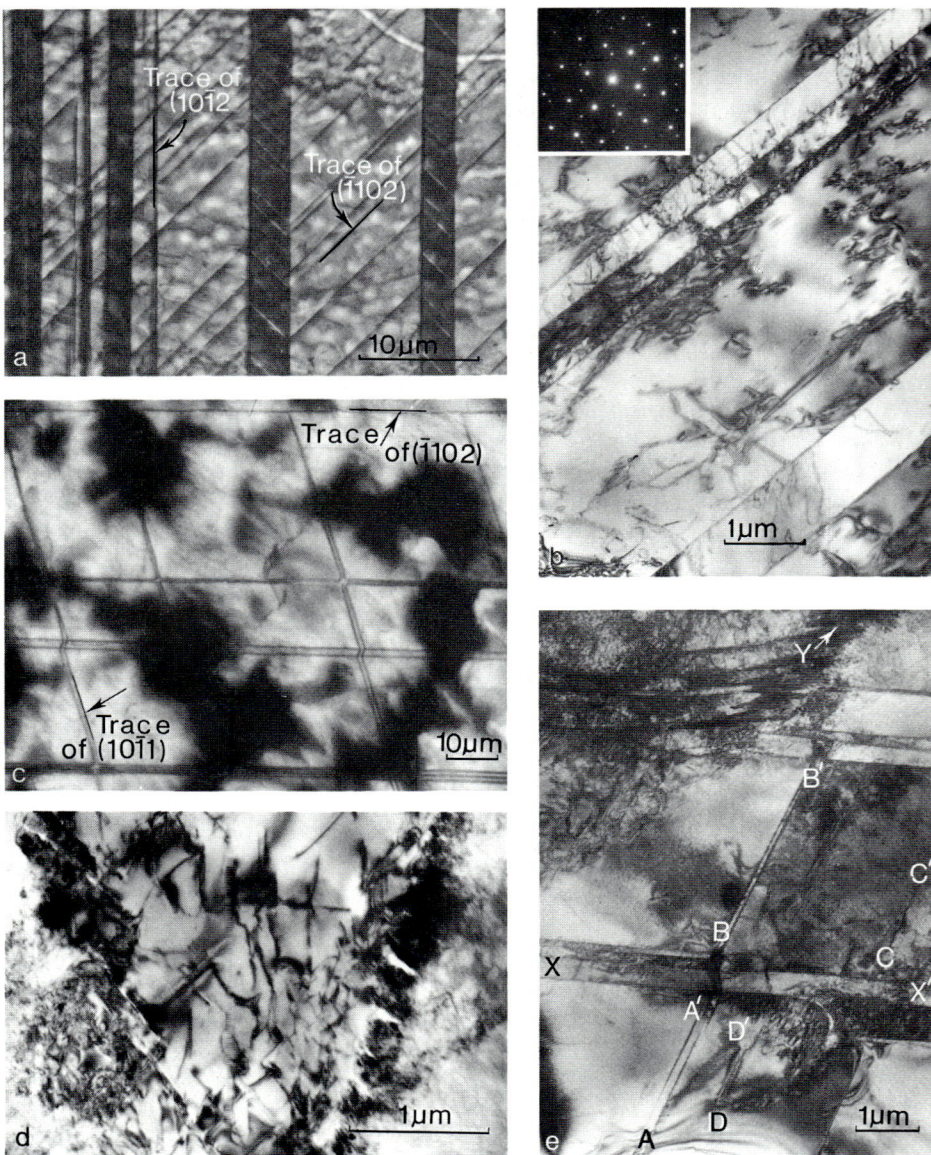

Fig. 5a–e. Microstructure in experimentally deformed Iceland spar VW 132. (a) Optical micrograph (partially crossed polars) showing primary and secondary e-twins. (b) Primary e-twins. (c) Optical micrograph showing e-twins and r-twins in an ion-thinned foil. (d) Fragmentation within r-deformation band. (e) e- and r-twins, with fringed wavy faults corresponding to traces of r-planes

ed and cracked and that the density of dislocations is so high that glide planes cannot identified. It is obvious that glide on $\{10\bar{1}1\}$ planes does not occur readily at low temperature; the tendency for twinning on $r=\{10\bar{1}1\}$ seems to support this. Occasionally a band of deformation (not twinning) corresponding to the trace of r planes is observed in specimens oriented for e-twinning. But in these cases the band is ill-defined and contains many tiny parallel microcracks which break the band up into little strained blocks whose nature is hard to determine (Fig. 5d), i.e. its structure is more closely analogous to a narrow fragmented kink band.

We are now studying the possibility that r-translation may occur by the movement of partial dislocations with Burgers vector of

$$|\tfrac{1}{2}(a_1+a_2+\tfrac{1}{3}c)|=4.04\,\text{Å},$$

so that the glide of these partials would create stacking faults with regard to the orientation of CO_3 groups. The existence of such partials could explain the low critical stresses for r-glide in calcite as compared, for example, to dolomite (Wenk and Shore, 1975). The features shown in Fig. 5d lend some weight to these ideas. If the only possible dislocation for r-translation was a perfect one with Burgers vector $|(a_1+a_2+\tfrac{1}{3}c)|=8.08\,\text{Å}$ then it would be hard to understand why translation does not occur in the direction of the shorter vector $a(|a|=4.99\,\text{Å})$ in the $\{10\bar{1}1\}$ plane. The passage of numerous partial dislocations could produce zones which were disordered with respect to CO_3. This might explain why significant deformation favoring the operation of an r-mechanism at low temperature creates intense strain and profuse microcracking, as illustrated in Fig. 5d. But proof of these ideas would have to come from quantitative contrast experiments for which we need better (i.e. less strained) material and then, possibly, recourse to comparison of dislocation images with computer-simulated images. We are currently following up this work using single crystals which have undergone minimal deformation.

References

Barber, D.J., Wenk, H.-R.: The microstructure of experimentally deformed limestones. J. Mat. Sci. **8**, 500–508 (1973).
Dietz, R.S.: Shatter cones in cryptoexplosion structures (meteorite impact?). J. Geol. **67**, 496–505 (1959).
Gansser, A.: The Insubric Line, a major geotectonic problem. Schweiz. Mineral. Petrog. Mitt. **48**, 123–144 (1968).
Gansser, A., Dal Vesco, E.: Beitrag zur Kenntnis der Metamorphose der alpinen Wurzelzone. Schweiz. Mineral. Petrog. Mitt. **42**, 153–168 (1962).
Griggs, D.T.: Hydrolytic weakening of quartz and other silicates. Geophys. J. **14**, 19–31 (1967).
Griggs, D.T., Turner, F.J., Heard, H.: Deformation of rocks at 500° to 800° C. Geol. Soc. Am. Mem. **79**, 41–48 (1960).
Hauser, J., Wenk, H.-R.: Optical properties of composite materials (submicroscopic twins, exsolution lamellae, solid solutions). Z. Krist., in press (1976).
Keith, R.E., Gilman, J.J.: Dislocation etch pits and plastic deformation in calcite. Acta Met. **8**, 1–10 (1960).

Müller, W.F., Wenk, H.-R., Thomas, G.: Structural variations in anorthite. Contrib. Mineral. Petrol. **34**, 304–314 (1972).
Paterson, M.S., Turner, F.J.: Experimental deformation of constrained crystals of calcite in extension. In: Experimental and natural rock deformation (ed. Paulitsch). Berlin-Heidelberg-New York: Springer 1970.
Turner, F.J., Griggs, D.T., Heard, H.C.: Experimental deformation of calcite crystals. Geol. Soc. Am. Bull. **65**, 833–934 (1954).
Wenk, H.-R.: The structure of the Bergell Alps. Eclogae Geol. Helv. **66**, 255–291 (1973).
Wenk, H.-R., Shore, J.: Preferred orientation in experimentally deformed dolomite. Contrib. Mineral. Petrol. **50**, 115–126.
Wenk, H.-R., Venkitasubramanyan, C.S., Baker, D.W.: Preferred orientation in experimentally deformed limestone. Contrib. Mineral. Petrol. **38**, 81–114 (1973).
Wenk, H.-R., Wenk, E., Wallace, J.H.: Metamorphic mineral assemblages in pelitic rocks of the Bergell Alps. Schweiz. Mineral. Petrog. Mitt. **54**, 507–554 (1974).

Chapter 6.6

Plasticity of Olivine in Peridotites

H.W. Green II

1. Introduction and Background

Prior to about 20 years ago peridotites were almost universally considered to have crystallized from an ultramafic magma to essentially their present microstructure. This mistaken impression persists today in that virtually all textbooks continue to classify these rocks as igneous. It was discovered, however, that many peridotites (when not badly altered to serpentine) display textures, structures, and preferred orientations strongly reminiscent of extensively deformed and recrystallized quartzites in metamorphic terrains. Carefully correlated optical microscopy studies of experimentally deformed and natural specimens have since confirmed that virtually all peridotites have undergone extensive straining, generally at temperatures sufficiently high for recrystallization to accompany the deformation (Raleigh, 1968; Carter and Ave'Lallemant, 1970; Ave'Lallemant and Carter, 1970; Raleigh and Kirby, 1970; Nicolas et al., 1971, 1973).

The concept of plate-tectonics (cf. Isaacs et al., 1968; Carter and Ave'Lallemant, 1970) has focused attention on these rocks as probably reflecting the plastic flow in the earth's mantle by which surficial plates are enabled to move about. About the same time the ion-bombardment thinning technique has enabled detailed transmission electron microscopy (TEM) studies of flow processes in peridotites to begin in several laboratories. Although this field is only a few years old, sufficient information has already been collected to confirm earlier inferences from optical work and to place important further constraints on the mechanisms of deformation, stress levels and strain-rates operating in the upper mantle, and on the modification of substructure produced by subsequent annealing or lower temperature deformation. In addition, discovery of an exsolved CO_2-rich fluid phase and its collection on grain boundaries during recrystallization places new chemical and physical constraints on models of the seismic properties and evolution of the upper mantle. This paper presents a review of these discoveries and points toward future profitable investigation. Detailed discussions will be restricted to olivine because the plasticity of the other phases of peridotites (which together generally make up less than 20–30% of the total volume) has been little studied.

2. Experimental Studies

The extensive optical studies which largely preceded and form a basis for the TEM investigations established the dominant deformation mechanisms in olivine, $(Mg,Fe)_2SiO_4$. As it is by far the most abundant phase in peridotites, to a first order approximation plasticity of peridotites is expected to reflect the plasticity of olivine. Raleigh (1968), using optical methods, found that the temperature dependence of slip on the major mechanisms in olivine is different, and he and Carter and Ave'Lallemant (1970) determined the P-T-$\dot{\varepsilon}$ fields in which the several mechanisms dominate in the deformation of polycrystalline aggregates (Fig. 1). The general results are that at lower temperatures and higher strain-rates, slip

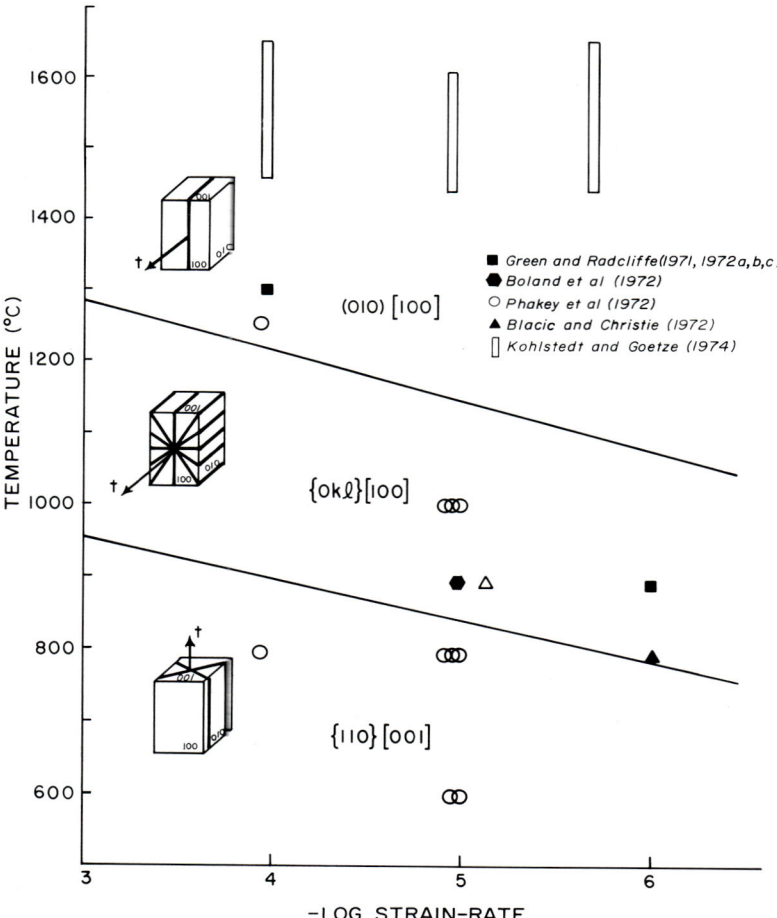

Fig. 1. Predominant slip mechanisms in experimentally deformed olivine as determined optically from deformation lamellae and crystal rotations. Experimental deformation conditions of specimens studied by TEM are indicated. Closed symbols are polycrystals, open symbols are single crystals. (After Carter and Ave'Lallemant, 1970)

on (100) and {110} in the [001] direction predominates, whereas at higher temperatures and lower strain-rates, {0kl} pencil glide parallel to [100] becomes progressively more important, and at the highest temperatures studied this [100] slip tends to become restricted to (010).

The "slip fields" of Fig. 1 are, of course, approximations; the different mechanisms would be expected to operate, even dominate, outside of these fields in suitably stressed single crystals and to function in a subordinate way even in polycrystals. These expectations are confirmed by the TEM studies. Phakey et al. (1972) reported on a large single crystal study, and Blacic and Christie (1973) compared a single crystal and a polycrystal. Although most (but not all) of the specimens of these studies were deformed in the nominal [001] slip field (Fig. 1), minor [100] slip was detected or suggested in all experiments at 800° C and above. (Surprisingly, rare perfect dislocations of the (100)[010] system were also recorded at 800° C despite the very large $|b|=10.2$ Å.) Even the single crystal specimens at 1000° C produced dominantly [001] slip. The substructure in these studies was dominated by abundant long, straight dislocation lines in the screw orientation with rare short, straight segments in edge orientations (Fig. 2a), indicating that the Peierls force in olivine is strong and that for both [001] and [100] Burgers vectors, edge dislocations are much more mobile than screws. Subordinate regions of tangled dislocations were found at 800–1000° C and were attributed to interference with the dominant [001] glide by minor slip parallel to [100] and to the extensive cross-slipping evident at 1000° C. No evidence for climb was found below 800° C, but its effects were distinctly present at 1000° C. The single experiment at 1250° C had its substructure apparently dominated by such effects, with a marked decrease in the flow stress and the general dislocation density (to $\sim 10^7$–10^8 cm^{-2} from $\sim 10^{10}$ cm^{-2}), and the appearance of subgrain formation and recrystallization. Boland et al. (1971) presented compatible qualitative results from a single specimen deformed at 10^{-5} sec^{-1}, 900° C.

Green and Radcliffe (1972a) and Phakey et al. (1972) identified the substructure responsible for optically visible deformation lamellae of the {0kl} [100] system. The lamellae are slip bands consisting of arrays of (presumably piled-up) parallel edge dislocations which are dominantly of the same sign and which lie in their slip planes—a structure essentially corresponding to the model proposed by Christie et al. (1964) to explain the optical properties of similar features in quartz. In polycrystalline material under these conditions, recrystallization begins at grain boundaries during the deformation and the new crystals themselves become deformed as the experiment continues (Fig. 2b).

At the highest experimental temperatures the optical studies are again confirmed (Green and Radcliffe, 1971, 1972b, 1972c; Kohlstedt and Goetze, 1974); recovery processes control the substructure, and in polycrystals [100] Burgers vectors greatly dominate over [001]'s and recrystallization accompanies the deformation. Green and Radcliffe (1972b, 1972c) found that when the temperature is high enough for climb to become important, deformation lamellae (slip bands) are much less abundant and the edge components of dislocation loops of the {0kl} [100] system climb into "picket fence" low angle kink-band boundaries closely paralleling (100), leaving the screw components of these loops extending across the bands. The screws are pinned to their edge components in the bound-

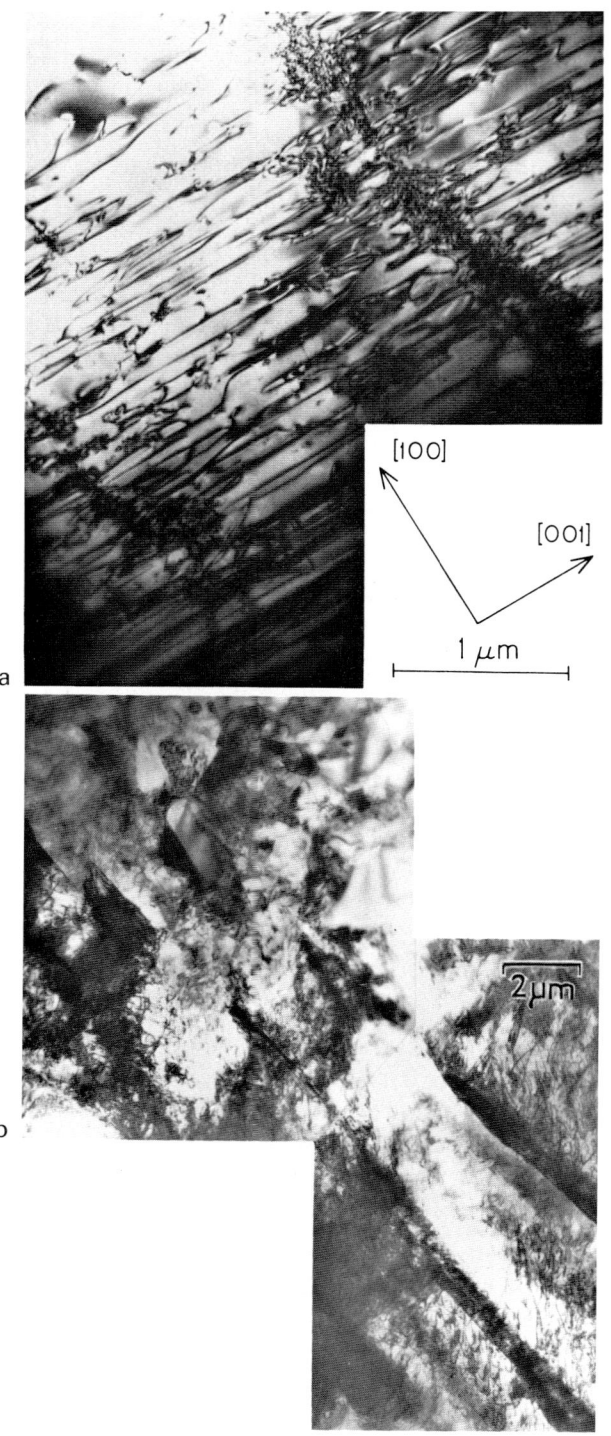

Fig. 2. (a) Substructure of olivine deformed at low temperatures or high strain rates (polycrystal crept at 800 °C). Screw dislocations are parallel to [100] and dense zones of dislocations are micro-kink boundaries subparallel to (001). (Blacic and Christie, 1973) 120 kV. (b) Slip bands (deformation lamellae) in olivine (900 °C, 10^{-6} sec^{-1}). Bands consist of arrays of edge dislocations with $b=[100]$ lying in their slip plane and spaced about 150 Å apart. Background dislocations have $b=[100]$ and $b=[001]$. Bands are partially recrystallized and new crystals are also deformed (Green and Radcliffe, 1972c) 800 kV. (c) Deformation (kink) bands in olivine deformed at high temperatures or low strain-rates (1 300 °C, 10^{-4} sec^{-1}). [100] screws pinned to their edge components in the band boundaries bow out within the bands (Green and Radcliffe, 1972b) 800 kV. (d) Syntectonic recrystallization in experiments results in polygonal crystals exhibiting substructural variation from lightly deformed to dislocation densities and configurations comparable to the parent crystals. 800 kV. (Green and Radcliffe, 1972b)

Plasticity of Olivine in Peridotites

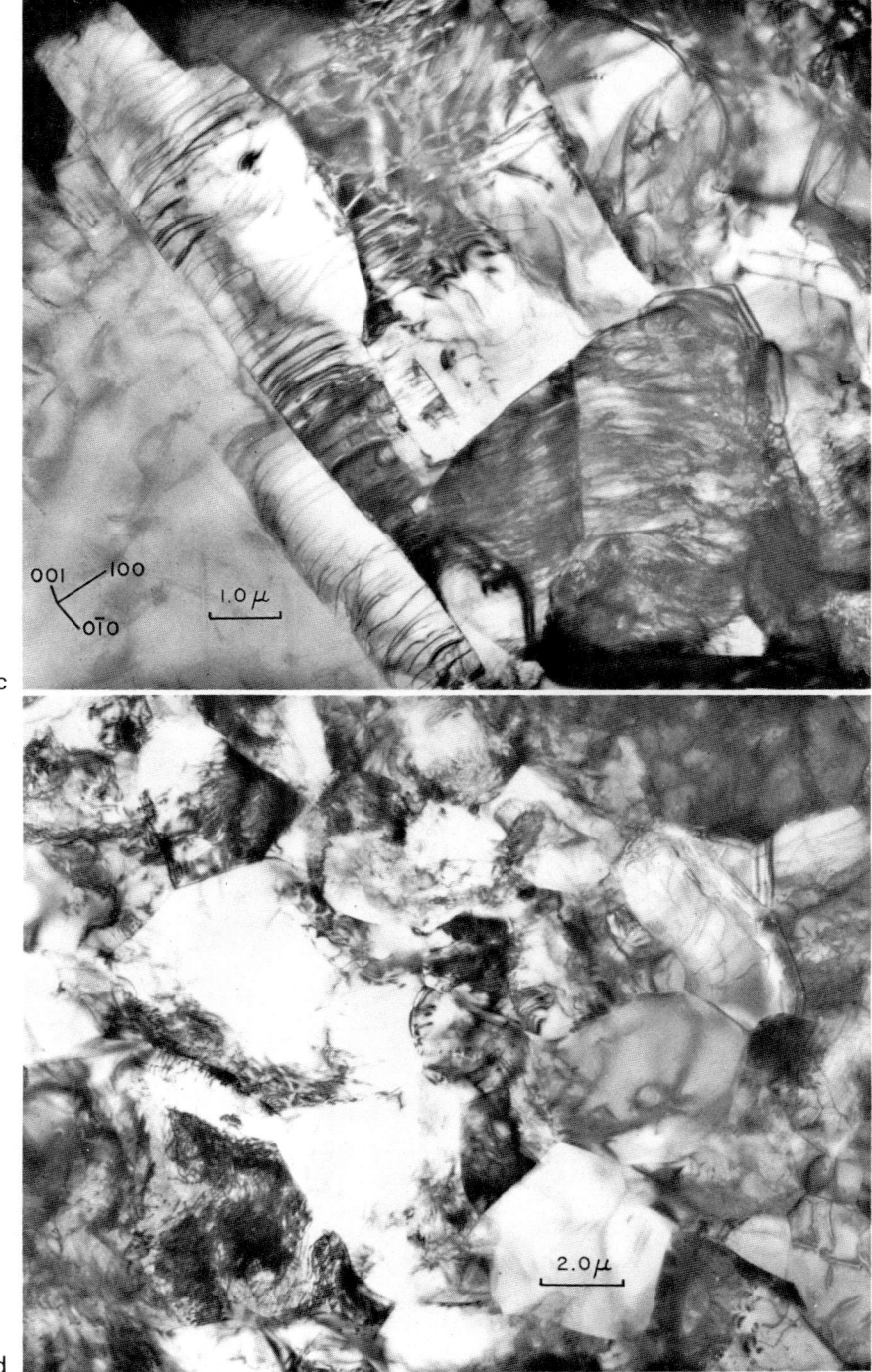

c

d

Fig. 2c and d

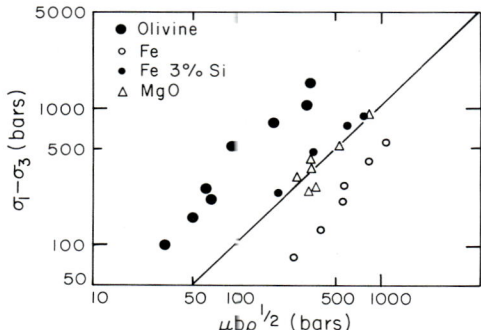

Fig. 3. Plot of applied differential stress $(\sigma_1-\sigma_3)$ versus internal stress $(\mu|b|\rho^{1/2})$ calculated from measured dislocation densities, with $\mu = 0.65 \times 10^6$ bars, and $|b|=5$Å (Kohlstedt and Goetze, 1974)

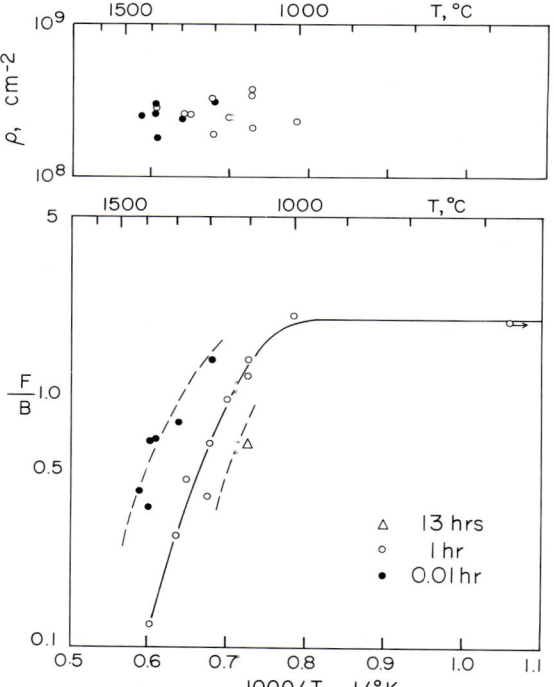

Fig. 4. Kinetics of annealing of naturally deformed olivine. Dislocation density (top) and ratio of free dislocations to those bound in low-angle boundaries (bottom). (Goetze and Kohlstedt, 1973)

aries and are commonly seen to bow out (Fig. 2c). Where kink bands end, a concentration of [100] screws is found lying approximately in the (001) plane (Green and Radcliffe, 1972, Fig. 2b). Syntectonic recrystallization is more prominent, with the younger crystals exhibiting substructural features similar to those in the parent grains (Fig. 2d). Kohlstedt and Goetze (1974) found that the free dislocation density in their single crystals was a function of differential stress alone (Fig. 3).

Goetze and Kohlstedt (1973) conducted annealing experiments on specimens of naturally deformed olivine to investigate the kinetics of recovery. They determined a lattice diffusion coefficient (tentatively identified with O^{-2}) from the collapse of sessile dislocation loops with $b=[001]$ lying near or in the (001) plane. The result was $D = 3 \times 10^4 \exp[(-135 \text{ kcal/mole})/RT]\text{cm}^2/\text{sec}$. The velocity of climb of free dislocations was found to be 10^4 times more rapid than that predicted by the diffusivity measured for sessile loops, yet the activation energy for both processes was found to be the same. It was proposed that the enhancement of climb rate for free dislocations was provided by pipe diffusion. The total density of dislocations changed little during annealing, indicating that annihilation was a small effect, but the organization into low-angle boundaries was a strong effect (Fig. 4). Radcliffe and Green (1973), in a parallel study, found trends very similar to those of Goetze and Kohlstedt (1973), but they found recovery rates to be less rapid. For example, Goetze and Kohlstedt (1973) at 1100° C found perceptible recovery in one hour and extensive recovery in 13 hours, whereas Radcliffe and Green (1973) could detect none after 24 hours. Similar disagreement was found at 1300° C (Radcliffe and Green, unpublished data). Despite this conflict, the experimental studies both show that fragments of the mantle (xenoliths) residing in a basaltic magma chamber (1100–1300° C) for times in excess of a few days would have a pronounced change produced in their olivine substructure.

3. Studies of Natural Peridotites

Peridotites in nature fall into three general categories: (a) igneous rocks with cumulus textures found in isolated layered intrusions and in the central portion of ophiolite complexes, (b) highly deformed metamorphic rocks tectonically emplaced onto continents or oceanic islands, and (c) xenoliths transported directly from the upper mantle in alkalic basalts and kimberlites. The igneous rocks of category (a) are fractional crystallization products of basic magmas, are generally undeformed, and have not been studied by TEM. The tectonites of category (b) have suffered varying degrees of structural modification during and after emplacement into the crust and therefore interpretation of their substructure is complex. They have been little studied to date and will be discussed last in this section. The xenoliths of category (c) potentially form our greatest source of chemical and (small-scale) structural information about the upper mantle. They constitute the deepest sampling of the earth available for direct study, and consequently have been investigated in greater detail than other peridotites. A number of significant differences in mineralogy, chemistry and texture separate xenoliths from basalts and those from kimberlites, and inasmuch as these differences may reflect differences in conditions beneath oceanic and continental regions, the two types of occurrence are separated in the discussion below and at the end we discuss the presence of fluid bubbles.

3.1 Xenoliths from Basalts

Peridotite and pyroxenite xenoliths in alkalic basaltic lava flows and tuff cones represent the freshest mantle material available for study. They are restricted to oceanic and tectonically active regions. These nodules lack the ubiquitous grain-boundary serpentinization of similar specimens from kimberlite diatremes and they tend to have angular facets on their surfaces rather than the well-rounded form of xenoliths from kimberlites. The pyroxenites and peridotites are sometimes found interlayered in the same specimen, indicating a host-dike relationship or primary layering in the mantle. Pyroxenites from some areas, notably Salt Lake Crater, Oahu, Hawaii, are garnetiferous and clearly represent material crystallized in the mantle. Peridotites from these localities fall into two classes: (a) igneous rocks with cumulus textures, and (b) metamorphic rocks with a variety of textures (Mercier and Nicolas, 1975). The latter type, with rare exceptions, lack garnet; the aluminous phase is spinel. Only spinel-bearing specimens from Salt Lake Crater (lherzolites and pyroxenites), Hualalai volcano, Hawaii (dunites), and Black Rock Summit (Lunar Crater), Nevada (lherzolites), have been investigated by TEM (Green and Radcliffe, 1971, 1972b, 1972c, 1975; Goetze and Kohlstedt, 1973; Kohlstedt and Vander Sande, 1973). In addition, dislocation etching studies have been performed on olivine specimens from San Carlos, Arizona (Kirby and Wegner, 1973; Kirby and Raleigh, 1973), and naturally decorated dislocations have been observed in olivine of igneous rock types from Mauna Kea, Hawaii and tectonites from San Quintín, Baja California (Zeuch, 1974).

The dislocation density in olivine of xenoliths from basalts is very much less than that in experimental specimens except for some single crystals deformed at very high temperatures and low stresses (Kohlstedt and Goetze, 1974). The lowest densities are found in those specimens classified as protogranular by Mercier and Nicolas (1974), who also demonstrated that these samples have experienced negligible strain. Except for broad kink bands in olivine, these coarse-grained rocks (1–5 mm) generally show no evidence of deformation in hand specimen or in the optical microscope (Fig. 5a). Some show a weak foliation defined by inequant crystals. The substructure is dominated by widely-spaced low angle (kink-band) boundaries, generally subparallel to (100), with a free dislocation density of $\sim 10^6$ cm^{-2} (Goetze and Kohlstedt, 1973) or less. The boundaries are almost exclusively "picket fence" arrays consisting primarily of dislocations of the same sign with [100] Burgers vectors, but also containing lesser numbers with Burgers vector of [001], [101] and [010] (see Goetze and Kohlstedt, 1973). Dislocation networks are very rare in these rocks and this probably reflects the small plastic strains and great preponderance of [100] Burgers vectors. Those specimens classified as porphyroclastic by Mercier and Nicolas (1974) contain remnants of large, highly deformed olivine crystals and less deformed orthopyroxene and clinopyroxene crystals set in a fine-grained matrix of recrystallized olivine (Fig. 5b). Their textures and substructure are strongly reminiscent of laboratory specimens deformed at the highest temperatures (Green and Radcliffe, 1972c); dislocation densities are less than those of experiments and kink bands are wider, but the substructures are qualitatively very similar. Kink bands with [100] screws extending across them are abundant and in the same crystal can occasionally be seen to grade laterally into regions where the distributed screws are less abundant and (001) walls appear (Fig. 6a). This structure grades further into regions where the (100) and (001) walls are equally developed defining long rectangular subgrains. Both sets of boundaries consist

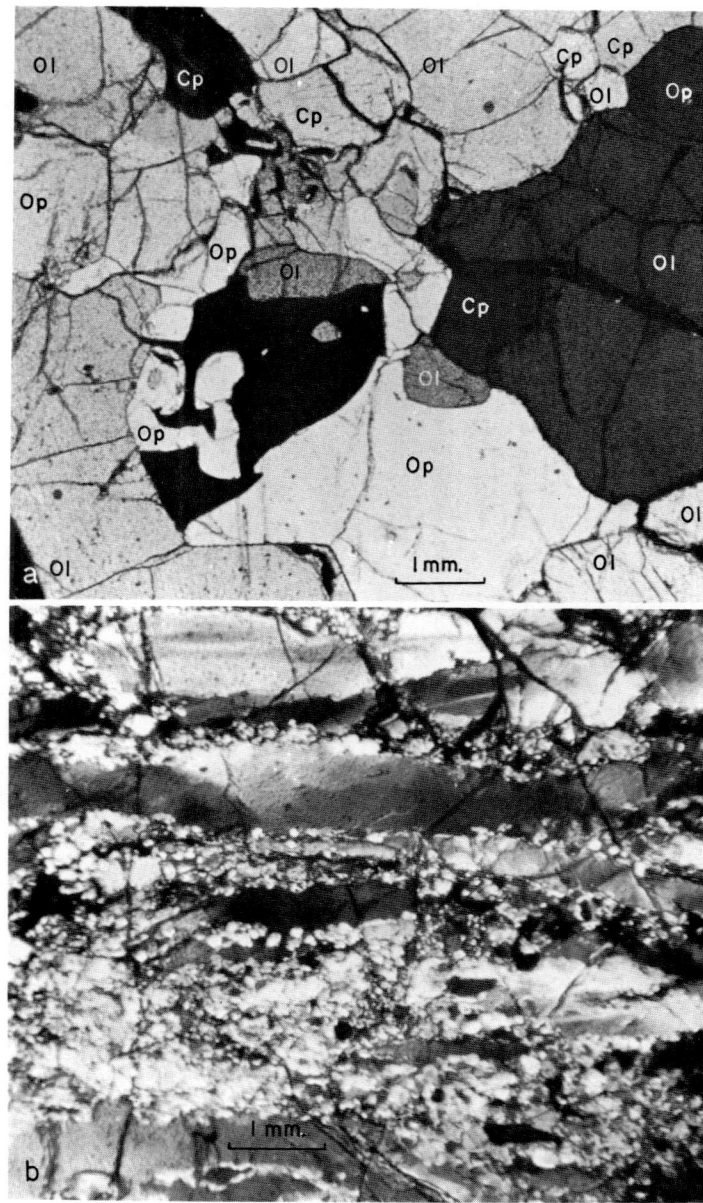

Fig. 5. (a) Optical micrograph of protogranular spinel lherzolite from Salt Lake Crater, Oahu, Hawaii. Ol, olivine; Op, orthopyroxene; Cp, clinopyroxene; black area, spinel. Complex intergrowths of spinel and pyroxenes such as seen here are characteristic of these specimens. No recrystallization is present and deformation is slight. (b) Optical micrograph of porphyroclastic lherzolite from Black Rock Summit, Nevada. Highly deformed olivine ribbons display abundant kink bands and subgrains in a matrix of fine-grained, recrystallized, olivine grains which also show deformation features. (Green and Radcliffe, 1972c)

Fig. 6. (a) Substructure of naturally deformed olivine (Porphyroclastic lherzolite from Black Rock Summit, Nevada). [100] screw dislocations extending across (100) kink bands (bottom) grade into region with development of (100) and (001) walls. 1 000 kV. (b) Olivine subgrains in Black Rock Summit lherzolite. (001) boundaries consist of parallel arrays of dislocations of b=[100] approximately in the screw orientation. Dark-field micrograph, 1 000 kV

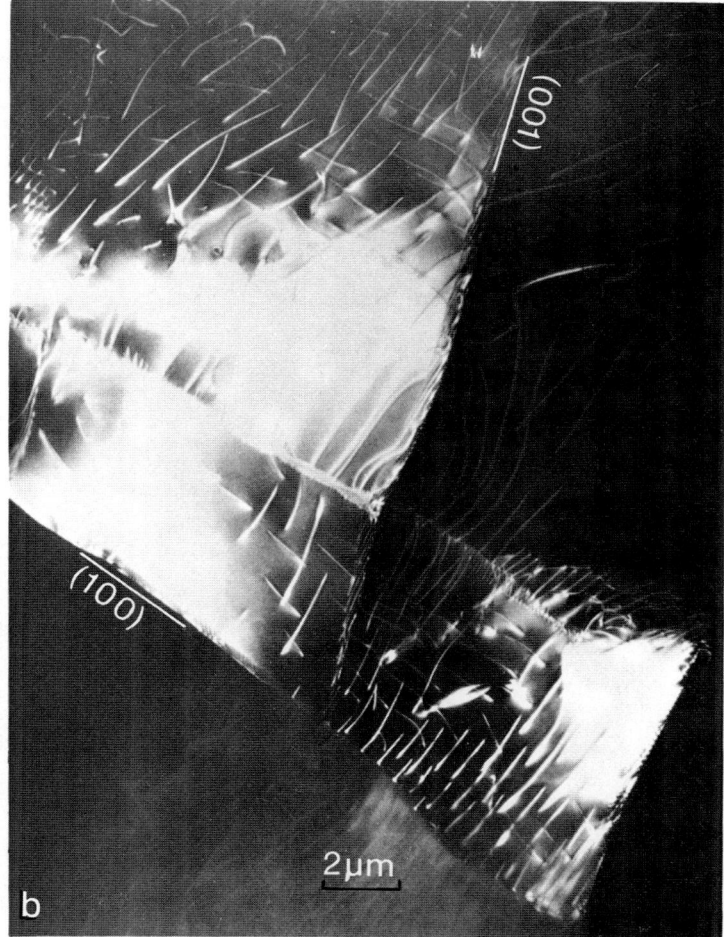

Fig. 6b

of dislocations dominantly of [100] Burgers vector; the (100) walls consist of edge dislocations and the (001) walls are defined by screw dislocations (Fig. 6b). Although these subgrains have the same general shape as those visible optically, they do not appear to be consistent with rotations determined across optically visible boundaries which show (001) boundaries to be tilt boundaries about [010]. Etching of the dislocations on these optically visible boundaries is consistent with the optical studies (Kirby and Raleigh, 1973). It is not yet understood how, if at all, these two types of (001) boundaries are related, and how the (001) tilt boundaries are formed. TEM has thus far shown no substructural evolution leading toward such boundaries.

3.2 Xenoliths from Kimberlites

Peridotite xenoliths from kimberlites, like those from basalts, fall into several textural classes. Boullier and Nicolas (1973, 1975) have identified four classes, two consisting of very coarse, little-deformed aggregates which are generally harzburgites (<5% clinopyroxene) and often spinel-bearing, and two classes consisting of deformed and partially recrystallized specimens which are generally lherzolites (>5% clinopyroxene) and exclusively garnet-bearing. These broad groupings coincide with the "granular" and "sheared" nodules which Boyd

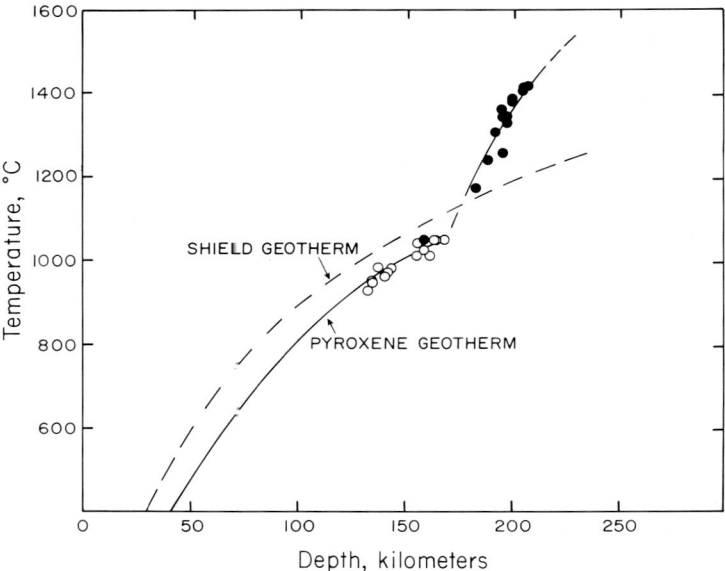

Fig. 7. Pyroxene geotherm for Lesotho. Data are from eight pipes. Closed circles represent "sheared" xenoliths, open circles represent "granular" nodules. Data presented are for "raw" Al_2O_3 values (Boyd, 1973)

(1973), and MacGregor (1974, 1975) have found to occupy different portions of the "pyroxene geotherm"; the granular, depleted, rocks being the shallower specimens and the sheared ones defining a deeper, steepened portion of the geotherm (Fig. 7). The more detailed classification of Boullier and Nicolas (1973, 1975) correlates even more remarkably with the distribution on the geotherm (Nixon et al., 1973; MacGregor, 1975). Green and Boullier (in preparation) have examined examples of these textural types from five South African localities and the following discussion principally summarizes their results.

3.2.1 Coarse Granular and Coarse Tabular Specimens

Although significant differences exist between these two classes on the optical scale (Boullier and Nicolas, 1973, 1974), no corresponding substructural differences have yet been found. The specimens of both classes have very low free dislocation densities ($\varrho \sim 10^6$ cm^{-2} or less) and subboundaries are widely spaced (Fig. 8a). These rocks on the electron microscope scale are closely similar to the protogranular spinel lherzolites from basalts, except that the kimberlite specimens always have serpentinized grain boundaries.

3.2.2 Porphyroclastic and Mosaic Specimens

These two textural types of Boullier and Nicolas (1973, 1974) grade from one to the other. The defining property of the porphyroclastic texture is that both porphyroclasts and recrystallized olivine are present. Specimens which have the olivine entirely recrystallized are called mosaic, even though they universally contain orthopyroxene, clinopyroxene and garnet porphyroclasts. As might be expected from the definitions, the substructures of these two types differ little except that only the younger, less deformed generation of olivine exists in mosaic specimens.

The substructure of both generations of crystals in these rocks stands in marked contrast to that of the coarse-grained nodules. The dislocation densities in olivine porphyroclasts can approach 10^9 cm^{-2} and in the matrix olivine the densities range from less than 10^5 cm^{-2} to 10^7 cm^{-2} (Fig. 8b). Furthermore, the organization of dislocations into low angle boundaries is much less pronounced than in any xenoliths from basalt and networks are more common. The percentage of [001] Burgers vectors is distinctly greater. This latter observation is in agreement with Olsen and Birkeland (1973; 1974), although their major interpretations and conclusions are in error (Green, 1974a).

3.3 Alpine Peridotites

The designation of Alpine peridotite applies to a variety of rocks which differ in their chemistry, structure and tectonic setting. Nicolas and Jackson (1972) have recognized two principal varieties: the harzburgite subtype which forms the basal portion of ophiolite complexes and is overlain by cumulate ultramafic and mafic rocks, and the lherzolite subtype which is less depleted in basaltic constituents and is not found in association with undoubted igneous rocks. The preliminary TEM studies published to date indicate that modification of mantle structures may occur by deformation or annealing in the crust, a result one would expect from the tectonic setting of these bodies (Boland *et al.*, 1971; Green and Radcliffe, 1971, 1972c; Goetze and Kohlstedt, 1973; Kohlstedt and Vander Sande, 1973).

Boland *et al.* (1971) examined the olivine substructure of a porphyroclastic dunite mylonite. They found the fine-grained matrix to have very low dislocation densities ($< 10^6$ cm^{-2}). The porphyroclasts are characterized by low-angle boundaries generally defined by a single "picket fence" array of dislocations, but networks are also observed. The sub-boundaries approximate low index planes and the dislocations parallel low index directions. The free dislocation density appears from their micrographs to be $\sim 10^6$–10^7 cm^{-2}.

Green and Radcliffe (1971, 1972c) found that harzburgite tectonites from Vourinos, Greece, contained the kink band and subgrain boundaries typical

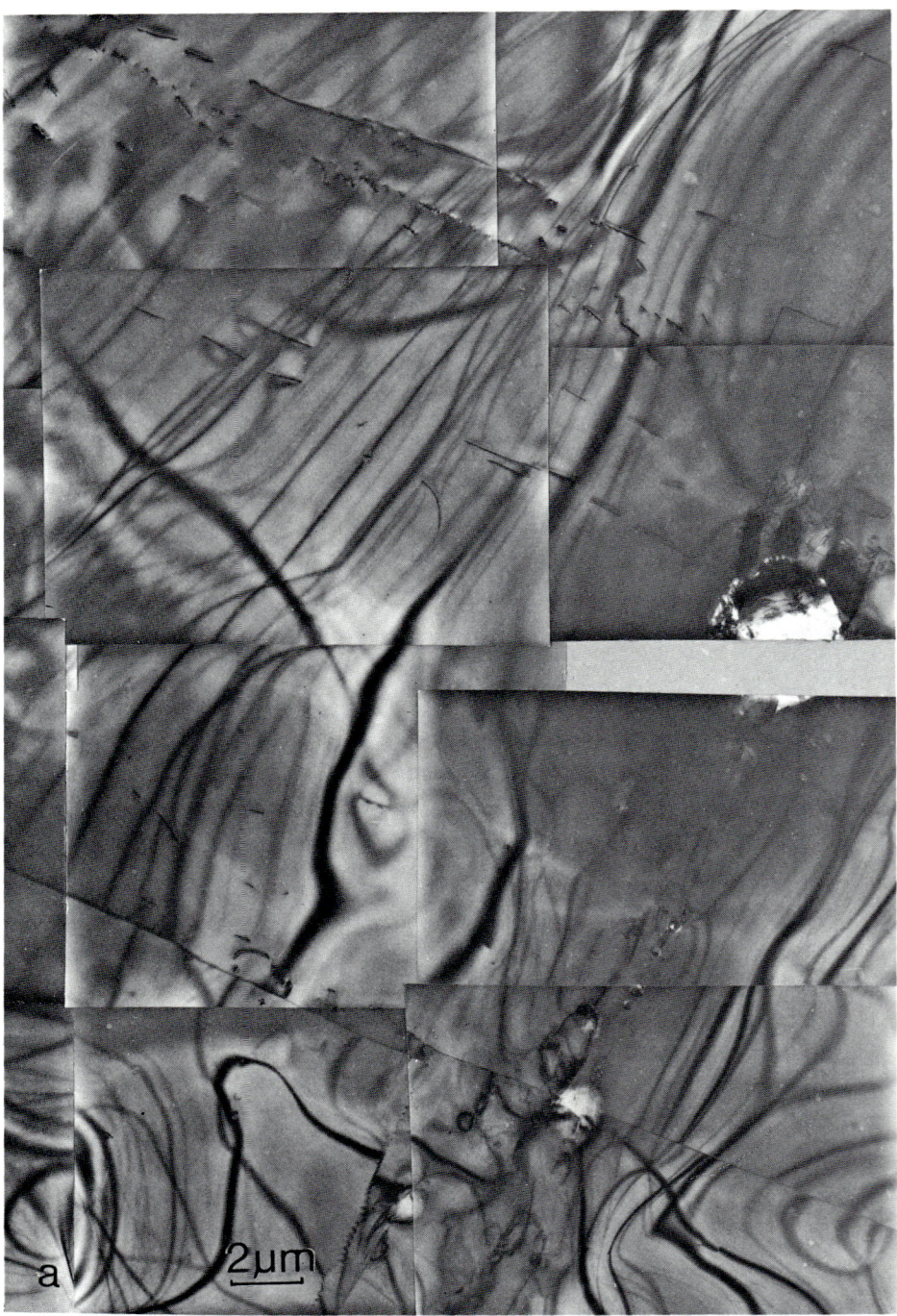

Fig. 8a

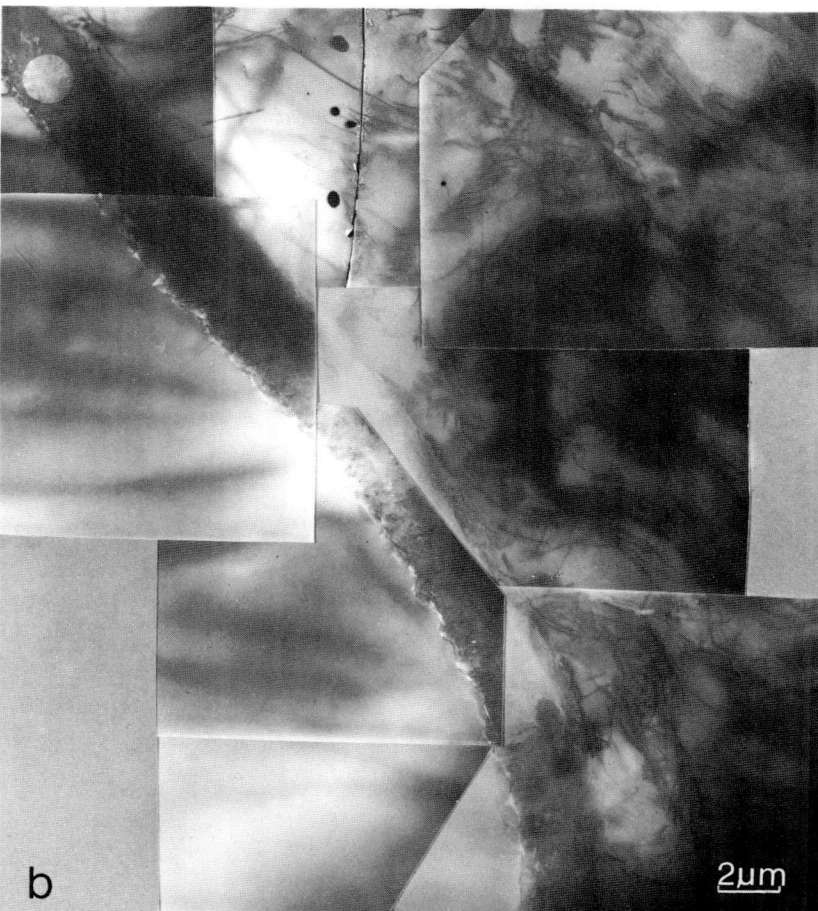

Fig. 8. (a) Olivine substructure of a course-grained, "granular" garnet harzburgite from the Monastery Mine, South Africa. Crystal is highly recovered except for later local deformation around large bubble (right). Other phases are even more highly recovered. Survey of more than $2 \times 10^3 \mu m^2$ of garnet revealed *no* free dislocations ($\varrho < 5 \times 10^4$ cm^{-2}). (b) Highly deformed olivine porphyroclast and recrystallized grain of a garnet lherzolite from the Kao pipe, Lesotho. Porphyroclast has dislocation density of $5-8 \times 10^8$ cm^{-2} whereas new crystal has 4 dislocations $= \sim \times 10^6$ cm^{-2}. Inclined grain boundary has serpentine "rind" < 1 μm thick. 3000 kV. (c) Substructure of harzburgite tectonite from Vourinos ophiolite complex, Greece. The high density of straight free dislocations along with simple low angle boundaries suggests minor low temperature deformation superposed on high temperature deformation. 1000 kV

of high temperature deformation, but in addition found a higher free dislocation density ($\sim 10^8$ cm^{-2}) suggesting a subsequent, lower temperature, deformation perhaps during emplacement in the crust (Fig. 8c). The similar substructure presented by the Mt. Albert material used by Goetze and Kohlstedt (1973) in their annealing experiments may have arisen in the same way.

Fig. 8c. Legend see page 457

Fig. 9a–c. Bubble precipitation on deformation features in olivine. (a) "Pearl necklace" decoration of isolated dislocations in porphyroclast from Thaba Putsoa pipe, Lesotho. Bubble images range down to ~ 100 Å where they disappear into the dislocation line image. No clear lower size limit could be established. 1000 kV. (b) Decoration of subboundaries in porphyroclast from Hualalai, Hawaii. Bubble images range from about 4000 Å (dark spots) abundant on only one boundary to ~ 100 Å describing "pearl necklaces" on the dislocations of both boundaries and the free dislocation at lower left. 1000 kV. (c) Collection of bubbles on grain boundary during recrystallization. Specimen of porphyroclastic lherzolite from Salt Lake Crater, Hawaii. The crystal on the left is growing into the porphyroclast on the right. The porphyroclast has extensive decoration of dislocations with small bubbles. As boundary has migrated, the small bubbles have been "swept up" and coalesced into the larger, lens-shaped bubbles on the boundary. 1000 kV. (Green and Radcliffe, 1974)

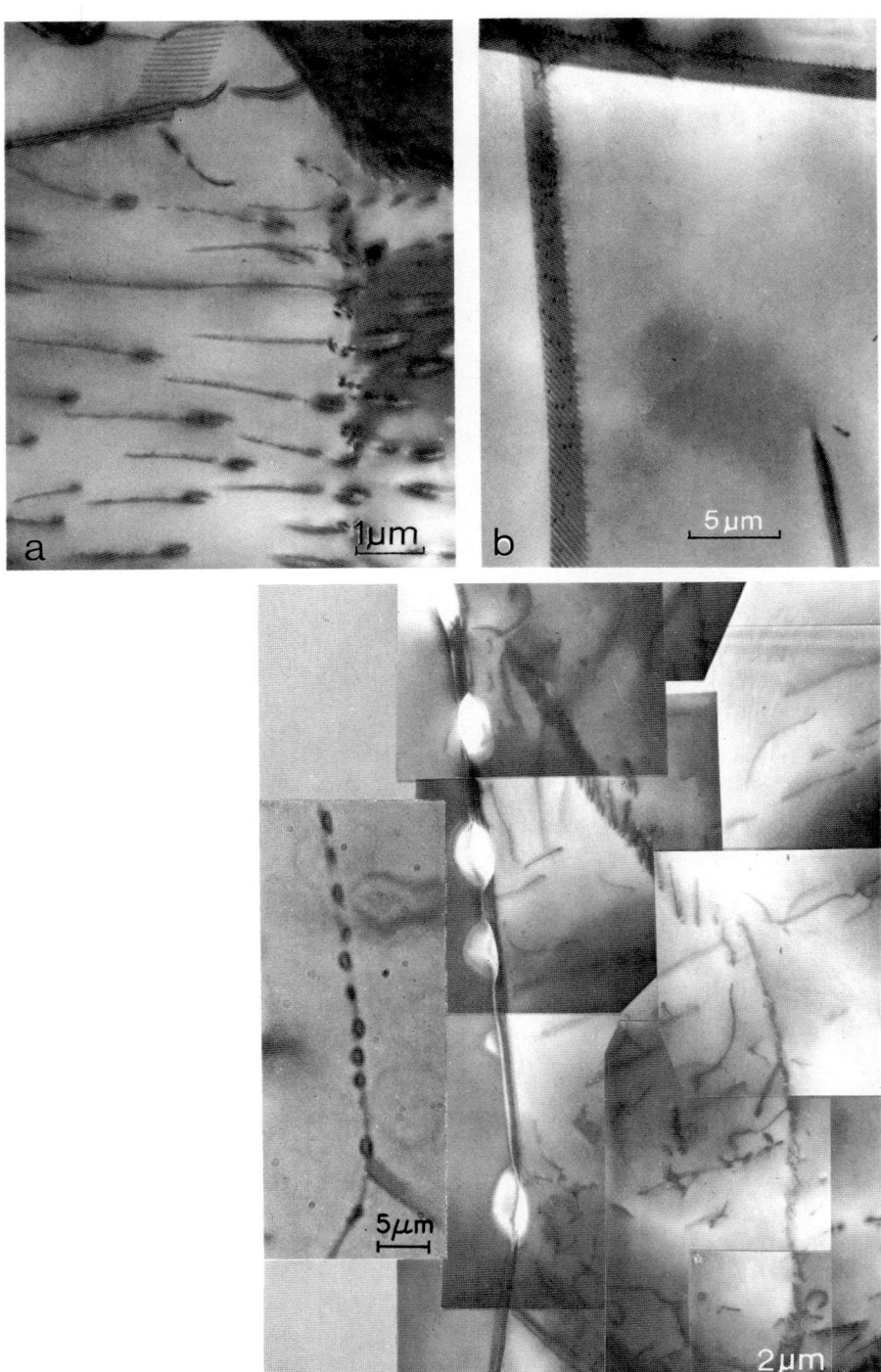

Fig. 9a–c

3.4 Bubbles

Roedder (1965) showed that large numbers of very small (≤ 5 μm) bubbles are usually present in peridotites. They are especially abundant in xenoliths from basalts where they usually occur on healed fractures and contain almost pure CO_2 at pressures as high as 5 kbar at eruptive temperatures. Similar bubbles, usually now empty, are also abundant in xenoliths from peridotites. Green (1972) and Green and Radcliffe (1975) found by TEM that submicroscopic bubbles exist in even greater quantities and that the submicroscopic and smaller microscopic bubbles are attached to crystal defects induced by deformation and exsolution and to grain boundaries. Bubbles with images as small as 100 Å have been found on isolated dislocations and subgrain boundaries in olivine (Fig. 9a, b), orthopyroxene and clinopyroxene and on the terminating edges of spinel and orthopyroxene exsolution lamellae in clinopyroxene. All of these observations are precise parallels to the well-documented precipitation (exsolution) of noble gas bubbles in irradiated metals, alloys and oxides. In the latter studies recrystallization after irradiation results in concentration of the gasses into larger bubbles on grain boundaries as the boundaries sweep up the previously precipitated volatiles. In xenoliths from basalts, where grain boundaries are often unaltered, similar bubble concentration onto the boundaries of new, recrystallized grains is observed (Fig. 9c). Green and Radcliffe (1975) conclude from a variety of lines of reasoning that the fluid inclusions in mantle rocks form by nucleation and growth of bubbles on crystal discontinuities; they are then collected onto grain boundaries during syntectonic recrystallization before incorporation into the basaltic or kimberlitic magmas that brought them to the surface. During passage to the surface, the external pressure release results in "punching out" of prismatic dislocations around the bubbles and fracturing to produce the bubble-decorated cracks described by Roedder (1965).

4. Discussion

Despite the wide variety of experimental and natural materials investigated, most of the observations are compatible with a rather simple picture of the variation of peridotite plasticity with physical conditions. The conclusions from previous optical studies are confirmed and strengthened, with the possible exception of a greater importance of [001] glide in olivine at higher temperatures. It is possible, however, that the transition to [100] slip is sensitive to H_2O content of the specimens analogous to that of the basal/prismatic slip transition in quartz (Griggs, 1967). The single crystals are almost certainly representative of dry olivine, but significant traces of H_2O might have remained in the polycrystalline specimens. The substructure of xenoliths from basalts appears to represent a continuation of the trends observed in experimental specimens. That is, as T increases or $\dot{\varepsilon}$ decreases diffusion-controlled processes increase in importance leading toward a subgrain structure of simple dislocation walls and increasing degrees of recrystallization.

In an effort to obtain more information, Goetze and Kohlstedt (1973) assumed that the observed olivine substructure in a Salt Lake Crater sample was a steady-state one, and argued that the external stress could be estimated by equating it to the back stress between dislocations. By this method they calculated a stress of 40 bars for the xenolith and estimated the value should be accurate within a factor of 3. Subsequently, Kohlstedt and Goetze (1974) determined the relationship between differential stress and dislocation density in a number of their single crystal experiments (Fig. 3), and found the relationship to be insensitive to temperature. A density of 10^6 cm^{-2} corresponded to about 100 bars—a result consistent with their calculation for the Salt Lake Crater specimen. Combining this result with their empirical σ-$\dot{\varepsilon}$ relations, we obtain for this xenolith a strain-rate of about 5×10^{-8} sec^{-1} for temperatures of 1400° C or 5×10^{-11} sec^{-1} for 1100° C. The latter temperature is about a maximum for Hawaiian samples determined by the pyroxene geotherm method (MacGregor and Basu, 1974). The much higher dislocation densities of tectonite xenoliths from kimberlites strongly suggests that their deformation has been at distinctly higher stresses and strain-rates than the other natural specimens. If one again applies the same methods of estimation, one obtains for the olivine porphyroclasts, $\sigma \sim 1000$ bars; $\dot{\varepsilon} \sim 10^{-5}$ sec^{-1} (1400° C) or $\dot{\varepsilon} \sim 10^{-8}$ sec^{-1} (1100° C), but for the small grains the results are comparable to the Salt Lake Crater specimen. Pyroxene geotherm temperatures for the porphyroclastic specimens of Green and Boullier (in preparation) are $\sim 1200°$ C. These estimated strain-rates are higher by 2–9 orders of magnitude than those consistent with asthenospheric flow deduced from lithospheric plate velocities (assuming an asthenosphere 50–100 km thick). The disagreement would be increased if these rocks have had a reduction in their free dislocation density, due to post-deformational recovery after incorporation into the magma, but would be reduced by perhaps a factor of 5 by the effect of pressure on the σ-$\dot{\varepsilon}$ relations. The latter effect cannot remove more than a small fraction of the discrepancy, and the existence of some essentially dislocation-free grains in these rocks (Figs. 8b and 9c) rules out the possibility of rapid deformation in the magma. On the other hand, the annealing studies summarized above demonstrate that substructures such as those found in Fig. 8b would be significantly altered within a few hours or days at mantle or magma temperatures, and hence we must be looking at material plucked "instantaneously" from its dynamic environment.

There appear to this writer only two avenues of escape from this apparent conflict between TEM and plate-tectonic deductions. The demonstration by Kohlstedt and Goetze (1974) that the theoretical relationship between stress and dislocation density holds well for their experimental specimens (the small divergence of their data from the theoretical relationship makes the TEM-plate tectonic discrepancy greater, not less) argues strongly against simple dismissal of their method. While one might defensibly argue that the conflict with sea-floor spreading rates of the Salt Lake Crater specimens lies within the uncertainty of the method, such an argument is untenable for the porphyroclastic specimens from kimberlites. Similarly, one could conclude as did Green and Radcliffe (1972b, c) that very low dislocation densities in Salt Lake Crater specimens are the result of static recovery and were not in equilibrium with a non-hydrostatic stress. Again, however, this argument is inapplicable to the tectonite xenoliths from kimberlites. One is left therefore with postulating that the flow reflected in these specimens was predominantly perpendicular to the earth's surface (in which case plate-tectonic constraints do not apply) or that something in the xenolith olivine makes the σ-ϱ relationship grossly different from that in the experiments of Kohlstedt and Goetze (1974). The first possibility implies convective motion and was used by Green and Gueguen (1974) in their model of mantle diapirism. The second possibility appears less likely, but it is known that in experimental deformation of quartz, introduction of H_2O into the structure can greatly reduce the flow stress and increase the dislocation density (Griggs, 1967). The presence of H_2O in these nodules makes such an effect a possibility, but the simple theory employed by Kohlstedt and Goetze (1974) suggests that the only effect should be minor, produced by reduction of the shear modulus.

The presence of dislocation substructures in naturally deformed olivine which are indistinguishable from those produced in the laboratory and the σ–$\dot\varepsilon$ relations discussed above emphasizes the conclusions from most of the studies reviewed here that flow in the specimens examined by TEM cannot have had a significant contribution from stress-directed diffusion (Nabarro-Herring) creep. Experimental specimens of peridotites, metals and ceramics deformed by climb-controlled slip processes display a power law relationship between stress and steady-state strain-rate (Kirby and Raleigh, 1973; Kohlstedt and Goetze, 1974) and hence the assumption of Newtonian viscosity ($\sigma \propto \dot\varepsilon$) usually made in geophysical models of mantle rheology is probably in error.

The discovery (Green and Radcliffe, 1975) that the CO_2-rich volatile phase in xenoliths is indigenous to them and not a contamination from the magma has important chemical and physical implications. The long-sought residence of the so-called incompatible elements (e.g. K, Rb, Ba, Sr, U, Th, and the light rare earth elements) may be in this phase, and therefore now concentrated onto grain boundaries. Preliminary electron microprobe investigation of grainboundary bubbles in the sample of Fig. 9c confirms the presence of Ba, Rb, U, Sr, and K (Green, 1974b). The extremely thin and regular serpentine film on grain boundaries of xenoliths from kimberlites suggests an *in-situ* origin and thus existence of an H_2O-bearing volatile phase on boundaries at high temperature. The presence of the volatiles as intragranular and intergranular bubbles should affect seismic velocities and attenuation and may weaken the rocks by promoting grain-boundary sliding (Green, 1972). The lack of a grain boundary melt phase in any of these xenoliths, and the steepened slope of the deeper portion of the pyroxene geotherms for kimberlite pipes are both in conflict with the partial melt model for the seismic high-attenuation, low-velocity zone.

References

Ave'Lallemant, H.G., Carter, N.L.: Syntectonic recrystallization of olivine and modes of flow in the upper mantle. Geol. Soc. Am. Bull. **81**, 2203–2220 (1970).

Blacic, J.D., Christie, J.M.: Dislocation substructure of experimentally deformed olivine. Contrib. Mineral. Petrol. **42**, 141–146 (1973).

Boland, J.N., McLaren, A.C., Hobbs, B.: Dislocations associated with optical features in naturally-deformed olivine. Contrib. Mineral. Petrol. **30**, 53–63 (1971).

Boullier, A.-M., Nicolas, A.: Textures and fabric of peridotite nodules in kimberlite. In: Lesotho Kimberlites (ed. P.H. Nixon), p. 57–66. Maseru: Lesotho National Development Corporation 1973.

Boullier, A.-M., Nicolas, A.: Classification of textures and fabrics of peridotite xenoliths from South African kimberlites. Phys. Chem. Earth **9** (1975).

Boyd, F.R.: A pyroxene geotherm. Geochim. Cosmochim. Acta **37**, 2533–2546 (1973).

Carter, N.L., Ave'Lallemant, H.G.: High-temperature flow of dunite and peridotite. Geol. Soc. Am. Bull. **81**, 2181–2202 (1970).

Christie, J.M., Griggs, D.T., Carter, N.L.: Experimental evidence of basal slip in quartz. J. Geol. **72**, 734–756 (1964).

Gillespie, P., McLaren, A.C., Boland, J.N.: Operating characteristics of an ion-bombardment apparatus for thinning non-metals for transmission electron microscopy. J. Mater. Sci. **6**, 87–89 (1971).

Goetze, C., Kohlstedt, D.L.: Laboratory study of dislocation climb and diffusion in olivine. J. Geophys. Res. **78**, 5961–5971 (1973).

Green, H.W.: A CO_2-charged asthenosphere. Nature Phys. Sci. **238**, 2–5 (1972).

Green, H.W.: Comments on paper by A. Olson and T. Birkeland: Electron microscope study of peridotite xenliths in kimberlites. Contrib. Mineral. Petrol. **46**, 69–72 (1974a).

Green, H.W.: Where are the incompatible elements in the upper mantle? (Abst.). EOS Trans. Am. Geophys. Union **55**, 1198 (1974b).

Green, H.W., Gueguen, Y.: Origin of kimberlite pipes by diapiric upwelling in the upper mantle. Nature Phys. Sci. **249**, 617–620 (1974).

Green, H.W., Radcliffe, S.V.: Deformation in the earth's upper mantle: electron petrography of peridotites. Jernkontorets Annaler **155**, 541–543 (1971).

Green, H.W., Radcliffe, S.V.: The nature of deformation lamellae in silicates. Geol. Soc. Am. Bull. **83**, 847–852 (1972a).

Green, H.W., Radcliffe, S.V.: Dislocation mechanisms in olivine and flow in the upper mantle. Earth Planet. Sci. Letters **15**, 239–247 (1972b).

Green, H.W., Radcliffe, S.V.: Deformation processes in the upper mantle. In: Flow and fracture of rocks (eds. H.C. Heard, I.Y. Barg, N.L. Carter and C.B. Raleigh). Geophys. Monogr. **16**, 139–156 (1972c).

Green, H.W., Radcliffe, S.V.: Fluid precipitates in rocks from the earth's mantle. Geol. Soc. Am. Bull. **86**, 846–852 (1975).

Griggs, D.T.: Hydrolytic weakening of quartz and other silicates. Geophys. J. **14**, 19–31 (1967).

Isaacs, B., Oliver, J., Sykes, L.R.: Seismology and the new global tectonics. J. Geophys. Res. **73**, 5855–5900 (1968).

Kirby, S.H., Raleigh, C.B.: Mechanisms of high-temperature, solid-state flow in minerals and ceramics and their bearing on the creep behavior of the mantle. Tectonophysics **19**, 165–194 (1973).

Kirby, S.H., Wegner, M.W.: Dislocation substructure of mantle-derived olivine as revealed by selective chemical etching. (Abst.) EOS Trans. Am. Geophys. Union **54**, 452 (1973).

Kohlstedt, D.L., Goetze, C.: Low-stress high-temperature creep in olivine single crystals. J. Geophys. Res. **79**, 2045–2051 (1974).

Kohlstedt, D.L., Vander Sande, J.B.: Transmission electron microscopy investigation of the defect microstructure of four natural orthopyroxenes. Contrib. Mineral. Petrol. **42**, 169–180 (1973).

MacGregor, I.D.: System MgO-Al_2O_3-SiO_2: solubility of Al_2O_3 in enstatite for spinel and garnet peridotite compositions. Am. Mineralogist **59**, 110–119 (1974).

MacGregor, I.D.: Petrologic and thermal structure of the upper mantle beneath South Africa in the Cretaceous. Phys. Chem. Earth **9**, 455–466 (1975).

MacGregor, I.D., Basu, A.: Thermal structure of the lithosphere a petrologic model. Science **185**, 1007–1011 (1974).

Mercier, J.-C., Nicolas, A.: Textures and fabrics of upper-mantle peridotites as illustrated by xenoliths from basalts. J. Petrol. **16**, 454–487 (1975).

Nicolas, A., Bouchez, J.L., Boudier, F., Mercier, J.-C.: Textures, structures and fabrics due to solid state flow in some European lherzolites. Tectonophysics **12**, 55–86 (1971).

Nicolas, A., Boudier, F., Boullier, A.-M.: Mechanisms of flow in naturally and experimentally deformed peridotites. Am. J. Sci. **273**, 853–876 (1973).

Nicolas, A., Jackson, E.D.: Répartition en deux provinces des péridotites des chaines Alpines longeant la Mediterranée: implications geotectoniques. Schweiz. Mineral. Petrog. Mitt. **53**, 385–401 (1972).

Nixon, P.H., Boyd, F.R., Boullier, A.-M.: The evidence of kimberlite and its inclusions on the constitution of the outer part of the earth. In: Lesotho Kimberlites (ed. P.H. Nixon), p. 312–318. Maseru: Lesotho National Development Corporation 1973.

Olsen, A., Birkeland, T.: Electron microscope study of peridotite xenoliths in kimberlites. Contrib. Mineral. Petrol. **42**, 147–157 (1973).

Olsen, A., Birkeland, T.: Electron microscope study of peridotite xenoliths in kimberlites: A reply. Contrib. Mineral. Petrol. **46**, 73–74 (1974).

Phakey, P., Dollinger, G., Christie, J.M.: Transmission electron microscopy of experimentally deformed olivine crystals. In: Flow and fracture of rocks (eds. H.C. Heard, I.Y. Borg, N.L. Carter and C.B. Raleigh). Geophys. Monogr. **16**, 117–138 (1972).

Radcliffe, S.V., Green, H.W.: Substructural changes during annealing of deformed olivine: Implications for xenolith tenure in basaltic magma (Abst.). EOS Trans. Am. Geophys. Union **54**, 453 (1973).

Raleigh, C.B.: Mechanisms of plastic deformation in olivine. J. Geophys. Res. **73**, 5391–5406 (1968).

Raleigh, C.B., Kirby, S.H.: Creep in the upper mantle. Mineral. Soc. Am. Spec. Papers **3**, 113–121 (1970).

Roedder, E.: Liquid CO_2 inclusions in olivine-bearing nodules and phenocrysts from basalts. Am. Mineralogist **50**, 1746–1782 (1965).

Zeuch, D.H.: Naturally decorated dislocations in olivine (Abst.). EOS Trans. Am. Geophys. Union **55**, 418–419 (1974).

CHAPTER 6.7

The Role of Crystal Defects in the Shear-induced Transformation of Orthoenstatite to Clinoenstatite

S.H. KIRBY

1. Introduction

Coe and Müller (1973) and Kirby and Coe (1974) have established that the structural mechanism of Coe (1970) is responsible for the transformation of orthoenstatite (OE) to clinoenstatite (CE) under shear stress. The Coe mechanism utilizes the concept of the OE structure as that of the CE structure "twinned" on a unit cell basis. The "twin plane" is (100) and thus the OE cell is comprised of two 9.1 Å thick subcell "twins" of CE. Coe selected an alternate CE cell so that when the OE structure is "untwinned" by shearing one of the OE subcell twins on (100) parallel to [001], the resultant orientation of a of CE (with respect to the sense of shear) is consistent with that observed. The b and c directions of the CE and its OE host are parallel. A specific macroscopic angle of shear of 13.3° is associated with the transformation mechanism and is consistent with observations (Kirby and Coe, 1974). I wish to amplify the results of Kirby and Coe (1974) and Coe and Kirby (1975) in regard to the role of partial dislocations and related stacking faults in the transformation.

2. Experimental

Right circular sample cylinders were prepared by coring from large single crystals of orthobronzite from Bamble, Norway. Most specimens were oriented with the core axis at 45° to a and c and were prepared as split cylinders by grinding and polishing to the diametrical (010) planes. The polished (010) surfaces were scribed with a diamond stylus to make possible the measurement of strain in the specimens. The half cylinders were separated by platinum foil and deformed in a solid medium deformation apparatus of the Griggs type at a temperature of 800 °C, confining pressure of 15 kilobars, and a strain rate of 1.07×10^{-4} sec^{-1}. The confining medium was talc in a sample assembly of standard design. Polished thin sections of the deformed samples were prepared in two orientations: (1) parallel to (010) at the diametrical scribed surface and (2) at ∼70° to a^* and containing b. After optical observations, 3 mm diameter discs were made from the thin sections, and special care was taken to establish the average shear strain in each disc from the angular rotation of scribe lines. The discs were thinned to electron-beam transparency by ion bombardment.

3. General Microscopy

Low strain specimens between crossed nicols show optically abundant (100) lamellar features with oblique extinction. Specimens strained beyond $\gamma \sim 13°$ appear to be optically pure CE with $Z \wedge c = 32°$ and much higher birefringence than OE in (010) sections. The principal optical directions Y and Z in OE and CE are normal to a common [010] axis.

The TEM structure of the deformed OE specimens is dominated by lamellar features parallel to (100) which increase in density and thickness with increasing strain below the transformation strain angle of 13.3° (Fig. 1a, b). Streaking in the diffraction pattern parallel to a^* accompanies the appearance of the lamellae. When the shear strain angle exceeds approximately six degrees, discrete CE spots can be accurately indexed with a^* and c in common with the OE host. Beyond the theoretical transformation shear angle of 13.3°, the OE spots disappear and pure CE patterns remain with little or no streaking (Fig. 1c, d). Numerous unit dislocations in the system (100) [001] are found in CE and increase in density with increasing shear strain (Coe and Kirby, 1975). Following Coe and Müller (1973) and Müller (1974), we can associate the lamellae in the low strain specimens with layers of CE in OE host which Müller (1974) identified by high resolution techniques in the experimentally deformed bronzite of Coe and Müller (1973). The fact that the lamellae occur only in low strain samples and disappear, along with the OE spots in the diffraction patterns, when the strain exceeds the transformation strain angle of $\gamma = 13.3°$ supports this identification.

In order to determine what role crystal defects play in the transformation, we can examine the interface between a CE lamellae and its OE host. Fig. 1e illustrates a CE lamella thickening from left to right as it crosses a bend contour where the Bragg condition is met for $g = (202)$. In the regions in which no overlap occurs between the top and bottom of the lamella, the contrast of the fringes and dislocations in the interfaces appears similar to that in the isolated stacking faults and partials in the bend contour. I conclude that the isolated stacking faults and partials are of the same type as those involved in the thickening of the CE lamella at the expense of its OE host. The average thickening in the a^* direction associated with each partial dislocation is 15 ± 5 Å, determined by thickness measurements and dislocation counts on long lamellae with many partials in the boundaries. I now turn to the characterization of these stacking faults and related partials which will enable us to identify the defect mechanism responsible for the transformation.

4. Defect Identification

Low-strain specimens were examined in a variety of two-beam diffraction conditions. It was found that those diffracting vectors which had large computed extinction distances gave poor contrast. This is to be expected since fringe spacing of stacking faults and image widths of dislocations vary with the extinction distance. For example, if the extinction distance is several times the maximum foil thickness that can be penetrated by the electron beam, then less than one

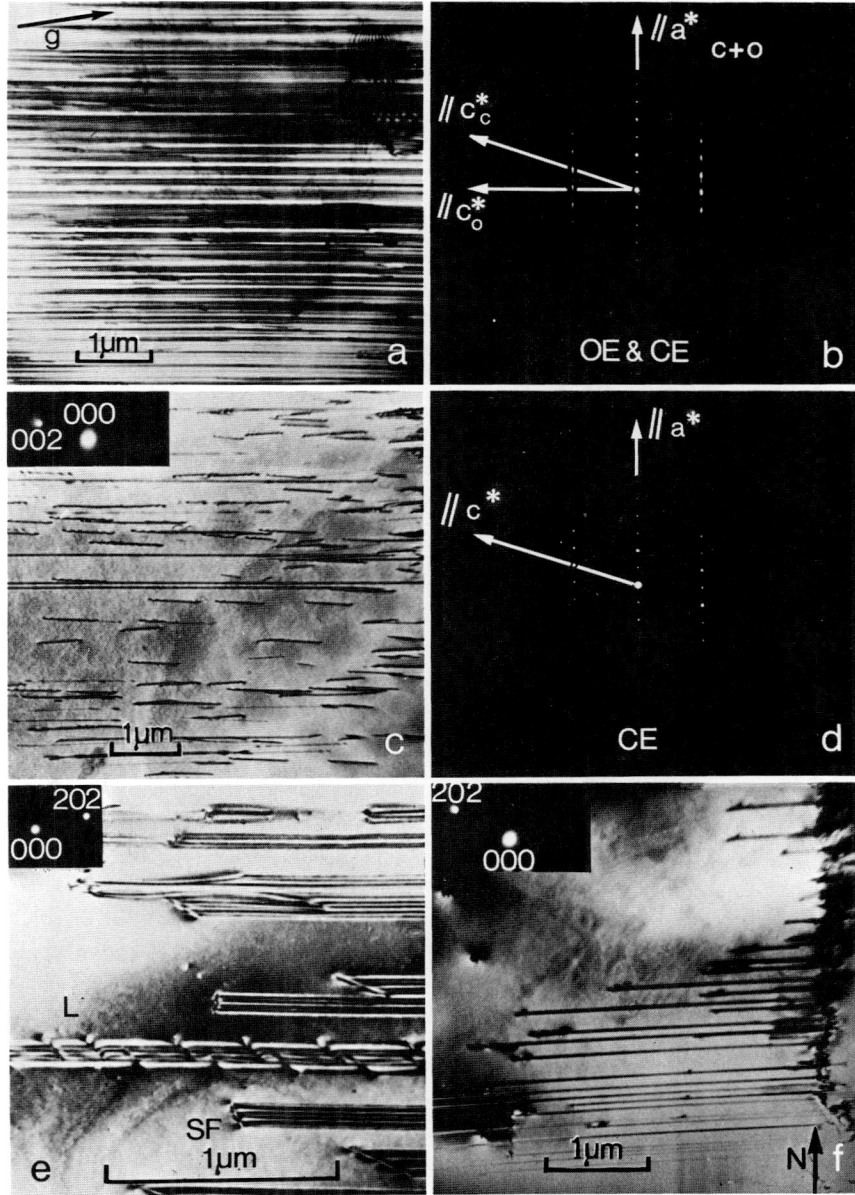

Fig. 1a–f. Bright field (BF) micrographs and associated diffraction patterns (SAD) from experimentally deformed orthobronzite with foil orientation approx. (010). The traces of (100) and the [001] direction are horizontal. (a, b) BF micrograph and SAD for a low strain ($\gamma \sim 9°$) specimen containing both OE and CE. (c, d) BF micrograph and SAD in the [010] zone for high strain specimen ($\gamma \sim 15°$) containing CE. (e) BF micrograph of CE lamella, L, stacking faults, SF, and associated partial dislocations in a low strain specimen. The lamella thickens from left to right. (f) BF micrograph of a (001) network, N, which has emmitted CE lamellae and associated partial dislocations

fringe can be expected in a stacking fault, which would result in relatively poor contrast. At the Bragg condition, image widths of dislocations are about half the extinction distance, ξ_g. If $\xi_g > 10000$ Å, then the dislocation image width is greater than 1/2 micron, which is a significant fraction of the dimensions of most micrographs. Dislocation contrast would then be difficult to distinguish from the normal variations of intensities in the field. In addition, McCormick (Chapter 2.4 of this volume) has shown that large extinction distances make dislocation images go out of contrast due to very small deviations from the Bragg condition. To illustrate the above points, it was found that $g=(102)$ produced essentially no stacking fault and dislocation contrast even though the [001] component of g is the same as for $g=(202)$, which produced strong contrast. $\xi_{202} = 1244$ Å and $\xi_{102} = 44800$ Å, which compare with a maximum foil thickness penetration of ~ 5000 Å for 120 kV electrons.

Eliminating these diffracting vectors with extinction distances greater than 10000 Å, I summarize the significant diffraction contrast experiments as follows: partial dislocation were of variable orientation in (100) but very frequently parallel to [001], [010] and $\langle 012 \rangle$. They were out of contrast for $g=(800)$, (020) and (060) and in contrast for $g=(202)$ and (302). Since the partials move in the glide plane (100), $b = X [0vw]$. Since the out-of-contrast observations indicate that $g \cdot b = 0$ for $g = (020)$ and (060), $v = 0$. Thus $b = X [001]$ where X is some fraction of the [001] cell parameter, c. The (100) stacking faults were also out of contrast for $g=(020)$ and (060) and in contrast for $g=(202)$ and (302). The out-of-contrast observations are consistent with $g \cdot R = 0$ for $R = N [001]$ where N is some fraction of the [001] cell parameter. The contrasts of the stacking faults and partial dislocations therefore indicate that $b \| R = N [001]$ with glide on (100).

4.1 Transformation Stacking Faults

We may now estimate the magnitude of the displacement vector, R, characterizing the transformation faults by interpreting the fringe contrast of the faults in light of two-beam dynamical theory at the Bragg condition (Van Landuyt et al., 1964; Amelinckx, 1970; Amelinckx and Van Landuyt, Chapter 2.3 of this volume). The transmitted and diffracted beam intensities, I_T and I_S, in a faulted crystal can be analytically represented as the sum of three distinct terms, $I_{T,S}^{(1)} + I_{T,S}^{(2)} + I_{T,S}^{(3)}$ (see Amelinckx, 1970). $I_{T,S}^{(1)}$ represents background and does not contribute to fringe contrast, $I_{T,S}^{(2)}$ gives rise to fringes of period $\xi_g/2$ symmetrical with the foil center in both BF and DF. The fringes are complementary between DF and BF and fringes are added at the sides of faults on crossing a thickness extinction contour. The third term $I_{T,S}^{(3)}$ produces fringes of period ξ_g, introduces asymmetry in the DF fringe intensities, and causes branching of fringes with increasing foil thickness. Fig. 2 illustrates the BF and DF contrast of stacking faults and partials in low strain OE samples. Fig. 2a–f show that the BF and DF fringes are complementary and that the DF fringes are qualitatively symmetrical for $g=(202)$. For the same diffracting vector, new fringes are added at the sides of stacking faults where they cross thickness extinction contours

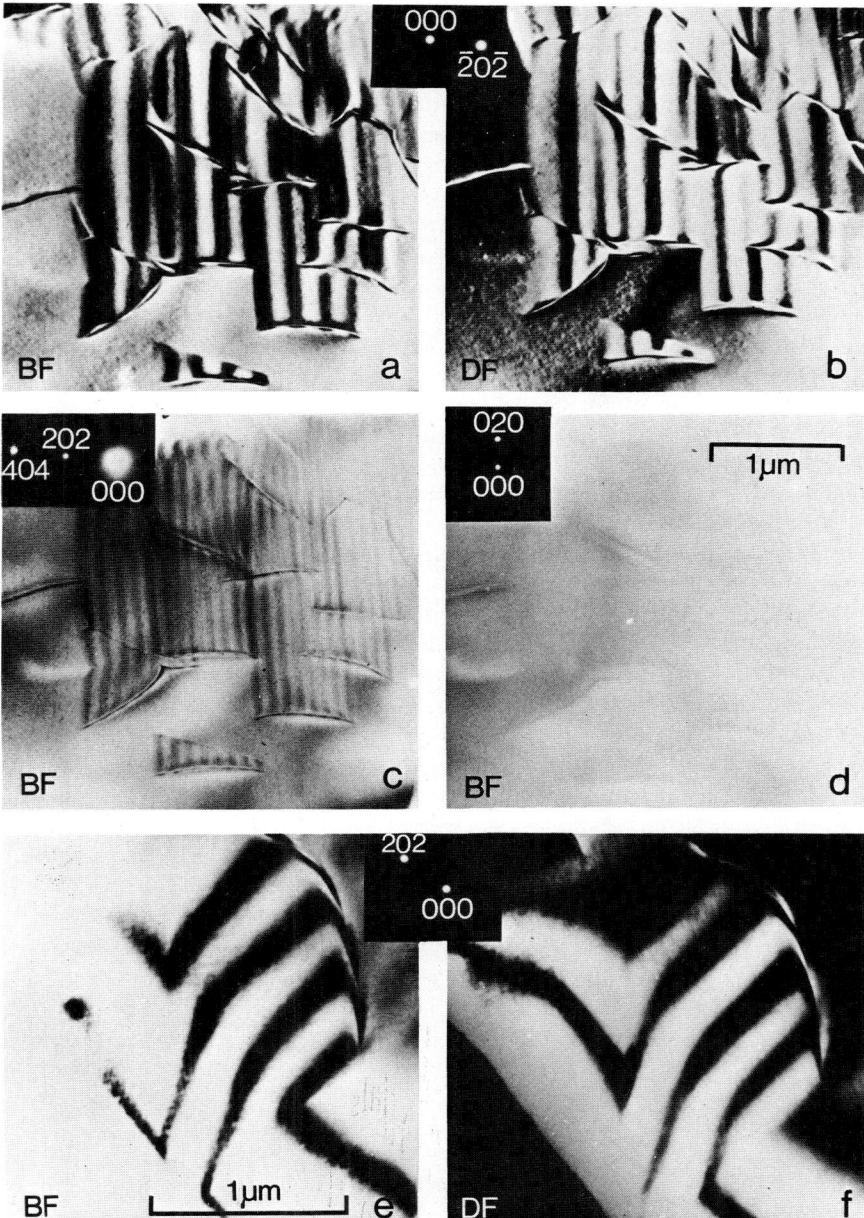

Fig. 2. (a–d) Micrographs from a low-strain bronzite specimen with foil orientation no. 2. The diffraction contrast of stacking faults and partial dislocations is shown for several diffraction conditions. The traces of (100) and [010] are approximately vertical. (e, f) BF-DF micrograph pair of a stacking fault in a foil thickening from the foil edge to the NE. New SF fringes (NE-trending fringes) are born at the thickness extinction contours (NW-trending contours)

(Fig. 2e). These and similar observations indicate the term $I_{T,S}^{(2)}$ dominates the contrast of the transformation stacking faults when $t \leq 4\xi_{202} \simeq 5000$ Å. If $\alpha = 2\pi \mathbf{g} \cdot \mathbf{R}$ is equal to $\pi \pm 2\pi m$, where m is an integer, $I_{T,S}^{(3)} = 0$ and $I_{T,S}^{(2)}$ are entirely responsible for the fringe contrast, independent of the foil thickness, t, and the extinction distance, ξ_g, and anomalous absorption length, ξ'_g. I attempted to test this possibility with $\mathbf{g} = (404)$ but was unable to eliminate the deviated (202) from significantly diffracting (Fig. 2c). The resulting multibeam image shows closely-spaced low amplitude fringes which could not be due to (404) or the deviated (202) alone. This prompted me to evaluate the range of values of α for which we would expect $I_{T,S}^{(2)}$ to dominate, given our diffracting conditions. If $\alpha \neq \pi \pm 2\pi m$ where m is an integer, then α, ξ_g, ξ'_g and t affect the relative contributions of the second and third terms of $I_{T,S}$. The smaller the t/ξ_g and $1/\xi'_g$, the greater the deviation of α from $\pi \pm 2\pi m$ may be and $I_{T,S}$ still dominates. Comparison of the attenuation of fringe contrast in Fig. 2 with computed stacking fault fringes suggests that the anomalous absorption parameter for $\mathbf{g} = (202)$ is likely to be in the range 10 ξ_{202} to 50 ξ_{202}. The thickness range over which the fringe character was observed is 0 to 4 ξ_{202}. Taking $t = 4\xi_{202}$ and $\xi'_{202} = 50\xi_{202}$, then $I_{T,S}^{(2)}$ and $I_{T,S}^{(3)}$ contribute about equally when $\sin\alpha/\sin^2(\alpha/2) \simeq 2$ (see Amelinckx, 1970). This implies that the deviation, Δ, of α from $\pi \pm 2\pi m$ must be less than $\sim 0.5\pi$ in order for $I_{T,S}^{(2)}$ to dominante $I_{T,S}$ and produce the observed fringe behavior. Computed profiles confirm this prediction. Since $\alpha = 2\pi \mathbf{g} \cdot \mathbf{R} = 2\pi \cdot 2 \cdot N$ for $\mathbf{g} = (202)$, then $N = (1 \pm 2m \pm < 0.5)/4$. Taking values of $m = 0, 1, 2, \ldots$, N takes on values $N = (1/4 \pm < 0.125)$, $(3/4 \pm < 0.125)$, $(5/4 \pm < 0.125)$, ..., where it is understood that N may range between the $\pm$ values except for $\Delta = 0$.

Coe's (1970) mechanism is consistent with the identified glide system, the measured angle of transformation shear, the orientation of CE, and with the observed average increase in thickness of CE lamellae associated with the transformation partials. It requires a total shear displacement of 0.83 $\mathbf{c}$ on the glide system (100) [001] in one OE subcell twin to produce two 9.1 Å thick layers of CE. This magnitude of N is in the range of values for $N = (3/4 \pm < 0.125)$ estimated from the stacking fault fringe character. In lieu of quantitative image matching, we can tentatively identify the transformation stacking fault displacement vector as $\mathbf{R} = 0.83$ [001] on the glide plane (100) with shear distributed over one of the 9.1 Å OE subcell twins where the sense of shear is consistent with Coe (1970) and Coe and Müller (1973).

4.2 Formation of Partial Dislocations

In isolated stacking faults, one of the bounding partials always shows strong contrast for $\mathbf{g} = (202)$, but the other is out of contrast (Fig. 2a, lower center). A pair of such strong and weak partials could be produced by dissociation of unit dislocations: [001] → 0.83 [001] + 0.17 [001], in which case $\mathbf{g} \cdot \mathbf{b}$ would be 1.66 and 0.34 respectively for $\mathbf{g} = (202)$. Isotropic theory predicts that partial dislocations are effectively out of contrast when $\mathbf{g} \cdot \mathbf{b} = 1/3$. Preliminary computer-simulated images using diffraction conditions of Fig. 2a with the partials separating the isolated stacking fault of type $b=0.83$ [001] and $b=0.17$ [001] (using the full anisotropic solutions) support this identification, since the former is in strong contrast, very similar to the experimental image of Fig. 2a and the latter is effectively out of contrast (J. McCormick, unpublished results). In addition, McCormick's simulations have shown that the strongly "beaded" character of the image changes to straight dark and bright lines when $|b|>c$, given the diffraction conditions of Fig. 2a. We can therefore eliminate values of $|R|$ and $|b|$ greater than c from the list of possible defects consistent with the stacking fault and "strong" partial dislocation contrast. The simulations of the strong partials cannot yet be considered as a quantitative match since the effects of variation ξ'_g and the sensitivity to deviation from the Bragg condition have not been systematically explored.

CE lamellae are commonly observed to propagate outward from $\sim(001)$ sub-boundaries (Fig. 1f) which, in the starting material, are composed of unit dislocation arrays. The partial dislocations which terminate these lamellae are always in strong contrast, consistent with $b=0.83$ [001]. This suggests that $b=$[001] unit dislocations have dissociated leaving the $b=0.17$ [001] partials in the complex region of the sub-boundaries. All the evidence is consistent with creation of CE lamellae 18.2 Å in thickness by such dissociation and subsequent motion of $b=0.83$ [001] transformation partials. The mechanism of thickening of lamellae as shear strain proceeds cannot be uniquely determined from present evidence. Either the unit dislocations multiply and dissociate, or a pair of $b=0.83$ [001] partials of opposite sign are created at lamellae boundaries and propagate in opposite directions. The observed thickening of CE lamellae in Fig. 1e (L) and in 1f suggests the former explanation.

In summary, the observed partial dislocation contrast is consistent with the dissociation of unit dislocations of $b=$[001] into transformation partial dislocations of $b=0.83$ [001] with strong contrast and into weak partials of $b=0.17$ [001] which are much less prominent and have no obvious relation to the transformation process.

5. Discussion

Many of the contrast properties of crystal defects in the experimentally-deformed orthobronzite reported here are nearly identical with those observed by Green and Radcliffe (1972), Kohlstedt and Vander Sande (1973), and Champness and Lorimer (1973, 1974) in natural orthopyroxenes of unknown or poorly known mechanical and thermal histories. In particular, the stacking fault contrast and the "strong" and "weak" partial dislocations observed by Kohlstedt and Vander Sande (1973) are strikingly similar to those described in this paper, suggesting that the defects may be the same. Their interpretation of the stacking fault contrast was that $R=\frac{1}{4}$ [001], apparently not recognizing the ambiguities discussed earlier which also admit identifications of $R=(1\pm 2\,m\pm\,<0.5)/4$ [001]. From the structural point of view, $R=\frac{1}{4}$ [001] is unlikely since it cannot produce a stable stacking of oxygens in the layers of tetrahedral chains. The $R=0.83$ [001] identified here, however, can restack the layers of tetrahedral chains to produce a structure which is known to be stable, i.e., clinoenstatite.

Ca-rich C2/c clinopyroxenes and P2$_1$/c clinopyroxenes such as CE are identical stacking polytypes. This suggests that if Ca-clinopyroxenes are produced coherently by exsolution from orthopyroxene, the shear associated with restacking should be nearly the same as in the transformation to clinoenstatite. Subsequent or simultaneous diffusion of Ca to the clinopyroxene then can complete the exsolution process (Champness and Lorimer, 1974). The close similarity between the clinoenstatite lamellae and associated partial dislocations described here and the coherent Ca-rich lamellae and associated partials described by Kohlstedt and Vander Sande (1973), Vander Sande and Kohlstedt (1974), and by Champness and Lorimer (1973, 1974) suggests that the 0.83 [001] shear identified here may be the same as in their exsolved Ca-rich clinopyroxenes. Thus non-hydrostatic stress may influence the rate of exsolution of Ca clinopyroxene in a way similar to the OE→CE transformation.

References

Amelinckx, S.: The study of planar interfaces by means of electron microscopy. In: Modern diffraction and imaging techniques in materials science (eds. Amelinckx, S., Gevers, R. Remaut, G., and Van Landuyt, J.), p. 257–294. Amsterdam: North Holland Publishing and American Elsevier 1970.

Champness, P.E., Lorimer, G.W.: Precipitation (exsolution) in an orthopyroxene. J. Mater. Sci. **8**, 467–474 (1973).

Champness, P.E., Lorimer, G.W.: A direct lattice-resolution study of precipitation (exsolution) in orthopyroxene. Phil. Mag. **30**, 357–365 (1974).

Coe, R.S.: Thermodynamic effect of shear stress on the ortho-clino inversion in enstatite and other coherent phase transition characterized by finite simple shear. Contrib. Mineral. Petrol. **26**, 247–264 (1970).

Coe, R.S., Kirby, S.H.: The orthoenstatite to clinoenstatite transformation by shearing and reversion by annealing: mechanism and potential applications. Contrib. Mineral. Petrol. **52**, 29–56 (1975).

Coe, R.S., Müller, W.F.: Crystallographic orientation of clinoenstatite produced by deformation of orthoenstatite. Science **180**, 64–44 (1973).

Green II, H.W., Radcliffe, V.: Deformation processes in the upper mantle. In: Flow and fracture of rocks (eds. Heard, H.C., Borg, I.Y., Carter, N.L., and Raleigh, C.B.). Am. Geophys. Union Monogr. **16**, 139–156 (1972).

Kirby, S.H., Coe, R.S.: The role of crystal defects in the enstatite inversion. Transactions, Am. Geophys. Union **55**, 419 (1974).

Kohlstedt, D.L., Vander Sande, J.B.: Transmission electron microscopy investigation of the defect microstructure of four natural orthopyroxenes. Contrib. Mineral. Petrol. **42**, 169 (1973).

Müller, W.F.: One dimensional lattice imaging of a deformation-induced lamellar intergrowth of orthoenstatite and clinoenstatite [(Mg, Fe) SiO$_3$]. Neues Jahrb. Mineral. Monatsh. H. **2**, 83–88 (1974).

Vander Sande, J.B., Kohlstedt, D.L.: A high-resolution electron microscopy study of exsolution lamellae in enstatite. Phil. Mag. **29**, 1041–1049 (1974).

Van Landuyt, J., Gevers, R., Amelinckx, S.: Fringe patterns at antiphase boundaries with $\alpha = \pi$ observed in the electron microscope. Phys. Stat. Solid. **7**, 519–546 (1964).

Section 7 Special Techniques and Applications

◄ SEM photograph of the surface of opal (Nr. 8163) from Hungary with crystallites of cristobalite. The diameter of the spheres is approximately 6000 Å. (Photograph by R. Wessicken)

CHAPTER 7.1

Amorphous Materials

M. L. RUDEE

1. Introduction

This paper describes methods for obtaining structural information about amorphous materials that are both promising and controversial. Studying the structure of an amorphous material might seem a contradiction in terms to one not familiar with the field, but it is a problem of long standing. In most investigations of amorphous materials — pure glassy SiO_2 being the archetypical subject — the first few nearest neighbor configurations can usually be determined with reasonable confidence from X-ray diffraction, density measurements, and various chemical techniques. The problem that remains then is to determine how far regular configurations extend. The two limiting-case models often used are the continuous random network (CRN) model, first put forward by Zachariasen (1932), and the connected microcrystallite model usually attributed to Valenkov and Porai-Koshits (1936). Both of these models were originally proposed for silica glass, and it is interesting that four decades later, the disagreement still persists: in quite recent papers Mozzi and Warren (1969) advocated the CRN model, while Konnert et al. (1973) favor more crystalline-like ordering. Both of these recent conflicting findings resulted from applying different new analytical techniques to the classical (Zernicke and Prins, 1927) radial density distribution function (RDF) method (see James, 1958, for a thorough introduction to this technique). The RDF method is not without its difficulties: extremely accurate diffraction data are needed to very high scattering angles, the RDF is averaged over a large sample and will not reveal fluctuations in the structure, and an RDF is not a unique description of a structure since Burger (1950) has shown that 230 crystallographic space groups yield only 24 unique symmetries even in their unaveraged RDF's (Patterson functions). Nevertheless, RDF analysis is at the very least a powerful starting place for structural analysis of amorphous materials, and any model must agree with observed diffraction patterns. This latter procedure is a relatively easy calculation using the Debye scattering equation (Warren, 1969) once some tentative atomic coordinates have been established.

Due to the apparent lack of agreement on the structure of glassy SiO_2 from RDF analyses alone, additional tools were sought when technological developments made a knowledge of the structure of amorphous semiconductors a problem of some urgency. What will be described below is the application of electron microscopy to the study of amorphous semiconductors, principally pure glassy

Ge. The two techniques used are dark-field and lattice-fringe electron microscopy, and these will be described in the next Section. In succeeding Sections, the application of these techniques first to carbon black and then to amorphous Ge will be reviewed. A discussion of carbon black is included because its study has been seminal to many developments in electron microscope techniques. Further, it has the unique attribute of being an amorphous material where agreement on the structure seems to exist.

Electron microscopy as a structural tool has its own difficulties and limitations and these will be discussed below. On the other hand it has the potential advantage of providing a real image without the ambiguity always inherent in traditional diffraction techniques. The image is a representation of a specific small volume of material, and thus it should be sensitive to fluctuations in any structure that would be lost in the averaging over a large volume typical of a diffraction pattern.

2. Experimental Techniques

The first technique used was dark field microscopy, a method long prominent in the study of defects in crystalline materials (see, for example, Hirsch et al., 1965). To form a dark field image, the incident beam is deflected off the optical axis so that scattered electrons, rather than transmitted electrons, pass through the objective aperture, see Fig. 1 b. If the specimen is crystalline, one single discrete diffracted beam is used, while with amorphous material that produces broad diffraction rings, the size of the objective aperture determines the region within the diffraction pattern that forms the image. A typical diffraction pattern from

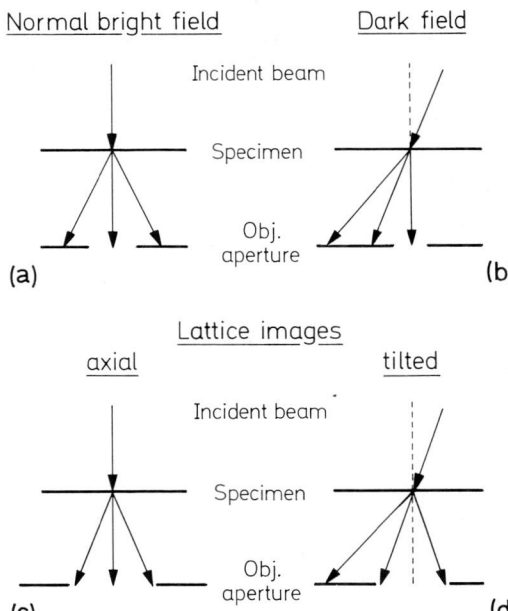

Fig. 1 a–d. Schematic representation of the following modes of electron microscope operation: (a) Normal bright field. (b) Dark field. (c) Axial beam lattice imaging. (d) Tilted beam lattice imaging

an amorphous material is shown in Fig. 2a; the position of an objective aperture for a dark field image is also illustrated. The simplest interpretation of dark field micrographs of amorphous material is that bright areas in the image are the projections of regions within the specimen that are sufficiently well ordered to diffract coherently, and that the size of the bright spot in the image is equivalent to the size of the ordered domain. Hence the presence of discrete bright spots in a dark field micrograph would favor an interpretation based on a structure containing ordered domains. As will be described in the next two Sections, this simple interpretation, despite its apparent success in carbon black, is not the whole story, and recent calculations stimulated by the interest in amorphous Ge indicate that the interpretation of dark field electron micrographs of glassy structures is by no means straightforward.

The next electron microscope technique to be applied to amorphous materials was lattice fringe imaging. In this technique an objective aperture of sufficient size is used so that it passes the transmitted beam and some of the diffracted intensity, see Fig. 1c. These are then recombined by the objective lens to form an image having fringes corresponding to the periodicities represented in the diffracted beams passed by the aperture. The theory of this kind of imaging was developed initially by Scherzer (1949), but seven years elapsed before Menter (1956) successfully resolved the 12 Å spaced $(20\bar{1})$ planes in copper and platinum phthalocyanine. As resolution improved, progressively smaller lattice spacings became the resolution standard for electron microscope manufacturers, until instruments have now been so refined that lattice images can be used for the study of the crystal structure, not just as a test of the instrument. The remarkable studies of atomic arrangements in crystalline oxides in Professor Cowley's laboratory are reviewed in Chapter 2.5 of this volume.

The theory of the formation of lattice fringe images is not without its subtleties. Space does not allow a complete review of all the electron optical problems involved but a brief outline is necessary. Lengthy discussions can be found in the work of Lenz (1971), Thon (1971), and Hawkes (1972).

The special feature of electron optics that complicates the interpretation of lattice fringe images is that spherical aberration, C_s, introduces a phase shift between the scattered and axial transmitted beam. This can be compensated for by defocussing the objective lens, Δf, with the result for weakly scattering objects, that an electron microscope objective lens acts as a spatial frequency filter. For a particular setting of Δf and the inherent C_s of the instrument, only certain d-spacings will be imaged and thus there will be an optimum defocus condition to form an image of any particular spacing. This property has been experimentally verified by Thon (1966) who analyzed through-focal series from specimens exhibiting a wide range of spatial frequencies. Optical diffraction patterns were made from the micrographs themselves, and the magnitude of the d-spacings actually present and their dependence on Δf revealed that the filtering action of the lens was in excellent agreement with the theoretical predictions. With axial illumination, then, it is necessary to "tune" the microscope by adjusting Δf to form an image of a particular d-spacing.

Most of the lattice-fringe studies of amorphous semiconductors published thus far have not used axial illumination, but have used instead an incident beam tilted so that the transmitted beam and a segment of the first diffraction ring make equal angles with the optical axis and pass through the objective aperture, see Fig. 1 d. The axial case theory does not apply, and since the observation of lattice fringes in amorphous Ge in the tilted illumination mode (Rudee and Howie, 1972) generated considerable interest, theoretical analyses of tilted beam images have been, and continue to be, performed.

Howie et al. (1973) found that the calculated tilted beam image of a small crystallite did not display the dependence on changes of focus that they observed in a through-focal series from amorphous Ge. These calculations, as well as those for the axial case, discussed above, have assumed that the illumination had negligible angular divergence. This assumption is often not justified, and Howie et al. (1973) showed that the effect of reasonable values of beam divergence would explain the dependence they observed of lattice fringe images on defocus. Additional results of image calculations will be discussed below.

3. Applications

3.1 Carbon Black

Carbon black has been used as a test specimen at almost every stage of electron microscope development, with the first application appearing in print (von Borries and Ruska, 1939) in the same year that the first commercial instrument reached the market. The early research was concerned with the hitherto unresolvable morphology of the carbon black particle, and the then new resolving power of the electron microscope demonstrated that particle sizes ranged from 200 Å to 5000 Å.

During this same period, the atomic arrangements within carbon black particles were under investigation by X-ray diffraction. Largely through the leadership of B.E. Warren, and later R.E. Franklin, it was recognized that the atoms in carbon black were in graphite-like layer planes, but that these planes were very limited in extent, and not stacked as regularly as in graphite. There existed, however, small regions within the particles that were sufficiently well-ordered to diffract coherently. These regions were called *domains*, and through X-ray diffraction their dimensions were found to be typically 15 Å to 50 Å. Heckman's (1964) review paper provides extensive citations to the early X-ray diffraction and electron microscope literature.

As electron microscope resolution improved, the research effort shifted to elucidation of the structure within the particle. Hall (1948) obtained axial illumination dark field micrographs by displacing the objective aperture, and showed that the layer planes within a particle tended to be parallel to the surface of the particle. By 1966, instrumentation had improved to the extent that tilted dark field micrographs could be obtained, and the resolution was thereby improved so that individual domains were resolved independently by Hess and Ban (1966) and by Rudee (1967). The domain sizes observed in the dark field micrographs agreed quite well with the X-ray diffraction results.

With further progress in electron microscope design, it became possible to use axial illumination lattice imaging and resolve the layer planes in carbon black, first in annealed (graphitized) material (Heidenreich et al., 1968), and shortly thereafter in unannealed, normal black (Ban and Hess, 1968). Subsequent electron microscope studies by Harling and Heckman (1969) and by Ban (1972) have led to a reevaluation of the structure of carbon black. It is now known that the layer planes are distorted but nearly continuous around the particle, analogous to a wavy onion, and that the domains detected in earlier X-ray diffraction and electron microscopy are occasional highly parallel segments within the larger, distorted layer structure. Fig. 2b, provided by Ban, is a micrograph showing the layer plane organization in unannealed carbon black. Ban (1972) suggests that carbon black is an example of the paracrystalline state of aggregation that was first hypothesized by Hoseman and Bagchi (1962).

To summarize, electron microscope images of the layer planes in carbon black suggest that a somewhat greater degree of order existed than was detected by either X-ray diffraction or dark field electron microscopy.

3.2 Amorphous Ge and Si

The sequence of the application of the electron microscope techniques to amorphous semiconductors paralleled the study of carbon black. Since the literature on the structure of amorphous semiconductors is huge, only the contributions of electron microscopy will be emphasized. See Moss (1974) for a discussion of all experimental techniques.

The first technique used was dark field electron microscopy and Rudee (1971) observed bright spots in amorphous Ge that were typically 14Å in diameter. This dimension agreed well with those X-ray studies that had been interpreted on the basis of a domain, rather than a network model, and these dark field electron microscope results were therefore interpreted by Rudee as supporting the presence of ordered domains. Subsequently, similar results were reported for Si (Rudee, 1972) and for Si, Ge, GeTe, and sputtered SiO_2 (Chaudhari et al., 1972).

The next method utilized was tilted lattice-fringe imaging and here small regions, again typically about 14Å diameter, were observed that exhibited quite regular fringes of about 3.3Å separation (Rudee and Howie, 1972); these fringes are illustrated in Fig. 2c. This observation was interpreted as support for the microcrystallite model.

As mentioned in the introduction, any successful model must also agree with the diffraction data as well as electron microscope observations, and Moss and Graczyk (1969) had clearly demonstrated that the diffraction pattern of amorphous Ge was incompatible with a model based on microcrystallites with Ge's equilibrium structure, diamond cubic. Rudee and Howie (1972) found that considerably better agreement was achieved by assuming that the structure contained small crystallites of hexagonal wurtzite, an allotropic modification of Ge and Si found in bulk specimens that had been subjected to high pressures.

Amorphous Materials 481

In subsequent work, Howie *et al.* (1973) showed that the observed fringe contrast was in quantitative agreement with theoretical predictions based on microcrystallites if account was taken of beam divergence, power supply ripple, and inelastic scattering. It was also shown in this paper that including connective

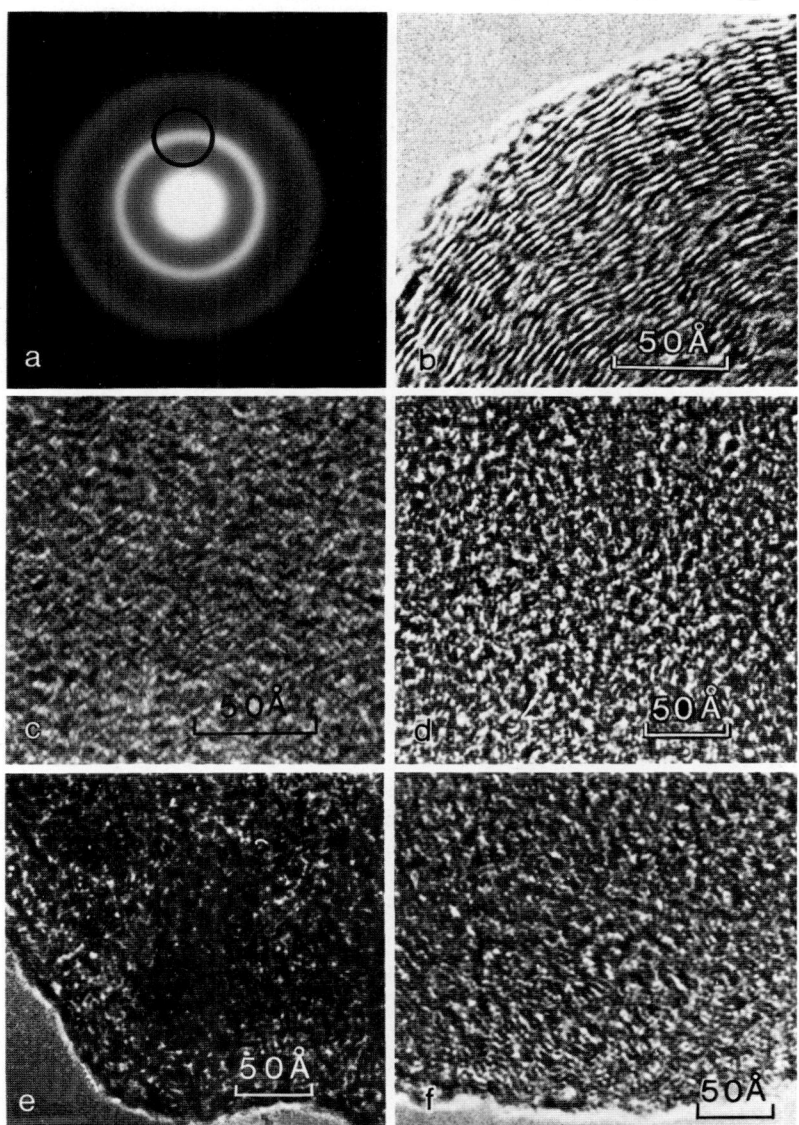

Fig. 2. (a) Diffraction pattern from amorphous Ge with the location of the objective aperture indicated for dark field. For lattice images, the central beam and part or all of some of the diffraction rings are included within the aperture. (b) Axial lattice fringe micrograph of carbon blacks from Ban. (c) Tilted lattice fringe micrograph of amorphous Ge. (d) Axial lattice fringe micrograph of amorphous Ge, (e) amorphous As, and (f) vitreous SiO_2 (From Gaskell and Howie, 1974)

boundary material between crystallites (Rudee, 1972) improved the agreement between the observed and calculated diffraction patterns.

The next question to be considered theoretically was what kind of lattice image would a random network produce? To find an answer, it is necessary to know the coordinates of all the atoms in a random network model. The first such set of coordinates published were for the 614 atom model of vitreous SiO_2 constructed by Bell and Dean (1972). Howie et al. (1973) calculated the tilted lattice fringe image using these coordinates, and the resulting theoretical micrographs did not produce anything like the regular fringes seen in the images of amorphous Ge.

The topology of a SiO_2 network would be different than in a network containing only four fold-coordinated atoms, so the calculated images using Bell and Dean's coordinates were not necessarily representative of Ge. Cochran (1973) used the coordinates of a 62 atom model of Ge that had been generated by Henderson, stacking these small clusters in order to obtain a model large enough to approach being representative of an electron microscope specimen. Cochran concluded that the random network model accounts qualitatively for much that is seen in the lattice fringe electron micrographs, but that the observed regions of highly regular fringes require the presence of at least a small proportion of crystallites.

These results still were not considered definitive because the coordinates used were subject to question: the consequences of stacking small clusters to form a large model introduced some uncertainty, and the diffraction pattern calculated from the 62 atom model did not agree well with experimental observation.

Due to the increasing interest in quantitative calculations pertaining to amorphous semiconductors, Polk measured the actual coordinates of his previously constructed 519 atom random network model (Polk, 1971; Polk and Boudreaux, 1973), and Chaudhari and Graczyk (1974) utilized these coordinates to calculate electron microscope images and other scattering properties. It was discovered that the Polk coordinates produced a good fit to the observed diffraction pattern. In additional studies of the diffracted intensity a marked anisotropy was discovered, i.e. in one direction the model produced a very strong, nearly crystalline-like diffraction maxima. When the atomic positions in the model were projected in a direction perpendicular to the direction of the strong diffraction maxima, it was discovered that the atoms formed layers spaced at the crystalline layer plane distance, and that these "quasi-planes" extended through the entire model. Chaudhari (private communication) has observed identical quasi-planes in another set of coordinates that have been developed by Steinhardt et al. (1974). Thus, it seems that tetrahedrally coordinated random networks are considerably less random than previously believed.

Chaudhari and Graczyk (1974) have calculated electron microscope images from the Polk coordinates and report that the results are very orientation dependent. If the quasi-planes are oriented for maximum contribution to the image, fairly regular fringes appear. Cochran (1974) evaluated these images from the Polk model as being "at most ill-defined fringes of average amplitude less than" the value observed by Rudee and Howie (1972), and advanced arguments (see also Cochran, 1973) that the probability of finding a properly oriented Polk cluster was too small to explain the observed fringes. In addition, Chaudhari and Graczyk used a rectangular aperture in their calculations, and Krivanek and Howie (1975) have used the same coordinates and have not found fringes when a circular aperture, the shape actually employed in experiments, was employed in the calculation.

Spaepen and Meyer (1974) used optical diffraction from a two-dimensional projection of the Polk coordinates to model the image formation in the electron microscope. Their experiment is not, however, an accurate analog of the electron microscope since the unscattered beam was not retained and the image is formed with rings and the center spot of the diffraction pattern. Their "micrograph" did exhibit blobs of contrast, but well-defined fringes were not apparent and the average spacing was less than the layer-plane distance in Ge.

The electron microscope calculations discussed thus far have been concerned with lattice fringe contrast, but questions also exist concerning the first technique used, dark field. Howie *et al.* (1973) noted that by using a microscope of the highest available resolution, the dark field spots observed were only 5 Å instead of the about 15 Å to 25 Å value reported earlier (Rudee, 1971, 1972; Chaudhari *et al.*, 1972). Chaudhari (1973) reported also observing similar smaller dark field images and used their small size as an argument against the micro-crystallite interpretation.

Krivanek and Howie (1975) have undertaken an extensive series of calculations of both dark field and tilted lattice-fringe contrast from both random network and microcrystallite coordinates in order to resolve the apparent discrepancies between theory and experiment.

In order to compare random networks and microcrystallite models of equivalent sizes, Krivanek and Howie stacked up to fifteen microcrystallites each of approximately 75 atoms. The center crystallite was oriented for maximum diffraction into the objective aperture. The constraints in the model were that no overlapping crystallites could be in registry and that space was filled. The initial results show that the dark field image of a single microcrystal of about 15 Å dimension produces a 5 Å image, in agreement with those obtained from high-resolution instruments, but the dark field images of a 64 Å thick array of microcrystals is essentially indistinguishable from the similarly sized Polk random network coordinates. Thus it appears that dark field images cannot distinguish between the two models.

The prospects appear more hopeful for lattice fringes. A single microcrystal produces an image in good agreement with the observed tilted beam micrographs, but adding surrounding randomly oriented microcrystals produces a loss in "fringiness" until only a mottle of blobs remains when the thickness reaches 64 Å. However, since they find that a random network does not produce fringes at any thickness, it does appear that tilted beam lattice fringe images of very thin films (≤ 30 Å) should be amenable to unambiguous interpretation. More sophisticated specimen-support techniques than the standard electron microscope support grids employed thus far will be required for stable, unsupported samples of such thin films.

These calculations by Krivanek and Howie present a paradox since neither model can explain the fringes that have actually been observed in films that are presumably thicker than the minimum thickness found necessary for fringes in the microcrystallite case. One possibility is that the films used have been thinner than suspected. Since neither special care has been taken in preparation nor independent thickness measurements made, an uncertainty in thickness sufficient to explain the fringes, about a factor of three, should be considered. Another possibility is that an unrecognized complication exists in interpreting the tilted beam technique. For instance, Howie *et al.* (1973) and Cochran (1973) suggest that some as yet unknown filtering action might exist in the electron microscope that selectively forms an image favoring fringes with a significantly more limited angular distribution around the optical axis than just the angle subtended by the objective aperture, and Cochran (1974) has put forward some tentative suggestions for a mechanism. Such filtering should be readily apparent in an optical diffraction pattern of a micrograph, but a pattern from Fig. 2c does not reveal sufficient filtering to explain the observed fringes. Nevertheless, unsuspected filtering mechanisms are a topic worthy of significant levels of theoretical and experimental investigation, and may lead to a deeper understanding of electron optics.

In order to reduce the experimental and theoretical uncertainties in the tilted beam method, Gaskell and Howie (1974) have utilized the axial configuration, Fig. 1c, and obtained lattice images in amorphous C, Ge, and As, vitreous SiO_2, and partially crystallized $Li-Al-SiO_2$ glass. Examples of their results are reproduced in Fig. 2d-f. The axial illumination technique demands more careful attention to instrument alignment and defocus than does the tilted configuration, but reduces the difficulty in correcting for astigmatism. Gaskell and Howie found that having the smallest possible projected filament image was necessary and accordingly used pointed-tip filaments. One especially noteworthy observation made by Gaskell and Howie was that fringes were visible in slowly evaporated amorphous C, but not in a rapidly prepared sample. This demonstrates that the fringes are related to the structure and are not artifacts as suggested by Berry and Doyle (1973). Further, a micrograph of partially crystallized $Li-Al-SiO_2$ glass revealed lattice fringes from both the "amorphous" matrix and the 50 Å crystallites of β-eucryptite indicating that imaging conditions for lattice planes in crystalline and amorphous regions are identical. The lattice fringes were more prominent in the thinner areas of the specimens, an observation that would be expected if the thickness dependence of Krivanek and Howie's calculations for tilted illumination conditions prove applicable to axial illumination.

Gaskell and Howie applied Cochran's (1973) criterion for fringe visibility to their observations and conclude that the likelihood of a random structure producing their results is about 10^{-11}.

The axial technique shows great promise and should be more amenable to theoretical analysis than is the tilted beam technique, although calculated images have not yet been published.

The model structures described thus far have been the continuous random network, the microcrystallite, and a mixture of the two. In response to the unsatisfactory explanation of the lattice fringe electron micrographs, three novel structural models have recently been proposed. The diamond and wurtzite structures differ in that in the latter the bonds between layer planes are rotated 60° with respect to the orientation in the former. Betteridge (1973) and Hodges (1974) consider the possibility of mixtures of many ratios of diamond-like and wurtzite-like player plane arrangements varying from plane to plane. Such varied stacking is found in the very complicated crystallography of SiC. They show that the mixed packing model improves somewhat the fit between the calculated and observed X-ray patterns. Gaskell and Howie (1974) have generated a model based on 14 atom clusters of diamond structure bounded by (111) faces of rotated bonds. Extention of this structure leads to encouraging agreement with the RDF. Weinstein (1974) proposes a mixture of clusters, 60% possessing Si III order and 40% having Clathrate II order, as a possible structure.

4. Conclusions

From this discussion, it appears that electron microscopy should play an important role in elucidating the structure of amorphous materials. It has already made a significant contribution to the understanding of carbon black. Although the state of understanding of amorphous semiconductors is much less satisfactory, the tempo and momentum of the activity in this field should produce the level of understanding necessary to make electron microscopy a definitive tool for structural analysis in the near future.

References

Ban, L.L.: Direct study of structural imperfections by high resolution electron microscopy. In: Surface and defect properties of solids, vol. 1 (eds. J.M. Thomas, M.W. Roberts), p. 54–94. London: The Chemical Society 1972.

Ban, L.L., Hess, W.M.: Microstructure of quasi-graphitic carbons. In: Proc. annual EMSA meeting (ed. C.J. Arceneaux), p. 256–257. Baton Rouge: Claitor's 1968.

Bell, R.J., Dean, P.: The structure vitreous silica: validity of the random network theory. Phil. Mag. **25**, 1381–1398 (1972).

Berry, M.V., Doyle, P.A.: Interpreting electron micrographs of amorphous solids. J. Phys. C **6**, L6–L10 (1973).

Betteridge, G.P.: A possible model of amorphous silicon and germanium. J. Phys. C **6**, L427–L432 (1973).

Borries, B. von, Ruska, E.: Eigenschaften der übermikroskopischen Abbildung. Naturwiss. **27**, 281–287 (1939).

Burger, M.J.: Tables of the characteristics of the vector representations of the 230 space groups. Acta Cryst. **3**, 465–471 (1950).

Chaudhari, P.: Oral presentation at the meeting of the American Physical Society in San Diego, March 1973.

Chaudhari, P., Graczyk, J.F.: Scattering properties of the Polk-Turnbull model for amorphous solids. In: Proc. 5th Int. Conf. on Amorphous and Liquid Semiconductors, vol. I, p. 59–68. London: Taylor and Francis 1974.

Chaudhari, P., Graczyk, J.F., Herd, S.R.: An electron microscope investigation of the structure of some amorphous materials. Phys. Stat. Sol. B **51**, 801–820 (1972).

Cochran, W.: Theory of electron micrographs of amorphous materials. Phys. Rev. B **8**, 623–629 (1973).

Cochran, W.: Scattering properties of amorphous structures. In: Tetrahedrally bonded amorphous semiconductors (eds. M.H. Brodsky, S. Kirkpatrick, D. Weaire), p. 177–187. New York: American Physical Society 1974.

Gaskell, P.H., Howie, A.: High resolution electron microscopy of amorphous semiconductors and oxide glasses. In: Proc. 12th Int. Conf. on the Physics of Semiconductors, p. 1076–1080. Stuttgart: B.G. Teubner 1974.
Hall, C.E.: Dark field electron microscopy. II. Studies of colloidal carbon. J. Appl. Phys. **19**, 271–277 (1948).
Harling, D.F., Heckman, F.A.: New information on carbon black fine structure. Materie Plastiche **35**, 80–84 (1969).
Hawkes, P.W.: Electron optics and electron microscopy, p. 124–162. London: Taylor and Francis 1972.
Heckman, F.A.: Microstructure of carbon black. Rubber Chem. Technol. **37**, 1254–1298 (1964).
Heidenreich, R.D., Hess, W.M., Ban, L.L.: A test object and criteria for high resolution electron microscopy. J. Appl. Cryst. **1**, 1–19 (1968).
Hess, W.H., Ban, L.L.: The microstructure of carbon black. In: Electron microscopy, vol. I (ed. R. Uyeda), p. 559–570. Tokyo: Maruzen 1966.
Hirsch, P.B., Howie, A., Nicholson, R.B., Pachley, D.W., Whelan, M.J.: Electron microscopy of thin crystals, p. 295–310. London: Butterworths 1965.
Hodges, C.H.: Structure factor of a random stacking sequence. Phil. Mag. **29**, 1221–1225 (1974).
Hosemann, R., Bagchi, S.N.: Direct analysis of diffraction by matter, p. 131–147, 302–353. Amsterdam: North Holland 1962.
Howie, A., Krivanek, O.L., Rudee, M.L.: Interpretation of electron micrographs and diffraction patterns of amorphous materials. Phil. Mag. **27**, 235–255 (1973).
James, R.W.: The optical principles of the diffraction of X-rays, p. 458–512. London: G. Bell 1958.
Konnert, J.H., Karle, J., Ferguson, G.A.: Crystalline ordering in silica and germania glasses. Science **179**, 177–173 (1973).
Krivanek, O.L., Howie A.: Kinematical theory of images from polycrystalline and random network structures. J. Appl. Cryst. **8**, 213–219 (1975).
Lenz, F.A.: Transfer of information in the electron microscope. In: Electron microscopy in materials science (eds. U. Valdré, A. Zichichi), p. 541–569. New York: Academic Press 1971.
Menter, J.W.: The direct study by electron microscopy of crystal lattices and their imperfections. Proc. Roy. Soc. (London), Ser. A**236**, 119–135 (1956).
Moss, S.C.: Prototype structures for amorphous semiconductors. In: Proc. 5th Int. Conf. on Amorphous and Liquid Semiconductors, vol. I, p. 17–30. London: Taylor and Frances 1974.
Moss, S.C., Graczyk, J.F.: Evidence of voids within the as-deposited structure of glassy silicon. Phys. Rev. Letters **23**, 1167–1171 (1969).
Mozzi, R.L., Warren, B.E.: The structure of vitreous silica. J. Appl. Cryst. **2**, 164–172 (1969).
Polk, D.E.: Structural model for amorphous silicon and germanium. J. Non-Cryst. Solids **5**, 365–376 (1971).
Polk, D.E., Boudreaux, D.S.: Tetrahedrally coordinated random-network structure. Phys. Rev. Letters **31**, 92–95 (1973).
Rudee, M.L.: A study of the domain structure of carbon black by both high-resolution dark-field electron microscopy and X-ray diffraction. Carbon **5**, 155–157 (1967).
Rudee, M.L.: The observation of ordered domains in amorphous Ge by dark-field electron microscopy. Phys. Stat. Sol. (b) **46**, K1–K3 (1971).
Rudee, M.L.: Dark-field electron microscopy of amorphous semiconductors. In: Electron microscopy and structure of materials (ed. G. Thomas), p. 1064–1073. Berkeley: University of California Press 1972.
Rudee, M.L., Howie, A.: The structure of amorphous Si and Ge. Phil. Mag. **25**, 1001–1007 (1972).
Scherzer, O.: The theoretical resolution limit of the electron microscope. J. Appl. Phys. **20**, 20–29 (1949).

Spaepen, F., Mayer, R.B.: Optical modeling of electron microscopy on amorphous tetrahedrally coordinated materials. J. Non-Cryst. Solids **13**, 440–446 (1974).

Steinhardt, P., Alben, R., Weaire, D.: Relaxed continuous random network models. I: Structural characteristics. J. Non-Cryst. Solids **15**, 199–214 (1974).

Thon, F.: On the defocussing dependence of phase contrast in electron microscopical images. Z. Naturforsch. **21a**, 476–478 (1966).

Thon, F.: Phase contrast electron microscopy. In: Electron microscopy in materials science (eds. U. Valdre, A. Zichichi), p. 571–625. New York: Academic Press 1971.

Valenkov, N., Porai-Koshits, E.: X-ray investigation of the glassy state. Z. Krist. **95**, 195–229 (1936).

Warren, B.E.: X-ray diffraction, p. 116–119. Reading: Addison-Wesley 1969.

Weinstein, F.C.: Applications of clathrates to the structure of amorphous Ge and Si. In: Proc. of the 5th Int. Conf. on Amorphous and Liquid Semiconductors, vol. I, p. 95–100. London: Taylor and Francis 1974.

Zachariasen, W.: The atomic arrangements in glass. J. Am. Chem. Soc. **54**, 3841–3851 (1932).

Zernicke, F., Prins, J.A.: Die Beugung von Röntgenstrahlen in Flüssigkeiten als Effekt der Molekülanordnung. Z. Physik **41**, 184–194 (1927).

Chapter 7.2

Signals Excited by the Scanning Beam

R. Blaschke

1. Introduction

The main part of this book deals with transmission electron microscopy of thin foils. In this chapter we illustrate very briefly with a few examples the possibilities of the scanning electron microscope (SEM). The electron beam excites electromagnetic waves and electron signals from a thin surface layer of a target in a wide energy range from fractions of one eV up to the kinematic energy of the incident electrons (Table 1). The signals not only vary in their energy but also the location of their excitation. Fig. 1 shows the different types of beam-excited signals and the conditions for their detection on bulk, semi-thin and ultra-thin specimens. In the transmission electron microscope (TEM), elastic scattering processes on ultra-thin specimens are used for high-resolution imaging and selected area diffraction (SAD). All the other excited signals require the scanning mode for imaging or analysis of small areas.

The first electron microscopes working with the scanning beam were constructed by Knoll (1935) for surface imaging and von Ardenne (1938) for transmission micrographs. Cosslett and Duncumb (1956) improved the conventional "microprobe" of Castaing and Guinier (1950) by introducing the scanning mode. But scanning microscopy became popular only after 1965 when the Cambridge "Stereoscan", developed by C.W. Oatley and co-workers in the Cavendish Laboratories of the University of Cambridge, between 1948 and 1965, became commercially available.

Newer developments of scanning beam instruments diverge strongly. Most manufactures strive for flexible instruments applicable to many kinds of operation modes using simultaneously all beam-excited signals under conditions ranging from observation of high-temperature processes to X-ray and cathodoluminescence (CL) microanalysis at temperatures of liquid He.

Other instruments are specialized either for high resolution, for contrast by energy selection of transmitted electrons, for work at low 1 kV accelerating voltages or for analysis of the Auger electron spectrum of a monolayer on the microscopic scale.

The SEM has so far mainly been used to photograph very small objects, making use of the large depth of focus and the moderate resolution. But in recent years the great potential of the SEM has become more obvious.

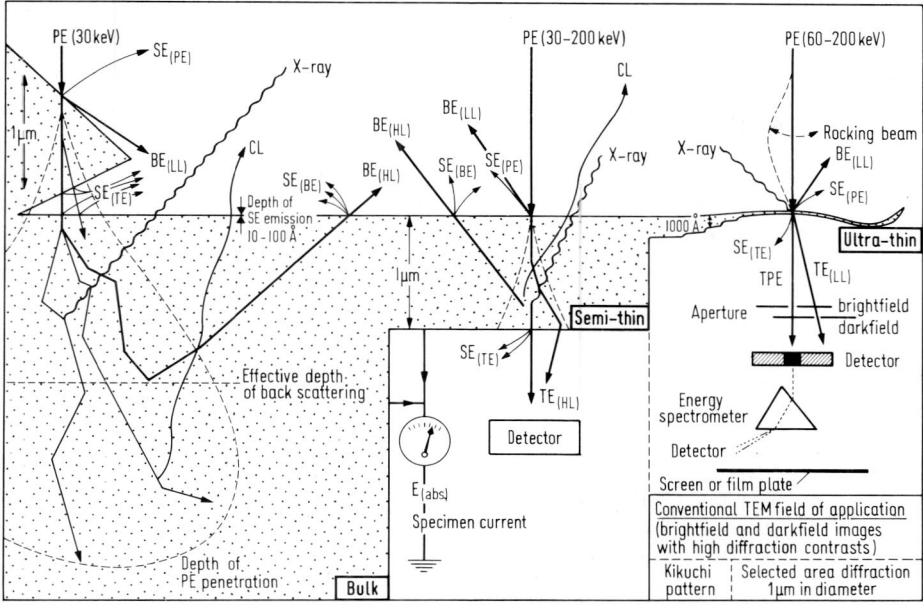

Fig. 1. Signals excited by the scanning electron beam and their relation to specimen morphology (cf. Table 1)

The rapidly expanding field of applications covers more and more areas of mineralogical interest and provides further information on the material which can often not be obtained with the conventional TEM in the ultra-thin foil.

2. Signals Excited by the Electron Beam

2.1 The Principle of the Scanning Mode

The scanning beam method consists in the coupling of the functions of two cathode ray tubes (CRT) of different dimensions. One of the CRT produces images in which the local brightness corresponds to the strength of the electron or photon signal which is excited by the fine-focused scanning beam of the other CRT (i.e. microscope tube) in a thin-surface layer of the investigated specimen. Technical details of the SEM are described in many handbooks (e.g. Oatley, 1972 or Reimer and Pfefferkorn, 1973). We will only briefly review the most important properties of the signals which are recorded in a SEM (compare also Fig. 1 and Table 1).

2.2 Back-scattered Electrons (BE)

The energy of back-scattered electrons varies from the energy of the primary electrons (PE) down to 50 eV. "Low loss" BE are the most important group

Table 1

Signal	Energy range	Depth of emission	Detector systems	Kinds of contrast or information	References
$SE_{(PE)}$ secondary electrons excited by primary E	0–50 eV (1–5 eV)	Au ≅ 10 Å C ≅ 100 Å	+230V biased collector grid + 12 kV accelerator electrode scintillator photomultiplier (after Everhart and Thornley, 1960)	High resolution topographic contrast with threedimensional impression if specimen is tilted (oblique view) Magnetic and electric field contrast	Schur, Schulte and Reimer (1967) Oatley (1972) Reimer and Pfefferkorn (1973)
$SE_{(BE)}$ secondary electrons excited by back scattered electrons	0–50 eV (1–5 eV)	Au ≅ 10 Å C ≅ 100 Å		Interfering contribution to total SE yield its fraction depends on PE energy	
$BE_{(LL)}$ "low loss" backscattered electrons	nearly $E_{(PE)}$	small (100 Å)(?)	1. grounded Al coated scintillator in low angle position relative to tilted specimen surface 2. solid state detector 3. energy selector	High resolution topographic contrast (threedimensional impression due to shadow effects) Channeling pattern and crystal orientation contrast Material contrast of substances with high Z (e.g. Au, Pt)	Wells (1972, 1974) Coates (1967, 1969) Booker (1970) Christenhusz (1968) Newbury (1974) Blaschke and Schur (1974)
$BE_{(HL)}$ "high loss" backscattered electrons	50 eV–$BE_{(LL)}$	≅ half of PE penetr. depth	−230V biased collector grid + 12 kV accelerator electrode scintillator + photomultiplier (collects all E. with energy > 230 eV)	Interfering contribution to total coefficient of backscattered electrons (reduce topographic and channelling contrast) Useful contribution to material contrast of substances with low Z (e.g. silicates)	Blaschke and Waltinger (1971) Blaschke and Schur (1974)

Signal	Energy	Resolution	Detector	Information	Reference
$E_{(abs)}$ absorbed electrons (specimen current)	(low)	—	microamperemeter	Material contrast with reduced topographic contrast limited by low resolution (used for electron beam calibration)	Macdonald (1971)
$E_{(Auger)}$ Auger electrons	10–3000 eV	30 Å or less	hemispherical or cylindrical mirror electron spectrometer	Electron energy spectra of very thin surface layers in the microscale (information about atomic number and chemical bond)	Yakowitz (1974) Yakowitz (1973)
X-ray$_{(char.)}$ characteristic X-ray radiation	depends on Z limited by $E_{(PE)}$	in the range of PE penetration 3 µm for silicates	1. energy dispersive Si (Li) detector 2. wavelength dispersive crystal spectrometer	X-ray spectra or element distribution images (Kossel pattern)	Hall and Höhling (1968)
X-rays$_{(cont.)}$ continuous X-ray radiation	$0-E_{(PE)}$			interfering signal (useful as standard for quantitative X-ray analysis of thin specimens)	
CL cathodoluminescence	1.7–3.1 eV (visual light) also IR and UV	differs widely in range	1. photomultiplier with lens and/or mirror (ellipsoidal mirror) (Hörl, 1972) 2. monochromator	Non-dispersive cathodoluminescence contrast or Luminescence spectra of selected areas	Knisely and Laabs (1973) Pfefferkorn and Blaschke (1974) Yoffe et al. (1973) Muir and Holt (1974)
TPE transmitted primary electrons (unscattered)	$= E_{(PE)}$	ultra-thin specimens	aperture with scintillator or solid state detector	Brightfield images with reduced Bragg diffraction contrast	Gentsch (1974) Blaschke et al. (1974b)
$TE_{(LL)}$ elastically scattered transmitted electrons "low loss"	$= E_{(PE)}$	ultra-thin specimens	1. ring aperture or ring detector 2. electron spectrometer	1. Darkfield images with reduced Bragg diffraction contrast 2. Convergent microdiffraction (200 Å in diameter)	Maher (1974)
$TE_{(HL)}$ inelastically scattered transmitted electrons "high loss"	$\ll E_{(PE)}$	semi-thin specimens	1. scintillator and PM 2. solid state detector	Transmission images of semi-thin specimens (resolution limited by top-bottom effect)	Reimer (1972) Blaschke (1972)

because they produce high-resolution topographic or material contrast and also channelling and crystal orientation contrast. Only in the case of materials with low atomic numbers (Z) can "high loss" BE produce useful additional material contrast.

Topographic contrast generated by BE is due to shadow effects of barriers in the ray path. Electrons which leave under flat trajectories produce the longest shadows. A barrier between the incident spot and the detector will be recorded as a shadow of a length proportional to its height. If the thickness of this barrier is smaller than the penetration depth of the back-scattered electrons, a fraction of them can arrive at the detector but not without loss of energy. It is essential to eliminate these in order to obtain sharp shadows. Specimens are usually tilted about 35° to the primary beam and a simple scintillation detector is placed at a low angle position (Wells, 1972; Blaschke and Schur, 1974). It would be advantageous to use an objective with an extremely short focal length and to tilt the specimen to very high angles (Wells, 1974).

We notice by comparing Figs. 2c and 3e with images produced by other signals that BE give the best three-dimensional representation of the morphology. This is because contrast corresponds to light optical illumination from the top side.

Geometrical and energetic conditions for *crystal orientation contrast* of the BE image are the same (Wells, 1972; Blaschke and Schur, 1974). In both cases, high energy electrons are required which have not lost any energy. Fig. 4 shows several examples. The changes in contrast are caused by a variation of the specimen-tilting in fractions of degrees. These effects were described by Christenhusz (1968) at about the same time channelling patterns were discovered by Coates (1967). The conditions for optimum crystal orientation or channelling contrast in BE mode are the following.

1. The detector system has to be set for low-loss back-scattered electrons which leave the crystal surface in low-angle trajectories..
2. The crystal surface has to be extremely smooth.
3. The surface layer a few hundred Å thick should have a low density of defects and lattice distortions.

Due to abrasion during mechanical polishing, points 2 and 3 can rarely be satisfied for minerals (Fig. 4d).

Since the surface conditions are very important, pseudo-orientation contrast can arise from different degrees of surface roughness in different crystallographic directions. This is illustrated for polycrystalline titanium in Fig. 4f.

The optimum *material contrast* is obtainable from electrons leaving the specimen surface in trajectories perpendicular to the surface plane. The local brightness of BE material contrast is proportional to the atomic number. In order to use a fine-focused primary beam it is advantageous to tilt the specimen to about 55 degrees. The image distortion caused by specimen-tilting can be corrected by calibrated changes in the Y-deflection of the primary beam. The detector is in a high-angle position above the specimen. Because of their high energy, BE are not influenced by weak magnetic or electric fields or slightly charged specimens.

2.3 Secondary Electrons (SE)

SE are generated in a very thin surface layer either by the incident PE or by BE. The first group ($SE_{(PE)}$) produces the high resolution topographic contrast while the second group ($SE_{(BE)}$) disturbs the signal. In order to eliminate the interfering fraction of BE-excited SE, two ways can be followed: Either we reduce the energy of incident electrons down to about 1 keV (i.e. the optimum electron energy for generating SE) but this causes a simultaneous loss in resolution due to an increased influence of the energy distribution of electrons emitted from the cathode. The second way consists in the reduction of specimen thickness down to much less than 1 μm and in increasing the accelerating voltage to a maximum (100 or 200 kV). In the latter case most of the high-energy electrons penetrate the thin specimen without back-scattering. Such SE images excited by 200 keV primary electrons in the surface of semi-thin specimens are very similar to Pt/C replicas in the TEM, enabling a resolution of about 20 Å.

Advantages of the low kinetic energy of secondary electrons are:
1. All excited SE are detectable by using a relatively low biased collector field of about +230 V.
2. SE are extractable from deeper holes in the specimen surface without shadow effects.
3. SE are influenced by magnetic and electric fields within the specimen surface layer. This can be used to obtain magnetic and electric field contrast.

The main disadvanges caused by the low energy of the SE are:
1. The topographic contrast of the SE signal does not correspond to the range of surface structure (i.e. the smallest particles on a specimen surface or sharp edges produce the highest SE yield, while coarse structures have a tendency to disappear). The contrast of the SE image becomes even more complex, if an energy-differentiated signal is used (Fig. 2b).
2. SE are sensitive to charging, which e.g. becomes visible as dark halos around charged particles. Nonconductors therefore ought to be coated with a conducting film (usually Au or C).

2.4 Absorbed Electrons ($E_{(abs)}$)

The number of absorbed electrons is given by the Equation

$$E_{(abs)} = PE - (SE + BE).$$

Under typical high-resolution conditions in the SEM, the specimen current is in the range of 10^{-11} amps. Such a beam current is usually too weak to obtain a passable $E_{(abs)}$ signal to noise ratio. In microprobes with a beam current in the range of 50 n amps, the absorbed electrons generate a strong signal which reflects atomic number contrast and is inverse to that of BE, as is illustrated in Fig. 2d. Resolution usually reaches 1 μm.

2.5 Auger Electrons and Characteristic X-rays

The inelastic impact of PE in the sample atoms produces energy transitions of electrons. Vacant positions are filled by electrons changing their energy level from outer to inner shells. During these transitions, energy is either released as emission of X-ray photons or, more rarely, it is transmitted to an electron which leaves the atom as an "Auger electron". The probability to generate Auger electrons increases with decreasing atomic number of the ionized atoms. For heavier elements it is much more probable to produce characteristic X-ray quanta than $E_{(Auger)}$. (About 1000 incident electrons excite 1 characteristic X-ray quantum.) $E_{(Auger)}$ interact with the nuclei and shell electrons of the target intensively resulting in an exit depth of at most 30 Å. Consequently $E_{(Auger)}$ spectra

▶

Fig. 2a–n. Spherical bodies recorded with different signals. (a–g) Iron sphere. (a) Secondary electrons (SE) image with superposed Y-modulated line scan demonstrating the dependence of the SE-yield from angle of the incident beam with the specimen surface using a linear amplifier system. (b) Differentiated SE signal showing better contrast in small details but loss of the three-dimensional impression of the sphere. (c) Back-scattered electron (BE) image with the BE detector on the left hand side (scan rotation = 90°). The three-dimensional appearance is very obvious. (d) Fe sphere recorded by the signal of specimen current ("absorbed electrons") showing complementary contrast to the SE image (a). (e–g) Recording of the same sphere using the emitted X-ray $Fe_{K\alpha}$-radiation. (e) Energy dispersive Si[Li]-detector (ED) in horizontal position of the left hand side. The linear (*lin*) and the logarithmic (*log*) calibrated scale of the Y-modulated $Fe_{K\alpha}$-signals demonstrate a high degree of independence of specimen morphology. (f) $Fe_{K\alpha}$-image recorded by ED detector in 40° position; angle between electron beam and detector axis is 50°. (g) $Fe_{K\alpha}$-image using a wavelength-dispersive (WD) crystal spectrometer positioned at 35° to the incident beam. The halo around the sphere is caused by X-rays which are excited by electrons back-scattered from the Al support in the direction of the Fe sphere. (h–j) Cathodoluminescence (CL) signals from minerals. (h) Sphere made of a single crystal of quartz recorded with the CL signal. Notice the luminescent lamellae parallel to the unit rhombohedron. Because the CL-detector is positioned in the upper right corner, radiation generated in the left part of the sphere must cross it in order to be detected. (i) CL image of a crystal plate cut perpendicular to the z-axis of lamellar quartz (specimen courtesy of Prof. H.-U. Bambauer) (Lehmann and Bambauer, 1973). (j) CL image of anatase. Of the three TiO_2 polymorphs only anatase shows cathodoluminescence (Pfefferkorn and Blaschke, 1974). This can be used to differentiate between rutile and anatase pigments in technical paints on the microscale. (k–n) Electron micrographs of flying ash consisting mainly of $SiO_2 + Al_2O_3$ glass spherules suspended on a thin organic film to illustrate advantages of scanning transmission electron microscopy. (k) Secondary electrons (SE); accelerating voltage 20 kV. (l) Conventional TEM 100 kV, bright field. (m) Scanning transmission 100 kV, bright field. (n) Scanning transmission 100 kV, dark field.

The comparison of images from the same specimen area illustrates the differing information obtainable from semi-thin objects. The SE image gives the best three-dimensional impression. In the TEM image only "low loss" electrons are recorded and this does not provide information from specimen areas with more than 1000 Å in thickness. The STEM method allows the detection of transmitted electrons with high energy losses, i.e. we can get transmission micrographs of specimens in the range of 1 μm thickness instead of <0.1 μm in the conventional TEM.

The extreme changes in specimen thickness generate a wide range of contrasts which are not covered by the linear amplifier system in the bright field geometry (m). In such cases the dark field image (DF) is more useful (n). (n) also shows the top-bottom-effect i.e. the border of the bigger sphere on the top side (1) is sharp, the border of the sphere lying underneath is diffuse

Signals Excited by the Scanning Beam 495

yield information about the atomic number and chemical bonds of very thin surface layers while X-rays penetrate micron-thick layers of the specimen. In order to eliminate $E_{(Auger)}$ and X-rays excited by back-scattered electrons it is advantageous to reduce the specimen thickness e.g. by ion beam etching. The X-ray detector most suitable for SEM's is the energy dispersive Si(Li) solid-state detector system recording simultaneously the full spectrum from 0.7 keV up

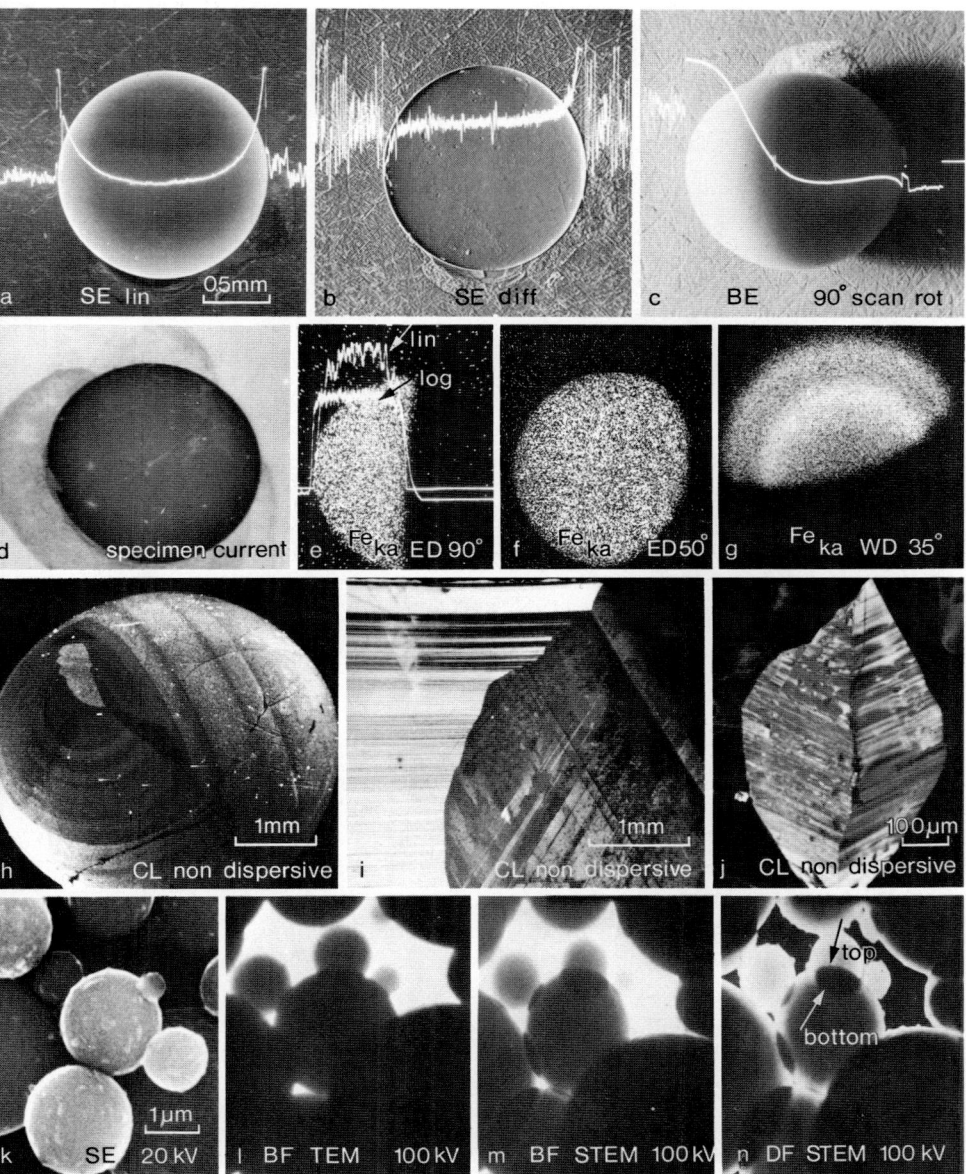

Fig. 2a–n

to the PE energy at high resolution SEM conditions (i.e. the beam current is in the range of 10^{-10} amps). Wavelength dispersive crystal spectrometers record only one peak of the emission spectrum and require a beam current of about 10^{-7} amp. [See articles by Lorimer and Cliff (Chapter 7.3); Suter et al. (Chapter 7.4) and König (Chapter 7.5) in this volume.]

Kossel Pattern. X-rays generated by a stationary electron beam within a single crystal propagate radially from the incident spot. Resulting from interactions with the outer shell electrons radiation is produced which is known as "Kossel radiation" and corresponds to Kikuchi lines in TEM. It is recorded on a film fixed either above a thick crystal or underneath a crystal less than 50 μm thin. Kossel pattern are useful to obtain lattice data of small magnetic crystals which cannot be analyzed in TEM or SEM with electromagnetic lenses. An informative review of Kossel pattern application is given by Yakowitz (1973) who advocates the use of specially designed Kossel generators instead of SEM or microprobe.

2.6 Cathodoluminescence (CL)

CL photons have the wavelength of visual light or of the neighboring IR and UV ranges. They get their low energy from radiative recombinations of electron-hole pairs generated within the specimen surface by electron bombardement.

The CL of many minerals (silicates, oxides, halogens) is generated by luminescence centers (i.e. impurities in a concentration of about 10^{-5}–10^{-3} foreign atoms per crystal atom). Impurities and crystal defects in quartz cause luminescence in fine lamellae (Fig. 2i). CL in anatase (Fig. 2j) seems to be the result of the lattice deformation (Pfefferkorn and Blaschke, 1974). Fine dispersed luminescent substances can be used to mark fine pores and fissures in mineralogical objects for CL micrographs in the SEM.

2.7 Transmitted Electrons (TE)

Transmitted electrons can be divided into transmitted primary electrons, elastically scattered electrons with low energy loss (LL) and inelastically scattered electrons with high energy loss (HL). The conventional TEM works on the basis of the elastic scattering processes (i.e. brightfield or darkfield images and selected area diffraction patterns). The scanning beam method (STEM) adds to the range of these TE applications:
1. Bragg contrast is reduced in bright field and dark field images of ultra-thin crystalline foils due to a higher aperture angle (about 10^{-2} rad) of the fine-focused beam compared to 10^{-4} rad in a conventional TEM.
2. Microarea diffraction can be obtained on an area as small as 200 Å in diameter by means of a slightly convergent stationary beam. Conventional selected area diffraction is limited to about 1 μm in diameter.
3. Microdiffraction can either be done using the rocking beam technique or the diffracted beam deflection method combined with electron energy selection.
4. Through STEM we can obtain additional surface images by SE and BE in TEM with scanning device. The resolution of SE images can be as good as 20 Å.

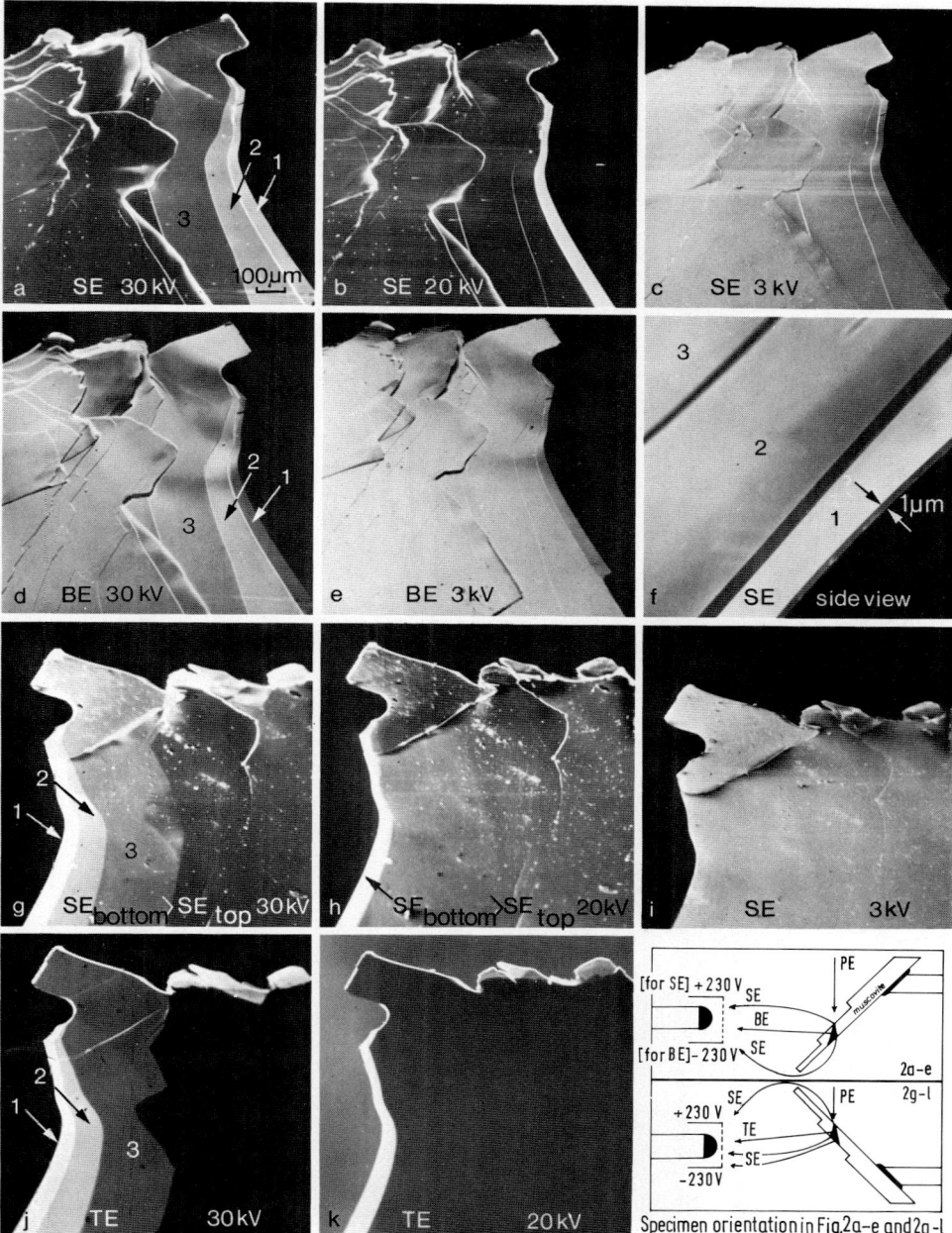

Fig. 3a–k. Influence of specimen orientation and PE energy on signals of SE, BE and TE of a cleavage plate of muscovite. Specifications are indicated on the micrographs

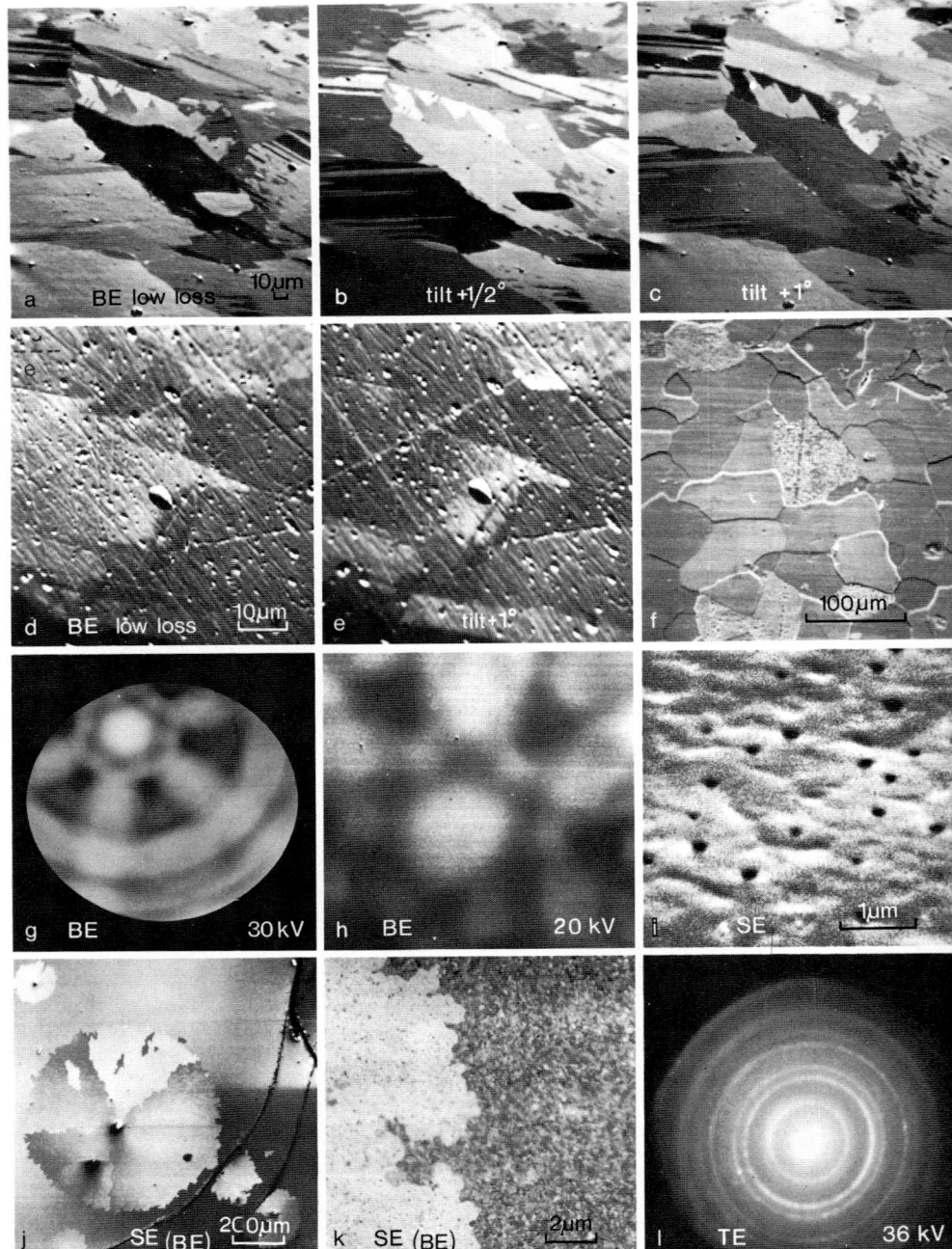

Fig. 4a–l

5. Semi-thin specimens cannot be penetrated by the electron beam in the conventional TEM. Transmission is greatly increased in the STEM and amounts in silicates to about 0.5–1 μm. Resolution in this case is limited by top-bottom-effects. The irradiated area increases because of scattering processes within the specimen ("electron diffusion").
   Semi-thin specimens display more truly structural features of the bulk sample than ultra-thin foils. Moreover, ion-thinned foils always contain large semi-thin areas.
6. A future field for the application of STEM will be *energy-loss analysis* of transmitted electrons after generating ionization processes within a 50 Å microarea of ultra-thin foils instead of X-ray or Auger microanalysis.

3. Some Applications of SEM Signals in Mineralogy

Most of the contributions in this volume explore with high-resolution TEM methods the structure of crystalline material and seemingly cover the whole range of research on the microstructure of minerals. In conclusion of our presentation of the scanning electron microscope we would like to present some examples to illustrate where this method has definite advantages over the conventional TEM.

The most important feature of the scanning beam method is the enormous depth of focus. The SEM is therefore the ideal instrument for problems in which sample morphology and composition are important. Fig. 2 a–n illustrate the range of useful beam-excited signals not obtainable in the conventional TEM (except l).

◀

Fig. 4a–l. Scanning beam interactions with crystal lattices. (a–c) BE images showing crystal orientation contrast of polycrystalline copper in an electrolytically polished section. The surface is cleaned by ion etching. The changes in contrast due to specimen tilting are in the range of fractions of degrees (courtesy of P. Schmidt, Münster). (d and e) Micrographs illustrating changes in crystal orientation contrast of *niccolite* (NiAs) on a mechanically polished section. (d) is tilted with respect to (e) by 1°. The shadows are caused by the micro-roughness of surface details. Geometrical and energetic conditions for crystal orientation contrast and for topographic contrast of the BE image are identical (Blaschke and Schur, 1974). (f) Pseudo-orientation contrast in SE image of ion-etched polycrystalline titanium. The variation in contrast is due to different degrees of surface micro-roughness. A change in the specimen tilting angle of more than 10° does not influence the relative contrast differences between the crystallites. Unfortunately it is not possible to observe ion etching *in situ*, because the ion beam produces a lot of electrons and ions which interfere with the signal. (g–l) Scanning beam interaction with gold crystal and foils. (g) Micro-channelling pattern from the surface layer of a single crystal of Au approximately normal to [111] (courtesy of Dr. Raith, Dortmund). (h) Channelling pattern of an epitaxial Au film evaporated on a muscovite crystal using the variations of beam incidence during scanning at low magnifications. (i) High magnification SE image of the Au film shown in (h). (j) Crystal orientation contrast in the SE (BE) image of a gold film demonstrating recrystallization from impurity centers. The differences in SE yield are generated by the fraction of SE excited by BE. (k) The border of the recrystallized area of (j) at higher magnification. (l) 30 kV rocking beam transmission diffraction pattern of a thin film taken from the face of the CRT on the SEM

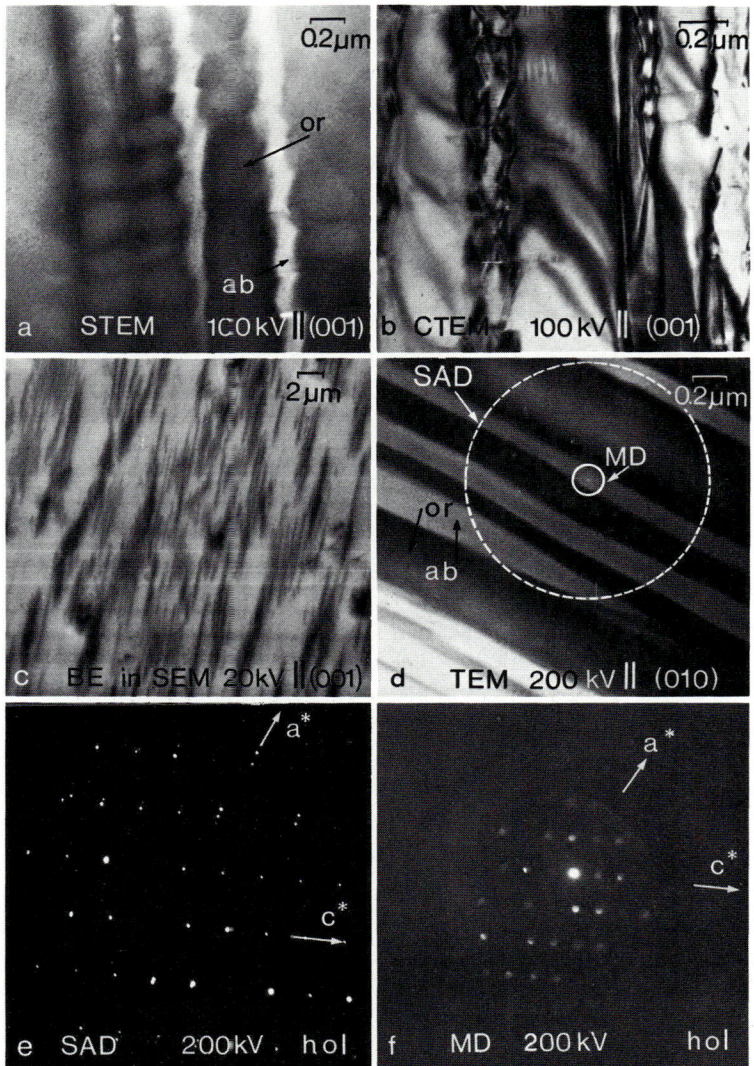

Fig. 5a–f. The use of fine-focused beam in an ultra-thin foil of moonstone (Blaschke *et al.*, 1974b). (a) Ion etched foil (001) in scanning transmission mode (STEM) at 100 kV. Bands of albite and orthoclase, both with different contrast, are indicated. Due to the strong convergence of the beam, the boundaries are not very sharp. (b) Nearly the same area as in (a) but under normal TEM conditions (i.e. objective aperture in the range of 10^{-4} rad.). Notice strong Bragg contrast which is absent in (a). (c) BE image of the same moonstone in polished section (001) using a special BE detector. Horizontal dark bands are albite lamellae. (d) 200 kV transmission mode TEM micrograph of ion-etched foil (010). (e) Selected area TEM diffraction pattern. The covered region is indicated by large circle in (d). The diagram contains the reciprocal lattice planes (*h0l*) of both, albite and orthoclase, superposed. (f) Convergent micro-diffraction (MD) pattern from the area indicated by small circle in (d) (about 500 Å). Notice that only albite diffractions are present

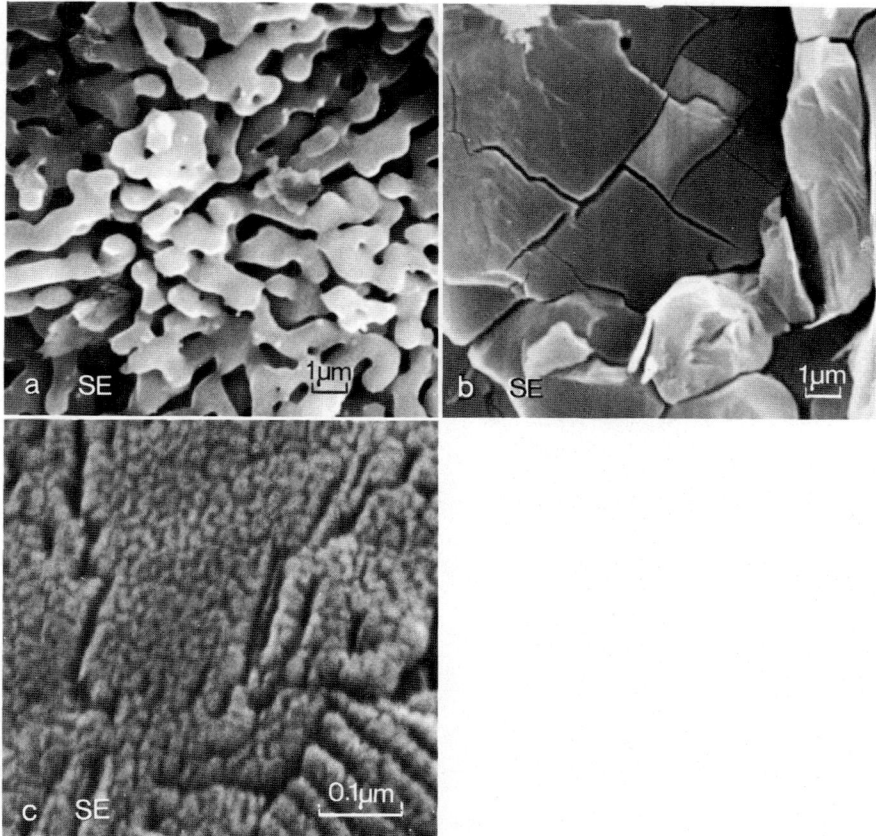

Fig. 6a–c. High temperature *in situ* experiments. Decomposition of limestone to CaO at 1000 °C in air (a) and under high vacuum conditions (b, c)

In Fig. 3 (muscovite flake) an attempt is make to compare scanning electron micrographs of an identical object observed with different electron signals and under different operating conditions. We conclude from Fig. 3 that:

1. 30 and 20 keV electrons penetrate layers of muscovite up to a thickness of about 1 μm (a, b, g, h, j, k).
2. The most real impression of the morphology of the particle is obtained by BE (e).
3. SE tend to overemphasize surface details such as small particles and crystal edges (a, b, g, h).
4. SE images, particularly at low accelerating voltages, are more disturbed by charging effects than BE images (c).
5. The full interpretation of electron micrographs requires knowledge of the used operating conditions.

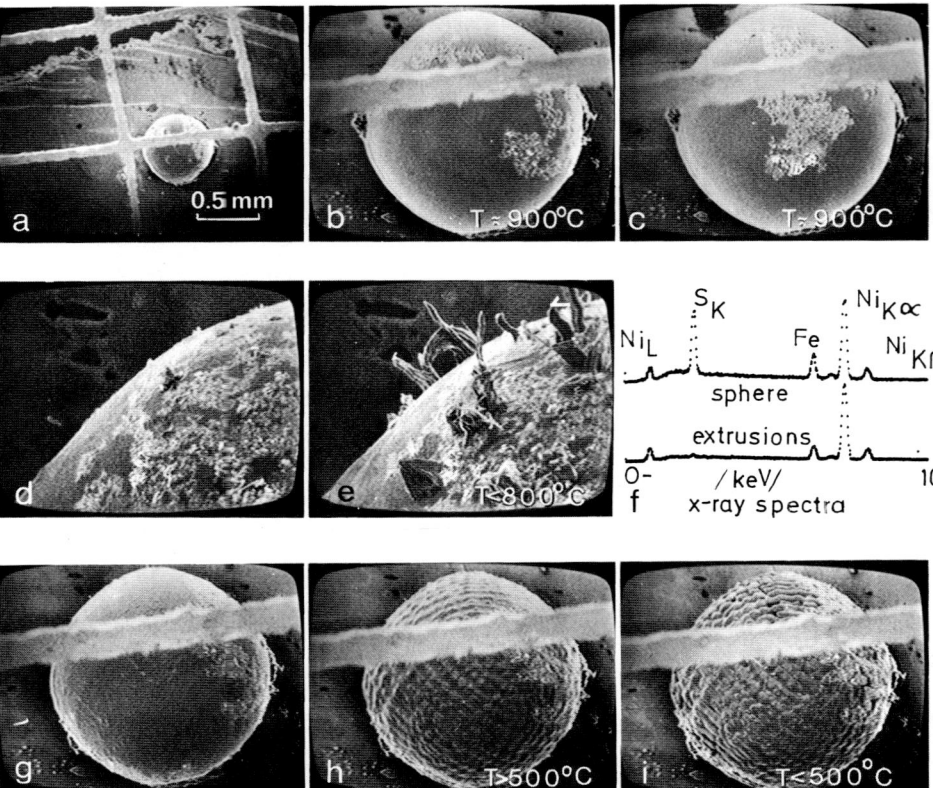

Fig. 7a–i. *Life observation* of crystal growth and melting processes at high temperatures in the SEM. All pictures except (f) are taken on the TV-monitor during replay of the magnetic tape. (a) General view into the specimen crucible. The low biased grid above the sample eliminates thermic electrons without noticeable disturbance of the SE signal. The sphere underneath the grid was produced by heating of a small amount of Ni_3S_2 powder (i.e. composition of heazlewoodite) on a MgO single crystal substrate in the SEM at 900 °C. (b, c) Same sphere at 900 °C. The rapid motion of NiO crystallites on the spherical surface demonstrates the low viscosity of Ni_3S_2 melt at this temperature. The morphology resembles drifting continents of the earth's crust which move at a velocity of about 1 mm/sec. (d, e) During slow cooling Ni crystals extrude. Velocity of growth is about 0.3 mm/sec. (f) Energy dispersive X-ray spectra of the sphere (top) and the extruded Ni-crystals (bottom). The Fe-peak is due to Fe_{K_α}-radiation excited on the surface of the pole-piece plate of the objective lens. (g, i) Changes in surface morphology during cooling can be observed. We attribute them to solidification (from g–h) and to a high/low phase transition of Ni_3S_2 (from h–i)

Applied to crystalline samples the scanning beam method of SEM or TEM with scanning device enlarges the analytical possibilities of electron microscopy in following points:
1. *Texture* of polycrystalline materials in polished sections can be imaged either by BE crystal orientation contrast (Fig. 4a–e) or SE pseudo orientation contrast after surface etching (Fig. 4f).

2. *Crystal orientation* and *lattice parameters* in polished sections or thin surface layers can be determined by electron channelling patterns in the microscale (rocking beam technique, Fig. 4g) or in the macroscale (Fig. 4h) and combined with morphological studies of surface microroughness (Fig. 4i) and crystalline composition (Fig. 4j–k).
3. Electron diffraction patterns are obtained from thin foils or small particles using the rocking beam method (Fig. 4e).
4. STEM bright field and dark field images of ultra-thin foils reduce Bragg contrast (Fig. 5a) in comparison with conventional TEM (Fig. 5b). The BE image controls the laminar composition by material contrast of the bulk specimen (Fig. 5c).
5. A convergent beam micro-diffraction pattern can be obtained on a very small area of less than 200 Å diameter. In contrast to this high spatial resolution, selected area diffraction in the conventional TEM mode has a better angular resolution (Fig. 5d–f).
   The spatial resolution is essential for lamellar crystals such as submicroscopic exsolution in alkalifeldspar ("moonstone").

Another advantage of the SEM is the large specimen chamber compared to the TEM. This permits the easy use of hot and cold stages within the microscope. Fig. 6 shows results from *in situ* experiments on a hot stage, illustrating the decomposition of limestone $CaCO_3$ to CaO at 1 000 °C in air (Fig. 6a) and under high vacuum in the microscope (Fig. 6b, c). In a study to investigate the structure of soft burnt limestone in the SEM with secondary electrons and using a hot stage we observed the formation of cleavage fractures (b) and not the structure typical of limestone burnt under atmospheric conditions (a). But a high-resolution micrograph with a LaB_6-cathode displays a similar texture on a much smaller scale (Blaschke *et al.*, 1974c).

An environmental cell to observe crystal growth by solid-gas and gas-gas reactions as well as by condensation from vapor under defined physical conditions in the SEM was developed by Maas (1974).

Combining *in situ* experiments with TV scan monitoring on magnetic tape, reactions can be followed and recorded directly. Fig. 7 illustrates such life observations of crystal growth and melting processes of Ni_3S_2, Ni and NiO. Similar experiments on geologically important systems could give information about mechanism and kinetics of chemical reactions and melting-crystallization phenomena, which could contribute to a better understanding of rock-forming processes.

References

Ardenne, M. von: Das Elektronenmikroskop, praktische Ausführung. Z. techn. Phys. **19**, 407–416 (1938).

Blaschke, R.: Einige anwendungstechnische Gesichtspunkte für Raster-Elektronenmikroskopie im Durchstrahlungsbetrieb. Beitr. elektronenmikroskop. Direktabb. Oberfl. **5**, 965–976 (1972).

Blaschke, R., Gentsch P., Grauer-Carstensen, E., Laves, F., Weber, L., Woodman, T.P.: Untersuchungen an Mondstein mit 2 Schillerflächen mittels PhEEM, REM, TEM, TEM(s) und Polarisationsmikroskop. Beitr. elektronenmirkoskop. Direktabb. Oberfl. **7**, 423–438 (1974b).

Blaschke, R., Münchberg, W., Obst, K.H., Warren, M.: Beobachtung der Entsäuerung von Kalkstein im Raster-EM. Beitr. elektronenmikroskop. Direktabb. Oberfl. **7**, 237–244 (1974c).

Blaschke, R., Schur, K.: Der Informationsgehalt des Rückstreubildes im Raster-Elektronenmikroskop. Beitr. elektronenmikroskop. Direktabb. Oberfl. **7**, 33–52 (1974a).

Blaschke, R., Waltinger, H.: Rasterelektronenmikroskopische Stereobildpaare von Anschliffen und ihre Bedeutung für die Gefügeanalyse. Beitr. elektronenmikroskop. Direktabb. Oberfl. **4/2**, 425–423 (1971).

Bond, E.F., Beresford, D., Haggis, G.H.: Improved cathodoluminescence microscopy. J. Microsc. **100**, 271–282 (1974).

Booker, G.R.: Electron channelling effects using the SEM. 3rd Ann. SEM Symp., p. 489–496, Chicago (1970).

Boyde, A.: Photogrammetry of stereo-pair SEM images using separate measurements from the two images. 7th Ann. SEM Symp., p. 101–108, Chicago (1974).

Castaing, R., Guinier, A.: Sur l'exploration et l'analyse élémentaire d'un échantillon par une sonde électronique. Proc. 1st Int. Congress on Electron Microscopy, p. 391–397, Paris (1950).

Christenhusz, R.: Zur Darstellbarkeit kristalliner Objekte in der Auflicht-Elektronenmikroskopie. Beitr. elektronenmikroskop. Direktabb. Oberfl. **1**, 67–77 (1968).

Coates, D.: Kikuchi-like reflection patterns obtained with the scanning electron microscope. Phil. Mag. **16**, 1179–1184 (1967).

Coates, D.: Pseudo-Kikuchi orientation analysis in the scanning electron microscopy. 2nd Ann. SEM Symp., 27–40, Chicago (1969).

Cosslett, V.E., Duncumb, P.: Micro-analysis by a flying-spot X-ray method. Nature **177**, 1172–1173 (1956).

Everhart, T.E., Thornlex, R.F.M.: Wide-band detector for micro-ampere low-energy electron currents. J Sci. Instrum. **37**, 246–248 (1960).

Gentsch, P.: Auflösung und Kontrast eines 100 keV-Transmissions-Elektronenmikroskops mit Rasterzusatz. Dissertation, Univ. Münster (Germany) (1974).

Hall, T.A., Höhling, H.J.: The application of microprobe analysis to biology. 5th Int. Congress on X-Ray Optics and Microanalysis (eds. G. Möllenstedt, K.H. Gaukler), p. 582–591 (1968).

Hörl, E.M., Mügschl, E.: Scanning Electron Microscopy of Metals using light emission. Proc. 5th Europ. Congress on Electron Microscopy, p. 502–507 (1972).

Knisely, R.N., Laabs, F.C.: Applications of cathodoluminescence in electron microprobe analysis. In: Microprobe analysis (ed. C.A. Andersen), p. 371–382. New York: John Wiley and Sons 1973.

Knoll, M.: Aufladepotential und Sekundäremission elektronenbestrahlter Körper. Z. techn. Phys. **16**, 467–475 (1935).

Lehmann, G., Bambauer, H.U.: Quarzkristalle und ihre Farben. Angew. Chem. **85**, 281–289 (1973).

Maas, A.: Direct observation and analysis of crystal growth processes in a scanning electron microscop. Proc. 8th Int. Congress on Electron Microscopy, Canberra, **1**, 162–163 (1974).

Macdonald, N.C.: Auger electron spectroscopy for scanning electron microscopy. 4th Ann. SEM Symp., p. 89–96, Chicago (1971).

Maher, D.M.: Scanning electron diffraction in a TEM and SEM operating in the transmission mode. 7th Ann. SEM Symp., p. 215–224, Chicago (1974).

Muir, M.D., Holt, D.B. Analytical Cathodoluminescence Mode SEM. 7th Ann. SEM Symp., p. 135–142, Chicago (1974).

Newbury, D.E.: The origin, detection, and uses of electron channelling contrast. 7th Ann. SEM Symp., p. 1047–1055, Chicago (1974).

Oatley, C.W.: The scanning electron microscope. Cambridge: Cambridge University Press 1972.

Pfefferkorn, G., Blaschke, R.: Nonconventional applications of non-dispersive cathodoluminescence. 7th Ann. SEM Symp., p. 144–149, Chicago (1974).
Reimer, L.: Physical limits in transmission scanning electron microscopy of thick specimens. 5th Ann. SEM Symp., p. 197–204, Chicago (1972).
Reimer, L., Pfefferkorn, G.: Raster-Elektronenmikroskopie. Berlin-Heidelberg-New York: Springer 1973.
Schur, K., Schulte, Chr., Reimer, L.: Auflösungsvermögen und Kontrast von Oberflächenstufen bei der Abbildung mit einem Raster-Elektronenmikroskop (Stereoscan). Z. Angew. Phys. **23**, 405–412 (1967).
Umland, F., Ritzkopf, M., Blaschke, R.: Korrosion an Auslaßventilen und Brennern von Dieselmotoren — Rasterelektronenmikroskopische Untersuchungen der Oberflächen. 7. Sitzung d. Arbeitskreises Rastermikroskopie DVM. p. 27–33 (1975).
Wells, O.C.: Explanation of the low-loss image in the SEM in terms of electron scattering theory. 5th Ann. SEM Symp., p. 169–176, Chicago (1972).
Wells, O.C.: Resolution of the topographic image in the SEM. 7th Ann. SEM Symp., p. 1–8 (1974).
Yakowitz, H.: A practical examination of the Kossel X-ray diffraction technique. In: Microsprobe analysis (ed. C.A. Andersen), p. 383–421. New York: John Wiley and Sons 1973.
Yakowitz, H.: X-Ray Microanalysis in Scanning Electron Microscopy. 7th Ann. SEM Symp., p. 1029–1042, Chicago (1974).
Yoffe, A.D., Howlett, K.J., Williams, P.M.: Cathodoluminescence studies in the SEM. 6th Ann. SEM Symp., p. 301–308, Chicago (1973).

Detailed bibliographies of scanning electron microscopy are published by:

Johari, O., Corvin, I. (eds.): Proc. of the Ann. Scanning Electron Microscope Symposium sponsored by IIT Research Institute Chicago (1968 ff.)
Pfefferkorn, G. (ed.): Beiträge zur elektronenmikroskopischen Direktabbildung von Oberflächen. Münster: Remy 1968 ff.
Wells, C.: Scanning electron microscopy. New York: McGraw-Hill Book Company 1974.

CHAPTER 7.3

Analytical Electron Microscopy of Minerals

G. W. LORIMER and G. CLIFF

1. Introduction

A multitude of analytical techniques which incorporate an electron optical column in the probe-forming and/or image-forming system have been developed during the last twenty years. These include the surface-sensitive technique of Auger electron spectroscopy, thermionic emission microscopy and photoelectron microscopy; the analysis of "bulk" specimens by electron probe microanalysis and the analysis of "thin" specimens by X-ray or electron spectrometry. Auger electron spectroscopy has been developed into a powerful tool for the chemical analysis of surface layers (Chang, 1973). It is particularly sensitive to the light elements C, H and O and has a spatial resolution for analysis of a few microns. Thermionic emission microscopy and photoelectron microscopy are useful techniques for studying phase distributions at a resolution in the micron range (Wegmann, 1972; Kinsman and Aaronson, 1972) but they cannot be classified as techniques of quantitative chemical analysis. Electron probe microanalysis is established as a standard tool for the analysis of bulk specimens, and it has made a very important contribution to mineralogy. The spatial resolution for analysis is a few microns with limits of detection, typically, 10–50 ppm. Procedures for carrying out quantitative analyses are well established and have been discussed by numerous authors (see, for example, Andersen, 1973; Heinrich, 1967).

There are two techniques which have been used successfully to obtain quantitative chemical information from thin specimens in the transmission electron microscope: X-ray spectrometry, the detection and analysis of the characteristic X-rays produced during the interaction of the incident beam with the specimen, and electron spectrometry, the detection and analysis of electrons which have lost energy while travelling through the specimen. Both techniques have been applied to scanning electron microscopes (STEM) and fixed beam (conventional) transmission electron microscopes.

2. Analysis Techniques

2.1 Transmission Electron Microscopy and X-ray Spectrometry

A prototype electron optical instrument (50 kV) *designed* to combine the functions of high resolution transmission electron microscopy and X-ray spectrometry,

EMMA-1 (Electron Microscope Microprobe Analyzer), was described by Duncumb in 1962 (Duncumb, 1962). Subsequent development of the instrument led to EMMA-3, a 100 kV instrument (Cooke and Duncumb, 1968; Cooke and Openshaw, 1969) and the commercial derivative EMMA-4 (Cook and Openshaw, 1970). When the instrument is used in its analytical mode the second condenser lens is switched off and an extra lens situated between the second condenser and the specimen, the mini lens, is activated to form a highly convergent beam at the specimen (Fig. 1). With a circular beam it is possible to obtain a probe size of less than $0.2\,\mu m$ and a probe current of 10 to 20 namps. X-rays produced in the specimen can be detected with either an energy dispersive detector or two fully-focusing spectrometers: both facilities have an X-ray take-off angle of 45°.

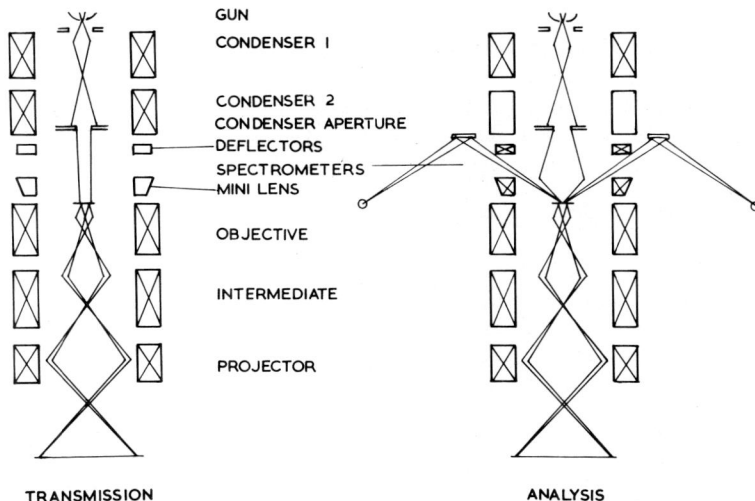

Fig. 1. Schematic ray diagram of the electron trajectories during the operation of an analytical electron microscope in either the microscopy of analysis modes

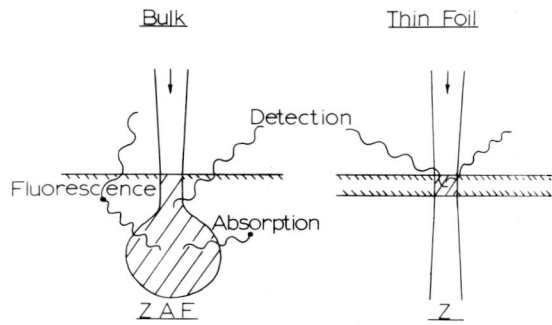

Fig. 2. Schematic electron distribution in a bulk sample and a thin ($\sim 1000\,\text{Å}$) foil. The probability of X-ray absorption or fluorescence is small in thin foil

While EMMA-4 was designed as a combined electron microscope-microprobe analyzer, most electron microscope manufacturers now provide microanalysis attachments/modifications to their basic instruments. To take full advantage of the increase in spatial resolution accompanying the use of a thin specimen (Fig. 2) a probe-forming facility must be incorporated into the instrument as well as a means of detecting X-rays. This can be done either by adding an extra lens immediately above the specimen or by using the pre-field of the objective lens.

2.2 Scanning Transmission Electron Microscopy (STEM)

A scanning transmission facility is a logical development of the provision of a probe-forming lens in the electron microscope. If scanning coils are inserted above the specimen and an electron detector is placed below, a scanning transmission image can be obtained. Advantages, as compared with conventional transmission electron microscopy, accrue when examining thick specimens due to the decrease in chromatic aberration and a decrease in damage rate. However the major bonus is that the STEM system provides a means of forming a small diameter probe which can be used to carry out microanalysis.

2.3 Electron Velocity Analysis (EVA)

Transmission electron microscopy and electron spectrometry are combined in EVA (Silcox and Vincent, 1972; Cundy et al., 1965; Edington and Hibbert, 1973). Electrons which pass through a thin foil may undergo several inelastic scattering processes, the most important of which are plasmon losses: the excitation of longitudinal oscillations of the electrons in the conduction band. These plasmon losses are, typically, a few hundred electron volts; their exact value depends on the atomic species. If the transmitted electron beam passes through an electron spectrometer, usually a Möllenstedt electrostatic analyzer placed below the viewing screen, the plasmon losses can be detected and separated into their constituent energies. As the electrons which pass through the analyzer are usually detected at the viewing screen (using a narrow slit) and the total magnification at the viewing screen is high (typically 20000 times), the potential spatial resolution for chemical analysis is high, typically 100 to 200 Å.

Quantitative analysis can be carried out by comparing the energy loss spectra with those of thin standards of known composition. The main limitation of the technique of EVA is that the spectra become increasingly complicated as the atomic number (Z) increases. At the present time direct quantitative analysis is restricted to elements of $Z < Si$; analysis of heavier elements in binary alloys can be carried out by monitoring the effect of the heavier element on the energy loss spectrum of the matrix ($Z < Si$).

2.4 Scanning Transmission Electron Microscopy and Electron Spectrometry

Crew et al. (1970), Crew (1972) have developed a scanning transmission electron microscope with a field emission source in which a fine – typically 5 Å diameter –

beam is scanned over a very thin specimen. The transmitted electrons are passed through a magnetic spectrometer and those which have suffered elastic or inelastic energy losses are detected. The inelastic signal is proportional to $Z^{1/3}$ and the elastic signal is proportional to $Z^{4/3}$; the ratio of the elastic to the inelastic signal is proportional to Z and is used to obtain "Z-contrast". Crew has been successful in obtaining micrographs of single heavy atoms or heavy atoms in organic molecules which have been dispersed on a thin carbon film.

While the point-to-point resolution which has been obtained using this technique is impressive, the chemical information which can be obtained is, at best, qualitative and the analysis of conventional "thin" specimens (specimens a few hundred or a few thousand Ångstroms thick) is not possible at the present time.

2.5 Scanning Electron Microscopy and X-ray Spectrometry

The development of the energy dispersive (solid state) X-ray detector has meant that chemical information can be acquired in the scanning electron microscope for a moderate additional cost and little, if any, modification in instrument design. If quantitative chemical analysis of bulk specimens is to be carried out, using either energy-dispersive or wavelength-dispersive X-ray detectors, it is necessary to follow the accepted procedures for quantitative microprobe analysis (quality of surface finish, preparation of standards, correction procedures, etc.).

If the scanning electron microscope is used to examine thin specimens, then the restrictions and simplifications outlined in Sections 3 and 4 are applicable (see, for example, the articles by Suter *et al.*, Chapter 7.4 and König, Chapter 7.5 of this volume).

3. X-ray Production in Thin Foils

The main advantages of the use of a thin specimen for microanalysis is that the spatial resolution for chemical analysis is dramatically improved, compared to a bulk specimen. The large bell-shaped region produced by the diffusion of the electron probe beneath the surface of the specimen, from which the majority of the X-rays are produced in a bulk sample (Fig. 2) is absent in a thin foil. At 100 kV the amount of diffusion of the electron beam in a sample 1000 to 3000 Å thick is limited and, to a first approximation, the beam diameter defines the activated volume from which X-rays are produced (Fig. 3a) (Coslett and Thomas, 1964a, b).

The main disadvantage of the use of thin specimens is that the activated volume is small compared to a thick ($>10~\mu m$) sample; X-ray counting rates are low and hence the potential accuracy of an analysis is inferior to that generally accepted for bulk specimens. To produce sufficient X-rays to carry out quantitative analyses with either crystal spectrometers or an energy dispersive detector 2–5 cm from a sample 1000 Å thick using a probe diameter of 1000 Å, a current density of approximately 20 amps cm^{-2} must be used. At these high-current densities some minerals, for example the plagioclase feldspars, suffer severe ioniza-

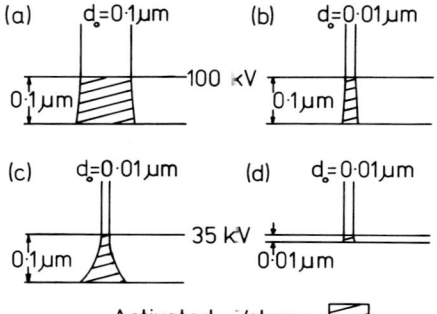

Fig. 3a–d. Schematic variation in activated volume with analysis probe diameter and accelerating voltage. (a) and (b) 100 kV, (c) and (d) 35 kV

tion and/or beam heating damage and the analysis results obtained must be interpreted with caution (Cliff et al., Chapter 4.10 of this volume).

The maximum current that can be focused into an analysis probe is given by

$$I_{max} = \frac{\pi^2}{4} \beta \Phi_0 \alpha^2 \tag{1}$$

where α is the semiangular aperture of the probe-forming lens, Φ_0 the Gaussian probe diameter and β the source brightness (in amps cm^{-2} ster^{-1}) (Bishop, 1974). α cannot be large due to spherical aberration which introduces a disc of confusion of diameter

$$\Phi_s = 2C_s \alpha^3 \tag{2}$$

where C_s is the spherical aberration coefficient of the lens. The final probe size is also limited by diffraction to form a disc of diameter

$$\Phi_d = 1.22 \frac{\lambda}{\alpha} \tag{3}$$

where λ is the electron wavelength. For a given probe diameter Φ the optimum aperture

$$\alpha_{opt} = \left(\frac{\Phi}{C_s}\right)^{\frac{1}{3}} \tag{4}$$

and the maximum probe current

$$I_{max} = \frac{3\pi^2 \beta}{16} \left(\frac{\Phi^{8/3}}{C_s^{2/3}} - \frac{4}{3}(1.22\lambda)^2\right). \tag{5}$$

With a tungsten hairpin filament β is approximately 1 amp × 10^5 cm^{-2} ster^{-1} with a source size of about 50 μm diameter. C_s for the probe-forming lens in

EMMA-4 is 3.5 cm and, with a suitable condenser aperture, usually 100 μm, it is possible to concentrate a current of 1 to 3×10^{-9} amps in a probe 800 to 1 200 Å diameter. Under these operating conditions sufficient X-rays are produced in a thin foil of approximately the same thickness as the beam diameter to carry out a quantitative analysis with either a crystal spectrometer or an energy dispersive detector. In order to improve on the spatial resolution for chemical analysis using an X-ray detector similar to those described above it is necessary to focus a higher current into a smaller diameter probe, either by decreasing C_s or increasing β. Modern transmission electron microscopes, in which the pre-field of the objective is used as a probe-forming lens, have C_s values of 1 to 2 mm and, with a conventional thermionic electron source, are theoretically capable of forming probes 250 to 450 Å diameter containing currents of 1 to 5×10^{-9} amps. The focal length of the lens is approximately the same as the value of C_s; a focal length of 1 to 2 mm can produce difficulties in optimizing specimen-detector geometry.

Improvements in β of approximately an order of magnitude can be obtained by using pointed filaments (Woolf and Joy, 1971) or LaB_6 filaments (Ahmed, 1971). Dramatic improvements in β, by a factor of 10^3 to 10^4, accompany the use of a field emission, rather than a thermionic emission, electron source (Joy, 1974). For the stable operation of a field emission source a vacuum of 10^{-10} to 10^{-12} torr is required in the gun chamber, an improvement of 10^6 to 10^8 over the minimum required for the successful operation of the conventional thermionic source.

The combination of a field emission source and a C_s value of 1 to 2 mm would enable a current of 1×10^{-9} amps to be focused into a spot approximately 20 Å in diameter. In a sample 1 000 Å thick there would be sufficient diffusion of the electron beam, even at an accelerating voltage of 100 kV, for the potential spatial resolution for chemical analysis to be measured in hundreds rather than tens of Ångstroms (Cosslett and Thomas, 1964a, b) (Fig. 3b). The most efficient shape of the activated volume is a cylinder whose diameter is equal to its height, i.e. a probe diameter equal to the foil thickness (Fig. 3a and d). If a β value of 5×10^7 amp cm^{-2} ster^{-1} could be obtained, in combination with a C_s value of 2 mm, it would be possible to focus an electron beam current of 10^{-6} amps into a probe 100 Å diameter, and the X-ray intensity from a sample 100 Å thick would then be the same as from a sample 1 000 Å thick with a probe current of 10^{-9} amps and a probe diameter of 1 000 Å; quantitative analysis would be possible with the X-ray detectors currently available. There are some practical difficulties in producing foils 100 Å thick from bulk specimens and at this current density severe ionization damage and beam-heating effects are to be expected.

The higher the electron voltage the shorter will be the diffusion distance of the incident electrons (for a given specimen thickness) and hence the better the spatial resolution for chemical analysis (Fig. 3). This point is a particularly important one when thin specimens are analyzed in conventional scanning electron microscopes, which are usually limited to voltages between 35 and 50 kV (Shimizu, 1972). Also the carbon contamination layer which forms on the surface of the sample within the analysis probe (see, for example, Figs. 5 and 6) has a very small effect on high energy (100 kV) electrons.

4. Quantitative Analysis of Thin Foils using X-ray Spectrometry

Quantitative microprobe analysis of bulk specimens is based on the comparison of the characteristic X-ray intensity from an element in the specimen with the characteristic X-ray intensity from the same element in a suitable standard, followed by the application of various correction procedures to account for the efficiency of X-ray production, detection, X-ray absorption and X-ray fluorescence (Fig. 2). In a specimen which is thin enough to be examined by conventional transmission electron microscopy (100 kV), the intensity of the characteristic X-rays produced cannot be compared with a bulk standard because the observed X-ray intensity is a function of foil thickness as well as of chemical composition. The large variations in foil thickness, which typify mineral specimens prepared by grinding or ion thinning, make it impractical to compare the absolute intensity of the characteristic X-rays from the specimen with thin film standards.

If the specimen is thin enough to carry out conventional 100 kV transmission electron microscopy, typically 1000 to 1500 Å for mineral specimens, it is transparent to most of the primary X-rays produced by the incident electron beam and X-ray absorption and fluorescence can, to a first approximation, be neglected.[1] Thus while the absolute characteristic X-ray intensity for any one element will be a function of specimen thickness, the characteristic X-ray intensity ratio for any two elements will be independent of thickness, i.e.

$$\frac{I_1}{I_2} = k \frac{c_1}{c_2} \qquad (6)$$

where I_1 and I_2 are the observed characteristic X-ray intensities, c_1 and c_2 are the weight fractions of the two elements and k is a factor which must be determined experimentally. If the bulk composition of the specimen is known and if the scale of the chemical inhomogeneities is a few microns, k can be determined by spreading the beam over an area that is thin but representative of the bulk composition. I_1 and I_2 [Eq. (6)] are monitored with the beam defocused and, as c_1 and c_2 are known, k can be calculated. The beam can then be focused into a small probe, I_1 and I_2 determined from a small area and the weight fraction ratio calculated using the experimentally determined k and Eq. (6) (Lorimer et al., 1972, 1973). If it is not possible to defocus the beam over an area that is thin but representative of the bulk composition, because of the scale of the inhomogeneities in the sample, as is often the case for minerals, k can be determined from an ion-thinned glass or mineral standard (e.g. Lorimer and Champness, 1973). During all of the above operations it is imperative to measure I_1 and I_2 simultaneously, for then the observed X-ray intensity ratio and, hence, the calculated weight fraction ratio, will be independent of any fluctuations in the probe current and of variations in specimen thickness.

---

[1] This assumption can be verified by plotting the intensity of one characteristic X-ray intensity vs a second characteristic X-ray intensity. A deviation from linearity indicates significant X-ray absorption or fluorescence and analyses must be carried out in thinner regions of the specimen.

An alternative to the crystal spectrometers is the energy-dispersive (usually silicon) detector (for a detailed comparison of the two see Jacobs, 1974; Hall, 1971). It does not have the wavelength discrimination of the crystal spectrometer and the peak to background ratio is inferior (hence its use for detecting trace elements is severely limited) and it is countrate-limited — approximately 3000–10000 cp sec maximum with equipment currently available. The energy dispersive X-ray detector has a distinct advantage over the crystal spectrometers: it can be used to record all of the characteristic X-ray intensities above $Z=11$ simultaneously (this range is being extended to $Z=6$ with windowless detectors). This not only speeds up the analysis, as compared to the spectrometers, but each pair of intensity ratios is independent of variations in the probe current. Eq. (6) can be used to convert these characteristic X-ray intensity ratios into weight-fraction ratios, assuming the relevant k_x's have been determined from suitable standards. The energy-dispersive detector is sufficiently stable that a given specimen will yield exactly the same "fingerprint" from one analysis to another, even if they are carried out days or weeks apart and by different operators. If the restriction of a thin sample is added — no absorption or fluorescence in the specimen — the relative efficiency of X-ray production for any two elements in the specimen will only depend on their weight-fraction ratio (this is an important difference from conventional microprobe analysis of bulk specimens where this ratio is also a function of the weight fraction of all the other elements present). X-rays produced in the thin specimen will be absorbed by the beryllium window in the detector and, in some instruments, by an additional plastic window at the interface between the detector and the electron-optical column. The calculation of the final observed X-ray intensity ratios from first principles, even with the above simplifications, is complicated. However the problem can be solved experimentally simply by obtaining X-ray intensity ratios from a number of specimens of known composition and then calculating the scaling factor k by which each ratio must be multiplied to give relevant weight-fraction ratios.

Fig. 4 is a plot of characteristic X-ray energy versus k_x, $\left(\dfrac{c_X}{c_{Si}} \cdot \dfrac{I_{Si}}{I_X}\right)$, which have been determined from thin mineral and alloy standards for which the chemical analyses were accurately known (Cliff and Lorimer, 1975). This calibration curve shows that the detection efficiency of the silicon energy-dispersive detector is a maximum at approximately $Z=14$ (Si). To obtain weight-fraction ratios (relative to Si) for lighter elements the observed peak heights or integrated peak counts (both with background subtracted) must be multiplied by a factor greater than one to compensate for the low efficiency of X-ray production and preferential absorption in the windows. For elements heavier than $Z=20$ (Ca) the observed X-ray intensity must again be scaled upwards. The error bars in Fig. 4 are $\pm \sigma$ values obtained from, typically, six standards. For heavier elements, $Z>20$, the error bars reflect the accuracy of the chemical analyses and the small — when compared to the microprobe analysis of bulk specimens — number of X-ray counts obtained (usually ten to twenty thousand for silicon). The primary cause for the large error bar on the k value for Na is the breakdown of the thin film criterion: in a thin foil of anorthite 1500 Å thick the average path for the X-rays to travel in the sample is 750 Å and 3.7% of the $Na_{K\alpha}$ X-rays

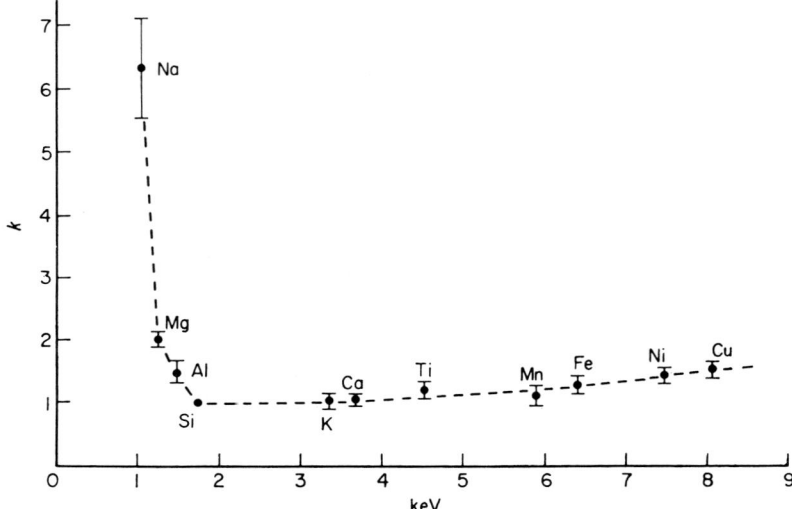

Fig. 4. Calibration curve of $k = \dfrac{c_X}{c_{Si}} \cdot \dfrac{I_{Si}}{I_X}$ vs $K\alpha$ X-ray energy for EMMA-4 fitted with an energy dispersive silicon detector (100 kV)

would be absorbed. In the same specimens Al and $Si_{K\alpha}$ absorption would be approximately 1.5% These absorption effects for the light elements are minimized by always measuring intensities relative to Si.

The major limitation of the technique, as outlined above, is that as no absolute X-ray intensities are recorded an independent determination of light element components, e.g. water, must be made before normalizing the measured weight fraction ratios to 100%. Assuming that this determination can be made, analysis results with an accuracy of 2 to 3% can be obtained as a matter of routine for counting times of 1 to 3 min.

Re-calibration need be carried out if the window material is changed or if it is the necessary to carry out the analysis at a different accelerating voltage. Although the shape of the calibration curve will be similar for all silicon detectors, the values of k will vary from instrument to instrument because of differences in the detectors themselves and in the details of the interface between the detector and the electron optical column; calibration curves must be determined for an individual instrument operating under specified conditions.

5. Examples of the Analysis of Minerals

5.1 Analysis of Small Crystals of Roggianite

The technique described in Section 4 has been used to determine the chemical composition of small particles of the natural mineral roggianite, which occurs as a fibrous aggregate.[2] The complex intergrowth of the fibres with other materials and their small size (0.1 μm diameter, 2–5 μm long) makes conventional microprobe or chemical analysis impossible. The analysis data from 5 individual fibres is given in Table 1.

The observed X-ray intensity ratios, X/Si, obtained in 200 seconds with an energy-dispersive detector and the necessary ancillary electronics, were converted into weight-fraction ratios using the calibration curve in Fig. 4.

Table 1

Fiber	Observed X-ray intensities (background subtracted)						Calculated oxide weight fractions				
	Na	Mg	Al	Si	K	Ca	Na_2O	Al_2O_3	SiO_2	K_2O	CaO
1	—	—	7580	19968	—	14001	—	23.1	43.2	—	32.0
2	105	—	7369	22391	453	14637	1.0	20.8	44.8	0.9	30.9
3	—	—	7382	19124	—	13928	—	23.3	42.7	—	32.9
4	—	—	9201	24425	—	15527	—	23.9	44.9	—	30.2
5	—	—	9701	24943	—	17861	—	23.6	43.1	—	32.6
Average	21	—	8247	22170	91	15191	0.2	22.9	43.7	0.2	31.7

5.2 Analysis of Ion-thinned Foils of Two Exsolved Pyroxenes

The two orthopyroxenes investigated were from the Stillwater complex, Montana (Hess, 1960), composition approximately $Mg_{0.81}Fe_{0.16}Ca_{0.03}SiO_3$ and from the Bushveld complex, South Africa (No SA722 of Atkins, 1969), composition $Mg_{0.78}Fe_{0.19}Ca_{0.03}SiO_3$. Both exhibit a lamellar structure parallel to (100) in the optical microscope. The lamellae are exsolved augite which nucleated heterogeneously, mainly at grain boundaries, and are structurally semi-coherent with the matrix. In addition to the large lamellae there are thin, coherent (100) lamellae and small, homogeneously-distributed, coherent plate-shaped precipitates on (100) (Fig. 5a). A precipitate-free-zone (PFZ) of the small plate-shaped precipitates was observed adjacent to both the semi-coherent and coherent lamellae. The microstructure of the Bushveld orthopyroxene is very similar to that of the Stillwater sample except that the average width and spacing of the augite lamellae is greater, and the coherent (100) platelets are often absent from the matrix between the augite lamellae, or, when present, are smaller than in the Stillwater sample (Fig. 6).

The object of the present investigation was to use the technique of analytical electron microscopy to determine the solute (Ca) distribution between the augite lamellae and to correlate this with the precipitate distribution. A series of analyses was carried out between the augite lamellae in both the Stillwater and Bushveld samples. Figs. 5a and 6 show typical areas after analysis (the contamination spots mark the position of the analysis probe). Fig. 5b shows the observed Ca/Si ratio between adjacent lamellae in the Stillwater sample; there is a depletion

---

[2] This investigation has been carried out in conjunction with Professor H.F.W. Taylor and Dr. J.A. Gard at the University of Aberdeen.

of calcium in the matrix adjacent to the augite lamellae and a high calcium-concentration in the central region between the lamellae. In the Bushveld sample an increase in calcium-concentration was not detected; the Ca/Si ratio was uniform between the lamellae and always at a lower value than that shown in Fig. 5b.

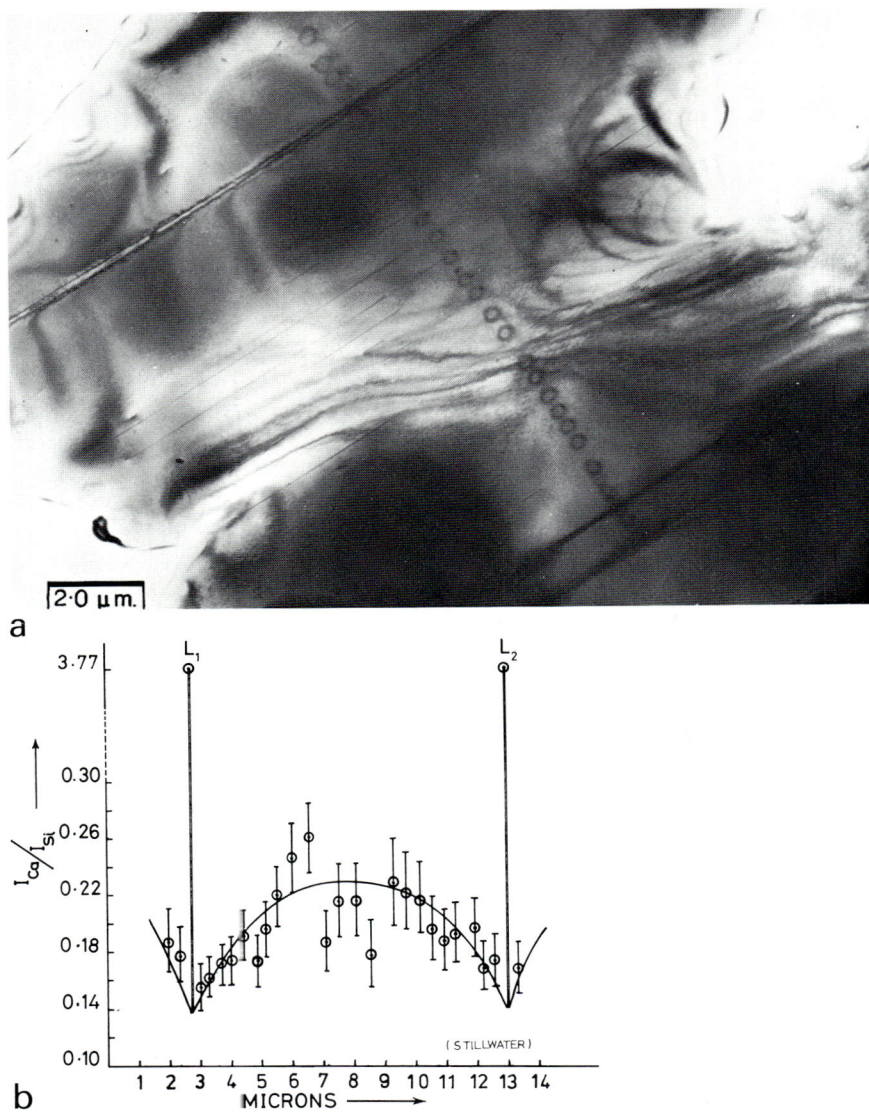

Fig. 5. (a) Electron micrograph of the Stillwater sample showing two thick augite lamellae with several thin lamellae and G.P. zones between them. The contamination spots mark the positions of the probe during analysis. (From Lorimer and Champness, 1973.) (b) Graph showing the variation in calcium concentration from the area in Fig. 6. (From Lorimer and Champness, 1973)

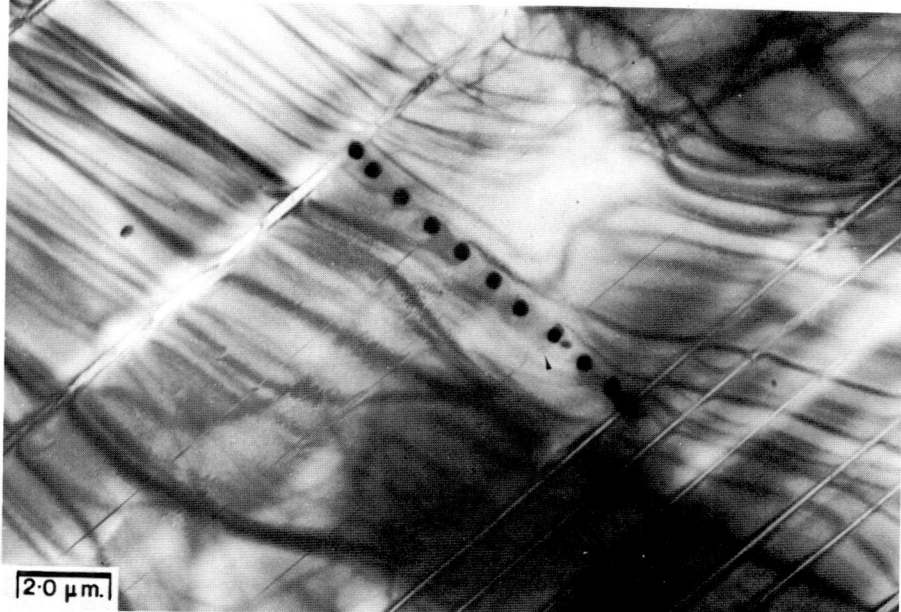

Fig. 6. Electron micrograph of the Bushveld sample showing several thick augite lamellae with several thin lamellae and G.P. zones between them. The contamination spots mark the position of the probe during analysis

In the Stillwater sample the calcium concentration is sufficiently high to have nucleated small coherent precipitates (G.P. zones) in the central region between all the lamellae, while in the Bushveld sample the calcium concentration is too low for nucleation of the G.P. zones to have occurred unless the augite lamellae are widely spaced. The calcium-concentration profiles, and the resulting precipitate distributions, are consistent with the slower cooling of the Bushveld, as compared to the Stillwater, sample.

6. Conclusions

The technique of analytical electron microscopy enables the correlation of microstructural features, as observed in the transmission electron microscope with a point-to-point resolution of less than 10 Å, and diffraction data from areas less than 1 μm in diameter with chemical analyses from areas 1000 to 2000 Å in diameter. This spatial resolution for chemical analysis is an improvement of approximately an order of magnitude over conventional microprobe analysis. It is not usually necessary to make corrections for X-ray absorption or fluorescence and, if an energy dispersive X-ray detector is used and the instrument calibrated using thin standards, quantitative analyses can be carried out very simply without referring to standards at the time of analysis. Because of the lower X-ray production rate in thin foils the analyses are less accurate than with conventional microprobe analysis for equivalent counting times. Electron-probe microanalysis is still very much tied to optical microscopy in terms of relation of analysis to microstructure and specimen preparation techniques. This is an advantage in terms of routine analysis when compared to the more complicated and less familiar techniques associated with transmission electron microscopy.

References

Ahmed, H.: The use of lanthanum hexaboride and composite boride cathodes in electron optical instruments. Proc. 25th Anniversary Meeting of EMAG, Inst. Physics, Cambridge, p. 30–32 (1971).

Andersen, C.A. (ed.): Microprobe analysis. London: John Wiley and Sons 1973.

Atkins, F.B.: Pyroxenes of the Bushveld intrusion, South Africa. J. Petrol. **10**, 222–249 (1969).

Bishop, H.E.: Recent instrumental developments in microanalysis. In: Advances in analysis of microstructural features by electron-beam techniques. Metals Society, London, p. 2–19 (1974).

Chang, C.C.: Auger electron spectroscopy. Surface Sci. **25**, 80–119 (1973).

Cliff, G., Lorimer, G.W.: The quantitative analysis of thin specimens. J. Microscopy. **103**, 203–207 (1975).

Cooke, C.J., Duncumb, P.: Performance analysis of a combined electron microscope and electron probe microanalyser "EMMA". Proc. 5th Int. Conf. on X-ray Optics and Microanalysis. Tübingen, p. 245–247 (1968).

Cooke, C.J., Openshaw, I.K.: A high resolution electron microscope with efficient X-ray microanalysis facilities. Proc. 4th Nat. Conf. on Electron Probe Microanalysis, Pasadena, p. 64–66 (1969).

Cooke, C.J., Openshaw, I.K.: Combined high resolution electron microscope and X-ray microanalysis. Proc. 28th annual EMSA Meeting, Baton Rouge, p. 552–553 (1970).

Cosslett, V.E., Thomas, R.N.: Multiple scattering of 5–30 keV electrons in evaporated metal films I: Total transmission and angular distribution. Brit. J. Appl. Phys. **15**, 883–907 (1964a).

Cosslett, V.E., Thomas, R.N.: Multiple scattering of 5–30 keV electrons in evaporated metal films II: Range-energy relations. Brit. J. Appl. Phys. **15**, 1283–1300 (1964b).

Crew, A.V.: Imaging of single atoms in scanning microscopy. Proc. 5th Europ. Congress on Electron Microscopy, Manchester, p. 640–642 (1972).

Crew, A.V., Wall, J., Langmore, T.: Visibility of single atoms. Science **168**, 1338–1340 (1970).

Cundy, S.L., Metherell, A.J.F., Whelan, M.J.: An energy-analysing electron microscope. J. Sci. Instr. **43**, 712–715 (1965).

Duncumb, P.: An electron-optical bench for microscopy, diffraction and X-ray microanalysis. Proc. 5th Int. Congress on Electron Microscopy. New York: Academic Press P.KK4 (1962).

Edington, J.W., Hibbert, G.: High-resolution microanalysis of aluminium solid solution alloys using combined electron microscopy and energy analysis. J. Microsc. **99**, 125–146 (1973).

Hall, T.A.: The microprobe assay of chemical elements. In: Physical techniques in biological research, 2nd edition, vol. 1A (ed. G. Oster), p. 157–275. New York: Academic Press 1971.

Heinrich, K.F.J. (ed.): Quantitative electron probe microanalysis. Nat. Bur. St. Spec. Publ. **289** (1967).

Hess, H.H.: Stillwater igneous complex, Montana. Geol. Soc. Am. Mem. **80** (1960).

Jacobs, M.H.: Energy and wavelength dispersive X-ray microanalysis. Advances in analysis of microstructural features by electron beam techniques. Metals Society, London, p. 80–118 (1974).

Joy, D.C.: The choice of electron sources for analysis. Advances in analysis of microstructural features by electron beam techniques. Metals Society, London, p. 20, 40 (1974).
Kinsman, K.R., Aaronson, H.J.: Application of thermionic emission electron microscopy to the study of phase transformations. In: Electron microscopy and structure of materials (ed. G. Thomas), p. 259–285. Berkeley: University of California Press 1972.
Lorimer, G.W., Champness, P.E.: Combined electron microscopy and analysis of an orthopyroxene. Am. Mineralogist **58**, 243–248 (1973).
Lorimer, G.W., Nasir, M.J., Nicholson, R.B., Nuttall, K., Ward, D.E., Webb, J.R.: The use of an analytical electron microscope (EMMA-4) to investigate solute concentrations in thin metal foils. In: Electron microscopy and structure of materials (ed. G. Thomas), p. 222–234. Berkeley: University of California Press 1972.
Lorimer, G.W., Razik, N.A., Cliff, G.: The use of the analytical electron microscope EMMA-4 to study the solute distribution in thin foils: some applications to metals and minerals. J. Microsc. **99**, 153–164 (1973).
Shimizu, R., Ikuta, T., Murata, K.: The Monte Carlo technique as applied to the fundamentals of EPMA and SEM. J. Appl. Phys. **43**, 4233–4249 (1972).
Silcox, J., Vincent, R.: Energy analysis and energy selection in electron microscopy and electron diffraction. In: Electron microscopy and structure of materials (ed. G. Thomas), p. 188–220. Berkeley: University of California Press 1972.
Wegmann, L.: Analytical Methods in Photoemission Electron Microscopy. In: Electron microscopy and structure of materials (ed. G. Thomas), p. 246–258. Berkeley: University of California Press 1972.
Woolf, R.J., Joy, D.C.: Quantitative assessment of pointed filaments as bright electron sources. Proc. 25th Anniversary Meeting of EMAG, Inst. Physics, Cambridge, p. 34–37 (1971).

CHAPTER 7.4

X-ray Microanalysis Using a Scanning Electron Microscope

M. SUTER, H.-U. NISSEN, R. WESSICKEN, and P. WIEDERKEHR

1. Introduction

Chemical analysis by electron microprobe methods has two principal limitations: the limits of detection in a homogeneous substance and the smallest analytical volume that can be achieved for specified conditions. The optimal spatial resolution for various heavy and light elements in microanalysis using block-shaped specimens is given by Reed (1973).

In the earth and materials sciences there is an increasing need to analyze the major elements of regions approximately 1 000 Å wide. This has been achieved for specimens of less than 1 000 Å thickness using a transmission electron microscope (for references, see Lorimer and Cliff, Chapter 7.3 of this volume).

This study describes an alternative method for the element analysis of very small areas in thin-foil specimens using X-rays generated by the electron beam of a commercial SEM and its application to an example in mineralogy, the exsolution lamellae of bytownite feldspar.

2. Description of the Method

Our method is a special line-scan procedure for quantitative microanalysis. The X-ray intensity is stored for each coordinate on the scanning line in a multichannel analyzer (MCA). This allows the collection of X-ray intensity through many repeated scans along the line and the summing up of these intensities for every coordinate. Several elements can thereby be assessed simultaneously.

Fig. 1 shows a schematic diagram of the instrumentation. Each time an X-ray quantum of one of the elements to be analyzed is received by the detector, the deflection voltage of the SEM is registered by the MCA. In this way the memory provides the intensity as a function of the deflection voltage which is proportional to the scanning line coordinate. (The MCA operates in the sampling mode and not in the usual pulse height analysis mode.) For the energy-analysis of the X-rays, single channel analyzers (SCA) are used. At the beginning of the experiment these SCA are adjusted to the spectral lines to be analyzed (in our example, the K-lines of Al, Si and Ca).

With several instrumental additions, the same arrangement can be used for the analysis in two dimensions using x- as well as y-deflection voltages.

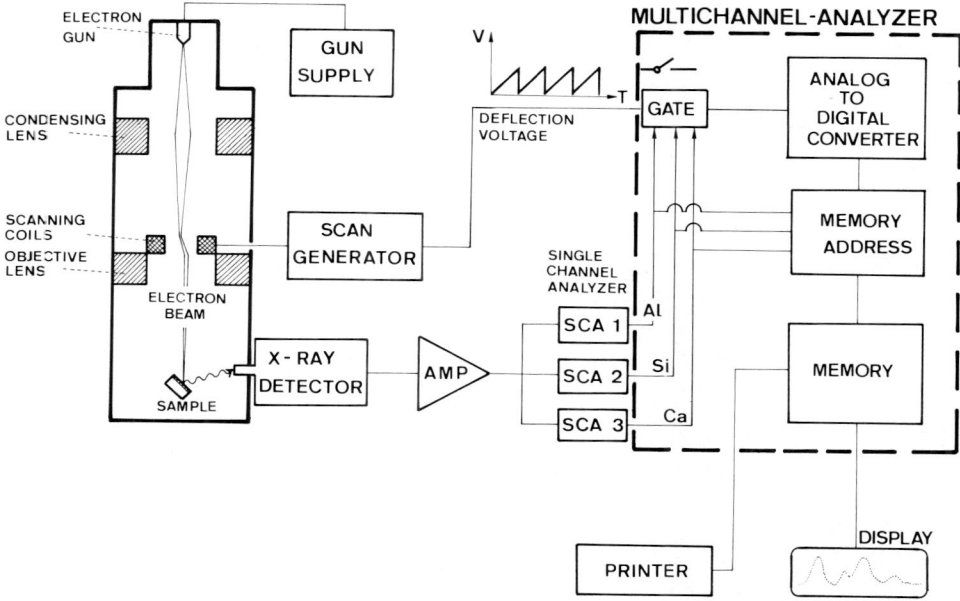

Fig. 1. Schematic diagram showing the instrumentation for simultaneous quantitative X-ray line scan analysis of several elements (Al, Si, Ca) using a SEM

A scanning electron microscope, manufactured by Cambridge Instruments Company, type Stereoscan Mark IIA, was used at an energy of 20 keV. The beam diameter was approximately 200–250 Å. The beam current was 1 namp. In most cases a line-scan length of 10 μm was used. The total length of the scanning line corresponded to 100 channels in the MCA, i.e. 1000 Å per channel. Each individual scan lasted 1–2 sec and each measurement lasted between 10 and 50 min. A Si(Li) solid state detector with 170 eV resolution at 5.9 keV was used.

3. Results

The bytownite specimen, from Roneval, Island of Harris, Scotland (Cliff et al., Chapter 4.10 of this volume) has a bulk composition determined by microprobe of approximately 69 mole % anorthite. The specimen contains calcic exsolution lamellae approximately 1000–3000 Å thick embedded in a more sodic matrix. The lamellar texture has been attributed by Nissen (1974) to two plagioclase phases. An argon-thinned specimen of the Roneval bytownite with thicknesses varying between several hundred and maximally 2000 Å was prepared perpendicular to the plane of the lamellae.

The scanning lines were positioned normal to the lamellae (Fig. 2a, b). Considerable intensity differences across the scanning line for each of the three elements analyzed were found (Fig. 3a). Since these changes are similar for all three elements, they cannot be due to composition, but must be due to strong variations in specimen thickness originating from different resistance of the lamellae and the matrix against the argon beam. The influence of thickness on counting rates is larger than that of compositional differences by roughly a factor of 10.

The distribution of anorthite contents along one scanning line calculated from the data in Fig. 3a is shown in Fig. 3b. A second analysis, parallel to the first one gave comparable anorthite contents (Fig. 3c).

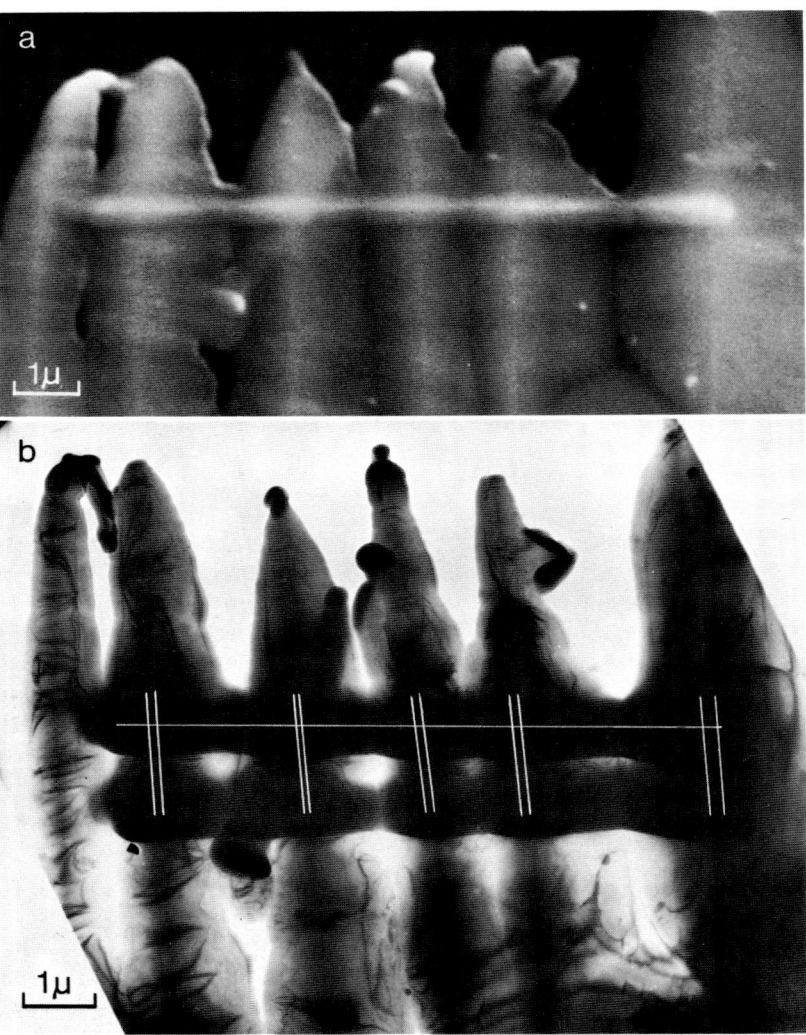

Fig. 2a and b. Electron micrographs of a bytownite specimen with lamellae showing analysis lines visible as stripes of contamination. (a) Scanning electron micrograph. Note that the plane of the specimen is inclined 45° to the image plane. One analysis line is visible. (b) Corresponding 100 kV transmission electron micrograph. The boundaries of the lamellae and one scanning line are marked by white lines

Fig. 3a–c. Analytical results for the region shown in Fig. 2. (a) X-ray intensities for K-lines of Si (above). Al (center) and Ca (below) along the upper of the two scanning lines shown in Fig. 2. (b) Calculated anorthite contents for bytownite with lamellae corresponding to Fig. 3a. Position of lamellae is indicated by dark bars. (c) Calculated anorthite contents for a line parallel to that of Fig. 3a and b. Note that for the thin areas of the specimen bigger fluctuations in calculated anorthite content occur as a result of larger statistical errors

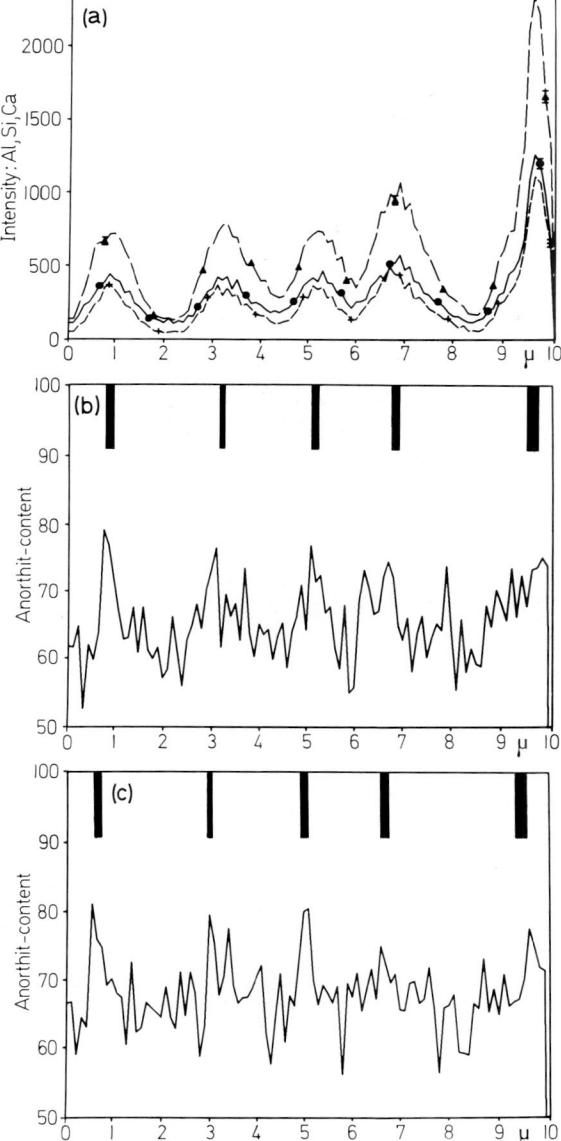

The anorthite contents were calculated from the raw data in the following way. First, concentration ratios for each point along the scanning line were calculated using the relation

$$\frac{I_1}{I_2} = k_{1,2} \cdot \frac{c_1}{c_2}$$

where c is used for the atomic concentration of any particular element, I for the X-ray intensity of that element and k for the proportionality factor which is a function of thickness and composition as well as beam energy. The physical effects which affect the above relation are primary excitation X-ray absorption, X-ray fluorescence, energy loss of the electron beam and the incidence angle of the electrons as well as the takeoff angle of the X-rays. The k-factors were determined from measurements in a non-exsolved, relatively homogeneous standard specimen with 67.2 mole% anorthite composition (Stewart et al., 1967) prepared in the same way as the specimen to be analyzed. These measurements were made under the same instrumental conditions as those in the bytownite as regards beam current, counting times and dimensions of the scanning line. Therefore the count rates, I, in the standard and the specimen to be analyzed can be compared. Since the count rate is, to a first approximation, a function of specimen thickness, the k-factors for all relative thicknesses of the standard specimen can be transferred to points of corresponding thickness (i.e. those with comparable count rates) in the specimen to be measured. An example of the k-factors ($k = 1/f$) obtained from the standard specimen is plotted versus Ca intensities in Fig. 4. This function has been approximated by an exponential function.

The anorthite contents were determined from the element ratios using any of the three relations

$$X_1 = \frac{3(c_{Al}/c_{Si}) - 1}{1 + c_{Al}/c_{Si}}, \tag{1}$$

$$X_2 = \frac{c_{Ca}/c_{Si}}{1 + c_{Ca}/c_{Si}}, \tag{2}$$

$$X_3 = (c_{Al}/c_{Ca} - 1)^{-1}, \tag{3}$$

where X is the molar anorthite component.

Hereby it was assumed that the stoichiometry of the plagioclases, composed of albite, $NaAlSi_3O_8$, anorthite, $CaAl_2Si_2O_8$ and minor potassium feldspar, $KAlSi_3O_8$, was perfect and that other minor elements were only present in negligible amounts. The weighted mean of the three anorthite calculations gave the final values shown in Fig. 3b, c. The calculated statistical relative errors for the anorthite contents at a particular point were approximately 3.5 mole% for relatively thick areas and 7 mole% for the thinnest areas. Within the statistical limits, X_1, X_2 and X_3 are consistent for each point measurement.

In Fig. 3b, c the lamellae can be recognized as peaks in calculated anorthite contents around 80 mole%. The average value for the matrix was calculated as 63.3 mole% (Fig. 3b) and 66.2 mole% (Fig. 3c). The curve in Fig. 3b appears shifted towards lower values by approximately 3 mole% compared to that of Fig. 3c. This can be explained by the fact that the takeoff angle in the analysis of Fig. 3b is larger than in the measurements of the standard while it is the same in the analysis giving a mean of 66.2 mole% for the matrix. The absorption for Al and Si becomes smaller when the takeoff angle increases. Therefore an apparent decrease of An content results. For Ca, the influence of absorption

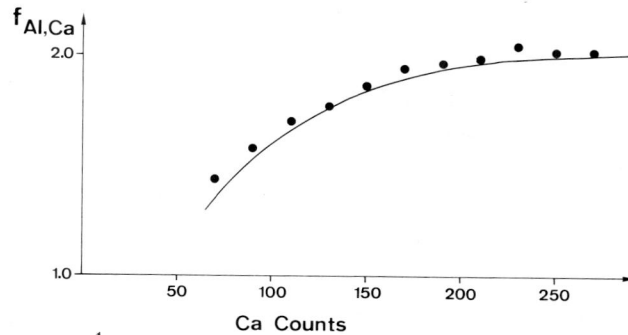

Fig. 4. Plot of calculated $f_{Al,Ca} = \dfrac{1}{k_{Al,Ca}}$ versus the Ca intensity (which is a measure of thickness) recorded within the wedge-shaped margin of the homogeneous labradorite specimen. The $f_{Al,Ca}$ values are means for intervals of Ca intensity comprising 20 counts each. The approximated curve has the form $f = f_0 [1 - a \cdot \exp(-b \cdot I_{Ca})]$

can be neglected. The maxima in anorthite content may be slightly higher than the calculated values as a consequence of the fact that the width of the lamellae is near the resolution limit and that the applied quantization of the multichannel analyzer was 1 000 Å per channel.

4. Discussion of the Method

The analytical method described here is particularly suitable for quantitative chemical analysis of small regions, as the data are stored in digital form. The simultaneous measurement of several elements at exactly the same point and for the same experimental conditions is advantageous. The effects of beam current variations can be eliminated by forming a mean over many scans. Also, the contamination is built up steadily in the procedure. Specimen damage due to heating is diminished at the same time. By comparing intensities in different areas of the specimen having the same compositions, an information about the relative specimen thickness can be obtained. Difficulties may arise from mechanical or electronic drift during measurement.

The microanalytical method described here should be applicable wherever an SEM and suitable ancillary electronics are available.

This work must be regarded as a preliminary study which shows that the technique described can resolve chemical inhomogeneities of approximately 1 000 Å diameter in silicates. The precision can certainly be improved.

References

Nissen, H.-U.: Exsolution phenomena in bytownite plagioclases. In: The feldspars (eds. W.S. MacKenzie and J. Zussman). Proc. NATO Adv. Study Inst. p. 491–521. Manchester: Univ. Press 1974.
Reed, S.J.B.: Principles of X-ray generation and quantitative analysis with the electron microprobe. In: Microprobe analysis (ed. C.A. Andersen), p. 59–81. New York: Wiley and Sons 1973.
Stewart, D.B., Walker, G.W., Wright, T.L., Fahey, J.J.: Physical properties of calcic labradorite from Lake County, Oregon. Am. Mineralogist **15**, 177–197 (1966).

CHAPTER 7.5

Quantitative X-ray Microanalysis of Thin Foils

R. KÖNIG

1. Introduction

As mentioned in the preceding articles (Lorimer and Cliff, Chapter 7.3; Suter et al., Chapter 7.4 of this volume), X-ray microanalysis of thin foils has the advantage of a better spatial analytical resolution and an improvement in absolute sensitivity compared to the microanalysis of bulk specimens. This is combined both in the STEM and TEM with the high optical resolving power of these instruments. But because of the low count rates in thin-film microanalysis, in some cases we come to the limits of this method. In order to take full advantage of the instrumental possibilities it is therefore necessary to optimize the instrument and its settings. The purpose of this paper is to supply the necessary physical background for this and for the quantitative evaluation of the measurements. In our derivations we use the K_α-line as an example. By changing the shell-dependent terms we can apply the equations also for other lines.

2. Interaction of the Electrons and the Thin Foil

2.1 Production of Characteristic X-rays

The number of K_α-photons produced in an infinitely thin layer of mass-thickness ds (s = density × thickness) is given as:

$$dn_1^{K_\alpha} = c_1 \cdot g_1^{K_\alpha}(E_0, E_K) \cdot ds \tag{1}$$

with

$$g_1^{K_\alpha} = \frac{N}{A_1} \cdot \omega_{K_\alpha} \cdot Q(E_0, E_K) \tag{1a}$$

the symbols are:

c_1 = weight fraction of the element 1 in the layer ($c_1 = 1 = 100\%$),
N = Avogadro's number,
$\omega_{K_\alpha} = K_\alpha$ fluorescence yield (see for example Bambynek et al., 1972 and Burhop, 1952),

$Q(E_0, E_K)$ = ionization cross-section for the K-shell (Bethe, 1930)

$$7.9 \times 10^{-20} \cdot \frac{\ln(E_0/E_K)}{E_0 \cdot E_K} \, [\text{cm}^2]$$

A_1 = atomic weight of element 1,
E_0 = impact electron energy in keV,
E_K = ionization energy of the K-shell in keV.

To arrive at the X-ray intensity produced in a foil of the thickness s_0 we cannot simply integrate Eq. (1) but have to consider the effects of elastic and inelastic scattering of electrons. In order to describe this in a simple manner we use the depth distribution function of X-ray production, which is defined as follows:

$$\varphi_1^{K_\alpha}(s) = \frac{dn_1^{K_\alpha}(s)}{dn_1^{K_\alpha}} \qquad (2)$$

with

$dn_1^{K_\alpha}(s) = K_\alpha$ X-ray intensity in a differential thin layer of thickness ds at the depth s in a bulk specimen,
$dn_1^{K_\alpha} \quad = K_\alpha$ X-ray intensity of the element 1 in a self supporting differential thin film of the same composition.

This function which was first measured and introduced by Castaing and Descamps, 1955 (see also Vignes and Dez, 1968, and Fig. 1) be adopted to our problem. We assume that the value of φ at $s=0$ (specimen surface) is determined mainly by back-scattered electrons with a sufficiently high energy to produce X-rays. The slope $m = d\varphi/ds$ of the $\varphi(s)$-curves is however mainly a result of forward scattering and the related increase of the mean path length of the electrons per element of mass thickness and secondly of the deacceleration producing a change in the cross-section.

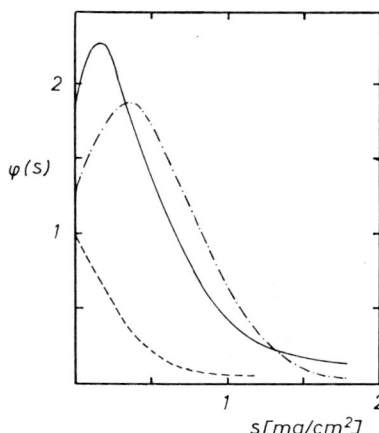

Fig. 1. Schematic diagram of $\varphi(s)$ —— curves for Au (——) and Al (—·—) at 30 kV accelerating voltage and $E_K \approx 10$ keV and curve for $\exp\{-\chi s\}$ ---- for $\chi = 2000$ cm^2/g

At a sufficiently small mass thickness s_0 of the foil (e.g. $s \leq 0.1$ mg/cm^2 and $E_0 = 30$ keV) $\varphi(s)$ can be approximated by

$$\varphi(s) = \varphi(0) + m \cdot s. \tag{3}$$

According to Eq. (2) $\varphi(s)$ for very thin films as are used in TEM and STEM is equal to 1 at $s = 0$ therefore

$$\varphi(s) = 1 + m \cdot s. \tag{3a}$$

m is essentially a function of the accelerating voltage U_0 and of the mean atomic number of the foil. It can be determined with the help of Monte-Carlo methods or can be directly measured on $\varphi(s)$ curves. A method to determine the energy dependence of m has been described by König (1972). The value of m is a function of the chemical composition and we can write:

$$m = c_i \cdot m_i(Z_i), \quad i = 1, 2, \ldots \tag{4}$$

Z equals the atomic number.

At $E_0 = 30$ keV the value of m for Al to Au ranges from 2 to 3 cm^2/mg and is largely independent of E_K for $2.5\, E_K \leq E_0$. In the high energy range ($E_0 \geq 100$ keV) m is smaller than 1 cm^2/mg for all atomic numbers. For the practical use in thin film analysis an accuracy of 20% for m is sufficient.

With the help of Eq. (3a) we can now integrate Eq. (1) over the mass-thickness s_0, regarding at the same time the absorption of X-rays in the layer at a take-off angle δ, defined as the angle between the detector axis and the specimen surface. With $\exp(-\chi s)$ expanded to $1 - \chi s$ for $\chi s < 0.2$ we have

$$n_1^{K\alpha}(s_0) = c_1 \cdot g_1^{K\alpha} \int_0^{s_0} (1 + m \cdot s)(1 - \chi_1 \cdot s)\, ds \, \frac{\text{photons}}{\text{electron}} \tag{5}$$

with

$$\chi_1 = \frac{\mu_1^{K\alpha}}{\sin \delta} \quad \text{and} \quad \mu_1^{K\alpha} = c_i \cdot \mu_{1i}^{K\alpha}, \quad i = 1, 2, \ldots.$$

$\mu_{1i}^{K\alpha}$ = mass-absorption coefficient of K_α-radiation of the element 1 in the element i.

In the following we set:

$$h(s) = s + (m/2 - \chi_1/2)s^2 - (\chi_1 m/3)s^3 \quad \text{for the integral in Eq. (5).} \tag{6}$$

2.2 Production of Bremsstrahlung in Thin Foils

A relation for the bremsstrahlung of thin films produced in an energy interval ΔE around E_{K_α} can be derived from a formula given by Albert (1974):

$$B^+(s) = 3.51 \times 10^{-1} \cdot \frac{Z_{\text{eff}}}{E_0 \cdot E_{K_\alpha}} \cdot \Delta E \cdot \text{s/g keV}^{-1} \text{ cm}^{-2} \frac{\text{photons}}{\text{electron}} \tag{7}$$

with

$$Z_{eff} = \sum_i c_i Z_i, \quad i = 1, 2, \ldots.$$

It should be noted, that the constant factor in Eq. (7) is strongly dependent on the angle between incident beam and detector axis.

3. X-ray Intensities as Measured by an Energy-dispersive Spectrometer

As has been described by Lorimer and Cliff (Chapter 7.3 of this volume) the use of an energy-dispersive detector is advantageous for simultaneous analysis of many elements and is often applied in mineralogic studies. A main advantage is stability of the detection probability W of the system over a long period of time. Therefore we introduce some properties of this detector system into our calculations. For crystal spectrometers minor modifications would be necessary. If W is the detection probability, $F/4\pi d^2$ the spatial angle, i the probe current and t the counting time, relation (5) can be rewritten to express the number of measured K_α-photons I_1 of element 1

$$I_1^{K_\alpha}(s) = 6.25 \times 10^{18} \cdot n_1^{K_\alpha}(s) \frac{F}{4\pi d^2} W(E_{K_\alpha}) \frac{i \cdot t}{amp \cdot sec}. \tag{8}$$

F is the effective detector area and d the distance detector-specimen. An approximation of this formula is given by Albert (1974):

$$I_1^{K_\alpha}(s) = 5 \times 10^{23} \cdot k_{20} \frac{\ln(E_0/E_K)}{E_0/E_K} c_1 \frac{W(E_{K_\alpha})}{Z_1^3} \frac{i \cdot t}{amp \cdot sec} \frac{s}{g/cm^2} \frac{F}{4\pi d^2} \tag{8a}$$

with

$k_{20} = Z_1/20, \quad 10 \leqq Z < 20$
$k_{20} = 1, \quad\quad\quad 20 \leqq Z < 42$

A corresponding expression for the bremsstrahlung we obtain from Eq. (7)

$$B(s) = 6.25 \times 10^{18} \frac{F}{4\pi d^2} \cdot W(E_{K_\alpha}) \cdot \frac{i \cdot t}{amp \cdot sec} B^+(s). \tag{9}$$

The number of photons registered in the channel with the highest count rate $P - B$ is important for the calculation of the peak to background ratio

$$P - B = p \cdot I_1^{K_\alpha} \frac{\Delta E}{\Delta E_H}. \tag{10}$$

The symbols are explained in Fig. 2 (see also Fig. 3).

It is interesting to determine the maximum value of the function $p \cdot \sqrt{\Delta E / \Delta E_H}$, derived from P/B considerations. This function has its maximum value at about $\Delta E = \Delta E_H$. If we therefore set the channel width ΔE equal to the half-width ΔE_H of the spectrometer we have the best conditions to detect very small signals.

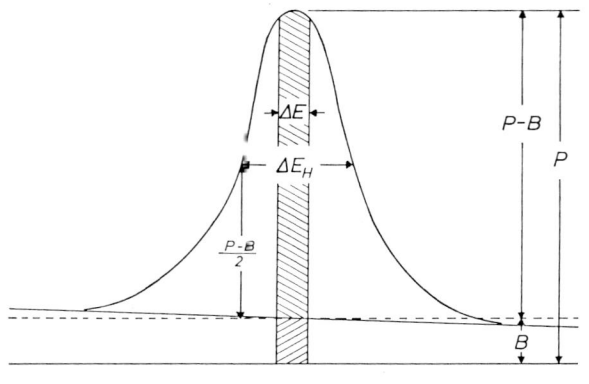

Fig. 2. Schematic diagram of a signal measured with an energy dispersive analyzer to illustrate the symbols used in the text

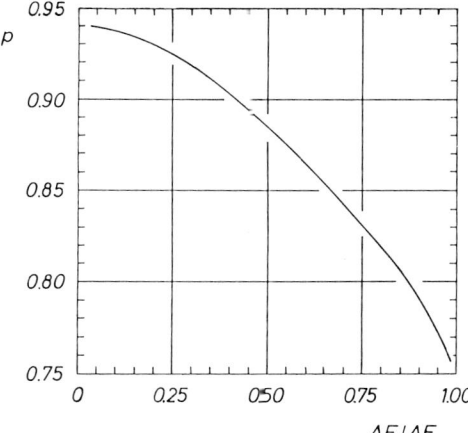

Fig. 3. Line-shape factor p for a Gaussian line-profile

4. Quantitative Analysis of Thin Films

To obtain a quantitative analysis of a thin foil we can proceed in two ways. Either we can compare the X-ray intensities of two elements by the use of a constant k with the weight fractions of the elements as was done by Lorimer and Cliff (Chapter 7.3 of this volume)

$$\frac{I_1}{I_2} = k \cdot \frac{c_1}{c_2} \tag{11}$$

or in the second approach we compare the intensity of one element in the thin film with the intensity of a thick pure element standard and use correction procedures to determine the weight fraction. This method is described extensively by König (1972) and is especially interesting for the application in the SEM.

Besides these methods which depend on ratios or external standards we could calculate element concentrations directly from X-ray intensities using Eq. (8). This is difficult however because not all physical constants are known with sufficient accuracy and because of uncertainties in the specimen thickness which varies considerably in ion-thinned foils (see for example Suter et al., Chapter 7.4). On the other hand if the chemical composition is known (such as for pure mineral inclusions like quartz which are present in many specimens) Eq. (8) can be used to calculate the thickness of the foil.

The factor k relating concentrations in a pair of elements can be calculated by combining Eqs. (8) and (11):

$$k = \frac{g_1^{K\alpha} \cdot h_1(s) \cdot W(E_1^{K\alpha})}{g_2^{K\alpha} \cdot h_2(s) \cdot W(E_2^{K\alpha})}. \tag{12}$$

It should be remembered that in this expression $h_1(s)$ and $h_2(s)$ are not constants but depend themselves on the chemical composition of the material and also on the foil thickness. For thin foils $h(s)$ can be approximated by a linear function $h_i(s) = h_i \cdot s$ (Fig. 4). Therefore the thickness cancels in Eq. (12) and we obtain:

$$k = \frac{g_1^{K\alpha} \cdot h_1 \cdot W(E_1^{K\alpha})}{g_2^{K\alpha} \cdot h_2 \cdot W(E_2^{K\alpha})}. \tag{12a}$$

In general h_i has a different value for each element. The ratio h_1/h_2 is therefore still dependent on the chemical composition and an iteration would be necessary to calculate chemical concentrations. For certain pairs of elements and their characteristic radiation and for sufficiently small s, h_1/h_2 is however close to unity. Then we can use the thin film asssumptions: neglection of X-ray absorption and electron-scattering.

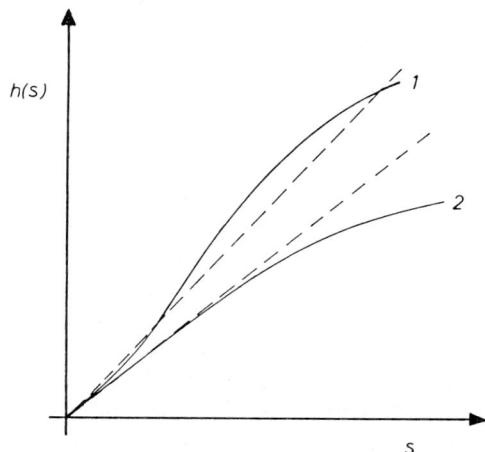

Fig. 4. Schematic diagram illustrating the effects of electron scattering and X-ray absorption as a function of mass thickness s. Linear approximations for small s are indicated. 1 shows the influence of high electron scattering, 2 the influence of low electron scattering and high absorption

The detection probability W is normally not specified by the manufacturers of detectors. It can be determined with limited accuracy (about 2–4%) by the use of the absolute intensity of thick pure element standards (Green and Cosslett, 1968; Albert, 1972). Between 2.5 keV and 12 keV W is close to 1 for energy-dispersive systems. If W is known Eq. (12a) is easy to evaluate and we can now make quantitative analysis under various experimental conditions without needing measurements of standards.

5. Detection Limits

The question of the minimum size of material which can be analyzed is of great importance. We assume that we use an electron microscope which is characterized by the following data:

a) gun brightness $\beta/\text{amp cm}^{-2}\,\text{sterad}^{-1} = r \cdot E_0/\text{keV}$
   ($r \approx 1.6 \times 10^3$ for a tungsten hairpin filament and $r \approx 1.6 \times 10^6$ for a Crewe-type field emission gun),

b) half-aperture angle

$$\alpha_0 = \alpha_{opt} \approx \frac{(i/\beta)^{1/8}}{C_s^{1/4}}, \qquad C_s = \text{spherical aberration constant [cm]}$$

of the last condensor.

With these conditions we can achieve a minimum beam diameter Φ_{min} which can be expressed as:

$$\Phi_{min} = \sqrt[4]{\frac{\pi C_s}{2}} (i/\beta)^{3/8}. \tag{13}$$

The mass-fraction of the element 1 irradiated in the thin foil by this electron beam is then given as:

$$M = c_1 \frac{\pi}{4} \Phi_{min}^2 \cdot s. \tag{14}$$

We want to detect localized masses. This limits the counting time if a beam or a specimen drift Ψ is present. We set

$$t = \Phi/(2 \cdot \Psi), \tag{15}$$

$\Psi/\text{cm} \cdot \text{sec} = $ beam drift.

In order to determine the minimum still detectable mass-fraction of the element 1 in the irradiated area, we calculate the minimum allowable mass-thickness s_{min}, applying the fundamental relation from counting statistics for the detection of the signal $P - B$ in respect to the background B:

$$P - B > 4\sqrt{B}. \tag{16}$$

(We use the factor 4 instead of 3 to have a little more scope for neglections.)

At this point we should remember that B in Eq. (16) is not only the bremsstrahlungsbackground of the specimen [compare $B(s)$ in Eq. (9)]. A considerable contribution to the background stems from bremsstrahlung produced by scattered electrons in parts of the electron microscope. This is despite careful shielding and screening measures, which are essential to obtain good results, in the same order of magnitude as the background produced in the foil. The factor f defines the quality of screening and is equal 1 for optimum shielding. Further we introduce a factor B^* which is a measure for the electron scattering in the specimen leading to production of bremsstrahlung in parts of the EM and relating the bremsstrahlung generated in the microscope with that produced in the specimen. For very thin foils B^* is independent of the specimen thickness ($B^* = 100$ if $s < 2.8 \times 10^{-9} Z_1^4/Z_{eff}$). Otherwise $B^* \approx 3.6 \times 10^{10} \cdot s Z_{eff}/Z_1^4$. We put in our calculations:

$$B \approx 10^{10} \cdot B^* \cdot f \cdot \frac{(E_0/E_K - 1)}{2} \cdot \frac{i \cdot t}{\text{amp} \cdot \text{sec}} \cdot \frac{F}{d^2} \cdot W \cdot \frac{\Delta E}{\text{keV}} \tag{17}$$

because the bremsstrahlung produced in thick targets is proportional to $(E_0/E) - 1$.

Using s_{min} determined in Eq. (16) by inserting Eq. (10) and Eq. (17) (for an explicit derivation see Albert, 1974) we get with Eq. (14) the smallest still detectable mass:

$$M_{min} \approx (C_s/\text{cm})^{3/8} \cdot (10^7/r)^{9/16} \cdot \sqrt{\frac{\Psi}{\text{cm/sec}} \cdot \frac{(Z_1/10)^2}{k_{20}}} \cdot \sqrt{\frac{\Delta E_H/\text{keV}}{F \cdot W/d^2}}$$

$$\times \sqrt[16]{\frac{i}{10^{-9} \text{ amp}}} \cdot 3.9 \cdot \sqrt{B^* \cdot f} \cdot \frac{(E_0/E_K)^{7/16}}{\ln(E_0/E_K)} \cdot \frac{\sqrt{(E_0/E_K) - 1}}{p \cdot \sqrt{\Delta E/\Delta E_H}} \cdot 10^{-19} \text{ g}. \tag{18}$$

M_{min} is minimum if we set $\Delta E = \Delta E_H$ and use an overvoltage ratio $E_0/E_K \approx 2.5$. Experimentally this is often difficult to achieve. For example in case of Al, Si, Mg and other light elements we have commonly $E_0/E_K \geq 20$ and if spectral resolution is needed, a ratio $\Delta E/\Delta E_H \approx 1/10$ is often used. This leads already to a deterioriation of the detection limit by a factor 6.5 or more. In the above calculations the absorption by the surrounding matrix and effects of electron-diffusion are not considered.

In Eq. (18) we notice the remarkable little influence of the beam current on M_{min}. This would mean that for example with a tungsten hairpin filament and a beam current of 10^{-12} amps we could achieve a spatial analytical resolution of ≤ 200 Å, but then the by-condition in Eq. (16) $B > 1$ is not fullfilled. In general we can say that for conventional guns, a beam current of about 10^{-10} amps is the lower limit.

Another interesting result of Eq. (18) is the interplay of electron optics and the geometrical dimensions of the detector, which is optimum if $C_s^{3/8} d/(r^{9/16} F^{1/2})$ is as small as possible.

6. Conclusions

In this article we have given essentially three results:

1. Formulae for the calculation of the expected characteristic photon emission in a thin film including electron scattering and X-ray absorption phenomena [Eqs. (8) and (8a)].

2. Relations for the use in quantitative chemical analysis of thin films [Eqs. (12) and (12a)].

3. Some comments about detection limits [Eq. (18)].

In conclusion it seems appropriate to illustrate with a mineral example how these formulae can be applied in order to obtain optimum results.

We chose plagioclase (Ca, Na)(Al, Si)AlSi$_2$O$_8$, which has been analyzed extensively by Cliff et al. (Chapter 4.10 of this volume). Our specimen is an intermediate plagioclase with fine lamellae spaced about 5000 Å. The foil is about 1000 Å thick, corresponding to a mass-thickness $s = 0.03$ mg/cm^2. We would like to investigate two questions:

1. Can Na be detected with a spatial resolution of <5000 Å and which are the most favorable conditions for this purpose using the system described below?

2. The An content of plagioclase can be calculated from either ratio c_{Al}/c_{Si} or c_{Ca}/c_{Si} as has been done by Cliff et al. (Chapter 4.10 of this volume). Assuming that the experimental conditions are as described below we want to determine values for h_i for Al, Si, Ca and also for Na in order to evaluate whether the thin film assumptions apply.

The electron optical system used in this example is equipped with a tungsten hairpin filament and has a C_s value of 2 cm. The detector system is characterized by the following data: $F = 0.12$ cm^2, $d = 5$ cm, $\delta = 30°$, $\Delta E_H = 0.15$ keV; W for Na is assumed to be 0.3, Al 0.6, Si 0.7 and Ca 1.0. The counting time will be 200 seconds for the Na detection we use $\Delta E/\Delta E_H = 1/10$ and $f = 1$.

A first consideration is the voltage selection. Low voltage and hence low overvoltage ratios such as are obtainable in a STEM is advantageous for reaching a low detection limit [Eq. (18)]. High voltages offer sometimes better spatial resolution due to less electron diffusion. In the following we check three different voltages 20, 30 and 100 keV.

First we use Eq. (18) to calculate M_{min} as a function of the beam drift Ψ, taking into account that $\sqrt[16]{i/10^{-9}}$ amp·sec ≈ 1 for all practical beam currents. The result is:

$$M_{min} = 9.0 \times 10^{-13} \sqrt{\Psi/\text{cm} \cdot \text{sec}^{-1}} \text{ g} \quad \text{for } E_0 = 20 \text{ keV}$$

$$M_{min} = 1.2 \times 10^{-12} \sqrt{\Psi/\text{cm} \cdot \text{sec}^{-1}} \text{ g} \quad \text{for } E_0 = 30 \text{ keV}$$

$$M_{min} = 2.7 \times 10^{-12} \sqrt{\Psi/\text{cm} \cdot \text{sec}^{-1}} \text{ g} \quad \text{for } E_0 = 100 \text{ keV}$$

or generally expressed $M_{min} = L \cdot \sqrt{\Psi}$. Now we assume that the beam drift Ψ is negligible compared to the size of the lamellae in the measuring time $t = 200$ sec. Therefore we replace $\sqrt{\Psi}$ by $\sqrt{\Phi/(4t)}$ [Eq. (15)]. From the condition that the irradiated mass of the element 1 (here Na) must be greater than the minimum still detectable mass we obtain an expression for the best attainable analytical resolution (beam or specimen drift neglected):

$$\Phi_{min} = 1.17 \cdot \sqrt[3]{\frac{L^2}{c_1^2 \cdot s^2 \cdot 4 \cdot t}}. \tag{19}$$

If we limit our measuring time by beam or specimen drift we obtain an equivalent expression:

$$\Phi_{min} = 1.13 \cdot \sqrt{\frac{L}{c_1 \cdot s}} \cdot \Psi^{1/4}. \tag{19a}$$

In our example we obtain the following data:

$\Phi_{min} = 1000$ Å for $E_0 = 20$ keV; $\quad \Phi_{min} = 1190$ Å for $E_0 = 30$ keV;

$\Phi_{min} = 2070$ Å for $E_0 = 100$ keV.

With Eq. (13) we see that we need a beam current of 6.9×10^{-10} amp in the case of 20 keV and 2.41×10^{-8} amp for 100 keV. The current density at the specimen is about an order of magnitude lower at 20 keV, therefore specimen contamination and beam damage are significantly reduced. 100 keV however is better in terms of counting statistics with 755 registered Na $K\alpha$-counts at 20 keV and 8100 registered Na $K\alpha$-counts at 100 keV.

Since the limiting factor is mainly radiation produced in parts of the EM it is essential to shield the signal from interfering noise by using a small detector aperture and to reduce X-ray production in parts of the microscope by shielding with low atomic number materials.

Next we calculate for 30 keV ($m = 2$ cm$^2 \cdot$ mg^{-1}) the h_i values to see whether the thin-film assumptions can be applied. Using Eq. (6) and setting $s = 0.03$ mg/cm^2 we obtain for

Ca $h(s) = s + (1 - 0.28) s^2 = 0.0306$, $\quad h_1 = 1.02$
Si $h(s) = s + (1 - 1.20) s^2 = 0.0298$, $\quad h_2 = 0.99$
Al $h(s) = s + (1 - 1.15) s^2 = 0.0299$, $\quad h_3 = 1.00$
Na $h(s) = s + (1 - 2.65) s^2 = 0.0285$, $\quad h_4 = 0.95$.

For 100 keV ($m = 0.5$ mg/cm^2) we get $h_{Ca} = 1.00$ and $h_{Na} = 0.93$. We see that in general the thin-film assumptions are fulfilled, only in the case of Na an intensity correction of 5% to 7% is needed. It should be noticed that for thicker foils, which can still be penetrated by the beam (especially in the STEM) the correction becomes much higher (0.06 mg/cm^2 foil: 15% for Na and 100 keV). Finally we calculate with Eqs. (1a)[1] and (12a) the correction factor k for two pairs of elements

---
[1] The following expression (Albert, 1972) was used for the calculation of $\omega_{K\alpha}$:

$$\omega_{K\alpha} = \frac{15}{17} \cdot \frac{Z^2}{1000} \cdot \left[1 - \frac{5}{6}\left(\frac{10}{Z} + \frac{Z}{100}\right)\right].$$

Al/Si and Si/Ca. The result is ($E_0 = 100$ keV)

$$k_{Si/Al} = 1.22 \quad k_{Si/Ca} = 0.93$$

which is in good agreement with the empirical data of Lorimer and Cliff (Chapter 7.3 of this volume).

References

Albert, L.: Photonen-Ausbeute, Signal/Untergrundverhältnis und Nachweisgrenze elektronenangeregter Kα-Linien. BEDO-Bd. **5** (1972).
Albert, L.: How thin layers and how low localized masses can be detected by means of electron probe microanalysis – A theoretical investigation. BEDO-Bd. **6** (1974).
Bambynek, W., Crasemann, E., Fink, R. W., Freund, H.-U., Mark, H., Swift, C. D., Price, R. E., Rao, P. V.: X-ray fluorescence yields, auger, and Coster-Kronig transition probabilities. Rev. Mod. Phys. **44**, 716–313 (1972).
Bethe, H.: Zur Theorie des Durchgangs schneller Korpuskularstrahlen durch Materie. Ann. Physik, (Leipzig) **5**, 325 (1930).
Burhop, E. H. S.: The Auger effect. Cambridge: Cambridge University Press 1952.
Castaing, R., Descamps, J.: Sur les bases physiques de l'analyse ponctuelle par spectrographie X. J. Phys. Radium **16**, 304–317 (1955).
Green, M., Cosslett, V. E.: Measurements of K, L and M shell X-ray production efficiencies. Brit. J. Appl. Phys. (J. Phys. D) Ser. 2, **1**, 425 (1968).
König, R.: Quantitative Röntgenmikroanalyse dünner Schichten am Rasterelektronenmikroskop. Diplomarbeit, Universität Karlsruhe (1972).
Vignes, A., Dez. G.: Distribution in depth of the primary X-ray emission in anticathodes of titanium and lead. Brit. J. Appl. Phys., Ser. 2, **1**, 1309 (1968).

CHAPTER 7.6

Particle Track Studies

I.D. HUTCHEON and J.D. MACDOUGALL

1. Introduction

Heavy charged particles produce damage trails (tracks) near the end of their range as they are brought to rest in dielectric materials (e.g. Fleischer et al., 1965). This discovery has been applied to problems in disciplines as diverse as anthropology and particle physics. In this short report we describe how electron microscopy is being used in studies of particle tracks, and illustrate with examples drawn mainly from our own work. For the interested reader, details of track formation, track sources, and experimental methods are extensively described in the literature (e.g. Fleischer et al., 1965; Lal, 1972; Lal et al., 1968; Macdougall et al., 1971; Price and Fleischer, 1971; Price et al., 1973a).

Although charged particle tracks were originally discovered by transmission electron microscopy (Silk and Barnes, 1959), virtually all of the subsequent studies prior to 1970 involved the observation of etched tracks with the optical microscope. The availability of lunar samples, with their high concentrations of radiation damage features, has provided the stimulus for the renewed use of the electron microscope in track studies. Since the moon is not shielded by either an atmosphere or an extended magnetic field, its surface has been continually bombarded by a stream of solar and galactic cosmic rays for more than 4 billion years. Much of the unique story preserved in the particle track records of lunar samples can only be deciphered with the high-resolution techniques of electron microscopy. In the discussion which follows, we have grouped our investigations into three categories: those which involve high-voltage transmission electron microscopy (HVEM); those which involve transmission electron microscopy of replicas; and those which involve scanning electron microscopy (SEM).

2. HVEM Observations

The high-voltage transmission electron microscope has been an essential tool in our studies of radiation effects in lunar samples. It has, for example, enabled us to see 80Å solar wind gas bubbles and extremely high densities of solar flare tracks ($\gtrsim 10^{11}/\text{cm}^2$) in micron-sized grains from the lunar soil and from the interior of lightly compacted breccias (Hutcheon et al., 1972a). HVEM observations of ion-thinned rock sections have revealed the microstructure of complex metaclastic lunar breccias, leading to a better understanding of lunar surface processes. The following discussion illustrates typical HVEM observations on two types of lunar samples, soils and breccias, and demonstrates the advantages of this technique for track studies of heavily irradiated material. Techniques of sample preparation are not discussed due to space limitations, but are thoroughly discussed in the literature.

The majority of lunar soil grains in the 0.1–5 μm size range is observed to be heavily irradiated (Fig. 1) containing $\gtrsim 10^{10}$ tracks/cm^2 which result from bombardment by low energy ($\lesssim 1$ MeV/amu) solar iron-group nuclei. Since such high densities are only produced during unshielded (i.e. surface) exposure of these grains to solar flares, we have made extensive HVEM observations of track density distributions in surface soil and drill core samples as a means of investigating lunar regolith dynamics. One of our most exciting discoveries was the observation of grains containing $\gtrsim 10^{11}$ tracks/cm^2 located at the bottom of the 3-m long Apollo 15 drill core (Phakey et al., 1972). Subsequent studies have shown that there is no systematic decrease in radiation damage in grains of this size range with depth in *any* of the lunar cores we have investigated. This suggests that at a given location the regolith is comprised of many layers, each of which was exposed on the surface for a period of $\gtrsim 4 \times 10^4$ years.

The extent of radiation damage is an indication of lunar soil "maturity". It was shown that soil particles may accumulate $\gtrsim 10^{12}$ tracks/cm^2 and begin to lose their crystalline structure due to the intense radiation damage (Phakey and Price, 1972). Grains containing the highest densities are observed to have rounded edges, suggestive of solar wind sputtering, and are characterized by the presence of a thin "skin" of intensely radiation damaged material (Fig. 2). The properties of these amorphous coatings have been extensively studied by Maurette and co-workers (Bibring et al., 1972). They have shown that the coating is a layer of metamict material, typically ~ 500 Å in thickness, produced by implantation of very low energy (~ 1 keV/amu) solar wind ions. By relating the thickness of the amorphous coating to the energy of the bombarding ions, Bibring et al. (1973) have shown that the mean energy of the solar wind probably has not changed over the past 500 million years.

Grains from the interior of some lunar breccias also contain high track densities, a memory of their pre-compaction exposure on the moon's surface. Similar features occur in gas-rich meteorites. The formation of these breccias must have been sufficiently gentle to cause no thermal or shock erasure of tracks. We have used the visibility of latent tracks (i.e. unetched tracks seen by HVEM), microstructure observations, and the presence or absence of etched tracks (see next Section) to classify lunar breccias on a scale of increasing metamorphic grade (Hutcheon et al., 1972a; Macdougall et al., 1973). Grains in loosely compacted soil breccias frequently have amorphous coatings and contain high densities of tracks showing strong contrast. These tracks can be chemically etched and are conveniently studied with TEM replicas. Higher on the metamorphic scale are breccias characterized by twinning and microfracturing, with abundant shock-induced glass. Tracks in such breccias are only faintly visible on dark field micrographs and cannot be etched. Metamorphism has annealed much of the damage that constitutes a track. Breccias of even higher metamorphic grade exhibit recrys-

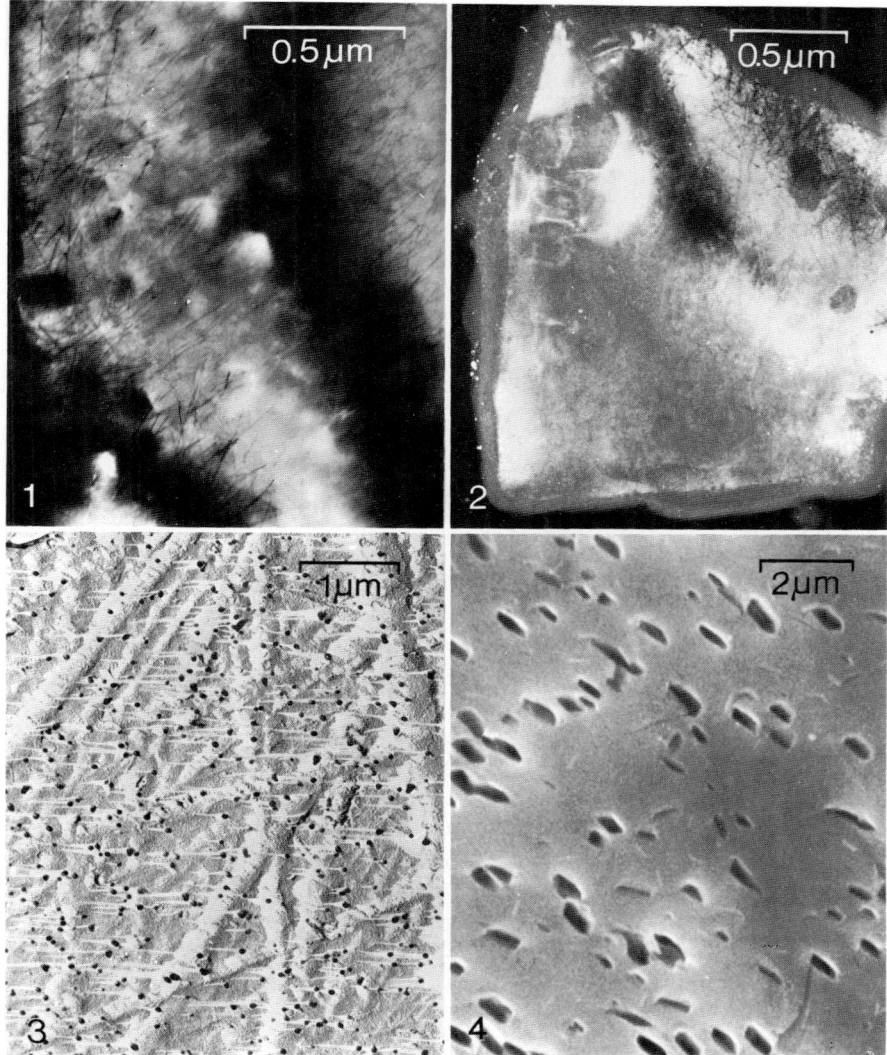

Fig. 1. Dark field micrograph of a small lunar soil grain. Tracks are visible as faint lines

Fig. 2. Dark field micrograph of a small grain from the lunar soil showing both tracks and an amorphous coating. (Photograph from Maurette and Price, 1975)

Fig. 3. Shadowed carbon replica of lunar feldspar grain etched 2 min in boiling 1:2 NaOH:H_2O. The density is 1.5×10^9 tracks/cm^2. In this photograph the tracks are black dots with light colored shadows; the large linear features crossing the figure are polishing scratches. (Photograph from Macdougall *et al.*, 1973)

Fig. 4. Scanning electron micrograph of an etched feldspar grain from a lunar rock

tallization features and retain no radiation damage features indicative of pre-compaction irradiation.

3. Replicas for TEM

As noted in the previous Section, tracks in materials that have not been severely heated or shocked can be chemically etched. Since studies of etched tracks are normally done on flat, polished surfaces, conventional shadowed carbon replicas for transmission electron microscopy (TEM) provide an excellent means for studying very high track densities. Tracks are easily visible as "flags" projecting upward toward the observer (Fig. 3). Using this method, we have been able to measure densities as high as 10^{11} tracks/cm^2. At such high densities, etching conditions must be very carefully controlled to ensure that track-hole diameters are not increased to the point where neighboring track-holes begin to overlap. Typically, lunar feldspar grains are etched for one to two minutes in a boiling 1:2 NaOH and water solution, a treatment which etches tracks to diameters of 300–500 Å. For lunar olivine crystals, similar results are obtained by etching for 10 min in boiling "WN" solution (see Krishnaswami et al., 1971, for recipe). Even for such short etching times, virtually all tracks present are etched and visible on the replica.

TEM replica studies, in addition to the HVEM observations discussed in the previous section, play an important role in our investigations of lunar soils and breccias. The main advantage of the replicas in this work is that much larger areas can be observed. Thus we can measure the track density gradient (which reflects the energy spectrum of the bombarding particles) across grains several hundred microns in diameter. These gradient measurements provide a glimpse of the history of our sun, since grains exposed on the lunar surface at some distant time in the past still retain the gradient produced during that exposure, providing they have subsequently been shielded by burial or inclusion within a large breccia. In our studies (Price et al., 1973b) we have noted that the *steepest* gradients observed in grains measured *in situ* in lunar breccias, in the Fayetteville gas-rich meteorite, in lunar soil grains collected from surface and buried locations during the Apollo 15 mission, and in a shielded grain at the bottom of a vug in lunar rock 15499 (Hutcheon et al., 1972b) are all similar to the gradient measured in glass from the Surveyor III camera, which resided on the moon during 1967–1970 (Crozaz and Walker, 1971; Fleischer et al., 1971; Price et al., 1971). This indicates that at various epochs in the past, extending from the last $\sim 10^5$ years to the time (thought to be at least 4 billion years ago) when gas-rich meteorites such as Fayetteville were assembled, the maximum slope in the solar flare spectra was similar to that determined from the Surveyor glass to be E^{-3}, where E is the energy per nucleon. This does not rule out the existence of time periods in which the flare spectra were sometimes shallower.

Different mineral phases exhibit different degrees of resistance to thermal annealing of particle tracks. Using replicas, we have surveyed a series of lunar breccias for the presence or absence of high track densities in the various constituent mineral phases. From data collected in this way, we have been able to put some limits on the thermal history of these breccias, based on laboratory-determined annealing temperatures (Macdougall et al., 1973). Replicas were invaluable in this study because they not only allowed us to survey representative areas but also provided the high resolution necessary to identify the very lightly etched tracks. In addition, the replicas faithfully reproduce grain-to-grain relationships in the breccia.

4. SEM Observations

The scanning electron microscope provides a further jump in the total area that can be conveniently studied on a single sample being examined for particle tracks, albeit at somewhat lower resolution than available with TEM. This capability has been particularly valuable in our measurements of *galactic* track density gradients in lunar rocks. In contrast to the *solar flare* tracks mentioned previously, which dominate in the outer few millimeters of most lunar rocks, the galactic particles are much more energetic and therefore penetrate to depths of many centimeters. Typically it is necessary to make measurements in a rock slice 5–10 cm in length. The large sample chamber of the SEM, coupled with the ability to change magnification rapidly and precisely measure distances, makes such measurements relatively simple. The possibility of scanning larger areas at lower magnification also makes the SEM a more convenient instrument than the TEM for measuring the generally lower densities (10^6-10^8 tracks/cm^2) of galactic cosmic ray tracks. Galactic cosmic ray tracks in a lunar rock as seen by SEM are shown in Fig. 4.

A second area in which we have made extensive use of the SEM is that of fission track dating of lunar rocks. In general, only rare accessory minerals such as whitlockite and zircon contain sufficient uranium to make this method possible. Because of the great value of lunar samples, it is not practical to grind up large amounts of rock and make conventional mineral separations. Using an energy-dispersive X-ray analyzer attached to the SEM we have found it relatively easy to locate such minerals simply by scanning rock sections for P and Zr rich phases. The minerals are relocated with the SEM for track counting after etching. Our fission track studies of lunar whitlockite have provided the first evidence for ^{244}Pu in the ancient moon (Hutcheon and Price, 1972) and, more recently, have indicated that lunar KREEP basalts may date back to $\sim$4.3 billion years (Haines et al., 1974).

5. Summary

The electron microscope is an invaluable instrument in the study of particle tracks, particularly in lunar samples. As outlined very briefly in this short paper, a myriad of problems can be addressed through the combined use of transmission and scanning electron microscopy, and conventional EM techniques such as surface replication. Advances in electron microscope technology will undoubtedly herald new discoveries in the particle track field.

References

Bibring, J.P., Chaumont, J., Comstock, G., Maurette, M., Meunier, M., Hernandez, R.: Solar wind and lunar wind microscopic effects in the lunar regolith. Lunar Science IV, 72–73. Houston: The Lunar Science Institute 1973.

Bibring, J.P., Duraud, J.P., Durrieu, L., Jouret, C., Maurette, M., Meunier, R.: Ultrathin amorphous coatings on lunar dust grains. Science 175, 753–755 (1972).

Crozaz, G., Walker, R.M.: Solar particle tracks in glass from Surveyor 3. Science 171, 1237–1239 (1971).

Fleischer, R.L., Hart, H.R., Jr., Comstock, G.M.: Very heavy solar cosmic rays: energy spectrum and implications for lunar erosion. Science 171, 1240–1242 (1971).

Fleischer, R.L., Price, P.B., Walker, R.M.: Solid-state track detectors: Applications to nuclear science and geophysics. Ann. Rev. Nucl. Sci. 15, 1–28 (1965).

Haines, E.L., Hutcheon, I.D., Weiss, J.R.: Excess fission tracks in Apennine Front KREEP basalts. Lunar Science V, 304–306. Houston: The Lunar Science Institue 1974.

Hutcheon, I.D., Braddy, D., Phakey, P.P., Price, P.B.: Study of solar flares, cosmic dust and lunar erosion with vesicular basalts. The Apollo 15 lunar samples, p. 412–414. Houston: The Lunar Science Institute 1972b.

Hutcheon, I.D., Phakey, P.P., Price, P.B.: Studies bearing on the history of lunar breccias. Proc. 3rd Lunar Sci. Conf. Geochim. Cosmochim. Acta Suppl. 3, 3, 2845–2865 (1972a).

Hutcheon, I.D., Price, P.B.: Plutonium-244 fission tracks: Evidence in a lunar rock 3.95 billion years old. Science 176, 909–911 (1972).

Krishnaswami, S., Lal, D., Prabhu, N., Tamhane, A.S.: Olivines: Revelation of tracks of charged particles. Science 174, 287–291 (1971).

Lal, D.: Hard-rock cosmic ray archaeology. Space Sci. Rev. 14, 3–102 (1972).

Lal, D., Murali, A.V., Rajan, R.S., Tamhane, A.S., Lorin, J.C., Pellas, P.: Techniques for proper revelation and viewing of etch tracks in meteoritic and terrestrial minerals. Earth Planet. Sci. Letters 5, 111–119 (1968).

Macdougall, D., Lal, D., Wilkening, L., Liang, S., Arrhenius, G., Tamhane, A.S.: Techniques for the study of fossil tracks in extraterrestrial and terrestrial samples I: Methods of high-contrast and high-resolution study. Geochem. J. (Japan) 5, 95–112 (1971).

Macdougall, D., Rajan, R.S., Hutcheon, I.D., Price, P.B.: Irradiation history and accretionary processes in lunar and meteoritic breccias. Proc. 4th Lunar Sci. Conf. Geochim. Cosmochim. Acta Suppl. 4, 3, 2319–2336 (1973).

Maurette, M., Price, P.B.: Electron microscopy of irradiation effects in space. Science 187, 121–129 (1975).

Phakey, P.P., Hutcheon, I.D., Rajan, R.S., Price, P.B.: Radiation effects in soils from five lunar missions. Proc. 3rd Lunar Sci. Conf. Geochim. Cosmochim. Acta, Suppl. 3, 3, 2905–2915 (1972).

Phakey, P.P., Price, P.B.: Extreme radiation damage in soil from Mare Fecunditatis. Earth Planet. Sci. Letters 13, 410–418 (1972).

Price, P.B., Chan, J.H., Hutcheon, I.D., Macdougall, D., Rajan, R.S., Shirk, E.K., Sullivan, J.D.: Low-energy heavy ions in the solar system. Proc. 4th Lunar Sci. Conf. Geochim. Cosmochim. Acta, Suppl. 4, 3, 2347–2361 (1973b).

Price, P.B., Fleischer, R.L.: Identification of energetic heavy nuclei with solid dielectric track detectors: Applications to astrophysical and planetary studies. Ann. Rev. Nucl. Sci. 21, 295–334 (1971).

Price, P.B., Hutcheon, I.D., Cowsik, R., Barber, D.J.: Enhanced emission of Fe nuclei in solar flares. Phys. Rev. Letters 26, 916–919 (1971).

Price, P.B., Lal, D., Tamhane, A.S., Perelygin, V.P.: Characteristics of tracks of ions of $14 \leq Z \leq 36$ in common rock silicates. Earth Planet. Sci. Letters 19, 377–395 (1973a).

Silk, E.C.H., Barnes, R.S.: Examination of fission fragment tracks with an electron microscope. Phil. Mag. 4, 970–971 (1959).

CHAPTER 7.7

Stony Meteorites

J.R. ASHWORTH and D.J. BARBER

1. Introduction

Meteorites are important as samples of parent-bodies formed early in the history of the Solar System, and later disrupted. Many stony meteorites have extremely complex structures. Particularly in "unequilibrated" chondrites and in gas-rich meteorites, much of the material is fine-grained and using optical methods can only be described as "unresolved matrix" or in similar terms. In a few selected meteorites we are studying the nature of such material and its relation to the coarser grains by HVEM (at 1 MV) of ion-thinned specimens. The extra penetration afforded at high voltages is essential because one cannot make significant areas of the inhomogeneous samples thin enough for study at 100 kV (minerals thin at different rates; some grains also fall out). In such material, it is often convenient to work in dark field, since this minimizes unwanted thickness-contrast, while a sufficient number of grains can be brought into diffraction contrast to illustrate grain morphology and intragranular substructures.

Although the literature of meteorites is very extensive and contains a great deal of data elucidated by optical petrography, the application of TEM methods to meteorites has been sparse. But our work benefits from the numerous electron microscope investigations which have been undertaken on lunar, terrestrial and experimentally-treated minerals. Since the detailed histories of meteorites are quite uncertain, deductions from observed microstructures have to be made at present in a comparative fashion.

The results reported here concern three areas of study: (1) shock stratigraphy and post-irradiation lithification of "gas-rich" meteorites; (2) characterization of the abundant deformation features in other stony meteorites; and (3) investigation of the fine-grained constituents of unequilibrated ordinary chondrites.

2. Gas-rich Meteorites

Most gas-rich meteorites are only loosely consolidated. They contain individual grains with histories of unshielded solar irradiation (see Pellas, 1972). The evidence suggests that the grains were irradiated under conditions very similar to those existing at the lunar surface, being presumably within asteroid regoliths (Macdougall et al., 1974).

Fig. 1. Clinoenstatite grains with a high density of stacking faults on (100), in shock-indurated Khor Temiki. Dark field

In *Khor Temiki*, a gas-rich achondrite which is poorly lithified and shows virtually unmodified regolithic phenomena, heavily shock-deformed grains are intimately associated with unshocked ones, some of which have high densities of solar cosmic-ray tracks (Ashworth and Barber, 1975a). These extremes of deformation history are characteristic of environments affected by impacts of varying intensity, as exemplified by lunar soil. In so far as Khor Temiki is lithified, shock is similarly an important factor. Veins of indurated material traverse the meteorite; samples of this material are found to consist essentially of clinoenstatite with a high density of (100) stacking faults (Fig. 1), formed *in situ* from the enstatite of the bulk meteorite by shock deformation and resultant heating. The thermal effect is probably predominant, since the clinoenstatite has features suggesting that it formed by inversion of the high-temperature polymorph protoenstatite: it is twinned on a fine scale, and in diffraction the a^* streaks arising from stacking faults have prominent diffuse intensity maxima at orthopyroxene reciprocal-lattice positions, an effect which allies this material to "slowly quenched" protoenstatite as studied by X-ray crystallographers (Reid *et al.*, 1974).

Glass may be important in bonding gas-rich meteorites, and is a major constituent of the gas-rich chondrite *Weston*. There are glassy fragments in Weston, and some interstitial glass may also predate lithification. Fig. 2 was taken in a sample known to have track-rich grains, revealed by a preliminary survey using scanning electron microscopy of lightly-etched material (cf. Macdougall *et al.*, 1974). This survey also reveals spatially inhomogeneous deformation, much

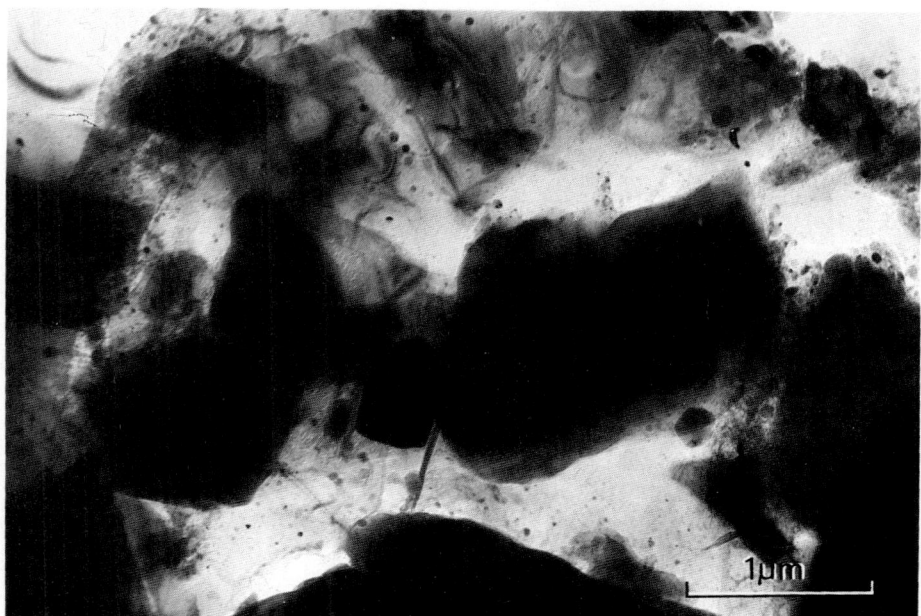

Fig. 2. Matrix of a gas-rich part of Weston, showing amorphous interstitial material with inclusions. Bright field

of which no doubt affected the individual grains prior to aggregation, but some of which may be related to lithification and glass production (cf. Christie et al., 1973).

3. Deformational Features in Olivine and Pyroxene

Olivine and calcium-poor pyroxene are the commonest minerals in chondrites. We have studied (Ashworth and Barber, 1975b) such meteorites with various shock histories. The metamorphosed chondrite *Olivenza* is lightly shocked; fractures are common in the olivine and are accompanied by slip dislocations which are mostly aligned along [001] and are often concentrated in slip bands. Fig. 3 shows such dislocations in the complex chondrite *Hedjaz*. This is more intensely deformed than Olivenza, and has dislocation tangles and loops closely similar to those observed in olivine experimentally deformed under the low temperature, high strain-rate regime characterized by $\{hk0\}$ [001] slip (Phakey et al., 1972). One optical effect of such fairly intense shock, known as mosaicism, is paralleled in the electron microscope by a "blocky" structure in which adjacent parts of a crystal are slightly misoriented across zones of high dislocation density. Such mosaicism is apparently due predominantly to fracturing, which is common, and most of the porosity of these lightly shocked meteorites resides in fractures.

In the very heavily shocked ureilite *Goalpara*, the thermal effect of shock has induced recrystallization of olivine. Substructures in the new grains (Fig. 4)

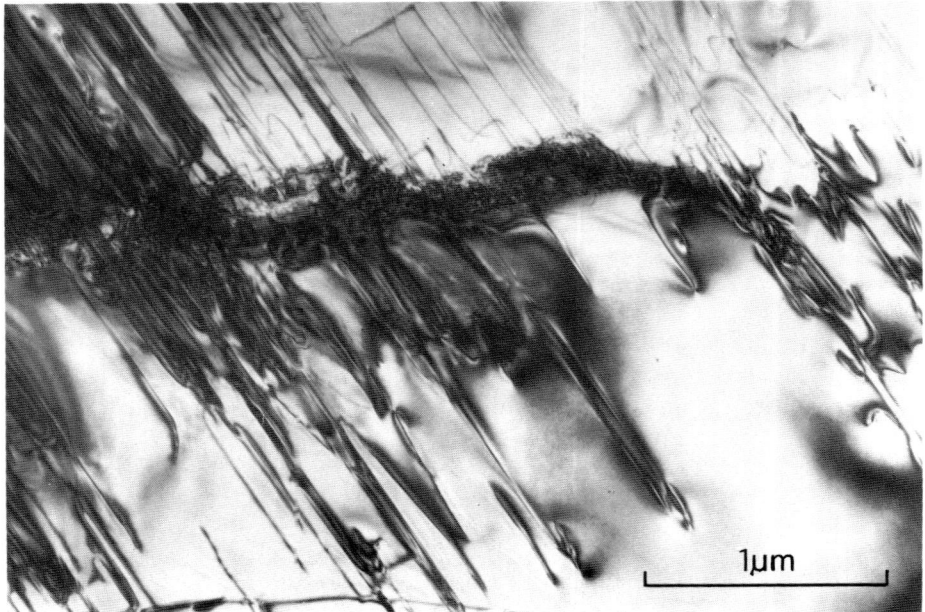

Fig. 3. Dislocations aligned predominantly along [001] in Hedjaz olivine. Bright field

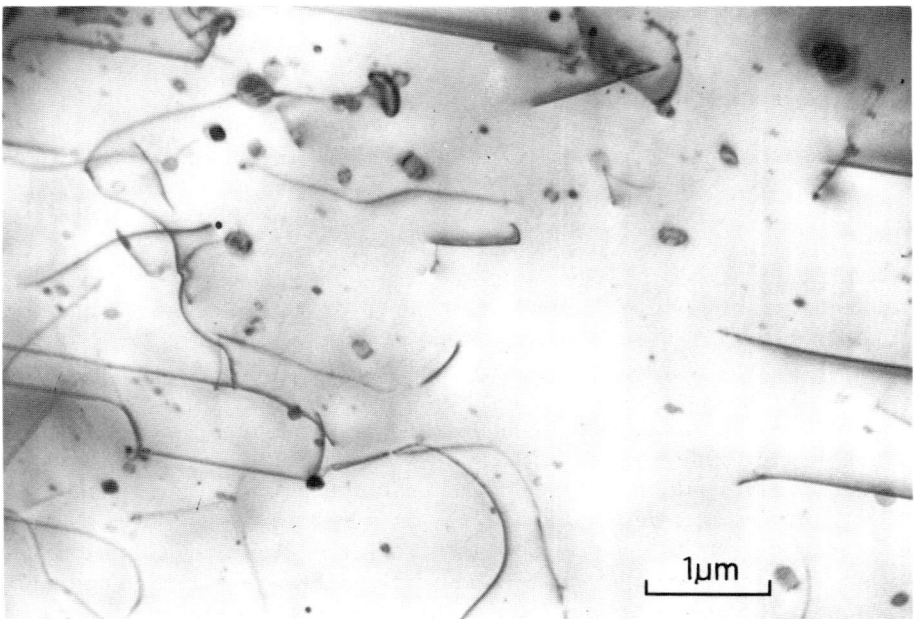

Fig. 4. Dislocation substructure and small precipitate particles in recrystallized olivine from Goalpara. Bright field

are closely similar to those produced by artificial annealing of dry olivine (Goetze and Kohlstedt, 1973).

Pyroxene in Hedjaz is also deformed with (100) stacking faults and high densities of dislocations lying out of the (100) plane (*cf.* Green and Radcliffe, 1972). Clinobronzite lamellae are present in this bronzite, so that the diffraction patterns are superficially similar to those described above from Khor Temiki. The proportion of clino-component, however, varies strongly within a given grain and, at least where subordinate in amount, it appears to be untwinned. Its association with heavily deformed orthopyroxene suggests that the clinobronzite formed directly by shear-induced inversion of the bronzite (*cf.* Kirby and Coe, 1974). The same association and the scarcity of micro-twinning distinguish this clinobronzite from that observed in the relatively undeformed primitive chondrite *Parnallee*, which is twinned. As in the other pyroxenes under discussion, a^* streaks indicate stacking disorder, but in Parnallee "orthopyroxene maxima" are lacking. Such clinopyroxene can possibly be attributed to rapid quenching of protopyroxene during high-temperature (chondrule-forming) processes prior to aggregation of the rock, as distinct from the "slow quenching" invoked above for the post-aggregation phenomena in Khor Temiki.

4. Fine-grained Constituents of Chondrites

Parnallee is an unequilibrated ordinary chondrite, LL3 in the nomenclature of Van Schmus and Wood (1967). It has not suffered the pervasive deformation seen in Hedjaz, though individual grains and chondrules have deformation features, as noted by Green *et al.*, (1970). The variation of deformation effects between grains probably reflects a history of repetitive fragmentation and reaggregation, as discussed by Dodd (1971). The matrix between the chondrules does not look clastic (Fig. 5). In contrast to the matrix of carbonaceous chondrites (Green *et al.*, 1970, 1971), it has low porosity and a "sintered" appearance. Grains of olivine and pyroxene are cemented by plagioclase. This structure indicates incipient metamorphism. It is more clearly seen in Hedjaz, where the matrix tends to be coarser (Fig. 6). Hedjaz is a breccia of which many parts have individual metamorphic histories (Kraut and Fredriksson, 1971); the coarsened matrix suggests that our sample has been heated more than Parnallee, so that it is petrologic type 4 in the nomenclature of Van Schmus and Wood (1967). This is consistent with our observation that the mesostasis within chondrules, though optically isotropic, is devitrified. In Parnallee, on the other hand, true glass survives in chondrules.

The low-temperature, partly cataclastic shock effects in Hedjaz presumably postdate the metamorphism, and probably date from disruption of the L-group parent body (*cf.* Pellas, 1972, p. 67; Carter *et al.*, 1968). It is uncertain whether the integrated structure of Parnallee matrix reflects moderately high temperature at the time of aggregation, or subsequent slight metamorphism; in either case it is probably a mild, optically cryptic manifestation of the metamorphic effects known in ordinary chondrites as a whole.

Fig. 5. Very fine-grained matrix in Parnallee. Dark field

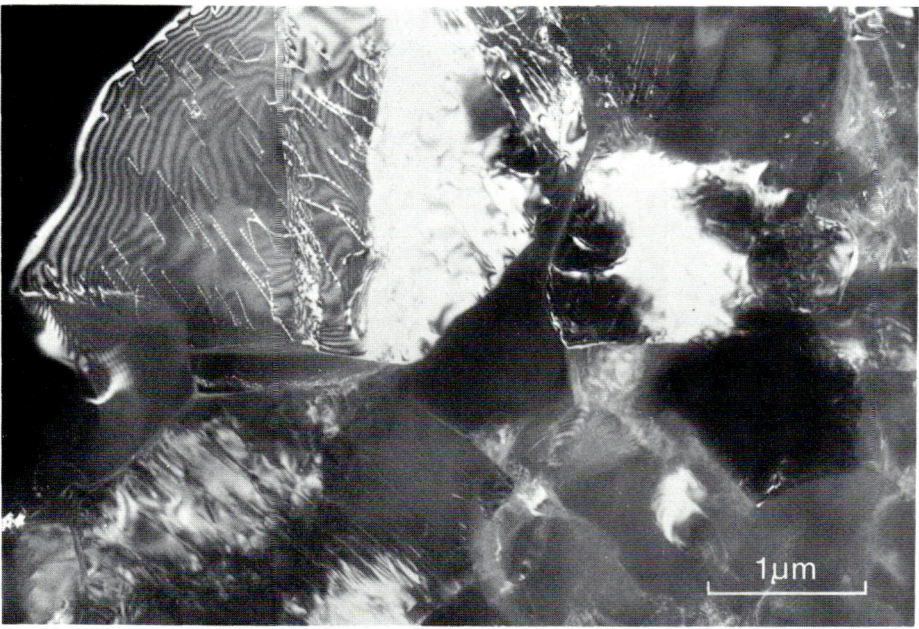

Fig. 6. Relatively coarse matrix in Hedjaz. Dark field

References

Ashworth, J.R., Barber, D.J.: Electron petrography of shock effects in a gas-rich enstatite-achondrite. Contrib. Mineral. Petrol. **49**, 149–162 (1975a).
Ashworth, J.R., Barber, D.J.: Electron petrography of shock-deformed olivine in stony meteorites. Earth Planet. Sci. Letters **27**, 43–50 (1975b).
Carter, N.L., Raleigh, C.B., DeCarli, P.S.: Deformation of olivine in stony meteorites. J. Geophys. Res. **73**, 5439–5461 (1968).
Christie, J.M., Griggs, D.T., Heuer, A.H., Nord, G.L., Jr., Radcliffe, S.V., Lally, J.S., Fisher, R.M.: Electron petrography of Apollo 14 and 15 breccias and shock-produced analogs. Proc. 4th Lunar Sci. Conf. Geochim. Cosmochim. Acta Suppl. 4, **1**, 365–382 (1973).
Dodd, R.T.: The petrology of chondrules in the Sharps meteorite. Contrib. Mineral. Petrol. **31**, 201–227 (1971).
Goetze, C., Kohlstedt, D.L.: Laboratory study of dislocation climb and diffusion in olivine. J. Geophys. Res. **78**, 5961–5971 (1973).
Green, H.W., II, Radcliffe, S.V.: Deformation processes in the upper mantle. In: Flow and fracture of rocks (eds. H.C. Heard, I.Y. Borg, N.L. Carter, C.B. Raleigh). Am. Geophys. U. Monograph. **16**, 139–156 (1972).
Green, H.W. II, Radcliffe, S.V., Heuer, A.H.: High voltage (800 kV) electron petrography of unequilibrated stony meteorites – Allende and Parnallee. Proc. 28th Ann. Mtg. Electron Microscopy Soc. of America, p. 490–491 (1970).
Green, H.W. II, Radcliffe, S.V., Heuer, A.H.: Allende meteorite: a high-voltage electron petrographic study. Science **172**, 936–939 (1971).
Kirby, S.H., Coe, R.S.,: The role of crystal defects in the enstatite inversion. Trans. Am. Geophys. U. **55**, 419 (1974).
Kraut, F., Fredriksson, K.: Hedjaz, an L-3, L-4, L-5 and L-6 chondrite. Meteorities **6**, 284 (1971).
Macdougall, D., Rajan, R.S., Price, P.B.: Gas-rich meteorites; possible evidence for origin on a regolith. Science **183**, 73–74 (1974).
Pellas, P.: Irradiation history of grain aggregates in ordinary chondrites. Possible clues to the advanced stages of accretion. In: From plasma to planet (Nobel Symposium 21) (ed. A. Elvius), p. 65–92. Stockholm: Almqvist and Wiksell 1972.
Phakey, P., Dollinger, G., Christie, J.: Transmission electron microscopy of experimentally deformed olivine crystals. In: Flow and fracture of rocks (eds. H.C. Heard, I.Y. Borg, N.L. Carter, C.B. Raleigh). Am. Geophys. U. Monograph. **16**, 117–138 (1972).
Reid, A.M., Williams, R.J., Takeda, H.: Coexisting bronzite and clinobronzite and the thermal evolution of the Steinbach meteorite. Earth Planet. Sci. Letters **22**, 67–74 (1974).
Van Schmus, W.R., Wood, J.A.: A chemical-petrologic classification for the chondritic meteorites. Geochim. Cosmochim. Acta **31**, 747–765 (1967).

CHAPTER 7.8

Microcracks in Crystalline Rocks

L. DENGLER

1. Introduction

The Scanning Electron Microscope (SEM) has become a useful tool in many areas of geologic endeavor. Its great depth of field makes it ideal for the study of rough topographies. In recent years it has been used to study fracture and microcracks in metals and ceramics (Majumdar, 1972) and is now being applied to these same topics in the study of rocks (Sprunt, 1973).

Microcracks are thought to play an important role in the physical behavior of rocks. They are believed to account for the non-linearity of stress-strain relationships at low stresses (Adams and Williamson, 1923; Walsh, 1965), affect electrical resistivity (Brace and Orange, 1968), thermal conductivity (Walsh and Decker, 1966) and wave propagation, causing the difference in velocities in saturated and unsaturated rocks at low pressure (Nur and Simmons, 1969) and anisotropy in rocks with oriented microcracks (Johnson and Wenk, 1974; Wang and Lin, 1974, Wang et al., 1975). The Griffith and Modified Griffith theories of failure are based on microcracks (Griffith, 1921, 1924; Brace, 1960).

Previous studies have attempted to observe microcracks both optically (Paulding, 1965) and with the SEM (Sprunt, 1973). The resolving power (2000–3000 Å) and limited depth of field, coupled with possible specimen damage during thin-section preparation, make optical results difficult to interpret. It is also difficult to distinguish cracks from grain boundaries. Sprunt's study used surfaces prepared by ion bombardment to minimize specimen damage although some artifacts — mounding and topographic steps due to differences in thinning rates — were reported.

This study differs from previous ones in that no sawing, grinding, polishing or bombardment of the sample surface occurred and therefore no artifacts were produced by sample preparation. Fracture surfaces from stressed and unstressed granodiorite were examined to study microcracks and pores. (The term "pore" refers to approximately equidimensional cavities or high-aspect ratio cavities, aspect ratio being defined as the ratio of minimum to maximum cavity dimensions. "Crack" refers to long narrow cavities where the length is at least ten times the width, or a low-aspect ratio cavity.).

A Coates and Welter Field Emission Scanning Electron Microscope was used in this study. Details of SEM operation and capabilities are available in several sources (Kimoto and Russ, 1969; Oatley et al., 1965). The resolving power

Microcracks in Crystalline Rocks 551

of the microscope is 100 Å with a continuous magnification range of 20 to 50000 diameters.

Samples studied were from the Climax stock of Area 15, Nevada Atomic Test Site, from the preblast coring for the Piledriver event. Details of this granodiorite may be found in Houser and Poole (1961). Unstressed specimen surfaces were obtained from Brazil test (tension) failure of virgin rock. Stressed samples were tested in a "triaxial" condition to failure at constant confining pressure. Specimen surfaces are fracture faces that were formed during the experiment. The only subsequent surface preparation in both stressed and unstressed rocks was treatment in an ultrasonic cleaner to remove dust particles and the evaporation of a 200 Å layer of 60% gold, 40% paladium alloy onto the specimen surface to provide a conducting layer for SEM operation. Mineral identification was made from characteristics such as cleavage, crystal form and fracture behavior and aided by an SEM study of isolated quartz and feldspar samples.

2. Observations

2.1 Unstressed Rock

The general appearance of the unstressed specimen surface was of a cohesive unit with no transgranular cracks and few grain boundary cracks. Quartz crystals

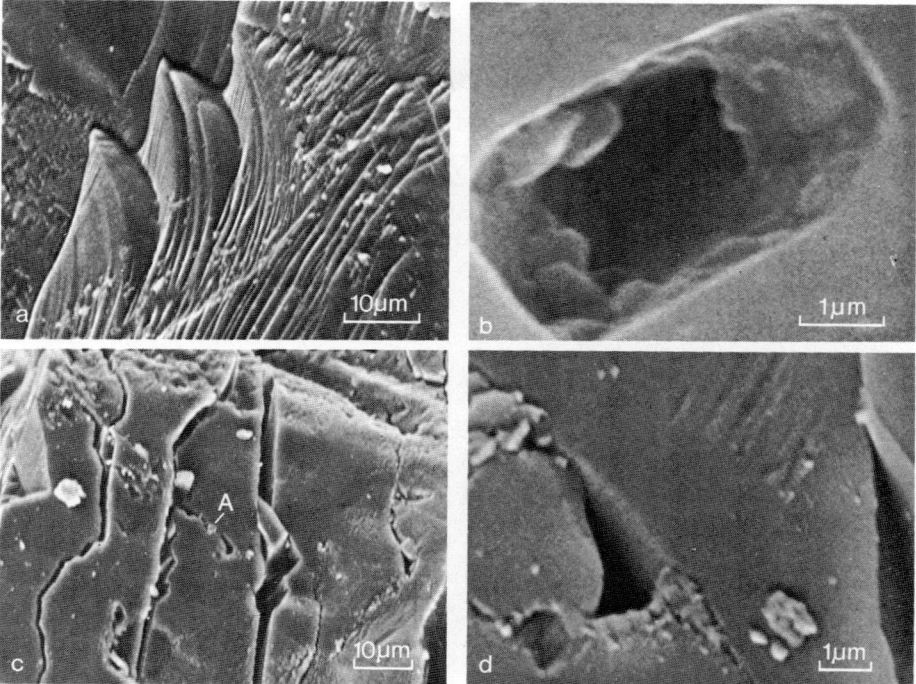

Fig. 1a–d. Quartz. (a) Unstressed rock, fracture surface. (b) Pore in unstressed quartz. (c) Transgranular fracture in stressed crystal. (d) Enlargement of A in (c). Fractured pore

were identified by conchoidal behavior and curved striations on fracture surfaces (Fig. 1a). No transgranular cracks were seen. A few apparently isolated pores with diameters on the order of microns were observed (Fig. 1b). They are possibly the remains of fluid inclusions. Plagioclase, recognized by cleavage and twin structures (Fig. 2a), showed greater density of pores than quartz. Pore diameters vary from less than 1 µm to 3 µm (Fig. 2b), and are smaller than those in

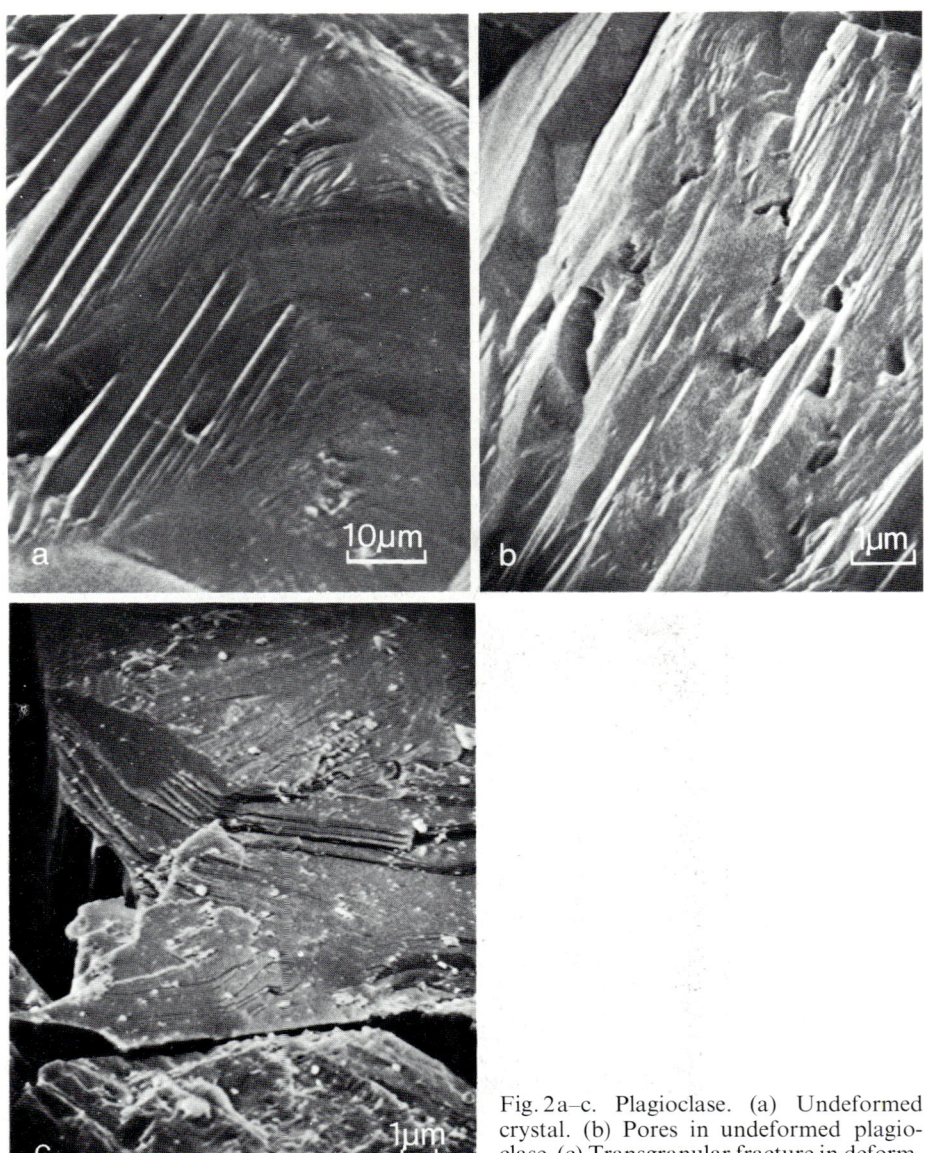

Fig. 2a–c. Plagioclase. (a) Undeformed crystal. (b) Pores in undeformed plagioclase. (c) Transgranular fracture in deformed plagioclase

quartz. Their shape is more angular than in quartz. Biotite grains were easily identified by cleavage plates. In the unstressed material these grains appear as undeformed banks of mica books (Fig. 3a). The cleavage bands appear tightly closed. The micas seem to be more loosely bound to their neighbors than quartz or feldspar crystals, with grain-boundary cracks commonly occurring in the unstressed material.

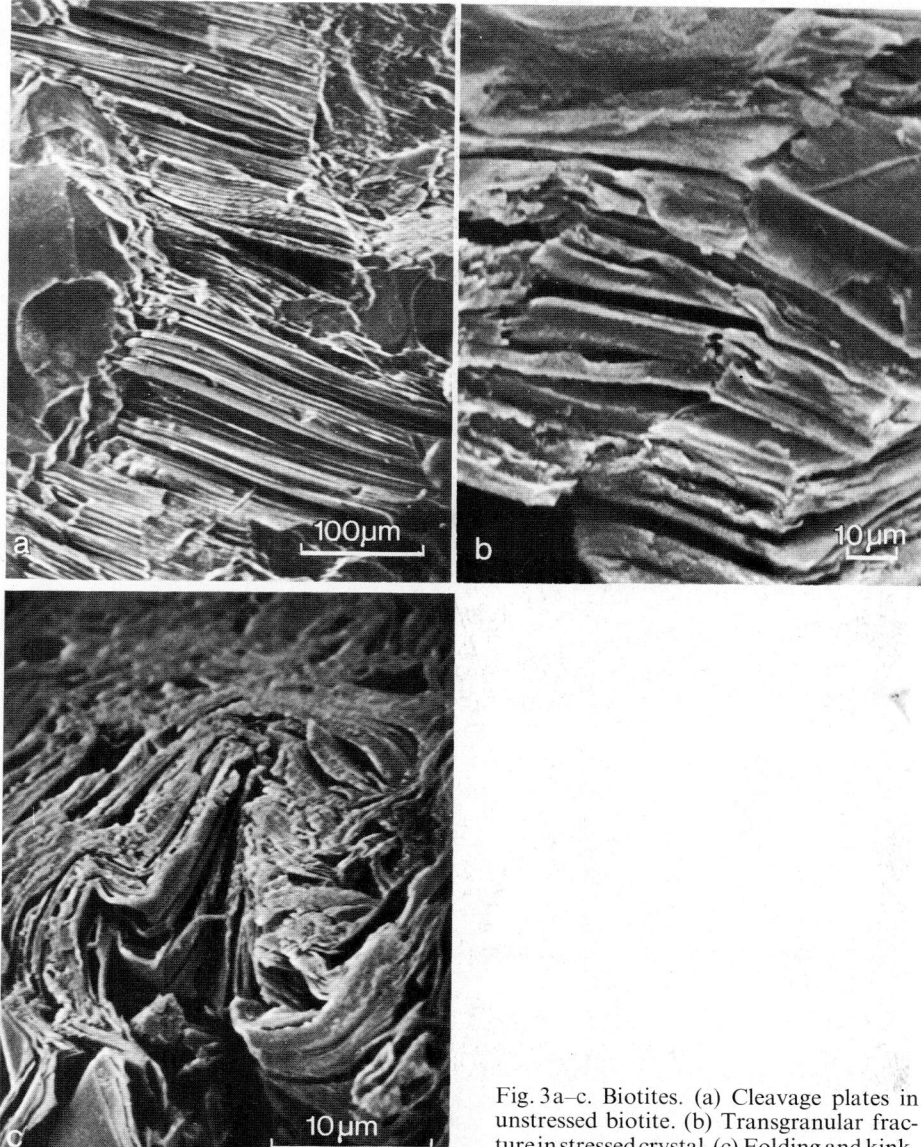

Fig. 3a–c. Biotites. (a) Cleavage plates in unstressed biotite. (b) Transgranular fracture in stressed crystal. (c) Folding and kinking in deformed biotite

554 L. Dengler:

2.2 Stressed Rock

In stressed samples there was an increased number of cracks in all samples studied. In quartz, transgranular cracks were common (Fig. 1 c) with preferred orientation parallel to the direction of maximum stress. Special attention was given to the role of pores in transgranular fracture. In general the pore features

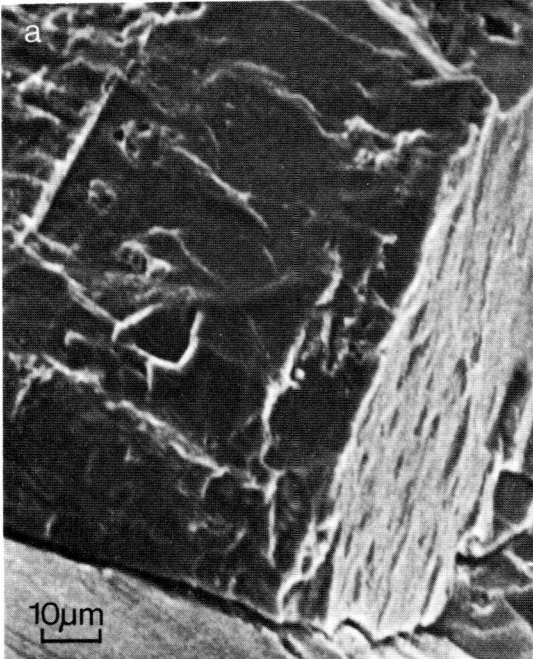

Fig. 4a and b. Grain boundary microcracks in stressed rock. (a) Feldspar-quartz, microcrack. (b) Fracture surface with predominance of grain boundary cracks

showed little or no fracturing. Only in grains having undergone catastrophic failure (numerous transgranular fractures extending the length of the grain) did the pores exhibit any fracturing at all. (Fig. 1d). They do not seem to play an important role in the initiation of fracture.

In plagioclase, extensive transgranular fracturing was also observed. These fractures generally followed cleavage planes, tending to join up in a series of right-angle steps traversing a grain. Pore structures exhibited no fractures even when located within 10 μm of a major transgranular fracture. The details of fracture in plagioclase seem to be controlled by preferred crystallographic directions such as cleavage and twin planes (Fig. 2c).

The affect of stress on biotite varied in the samples studied. In some samples both seemingly undeformed and fractured micas could be observed. An optical investigation of the same samples showed some degree of kink banding in all stressed micas, a stress affect which may not be observable on the SEM. The SEM study showed grain boundary cracks in all the micas, frequent opening of cleavage plates, transgranular cracks across the cleavage plates and extensive folding — especially in samples tested under high confining pressure (Fig. 3b,c).

Although extensive transgranular cracks were observed in all mineral constituents, stressed samples showed a predominance of grain boundary fracture (Fig. 4a). These fractures often extended the length of ten or more grains with widths of up to 10 μm. Length may be a function of confining pressure. Grain-boundary fractures occurred most readily between dissimilar grains (i.e. quartz — feldspar, feldspar — mica, etc.) as in Fig. 4b.

The study of fracture surfaces by SEM yields valuable information about the actual shape and size of microcracks in rocks. A number of the topics presented above are certainly worthy of further exploration — cause of pore structures and their behavior under stress, the role of confining pressure in failure, the initiation of microcrack growth through the study of samples unloaded prior to failure — to mention a few. Tensile and compressive stages have been developed for the SEM to observe the actual deformation of materials in a dynamic experiment. At present such an experiment can only be performed under high vacuum conditions, so the effect of pressure cannot be studied.

References

Adams, L.H., Williamson, E.D.: The compressibility of minerals and rocks at high pressure. J. Franklin Inst. **195**, 475–529 (1923).
Brace, W.F.: An extension of the Griffith theory of fracture to rocks. J. Geophys. Res. **65**, 3477–3480 (1960).
Brace, W.F., Orange, A.S.: Further studies of the effects of pressure on electrical resistivity of rocks. J. Geophys. Res. **73**, 5407–5420 (1968).
Griffith, A.A.: The phenomena of rupture and flow in solids. Phil. Trans. Roy. Soc. London Ser. A **221**, 163–197 (1921).
Griffith, A.A.: The theory of rupture. Proc. 1st Int. Congress Appl. Mech. Delft, p. 55–63 (1924).
Houser, F.N., Poole, F.G.: Age relations of the Climax composite stock, NTS, Nye Co., Nevada. U.S. Geol. Surv. Tech. Letter, Area 15-1 (1961).

Johnson, L.R., Wenk, H.-R.: Anisotropy of physical properties in metamorphic rocks. Tectonophysics **23**, 79–98 (1974).

Kimoto, S., Russ, J.C.: The characteristics and applications of the scanning electron microscope. Am. Sci. **57**, 112–124 (1969).

Majumdar, A.J.: The application of scanning electron microscopy to textural studies. Proc. Brit. Ceram. Soc., No. 20., 43–69 (1972).

Nur, A., Simmons, G.: The effect of saturation on velocity in low porosity rocks. Earth and Planet. Sci. Letters **7**, 183–193 (1969).

Oatley, C.W., Nixon, W.C., Pease, R.F.W.: Scanning electron microscopy. Advan. Electron. Electron Phys. **21**, 181–247 (1965).

Paulding, B.W.: Crack growth during brittle fracture in compression. PhD Thesis, Mass. Institute of Technology (1965).

Sprunt, E.S.: Scanning electron microscope study of cracks and pores in crystalline rocks. Master's Thesis, Mass. Institute of Technology (1973).

Walsh, J.B.: The effect of cracks on the compressibility of rock. J. Geophys. Res. **70**, 381–389 (1965).

Walsh, J.B., Decker, E.R.: Effect of pressure and saturating fluid on the thermal conductivity of compact rock. J. Geophys. Res. **71**, 3053–3061 (1966).

Wang, C.-Y., Lin, W.: Velocity ratios for rocks with oriented microcracks. Nature **248**, 579–580 (1974).

Wang, C.-Y., Lin, W., Wenk, H.-R.: The effects of water and pressure on velocities of elastic waves in a foliated rock. J. Geophys. Res. **80**, 1065–1069 (1975).

Subject Index

Entries in *italics* denote names of minerals; page numbers in **bold face** refer to electron photomicrographs.

Abbe Theory   123
absorption   32, 119, 493, 528
acmite $NaFeSi_2O_6$ 412
age determination   12, 540
albite $NaAlSi_3O_8$ 7, **193**, 249, 286
alkali feldspars
   exsolution   7, **98**, **109**, 189, **193**, 249, **500**
   polymorphism   248, 285
Al_2O_3 146, 404, **406**
α-β transitions   282
amorphous materials   385, 476, **481**, 545
amphiboles   172, 189, **238**
amplitude contrast   22, 44, 108
analytical EM   244, 258, 494, 506, 520, 526
anatase TiO_2 495
anisotropy effects on contrast   42, 113, 382
anomalous absorption length   32, 80, 119
anorthite $CaAl_2Si_2O_8$ 10, 46, 59, **287**, **345**
anthophyllite $(Mg,Fe)_7[OH/Si_4O_{11}]_2$ 189, **242**
antiphase boundary (see APB)
APB
   conservative   73
   definition   9, 72, 278
   non-conservative   73
   periodic   74, **267**, 292, 332
   variation in size   9, 285, 288, 290
APB's
   anorthite (b)   9, **287**
   anorthite (c)   10, 46, **289**, **345**
   bytownite (b)   **267**
   cristobalite   284
   hexacelsian   293, **355**
   labradorite   292, **332**
   leucite   **107**
   pyroxenes   **2**, 272, 283, 284
   pyrrhotite   **105**, **137**, 304

rutile   **92**, 310
scapolite   **284**, 294
spinel   37, **281**
tealite   **105**, **107**
wenkite   294, **369**
As, amorphous   **481**
astigmatism   19
Au   **498**
Auger electrons   494, 506
augite $(Ca,Mg,Fe)SiO_3$ **2**, 184, **186**, 222, 230, 235, 272, **283**, 516

$BaAl_2Si_2O_8$ 354
baddeleyite ZrO_2 **297**
$BaFe_{12}O_{19}$ 74
$BaNa(NbO_3)_5$ **69**
bastnaesite $Ce[F/CO_3]$ 108
beam damage   20, 259, 309, **345**, **366**, **381**, 430, 509
beam heating   163, 309, 345
bend contours   31
bending   376
bibliography
   SEM   489, 505
   TEM   49
biotite $K(Mg,Fe)_3[(OH,F)_2/AlSi_3O_{10}]$ **553**
boundary (see fault, interface)
boundary energy   215, 252
boundary plane of exsolution lamellae
   alkalifeldspar   192, 252
   plagioclase   194, 199
Bragg contours   31, 496
Brazil twinning   283, 383
bremsstrahlung   528
bright field E.M.   22
brittle deformation   550
bronzite (see enstatite)
bubbles   373, 426, **459**

Burgers vector of dislocations, determination 41, 45, 119, 376
  calcite 441
  corundum 405
  ilmenite-hematite 214
  olivine 394, 445
  pyroxenes 226, 390, 471
  quartz 119, 382, 412
  spinel 45
  wollastonite 331
bytownite (Ca, Na)[(Al,Si)Si$_2$O$_8$] **197**, 260, **267**, **522**

C, amorphous **481**
calcite CaCO$_3$ 428, **433**
camera constant 29, 53
CaO **501**
carbon film 147
carbonate rocks **428**
carborundum SiC **166**, 278
cathode ray tubes (CRT) 439
cathodoluminescence 496
celsian BaAl$_2$Si$_2$O$_8$ 293, 354
channelling patterns 492
charging 147, 493
chemical analyses 244, 258, 494, 506, 520, 526
chemical fault 73, 108
chromatic aberration 19, 125
cine photography 164, 282, 502
clay minerals 5
climb of dislocations 11, 378, 394, 404, 412, 429
clinoenstatite MgSiO$_3$ **297**, **319**, 391, **467**, **544**
clinopyroxenes (see augite, clinoenstatite, diopside, pigeonite)
Co 97
coarsening 182, 205
Coble diffusion 376
coherent precipitates 43, 98, 177, 214, 254
coherent scattering 27
cold-working 11, 378, 422, 428, 436
colemanite Ca[B$_3$O$_4$(OH)$_3$]·H$_2$O 277
complex faults 74
compositional fault 73, 108
computer methods 45, 113, 128
conservative motion of dislocations 10, 378
contrast
  dislocations 41, 113, 379, 470
  faulted crystal 32, 83, 96, 320, 466
  perfect crystal 30, 78
  topographic 492
cooling history 9, 182, 255, 270
core of dislocation 377

correspondence matrix 296
corundum Al$_2$O$_3$ **146**, 404, **406**
cosmic radiation 537
cracks 550
cracks healed 145, **437**
creep 378, 404
cristobalite SiO$_2$ **284**, **474**
cross-slip 378, 394, 412
cryptoperthite 7, **98**, 189, **193**, **249**, **500**
crystal spectrometers 490, 512
crystallographic shear (CS) 73, 300, 311
Cu **498**
cummingtonite (see grunerite)

damage
  abrasion 145
  cosmic radiation 538
  electron beam 20, 259, 309, **345**, **366**, **381**, 430, 509
  ion thinning 149
dark field E.M. 22
  amorphous substances 477
  high resolution 134, 137
Dauphiné twinning 283, 383
defects
  linear (see dislocations)
  planar (see APB, inversion boundary, stacking fault, twin)
  small volume 43
defocus 126
deformation
  amphiboles 245
  biotite **553**
  calcite 428, **433**
  enstatite 387, **391**, **465**
  olivine **392**, **443**
  plagioclase **552**
  quartz **382**, **410**, **419**
deformation lamellae
  olivine 394, 445
  quartz 383, 387, 410
dehydration 59
density of dislocations 397, 407, 412, 422, 432, 450
depth of focus (SEM) 499
detection limits, microanalysis 532
detectors 490, 512, 529
deviation parameter 31, 118
diaplectic glass (thetomorphic glass) 387, 398
diffraction of electrons 23, 52, 316, 503
diffraction patterns, calibration 53
  indexing 55
diffusion 174, 248, 449
diopside (Ca,Mg)SiO$_3$ **391**
dipole of dislocation 394, **405**, 429, **437**

Subject Index

dislocation contrast 41, 113, 379
   multiplication 383, 404
   theory 376, 397, 404, 410
dislocations
   alkalifeldspar 254
   calcite 428
   corundum 404
   hematite-ilmenite 214
   olivine 293, 443
   pyroxenes 221, **223**, **391**, **470**
   quartz 113, **372**, **381**, **410**, **419**
   spinel 27, 42, **46**
disorder 69, 104, 277, 304, 326, 361
   effects on diffraction pattern **60**, **326**, **349**
displacement vector
   anorthite 9, 288
   definition and determination 35, 72, 95
   hexacelsian 358
   pyroxenes 231, 283, 466
   rutile 311
   wenkite 369
   wollastonite 328
displacive transformation 106, 274
dolomite $(Ca,Mg)CO_3$ 441
domain boundaries, definition and contrast 76, 84, 96
domain structures, origin 68
dynamic experiments 164, 503
dynamical theory 31, 45, 78, 113, 124

edge dislocation 376
edge trapping 405
elastic scattering 18, 527
electron diffraction 23, 52, 503
   camera constant 29, 53
   extinctions 59, 331
   forbidden reflections 59, 380
   micro-area diffraction 496
   multiple diffraction 59
   selected area diffraction (SAD) 22, 25, 53, 496
   streaks **60**, **326**, **349**, **423**
   versus X-ray diffraction 4, 52
electron spectrometry 506
electron velocity analysis 508
enantiomorphism, contrast 40, 77, 81, 84, 99
   cristobalite 282
   quartz 283
   spinel 39, 41, 282
energy-dispersive detector 490, 512, 529
energy-loss analysis (STEM) 499
enstatite $MgSiO_3$ **188**, **230**, **235**, **297**, **319**, **391**, **516**, **544**
environmental cells 161
epitaxy 209

etching solutions 154, 540
evaporation in electron beam 259
Ewald sphere 25, 87
excitation error, s 23, 34, 76, 118
experimental deformation
   calcite 428, 439
   corundum 404
   enstatite 465
   olivine 393, **444**
   quartz **382**, **410**
exsolution
   alkali feldspar 7, **98**, **107**, **189**, **248**, **500**
   amphiboles **172**, **189**, **238**
   bytownite **198**, **260**, **266**, **522**
   labradorite **109**, **196**, **261**, **332**
   peristerite **194**, **261**
   pyroxene **2**, **182**, **220**, **228**, **234**
   spinel **25**
exsolution lamellae 98, 174
extinction distance $1/s$ 31, 117
extinctions in diffraction patterns 59, 331

failure 550
fault
   chemical (compositional) 73, 108
   complex 74
   double 72
   extrinsic 37, 72, 95
   geometrical 73
   intrinsic 37, 72, 95
   overlapping 39
   sequential 108
   single 72
Fe **495**
Fe_2O_3 **211**, **215**
Fe_3O_4 **211**, **282**
feldspars (see specific feldspar minerals)
ferrite 25, **26**, **37**, **41**, **74**, **211**, **281**
field emission for
   high resolution 132
   microanalysis 511
filaments
   high resolution 132
   microanalysis 511
   SEM 503
fission tracks 12, 537
flow law, olivine 396, 462
flow-stress curve, corundum 407
fluid inclusions 373, 426, **459**
flying ash **495**
foil thickness 34
forbidden reflections 59, 380
Fourier transform 5, 124
foshagite $Ca_4[(OH)_2/Si_3O_9]$ 59
fracture 550
fragments, preparation 147
Friedel's law 46, 81, 99, 281

fringe profiles
  α fringes 47, 89
  δ fringes 91
  mixed fringes 93, **320**, **327**
fringes, lattice 10, 22, 103, 109, 123, **235**, **243**, **267**, **291**, **297**, **320**, **327**, **332**, **351**, **483**

Ge, amorphous **481**
gedrite (Mg,Fe)$_{6-5}$Al$_{1-2}$[OH/(Al,Si)-Si$_3$O$_{11}$]$_2$ 172, 189, **243**
geological interpretation
  alkali feldspar exsolution 255
  anorthite APB's 9, 288, 290, 345
  calcite dislocation microstructure 438
  olivine dislocation microstructure 443
  pyroxene microstructure 182, 285
  quartz dislocation microstructure 424
glass 385, **476**, **481**, **545**
glide plane 376
gold **498**
goniometric specimen stage 55
G.P. (Guinier-Preston) zones **188**, **517**
grain-boundary diffusion 375
grinding 147
growth of precipitates 179, 241, 281
grunerite (Fe,Mg)$_7$[OH/Si$_4$O$_{11}$]$_2$ 189, **240**

hausmannite Mn$_3$O$_4$ 282
healed cracks 145, **437**
heating experiments 292, 309, 345, 503
heating stage 158, 503
heazlewoodite Ni$_3$S$_2$ **502**
hematite Fe$_2$O$_3$ **211**, **215**
hexacelsian BaAl$_2$Si$_2$O$_8$ 293, **355**
high order g bright field E.M 48
high resolution E.M. **5**, **123**, **137**, **304**, **319**, **338**, **478**
high-voltage E.M. 20, 44
H-Nb$_2$O$_5$ **133**
homogenization
  alkali feldspar 189, 286
  anorthite APB's 292
  bytownite exsolution 199
  intermediate plagioclase APB's 293
hornblende 189, **240**
hot stage 158, 503
hot-working 11, 378, 424
Howie-Whelan equations 80 114
hydration 164
hydrolytic weakening 384, 399, 410, 426, 462
identification of phases 25, **30**
idocrase Ca$_{10}$(Mg,Fe)$_2$Al$_4$[(OH)$_4$/(SiO$_4$)$_5$/(Si$_2$O$_7$)$_2$] **131**
ilmenite FeTiO$_3$ **215**
image contrast (see contrast)

image formation 22, 123
image matching 113, 127
inclusion 412, 426, **460**
inelastic scattering 18, 27
in situ experiments 163, 180, 210, 282, 292, 295, 309, 345, 503
intensity of
  electrons 24, 62, 80, 123
  X-rays 510, 526
interface
  combination of 99
  glissile 276, 294
  identification of 100
  interphase (see also exsolution lamellae, polytypes) 97
  intrinsic lextrinsic 37, 72, 95
  inversion **39**, 77, 81, 84, 99, **282**
  nature of 95
  translation (see also APB, stacking fault) 71, 95, 282
  twin (see also twins) 75, 96
  types 95
interface dislocations 27, **214**, **226**, **234**, 252, **269**
interference fringes 320
intermediate plagioclase 194, **266**, 292, **332**
interphase interfaces (see also exsolution lamellae, polytypes) 97
interstitials 69
inversion boundaries **39**, 77, 84, 99, **282**
invisibility criteria 34, 119, 380
ion beam thinning 148

Jahn-Teller effect 69, 282
jogs in dislocations 394

*k*aolinite Al$_2$[(OH)$_4$/Si$_2$O$_5$] **64**
KCl **63**
Kikuchi pattern 27, **30**, 165, 198, 258
kinematical theory 23, 30
kinetics of recovery 448
kinks 376, 555
Kossel pattern 496

labradorite (Ca,Na)[(Al,Si)AlSi$_2$O$_8$] **109**, 196, **261**, 332, 534
lattice fringes
  amorphous material **483**
  amphiboles **243**
  anorthite **10**, **291**
  hexacelsian **357**
  intermediate plagioclase **267**, **332**
  pyroxenes **235**, **297**, **320**
Laue zones 54
layered structures 108
ledges (growth steps) 234
leucite K[AlSi$_2$O$_6$] **78**, **107**

Subject Index

$LiFe_5O_8$ 25, 37, 41, 280
limestone 428, 501
line of dislocation 376
linear defects (see dislocations)
loops of dislocations 376, 394, 410, 419, 429, 437
low angle boundaries 11, 372, 378, 388, 419, 428, 443

magnetite Fe_3O_4 211, 282
mantle of earth 393, 443
marble 429
martensic transition 245, 274, 294, 331, 376, 387, 465
martensite 11, 212
metals versus minerals 174, 376, 398
metamorphism 266, 387, 420, 429, 538, 547
meteorites 12, 537, 543
mica 5, 497, 553
microanalysis 244, 258, 494, 506, 520, 526
  minimum mass 532
  spatial resolution 511, 520, 532
micro-area diffraction 496
microcline $KAlSi_3O_8$ 7, 249, 286
microcracks 436, 550
microcrystallite model for amorphous substances 476
misfit dislocations 27, 214, 226, 234, 252, 269, 412
mixed layer compounds 74
modulated structures
  alkali feldspar 191
  bytownite 199
  labradorite 196
  peristerite 195
  pyroxenes 184, 221
Moiré fringes 60
molybdenite MoS_2 96, 109
moonstone (see also perthite) 98, 107, 190, 249, 500
Mn_3O_4 282
muscovite $KAl_2[(OH,F)_2/AlSi_3O_{10}]$ 498
mylonites 387, 422

Nabarro-Herring creep 376, 462
n-beam high resolution method 44, 123, 137, 304, 319, 338, 478
Nb_2O_5-H 133
$2Nb_2O_5 \cdot 7WO_3$ 131
network of dislocations 216, 373, 378, 394, 419, 428
Ni 502
NiAs 104, 498
Ni_4Mo 70
NiO 502
Ni_3S_2 502

niccolite NiAs 104, 498
non-conventional techniques 48
non-stoichiometry 73, 104, 137, 299, 304, 310
nucleation
  heterogeneous 176, 220, 239, 360
  homogeneous 175, 241, 282

offretite $(Ca,Na,K)_2[Al_3Si_9O_{24}] \cdot 9H_2O$ 61
okenite $Ca_3[Si_6O_{12}(OH)_6] \cdot 3H_2O$ 61
olivine Mg_2SiO_4 395, 443, 546
omphacite $(Ca,Na)(Mg,Fe,Al)[Si_2O_6]$ 284, 285
opal $SiO_2 \cdot nH_2O$ 475
optimum defocus 126
order
  electric 69
  interstitial 69
  magnetic 69
  positional 69, 277
  stacking 108, 277, 326
  substitutional 277
  vacancies 69, 104, 137, 304
ordering transitions 274
ordering twins 104, 248, 285
orientation contrast, SEM 492
orientation variant 70
orthoclase $KAlSi_3O_8$ 6, 249, 286, 500
orthoenstatite (see enstatite)
orthopyroxene (see enstatite)

paracelsian $BaAl_2Si_2O_8$ 358
parawollastonite $CaSiO_3$ 326
partial dislocations 42, 378
  calcite 441
  ilmenite-hematite 216
  molybdenite 96, 110
  pyroxenes 223, 390, 392, 470
  spinel 48
  tealite 106
  wollastonite 331
partial occupancy 363
particle track 12, 537, 544
$PbSnS_2$ 105, 107
Peierls force 383, 394, 445
pencil glide 445
penetration of electron beam 21, 490
perfect crystal 30, 78
peridotites 392, 443
periodic faults
  APB's 74, 266, 341
  stacking faults 108, 327
peristerite 196, 261
perthite 7, 98, 109, 189, 193, 249, 500
phase contrast 20
phase transformation (see transformation)
phase object approximation 125

pigeonite (Mg, Fe, Ca)SiO$_3$ **2**, **185**, **222**, **230**, **272**, 283
plagioclase (see albite, anorthite, bytownite, labradorite, peristerite)
  phase diagram 199
planar defects (see also APB's, inversion boundaries, exsolution lamellae stacking faults, twins) 35, 68, 174, 274
polymorphism, general discussion 68, 274
  alkali feldspar **249**, 285
  anorthite **9**, **286**, **345**
  baddeleyite **295**
  paracelsian-hexacelsian **354**
  pyroxenes 182, **283**, **319**, **465**
  silica minerals **282**
  spinel **280**
polytypism, general discussion 59, 73, 97
  enstatite-clinoenstatite **293**, **319**, **465**
  pyrrhotite **138**, **304**
  wollastonite **324**
pores **437**
precipitation (see exsolution)
preferred orientation 63, **421**, **424**, **432**
projected charge approximation 127
pseudosymmetry 324
pyroxenes (see also specific pyroxene minerals)
  deformation **389**, **465**, **54**
  exsolution **2**, **9**, **182**, **220**, **228**
  microanalysis **516**
  polymorphism **2**, 182, **272**, **283**, **319**, **465**
pyrrhotite Fe$_{1-x}$S 69, **104**, **137**, 299, **304**

quartz SiO$_2$
  deformation 113, **372**, **381**, **410**, **419**
  luminescence **495**
  natural **372**, **386**, **419**
  polymorphism 283
  synthetic **383**, **410**
quartzite **372**, **386**, **419**

radial density distribution function for amorphous substances 476
random network model for amorphous substances 476
rates of transformation 180
reciprocal lattice vector g, definition 23
reconstructive transitions 274
recovery of dislocations 11, **378**
  calcite **436**
  olivine **394**, **445**
  quartz **381**, **410**, **424**
recrystallization 12, **376**, **448** 547
  calcite **436**
  quartz 384, **422**
reduction, *hematite* 210

reflection twins 76
refraction of electrons 62
replica methods for tracks 540
resolution limits
  microanalysis 506, 511, 520
  SEM 499
  TEM 18, 124
response matrix 82, 87, 99
rocking beam technique 496
roentgenite Ca$_2$Ce$_3$[F$_2$/(CO$_3$)$_3$] **109**
roggianite NaCa$_6$Al$_9$Si$_{13}$O$_{46}$·20H$_2$O 56, **514**
rotation twins 76
rutile **69**, 74, **92**, 206, 300, 310
  doped with Cr^{3+} 315
  doped with Fe^{2+} 317
  doped with Ga^{3+} 317

sanidine KAlSi$_3$O$_8$ **249**, 285
satellites **191**, **267**, **332**
scanning electron microscope (SEM) 488
  absorbed electrons 493
  application 499, 541, 550
  Auger electrons 494
  backscattered electrons 489
  cathodoluminescence 496
  microanalysis 494, 520, 526
  microdiffraction 496
  references 505
  STEM 496
scapolite (NaCl·NaAlSi$_3$O$_8$-CaCO$_3$·CaAl$_2$Si$_2$O$_8$) **284**, 294
schiller **194**, **332**
screw dislocations 377
selected area diffraction (SAD) 22, 25, 53, 496
shear structure 73, 300, 311
shear-transformation 245, 274, 294, 331, 376, 387, 465
shock 375, 387, **392**, 397, 398, **435**, 547
Si 30, 34, 48
SiC **166**, 278
silica minerals (see cristobalite, quartz, tridymite)
Si$_3$N$_4$ 152
SiO$_2$ amorphous **474**, **481**
slip bands 376
slip systems
  calcite 428
  olivine 394, 444
  quartz 383, 412
SnO$_2$ **206**
solar radiation 12, **537**, 543
spatial resolution
  electron probe 506
  microanalysis 511, 520, 532

Subject Index

specimen preparation
  chemical thinning 153
  crushing fragments 147
  etching 154, 540
  impregnation 148
  ion beam thinning 148
  SEM 492, 550
  thin sections 147
specimen stages 155
  cold 160
  environmental 161, 210
  goniometric 55, 155
  hot 158
  straining 160
  tilting 157
spectrometers 490, 512, 529
spinel AB_2O_4 25, **26**, **37**, **41**, 74, **281**
spinodal decomposition 178, 192, 205, 276
sphalerite ZnS 97, 101
spherical aberration 19
  high resolution 126
  microanalysis 510, 532
$SrAl_2Si_2O_8$ 293
$SrSiO_3$ **61**
stacking fault
  contrast 84
  definition 72, 379
  periodic **108**, **329**
stacking faults
  bastnaesite **108**
  calcite 441
  molybdenite **96**, 110
  pyrrhoxenes 392, 466
  pyrrhotite 105
  TaC **38**
  wollastonite **329**
stacking order 59, 73, 108, 278, 313, 324
stages (see specimen stages)
steel **11**, **92**
STEM 496, 508
stereo microscopy 165, 216
stoichiometry 73, 104, 137, 299, 304, 310
strain contrast 43, 334, 379
strain energy 194, 248
streaks in diffraction patterns 60, **326**, **349**, **423**
stress-strain curve 405, 410
structure determination (see also high resolution) 5, 66, **123**, **137**, **304**, **319**, **338**, **478**
  rules and prescriptions 130
structure factor 24, 62, 124
structure factor contrast 22, 44, 108
subgrain boundaries **11**, **372**, **378**, **419**, **428**, **437**, **443**
superreflections 57, 286, 326, 345, 355, 364

superstructure
  anorthite **286**, **345**
  intermediate plagioclase 292, **332**
  pyrrhotite **137**, **304**
  spinel **25**, **37**, 74, **281**
  wenkite **363**
surface energy 229
surface morphology 550
symmetry change 70, 279
synchisite $CaCe[F/(CO_3)_2]$ 108
syntectonic recrystallization 448

TaC **38**
tacharanite $(Ca,Mg,Al)(Si,Al)O_3 \cdot H_2O$ **57**
tangles of dislocations **383**, **422**
taellite $PbSnS_2$ **105**, **107**
tectonic deformation 424, 432
texture by diffraction 63, 424
thermionic emission 506
thetomorphic glass 387, 398
thickness fringes 31
thin-sections 147
Ti **498**
$Ti_2Nb_{10}O_{27}$ **133**
$Ti_2Nb_{10}O_{29}$ **129**
TiO_2 **69**, 74, **92**, **206**, 300, 310, **495**
topographic contrast 492, 493, 550
topotactic reactions 62
tourmaline $Na(Mg,Fe,Al)_3Al_6[(OH)/(BO)_3/Si_6O_{18}]$ **6**
tracks **12**, **537**, **544**
transformation
  displacive 68, 274
  ferroelectric 280
  heterogeneous 275
  homogeneous 275
  martensitic (see shear)
  ordering 68, 274
  reconstructive 274
  shear 245, 274, 294, 331, 376, 387, 465
transition (see transformation)
transitional anorthite 266
translation boundaries (see also APB, stacking fault) 71, 95, 282
translation variants 71
transmission function 123
tridymite SiO_2 282
troilite FeS 304
TTT diagram 180
tweed structure (see also modulated structure) 44, 195, 199, 221
twin 70
  mechanical 376, 439
  periodic **193**, **253**, **298**, **319**, **465**
  reciprocal 390

twin interfaces   75, 96
twinning
  alkali feldspar   7, **193**, **253**, 285
  calcite   **429**, **439**
  leucite   **107**
  pyroxenes   **231**, **298**, **319**, **389**, **465**
  pyrrhotite   **105**
  quartz   283, 383
  wenkite   363
twinning partials   378
two-beam dynamical theory   31, 78, 113, 397
"two-dimensional" structures   60

unit cell   54

vacancies   69, 137, 299, 304
vacuum matrix   88
Verwey ordering   282
vesuvianite (see idocrase)
Video tape recording   164, 503
visibility criteria   34, 119, 380
vitrification   388, 389
volume diffusion (Nabarro-Herring)   376, 462

water weakening   384, 399, 410, 426, 462
wavelength   19, 53
weak beam dark field technique (TEM)   48
wenkite $Ba_4Ca_6(Al,Si)_{20}O_{39}(OH)_2(SO_4)_3$ 294, **361**
whitlockite $Ca_3[PO_4]_2$ 12, 541
wollastonite $CaSiO_3$ 299, **324**, 390
work-hardening   378, 404, 410
wüstite FeO   299
wurtzite ZnS   97, **101**, 278

xenoliths   449
xonotlite $Ca_6[(OH)_2/Si_6O_{17}]$ 59
X-ray
  microanalysis   244, 258, 506, 520, 526
  production   494, 506, 526
X-rays versus electrons   4, 52

zeolite   61
zinc hydroxysulphate   **60**
zincblende ZnS   97, **101**, 278
zircon $ZrSiO_4$ 541
ZnS   97, **101**, 278
Zr_3Al_3 **93**
ZrO_2 **297**

Minerals and Rocks

As from volume 10 the series formerly titled Minerals, Rocks and Inorganic Materials is continued under the new title.

Editor-in-Chief: P.J. Wyllie
Editors: W. von Engelhardt, T. Hahn

Vol. 1: W.G. ERNST
Amphiboles
Crystal Chemistry, Phase Relations and Occurrence
With 59 figures
X, 125 pages. 1968
Subseries: Experimental Mineralogy

Vol. 2: E. HANSEN
Strain Facies
With 78 figures. 21 plates
X, 208 pages. 1971

Vol. 3: B.R. DOE
Lead Isotopes
With 24 figures
IX, 137 pages. 1970
Subseries: Isotopes in Geology

Vol. 4: O. BRAITSCH
Salt Deposits — Their Origin and Composition
Translated from the German edition by P.J. Burek and A.E.M. Nairn in consultation with A.G. Herrmann and R. Evans
With 47 figures
XIV, 297 pages. 1971

Vol. 5: G. FAURE, J.L. POWELL
Strontium Isotope Geology
With 51 figures
IX, 188 pages. 1972
Subseries: Isotopes in Geology

Vol. 6: F. LIPPMANN
Sedimentary Carbonate Minerals
With 54 figures
VI, 228 pages. 1973

Vol. 7: A. RITTMANN
Stable Mineral Assemblages of Igneous Rocks
A Method of Calculation
With contributions by V. Gottini, W. Hewers, H. Pichler, R. Stengelin
With 85 figures
XIV, 262 pages. 1973

Vol. 8: S.K. SAXENA
Thermodynamics of Rock-Forming Crystalline Solutions
With 67 figures
XII, 188 pages. 1973

Vol. 9: J. HOEFS
Stable Isotope Geochemistry
With 37 figures
IX, 140 pages. 1973

Vol. 10: J.T. WASSON
Meteorites
Classification and Properties
With 70 figures
X, 316 pages. 1974

**Vol. 11:
W. SMYKATZ-KLOSS**
Differential Thermal Analysis
Application and Results in Mineralogy
With 82 figures. 36 tables
XIV, 185 pages. 1974

Crystal Chemistry of Non-Metallic Materials

Editor: R. Roy
This new independent series was originally intended to appear as a subseries Crystal Chemistry within the former series Minerals, Rocks and Inorganic Materials as announced.

Vols. 1 and 3 of the new series CRYSTAL CHEMISTRY OF NON-METALLIC MATERIALS are in preparation

Vol. 1: R. ROY, R.E. NEWNHAM
Principles of Crystal Chemistry

Vol. 2: R.E. NEWNHAM
Structure-Property Relations
With 92 figures
IX, 234 pages. 1975

Vol. 3: O. MULLER, R. ROY
The Major Binary Structural Families

Vol. 4: O. MULLER R. ROY
The Major Ternary Structural Families
With 46 figures
IX, 487 pages. 1974

Springer-Verlag
Berlin
Heidelberg
New York

Contributions to Mineralogy and Petrology

In Cooperation with the International Mineralogical Association (I.M.A.)

Editors in Chief:
C.W. Correns,
I.S.E. Carmichael
Managing Editor:
J. Hoefs

The journal publishes contributions in geochemistry, including isotope geology; the petrology and genesis of igneous, metamorphic, and sedimentary rocks; experimental petrology and mineralogy; and the distribution as well as the significance of elements and their isotopes in rocks. In addition to original investigations, review articles are also included.

Sample copies as well as subscription and back volume information available upon request.

Please address:
Springer-Verlag, Promotion Department 4021
D-1000 Berlin 33, Heidelberger Platz 3

or

Springer-Verlag New York Inc.
Promotion Department, 175 Fifth Avenue
New York, NY 10010, USA

Springer-Verlag
Berlin Heidelberg New York

QE
369
M5
E35

Electron microscopy in mineralogy

QE
369
M5
E35

DATE DUE | BORROWER'S NAME

010460896
77 12 22
MELANSEL KEVIN R

DEC 22 1978
MAY 2 5 1979

JUL 12 1976